Fairy Tales About Textbook Reading

Fairy Tale 1. I need to understand something before I am able to read it. It makes no sense to read new material before somebody explains it to me.

Fairy Tale 2. The most effective time to read the exam material is the night before. It will be fresh in my mind.

Fairy Tale 3. It is stupid to underline and highlight material that I read since it clutters up the text and my brain can't process it efficiently. Furthermore, I get less money when I sell it.

Fairy Tale 4. Albert Einstein came up with his great ideas because he did not "mess up" his brain with reading complicated text in physics and math.

Fairy Tale 5. A good engineering student sits down and reads the assigned text once (after the class). S/he avoids the graphs and figures since they are commonly very distracting, taking up too much time to understand.

Fairy Tale 6. It is most efficient to memorize words that I don't understand so I can ask the teacher the next time before class. Looking up the meaning of the words myself would be too time consuming.

Fairy Tale 7. Research has shown that summarizing what I read using my own words and in mind-maps is dangerous. Not using the writer's own words will make it impossible for me to really comprehend what s/he explains. I will also fail those questions on the exam.

The Creative Mind

Genius is one percent inspiration and ninety-nine percent perspiration. . . . Opportunity is missed by most people because it is dressed in overalls and looks like work." (Thomas Edison)

A creative person requires both convergent and divergent thinking capabilities, each associated with one side of our brain.

Left Brain Features
Uses logic, rational
Detail oriented
One-at-a-time processing
Looks at a detail
Knows "how"
Facts "rule"
Focus on present and past
Loves math and science
Acts systematically, linearly
Remembers numbers, names
Follows important rules, laws
Relies on words, facts
Forms strategies
Thinks convergently
Operates cautiously
Set to be objective

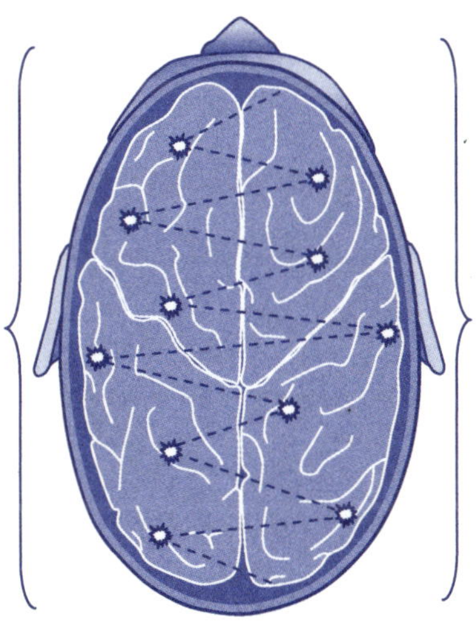

Right Brain Features
Uses feelings, intuition
Big-picture oriented
All-at-once processing
Looks at the whole
Discovers "why"
Imagination "rules"
Focus on present and future
Loves philosophy
Acts casually, randomly
Remembers places, faces
Follows common sense
Relies on images, emotions
Forms possibilities
Thinks divergently
Operates adventurously
Willing to be subjective

Creative ideas come to most people when their minds are involved in an unrelated activity such as drawing or showering in the morning. As the brain works 24/7 it operates at full throttle when we sleep, working on problems similar to a fermentation process. We must keep in mind, though, that the brain must have available all the up-to-date knowledge to access and "play" with creating infinite number of combinations.

A creative engineer is: *(a) inquisitive, (b) prepared to fail, (c) willing to jump into "the swift flowing river," (d) able to daydream, (e) "different", and (f) a person who says "I can do better."*

My Productive Study Environment

When you play, play! When you dance, dance! When you study, study!

Everybody has a different "best" method to learn. Nevertheless, for every brain to function effectively it needs to feel relaxed, ready to accept and process new information or solve problems. What is the time during the day when your brain feels awake, is most creative, or "computes" with the least number of errors? If you don't know the answer, you need to test it. Where is the best environment for me to read, to do math homework, to design, work with the study team, etc.? Each place has a different atmosphere, providing different stimuli to the brain.

Here is a set of key features of a productive study environment:

1. Are TV, phone, games, loud music/conversations, etc. off?
2. Will nobody interrupt me for at least 30 minutes at a time?
3. Is this a place where I study several times/week?
4. Are the temperature and the air(flow) comfortable?
5. Are the seat adjustable and the lighting sufficient?
6. Can I bring my study tools (highlighter, eraser, dictionary, calculator, . . .)?
7. Is there a place close by to take a quick break and eat a snack?

Look for another spot if one answer is No!

The Learning Journal

What I hear, I forget/What I see, I remember/What I write and draw, I understand!

Creating a learning journal, using your own words and sketches, is a powerful means to establishing a deep understanding about what you have learned every week. The final class journal is a powerful record that documents the progress of your learning. Professionally written journals serve as a portfolio to take to an employment interview.

Example topic questions for weekly entries:

1. What is the topic of the week?
2. What happened in class? (Lecture topics, in-class group work, new assignments, comments, quizzes, interesting question by a student, . . .)
3. What happened outside the class that relates to the topic? (news article, TV, group meeting, on-line search, interesting assignment, ...)
4. What were the two most interesting new things I learned and why?
5. What was the toughest subject matter to understand? Which technique helped to overcome the tough spot?
6. What are meanings of the 5 most curious new words of the week?
7. Which figure(s) or table(s) has most important information (add a copy here)?
8. What questions arose during pre-reading of the text?
9. If I had known that . . . when I worked on/read . . . I would have been able to . . .

It is an excellent habit to carry a small sketchbook with an attached small pencil. Modern alternatives are iPad, SMS, phone message to oneself, . . . The purpose is to collect ideas and thoughts anytime during the week to be transcribed into the Learning Journal at the weekend.

Construction Equipment and Methods
Planning, Innovation, Safety

Leonhard E. Bernold
School of Civil and Environmental Engineering
The University of New South Wales, Sydney, Australia

WILEY

VP & EXECUTIVE PUBLISHER	Don Fowley
SENIOR EDITOR & PRODUCT DESIGNER	Jennifer Welter
ASSISTANT EDITOR	Samantha Mandel
EDITORIAL ASSISTANT	Hope Ellis
EXECUTIVE MARKETING MANAGER	Christopher Ruel
MARKETING ASSISTANT	Marissa Carroll
PRODUCTION MANAGER	Janis Soo
ASSISTANT PRODUCTION EDITOR	Elaine S. Chew/Yee Lyn Song
PHOTO EDITOR	Sheena Goldstein
MEDIA SPECIALIST	Andre Legaspi
COVER DESIGNER	Kenji Ngieng
COVER PHOTO CREDIT	© Paul Springett A/Alamy Limited

This book was set in 9.5/12 Palatino by Thomson Digital and printed and bound by Courier Kendallville, Inc. The cover was printed by Courier Kendallville, Inc.

This book is printed on acid-free paper. ♾

Founded in 1807, John Wiley & Sons, Inc. has been a valued source of knowledge and understanding for more than 200 years, helping people around the world meet their needs and fulfill their aspirations. Our company is built on a foundation of principles that include responsibility to the communities we serve and where we live and work. In 2008, we launched a Corporate Citizenship Initiative, a global effort to address the environmental, social, economic, and ethical challenges we face in our business. Among the issues we are addressing are carbon impact, paper specifications and procurement, ethical conduct within our business and among our vendors, and community and charitable support. For more information, please visit our website: www.wiley.com/go/citizenship.

Library of Congress Cataloging-in-Publication Data

Bernold, Leonhard E. (Leonhard Emil), 1952–
Managing construction equipment / Leonhard E. Bernold.
p. cm.
Includes bibliographical references and index.
ISBN 978-0-470-16986-5 (hardback)
1. Hoisting machinery. 2. Earthmoving machinery. 3. Building—Equipment and supplies. I. Title.
TJ1350.B44 2013
624.068'2—dc23

2012020567

Printed in the United States of America

10 9 8 7 6 5 4 3 2 1

I dedicate this book to my family for their continued love and encouragement. My gratitude goes towards my wife Marilyn and my beautiful daughters Sarah and Elizabeth.

Preface

This textbook on construction equipment and methods "feels" different than what is presently available since it presents subjects in a multidimensional manner while, at the same time, engaging the students in developing their personal learning preferences. It presents construction knowledge not as a list of facts but rather as a web linking principles of mechanics, management, historic innovations, engine power, materials, computers, electronics, safety, and highly skilled human resources. Special effort is made to make the material come alive by adding short stories about key people and providing opportunities for creative and innovative minds to earn credits towards their final grades if their instructor includes the book's activities in class assignments.

Tapestry of Learning

Deep learning is a complex process in which a teacher both enables and facilitates. I consider myself a coach who demonstrates expertise, provides formative feedback, forms the students throughout the semester but most importantly encourages them to develop the professional learning skills that are necessary for success. Equally important, I do not grade on a curve as a step toward encouraging teamwork, in which everybody can win if s/he is meeting the level of excellence set by the coach.

A textbook does not have to be dry and sterile. This text engages the reader in a conversation, with the intent of stimulating interaction with the teacher. It presents small problems that challenge the student to follow the thoughts of an experienced engineer just as an apprentice looks over the shoulder of his or her master. Equally important, each reader is recognized as having unique learning strengths, as illustrated by David Kolb's learning model in Figure P.1. Various methods and techniques are employed in each chapter to let all readers apply their individual learning strengths.

What is the feeling one gets when one really "gets" new material? One understands the theory behind it, knows how to apply it to solve problems, knows why it is important, and, finally, has an instinctive sense of what would happen if critical variables were changed. In fact, one has visited all four quadrants of Kolb's circle, while scaling Bloom's learning pyramid (second part of Figure P.1), beginning with memorizing the facts and ending with being able to apply the new knowledge creatively. The teacher leads the students through the process of acquiring knowledge via a personal guided inquiry into the subject.

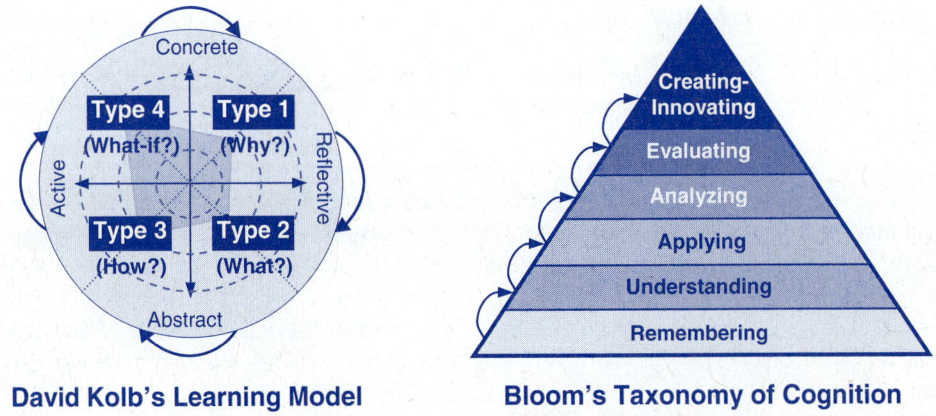

Figure P.1 Applied Learning Models

Generic Structure of a Class Journal

A. Executive Summary
B. Table of Contents
C. Weekly Journal entries with the following five topics:

1. Weekly Review: **WHAT**? happened in class? **What** did we have to do?
2. Personal discussion: Reflections about each main subject of the week:
   - **WHY?** - are the new principles and methods important?
   - **HOW?** - does the new material relate to what I already know?
   - **WHAT?** - are real-world problems that I can solve now?
   - **WHAT?** - helped me most to understand?
3. Special Journaling Questions of the Chapter (assigned by teacher)
4. New Terms: The meaning of 10 new (to me) terms in construction
5. My Favorite Sketch-Figure-Picture that communicated most effectively.

D. SUPPLEMENTARY Material: Personal observations, evaluations, poems, newspaper articles. PowerPoint, website, etc.

Most students have little knowledge of the historical tradition and background of the technologies they use almost every day. To fill this vacuum, short sidebars are added to the main text in order to encourage further inquiry. In the same vein, chapter reviews include a variety to avenues for the students to move through Kolb's learning quadrants and climb Bloom's pyramid of understanding. Crossword puzzles allow students to test their knowledge of new terms and terminologies, traditional homework offers means to practice new concepts, while journaling and open-ended problems challenge them to think creatively, reflecting on how the new material increases their ability to solve new problems. The "lost art" of journaling is given an opportunity for a comeback. There is no doubt that the most prominent engineer "journaler" worked in the sixteenth century. Who does not admire Leonardo da Vinci's sketches of lifting cranes and flying machines, and the text and calculations that accompany them explaining how and why the proposed system would work?

Many of the skills that students practice while journaling are among those suggested by the Accreditation Board for Engineering and Technology (ABET): communication, analytic skills, information literacy, lifelong learning, knowledge of contemporary issues, holistic approach to engineering problems, and reflective thinking about professional issues. One of a teacher's most satisfying experiences is to observe students, after first struggling during a semester, present their journals at semester's end with pride and a smile, and say, "This was the toughest thing I did in school, but it will guide my future professional life." This and similar statements I have heard many times.

Presentation Strategy

The material in *Construction Equipment and Methods* is presented using two integrated media: the traditional textbook and the supporting website, the latter offering a large variety of supplementary materials for students to explore, if they wish. At the same time, the teacher is helped to find ways to include the web material into their teaching.

Some of the chapters contain features called header problems and/or worked-out problems, positioned at "pressure points." Both features are nontraditional in being that they are embedded in real-world situations that can occur during a construction project. Rather than providing straightforward solutions, they guide readers through problem-solving exercises, requiring them to make assumptions and collect and analyze data; the problems conclude with a discussion of the results, adding a qualitative component and final recommendations to make to a "virtual boss."

Each chapter ends with a chapter review and bibliography. An extensive Glossary is on the book's website. The chapter review generally has four components: (1) the Equipuzzle, (2) journal questions, (3) traditional homework problems, and (4) open-ended problems. The journal questions are tailored to each chapter and can be selected by the teacher to amplify the generic outline of the journal structure discussed earlier.

Book Contents

Not all equipment used today can be covered in a 600-page textbook. While the most common (= available for students to observe) are included in this text, many of the principles discussed apply also to equipment that is omitted. By encouraging students to become information literate, it is hoped that they will have few problems in "connecting the dots" and finding the relevant information when they need it.

Construction Equipment and Methods covers a wide spectrum of topics in 11 chapters:

Chapter 1, HUMAN INGENUITY BUILDS MIRACLES, introduces the basic underpinnings supporting the effective deployment of construction technology: (1) simple machines, (2) skillful people, and (3) capital to deploy all the assets necessary to complete the projects. All modern equipment is still based on five simple machines: (1) lever, (2) inclined plane, (3) block-and-tackle pulleys, (4) wheel and axle, and (5) screw. At the center are competent humans who can turn technology into sophisticated systems able to perform the needed tasks safely, with highest quality and least waste.

Chapter 2, COSTING THE OPERATION OF EQUIPMENT, reviews the many cost components and the methods for calculating hourly as well as unit costs of doing work. In a competitive marketplace, establishing a winning bid requires a skillful synergism of competencies in the field operation, advanced technology, and sound cost accounting. The goal of this chapter is to lay out how these are connected and to sketch basic principles that will be revisited throughout the book.

Chapter 3, CREATING AND DEPLOYING MASSIVE POWER SAFELY, expands the static description of equipment use in Chapter 1 by addressing the powerful dynamics that let them work efficiently and safely. It first takes a look at the history and elements of the powertrain that begin with a power source and ends with an articulator that produces work mostly by applying force. Today's operators control large equipment by moving tiny levers, foot pedals, switches, or joysticks while depending on Global Positioning Systems (GPS) or electronic sensors for guidance. Technology is not without risk, although such developments have generally improved work safety. Nevertheless, only an awareness of the many health and safety risks faced by managers, operators, drivers and workers, coupled with a solid safety program, will prevent accidents.

Chapter 4, THE OMNIPRESENT BACKHOES, pays tribute to a unique piece of equipment that is used for a wide variety of tasks, making it the workhorse of global construction. Its versatility led to many add-ons that not only expand its use but also challenge the basic mechanisms and skills of the operator. The chapter highlights the meaning of motion-skill in operation, but also addresses the many planning skills that are required. Examples are the call-before-you-dig law (One-Call), the use of electronic technology, and accident avoidance measures during trenching.

Chapter 5, DOZERS, SCRAPERS, AND GRADERS, covers equipment that originated when people started to settle down and cultivate the land. More than for other types of equipment, the focus of developmental progress has been the challenge of integrating large amounts of power with "earth-shaping" tools requiring traction between ground. This chapter shows the many innovative developments instrumental in today's state of the art. The author's absolute favorite is the scraper, which had a fascinating "life" after James Porteous (a Scottish immigrant from Fresno, California) invented the first efficient system in the 1880s. Equally fascinating is the integration of GPS and laser-based technology that transforms these huge machines into extremely precise and highly productive devices.

Chapter 6, TRUCKS, LOADERS AND THE QUARRY OPERATION, highlights in part a construction workhorse that is under-appreciated—trucks. Either as on-road or off-road vehicles, they transport virtually all essential construction material, equipment and tools and move mountains of earth and rock as fill for roads or to create large dams. The proper rigging of loads is as important as the use of GPS satellites to ensure their efficient and safe deployment. Off-road haulers need not adhere to federal laws on weight limits and have turned into mighty vehicles with extremely sophisticated electronics and ergonomically designed operator cabins. This chapter reviews the historical background of this important piece of equipment before studying its use within a fleet that is "served" by loaders and excavators. The later part of the chapter introduces an interesting application area where the equipment addressed in this chapter are needed, the rock quarry. For this, the many different crusher types, conveying systems and aggregate processing technologies are introduced. Worked-out problems lead the reader through the process of designing a quarry.

Chapter 7, CRANES—GENTLE GIANTS IN CONSTRUCTION, is yet another piece of equipment tracing its family tree back to the times when humans settled down in the flat areas of the Middle East and China. First used for lifting water buckets from wells, no Greek temple could have been erected without them, no cathedrals and no Empire State Building erected. Even today's cranes are comprised of several simple mechanisms: the wheel and axle winch, the block-and-tackle, lever with counterweights, the screw, the sheave, and lightweight truss. Equally important, but not as simple, are the sophisticated wire ropes lifting a large load with a few tiny wires. As spectacular accident reports demonstrate, cranes can be extremely hazardous if they are not competently operated and maintained. This chapter reviews the crane's vast family tree, reaching all the way to today's International Space Station, and discusses the many innovations and unique applications before looking at the newest electronics employed to make them safer and more productive.

Chapter 8, CREATING AND SECURING A DEEP DIG, describes the large number of technologies and methods needed to excavate, dewater, and stabilize the sides and bottom of a dig. What typically necessitates this kind of construction is the cost of land in urban areas, together with zoning laws restricting the height of a building. The non-visible underground space is thus turned into rentable areas for cars, shops or storage. This chapter discusses the many methods of installing retaining walls, wall shoring, dewatering, earth anchoring, and grouting. A fascinating device covered in this chapter is the centrifugal water pump, which is instrumental in dewatering excavation pits or circulating slurry. Its ancestry goes back to irrigation needs in Mesopotamia, and efforts to mine the lucrative salt in ancient China. Only physicists like Pascal and innovators of the twentieth century, however, made the pump into the highly efficient apparatus it is today.

Chapter 9, THE MANY WAYS TO CONSTRUCT DEEP FOUNDATIONS, emphasizes that the foundations for any kind of structure are increasingly important as larger and heavier structures have to be built in areas with ever poorer ground conditions. Again, the reason is the growth of the urban metropolis located near lakes, ocean or rivers, and rising land prices. Not surprisingly, this is the area that has undoubtedly produced the largest number of innovations over the last 30 years, from the field of strong piling, to in-situ soil improvement methods such as jet grouting and soil mixing.

Chapter 10, FORMS AND TEMPORARY STRUCTURES FOR MATERIAL AND LABORERS, introduces the critical underappreciated set of mechanical systems required to build large structures, namely shoring, formwork, and scaffolding. While enjoying a long history of use and development, these systems have improved dramatically in the last 25 years. No one could have predicted 30 years ago the rapid growth of aerial platforms allowing workers to reach places that previously required elaborate scaffolding. Mechanized self-climbing and slipping formwork, for vertical and horizontal use, have revolutionized the speed and quality of concrete placement and molding. Formwork systems for elevated decks, tall walls and columns can be put together from prefabricated but flexible elements, moved by cranes as large structures and supported by sensor-equipped tower shores. Reshoring is made simpler by innovative mechanisms that allow lowering of the forms while leaving the shore in place. Special attention is paid to the danger points of scaffolds designed to support both workers and materials.

Chapter 11, BRIDGE-BUILDING TECHNOLOGIES, presents the fascinating topic of building bridges that link communities and countries. Not only are bridges one of the outstanding contributions of construction, but their erection requires most of the equipment and methods discussed in the preceding chapters. Still, bridge construction has developed its own technologies and methods with many innovations addressing unique demands and circumstances. While suspension bridges have achieved an amazing central span of 2 km (1.24 miles) and cable-stayed bridges have passed 1.2 km (0.75 miles), the majority of bridges are built to relieve traffic congestion in downtown areas. In these circumstances, the small space available for construction and the required speed has led to many unique technologies. In fact, more than one method is commonly used, each providing unique qualities that fit the varying challenges along the centerline of a bridge. Using many drawings, this text presents a wide variety of bridge building technologies after presenting the know-how for constructing the most common structural elements, such as post-tensioned box-girders, expansion joints, and suspension as well as stay-cable systems.

The Book Site

The book is accompanied by a website for both students and instructor accessible by visiting www.wiley.com/college/bernold. It offers a large amount of additional information for the student and instructor to pursue specific topics in more depth. In each chapter, visual tags in the text alert the reader to the availability of more information on the booksite. The associated code can be used to find the identified document. For example, the first tag in Chapter 3 appears as follows:

S3.1 Heavy-Duty Diesel
Engine Frictional Losses, U.S.

Army Research

This tag translates into: More about diesel engines can be found in the Supplemental materials folder under Chapter 3. The same code is used to identify the associated file.

A second type of tag indicates the availability of more visual material such as photos or brochures. One such example is:

VG4.1 Excavator Tracks

This tells the reader that visuals about excavator tracks can be found inside the Virtual Graphics folder under Chapter 4. Again, the files match the tag label.

Listing URLs in a textbook can be risky, given the frequency with which they change. But whenever a relatively "stable" website is relevant, a direct link is given. One example from Chapter 1:

For information about the U.S. Interstate System, see http://www.fhwa.dot.gov/interstate/homepage.cfm/.

Instructor Companion Site

The instructor companion site includes everything on the student companion site, but is password-protected and also contains the following:

- Solutions Manual
- Illustrations from the text
- Visually oriented PowerPoint Lecture Slides for selected topics

Instructors should visit the website at www.wiley.com/college/bernold to register for access to the password-protected portion of the site.

Weaving a Knowledge Web

Several important topics are woven into the structure of the eleven chapters of the text, thus fostering the creation of a knowledge web with ever more links to nodes previously created in students' minds. The goal of creating these integrated linkages is to reinforce already existing knowledge nodes while strengthening the "depth" of new experiences. Examples are the historical views back to 4000 BC, and references estimating, physics, material behavior, construction safety and health, mechanics, surveying, hydraulics, etc. The following matrix shows where some of the main concepts appear and re-appear to be discussed in different ways, emphasizing their importance and strengthening the students' understanding.

Acknowledgments

I would like to acknowledge the many personal friends who helped and contributed in many ways and my former students whose years of work have flowed into the content of the book. A special thank you goes to my courageous friend Willi Immer, a Lecturer in Quality Management, who took time to proofread the draft material. Last but not least, I extend my thanks to James Worrell, a crane expert with many years of experience—the chapter on cranes would not have been possible without his input.

#	Equipments & Methods	Chap. 1 Building Miracles	Chap. 2 Costing	Chap. 3 Power-Safety	Chap. 4 Backhoes	Chap. 5 Dozers, Scrapers, Graders	Chap. 6 Trucks, Loaders, Quarry	Chap. 7 Cranes, Ropes	Chap. 8 Creating Deep Digs	Chap. 9 Deep Foundations	Chap. 10 Temporary Structures	Chap. 11 Building Bridges
1	Accident Prevention	x		x	x	x	x	x	x	x	x	x
2	Actuators	x		x	x	x	x	x	x	x	x	x
3	Cost & Economics	x	x	x	x	x	x	x	x	x	x	x
4	Force Modeling	x		x	x	x	x	x	x	x	x	x
5	History of Technology	x		x	x	x	x	x	x	x	x	x
6	Mechatronics	x	x	x	x	x	x	x	x	x	x	x
7	Plan for Safety	x		x	x	x	x	x	x	x	x	x
8	Productivity	x	x	x	x	x	x	x	x	x	x	x
9	Craning	x		x	x			x	x	x	x	x
10	Excavation	x	x	x	x	x	x			x		x
11	Health			x	x	x	x	x	x	x	x	x
12	Opera. Skill-Training		x	x	x	x	x	x	x	x	x	x
13	OSHA Rules			x	x	x		x	x	x	x	x
14	Resistance-Friction			x	x	x	x	x		x		x
15	Tower Crane			x			x	x	x		x	x
16	Diesel Engine		x	x	x	x	x			x		
17	Hydraulic Power		x	x	x	x			x		x	x
18	Rimpull			x		x	x					
19	Cycle Time				x	x	x	x	x	x	x	x
20	Soil Behavior				x	x			x	x	x	x
21	Surveying				x	x	x	x				x
22	Shoring				x				x	x	x	x
23	Trucking				x		x	x				x
24	Concrete CIP						x	x	x	x	x	x
25	Concrete-Precast						x			x		x
26	Form-Falsework								x	x	x	x
27	Crawler Crane							x		x		x
28	Anchoring									x		x
29	Drilling									x		x
30	Drill Rig									x		x
31	Piling									x		x
32	Pumping									x		x
33	Truck Crane							x				x
34	Concrete-Post-Tens.								x		x	x
35	Pavers		x									
36	Maintenance		x									

Brief Contents

Contents

chapter one

Human Ingenuity Builds Miracles

One machine can do the work of fifty ordinary men. No machine can do the work of one extraordinary man.

—Elbert Hubbard

I use technology to work more professionally, but especially to finish work quickly.

—Wayne Hemingway

It is only when they go wrong that machines remind you how powerful they are.

—Clive James

In essence, the main function of construction equipment is to move, lift, assemble, dig, load, pour, compact, and place bulk material, parts, and prefabricated elements at a desired location. The goal is to employ the equipment and tools handled by skilled operators to maximize performance, while maintaining the highest standards of safety and quality. The underpinnings that serve as enablers are basic physical principles allowing us to apply and magnify forces that are required to accomplish the desired tasks. Since 4000 BC, when construction technology was first purposely being applied, five rather simple but extremely powerful machines have made tremendous contributions, albeit mostly hidden to the eye of the unaware. These classical machines are *(a) Lever, (b) inclined plane, (c) block-and-tackle, (d) wheel and axle,* and *(e) screw*. The enabling concepts of the methods on which today's technologies depend are basically the same as those 2,000 years ago, differing only in the application of power, which of course has drastically changed. Raw human power was first augmented with animal power before it was harvested from rushing water turning large wheels. All that changed with the arrival of the steam followed by the combustion engine. Parallel to the development of new power sources was the emergence of the banking system and the introduction of double-entry book keeping by the Medici in fourteenth-century Italy, granting loans to individual entrepreneurs. Now, innovative builders were able to buy equipment necessary to do business with powered machinery.

The first three chapters address generic issues before covering the essential pieces of equipment and methods. Specifically, Chapter 1 highlights the importance of classical machines and shows how construction technology advances a society. Chapter 2 examines the significance of capital and money flow in the deployment of construction equipment. Finally, Chapter 3 discusses the principles of creating and controlling power while keeping humans safe.

1.1 Each Construction Equipment Is an Assembly of Classical Machines

A Short Story About an Amazing Record

The speed record for construction of tall buildings is held by the 103-story Empire State Building (see Figure 1.1). General contractors Starrett Brothers, Inc. and Eken, Inc. designed and purchased all new equipment, which was fit to the building. The contractor calculated that this would cost less than renting secondhand equipment and would speed up construction because the individual pieces could be tailored for the job. For example, 16 electrically-driven guy derrick cranes of different sizes, all with automatic hoists, having 10 to 15 tons' capacity, were needed to hoist the steel bundles from the ground and to spread out the pieces on the floor. Many mechanical hoists moved all the other material, such as brick, lumber, concrete, etc., to the different levels.

Steel girders and columns, some of them partially preassembled, arrived on site on trucks after they had traveled by train from Pittsburgh to New Jersey. Each piece carried a tag with its location in the frame and the number of the correct crane derrick. Eighty hours after they had left the furnace, they were already riveted together. Not surprising, the largest building at that time was completed in less than 15 months, 3 months ahead of schedule and under budget. It was 1931, the time of the Great Depression in the U.S. when a worker made $15/day. Seven people died during construction despite the 50-man safety crew that supervised the site.

For more information about the Empire State Building (ESB), visit:
http://www.skyscraper.org/home.htm
http://www.esbnyc.com/esb_story_historical_timeline.asp
http://digitalgallery.nypl.org/nypldigital/dgkeysearchresult.cfm?parent_id=131032

1.1.1 THE LEVER IN ACTION

Without question, the lever is one of the earliest tools used to apply the large forces caused by the multiplier effect resulting from the difference in distance from the fulcrum or its pivot point. The crowbar is an interesting tool to demonstrate how the lever action is so crucial in construction, even today. Assume that we are trying to pull a nail out of a plank that has been fastened to wooden beam underneath as shown in Figure 1.2a. The notched wedge-shaped end of the bar has been used to grab the head of the nail ready to be pulled.

The initial force $\mathbf{F_P}$ that has to be created is necessary to overcome the static friction force that holds the nail inside the two pieces of wood. Sir Isaac Newton taught that the forces on an object that is not in motion have to be in equilibrium. As a result, the moment in $\mathbf{P_1}$ created by $\mathbf{F_1}$ has to be equal the moment created by $\mathbf{F_{P1}}$. Thus we can write $\mathbf{F_1} * \mathbf{l_A} = \mathbf{F_{P1}} * \mathbf{l_{N1}}$.

Figure 1.1 Empire State Building

Now we are ready to figure out how much pulling force we can create:

$$\mathbf{F_{P1}} = \mathbf{F_1} * (\mathbf{l_A}/\mathbf{l_{N1}})$$

Now we recognize that the force that we can generate from a perpendicular base force changes linearly based on the distances between its point of attack and the pivot point. A larger distance to the driving force will increase the lever force, while the inverse is true for increasing the distance between pivot point and levered force.

Another extremely important tool used for constructing the Empire State Building was the wrench. A simple wrench, depicted in Figure 1.3, is made of one piece of steel with a U-shaped opening that grips two opposite faces of a nut. The end is at an angle to the longitudinal axis of the handle to allow for a greater range of

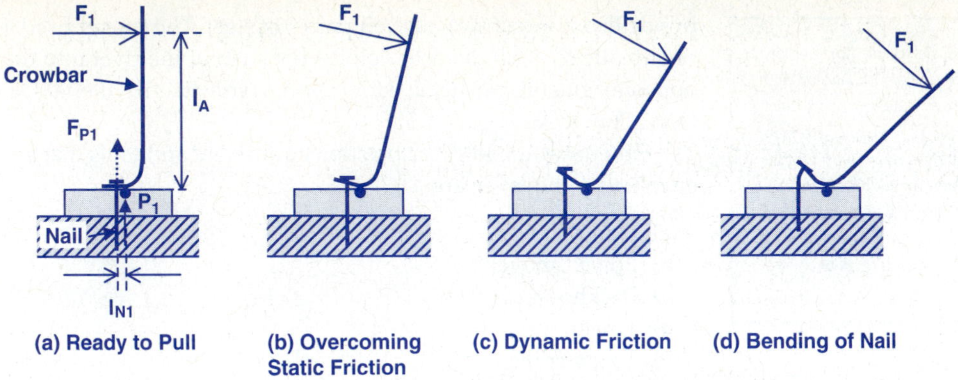

Figure 1.2 Lever Transforms a Small Pushing Force into a Large Pulling Force

movement when the wrench is flipped. The pivot in this case is in the center of the nut or bolt that is being tightened. Also, instead of just one force there are two forces that drive the nut.

However, the most prominent lever-type equipment that could be seen everywhere during the construction were the 16 derricks (Figure 1.4).

The two main parts of the derrick, the mast and the boom, are both levers with the pivot on the bottom end. Each has two additional forces that act on it at the upper end. The guy lines hold the mast in place while the boom line, also attached to the mast, lowers and raises the boom. Finally, the boom holds the hoist line pulleys and with it the load that is being lifted. As we will learn in the chapter about cranes, this type of equipment has an immensely rich history contributing many great achievements.

After a piece of steel was set in place by the erection crew, the riveting crew started. Steel rivets, looking like small mushrooms, had to be heated in an oven until they became soft before one by one they could pushed through the

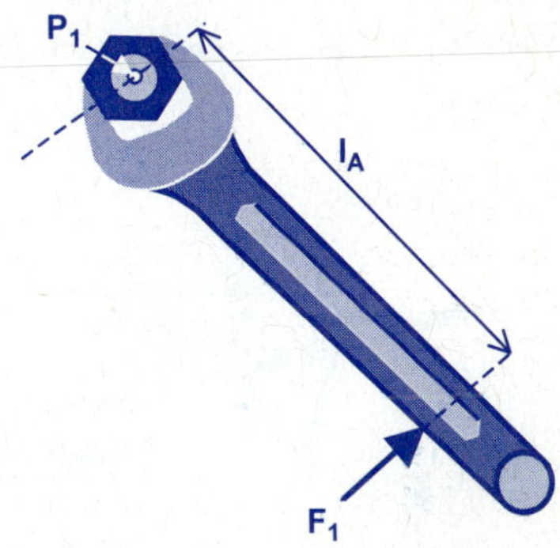

Figure 1.3 Open-End Wrench

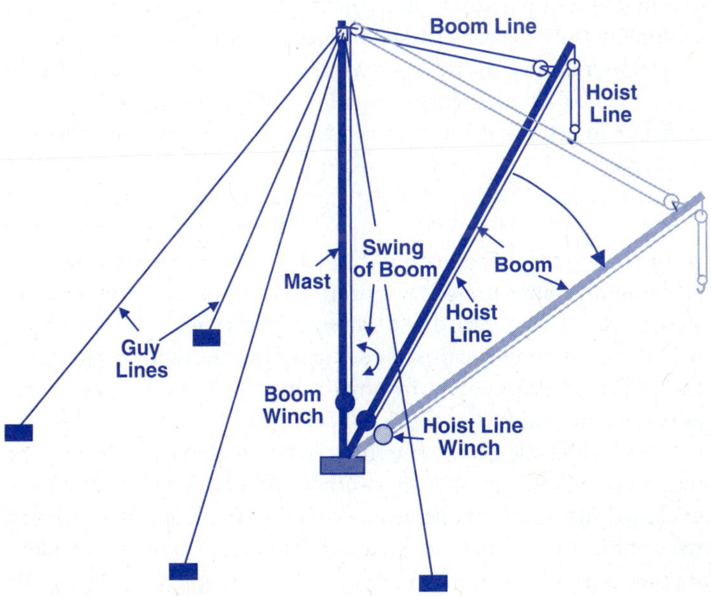

Figure 1.4 Guy Derrick Crane with Electric Winches

pre-drilled holes of two aligned pieces of steel. The riveter used a pneumatic rivet hammer to deform the stem of the rivet into the opposite mushroom head. How would you handle the 600 °C (1,112 °F) rivets (Figure 1.5)?

> The four-man riveting crew consisted of: (a) heater, (b) catcher, (c) bucker-up, and (d) riveter. The heater kept enough rivets hot inside a small oven and grabbed them one by one when ready using a 3-ft (0.9m) tong. He then tossed it to the catcher some distance away, who caught the red-hot rivet with a bucket. The catcher had a second pair of tongs to insert the rivet into the aligned holes. How would you transport a small piece of steel with a temperature of 600 °C (1,112 °F) from the small oven to a column 9 m (30 ft) away?

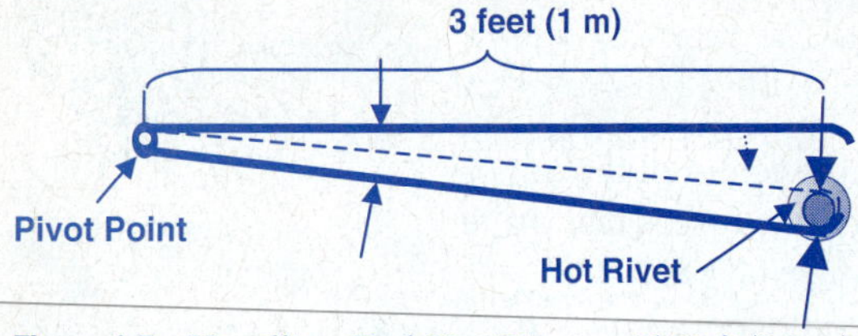

Figure 1.5 Handling Red-Hot Rivets with 3-ft (0.9 m) Tong

Other examples of construction equipment and tools that are enabled by the lever are:

(a) Hand shovel
(b) Power shovel
(c) Backhoe excavator
(d) Tower crane
(e) Truck crane
(f) Crawler crane
(g) Wheelbarrow

See Figure 1.6 for an example of synergy between the lever and wheel.

Figure 1.6 Synergizing Lever and Wheel

1.1.2 THE INCLINED PLANE

The inclined plane does not move, but provides an important mechanism to scale heights or to descend vertical precipices without a vertical free-fall. In construction, its most important applications, even today, are in the form of ramps and chutes. The ramp allows wheeled vehicles to reach a higher level in a building while traveling a solid surface with a much reduced power requirement. Chutes provide a means to transfer construction material to a lower level in a controlled manner without destroying it due to a vertical drop.

1.1.3 THE BLOCK-AND-TACKLE PULLEYS

It was the giant of applied physics and engineering, Archimedes of Syracuse (287–212 BC), who formulated the mechanical advantage one can get from using several pulleys aligned on one axis in a block arrangement. Even before Archimedes, the power of such a mechanism had been used but not really understood. The Romans could not have erected all their statues, harbor walls, or sailed their large merchant ships without the block-and-tackle. Construction is certainly a thankful beneficiary of Archimedes' findings even today.

> Behind the Indiana limestone and granite cladding brick walls had to be built, needing 10 million bricks. The old method of using wheelbarrows to distribute them to the masons was totally impractical. Since the brick walls were not visible when complete, surface damage to an individual brick did not matter as long as it was structurally intact. Thus, Starrett Brothers and Eken installed chutes that led from the street level down into the basement feeding into hoppers. The bricks were dumped directly from the trucks into the chutes. When needed, the bricks were released from the hopper into rail carts, pushed to the elevator, and delivered directly at the appropriate floor to the masons who needed bricks.

A single pulley is a simple machine that redirects force while losing a little energy due to friction between the static center pin and the turning pulley (wheel). The mechanical advantage comes to play when the rope is wound around two blocks as in Figure 1.7. Having two blocks connected by a continuous rope is called a tackle. In the shown arrangement, the pulling force needed to hoist a load of $\mathbf{F_{Load}}$ is $\frac{1}{4}\mathbf{F_{Load}}$. By adding more lines, the required force can be further reduced.

A critical element, which is often forgotten, is the rope made of twisted natural or synthetic fibers. The Egyptians in 4,000 BC made ropes out of flax or papyrus fibers that allowed thousands of laborers to pull large blocks of stone on sleds (Figure 1.8).

The time between 1849 and 1889 was the heyday for wire ropes, which now are so critical both in designing and building structures such as suspension- or cable-stay bridges, post-tensioning, and cranes. High-strength steel wires were twisted into strands that were then laid around a core into the rope. The biggest demand for wire ropes at that time came from cable car operators in San Francisco and other cities in the United States.

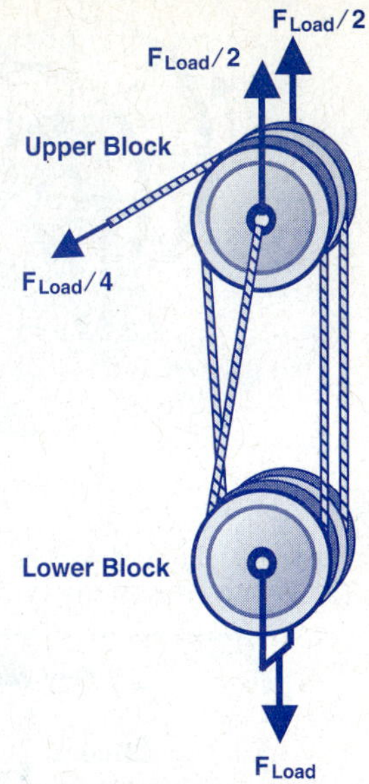

Figure 1.7 A Four-Line Tackle

1.1.4 THE WHEEL AND AXLE

This simple machine can be found everywhere today, as gears in watches and steering wheels in cars, as screwdrivers and drills in the toolbox, or as the powered wheel of a motorcycle. Again, this mechanism is able to amplify a force $\mathbf{F_W}$ that is applied to the outer rim of a wheel at distance of $\mathbf{r_W}$ from its center, as shown in Figure 1.9. A smaller-diameter wheel or axle, with a radius $\mathbf{r_A}$, is rigidly attached to the center of the larger wheel. At the rim of the axle, a much larger force $\mathbf{F_A}$ develops and can be used for work such as lifting water buckets from deep wells or drilling into wood. To calculate the amount of force amplification that can be achieved from a particular machine, we can again use the equilibrium equation for the fulcrum of the system:

$$\mathbf{F_W} * \mathbf{r_W} = \mathbf{F_A} * \mathbf{r_A} \text{ or } \mathbf{F_A} = \mathbf{F_W} * (\mathbf{r_W}/\mathbf{r_A})$$

As with the lever, the augmentation ratio is in fact the ratio of the distances from the fulcrum to the vector of the force perpendicular to the radius. The larger the difference between the radius of wheel and axle, the bigger its multiplication factor.

The Empire State Building was a major user of wheel-and-axle machines, mostly to lift heavy steel beams to the raising building. The 16 derricks depended on levers, guys to support the mast, and block-and-tackle to operate the boom and the hoist. But they needed a machine to operate the lines for the boom and hoist. Figures 1.10 and 1.11 present the first electrically driven winches with automatic brake systems. Hand-cranks had been replaced with a geared motor that could be controlled via a lever. Connected to the axle or drum, the motor either pulled the line in or let it out. This same winch configuration can be found at today's tower cranes, while mobile cranes, powered by diesel engines, mostly employ hydraulic motors.

For a picture from the ESB: http://digitalgallery.nypl.org/nypldigital/dgkeysearchresult.cfm?parent_id=131032 Image ID: 79853?

A slightly different application of the wheel and axle was the extremely effective small rail carts that were used to transport heavy material other than steel beams

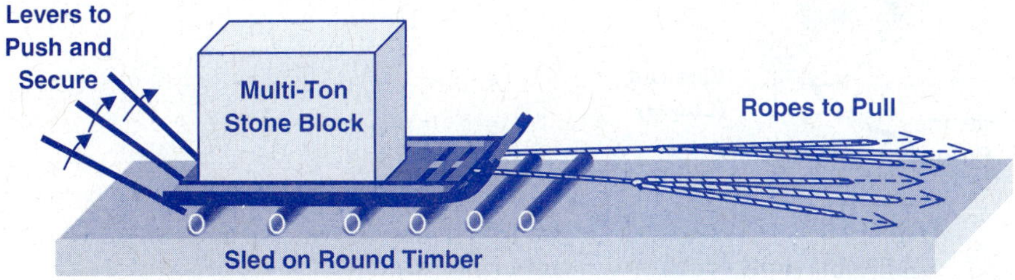

Figure 1.8 Ropes and Levers Used for Egyptian Construction 6,000 Year Ago

and columns (Figure 1.12). While those used for the Empire State Building were pushed by the laborers, the same carts were available with a hand-crank or a motor that powered the axle. Different from Figure 1.9, the small-diameter axle receives the initial force, which is reduced to a smaller force between the rim of the wheel and the rail. If there is sufficient friction available, the cart will start moving in the opposite direction from the force vector, turning the rotation into a linear motion with a velocity related to the radius of the wheel and the rotational speed with which the drive axle is turning.

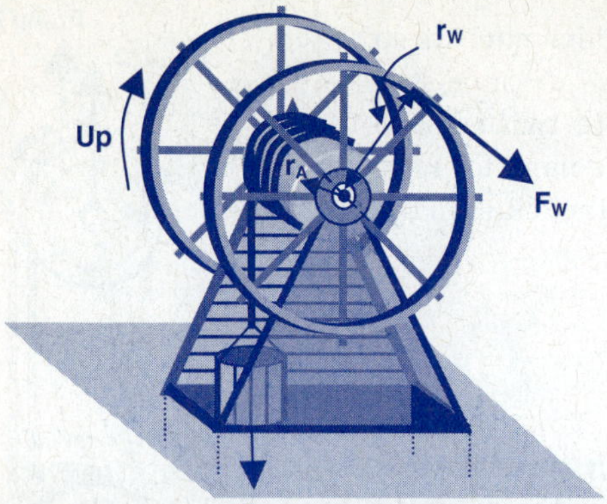

Figure 1.9 First Main Use of Wheel-and-Axle to Lift Water from the Well

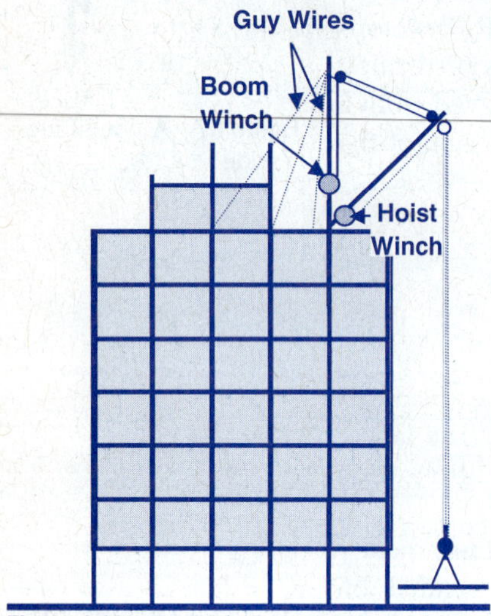

Figure 1.10 Derrick Cranes Depend on Wheel-and-Axle

If we assume that the axle is turning once per second, or 60 rounds-per-minute (RPM), then the axle would travel on its circumference per second: $\mathbf{Travel - Distance_A/sec = D_A * \pi/sec}$. As the larger wheel also turns 1 time per second, we get: $\mathbf{Travel - Distance_W/sec = D_W * \pi/sec}$.

As we observe, the speed relationships between axle and wheel are opposite to that of the forces. Here, larger diameter or radius results in a linear increase of distance traveled. Thus, an increase of force at a wheel-and-axle machine goes hand in hand with a reduction in travel distance, while the speed of rotation or RPM naturally stays the same.

1.1.5 THE SCREW

A screw converts a rotational into a linear motion or a torque into a linear force as its helical thread works itself into a base material, such as a cork or wood. If we unravel a screw, we create a long wedge, as depicted in Figure 1.13.

A torque $\mathbf{T_R}$ must be applied on the bolt or the nut to offset the thread resistance due to the friction coefficient $\boldsymbol{\mu}$ and the force needed to overcome the incline $\boldsymbol{\alpha}$. As we learned in 1.1.1, it's the lever or, in this case, a wrench, that will allow us to apply a large torque using a small force **F** at the end of a lever arm (see Figure 1.3). Using an appropriately sized wrench with a lever of length $\mathbf{r_W}$, we can create a torque $\mathbf{T_W = r_W * F}$. If $\mathbf{T_W}$ is larger than $\mathbf{T_R}$, the torque needed to overcome all the resistances, the bolt will turn and move linearly along its center axle.

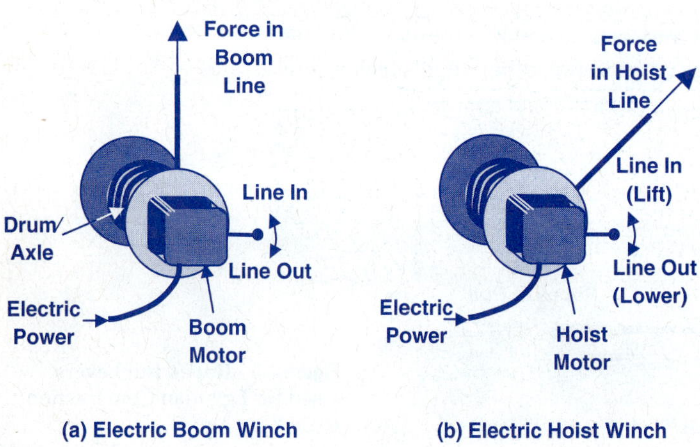

Figure 1.11 Powered Winch Drums Used by Cranes

Bolts and screws are abundantly used in construction. Unlike rivets, bolts allow a nondestructive separation after steel pieces have been fastened. Depicted in Figure 1.14a, bolts are externally threaded cylinders with a large head on one end. The pitch is the distance between two threads measured parallel to the central axis of the bolt. A die cuts a thread on a preformed cylindrical rod, creating what is called the male threaded piece of the bolt. Bolts are made of various lengths and diameters to be inserted into pre-drilled and aligned holes of two or more overlapping pieces that have to be joined. A washer, serving as a distributor of the tension force, and the nut are twisted onto the free threaded end of the bolt and tightened, thus creating tension in the bolt and friction between the overlapping pieces. Sometimes, special calibrated wrenches are used to apply a specific amount of torque to the nut, since under-tensioning could cause a weak connection and over-tensioning could damage the bolt. Sometimes, it is impossible to apply a nut. In such a case, one of the pieces to be joined (made of steel, wood, plastic, aluminum, etc.) will be prepared to serve as the female part of the bolt, replacing the nut. At the appropriate place, a hole will be drilled beforehand, tapped with the special tool called a tap. Tapping a prepared hole means cutting the threads into the walls of the pre-drilled hole.

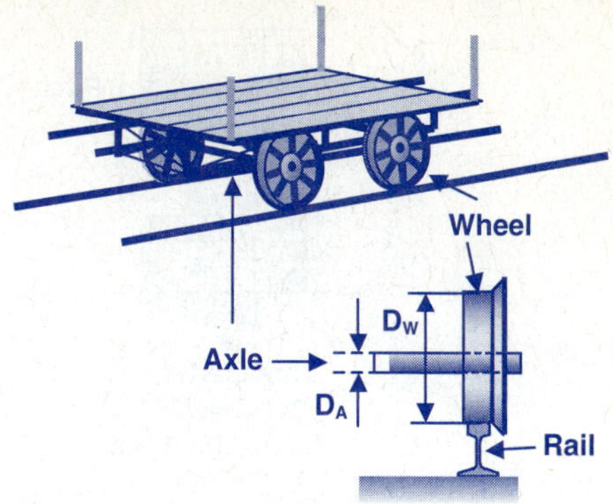

Figure 1.12 Rail Cart on Steel Wheels

Small-gauge rail tracks were built around the construction site to move materials quickly on rail carts. The carts fit into specially built hoists; hence they were able to reach each floor, where turntables and snaking tracks allowed delivery of material without re-handling.

The lead-, translation-, or jack-screw can be found in many places in construction, used for many different purposes and designed to turn rotational motion of the screw into a liner motion of a free-moving nut. Figure 1.14b presents one application where a solid shaft with a specially cut thread stays and turns in place held by two end-blocks with bushings or roller-bearings. When the nut is prevented from rotating, the threads force it to move up or down depending of the direction of the shaft rotation. Instead of a simple bolt, ball assemblies can further reduce friction.

Some simple applications of the leadscrew are vises, c-clamps, and even the adjustable- or pipe-wrench. Those are appropriately called ball-screws. Shown in Figure 1.14b is the Acme thread where the walls are cut in a 29% angle with a flat valley and apex. These screws are designed to have minimal friction between the female and male threads. An example of a lead-screw that everybody has encountered is the jackscrew used to lift cars and other heavy

$T_W = r_W * F$ (Torque resulting from wrench)
Resistant torque $\mathbf{T_R}$ from bolt friction =
$T_r = Q((2 * \pi * \mu * r_s + p)/(2 * \pi * r_s - \mu * p)) * r_s$
Since $T_W = T_R$ we can calculate:
$F = Q((2 * \pi * \mu * r_s + p)/(2 * \pi * r_s - \mu * p)) * (r_s/r_W)$

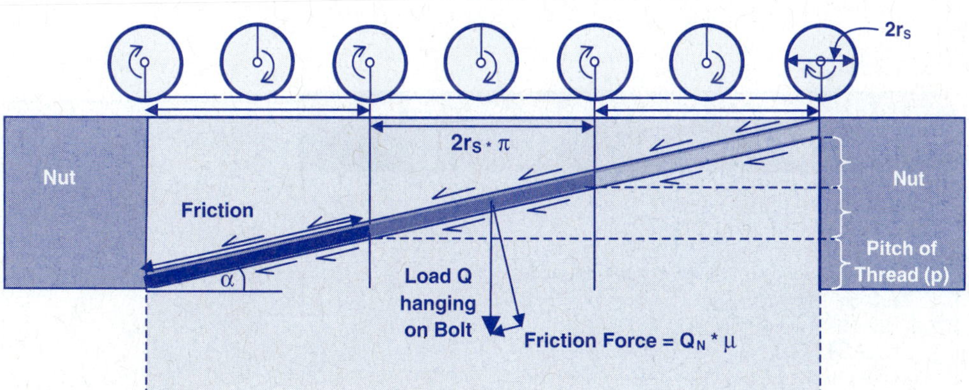

Figure 1.13 Model of Bolt Thread Moving Up an Incline Inside a Nut (not to scale)

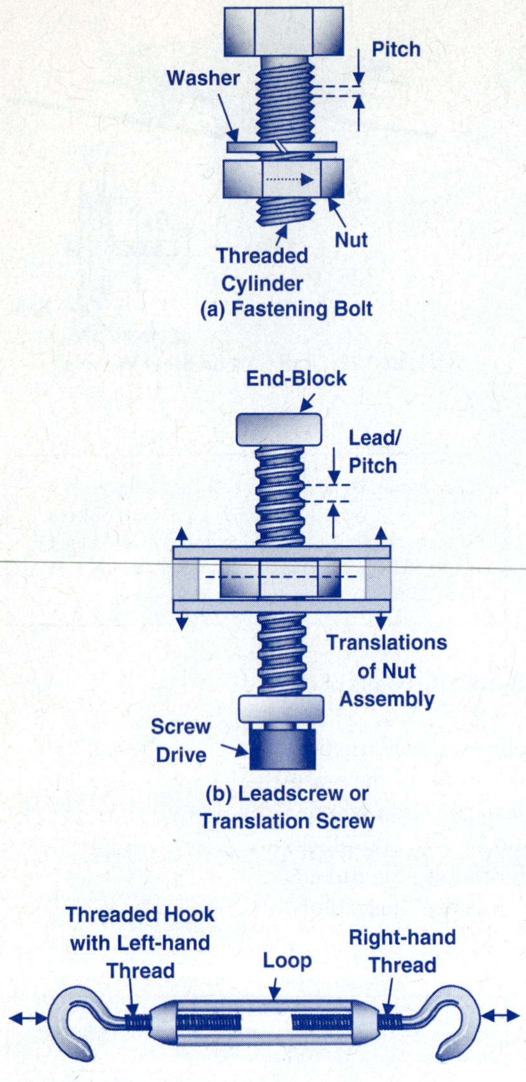

Figure 1.14 Applications of the Screw

weights. Two scissor mechanisms are pinned into a four-joint mechanism, and a lead-screw is used to connect two opposite joints, one of which contains the nut. By turning the screw, the two ends are moving closer to each other thus forcing the other, two joints farther apart.

An important screw mechanism that sped up the relocations of one of the 16 derrick cranes on the Empire State Building was the turnbuckle, depicted in Figure 1.14c. Its main application is in adjusting the length of tightening ropes, cables, wires, or rods and creating large tension forces. The key is the loop that connects two hooks or eyelets threaded in opposite direction. Turning the loop with the help of a lever or wrench and preventing the hook from turning will cause both threaded ends to get closer or separate more. While turnbuckles are used to tighten guy wires of derricks or cross-bracings of steel structure, applications on final structures include the adjustment of suspension cables of suspension bridges.

1.2 The Need for Construction Equipment and Methods

1.2.1 PUBLIC WORKS

Without a doubt, the largest users of construction equipment are public works worldwide. The most impressive example is the U.S. Interstate Highway system, which at present has a total length of approximately 47,000 miles (75,600 kilometers), making it not only the largest highway system in the world, but also the largest infrastructure and the largest public works project in history. As the list below of infrastructure and public works facilities demonstrates, the range of work that requires the mobilization of a wide variety of construction equipment is astonishingly large. Of course, the public expects that the equipment will be managed effectively, providing the best

Infrastructure and Public Works Projects Involving Construction Equipment

Transportation

- Roads highways (bridges, tunnels, intersections, etc.)
- Railways (bridges, tunnels, stations, rail yards, etc.)
- Canals and waterways
- Seaports and lighthouses
- Airports, including air navigational systems
- Mass transit systems (rails, subways, tramways)
- Ferry terminals with maintenance shops

Energy

- Electrical power network with generation plants, dams, electric grid, distribution substations, etc.
- Natural gas pipelines with storage facilities
- Petroleum supply with refineries, pipelines, terminals
- Coal supply with loading and storing facilities
- Steam or hot water production and distribution

Water

- Drinking water supply with structures for collection, treatment, storage, and distribution
- Sewage collection and treatment facilities
- Rain water drainage network
- Large irrigation systems with dams, pumping stations, canals, and reservoirs
- Flood control depending on dikes, levees, pumping stations, and floodgates

Communication

- Postal service necessitating sorting facilities
- Telephone networks with buried cables
- Mobile phone networks needing transmission towers
- Television and radio stations
- Cable television needing buried communication lines
- Internet backbone with buried high-speed cables
- Communication satellites with ground stations
- Undersea cables

Waste Management

- Municipal garbage and recyclables collection
- Solid waste landfills
- Solid waste incinerators
- Materials recovery facilities
- Hazardous waste disposal facilities

Military

- Barracks and training facilities
- Headquarters
- Airfields
- Storage facilities for military equipment
- Port installations

Public Works (Infrastructure + . . .)

- Governmental buildings, courts, and prisons
- Schools, libraries, and universities
- Hospitals, health, and recreational facilities
- Parks, museums, and conservatories
- Bicycle paths, pedestrian walkways, etc.

value possible for the taxpayers' money. This mandate feeds right back to the tenet of this book, which is to understand the principles necessary for deploying construction machinery in the most efficient manner. For now, let us review some of the key aspects of the largest infrastructure project.

On June 29, 1956, President Dwight Eisenhower signed the Federal-Aid Highway Act of 1956. According to Eisenhower, this effort would eliminate unsafe roads, inefficient routes, traffic jams, and all of the other things that got in the way of "speedy, safe transcontinental travel." The initial cost estimate for the 41,000 miles (66,000 km) system was $25 billion and 12 years to build; it ended up taking 35 years and costing $114 billion (adjusted for inflation). Kansas, Missouri, and Pennsylvania all claim that it was the first to start paving after the act was signed. As Figure 1.15a

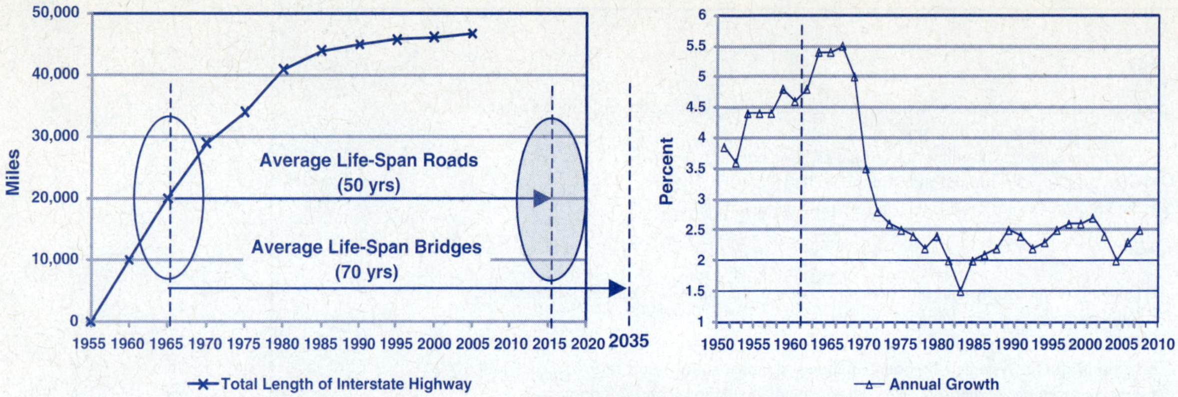

(a) Growth and Life Span of U.S. Interstate Highway **(b) Average Real Growth in U.S. Public Works Investment**

Figure 1.15 History of U.S. Investment in Public Works
Source: U.S. Dept. of Commerce, Bureau of Economic Analysis

depicts, the growth measured by miles of completed highway was spectacular until 1980 when its intended length of 41,000 miles was reached. The Interstate Highway System touches the life of every American. Even groceries find their way into stores via highway trucks. As of 2010, the system has a total length of 47,182 miles (75,932 km).

For information about the US Interstate System, see http://www.fhwa.dot.gov/interstate/homepage.cfm/.

In order to put the sheer size of this enormous project in perspective, here are some comparative values:

- ✓ The amount of concrete used would build 80 Hoover dams.
- ✓ The amount of concrete used would provide six sidewalks to the moon.
- ✓ The Portland cement used would fill three trains of cement hopper cars with their locomotives in New York City and their cabooses in San Francisco.
- ✓ In 2010 the Interstate System had over 600,000 bridges.

Bridges and roads experience different loading and environmental conditions (e.g., deicing salt) impacting their level of yearly deterioration. Of course, a lot also depends on the level of maintenance and repair efforts relative to the wear and tear. Thus, the assumed average life spans for highway bridges of 70 and roads of 50 years is a very rough benchmark. Nevertheless, Figure 1.15a shows that in the next 5 to 20 years, 30,000 miles of highway and possibly 300,000 bridges will need to be replaced or to undergo major overhauls.

Reviewing the past spending for U.S. public works projects presented in Figure 1.15b, we realize that the rapid expansion of the interstate highway system in fact was paralleled by its growth from 4.4% in 1956 to 5.5% in 1964. With its drop to 1.5% in 1984, the number of new highways built also dropped. What can we learn from these observations? One lesson is that massive new investments are needed to keep a highway system in service after its life span is up.

The highways are not the only public works that needs the construction industry. A second major public works that is reaching the end of service is the nation's water supply and sewer systems, major portions of which were built in the early twentieth century. The American Society of Civil Engineers (ASCE) believes that every day the United States loses 7 billion gallons of clean drinking water due to leaks in water pipes. Leaking sewer pipes are a major health hazard, as untreated sewage can find its way into sources of drinking water. It is expected that building and replacing water and sewage lines will cost

$660 billion to $1.1 trillion over the next 20 years. As Figure 1.15a shows, this time frame coincides with the end of life of major sections of the Interstate highway.

Figure 1.16 depicts the real world of replacing old pipelines with a new network. As shown, the contractor had to dig very careful around a series of utilities (e.g., gas, water, electrical, drainage, communication). Unfortunately, the location of the older lines was never documented, which increases the risk that one of them may be accidentally damaged during excavation. What does this tell us? For one thing, renewing the underground infrastructure will not be as straightforward as it was a century ago, when most of it was first put in place.

Figure 1.16 Installing New Pipe Underneath Existing Utilities

On a global level, the Organization of Economic Co-operation and Development (OECD) wrote in its publication *Infrastructure to 2030* (2007) that worldwide, the annual cost for communication, transportation, electricity, and water will likely average $ 1.5 trillion in today's money.

The ASCE has taken on the challenge to grade most of the major public works in the United States every two or three years. The summary is then published in a report called "Report Card for America's Infrastructure." It not only grades each system, but also estimates the cost necessary to bring it to an acceptable grade such as an A or a B. Table 1.1 presents summaries from the 2009 report for some of the infrastructure systems that were covered.

Overall, the grade point average of America's infrastructure was a D, with $2.2 trillion of investment need for the next 5 years, which works out to $440 billion/year.

Table 1.1 Excerpts from ASCE's Report Card 2009 (*www.asce.org/reportcard*)

Examples	Key Assertions by ASCE's 2009 Report Card	Projected Investment Needs
Drinking Water *Grade* D^-	Large sections of America's drinking water system have passed the end of their service lives. They need to be replaced or upgraded to comply with existing and future federal water regulations. Leaking pipes lose an estimated 7 billion gallons of clean drinking water a day.	At least $11 billion annually is needed to replace and upgrade existing lines. Further money is required to satisfy the growing demand for water.
Wastewater *Grade* D^-	According to the Environmental Protection Agency (EPA), sanitary sewer overflows, caused by blocked or broken pipes, result in the release of as much as 10 billion gallons of raw sewage yearly.	$390 billion over the next 20 years to update or replace existing systems and build new ones to meet demand.
Bridges *Grade C*	According to the U.S. Department of Transportation, of the 600,905 bridges across the country as of December 2008, 72,868 (12.1%) were categorized as structurally deficient, and 89,024 (14.8%) were categorized as functionally obsolete.	$17 billion to substantially improve current bridge conditions. Currently, only $10.5 billion is spent annually on the construction and maintenance of bridges.
Rail *Grade* C^-	A significant potential increase in passenger rail demand will add to the freight railroad capacity challenges.	More than $200 billion through 2035 to pay for the anticipated growth.
Roads *Grade* D^-	Americans spend 4.2 billion hours a year stuck in traffic at a cost to the economy of $78.2 billion, or $710 per motorist. Poor road conditions cost motorists $67 billion a year in repairs and extra operating costs.	$186 billion annually to substantially improve conditions. The current $70.3 billion per year is well below estimated needs.
Schools *Grade D*	Spending on the nation's schools grew from $17 billion in 1998 to a peak of $29 billion in 2004. However, by 2007 spending fell to $20.28 billion.	The National Education Association estimates that the nation's schools need $322 billion just for repairs.
Energy *Grade* D^+	Demand for electricity has grown by 25% since 1990. Public and government opposition and difficulty in the permitting processes have restricted modernization.	$1.5 trillion by 2030.
Levees *Grade* D^-	The United States has 100,000 miles of levees. Many are more than 50 years old and were originally built to protect crops from flooding. With an increase in development behind these levees, the risk to public health and safety from failure has increased.	$100 billion to repair and rehabilitate the nation's levees.

1.2.2 EXTREME CONSTRUCTION—MEGAPROJECTS

Tall Buildings. The rising cost of energy and pollution as well as the global aspirations of up-and-coming countries will lead the construction industry further to push the range of the doable. New materials and more sophisticated equipment will lead the way to what we now may call extreme construction. The Empire State Building, at 1,454 ft (443.2 m), was the tallest building of the world until 1972, when the World Trade Center was completed. Today, a new record holder may keep the title for only two or three years. Today, the record belongs to the Burj in Dubai, which is 2,300 ft (708 m) high. The next aspirant might be a 3,284 ft (1,010 m) high tower in Kuwait.

Multifunctional Underground Space. Amsterdam is well known for its canals that were built in the seventeenth century. The thick layers of clay and sand create a difficult foundation to build on. Recent ideas are to excavate the areas below the canals and replace the weight of the watertight clay with a large underground building with a canal on top. Figure 1.17 shows the concept.

The construction of such a large space below water and without space for a large pit to dig will be a great challenge. Applying an open-cut method, sections of the canal would have to be drained and deep vertical sidewalls would have to be dug first. Those walls would keep the clay and sand from falling into the construction pit during excavation. Large amount of soil material would have to be removed and transported through the middle of the city to a fill area. After the top floors are done the canal floor can be sealed and the water can flow back in.

Tunnels. The new Swiss tunnel 7,546 ft (2.3 km) below the highest mountains of the Alps will become the longest tunnel in the world, 35 miles (56 km) long. It will reduce the train travel time between Zurich and Milan by one hour. More importantly, freight trains will pick up highway trucks on either side and transport them at 155 mph (250 km/hr) through the base of the mountains. One of the main goals of the project is to drastically reduce the exhaust fumes created by the heavy trucks traveling between the south and the north of Europe. Thus, the air quality in major parts of Switzerland will be improved and trees that have suffered from the toxic fumes will regain their strength. Of course, rail transport through the base of the Alps will also drastically increase the energy efficiency of moving cargo overall.

Over the last 25 years, tunnel construction has involved many innovations in equipment, methods, and materials. Important examples are various grouts needed to prevent the water pressure inside the mountain from breaking into the tunnel. Thus, underground work will certainly produce new record-breaking feats in the future.

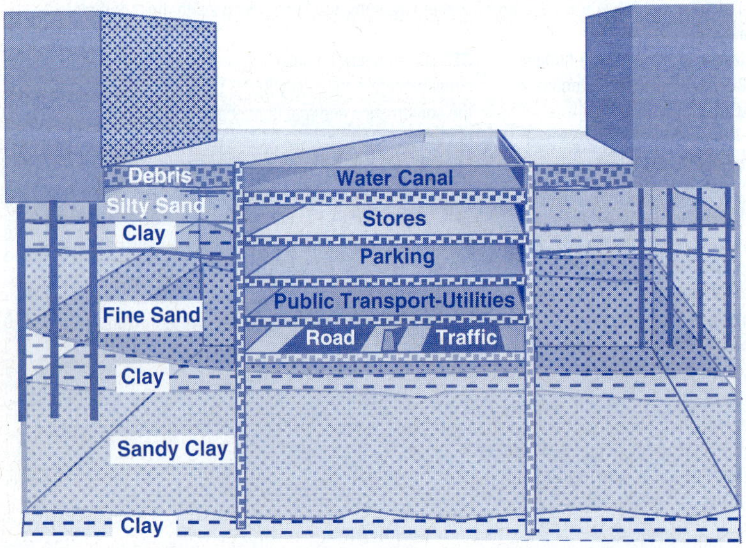

Figure 1.17 Concept of a Multipurpose Underground Space for Amsterdam

Figure 1.18 The Submerged Floating Tunnel Concept

On the other hand, tunnels do not always run through mountains. Figure 1.18 presents the basic concept of the submerged floating or suspended tunnel anchored to the bottom. Rather than tunneling below the water or placing a tube on the bottom of the water, a tunnel is created that has enough buoyancy to make it float to the surface. In order to keep it safely below the largest ships that may pass over it, the tunnel is held in place via cables attached to large weights placed beneath. This approach allows the water pressure on the tunnel to be kept at a minimum level, just deep enough to avoid a collision with a ship, even in areas with a high water depth.

Bridges. These structures, linking people and countries, will also continue to push their limits. The record, with an open-span length of 6,464 ft (1,991 m), is still the Japanese suspension bridge Akashi-Kaikyo, finished in 1998. While this is an amazing record, it will not hold. A new competitor is the cable-stayed bridge, with several examples passing the 3,280 ft (1 km) mark. In 2009 the $1.6 billion Incheon bridge was opened to traffic (Figure 1.19). While the main span of the cable-stayed bridge is "only" 2,623 ft (800 m), the two access bridges combine to a total length of 7.65 miles (12.3 km).

While the main span was built in steel, the access sections were constructed of concrete that was cast on shore and heaved into place by gigantic barge-cranes or cantilevered derrick cranes.

Figure 1.19 Incheon Cable-Stayed Bridge with 74 m (242 ft) Clearance

1.3 Chapter One Review

JOURNALING QUESTIONS

(1) Present your thoughts about how machines, ingenuity, and capital are linked when it comes to construction technology. Do you know of a good case that illustrates this linkage? Are you aware of a situation where the link was missing? Explain.

(2) In the European Middle Ages, castles and cathedrals were the greatest construction projects. Find a particularly interesting example, discuss how simple machines were (most probably) used, and sketch how they were applied.

(3) The switch from human and animal power to steam and combustion engines had drastic effects on society and the way the economy operated. Discuss some of the most important early impacts both negative and positive (e.g., in eighteenth-century England).

OPEN-ENDED QUESTION

(1) As you look around on construction sites you realize that lifting and moving material, elements, assemblies, and so forth is a central theme. Looking back in history, you realize that the lifting technologies available at a particular time period dictated the amount of human power needed to build large structures. Reflect on the effect that simple machines had on construction methods over time (from Egyptians until today). Predict future developments in technologies related to lifting and moving objects in construction.

BIBLIOGRAPHY

Ball, N. "Circular Saws and the History of Technology." *Bull. Assoc. Preservation Tech.* 7, no. 3 (1975).

Ball, N. R. "Observations on a Contemporary Account of Naval Block-Making Machinery." *Bull. Assoc. Preservation Tech.* 9, no. 1 (1977).

Barley, S. R. "What Can We Learn from the History of Technology?" *J. Eng. Technol. Mgmt.* 15 (1998).

Cavanagh, T. "Balloon Houses: The Original Aspects of Conventional Wood-Frame Construction Re-Examined," *J. Arch. Edu.* 51, no. 1 (1997).

Centre for the Study of Living Standards, "Productivity Trends in the Construction Sector in Canada: A Case of Lagging Technical Progress." *Rep. Centre Study of Living Stand.* Ottawa, Ontario, 2001.

Coulton, J. J. "Lifting in Early Greek Architecture." *J. Hellenic Stud.* 94 (1974).

Curtis, J. O. "The Introduction of the Circular Saw in the Early 19th Century." *Bull. Assoc. Preservation Tech.* 5, no. 2 (1973).

Dobres, M. A. "Gender and Prehistoric Technology: On the Social Agency of Technical Strategies." *World Archaeology*, 27, no. 1 (1995).

Gomory, R. "Technology Development, *Science.* n.s., 220, no. 4597 (1983).

Gottmann, J. "Why the Skyscraper?" *Geographical Review* 56, no. 2 (1966).

Gutman, M. "Building the Empire State, A Review." *J. Soc. Arch. Historians* 59, no. 2 (2000).

Heizer, R. F. "Ancient Heavy Transport, Methods, and Achievements." *Science*, n.s. 153, no. 3738 (1966).

Inkster, I. "Technology in World History Cultures of Constraint and Innovation, Emulation, and Technology Transfers." *Comp. Tech. Transfer Soc.* 5, no. 2 (2007).

Jackson, D. C. "The Origins of Engineering." *Science*, n.s. 78, no. 2035 (1933).

Kleinschmidt, H. E. "Evolution of the Wheel." *Scientific Monthly* 59, no. 4 (1944).

Lee, D. "Science, Philosophy, and Technology in the Greco-Roman World II." *Greece & Rome*, 2nd ser., 20, no. 2 (1973).

Li, Y., Y. W. Mai, L. Ye. "Sisal Fibre and Its Composites: A Review of Recent Developments." *Composites Sci. and Tech.* 60 (2000).

McNabb, D. E., and F. T. Sepic. "Culture, Climate, and Total Quality Management: Measuring Readiness for Change." *Pub. Prod. & Mgmt. Rev.* 18, no. 4 (1995).

Organization of Economic Co-operation and Development (OECD). *Infrastructure to 2030.* Paris: OECD Publications, 2007. http://www.oecd.org/dataoecd/61/27/40953164.pdf/.

Peters, T. F. "Technological Thought Is Design's Operative Method." *Perspecta* 31 (2000).

Peterson, C. E. "The Technology of Early American Building (TEAB)." *Newsletter Assoc. Preserv. Tech.* 1, no. 1 (1969).

Peterson, C. E. "Inventing the I-Beam: Richard Turner, Cooper & Hewitt and Others." *Bull. Assoc. Preservation Tech.* 12, no. 4 (1980).

Prager F. D. "Brunelleschi's Inventions and the Renewal of Roman Masonry Work." *Osiris* 9 (1950).

Shaw, J. W. "A Double-Sheaved Pulley Block from Kenchreai." *Hesperia* 36, no. 4 (1967).

Stephenson, C. "In Praise of Medieval Tinkers." *J. Econ. Hist.* 8, no. 1 (1948).

Tuan, Y. F. "Cultural Pluralism and Technology." *Geographical Rev.* 79, no. 3 (1989).

U.S. Bureau of Labor Statistics, Consumer Price Indexes for 16 Countries, 1950–2008, U.S. Dept. of Labor, Div. Intern. Labor Comp., 2009.

Wilson, A. "Machines, Power and the Ancient Economy." *J. Rom. Stud.* 92 (2002).

chapter two

Costing the Operation of Equipment

Inflation is when you pay fifteen dollars for the ten-dollar haircut you used to get for five dollars when you had hair.

—Sam Ewing

A bank is a place that will lend you money if you can prove that you don't need it.

—Bob Hope

A cost is commonly associated with a sacrifice of money, time, or other valuables that are accepted as payment. In the case of purchasing a new piece of equipment, its cost to the purchaser is dictated by the sales price that the seller has agreed to. During operations, if a newly purchased piece of equipment leads to further expenses or expenditures, these are referred to as operating costs. The process does not end there. More cost is accumulated when the machine is maintained in the equipment yard and broken parts have to be replaced. The goal of this chapter is to find an answer to following question: How much does the contractor have to charge the customer for the use of a piece of equipment? Only an accurate response will guarantee the survival of the company, as all the ensuing costs have to be paid while making a profit and providing a cover for bad days.

Table 2.1 Main Topics Covered in This Chapter

Cost Planning & Control		Financial Planning		
Cost Factors	**Methods of Analysis**	**Methods & Tools**	**Assets and Equity**	**Key External Factors**
✓ Credit Account	✓ Cost Tradeoff	✓ Cash-Flow Analysis	✓ Cash	✓ Inflation
✓ Economic Life	✓ Time Tradeoff	✓ Accounting	✓ Fixed Capital	✓ Investment Tax Credits (ITC)
✓ Ownership Cost	✓ Crash Duration	✓ Amortization	✓ Bank Credit	✓ Internal Revenue Service (IRS)
✓ Direct/Indirect Cost	✓ Sampling Method	✓ Annuity (Uniform)	✓ Account	✓ Interest Rate
✓ Equipment Tax	✓ Inspection Scheme	✓ Present Value (PV)		✓ Tax Rate
✓ Functional Obsolescence	✓ Record Keeping	✓ Minimal-Rate-of-Return (MRR)		
✓ Operating Cost	✓ Straight-Line Depreciation	✓ Cumulative Cost Index		
✓ Filters, Oil, and Grease (FOG)	✓ RS Means Rental Rates	✓ Time-Value-of-Money		
✓ Preventive Maintenance	✓ Blue Book Cost Data	✓ Tax Depreciation		
✓ Overhead	✓ Sensitivity Analysis	✓ Declining Balanced Depreciation		
✓ Replacement-Salvage Value	✓ Value Depreciation	✓ Buying-Owning		
✓ Operating Factors		✓ Leasing		
✓ Reliability-Centered Maintenance		✓ Renting		
✓ Tire Cost				
✓ Hourly Operating Cost				
✓ Condition-Based Maintenance				
✓ Maintenance & Repair Cost				
✓ Operator Skill				
✓ Job Efficiency				

2.1 The Economics of Deploying Machinery

We learned in Chapter 1 that the use of appropriate machines boosts the productivity of construction. Throughout history, limited human or animal muscle power has gradually been replaced with engines driving large mechanical systems. Looking around today's construction sites in developed countries, one finds many examples where smart and innovative systems offer totally new capabilities to do things accurately, in less time, with fewer people, and with fewer safety risks. As depicted in Figure 2.1a, the situation in developing countries does not favor the use of highly productive equipment such as a crane to move mortar to the masons working on scaffolds.

Paying the Bills Through History

Barter: Direct exchange of goods for goods between individuals, groups, or even countries after an acceptable rate has been bargained. Used when no trusted currency is available (e.g., the communist countries during the Cold War).

Money: Valuable objects that are used for making payments. The word derives from the Roman temple to the goddess Juno Moneta where for 400 years silver coins were minted. Examples of primitive money include beads, hoes, ivory, oxen, rice, salt, silk, nails, and vodka.

Dollar: The Taler coin was first minted during the 16 century in Joachimstal in Bohemia. The word *dollar* is used in Section 9 of Article 1 of the U.S. Constitution and referred to a Spanish silver coin, later called the peso, worth 8 Spanish reales that was equivalent to the German Taler.

Cash: Form of a currency that can be easily stored (in a *cassa* or box), transported and used for payment. Includes banknotes and coins.

Banknote: (bill, paper money) Contract where the issuer (e.g., bank) makes an unconditional promise that it will pay a certain amount to the bearer on demand.

Continentals: Paper currency in dollars issued in 1775 by the Continental Congress lacking coinage to pay for the army. By the end of the war the *Continentals* had fallen to one-thousandth of their nominal value.

The principles that lead to different optimal solution are tradeoffs between various ways to accomplish the same task, such as mixing and supplying mortar to brick layers. For the same kind of work, there are different factor costs that the contractor has to pay for the necessary resources.

Figure 2.2 illustrates the two options for mixing/transporting mortar and sand that were delivered by truck to the construction site. The graphs along the time vs. cost axis identify two kinds of tradeoffs, economic and time. These tradeoffs, marked with diamonds and crosses respectively, represent situations where the two methods, human-powered and mechanized, are considered equal for one of the two factors time or cost. To be more specific, the point labeled ❶ represents the total cost and the time needed to mix and deliver a truckload of mortar sand, consisting of the cost for a crane with operator, one mixer, and a laborer finishing in time T_1. Naturally, compared to a crane a human is very limited in the load he can carry and how fast he can walk up steep steps, and so forth. This fact is highlighted by the position T_2 indicating the time it takes for two laborers to do the same work. When comparing points ❶ and ❷ one realizes the effect of replacing human labor with machines in shorter time to finish and higher cost.

Of course, adding more laborers will reduce the time needed, while adding a second crane will have little impact on the time. The result of increasing the crew size and the effect of the labor wage is shown on the left-hand graph, crew cost. The slope α represents higher cost per laborer from low to high. Let's assume that the maximum number of laborers we could hire is 16, each working at the same speed as the others. First we

(a) Human-Powered Hauling and Mixing Mortar

(b) Mechanized Approach (Mixer + Crane)

Figure 2.1 Two Solutions to Producing Mortar

recognize that the 16-person crew is able to finish much quicker in time T_3, but still a little bit longer than with crane and mixer, as indicated by ❸ for less cost. While the cost of the crew increases linearly, based on the base labor rate, the reduction in time follows a curve that is getting closer to a so-called crash-duration, a duration that cannot be lowered by adding more people. In fact, employing too many laborers can result in a situation where they get locked in place like heavy traffic when there is no red light. The smaller marginal time reductions mirror the fact that the laborers are being continuously slowed down by

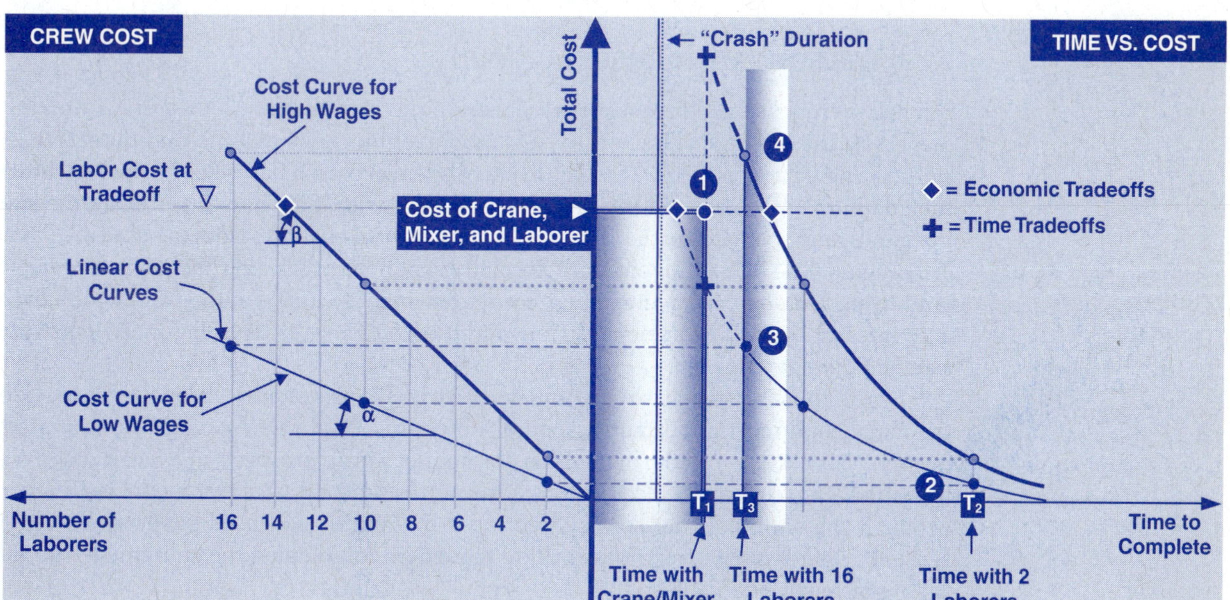

Figure 2.2 Comparison of Mortar Supply Methods Based on Time and Cost

queues when loading, unloading or at narrow spots on the walkway. Small delays caused by inattentiveness by one will create a ripple effect for all the others in line. The line for time tradeoffs between the human-powered method and the crane operation lies to the left of time T_4, pointing out that the human will never reach its speed.

Finally, the effect of raising labor costs on the economic desirability of buying or renting equipment is self-evident. As the slope of the cost-curve changes from α to β, the total crew cost surpasses the cost for the crane-mixer option, eventually reaching point ❹. At this high labor wage, the crew of 16 is not only slower but, in addition, more expensive. The economic tradeoff is reached with the slope β, since the cost of 16 laborers matches the cost for the crane option. Of course, substituting machines for people cannot be based only on this simple economic analysis. For example, when the old Aswan dam in Egypt was constructed, the contractor had opted to import a large rock crusher powered by a diesel engine. As the first equipment parts started to break, the entire operation came to a standstill as the unique replacement parts were shipped from Europe and up the Nile. Eventually, the crusher was abandoned and replaced by thousands of Bedouin families who broke the rocks with hammers. The ripple effect of the local people being able to earn money was the establishment of a vibrant community with farmers, schools, and medical care.

2.2 Cash, Credit, and Capital

The saying "you need money to make money" is certainly true when it comes to construction equipment, which is commonly built by equipment fabricators and not by contractors. Even renting machinery requires capital, as the rent and the other operational expenses have to be taken care off before the contractor is paid by the owner. Failure to fully account for all the costs that ensue when owning and renting equipment will lead to low bids, caused by underestimating the cost of machinery, and to unsustainable losses. Let's try to understand the basic concept of equipment costing, which, like classical machines, is fundamental in managing construction equipment.

2.2.1 THE EQUIPMENT CREDIT ACCOUNT

Asphalt pavers are good examples to illustrate the money flow as they are commonly owned by the paving contractor, say QualityPave, Inc. Let's assume that the purchase price of a new paver is $78,000 and the average overall cost per day is $680, which includes fuel, daily maintenance and repairs, transportation with a lowboy to and from the site, insurance, and so forth. Furthermore, the average job takes 7 days from the time the paver is delivered. The bills or invoices are normally sent out 1 week after the job is completed (and inspected), and the average time for the payment to arrive in the contractor's bank account, sent by the Department of Transportation (DOT), is 45 days. Figure 2.3 maps out a possible money flow.

Before any large purchase QualityPave has to make some key decisions about the size, the make, and, most importantly, should it buy a used or a new paver. One important factor in this judgment is the owner's assessment of future prospects in asphalt paving. If the outlook is positive, the acquisition of the newest, most productive model is advantageous. If the outlook is not so rosy, renting or buying a good used paver makes more sense. Further strategic decisions include the length for keeping the equipment and the timing of major overhauls.

As Figure 2.3 shows, QualityPave made a decision at $T_0 = 0$ to purchase a paver for $78,000, expecting to keep it for 6 years, at which time it would be sold for about $12,000.

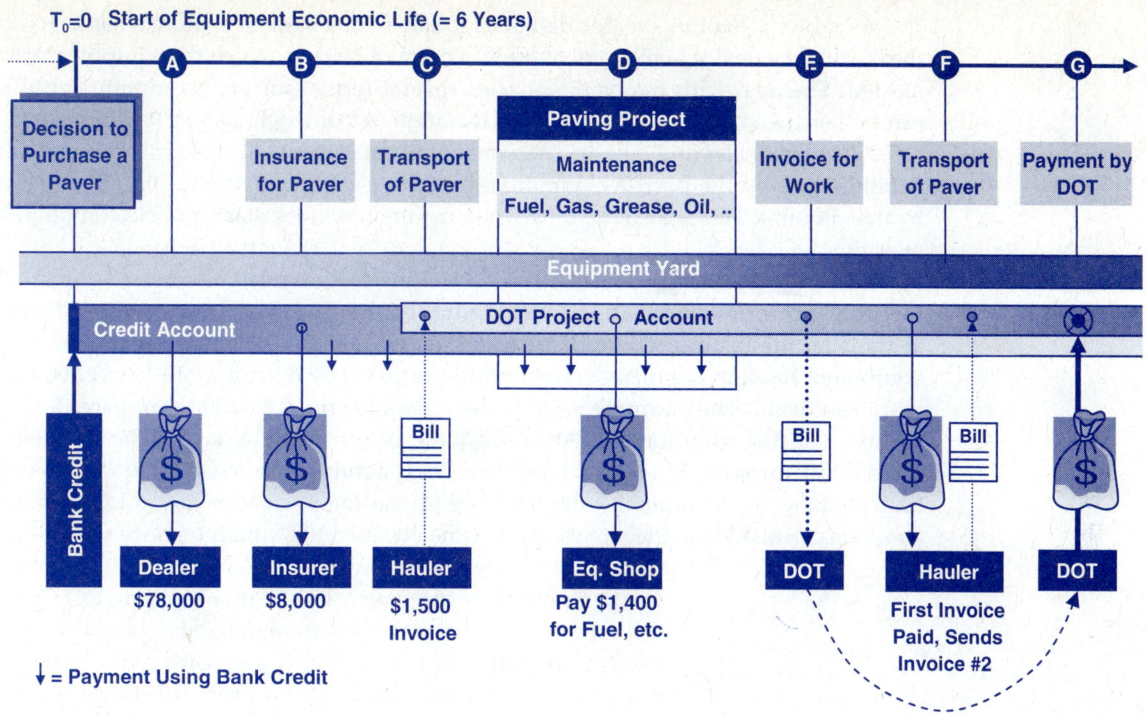

Figure 2.3 The Flow of Cash In and Out of a Bank Account

Naturally, QualityPave needs additional funds to pay the dealer, the insurance company, and so forth. The question now is, Where does the money come from? Does the company have sufficient savings from previous profits to finance the paver for 6 years, turning liquid funds into fixed capital? Is it even wise to tie down its own liquidity instead of borrowing the needed money from a bank or other lending institution?

Fixed Capital: These are longer-term (fixed) investments of a company into equipment, new technology, buildings, shops, or computers that are necessary to operate.

Working Capital: These are the funds used by a company to operate the business on a daily basis, including inventories or work-in-progress (WIP) products. It also includes the liquid assets needed to pay the bills for material, rentals, fuel, salaries, etc.

Assets: List of valuables that a company owns for the purpose of executing its business.

Liabilities: A company's financial obligations to other parties (companies, government, persons) used to fund part of the assets.

Bank Credit: A maximum amount of funds made available by a lending institution to allow a borrower to finance a specific undertaking. Borrower has to provide collateral, which will go over to the lender in case the credit cannot be paid back at the end of its term.

Account: A virtual piggy bank at a commercial institution or held within the company that records all financial transactions/cash-flow between its "owner" and other entities (e.g., DOT Project Blacktop).

Interest: A fee paid by the borrower to the lender for the use of loaned money.
Simple interest is calculated only on the principal (amount) that has been borrowed. The common time period for interest is one year.

Compounded interest is used when a borrower does not pay the interest until the entire loan is due and has to be paid back (e.g., after 4 years). He pays interest on the interest, or compound interest.

As shown, QualityPave decided to negotiate with a bank to acquire a bank credit for the life of the paver. As collateral, it offers a piece of land on which the equipment yard is located. The deal with the bank specifies several terms, such as maximum funds that can be borrowed, the interest rate, and duration. At time **A**, QualityPave transfers the $78,000 to the dealer and takes possession of the equipment. At this time, the company's liabilities rise by the borrowed amount and its assets (and fixed capital) also increase by that amount. This is also the time when the interest clock starts to tick, compounding over time.

Events **B**, **C**, and **F** illustrate why the ceiling of the bank credit must be larger than the purchase price ($78,000). For one thing, before the paver can even be deployed at the site, the insurance company wants the $8,000 in its bank account. The heavy equipment hauling company transporting the paver will send a bill for $1,500 before work can begin. The equipment shop also needs to spend $1,400 to prepare the paver (grease, oil, fuel, etc.) for use. At time **D**, the paver is finally laying the new asphalt for a DOT project, burning diesel fuel and natural gas to heat up the asphalt, delivered by the equipment shop, before spreading it on the ground. Every other day, a machinist visits the site to spend time on preventive maintenance. As shown in Figure 2.3, the money to pay for these costs does not come from the bank credit of the paver, but from a separate DOT project account that has been opened to manage its cash flow. Such a project account is commonly also a credit account at a bank established solely for financing the project in times when the expenses exceed the income. This certainly is the case between **D**, when the work is being done, and **G**, when DOT transfers the money for the work. Of course, QualityPave can only send the invoice after the work is completed and signed off by DOT's project manager at time **E**. DOT's payment feeds into both the equipment credit and the project account. Of course, the amount received from DOT covers only a small portion of the purchasing price of the paver, since it will create revenue streams for the 6 years of its economic life. The amount dedicated to the project account is needed to pay all the other expenses, such as diesel fuel, natural gas, transportation, and so on.

Inflation: This is a measure that indicates how much the price for a selected basket of goods has increased as a percent. Because of inflation, the same amount of money (e.g., $100) will allow one to buy fewer goods.

Amortization: An amortized loan requires the borrower to pay back portions of the loan in regular intervals and interest on the remaining amount. Thus, each payment consists of interest and payments on the original loan.

Supplemental Material S2.1 Consumer Price Index

2.2.2 BANK'S PAYBACK SCHEDULE

It is common that bank credits have to be amortized with an interest rate that reflects the risk of the loan, the cost of managing the credit account, and the expected rate of inflation. Figure 2.4 offers a graphical representation of a possible repayment schedule for the loan with a yearly interest of 9%.

Figure 2.4 tells us that the yearly repayment is a constant 1/6 of the $78,000 (= $13,000) starting at the end of the first year. The bank charges 9% interest for the amount that is still outstanding. Thus, the interest payments will change every year, starting with 9% of $78,000 = $7,020 at the end of the first year, and 9% of $65,000 = $5,850 at the end of the second.

It is apparent that a bank loaning different amounts to hundreds of companies will end up with an erratic money inflow, changing monthly and requiring it to constantly recalculate the amount of new credits it will be able to issue. To eliminate those variations, banks use annuities, which are fixed amounts of payments payable at regular intervals during

Annuity: Indicates a series of equal payments or receipts that occur at evenly spaced intervals. Examples are monthly car loan or home mortgage payments.

Time Value of Money: Inflation will reduce the value of money over time since today's $100 will buy fewer goods in a year. Considering the time value of money in an investment analysis means to account for its increase or decrease over time (e.g., foreign exchange).

Present Value (PV): The PV represents the present value of future income and expenses considering the time value of money. A discount rate is selected, functioning like an interest rate. Thus future money streams are discounted.

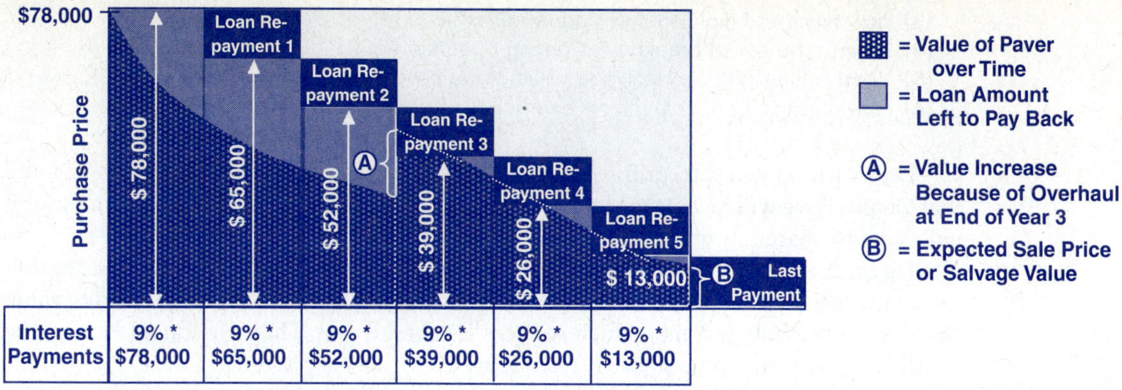

Figure 2.4 Simplified Payment Schedule for the Bank Credit

the entire duration of the loan. The function that can be used to calculate the uniform annuity looks as follows:

Supplemental Material S2.1 Consumer Price Index 1950–2009

$$\mathbf{A} = \mathbf{P} * \left(\frac{\mathrm{i}}{1 - \frac{1}{(1+\mathrm{i})^{\mathrm{n}}}} \right)$$

where

$\mathbf{i} =$ the interest rate per period
$\mathbf{n} =$ the number of periods
$\mathbf{P} =$ the principal loan amount (or present value)
$\mathbf{A} =$ the periodic payment in an annuity (the amortized payment)

For QualityPave this means that the bank will require the following fixed amount per year:

$$\mathbf{A} = \$78{,}000 * \left(\frac{0.09}{1 - \frac{1}{(1+0.09)^{6}}} \right) = \$78{,}000 * 0.2229 = \$17{,}386/\text{year}$$

Over the 6-year life of the loan QualityPave would thus pay interest in the amount of = (6 ∗ $17,386 – $78,000) = $26,316 or an average of $4,386/year (= 5.6% of initial loan).

As a result, the cost for borrowing the capital is $26,316, not including eventual fees that a bank may charge for initiating the loan.

2.2.3 RECOVERING BORROWED MONEY

As the bank loan needs to be repaid month by month as an annuity, QualityPave needs to recover its investment into the paver by charging for its use. We learned from Figure 2.3 that at time **G** a portion of DOT's payment for the work performed flows into the credit account that had been opened for the purchase of the paver. Commonly, such charge rates are based on time, whether hourly or daily, while the money to be paid back to the bank is fixed over time. Setting a proper rate, for example $x/hour, is extremely tricky as this has be done before one knows how well the paver will perform. Some of the questions that the QualityPave has to answer or "guess" are:

(1) How many projects per month will we have?
(2) What will the production rate of the paver be (e.g., ft^2 (m^2)/hour)?

(3) How much will maintenance and repair cost?
(4) What is the risk of breakdowns during operation?
(5) What is the residual or salvage value of the paver after 6 years?
(6) How much will our competitors charge?

Let's assume that the equipment manager expects that the weekly average of hours that QualityPave will be able to charge for the use of the paver is 22. What is the amount it will have to charge in order to recover the cost of borrowing money for 9%?

Figure 2.5 presents the cash flow of the situation at hand showing that besides that expenditure to pay the purchase price we will have to consider the expected salvage value since this represents income, which reduces its overall cost. Thus, the calculation of the yearly annuities has to be done in two steps.

Step 1: The annuity of the purchase can be computed, using the same uniform formula as above. Thus we get:

$$A_1 = \$78{,}000 * \left(\frac{0.09}{1 - \frac{1}{(1+0.09)^6}} \right) = \$78{,}000 * 0.2229 = \$17{,}386/\text{year}$$

Step 2: The annuity of the income from selling the paver in 6 years can be computed in two steps. First we calculate its present value, which we subsequently spread out over the 6 years.

Present Value (**PV**) of \$12,000 considering 9% interest over 6 years:

$$\mathbf{PV} = \mathbf{F} * \left(\frac{\mathbf{1}}{(\mathbf{1}+\mathbf{i})^{\mathbf{n}}} \right)$$

S2.2 Compounding
S2.3 Annuity Calculator

where

i = the interest rate per period
n = the number of periods
PV = present value
F = future payment

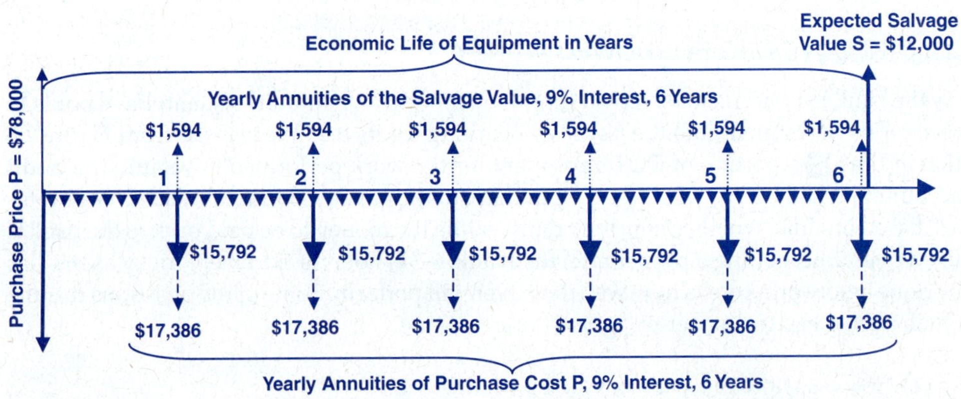

Figure 2.5 Monthly and Yearly Cash Flow Necessary to Recover the Capital Investment

$$\mathbf{PV} = \$12{,}000 * \left(\frac{1}{(1+0.09)^6}\right) = \$12{,}000 * 0.596 = \$7{,}152$$

$$\mathbf{A_2} = \$7{,}152 * \left(\frac{0.09}{1 - \dfrac{1}{(1+0.09)^6}}\right) = \$7{,}152 * 0.2229 = \$1{,}594/\text{year}$$

Total annuity $\mathbf{A_T} = \mathbf{A_1} - \mathbf{A_2} = \$17{,}386/\text{year} - \$1{,}594/\text{year} = \underline{\$15{,}792/\text{year}}$

Average use of paver = 22 hours/week * 52 weeks/year = 1,144 hours/year

Hourly charge for the equipment to recover the investment = \$15,792/1,114 hrs = $\underline{\$13.80\text{ hr}}$

We now recognize that the \$17,386/year required by the pay back per year is larger than the \$15,792/year that QualityPave is charging for the use of the paver. How can that be? The answer is given by the sale at the end. QualityPave will receive \$12,000 from selling the paver after 6 years with a PV of \$7,152. However, the bank interest charges do not consider the sale since it does not want to take the risk of misjudging the actual salvage value after 6 years. Hence, similar to the bank annuity we can compute the total amount of interest:

Cumulative interest to be charged for its use = (6 * \$15,792 – \$78,000) = \$16,752

Earlier we calculated the total interest over 6 years that the bank charges QualityPave as \$26,316. It can't come as a big surprise when the difference (\$26,316 – \$16,752) = \$9,564/6 = \$1,594 equal the annuities of the expected salvage value at the end of 6 years.

For the company, the charges for borrowing money from the bank are its financial cost that it has to pay to get into the paving business with a piece of equipment. However, in order to calculate the total costs that need to be billed to a customer when that piece of equipment is being employed on a job, the planner has to consider a long list of costly factors. Let's start with a clean sheet of paper.

2.3 Costing Construction Equipment

The economic success of a company that decides to own equipment depends on the knowledge about its total cost. We just learned that one cost item is the interest that has to be paid for borrowed money. In addition, the loan itself has to be paid back month by month over the economic life of the equipment, requiring the contractor to charge the customer for its use, thus creating sufficient cash in-flow. Another incentive to know the equipment cost is to establish competitive charge rates in order to win a sufficient number of bids to keep the equipment working. It is easy to understand that rates that are too high will lead to fewer projects, which will in turn result in higher charge rates as the machine is used less.

The total costs to operate construction equipment are arranged in two categories, ownership and operating cost.

(a) Ownership Costs: The contractor incurs these costs even without ever using the equipment on a project. In other words, ownership costs have to be borne even if the equipment stays in the yard. They include

(1) Value depreciation
(2) Interest on capital

(3) Storage facility
(4) Insurance (accident, theft, fire, flood, or other natural disasters)
(5) Machine tax
(6) Functional obsolescence

(b) **Operating Costs:** These costs are a direct result of the machine being used or operated. They include

(1) Fuel
(2) Filters, oil, and grease (FOG)
(3) Preventive maintenance
(4) Repairs
(5) Tires (for wheeled equipment)

(c) **Equipment Overhead Cost:** The costs for owning and operating do not consider the effort necessary to get the equipment used or rented. Advertisements for rent and administrative work for processing invoices or paying the bills need to be paid for. How about the building and the computers needed to do the accounting of all the cost data needed for the preventive maintenance?

The following section will discuss in more detail the many factors that have to be considered when computing the total cost of equipment. In addition, examples are used to explain the mechanisms for establishing charge rates.

2.4 Computing Ownership Costs

2.4.1 VALUE DEPRECIATION

Any piece of equipment loses value measured by the difference in the amount of money that has to be paid at the purchase and the amount that can be received during a sale. The loss of equipment value is referred to as depreciation, which is a major portion of the ownership cost.

Let's use QualityPave's purchase of the paver for $78,000 as an example. The company expects that it will be able to sell the unit after 6 years for $12,000. Thus, the total loss in value can be calculated:

$$\text{Total equipment depreciation} = \$78{,}000 - \$12{,}000 = \$66{,}000$$

In order for the company to recover the depreciation cost evenly, we should spread it over its lifetime. Figure 2.6 presents graphically the mechanics of the depreciation process.

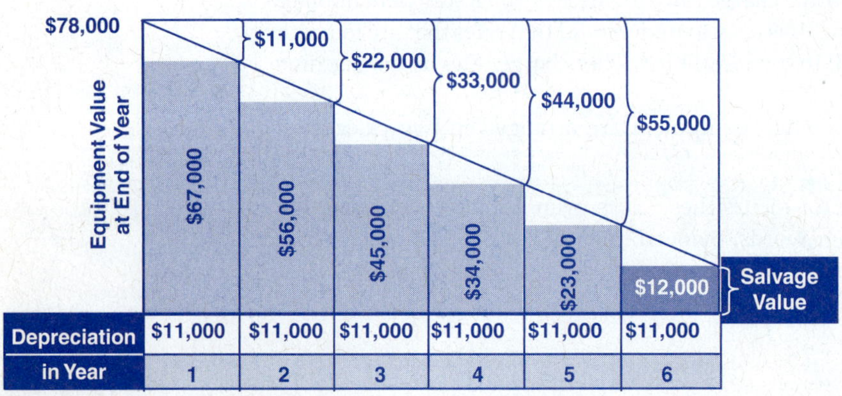

Figure 2.6 Straight-Line Depreciation of Paver

As shown, the yearly depreciation follows a straight line between the purchase and the time of sale and is therefore called the straight-line depreciation. Later, when we cover tax structure, we will meet some different methods. The linear approach leads to a constant \$11,000 (\$66,000/6 years) of yearly loss resulting in an ever-decreasing value of the equipment at the end of a year.

One open question is: how is the contractor able to predict the salvage value of the equipment at the end of its economic life? The problem is to forecast what somebody is willing to pay in 6 years. Some of the main parameters to consider are

(1) Yearly inflation during the six years
(2) Supply and demand at the time of sale (economic down- or up turns)
(3) Physical deterioration of the machine (exposure to environment, quality of operators)
(4) Quality of maintenance and repairs
(5) Technological improvements of similar equipment during the six years

One approach is to use the advertisements for used equipment in sales magazines such as the *Green Guide Volumes I and II, Handbook of New and Used Construction Equipment Values, Equipment Watch*. Naturally, such data only shows today's asking prices and not the actual sale price of the equipment in 6 years.

Since the accuracy of the forecasts will increase over time, each factor could be subjectively established every year. Of course, this would result in a constantly changing salvage value and, as a result, in a changing ownership cost, making it estimating new jobs very confusing as the rates constantly change.

In summary, establishing a true salvage value is very important to computing actual ownership cost and requires an experienced equipment manager, supported with a solid data base about the past facts relative to physical deterioration, quality of maintenance, and so on.

2.4.2 INTEREST ON CAPITAL

Any person or business investing money is expecting a return on the money that at least covers inflation and reflects the risk of the investment. In section 2.2.2 we assumed that QualityPave would buy the paver with a funds from a bank credit, which it has to amortize at an interest rate that has been negotiated with the bank. Of course, if QualityPave has sufficient liquidity on a long-term basis, it can decide to finance the purchase internally or instead to put it in a long-term savings account, or to buy bonds or stocks. It is common for a company to set a minimum attractive rate-of-return (MARR) defining the lowest interest rate it will charge money that it invests on new ventures such as buying equipment. If a certain investment option does not generate the required MARR, it will not be chosen. Naturally, for risky options the company will increase the interest rate beyond the MARR.

There are two main methods for calculating the financing expenses to be included in the ownership cost.

Method 1: In section 2.2.2 we calculated how much the bank charges QualityPave for borrowing the \$78,000 amortized with a 9% bank interest to purchase the paver expecting a 6-year life. We used the annuity formula and computed

$$\mathbf{A} = \$78{,}000 * \left(\frac{0.09}{1 - \frac{1}{(1+0.09)^6}} \right) = \$78{,}000 * 0.2229 = \$17{,}386/\text{year}$$

Over the 6-year life of the loan, QualityPave would have to pay \$26,316 in total interest. Thus, interest cost/year = \$4,386/year (= 5.6% of initial loan).

Method 2: A simpler method is sometimes used by contractors who self-finance the purchase. It is based on charging a desired interest rate on an average annual investment (AAI) applying the straight-line value depreciation method:

$$\mathbf{AAI} = \frac{(\mathbf{n}+\mathbf{1}) * \mathbf{P} + (\mathbf{n}-\mathbf{1}) * \mathbf{S}}{\mathbf{2} * n}$$

where

n = the number of periods
P = the principal loan amount
S = Salvage Value

If we assume that QualityPave intends to use its MARR = 9% for calculating the annual financing charge for the paver, we compute

$$\textbf{Annual Interest} \text{ (for P = \$78,000, S = \$12,000 and n = 6)} = \mathbf{MARR} * \mathbf{AAI}$$

$$= 0.09 * \text{AAI} = 0.09\left[\frac{(6+1) * \$78{,}000 + (6-1) * \$12{,}000}{2 * 6}\right]$$

$$\boxed{\text{Interest} = 0.09 * \$50{,}500 = \$4{,}545/\text{year}}$$

One observes that Method 2 results in a slightly higher interest, \$4,545/year compared to \$4,386/year. The reason is that the bank requires a yearly amortization of the entire purchase cost while the AAI amortizes only the difference between the purchase and the salvage value. As a result, the bank will charge less interest on average as QualityPave pays back the loan much more quickly, resulting in lower interest charges on the remaining loan.

2.4.3 STORAGE FACILITY, INSURANCE, EQUIPMENT TAX

These costs can only be included in the ownership cost if they are paid *specifically* for the machine regardless if the unit is *used or not*. In other words, property tax, license fees, or general insurance and general storage, as well as jobsite security, inspection fees, or highway permits, cannot be included.

2.4.4 FUNCTIONAL OBSOLESCENCE

This last component of ownership cost accounts for the loss of value due to changes that are not related to the physical deterioration of the equipment. Examples include technological improvements that make a replacement machine superior, new regulations about noise or exhaust that will require modifications in the future, or a major accident caused by the same kind of machine.

Similar to predicting the residual value at the end of its economic life, assessing the obsolescence cost requires experience. One approach is the substitution principle. Since improvements in technology will cause a company to replace its old piece of equipment with a newer and superior one, the functional value of an older piece will diminish year by year. Therefore, the question can be formulated as follows: What is the replacement value of a virtual unit that would provide the same performance as the original piece of equipment?

Figure 2.7 shows the calculation for functional obsolescence that is based on the loss in value of the paver, starting with its purchase price of \$78,000. Over its economic life,

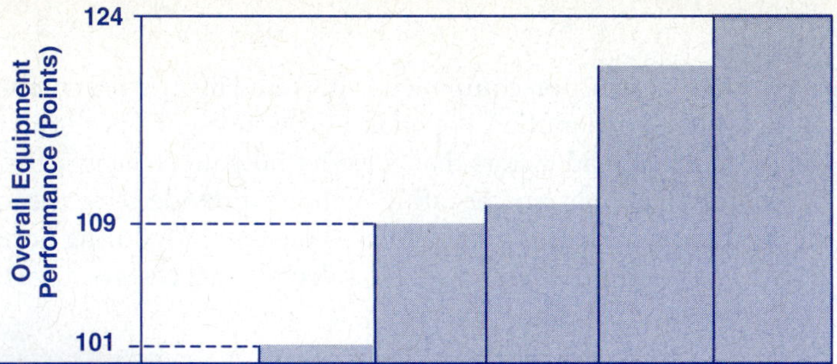

Model Year	1	2	3	4	5	6
Price (0% Inflation)	$78,000	$78,500	$81,600	$82,000	$88,400	$90,300
Tot. Performance Points	100	101	109	110	120	124
Funct. Replacement Value	$78,000	$77,700	$74,900	$74,500	$73,700	$72,800
Functional Obsolescence	$0	$300	$3,100	$3,500	$4,300	$5,200

Figure 2.7 Functional Obsolescence-Based Replacement Value

comparable paver models are being introduced on the market each year and sold for higher prices, as shown in the second row (corrected for inflation). The higher prices go along with performance improvements on a point scale. The fourth row lists the costs that would be incurred if QualityPave replaced its old paver with a virtual new paver that would also have a 100-point performance (like the one they have bought). One can easily see that the replacement values decrease every year to $72,800, resulting in an increase of functional obsolescence, over 6 years, of $5,200.

In the place of performance points, one might use the costs related to performance. Let's say that 100 points is the equivalent of paving on average 300 ft (100 m) per hour bringing a profit of $200. In that case, 124 points would result in a new paver laying 372 ft (124 m) per hour. Assuming the profit is based on the length of paved road, the new profit = $248 for 1 hour of paver operation. Since the new paver costs $12,300 more ($2,050/year), the yearly capital interest and the value depreciation will increase as well. Assuming it will also be deployed for 1,144 hours/year, the profit might shrink by $20 to $228. Thus, QualityPave forgoes $28/hour in profit due solely to higher production of the newer machine. For the entire year this would add up to $32,032 (28 ∗ $1,144), approximately 6 times the amount in Figure 2.7 ($5,200).

In summary, adding obsolescence to ownership costs accounts for the lost opportunity to benefit from higher production and thus higher profits. In that sense the costs of functional obsolescence are opportunity costs. Increasing the cost by the opportunity cost will result in higher charges, higher income, and higher profit. Of course, it would be prudent to save some of that amount, since the replacement machine will be more expensive to buy.

2.5 Computing Operating Costs

Operating costs result from using the equipment. They are not only linked to the hours of use, but also to the physical obstructions that have to be overcome by the powered machine. In other words, the fuel and maintenance needs of a bulldozer working in a sandy beach region will be significantly different from one that is working in a steep hillside with hard sandstone. At the same time, an experienced operator will be able to minimize the abuse that the equipment experiences, thus minimize its operating cost. Let's review the main contributors to the operating cost.

2.5.1 FUEL

With few exceptions, construction equipment is powered by internal combustion engines using diesel fuel. Fuel consumption depends on the horsepower-rating or size of the engine as well as fuel and load factors that consider how much energy the engine has to produce. Of course, the horsepower-rating will be provided by the company that fabricates the machine as well as the average fuel consumption during normal operation.

The following base-formula can be used to calculate fuel costs of a particular piece of equipment to be deployed at a new jobsite:

$$\textbf{Fuel Cost (\$)/hr} = \textbf{Horsepower (bhp)} \times \textbf{Fuel Cost/Gallon (\$/gal)} \times \textbf{Fuel Factor (gal/bhp-hr)}$$

$$\textbf{Fuel} \text{ Cost (\$)/hr} = \text{Horsepower (bhp)} \times \text{Fuel Cost/Liter (\$/l)} \times \text{Fuel Factor (l/bhp-hr)}$$

where

$$\begin{aligned} 1\ \text{hp} &= 33{,}000\ \text{ft}\cdot\text{lbf/min} = 746\ \text{Watt} \\ 1\ \text{gal} &= 3.78\ \text{liters} \\ \text{bhp-hr} &= \text{brake horsepower-hour} \\ \text{bhp} &= \text{a measure of an engine's horsepower without the loss in power caused by} \\ &\quad\ \text{the gearbox, alternator, and other auxiliary components} \end{aligned}$$

While the first two factors are fixed after a particular piece of equipment has been selected, the fuel factor varies according to the project conditions, operator skill, and operational setup. The following function can be used:

$$\textbf{Fuel Factor (gal/bhp} - \textbf{hr)} = \frac{\textbf{Operating Factor (OF)} * \textbf{Fuel Consumption (lbs/bhp} - \textbf{hr)}}{\textbf{Fuel lbs/gallon}}$$

The fuel consumption is given for an average operating factor. The operating factor itself indicates the percentage of the working time the engine works using 100% of its brake-horsepower per hour. The average fuel consumption of diesel engines can be assumed as:

$$\begin{aligned} &\textbf{Average diesel fuel consumption} = \textbf{0.36 lbs per bhp} - \textbf{hr (0.16 kg/bhp} - \textbf{hr)} \\ &\textbf{Weight of diesel fuel} = \textbf{7 lbs/gal (0.85 kg/l)} \end{aligned}$$

Assuming an operating factor of 60%, we can calculate the fuel factor as

$$\text{Fuel factor (gal/bhp-hr)} = \frac{0.6 * 0.36\ \text{lbs/bhp-hr}}{7\ \text{lbs/gal}} = 0.03\ \text{gal/bhp-hr}$$

$$\text{Fuel factor (l/bhp-hr)} = \frac{0.6 * 0.16\ \text{kg/bhp-hr}}{0.85\ \text{kg/l}} = 0.113\text{l/bhp-hr}$$

The variable in the fuel factor function is the operating factor, which has to reflect many situational conditions affecting the engines fuel consumption per hour. These include:

- **Loading Conditions:** Forces and obstacles that the equipment has to overcome relative to its size (low, average, severe).
- **Operator Skill:** A measure describing the smoothness of operation, efficiency of job layout, adherence to safe procedures, etc.
- **Job efficiency:** A ratio that describes how well a job is running compared to an average job performance. An idle-free operation is highly efficient, leading to a job duration that is lower than average.

Figure 2.8 Wheeled Backhoe Excavator

Since each condition is independent of the others, we can multiply numerical values that represent the situation in each of the three factor areas. Let's apply this concept to find the fuel cost for a wheeled excavator such as the one shown in Figure 2.8.

Worked-Out Problem 2.1/Expected Fuel Cost per Hour

The company that you are work for, Brenner, Inc., is planning to rent wheeled backhoe excavator X to work on a new job on the west coast of the U.S. to start in 3 months. Your supervisor, Jack Bradley, asks you to calculate the fuel cost for an average 8-hr day. He gives you key data about the excavator and the job:

- The diesel engine of the excavator type X has 156 hp.
- The operator is average.
- The loading conditions will be rough since the job is in the mountains.
- Jack will personally set up and monitor the job, which guarantees good job efficiency.

Background and Assumptions:

Brenner, Inc., has been extremely diligent in collecting data about the performance factors related to the type of excavator to be rented. Table 2.2 shows the information that the company has on file:

Table 2.2 Operating Factors for Wheeled Backhoe X

	Loading Conditions				Operator Skill				Job Efficiency			
	Easy	Average	Harsh	Extreme	Poor	Average	Good	Excellent	Poor	Average	Good	Excellent
Factor Value	0.5	0.6	0.7	0.8	1.2	1.0	0.9	0.8	1.4	1.0	0.95	0.75

To figure out the cost of diesel fuel during the operation, you consult the price history of diesel fuel on the West Coast, which has been fairly stable for the last 6 months, hovering around \$3.1/gallon (\$0.82/l) with a slight upward trend.

Calculations:

The basic function to calculate fuel cost for backhoe X:

$$\textbf{Fuel Cost/hr} = \textbf{Horsepower} * \textbf{Fuel Cost/Gallon (Liter)} * \textbf{Fuel Factor}$$

Horsepower = 156 hp

Fuel cost = $3.1/gal ($0.82/l)

$$\text{Fuel Factor} = \frac{\textbf{Operating Factor} * 0.36\,\text{lbs/bhp-hr}}{7\,\text{lbs/gal}} \; (\text{OF} * 0.1941\,\text{l/bhp-hr})$$

The operating factor has to consider three main conditions: (1) loading, (2) operator skill, and (3) job efficiency. Utilizing the company's database, we can create the matrix shown in Table 2.3:

Table 2.3 Two-Factor Matrix to Establish the Operating Factor

			Loading Conditions			
			Easy	Average	Harsh	Extreme
		Factor Value	0.5	0.6	0.7	0.8
Operator Skill	Poor	1.2	0.60	0.72	0.84	0.96
	Average	1.0	0.50	0.60	0.70	0.80
	Good	0.9	0.45	0.54	0.63	0.72
	Excellent	0.8	0.40	0.48	0.56	0.64

As shown, the result of combining the loading conditions with the operator skill is 0.70. The final operating factor can be established by multiplying:

$$\textbf{Operating Factor} = 0.70 * 0.95 = 0.66$$

$$\text{Fuel Factor} = \frac{0.66 * 0.36\,\text{lbs/bhp-hr}}{7\,\text{lbs/gal}} = 0.034\ \text{gal/bhp-hr}\ (0.128\,\text{l/bhp-hr})$$

$$\text{Fuel Cost/hr} = 156\,\text{hp} * \$3.1/\text{gal} * 0.034\,\text{gal/bhp-hr} = \$16.44/\text{hr}$$

$$= 156\,\text{hp} * \$0.82/\text{l} * 0.129\ \text{l/bhp-hr} = \$16.44/\text{hr}$$

$$\text{Fuel cost/day} = 8\,\text{hrs/day} * \$16.44/\text{hr} = \$132/\text{day}$$

Discussion: The outcome of this calculation depends heavily on the assumptions made by Jack. For example, should the job efficiency be poor, the production factor will increase from 0.66 to 0.70 * 1.4 = 0.98.

As a consequence, the fuel cost/day will soar to 8 hrs/day * 156 hp * $3.1/gal * 0.05 gal/bhp-hr = $193.4/day, a 46% increase in fuel cost (8 hrs/day * 156 hp * $0.82/l * 0.189 l/bhp-hr).

As a consequence, it is important to remind Jack to monitor closely his assumptions when the job begins, or the premises of the fuel calculations may become totally obsolete.

2.5.2 FILTERS, OIL AND GREASE (FOG)

Expenses for FOG, similar to fuel, vary with machine size and operational factors. As with fuel, higher demands on the engine will result in higher FOG costs. Thus, it makes sense to link their calculation to the fuel consumption, which has already been computed.

FOG costs covers a wide variety of items that result from routinely servicing the equipment:

- Salary for the servicing mechanics
- Service and fuel trucks, tools, and equipment
- Cost of filters, oil, and grease
- Special supplies
- Handling and disposal of hazardous material and oil
- Cost of oil analysis to assess the health of the engine

The following function can be used:

FOG cost/hr = FOG Factor * Fuel Cost/hr

As one would expect, the FOG factors are closely linked to the type of equipment. As the operational factors, each company should collect its own factor data. A start might be the U.S. Army Corps Engineers' predetermined values published in "Construction Equipment Ownership and Operating Expense Schedule" (EP 1110-1-8, Volume 10, November 2009), which includes the following five values:

S2.4 EP 1110-1-8 Army Corps Cost Calculations

	FOG Factor
1. Asphalt paver	0.119
2. Hydraulic truck cranes over 26 tons	0.110
3. Compactors	0.102
4. Wheeled hydraulic backhoe excavator	0.141
5. Crawler loader or backhoe	0.524

If we adopt the value recommended by the Army Corps of Engineers, we can calculate the FOG cost for the wheeled backhoe excavator X to be deployed on the US West Coast as:

Wheeled backhoe FOG cost/hr = FOG Factor * Fuel Cost/hr = 0.141 * \$16.44 = \$2.32/hr

Wheeled backhoe FOG cost/day = 8 hrs/day * \$2.32/hr = \$18.56/day

2.5.3 MAINTENANCE AND REPAIR COST (M&RC)

As with fuel, the cost for repairs increases with the load that is put on the equipment, the skill of the operator, and its age in terms of hours of use. Generally, the older the unit, the higher the repair costs that have to be expected. Repairs can be done in the field or in the shop by qualified mechanics and the necessary support equipment such as lifting cranes, repair truck, tools, and so on. Most often repairs include replacement parts that are held in the shop inventory or have to be shipped overnight from either the regional or the central warehouse by truck or airplane. Thus, repair costs include again a wide range of cost items, such as salary, service trucks, special equipment, tools and replacement parts, shop facility, freight, inventory, and shop overhead.

Texts on construction equipment commonly focus only on repair costs. This book considers repair as reactive maintenance, alluding to the fact that the repair could have been avoided with proactive actions. A better term for proactive is *preventive maintenance*, an investment to reduce the risks of breakdowns during operation, possibly

Equipment Maintenance Strategies

Corrective Maintenance or Wait 'til Fail
Equipment is operated until it breaks and the failed parts repaired. Often leads to downtime and costly ripple effects slowing down the entire operation. Cost of repair can be excessive.

Preventive Maintenance (PM)
Proactive and systematic service to prevent failures of equipment when in operation.

Planned Maintenance: Scheduled service of all key items separately, according to manufacturers' recommendations or part histories. Examinations can be planned based on hours-of-operation, calendar dates, etc.

Condition-Based Maintenance (CBM): Preventive service of the equipment based on information from examining indicators that reliably predict a failure to be imminent. Requires the installation of a condition-monitoring system that measures temperature, particles in motor oil, etc.

Reliability-Centered Maintenance (RCM): Comprehensive approach to establish the safe minimum level of maintenance with minimal failure risks. User-defined level of operation is achieved using a holistic view of the equipment to include modifications in operational procedures. Starting point should be the establishment of the *Evaluation Criteria for RCM Processes* (SAE JA1011).

resulting in costly consequences not only in terms of the high cost of special repairs and freight cost for express parts shipments but also the cost of delaying the entire operation. For example, a broken-down loader or dozer might have to be replaced with an inefficient substitute.

As the inserts show, there are many different strategies to maintain equipment. Data has shown that preventive maintenance leads to lower repair costs, higher productivity of the machine, and a longer service-life for the unit. Naturally, having an excellent service record to go along with the sale of a piece of equipment will increase its salvage or residual value.

In contrast, implementing, for example, a condition-based-maintenance (CBM) program will cost money from the beginning. A condition-monitoring system will require the sampling of the machine and the analysis of the samples in a laboratory, which might be done by a service company. It also requires the use of electronics and training of the operators and managers.

Z. W. Mitchell (1998) presented an interesting statistical analysis of maintenance and repair (M&R) data from several contractors owning and operating a wide variety of equipment (e.g., dozers, loaders, scraper, backhoes). Due to the availability of detailed data, he was able to relate the cost of M&R to the hours of operation, not the age of the equipment. The data analysis is also supported with important qualitative information concerning the severity of operational loads that each unit experienced. The outcome of the study was a generic function for computing the cumulative expenses for maintenance and repair as a function of hours of use. Introducing CCI (cumulative cost index), he set up the following equation:

$$\sum^t (P_t + L_t + O_t) = PP_o(CCI_t - 1)$$

where

$CCI_t =$ cumulative cost index at time t
$P_t =$ cost of parts at time t
$L_t =$ cost of labor at time t
$O_t =$ other maintenance costs at time t
$PP_o =$ original purchase price of the equipment

The crucial cost index function can be established via a regression analysis of maintenance records of similar equipment. It was shown that the best-fitting function for the many pieces that were analyzed is of the form

$$CCI = 1 + \beta_1 x + \beta_2 x^2$$

where

$CCI =$ cumulative cost index
$x =$ cumulative hours of use/1000
$\beta_1, \beta_2 =$ coefficients determined via data regression

It was found that the coefficients β_1 and β_2 varied, based on the operational loads put on the equipment and the level of preventive maintenance that was provided. High β_1s,

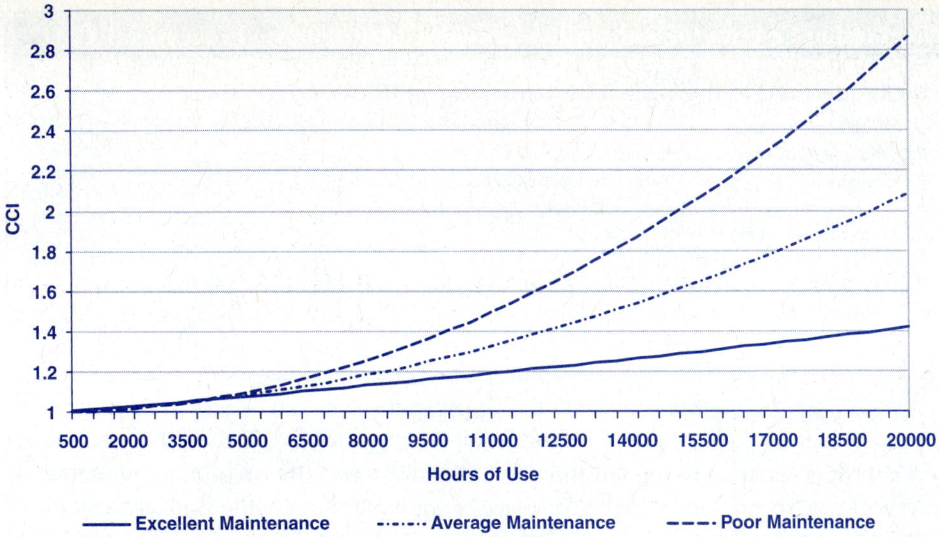

Figure 2.9 Cumulative Cost Indices as a Function of Use and Maintenance Effort

indicating expenditures for up-front maintenance, were accompanied by low β_2s, and vice versa. Figure 2.9 presents three generic curves for CCI, indicating the effect of maintenance, that apply Mitchell's study.

The three CCI curves clearly show the effect of maintenance on the long-term M&RC. As shown, until approximately 4000 hours of use, the solid line indicating excellent maintenance lies above the two other lines, indicating that more money is spent early on. However, beyond 10,000 hours, the curves for poor maintenance turn drastically upward. Still, these three curves represent only an average loading factor. Operating the same equipment in lighter or harsher job environments will affect the M&RC. Figure 2.10 depicts the result of deploying the same equipment in differing operational conditions.

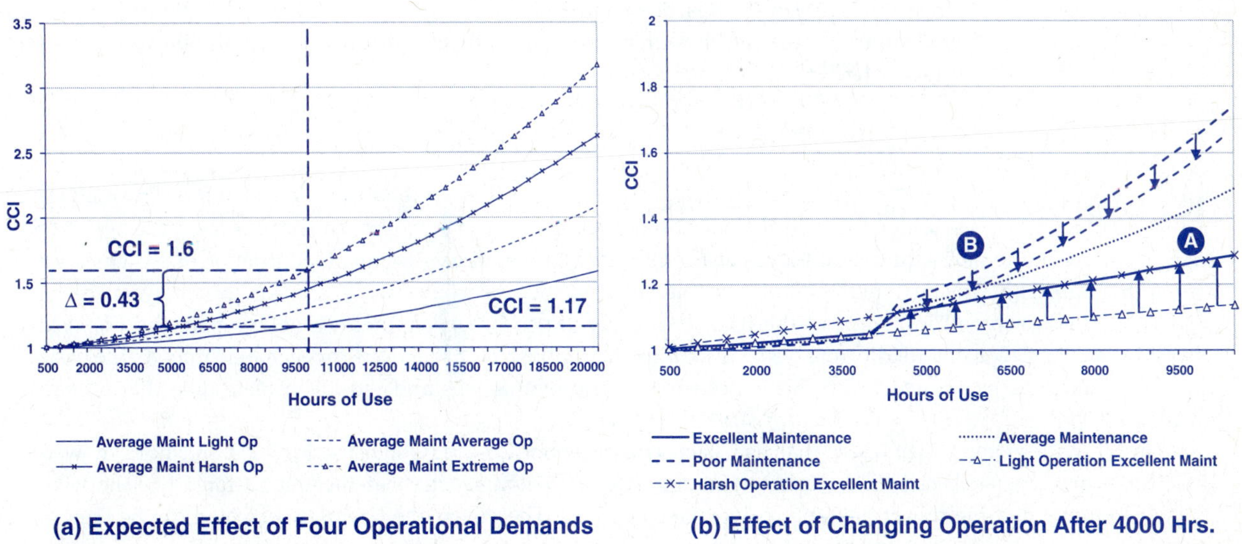

Figure 2.10 CCI Functions for Equipment with Average Maintenance

Evaluation Criteria for Reliability-Centered-Maintenance Processes (RCM)

Eight questions as the basis for establishing an RCM

1. What is the item supposed to do?
2. What are its associated performance standards?
3. In what ways can it fail to provide the required functions?
4. What are the events that cause each failure?
5. What happens when each failure occurs?
6. In what way does each failure matter?
7. What systematic tasks can be performed proactively to prevent, or to diminish to a satisfactory degree, the consequences of the failure?
8. What must be done if a suitable preventive task cannot be found?

The reader must be reminded that the CCI graphs are either based on the statistical analysis of past records for a particular piece of equipment or are the expected values set up by an experienced manager. In the latter case, the assumed values for the coefficients are updated every 500 hours when the actual sums for the M&RC become available. Figure 2.10a highlights that after 10,000 hours, the differences in the CCI could be significant, in this example 0.43, which means $43,000 of cumulative M&RC for a machine that had cost $100,000.

Figure 2.10b depicts a realistic situation where at 10,000 hours one piece of equipment is being moved from one operational level to another. The situation **A** indicates the inconsistency when changing from one CCI curve, Light Operation with Excellent Maintenance, to the curve for Harsh Operation with Excellent Maintenance. This abrupt jump from one to the other fails to consider that a piece of equipment has a memory. In other words, if a piece has experienced light loads it will have less wear and tear, and thus a lower risk of breaking down, when it moves to the harsh environment. Of course, harsher environments will demand more expenditures for maintenance than light ones, which should result in a steeper CCI after 4,000 hrs. Scenario **B** shows how the memory of the equipment can be considered. As indicated by the downward arrows, the CCI curve for harsh environments has been translated downward. The logic for the translation of the curve is the fact that each curve is cumulative. In other words, since the equipment has been working with light loads until 4,000 hours, the harsh environment and the higher maintenance costs will only gradually show their effect on the M&RC.

Worked-Out Problem 2.2/ Hourly Cost for Maintenance and Repair

Gregg Hunter is the new Equipment Manager for Clarkson, Inc., a heavy-highway contractor that owns a lot of equipment. The previous person holding the job did not keep complete records of every piece of equipment and even had lost some of the records entirely. Since Gregg planned to make repair and disposal decisions for the fleet using M&RC forecasts, he planned to establish cost functions for each machine. As the basis for those functions he wanted to use his experience from previous jobs and the incomplete data that he had available.

He picked a medium-size dozer that had been bought about $2^1/_2$ years ago for $275,000 and now showed 4,600 hours of use. It had been used for light work early on for 2,000 hours. After that, the unit operated 1,500 hours in harsh conditions and the rest of the time in average conditions. The matrix the Gregg had created can be seen in Table 2.4.

Table 2.4 CCI Function Coefficients for $275,000 Dozer with 4,600 Hours of Use

	Maintenance Levels					
	Poor		Average		Excellent	
	a	b	a	b	a	b
Light	0.0006	0.0022	0.006	0.0011	0.012	0.00011
Average	0.0008	0.0044	0.008	0.0022	0.016	0.00022
Harsh	0.0012	0.0066	0.012	0.0033	0.024	0.00033
Extreme	0.0016	0.0088	0.016	0.0044	0.032	0.00044

Gregg is asking you to check how close his functions match the available records. In the record, the following general M&RC data are found (no exact accounting of all that was included):

When the dozer was moved to harsh work, M&RC/hr was $2.85/hr.
At the end of that period M&RC was = $17,820
Today, the record shows cumulative M&RC = $26,650 and unit cost =$9.25/hr

Background and Assumptions:

Gregg is using the function

$$CCI = 1 + \beta_1 x + \beta_2 x^2$$

and sets $\beta_1 = a$ and $\beta_2 = b$. Based on the definition of CCI: x = Hours of use/1000.

Since the dozer was moved from different jobs, for which the operating factor is given, the coefficients can be adapted accordingly.

The biggest question is the level of maintenance that the dozer received. It might be possible to make an educated guess at the end of the first deployment phase, light work, and to confirm it at the end of the second and third (Table 2.5). However, this means that we have to assume that the maintenance level was kept the same at all times.

Calculations:

$x_1 = 2{,}000$ Hours of use/1,000 = 2 $\quad 0 - x_1$: Light work
$x_2 = 3{,}500$ Hours of use/1,000 = 3.5 $\quad x_1 - x_2$: Harsh work
$x_3 = 4{,}600$ Hours of use/1,000 = 4.6 $\quad x_2 - x_3$: Average work

We can use a spreadsheet that allows us to set up a matrix quickly for the three maintenance cases related to the light work between time 0 and 2000 hours. A cell function looks as follows:

$$= 1 + (\$C15 * E\$3 + (\$C15 * \$C15) * E\$4)$$

Here C15 contains x (= Hours/1000), E3 stores the coefficient a, and E4 the coefficient b.

Table 2.5 CCI Values for Dozer in Light Work

	Maintenance		
	Poor	Average	Excellent
a =	0.00060	0.00600	0.01200
b =	0.00220	0.00110	0.00011
500 Hours	1.00085	0.00328	1.00603
1000 Hours	1.00280	1.00710	1.01211
1500 Hours	1.00585	1.01148	1.01825
2000 Hours	1.01000	1.01640	1.02444

The cumulative and the average M&RC/hr after 2000 hours for the three maintenance levels are

$$\text{M\&RC(poor)} = (1.01 - 1.0) * \$275,000 = \$2,750: \quad \text{M\&RC/hour} = \$2,750/2,000 = \$1.375/\text{hr}$$
$$\text{M\&RC(average)} = (1.01640 - 1.0) * \$275,000 = \$4,510: \quad \text{M\&RC/hour} = \$2.255/\text{hr}$$
$$\text{M\&RC(excellent)} = (1.02444 - 1.0) * \$275,000 = \$6,721: \quad \text{M\&RC/hour} = \$3.3605/\text{hr}$$

From the records, we learn that the M&RC/hr was $2.85/hr, an amount that does not fit any of the averages. This may not be an accident, since the calculated costs represent the average over the entire 2000 hours but the data from the record really only considers the costs for the last 100 –200 hours of use. Let's use the CCI function to figure out the hourly M&RC for the last 500 hrs. The ΔCCI values for the three cases can be calculated by subtracting the CCI for 2,000 hrs from the one at 1,500 hrs (e.g., 1.01 – 1.00585 = 0.00415).

Poor: M&RC/hour = (0.00415 ∗ $275,000)/500 hr = $2.2825/hr
Average: M&RC/hour = (0.00492 ∗ $275,000)/500 hr = $2.706/hr
Excellent: M&RC/hour = (0.00619 ∗ $275,000)/500 hr = $3.4045/hr

Now, it is apparent that $2.85/hr is much closer to the $2.706/hr than all the other values.

Thus, we might assume that the dozer most likely received an average amount of maintenance. Let's see if we can confirm this prediction by adding the information about the second phase.

The CCI values for a dozer working in a harsh environment, depicted in Table 2.6, do not apply for the first 2000 hrs in light work. The solution uses the data from Table 2.5 for the first 2000 hrs before adding the increases in CCIs due to the harsh work environment starting with 2500 hrs. By combining the two data sets, we acknowledge that the harsher work will result in higher maintenance and repair costs, especially when poorly maintained. However, the resulting values that hover around 1.063 for all three maintenance schemes will not allow us to identify which is responsible for the $17,800 of total M&RC. The consequences of average maintenance are that its cumulative costs are now $16,950 at its lowest, while poor maintenance is $18,220 at its highest.

Let's hope that the final hours will allow us to answer Gregg's question. For this, we have to add the 1,100 hours in average work.

Figure 2.11 shows how poor maintenance would cause a comparatively rapid increase in hourly costs (e.g., to $11.71/hr after 4500 hrs), while higher early costs for excellent maintenance are starting to pay dividends in that the hourly cost stays significantly lower (= $4.97/hr) due to lower repair costs.

Comparing the present cumulative M&RC of $26,650 and the latest unit cost of $9.25/hr seems to confirm the first guess, at 2,000 hrs, that the dozer was maintained at about an average level. Clearly, the high hourly cost during the last 200–300 hrs provides evidence that the unit did not receive excellent maintenance. Similarly, the present values lie significantly lower than would be expected if a poor maintenance scheme had been followed (see Table 2.6).

Table 2.6 CCI Values for 3500 Hours (Harsh, Light and Harsh Work)

	Poor	Average	Excellent		From Table 2.5		
a =	0.0012	0.012	0.024				
b =	0.0066	0.0033	0.00033				
500 Hours	1.002250	1.006825	1.012083	Light Work	1.00085	1.00328	1.00603
1000 Hours	1.007800	1.015300	1.024330	Light Work	1.00280	1.00710	1.01211
1500 Hours	1.016650	1.025425	1.036743	Light Work	1.00585	1.01148	1.01825
2000 Hours	1.028800	1.037200	1.049320	Light Work	1.01000	1.01640	1.02444
2500 Hours	1.044250	1.050625	1.062063	Harsh	1.02545	1.02983	1.03718
3000 Hours	1.063000	1.065700	1.074970	Harsh	1.04420	1.04490	1.05009
3500 Hours	1.085050	1.082425	1.088043	Harsh	1.06625	1.06163	1.06316

Good Habits for Preventive Maintenance of Large Construction Equipment

1. **Pre-planned service** of key parts such as filters, grease boxes, tires, etc. Cleaning of important parts.
2. **Sampling and analysis** of the contaminants in the **oil** (ferrographic or spectrographic). This measures the wear on the components in the transmission and may even reveal the reason for the wear.
3. **Routine inspections** twice/day (by trained operator or mechanic) to spot simple problems such as oil leaks, worn belts and tires, corrosion, overheating, etc.
4. **Scheduling of preventive maintenance** tasks at convenient times so as not to slow down the sitework.
5. **Keeping** detailed computerized **records** for every major piece of equipment with time, duration, and cost of service and repairs, wear of parts, etc. Effect of downtime on the entire operation should be recorded, too.
6. **Information-based management of repairs**, rebuilds, or sale of each unit. Use of records and sample analysis to maximize the value of the equipment.
7. **Education and training** of managers and operators on the value and the process of preventive maintenance.

Discussion of Results

The CCI function coefficients that Gregg chose for the dozer mirror the available data for M&RC over its past 4,600 hrs. While the early data was not distinctly indicative of any maintenance level, the latest data match that of average maintenance but with a clear tendency toward the data for poor maintenance, especially the presently relative high unit cost ($9.25/hr compared to $7.95/hr for the average maintenance mode).

Recommendations

From the analysis of the present data, in combination with Gregg's experience with using cumulative cost functions to predict the future M&RC cost, it appears that the company would benefit greatly in terms of limiting the future cost by immediately implementing a condition-based preventive maintenance system for the dozer. While past neglect cannot be immediately rectified, future breakdowns and downtimes can be minimized by knowing the risk of part failures. To install a condition-monitoring system, one of several service companies can be hired, and the data analysis can be highly automated. After 1 or 2 years, the benefits should already be measurable in terms of reduced direct repair costs and reduced fleet costs because of less downtime. In order to

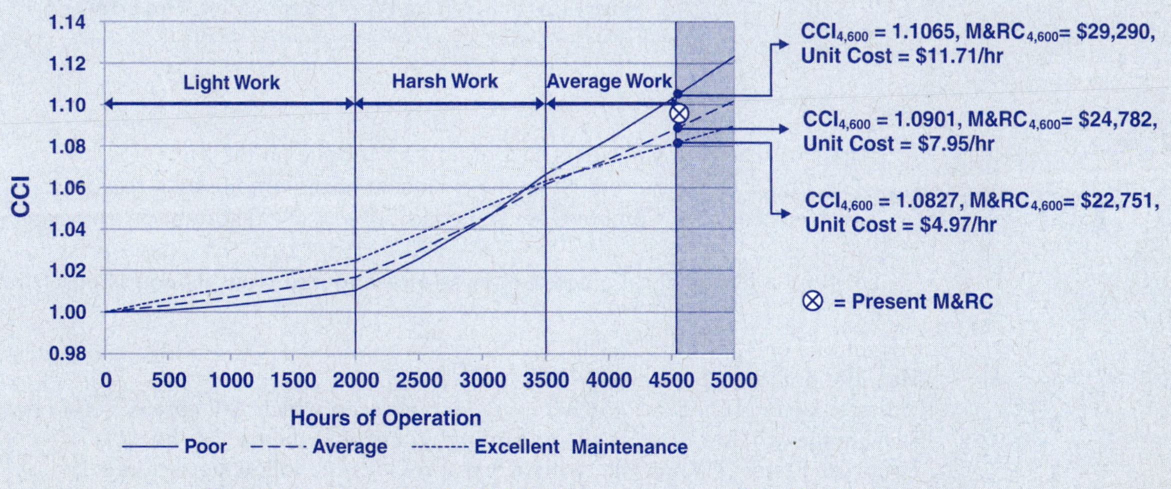

Figure 2.11 Comparing M&RC for Poor, Average, and Excellent Maintenance

create the basis for an economic comparison (before and after), it is also recommended that Gregg update the records with automated data acquisition for maintenance and repair for each unit. Furthermore, he should follow up any downtime to inquire about the indirect cost caused by lower productivity of the fleet to which the down equipment belonged.

Finally, Gregg is encouraged to suggest that the company launch a special educational program that is based on "Good Habits of Preventive Maintenance of Large Construction Equipment."

2.5.4 TIRE COST

Wheeled equipment travels on air-pressured rubber tires that experience slow abrasion and other damage, depending on the cumulative amount of stresses at the point of contact with the traveling surface. In other words, tires wear during their use and need to be either recapped or replaced. The life of tires is closely related to the way they are treated. For example, keeping the tire pressure too low will add to fuel costs and will create excessive heat that eventually leads the tire to blow out at the side. Similarly, overloading and obstacles on the road can cause damage to tires.

The life of tires is generally shorter than the service life of the equipment. Thus, tire costs, which can be substantial, are treated separately from the equipment itself. For that purpose, the purchase price of wheeled equipment, used for owning cost calculations, does not include the cost of the new tires.

Today, contractors are able to extend tire life using retreading, also called recapping. This is cheaper than buying a new tire of equal quality.

The single most important factor linked to tire cost is the expected service life for its level of use and abuse.

The Army Corps of Engineers (2009) recommends the following function to compute the tire cost/hr:

$$\textbf{Tire Cost/hr} = \frac{\textbf{Tire Cost Factor} * \textbf{Tire Cost}}{\textbf{Tire Life Factor} * \textbf{Tire Wear Factor} * \textbf{Max. Tire Life}}$$

where

Tire Cost Factor = Considers cost of a recap if done on the tire
Tire Life Factor = Accounts for the extension of the tire life by a recap
Tire Wear Factor = Combines several conditions of use that impact tire wear

Let's use a hypothetical wheel loader as a case to figure out its hourly cost/tire:

Cost of one of 4 tires = \$2,450
Max. life of a tire = 3,000 hrs
Tire cost factor (TC) = 1.65 (there will be one recap which will cost 65% of a new tire)
Tire life factor (TL) = 1.8 (the recapping will extend the tire life by 80%)
Tire wear Factor (TW) = The Army Corps provides the following factors:
Front Tire: Average work load = 0.83, **severe work load** = 0.25
Drive Tire: **Average work load** = 0.54, **severe work load** = 0.15

$$\text{Tire Cost/hr for a drive tire with an average operational load} = \frac{TC * \$2,450}{TL * TW * \text{Tire Life}} = \frac{1.65 * \$2,450}{1.8 * 0.54 * 3,000\text{ hrs}} = \frac{\$4,042}{2,916\text{ hr}} = \$1.39/\text{hr}$$

$$\text{Tire Cost/hr for a drive tire with a severe load} = \frac{\$4,042}{1.8 * 0.15 * 3,000\text{ hrs}} = \frac{\$4,042}{810\text{ hrs}} = \$4.99/\text{hr}$$

Wear drastically reduces the useful life of the tire, which, even with recap, is reduced from 2,916 hours to 810 hours.

Of course, these calculations leave out the cost of money and the increase in prices for tires over the service life of the equipment. The following will quickly revisit the approach to calculate the hourly cost considering a 9% interest rate for the capital that is needed to pay for the tires.

Worked-Out Problem 2.3/ The Cost of Loader Tires

Gregg Hunter from Clarkson, Inc. requests that we help him in calculating the cost of 4 loader tires that are being used in an average work environment. One new tire sells right now for \$3,150 and has a maximum life of 3,200 hrs. Each tire will be recapped one time at 50% cost of a new one, extending its life by 75%. He expects that the prices of new tires will increase at about 5% per year. The expected service life of the loader is 15,000 hrs over 6 years (use = 2,500 hrs/year). The interest rate for the bank credit for purchasing the tires is 9%.

Base Data/Assumptions

Tire life = 3,200 hrs
TC = 1.5
TL = 1.75
TW front tire = 0.83
TW drive tire = 0.54

The loader will be sold with the old tires.

Service Life:

Front tires: 1.75 * 0.83 * 3,200 hrs = 4,648 hrs = 1.9 years
Recap after 2,656 hrs = 1.06 years
Drive tires = 1.75 * 0.54 * 3,200 hrs = 3,024 hrs = 1.2 years
Recap after 1,728 hrs = 0.7 years

Calculations

Cost of Front Tires:

Recap 1 = 0.5 * \$3,150 = \$1,575/recap
Cost after 1.06 years = 2 tires * \$1,575 = \$3,150
Set 2 after 2 years = \$3,150 * $(1 + i)^2$ = \$3,150 * $(1 + 0.05)^2$ = \$3,473/tire
Cost after 1.9 years = 2 new tires * \$3,473 = \$6,946
Recap 2 = 0.5 * \$3,473 = \$1,736/recap
Cost after 2.96 years = 2 tires * \$1,736 = \$3,472
Set 3 year 4 = \$3,150 * $(1 + .05)^4$ = \$3,829/tire

Monthly Interest Rate

Compounded Interest = $(1+i)^n$
where i = interest in n periods
Equivalent monthly interest compounded to a 5% yearly interest:
$5\% = (1+i)^{12}$, i = 0.407%

Cost after 3.86 years = 2 new tires
* \$3,829 = \$7,858
Recap 3 cost = 0.50 * \$3,829 = \$1,914/tire

Cost of Drive Tires:

Recap 1 after 0.7 year = \$1,575/recap
Set 2 after 1.2 years = (1.05 + 2 * 0.00407) * \$3,150 = \$3,333/tire (\$6,666)
Recap 2 after 1.9 years = 0.5 * \$3,333 = \$1,667/recap
Set 3 after 2.4 years = $((1.05)^2 + 4 * 0.00407)$ * \$3,150 = \$3,524/tire (\$7,048)
Recap 3 after 3.1 years = 0.5 * \$3,524 = \$1,762/recap
Set 4 after 3.6 years = $((1.05)^3 + 6 * 0.00407)$ * \$3,150 = \$3,723/tire (\$7,446)
Recap 4 after 4.3 years = 0.5 * \$3,723 = \$1,862/recap
Set 5 after 4.8 years = $((1.05)^5 - 2 * 0.00407)$ * \$3,150 = \$3,995/tire (\$7,990)
Recap 5 after 5.5 years = 0.5 * \$3,995 = \$1,997/recap

Flow Model of the Cash Spent in Figure 2.12

Total cost for purchasing new tires and 1 recap without inflation (to today's prices) = \$75,600
Total cost taking into account 5% inflation per year = \$84,530
Cost per hour without considering time-value-of-money = \$84,530/15,000 hrs = \$5.63/hr

Present Value and Annuities: Table 2.7 shows the result of the spreadsheet used to calculate the Present Value (PV) of each expense, using a yearly interest rate of 9%. Each PV is subsequently spread out evenly over the 6 years of service life.

The spreadsheet shows a total PV of \$68,181 and a yearly annuity of \$15,199. Finally, the cost per hour is computed by dividing the annuity by hours used yearly:

\$15,199/2,500 hrs = \$6.08/hr

Discussion of Results

The difference in tire life for front and drive tires doubles the number of times that the loaded needs to serviced for tire changes. On the other hand, changing all four tires at the same time by either using one set beyond its desired use or by taking one set off early will either increase the risk or add cost.

The inclusion of costing the 9% interest for the money borrowed for buying new or recapping old tires has a significant effect on the hourly cost, raising it from \$5.63/hr to \$6.08/hr (8.1%). As tires are a significant element of the overall equipment cost, neglecting to include time value of money would create a considerable gap in assigning the effective cost for using the loader (see Figure 2.12 and Table 2.7).

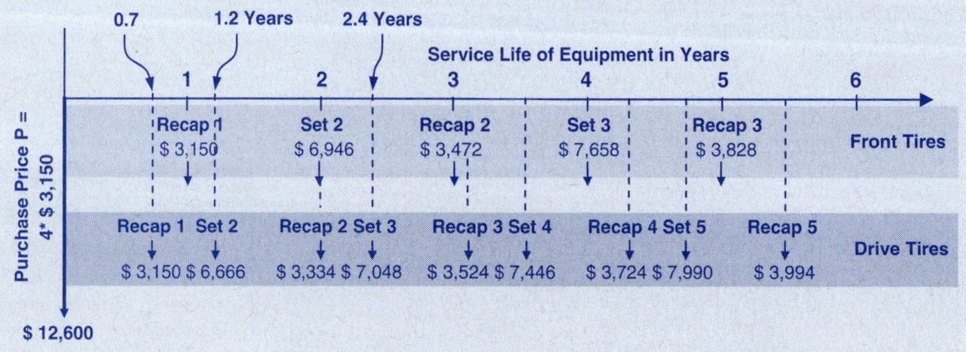

Figure 2.12 Time-Based Expenses for Tires

Table 2.7 Calculation of Tire Cost/Hr Considering Time Value of Money (9%)

$$PV = 3{,}150 * \left(\frac{1}{(1 + 0.09)^{1}} \right)$$

	4 Tires	Recap D1	Recap F1	Set D2	Set F2	Recap D2	Set D3	Recap F2	Decap D3	Set D4	Set F3	Recap D4	Set D5	Recap F3	Recap D5	Total
Expense	12600	3,150	3,150	6,666	6,946	3,334	7,048	3,472	3,524	7,446	7,658	3,724	7,990	3,828	3,994	$84,530
Time in Years	0	0.7	1	1.2	2	2	2.4	3	3.1	3.6	4	4.3	4.8	5	5.5	
Present Value	12600.0	2974.1	2889.9	6115.6	5846.3	2806.2	5932.2	2681.0	2721.2	5274.9	5425.1	2638.2	5193.0	2487.9	2595.8	$68,181
Annuity (Year)	2808.8	663.0	644.2	1363.3	1303.3	625.5	1322.4	597.7	606.6	1175.9	1209.4	588.1	1157.6	554.6	578.7	$15,199
Cost/Hr	1.124	0.265	0.258	0.545	0.521	0.250	0.529	0.239	0.243	0.470	0.484	0.235	0.463	0.222	0.231	$6.08

$$A = PV * \left(\frac{0.09}{1 - \frac{1}{(1+0.09)^{6}}} \right)$$

2.6 Equipment Overhead

So far, we have considered costs that are the consequence of owning and operating a piece of equipment. However, having enough money to cover these expenses does not mean that the machinery will automatically be put to work resulting in bills to the customer covering our expenses. Also, we have not covered the salaries for the accountants or for the marketing department charged to find work for the loaders, dozers, or paver.

Generally, overhead costs cover all the expenses that cannot be precisely linked to a source. For example, compared to the cost of fuel for a dozer, the light for the business office is impossible to divvy up accurately. Overhead, as the term implies, includes all the cost that are over the equipment's head.

In construction, we consider three types of overhead: (a) general overhead, (b) project overhead, and (c) equipment overhead. Usually, such difficulties in assessing costs are distributed as a percent of a base number, such as direct cost (e.g., owning and operating cost). The accounting of overhead can become quite tricky in that, for example, the cost for the land and building of the repair shop should be used to ascertain repair cost and thus cannot be considered overhead. On the other hand, base telephone charges, utilities, landscaping cost, and so on need to be covered somehow.

One has to separate between renting the equipment or using it on a project that was acquired through the low bid process, in that the low bid has to include a summary overhead. In other words, there is no means to charge overhead for a piece of equipment since it has to be included in the general overhead.

In summary, equipment overhead is commonly charged as a percent of owning and operating cost (e.g., 25%). However, this overhead is only applicable where it is rented out independently of a project.

2.7 Reducing Income Taxes with the Help of Tax Depreciation

2.7.1 WHAT IS *TAX DEPRECIATION?*

S2.5 IRS How to Depreciate Property
S2.6 IRS Instructions for Business

Income taxes paid by a business are calculated by multiplying the difference between expenses + tax deductions by the total revenues and tax rate(s). Expenses are assets that have been expended or used up in the process of producing revenue. In construction, the most common means of expending (liquid) assets is by paying with cash or checks for material, tools, laborers, and so on. In theory, equipment would only qualify as a normal expense in the year it is purchased. But, of course, the life of a piece of equipment extends beyond one year. For this reason, the Internal Revenue Service (IRS) developed rules for what is called tax depreciation, which is different from economic or value depreciation, which we covered in section 2.4.1. In IRS Publication 946, "How to Depreciate Property," the IRS stipulates not only what can be depreciated for tax purposes, but also the "tax" life of the property (e.g., 5 years for trucks). During this life span, a company can declare a percentage of its cost as depreciation expense since the value of the truck is used up by that percentage. During certain periods, as in an economic downturn, the government may implement so-called accelerated depreciation laws in order to stimulate the economy. In the 2008 Economic Stimulus Act, the government provided a special depreciation allowance whereby the cost of certain equipment or leases could be depreciated within one year. In order to benefit from such depreciation expenses, the company must (1) maintain depreciation records and financial statements and (2) maintain depreciation records for the tax return based on the IRS rules.

2.7.2 CALCULATING DEPRECIATION EXPENSES

QualityPave, Inc bought the paver for $78,000 in total, paid with the help of a bank credit. As mentioned above, the IRS allows the yearly income statement to include a portion of the purchasing cost as an expense, which will reduce the amount of taxes QualityPave has to pay.

The U.S. Internal Revenue Service (IRS) has devised three tax depreciation methods for a contractor to choose from: (1) Straight Line, (2) Declining Balance, and (3) Sum-of-Years' Digits Method.

IRS Publication 946 defines them as follows:

Straight-Line Method of Tax Depreciation:

This method lets you deduct the same amount of depreciation each year over the useful life of the property. To calculate your deduction, first determine the adjusted basis, salvage value, and estimated useful life of your property. Subtract the salvage value, if any, from the adjusted basis. The balance is the total depreciation you can take over the useful life of the property. Divide the balance by the number of years in the useful life. This gives you your yearly depreciation deduction.

Declining-Balance Method of Tax Depreciation:

An accelerated method to depreciate property. The General Depreciation System (GDS) and Modified Accelerated Cost Recovery System (MACRS) uses the 150% and 200% declining balance methods for certain types of property.

A depreciation rate (percentage) is determined by dividing the declining balance percentage by the recovery period for the property.

As the following worked-out problem will demonstrate, each method will allow exactly the same amount of deprecation over its tax life, 5 years for major construction equipment. If the eventual sales price is higher than the assumed salvage value, the contractor would have to pay income tax on the difference. In other words, tax collection agencies around the world will get their due whatever method or assumption the contractor makes.

Worked-Out Problem 2.4/ Yearly Tax Savings Due to Depreciation

Erica Perl is the asset manager for QualityPave and would like to know which one of the three depreciation methods she should pick for the $78,000 paver she just added to the list of equipment assets. She understands that the company expects an excellent business year with a high profit margin, which will slow a little bit in the second, and certainly drop briskly for the third and fourth year. No prognosis can be made for the fifth year at this point. QualityPave's tax rate is 29%.

Overview

We need to recognize that the tax life of the equipment is only 5 years, mandated by the Tax Office (Internal Revenue Service in the U.S.), and not the 6 years of service-life that QualityPave has decided on for economic reasons.

Figure 2.13 compares the progress of depreciation over the 5 years. We decide to use the same salvage value as after 6 years, when the paver will be sold for 15% of its purchase price (= expected salvage value). During that year, no depreciation for tax savings will be possible but eventual sales gain will be taxed as income, loss will be recognized as cost.

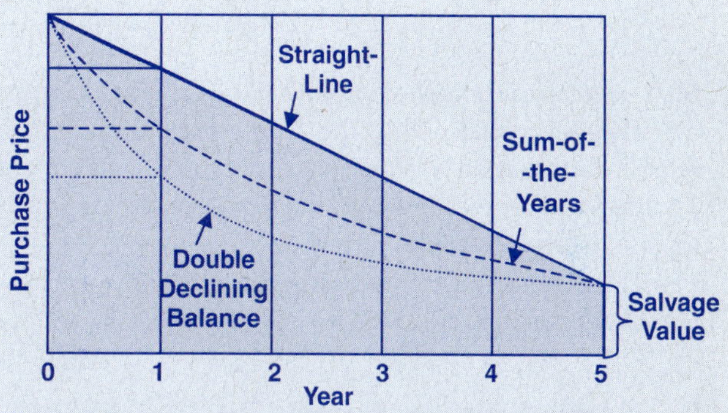

Figure 2.13 Equipment Tax Depreciation Purposes

Salvage value = 15% of $78,000 = $11,700

Straight-Line Depreciation: This is the simplest and most-often-used method. The straight line connects the initial cost with the estimated salvage value. Each year, 1/5 of the difference will be depreciated. Hence, the yearly depreciation expense for tax purposes is

$$D_{Str} = (\$78{,}000 - \$11{,}700)/5 = \$13{,}260/\text{year}$$

Since this amount will be considered as expense for the year, for each of the 5 years the revenue will be reduced by this amount and with it the net profit that is the basis for calculating taxes. Assuming that QualityPave will have a revenue that is larger than the $13,260, the straight-line depreciation will allow it to save:

Yearly tax savings due to depreciation = 29% * $13,260 = <u>$3,845.4</u>
Total tax savings = 5 * $3,845.4 = <u>$19,227</u>

Double-Declining-Balance Method (DDBM): Figure 2.13 shows that this method accelerates the depreciation during the first years when compared to the straight-line method. In fact, the name comes from doubling the straight-line depreciation for the first year, which is 20% or 1/5 of the total amount. However, for the following years the 40% (double 20%) is only applicable to the amount that has not been depreciated yet.

As depicted in Table 2.8, the DDBM begins with depreciating the full purchase price without considering the salvage value. In fact, it anticipates that value to be less than 100% – 86.5% (accumulated in year 4) or 13.5%. Since QualityPave anticipates selling the paver for $11,700, no depreciation expenses will be recognized in the fifth year, because the depreciable value of the paver cannot fall below $11,700. In fact, as Table 2.8 indicates, the salvage value was already reached at the end of the 4th year, causing a correction. Instead of the 40%, which would have led to a value below the sales price of the paver, the depreciation rate had to be reduced to 31% so that exactly $11,700 of the end of year value would result.

As mentioned earlier the total savings are the same as with the straight-line method, namely $19,227. It is apparent, however, that the savings in the first two years lies drastically above those from the straight-line method.

Table 2.8 Depreciation and Tax Savings Using DDBM

Depr. Value - Beg. of Year	Depr. Rate	Depreciation Expense	Accumulated Depreciation	Accumul. Depr. in %	Depr. Value - End of Year	Tax Savings
$78,000	40%	$31,200	$31,200	40%	$46,800	.29 * 31,200 = $9,048
$46,800	40%	$18,720	$49,920	64%	$28,080	.29 * 18,720 = $5,429
$28,080	40%	$11,232	$61,152	78.4%	$16,848	.29 * 11,232 = $3,257
$16,848	40%	$5,148	$66,300	86.5% 85%	$11,700	.29 * 5,148 = $1,493
$10,109	0%	$0	$66,300		$11,700	0

Total Savings = $19,227

Sum-of-Years' Digits Method (SOYD): This is again an accelerated method of depreciation.

The base for calculating the depreciation for the year is the sum of the digits of those years during which the tax-related depreciation is taking place. For our 5-year paver this amount would be calculated as: 1 + 2 + 3 + 4 + 5 = 15. To assess the depreciation rate, the sequence of year number is reversed and divided by 15. Table 2.9 shows how this method is applied to the paver.

The base for the depreciation considers, like the straight-line, the salvage value. In the case of our paver, the SOYD depreciates only $78,000 − $11,700 = $66,300.

Recommendation for Erica Perl

The double-declining-balance method provides by far the most in terms of tax saving for the first year. DCBM considers $31,200 as expense compared to $22,100 of the SOYD or $13,260 when the Straight-Line Method is used. For the second year, DDBM is still highest but is lower thereafter. In fact, in year 5, DDBM will account for $0 in depreciation expenses. SOYD is clearly a compromise between the straight-line and DDBM in that it accelerates at the beginning but shows expenses every year in a diminishing amount. For example, in year 5 the depreciation expenses of SOYD are $4,420 compared to the $0 of the DDBM or the $13,260 of the Straight-Line Method.

Based on projections for excellent business during the first year with a continued slowdown thereafter, it seems most beneficial for QualityPave to adopt the DDBM for the paver.

In times of high unemployment, the government may act to create incentives for companies to invest in equipment, which in turn would benefit the equipment manufacturing sector, which would have to hire more employees to satisfy the higher demand. The so-called Investment Tax Credits (ITC) are a quick and effective means to address this problem. For example, the Economic Recovery Tax Act of 1981 instituted a 10% Investment Tax Credit to spur purchases of equipment. This one-time credit, applied in the year of the acquisition, can be used to reduce the tax itself. In other words, while the depreciation expenses are pre-tax reductions, tax credits can be subtracted directly from normal taxes. For QualityPave this would mean that in a year when the government provides 10% ITCs, the purchase of the paver would allow $7,800 to be subtracted from the taxes that already include the tax savings from the depreciation in that year.

Table 2.9 Depreciation and Tax Savings Using SOYD

Base of Calculation	Depr. Rate	Depreciation Expense	Accumul. Depr.	Value - End of Year	Tax Savings
$66,300	5/15	$66,300 * 5/15 = $22,100	$22,100	$55,900	.29 * 22,100 = $6,409
$66,300	4/15	$66,300 * 4/15 = $17,680	$39,780	$38,220	.29 * 17,680 = $5,127
$66,300	3/15	$66,300 * 3/15 = $13,260	$53,040	$24,960	.29 * 13,260 = $3,845
$66,300	2/15	$66,300 * 2/15 = $ 8,840	$61,880	$16,120	.29 * 8,840 = $2,564
$66,300	1/15	$66,300 * 1/15 = $ 4,420	$66,300	$11,700	.29 * 4,420 = $1,282

Total Savings = $19,227

2.7.3 HOW MUCH SHOULD WE CHARGE FOR ITS USE?

So far, we have tallied up all the costs that result from owning and operating a piece of construction equipment. But eventually we have to figure out a way to get reimbursed for all the costs and to earn the profit necessary to keep the business moving ahead. In a first step towards this end, we have to calculate the average cost of operation. This is not straightforward, as different costs arise at different times during the life of the equipment, 6 years for the paver. A common approach to accomplish this is to spread all the cost evenly over the years and divide the total for one year by the expected hours of use.

Worked-Out Problem 2.5/ Hourly Cost of Operating an Asphalt Paver

Fred Wilson is QualityPave's manager of estimating. He is working on a new bid and needs to know the cost per hour of the brand-new paver (see Figure 2.14) for which he has no records from previous jobs. He asks us to give him some input, because he has nobody who is available and qualified to calculate such a rate. Fred wants us to use an average inflation rate of 4% for the 6 years of service life, and we should assume that the paver will be productively operating for 1,400 hours per year.

Figure 2.14 Small Asphalt Paver

Cost Figures and Assumptions

There are several cost factors that we have worked out but several others we have to assume based on similar equipment or manufacturer's specifications. The following cost figures will be used:

1. Purchase price $78,000
2. Salvage after 6 years $11,700
3. Cost of major overhaul at end of year 3 $21,000
4. Yearly annuity payment $17,386
5. Tax savings according to DDBM total $19,227
6. Insurance, taxes, and permits/year = 10% of purchase price
7. Fuel per hour $ 14
8. Oil, lubricants, filters, battery, cleaners, etc. per hour $ 8
9. Cost of a set of new tires lasting 2,800 hours $ 4,800
10. Expected scheduled maintenance cost per year:

Year 1	$3,200	Year 3	$4,500	Year 5	$4,500
Year 2	$3,900	Year 4	$3,900	Year 6	$7,000

Present Value of a Future Sum

$$PV = \frac{FV}{(1+i)^n}$$

PV = Present value, or value a time = 0
FV = Future value, or value at time = n

Uniform Annuity of Present Value

$$A = PV * \left(\frac{i}{1 - \frac{1}{(1+i)^n}} \right)$$

A = Uniform annuity of PV

Uniform Annuity of a Future Value

$$A = FV * \left(1 - \frac{i}{(1+i)^n - 1} \right)$$

i = interest rate per period
n = number of periods

Approach

Our goal is to come up with an hourly cost that can be uniformly utilized during the entire life of the paver. As the costs change every year, for example the maintenance cost, it is necessary to spread all the cost items, including the salvage value, evenly over the 6 years of use. Subsequently, we can divide the cost-per-year by the hours of operation for a year, resulting in the cost per hour. Finally, we can just add those costs that are already given as per hour.

Step 1: Cash-Flow Diagram (for Non-Hourly Costs and Revenues) as shown in Figure 2.15.

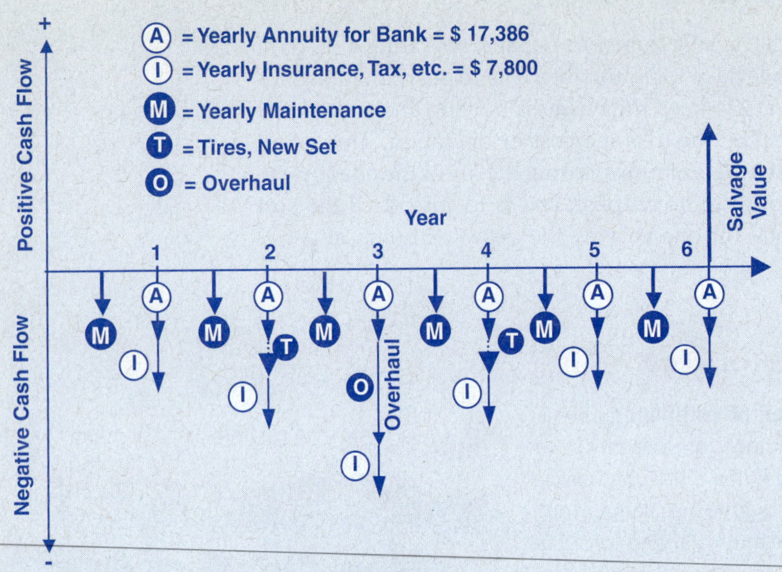

Figure 2.15 The Irregular Yearly Cash Flow of the Paver

Step 2: Turning a Singular Non-Uniform Cash Transaction into a Uniform Series

We learned that banks want their loans amortized but with a uniform cash flow for the time of the loan. For that purpose, they use annuities, which are a mix of payback and interest. The same approach can be used to transfer a single cash transaction into a uniform annuity over a period of 6 years. Let's start with the salvage value of $11,700, which will be generated at the end of the 6th year. Without consideration of the time value of money, we could simply divide that amount by 6 years to get its yearly equivalent of $1,950. However, we have to consider a yearly inflation rate of 4%, reducing the real value of the sale at present: $PV = \$11{,}700/1.04^6 = \$11{,}700/1.265 = \$9{,}250$.

The annual equivalent of the salvage value is

$$A = \$11,700 * (0.04/((1+0.04)^6 - 1) = 0.1508 * \$11,700 = \$1,764$$

The **overhaul** after 3 years cannot be handled the same way, as the cash payment takes place right in the middle. The simplest way is to calculate its present or its future value and from that compute the annuity. In other words, it requires two independent steps.

PV of $21,000 overhaul cost to be paid at the end of year 3 considering 4% inflation:

$$PV = \$21{,}000 * (1/((1+0.04)^3) = \$18{,}669$$

$$\text{A of a PV} = \$18{,}669 \text{ for a period of 6 years and an } i = 4\%$$

$$A = \$18{,}669 * (0.04/(1-(1/(1+0.04)^6) = 0.1908 * \$18{,}669 = \$3{,}562 \text{ (overhaul)}$$

Thus, the annual equivalent of a one-time charge of $21,000 is an annual charge of $3,562.

Two sets of tires will be needed (it will be sold with old tires). The simplest way is again to calculate the two PVs, sum them and come up with the annual equivalent.

$$\text{PV of } \$4{,}800 \text{ FV with } n = 2 \text{ and } i = 4\%, PV = \$4{,}800 * (1/((1+0.04)^2)) = \$4{,}439$$

$$\text{PV of } \$4{,}800 \text{ FV with } n = 4 \text{ and } i = 4\%, PV = \$4{,}800 * (1/((1+0.04)^4)) = \$4{,}103$$

$$\text{A of a PV } \$8,542 \text{ for a period of 6 years and } i = 4\%$$

$$A = \$8,542 * (0.04/(1-(1/(1+0.04)^6) = 0.1908 * \$8{,}542 = \$1{,}630$$

The annual equivalent of two new sets of tires is an annual charge of $1,630.

Each year, a different amount of maintenance is expected, reflecting its years in operation.

Indicated in the cash flow diagram is the fact that maintenance is done across the year and its costs do not show up at its end. The effect of having cost spread across the year is that the cash expense is recognized in the middle of the year.

It is apparent that the calculation of PVs for the maintenance expenses cannot use one year as a period (as the costs occur in the middle of a year). As a consequence, we will have to switch to a 6-month or 1/2-year period despite the fact that the inflation rate is based on the year. What we have to do is to turn the 4% inflation into its half-year equivalent. We can accomplish this by calculating the compounding effect over two periods with its future interest being 4 (without dimension) and i=1%.

$$i = 4 * (0.01/((1+0.01)^2 - 1)) = 1.99$$

Now we have the means to switch to the 6 months period with i = 1.99:

PV of \$3,200 FV with n = 1 and i = 1.99%, $PV = \$3,200 * (1/((1+0.0199)^1)) = \$3,138$
PV of \$3,900 FV with n = 2 and i = 1.99%, $PV = \$3,900 * (1/((1+0.0199)^3)) = \$3,676$
PV of \$4,500 FV with n = 3 and i = 1.99%, $PV = \$4,500 * (1/((1+0.0199)^5)) = \$4,078$
PV of \$3,900 FV with n = 4 and i = 1.99%, $PV = \$3,900 * (1/((1+0.0199)^7)) = \$3,393$
PV of \$4,500 FV with n = 5 and i = 1.99%, $PV = \$4,500 * (1/((1+0.0199)^9)) = \$3,762$
PV of \$7,000 FV with n = 6 and i = 1.99%, $PV = \$7,000 * (1/((1+0.0199)^{11})) = \$5,624$

Total PV for yearly maintenance = \$23,671

From the PV we are now able to establish its uniform distribution over 6 year with i=4%:

$$A = \$23,671 * (0.04/(1-(1/(1+0.04)^6)) = 0.1908 * \$23,671 = \$4,517$$

The annual equivalent of 6 years of maintenance is \$4,517.

Step 3: Summary of Uniform Cash Flows

Figure 2.16 summarizes the effect of creating uniform annuities from the irregular outlays and incomes during the service life of the paver. Now we are ready to calculate the total cost per year.

Summation of annual cost and salvage income:

Annuity of bank loan	\$17,386
Insurance, tax, permits, etc.	\$7,800
Maintenance	\$4,520
2 sets of new tires	\$1,629
Overhaul	\$3,562
Salvage income	(\$1,764)
Total cost per year =	\$33,133

Now we are ready to come up with the answer to the original question, cost per hour of use.

Step 4: Calculation of Hourly Operating Cost

This final step will decide the minimal hourly cost of the paver to be used for a new bid. To establish its true cost is important for financial survival as the time of productive operation is the only way to recover the capital outlay of the

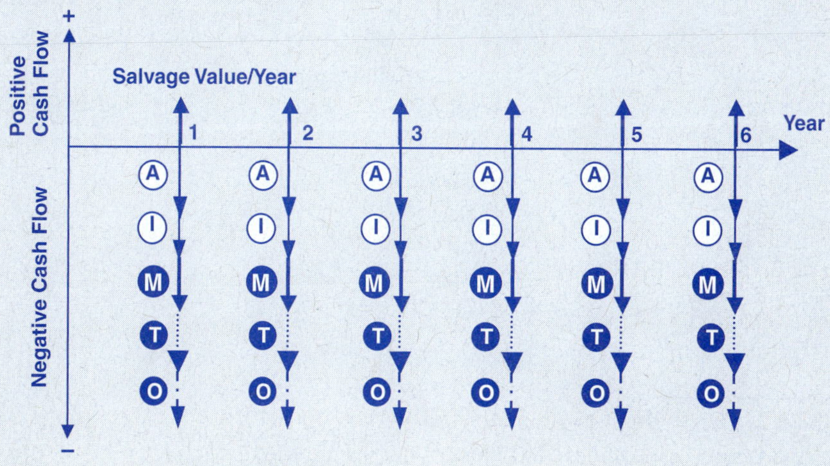

Figure 2.16 Annual Diagram of All Uniform Cash Flow Streams

company. In other words, QualityPave is unable to charge ownership and operating costs when the paver sits idle in the maintenance shop or is being transported to another site.

Based on Fred Wilson's estimate, the new paver will be laying down asphalt for 1,400 hours of the year. This prediction allows us now to compute the total cost per hour.

Cost per hour based on annual cost = $33,133/1,400 =	$23.7
Cost of fuel per hour ($2.50/gal)	$14.0
Cost of oil, lubricants, filters, etc.	$8.0
Total cost per hour =	**$45.7**

Discussion of Results

It is apparent that the accuracy of the calculated cost per hour rests on the accuracy of several assumptions such as the salvage value, the price of fuel, and the cost of the overhaul. On the other hand, two large portions of the yearly costs, $17,386 and $7,800 for loan and insurance, can be considered fixed accounting for $25,186/1,400 = $18.

Another critical assumption is the number of hours that the paver will be used per year. Figure 2.17 presents the result of a sensitivity analysis that focuses on the price of fuel and the number of hours used.

The cause for the convex curve can be found in the fact that a portion of the total cost is fixed (e.g., annuity for loan) while others are variable and change linearly with the number of hours used (e.g., fuel cost). As expected, the cost per hour is extremely sensitive to both variables. For example, reducing the hours used from 1,400 to 1,000 hours will increase the cost by 21% from $45.7/hour to $55.1/hour (3). The price of fuel has an equally drastic effect, as the paver uses 5.6 gallons per hour of use. Indicated by (1) is the 18% rise from $45.7/hr to $54.1/hr. The sensitivity to changing fuel prices is slightly smaller for lower hours used. For example, the increase at (4) is "only" 15%, from $55.1/hr to $63.5/hr.

What is not included yet are the so-called indirect costs. As the terms specifies, these cost are not a function of operating the paver. A good example is the overhead needed to cover the expense of running the main office (e.g., cost of estimating bids that are not successful), which commonly is assessed at 5%. Another such factor is *profit* (e.g., 15%) that could be charged directly to the equipment rather than the general overhead. This certainly can be used when the paver is rented out to other contractors.

In contrast, the tax savings of $19,227 could be considered as a cost reduction of approximately $2.3/hr.

The final results of calculating cost per hour with the given assumption are:

Direct cost	$45.7/hr
Indirect cost—Overhead & Profit 20%	$9.1/hr
Total hourly cost (gross)	$54.8/hr
Average Tax Savings ($19,227/6)	($2.3/hr)
Total hourly cost(net)	$52.5/hr

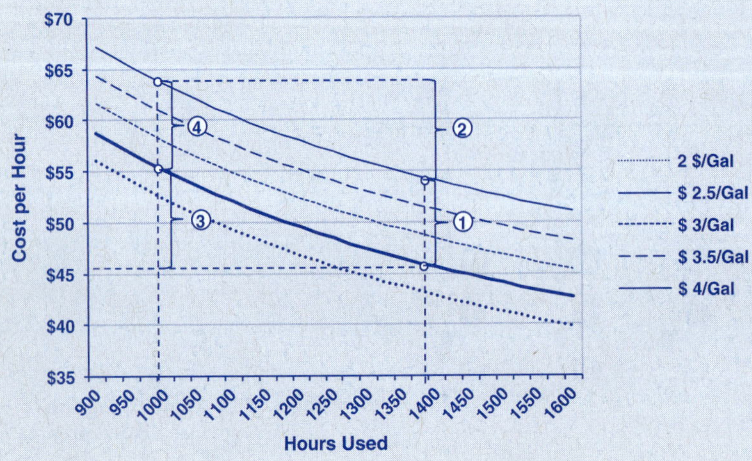

Figure 2.17 Sensitivity Analysis of Hours Used and Price of Fuel

Conclusion

Which of these numbers will be used depends on company policy and accounting method. For example, one policy could be that tax savings from equipment are used for specific strategic investments and not to reduce hourly cost. In the same vein, the profit margin might be changed in accordance with supply and demand.

Finally, it is self-evident that QualityPave could adjust the hourly cost rate for the paver every year or even every month. The problems that arise, though, are in estimation. What should the estimators use for their bids when the numbers constantly change? What is most baffling is the inverse relationship of low demand and hourly cost. As the sensitivity analysis showed, reduced hours of use, as happens in time of low demand, will cause the cost per hour to go up. This is certainly not what the estimator likes to see, higher estimates during times of high competition when everybody will try to keep their estimates as low as possible. In effect, the strategy should be to inversely link the cost per hour to the expected business situation. The question would still remain, by how much?

It is therefore most opportune to select one of the cost figures above most fitting to the company policy, and keep it fixed while adjusting the overall profit margin flexibly to respond to the supply-demand cycle.

2.8 Buy, Lease, or Rent?

Our cost calculations so far have assumed that the contractor always owns a piece of equipment. Counting the many equipment rental places across the United States, one knows that this can't always be true. How does one figure out what is the best strategy, buying, leasing, or renting?

2.8.1 QUALITATIVE COMPARISON OF THE OPTIONS

Owning: There is no question that owning a piece of equipment provides many advantages when there is a lot of work available and one does not have depend on rental companies that might not have the best-fitting machine available. The second key issue relates to the availability of capital. Especially small or young companies may not even have access to bank credit.

If the new equipment can be financed with capital from the company's equity, buying provides many benefits for the contractor. First, it helps in managing taxes for the next 5 years by allowing the choice among three different depreciation strategies. When a bank credit is used, liquid equity can be built up that will feed back into the cash flow stream, helping in the payment of bills before the contractor gets paid himself.

From an operational point of view, the contractor has the opportunity to train an operator to become highly efficient thanks to knowing the machine inside out. In the same vein, through a fine-tuned preventive maintenance scheme the efficiency, cost, life, and the value of the equipment can be kept at a high level.

Leasing: This is the alternative for a company that expects a lot of work for the next years but lacks capital or collateral to qualify for a bank credit. A lessor (leasing company) will purchase the desired piece of equipment and lease it to the contractor for a monthly payment that covers the ownership cost, minus the tax savings, as an annuity. The operating costs have to be borne by the lessee directly. Much as with owning, a big pro is that one has the resource available for a longer time. Also, operating costs can be controlled by preventive maintenance and the training of a very skillful operator. What might also turn out to be a pro is the riskless switch to a new piece of equipment at the end of its life. As the lessor owns the equipment he will find a way to sell it. This may turn into a trade-in that allows the contractor to move up to a more productive machine and shortcutting the loss due to functional obsolescence, discussed earlier.

Table 2.10 Matrix Linking Background Conditions to Options

Conditions	Buy	Lease	Rent
Sufficient cash/capital available	X		
Taxable income	X		
Lots of work for equipment for next years	X	X	
Excellent equipment operator available	X	X	
Preventive maintenance program implemented	X	X	
Interest in buying equipment only after it is tested		X	
Short on cash and collaterals		X	X
Desire to use most sophisticated equipment		X	X
Need to build up equipment on jobs quickly			X
Lack of experience with maintenance program			X
Only intermittent use for equipment			X

Renting: Generally, the hourly cost for renting is the highest for the three options. Why would anyone rent, you may ask? The key factor is the usage per year. If a company does not have sufficient work to keep the equipment busy, the ownership cost will have to be spread between fewer hours, thus pushing the hourly cost up, possibly far above the rental cost. When renting, the ownership and the long-term operating costs (e.g., maintenance) are paid by the rental company. In order to respond to breakdowns, most rental companies have 24/7 assistance. While the availability of equipment types for renting might be limited, larger outfits may provide an opportunity to rent just the right machine for a certain job.

Table 2.10 summarizes some of the main circumstances that favor buy-lease-rent options.

2.8.2 INFORMATION SOURCES FOR QUANTITATIVE DECISION MAKING

How should a contractor needing a piece of equipment make the most economically beneficial decision, considering business circumstances? The short answer is: By using as much unbiased information as possible. It is only natural that any equipment dealer, rental, or leasing company will publish data and information that puts their solutions in the best light for any situation. Thus, there is a need of data sources that have no interest in pushing one solution over another. Especially in need for assistance are contractors who venture into a new area of construction where they have little experience with equipment costs. What is an acceptable purchase or sales price for a new or used piece of machinery? What is a reasonable rental rate in an area? What is the effect of equipment age on the price of a piece of equipment? Finally, how does the equipment charge/hour compare to standard rates for a certain piece of equipment? These are only some of the questions that a comprehensive comparison needs to address.

Over the last 60 years, several services have emerged that offer data that meet the needs of a contractor making a decision related to a piece of equipment. Some companies offer up-to-date prices of equipment being purchased and sold, some show nationwide rental rates, while others have both.

The Blue Book: A first example is the "Rental Rate Blue Book for Construction Equipment," published by EquipmentWatch, San Jose (www.equipmentwatch.com). The sources of the data are rental companies located all across the United States. Table 2.11 presents some sample cost figures (not up-to-date) that are available in the Blue Book. These rates are averages that can be adapted to locations in the United States with the help of indexes that are provided. For example, rates in larger cities are lower than the average as more competition tends to push the hourly rates downward (see next section). The cost

Table 2.11 Examples of Average Equipment Rental Rates as a Function of Class and Age

Class		Age	
		5 Years	15 Years
Hydraulic Backhoe Excavators—Crawler-Mounted 			
Caterpillar 312C L, 0.68 cy, 90HP	$/hr	60.40	58.50
Caterpillar 325C L, 1.5 cy, 186HP	$/hr	117.50	113.70
Deere 120C, 0.79 cy, 89HP	$/hr	62.00	60.70
Hitachi ZAXIS 270LC-3, 1.75 cy, 188HP	$/hr	115.50	112.00
Komatsu PC120-6, 0.63 cy, 89HP	$/hr	57.60	55.80
Komatsu PC270LC-8, 1.85 cy, 187HP	$/hr	135.30	121.40
Wheel Loader: Diesel, 4WD, Articulated 			
Case 21D, 0.92 cy, 51 HP	$/hr	27.40	26.80
Komatsu WA250-5, 3 cy, 134 HP	$/hr	50.30	49.20
Caterpillar IT28G, 2.35 cy, 143 HP	$/hr	52.00	50.90
Wheel Loader-Backhoe 			
Deere 310G, Diesel, 1 cy, 70 HP, WD, Extendable backhoe stick, Digging	$/hr	34.60	33.90
Case 580 Super M Series 2, 2WD, 1 cy, 90 HP, Extendable backhoe stick, Digging Depth < 15'	$/hr	40.50	39.60
Case 580 Super M Series 2, 4WD, 1 cy, 90 HP, Extendable backhoe stick, Digging Depth < 15'	$/hr	42.30	41.50
Crawler-Mounted Asphalt Paver			
Lee-Boy 1000F, Diesel, 37 HP, Screed Model: Extendable	$/hr	43.40	42.40
Caterpillar AP-655C, Diesel, 158 HP, Screed Model: 8-16B Extend-A-Mat	$/hr	237.80	231.40

figures do not include the salary for operators as well as the contractor's overhead and profit. As we learned earlier, ownership, maintenance, and repairs are all included.

The class or size of the equipment is closely related to the HP of the engine. Reviewing the table for the backhoe excavator, it is clear that the prizes for units with similar HP and bucket capacity in cubic yards (cy) tend to cluster together. For example, a 5-year-old *Caterpillar 312C-0.68* cy (0.52 m^3)-90 HP, a *Deere 120C*-0.79 cy (0.6 m^3)-89 HP, or a *Komatsu PC120*-0.63 cy (0.48 m^3)-89 HP cost $60.40/hr, $62.00/hr, and $57.60/hr respectively. The Deere 120C costs $2–$4 more than the others but also has the largest bucket capacity.

The last example covers two classes of pavers. In Worked-Out Problem 2.5 we calculated that the direct cost was $45.70/hr. According to Table 2.11, this cost figure is close to the $43.40/hr of the 5-year-old *Lee-Boy 1000F, Diesel, 37 HP, Screed Model: Extendable*. In contrast, the $237.80/hr *Caterpillar AP-655C, Diesel, 158 HP, Screed Model: 8-16B Extend-A-Mat* clearly indicates that the 158 HP engine is powering a much larger and more expensive piece of equipment than the 37 HP Lee-Boy.

RSMeans: Published by Reed Construction Data, RSMeans offers a series of specialized for building, heavy-highway, electrical, mechanical, construction etc. In addition to selected equipment, RSMeans also collects cost data on material and labor for the most important building trades.

Figure 2.18 is extracted from of the RSMeans Heavy Construction Cost Data Book. As implied in the name, it covers operations that are related to building highways including bridges—all requiring heavy machinery. As can be expected, earthmoving equipment and cranes are covered as well as pile-driving machinery needed to build bridge foundations. The page provides also some basic guidance: *"The selection of earthmoving equipment depends upon the type and quantity of material, moisture content, ..."*

Reference Tables

General Requirements R0154 Construction Aids

R015433-10 Contractor Equipment

Rental Rates shown elsewhere in the book pertain to late model high quality machines in excellent working condition, rented from equipment dealers. Rental rates from contractors may be substantially lower than the rental rates from equipment dealers depending upon economic conditions; for older, less productive machines, reduce rates by a maximum of 15%. Any overtime must be added to the base rates. For shift work, rates are lower. Usual rule of thumb is 150% of one shift rate for two shifts; 200% for three shifts.

For periods of less than one week, operated equipment is usually more economical to rent than renting bare equipment and hiring an operator.

Costs to move equipment to a job site (mobilization) or from a job site (demobilization) are not included in rental rates, nor in any Equipment costs on any Unit Price line items or crew listings. These costs can be found elsewhere. If a piece of equipment is already at a job site, it is not appropriate to utilize mob/demob costs in an estimate again.

Rental rates vary throughout the country with larger cities generally having lower rates. Lease plans for new equipment are available for periods in excess of six months with a percentage of payments applying toward purchase.

Rental rates can also be treated as reimbursement costs for contractor-owned equipment. Owned equipment costs include depreciation, loan payments, interest, taxes, insurance, storage, and major repairs.

Monthly rental rates vary from 2% to 5% of the cost of the equipment depending on the anticipated life of the equipment and its wearing parts. Weekly rates are about 1/3 the monthly rates and daily rental rates about 1/3 the weekly rate.

The hourly operating costs for each piece of equipment include costs to the user such as fuel, oil, lubrication, normal expendables for the equipment, and a percentage of mechanic's wages chargeable to maintenance. The hourly operating costs listed do not include the operator's wages.

The daily cost for equipment used in the standard crews is figured by dividing the weekly rate by five, then adding eight times the hourly operating cost to give the total daily equipment cost, not including the operator. This figure is in the right hand column of the Equipment listings under Equipment Cost/Day.

Pile Driving rates shown for pile hammer and extractor do not include leads, crane, boiler or compressor. Vibratory pile driving requires an added field specialist during set-up and pile driving operation for the electric model. The hydraulic model requires a field specialist for set-up only. Up to 125 reuses of sheet piling are possible using vibratory drivers. For normal conditions, crane capacity for hammer type and size are as follows.

Crane Capacity	Hammer Type and Size		
	Air or Steam	Diesel	Vibratory
25 ton	to 8,750 ft.-lb.		70 H.P.
40 ton	15,000 ft.-lb.	to 32,000 ft.-lb.	170 H.P.
60 ton	25,000 ft.-lb.		300 H.P.
100 ton		112,000 ft.-lb.	

Cranes should be specified for the job by size, building and site characteristics, availability, performance characteristics, and duration of time required.

Backhoes & Shovels rent for about the same as equivalent size cranes but maintenance and operating expense is higher. Crane operators rate must be adjusted for high boom heights. Average adjustments: for 150′ boom add 2% per hour; over 185′, add 4% per hour; over 210′, add 6% per hour; over 250′, add 8% per hour and over 295′, add 12% per hour.

Tower Cranes of the climbing or static type have jibs from 50′ to 200′ and capacities at maximum reach range from 4,000 to 14,000 pounds. Lifting capacities increase up to maximum load as the hook radius decreases.

Typical rental rates, based on purchase price are about 2% to 3% per month.

Erection and dismantling runs between 500 and 2000 labor hours. Climbing operation takes 10 labor hours per 20′ climb. Crane dead time is about 5 hours per 40′ climb. If crane is bolted to side of the building add cost of ties and extra mast sections. Climbing cranes have from 80′ to 180′ of mast while static cranes have 80′ to 800′ of mast.

Truck Cranes can be converted to tower cranes by using tower attachments. Mast heights over 400′ have been used.

A single 100′ high material **Hoist and Tower** can be erected and dismantled in about 400 labor hours; a double 100′ high hoist and tower in about 600 labor hours. Erection times for additional heights are 3 and 4 labor hours per vertical foot respectively up to 150′, and 4 to 5 labor hours per vertical foot over 150′ high. A 40′ high portable Buck hoist takes about 160 labor hours to erect and dismantle. Additional heights take 2 labor hours per vertical foot to 80′ and 3 labor hours per vertical foot for the next 100′. Most material hoists do not meet local code requirements for carrying personnel.

A 150′ high **Personnel Hoist** requires about 500 to 800 labor hours to erect and dismantle. Budget erection time at 5 labor hours per vertical foot for all trades. Local code requirements or labor scarcity requiring overtime can add up to 50% to any of the above erection costs.

Earthmoving Equipment: The selection of earthmoving equipment depends upon the type and quantity of material, moisture content, haul distance, haul road, time available, and equipment available. Short haul cut and fill operations may require dozers only, while another operation may require excavators, a fleet of trucks, and spreading and compaction equipment. Stockpiled material and granular material are easily excavated with front end loaders. Scrapers are most economically used with hauls between 300′ and 1-1/2 miles if adequate haul roads can be maintained. Shovels are often used for blasted rock and any material where a vertical face of 8′ or more can be excavated. Special conditions may dictate the use of draglines, clamshells, or backhoes. Spreading and compaction equipment must be matched to the soil characteristics, the compaction required and the rate the fill is being supplied.

Source: Copyright RSMeans Heavy Construction Cost Data Book 2012

Figure 2.18 General Requirements Related to Contractor Equipment

Table 2.12 RSMeans Rental Rates for Earthwork Equipment

01 54 | Construction Aids								
01 54 33 | Equipment Rental			UNIT	HOURLY OPER. COST	RENT PER DAY	RENT PER WEEK	RENT PER MONTH	EQUIPMENT COST/DAY
20	0010	**EARTHWORK EQUIPMENT RENTAL** without operators R015433 -10						
	0040	Aggregate spreader, push type 8' to 12' wide	Ea.	2.70	25.50	76	228	36.80
	0045	Tailgate type, 8' wide		2.55	33	99	297	40.20
	0055	Earth auger, truck-mounted, for fence & sign posts, utility poles		15.30	470	1,405	4,225	403.40
	0060	For borings and monitoring wells		41.30	645	1,940	5,825	718.40
	0070	Portable, trailer mounted		2.75	31.50	95	285	41
	0075	Truck-mounted, for caissons, water wells		89.05	2,875	8,645	25,900	2,441
	0080	Horizontal boring machine, 12" to 36" diameter, 45 H.P.		22.20	192	575	1,725	292.60
	0090	12" to 48" diameter, 65 H.P.		30.95	335	1,000	3,000	447.60
	0095	Auger, for fence posts, gas engine, hand held		.50	6	18	54	7.60
	0100	Excavator, diesel hydraulic, crawler mounted, 1/2 C.Y. capacity		18.80	370	1,105	3,325	371.40
	0120	5/8 C.Y. capacity		28.60	540	1,615	4,850	551.80
	0140	3/4 C.Y. capacity		34.90	615	1,840	5,525	647.20
	0150	1 C.Y. capacity	↓	42.70	680	2,035	6,100	748.60

Source: Copyright RSMeans Heavy Construction Cost Data Book 2012

Table 2.12 tabulates the total operating cost for renting some example pieces of machinery needed for soil boring and excavation. For example, RSMeans' 12-digit line number 01 54 33.20 0075 identifies a piece of truck-mounted boring equipment to drill caissons and water wells. Here, the operating cost/hour is kept separate from the rental cost, which is given per day, per week, and even per month. In Figure 2.18 we read: "*The hourly operating costs for each piece of equipment include costs to the user such as fuel, oil, lubrication, normal expendables for the equipment, and a percentage of mechanic's wages chargeable to maintenance. The hourly operating costs listed do not include the operator's wages.*" The daily costs are based on a weekly rental duration. It is established by: "*. . . dividing the weekly rate by five, then adding eight time the hourly operating cost.*" For example:

$$\text{Daily cost of } 01\,54\,33.20\,0075 = 8,645 \text{ per week}/5 \text{ days} + 8 \text{ hrs} * \$89.05/\text{hr}$$

$$= (\$1,729 + \$712)/\text{day}$$

Unlike the Blue Book, RSMeans does not classify backhoe excavators using manufacturers and HP but rather by the size of the bucket they use. Of course, larger buckets require larger engines delivering more power. Let's see how the two data sources compare by computing hourly operating costs, which should match the hourly rental cost based on the Blue Book.

Costs for 01 54 33.20 0150, 1 cuyd (0.76455 m^3) capacity crawler excavator:

Operator cost = \$42.70/hr
Daily Rental = \$680/day
Weekly rental = \$2,035/week

(See Figure 2.17: *Costs to move equipment to a job site (mobilization) or from job site (demobilization) are not included in rental rates, nor in any Equipment costs . . .*)

Daily rental cost based on week's rent		= \$407/day
Hourly rental cost	= \$407/8 hrs	= \$50.95/hr
Hourly rental based on a day's rent	= \$680/8 hrs	= \$85.0/hr
Total hourly cost with 1 day's rent	= \$42.70/hr +\$85.00/hr	= \$127.7/hr

Table 2.11 does not list a 1 cy excavator, but one with 0.8 cy costing \$62/hr and 1.5 cy for \$117.5/hr. Through a simple interpolation, we find that the cost is expected to be \$77.50/hr.

As we can see, the RSMeans hourly cost is slightly higher than the interpolated value. If we assume that the backhoe is rented longer-term, such as for a month, the hourly rental cost would be reduced to \$6,100/(4 weeks * 5 days * 8 hrs) = \$38.1/hr, which will result in a total hourly cost of \$81/hr. The remaining difference is not surprising, in that RSMeans separates the maintenance and repair cost as well as fuel cost, not leaving it to the rental company to cover. Furthermore, since both use averages adjusted for the location of the rental, differences may even out eventually. Let's investigate the effect of locations on the cost of the backhoe based on RSMeans data.

Anybody who has traveled to different regions in the United States has become aware of the differences in food and gas prices influenced by state taxes as well as supply and demand. The same holds true for construction material and installation, which includes equipment.

For the crawler excavator for which we came up with an average rate of \$100.20/hr, the adjustment factor based on location creates large differences. Here are some example adjustments that are based on Table 2.13.

Hourly cost of a crawler excavator rented in:

San Jose, California	$1.351 * \$93.65/\text{hr} = \$126.52/\text{hr}$
Grand Junction, CO	$0.796 * \$93.65/\text{hr} = \$\ 74.54/\text{hr}$
Washington, DC	$0.928 * \$93.65/\text{hr} = \$\ 86.91/\text{hr}$
Panama City, FL	$0.605 * \$93.65/\text{hr} = \$\ 56.66/\text{hr}$

Table 2.13 Adjustment Factor Based on Location

Location Factors

Costs shown in RSMeans cost data publications are based on national averages for materials and installation. To adjust these costs to a specific location, simply multiply the base cost by the factor and divide by 100 for that city. The data is arranged alphabetically by state and postal zip code numbers. For a city not listed, use the factor for a nearby city with similar economic characteristics.

STATE/ZIP	CITY	MAT.	INST.	TOTAL
CALIFORNIA (CONT'D)				
951	San Jose	103.2	135.1	117.3
952	Stockton	100.7	117.2	108.0
953	Modesto	100.7	116.9	107.9
954	Santa Rosa	100.3	135.0	115.6
955	Eureka	101.6	114.2	107.2
959	Marysville	100.9	114.4	106.8
960	Redding	102.8	114.4	107.9
961	Susanville	101.5	114.4	107.2
COLORADO				
800-802	Denver	101.7	84.7	94.2
803	Boulder	98.6	82.9	91.7
804	Golden	100.7	81.7	92.3
805	Fort Collins	102.6	79.1	92.2
806	Greeley	100.0	78.2	90.4
807	Fort Morgan	99.3	82.0	91.7
808-809	Colorado Springs	101.6	81.7	92.8
810	Pueblo	100.4	80.8	91.8
811	Alamosa	101.5	76.7	90.6
812	Salida	101.3	78.3	91.2
813	Durango	101.9	79.6	92.1
814	Montrose	100.8	79.3	91.3
815	Grand Junction	104.2	79.6	93.4
816	Glenwood Springs	101.6	82.2	93.1

STATE/ZIP	CITY	MAT.	INST.	TOTAL
CONNECTICUT				
060	New Britain	101.1	118.6	108.8
061	Hartford	102.0	118.7	109.4
062	Willimantic	101.7	118.4	109.1
063	New London	97.6	118.6	106.9
064	Meriden	99.7	118.6	108.0
065	New Haven	102.8	118.6	109.8
066	Bridgeport	102.2	118.7	109.5
067	Waterbury	101.8	118.6	109.2
068	Norwalk	101.8	126.3	112.6
069	Stamford	101.9	126.3	112.7
D.C.				
200-205	Washington	102.4	92.8	98.2
DELAWARE				
197	Newark	99.4	107.9	103.2
198	Wilmington	99.1	108.0	103.0
199	Dover	99.5	107.9	103.2
FLORIDA				
320,322	Jacksonville	98.7	69.4	85.7
321	Daytona Beach	98.8	75.2	88.4
323	Tallahassee	99.4	59.3	81.7
324	Panama City	100.1	60.5	82.7
325	Pensacola	102.6	64.1	85.6

Source: Copyright RSMeans Heavy Construction Cost Data Book 2012

2.9 Chapter Two Review

THE EQUIPUZZLE

The letters in the circled boxes can be combined to identify a formal method used to spread out the cost of an expensive asset:

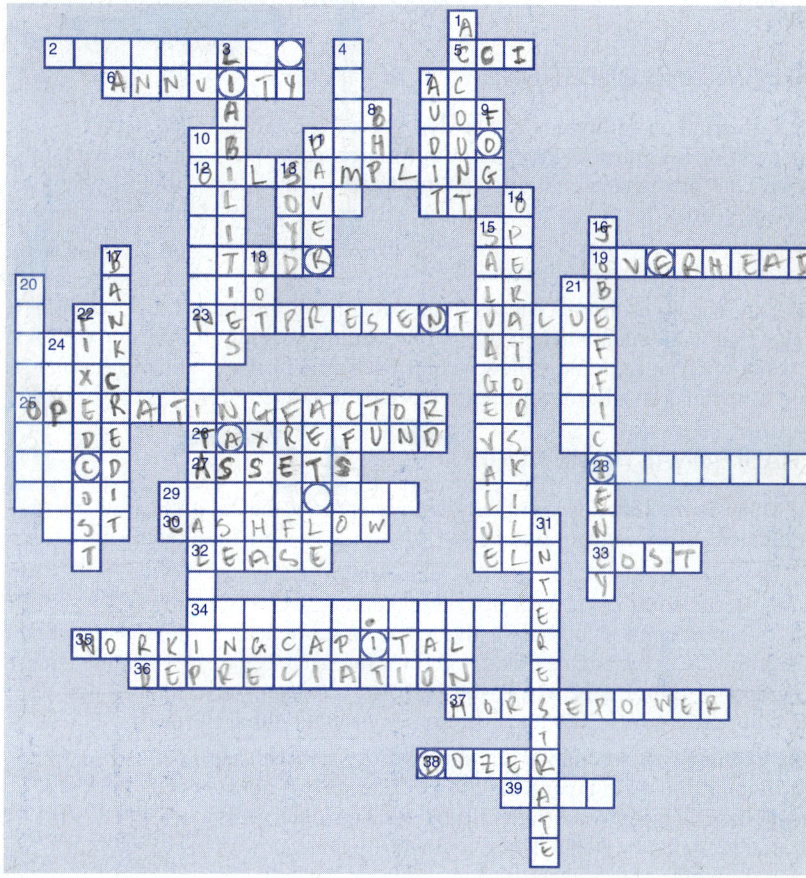

Across

2. Calculated by subtracting the depreciations or amortization from the acquisition cost of a tangible asset (two words)
5. Short for Cumulative Cost Index
6. Uniform or fixed amounts of payments at regular intervals that consider the time-value of money
12. A method to proactively assess the health of engine and transmission in terms of wear and tear (two words)
19. A group of cost items that are not directly associated with the construction of a billable work item (two words)
23. NPV (three words)
25. A rate that considers the combined operational situations of a job such as hard soil and tight work space (two factors)
26. An amount that can be subtracted from the taxes to be paid in one year (two words)
27. List of the economic value of all that is owned by a company
28. A financial charge on individuals or property owners by a state to support its functions (two words)
29. Causes the loss of purchasing power
30. The stream of liquid assets in and out of a financial portfolio (two words)
32. A real-estate rental agreement
33. The value of money that one has to give up in order acquire resources or to produce an output
34. Results in an uniform amount of depreciation
35. The assets available to conduct the daily operation of a company (also referred to as current assets) (two words)
36. The state of becoming outdated and decreasing in usefulness
37. An old measure, defined by James Watt, describing the amount of energy being created or used per unit of time
38. Short for bulldozer or crawler dozer
39. Funds that used to be stored in a cassa

Down

1. A virtual space in which to post a financial transaction
3. Name for all the financial obligations of a company toward borrowers
4. That what is left after all the costs of operation are subtracted from the revenue
7. A formal review of the accounts
8. Short for brake horsepower
9. Short for filters, oil, and grease
10. Allows the comparison of different investment alternatives when making an economic decision (three words)
11. Equipment that spreads asphalt or concrete onto a prepared road base
13. Short for Sum-of-Years' Digits depreciation method.
14. A ratio that describes the proficiency of an operator in performing a task (two words)
15. Another term for residual value (two words)

16. A measure that describes how smoothly a job is being done (two words)
17. The amount of funds a bank is willing to loan an individual or business (two words)
18. Short for Department of Transportation
20. Another word for shareholders equity (two words)
21. An economic plan of future cost and revenues
22. That portion of the total cost that does not vary with the amount of construction going on (two words)
24. Expenditure resulting in the outflow of a (liquid) asset
31. The percentage of money that a borrower has to pay a lender (two words)

JOURNALING QUESTIONS

(1) Construction equipment can be depreciated for value/business or tax purposes. Explain the differences in purpose. One of them helps the company to save taxes for several years after the purchase of a new piece of equipment. How is that possible when only business expenses can be used to calculate income?

(2) Functional obsolescence leads to the loss of equipment value that is independent of wear and tear on the machinery. Show on an example of how a contractor might calculate functional obsolescence using replacement value.

(3) Repair is sometimes referred to as reactive maintenance. Explain the concept behind this statement. What are the main factors that influence the repair cost of a piece of equipment over its service life?

(4) Owning a piece of equipment, such as the 16 cranes used in the steel erection of the Empire Steel Building, can have multiple benefits over renting equipment. Please address this statement both from the technical and the financial point of view.

TRADITIONAL HOMEWORK

(1) What is the monthly compounded interest rate that is equivalent to an annual rate of 10%?

(2) You are trying to calculate the fuel cost for a 150-ton lattice boom truck crane with two diesel engines, a 290-hp for the crane, and a 440-hp for the truck. Assume a fuel factor for the crane engine of 0.024 and for the truck 0.005. The cost of diesel is $2.94/gal ($0.78/l) for off-road and $3.42 for on-road use. What are the hourly costs for either?

(3) A company just bought a brand-new dozer for $290,000 and would like to compare the M&RC after 15,000 hours using the CCI factor coefficient in Table 2.6. The first 5 years the dozer will do harsh work, followed by 5 years of average work, and 5 years of light work. What will the cumulative cost be at the end of 15,000 hours for both average and excellent maintenance. What will be the hourly M&RC at that time (not the average over 15,000 hours)?

(4) Assume that the 4 larger cranes used for the Empire State Building cost $66,000 each and the 12 smaller ones $38,000 in 2009. The service life was 18 months at which time they were sold with a salvage value about 20% of the purchase price, and the interest rate for commercial loans at that time was 4.5%.

(a) How much was the monthly annuity that Starrett Brothers, Inc. and Eken, Inc had to pay if the life of the bank loan was also 18 months?

(b) How much would they be able to save in monthly payments if the bank amortized only the difference between the purchase price and the salvage value, covering the rest of the purchase price with a simple loan?

(5) Assume that the ownership cost of the cranes consisted of straight-line depreciation over 18 months with 4.5% interest and annual insurance cost of 15% of its purchase price.

The average operating cost per week consisted of maintenance, 0.5% of purchase price, and electrical power need 0.4% of purchase price. Finally, we need to add 10% overhead for the safety brigade that was constantly monitoring the craning operation and 5% general overhead. How much was the cost for all cranes per floor when the steel gangs erected an average of 4.5 floors per week?

(6) RSMeans lists an 80-HP wheel backhoe-loader with a 1–$1^1/_4$ cy (0.76–0.95 (m^3) capacity bucket with a daily rental price of $320, weekly $960 and monthly $2,875. The operating cost is $24.45/hr.

(a) Calculate the total hourly rate for the backhoe-loader (using the RSMeans method).

(b) Compute an hourly rate based on the Blue Book for an equivalent machine.

(c) Explain the difference, if any.

OPEN-ENDED QUESTION

(1) As you look around on construction sites, you realize that lifting and moving material, elements, assemblies, and so on is a central theme. Looking back in history, you realize that lifting technologies available at a particular time period dictated the amount of human power needed to build large structures. Reflect on the effect that simple machines had on construction methods over time (from Egyptians until today). Predict future developments in technologies related to lifting and moving objects in construction.

BIBLIOGRAPHY

Army Corps of Engineers. *Construction Equipment Ownership and Operating Expense Schedule*. EP 1110-1-8, vol 1. Washington, DC: DoD, 2009.

Bhurisith, I., and A. Touran. "Case Study of Obsolescence and Equipment Productivity, Technical Note," *Constr. Engrg. Mgmt.* 128, no. 4 (2002).

Fan, H., S. AbouRizk, H. Kim, and O. Zaïane. "Assessing Residual Value of Heavy Construction Equipment Using Predictive Data Mining Model." *J. Constr. Engrg. Mgmt.* 22, no. 3 (2008).

IRS, *How to Depreciate Property*. U.S. Department of the Treasury, Publication 946 Cat. No. 13081F, 2009.

Jaafari, A. and Mateffy, V. K., "Realistic Model for Equipment Replacement," *J. Constr. Engrg. Mgmt.*, 116 (3), 1990.

Lucko, G., M.C. Vorster, and C.M. Anderson-Cook. "Unknown Element of Owning Costs—Impact of Residual Value." *J. Const.* 133, no. 1 (2007).

Mitchell, Z. W. "A Statistical Analysis of Construction Equipment Repair Costs Using Field Data & the Cumulative Cost Model." Ph.D. Dissertation, Virginia Polytechnic Inst. State Univ., Blacksburg, VA, 1998.

Peterson, S. J. *Construction Accounting and Financial Management*. Upper Saddle River, NJ: Prentice-Hall, 2008.

RSMeans. *Heavy Construction Cost Data*. Kingston, MA: Reed Construction Data Company.

SAE (Society of Automotive Engineers). *Evaluation Criteria for Reliability-Centered Maintenance (RCM) Processes*, SAE Standard, Product Code: JA1011. Warrendale, PA, 2009.

Tatum, C.B., M. Vorster, and M.G. Klingler. "Systems Analysis of Technical Advancement in Earthmoving Equipment." *J. Constr. Engrg. Mgmt.* 132, no. 9 (2006).

chapter three

Safely Deploying Massive Power

The majority of the construction technologies discussed in the later part of this book depend on powered machines and mechanisms exerting large forces. Even scaffolds, framework, and formwork have become mobile systems requiring hydraulic power for operation. With the rise of size and complexity came the need for better controls that led to a totally new method of communication between the equipment and the operator and also between components of the machines themselves. Construction equipment operators maneuver huge pieces of equipment by moving levers, foot pedals, throw switches, or joysticks, and use Global Positioning System (GPS) technology as spatial guidance. On the other hand, the risk of health problems and injuries faced by operators and workers has received increased attention from manufacturers and governmental agencies. After introducing many concepts related to powering modern equipment, we will review what is being done to prevent accidents and to protect the human body (see Table 3.1).

Table 3.1 Main Topics Covered in This Chapter

Power & Transmission & Actuation		Strains on Operator's Body	
Critical Mechanisms	**Physics**	**Hazards**	**Rules and Measures**
✓ Powertrain	✓ Thermal Expansion—Gas	✓ Lower Back Pain	✓ Ergonomics
✓ Internal Combustion Engine	✓ Electricity	✓ Lifting Excessive Weight	✓ Seat Damping
✓ Electric Motor—Generator	✓ Chemical Reaction	✓ Falls—High Jumps	✓ Limited Time Exposure
✓ Muscle Contraction	✓ Friction Losses	✓ Whole-Body Vibration	✓ Permissible Exposure Limit
✓ Steam Cylinder	✓ Pressure—Volume	✓ Hands-Arms Vibration	✓ Hearing Protection
✓ Driveshaft-Gear-Coupler-Chain	✓ Energy—Work-Horsepower	✓ Noise—Hearing Loss	✓ Machinery Noise Reduction
✓ Pump—Pipe—Hose	✓ Power Losses	✓ Heat Stress—Cold	✓ Metabolism
✓ Hydraulic—Pneumatic—Steam Cylinder/ Motor	✓ Rolling Resistance	✓ Lumbar Vertebrae	✓ Heat Index
✓ Block-and-Tackle	✓ Aerodynamic Drag	✓ Fibrous Disks	✓ Cooling-Off Periods
✓ Treadwheel	✓ Fluid Statics—Hydrostatic Pressure (Pascal)	✓ Pinching Spinal Nerve	✓ Cold-Protection Clothing
✓ Horsepower	✓ Analog-Digital Conversion	✓ Tendinitis	✓ Biomechanics
✓ Winch—Rope—Sheave	✓ Electronics	✓ Body Resonance	✓ INDG242 (rev1)
✓ Diesel Engine—Two Stroke	✓ Mechatronics		✓ OSHA 29 CFR 1910.95
✓ Locomotion (Wheel-Crawl)	✓ Computer Control		✓ ISO 2631-1: 1997
✓ Steering-Braking			✓ ISO 2204
✓ Gear Ratio—Gear Box			✓ ANSI S12.13
✓ Hydraulic Pressure Pump			✓ NIOSH TIC-2 No. 20034733
✓ Proportional Electro Valves			✓ OSHA TED 01-00-015: 4
✓ Piston Rod			
✓ Joystick Control			
✓ Hydraulic Hoses			

3.1 The Mechanical Powertrain

Creating any kind of mechanical motion within earth's gravitational field requires energy that can be easily created and transmitted to the location of use. Appropriately, the word *energy* is borrowed from the Greek word "energeia," meaning activity or operation. For example, a "hand-operated" shovel has to transform a person's muscle energy into a cutting force at the edge of the blade when digging. We saw in Chapter 1 how simple machines are extremely efficient in translating and altering forces. Hence it is no surprise that they make excellent mechanisms to transmit energy from the point of creation to where it is needed with minimal losses. Borrowing from the concept of a train that has its source of energy at one end, the assembly of parts distributing power in construction equipment is called the powertrain.

Figure 3.1 models the different components of the "train" indicating also the main sources of power losses. The list of energy sources in ❶ shows the predominant fuels tapped to energize a motor, including the ATP (adenosine triphosphate) fueling the human muscle. William Rankine showed how chemical fuel, such as coal, can be converted into work. For example, heat turns water into a vapor that takes up much more space. If the volume of the vapor is contained, the vapor can be pressurized, which subsequently can be turned into a mechanical force F (= Area * Pressure) pushing a piston inside a cylinder. The system that turns energy into a mechanical motion, such as the turning of a shaft, is called the motor or engine ❷. The potential and kinetic energy of flowing water have long been harvested using various water wheels to drive rock-breaking hammers, irrigation pumps, flour or steel mills, or pumps necessary to keep coal or ore mines dry. As one watches a construction site today, internal combustion engines and electrical motors are the dominant sources of power. Although one of the key laws in physics states that energy can be transformed into other forms keeping the scalar amount, only part of the energy that feeds such motors will actually produce work. Why is that? It's because all the moving parts along the powertrain demand a "tax" commonly referred to as energy loss. In fact, the transmission ❸ is a major contributor to this energy loss as a result of mechanical friction in bearings, chains, pins, and hydraulic friction inside pipes and hoses. Losses also occur in gears that are needed to change speed and torque, block-and-tackle pulleys, idlers, valves, and so forth.

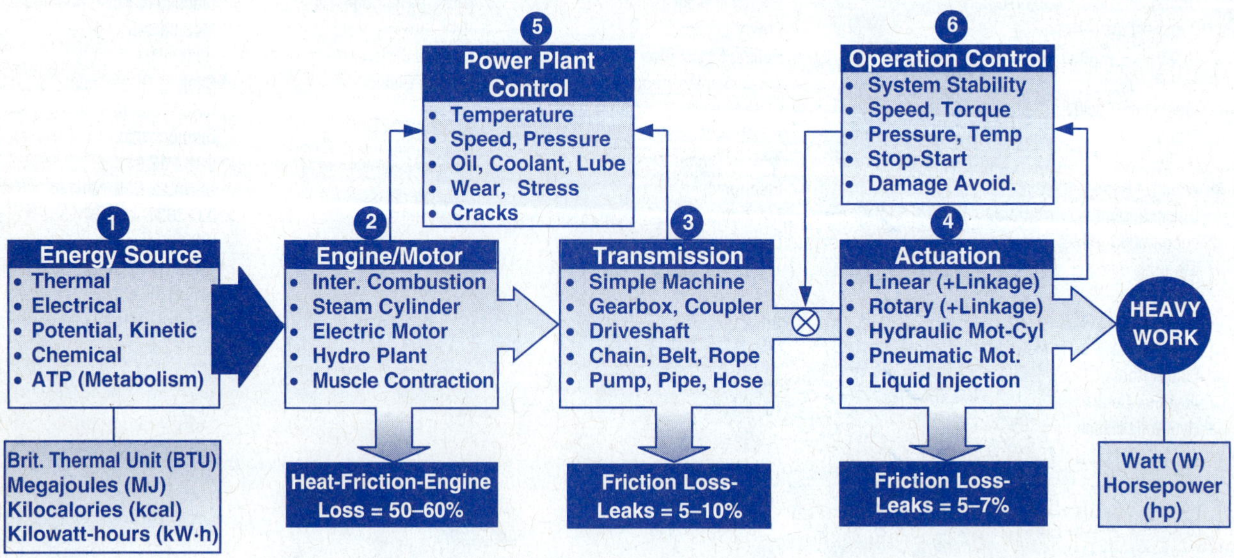

Figure 3.1 Powertrain with Control Mechanisms

The "front rail car" of the powertrain is the actuator ❹, a mechanical device that converts energy into actions or motions to execute the desired work. Such actuators include hydraulic motors and cylinders that move telescoping crane booms or lift heavy objects off the ground. In olden times, continuous wheels or chains lifted buckets full of water or lifted and released heavy hammers to ram wooden piles into the soft ground. Unfortunately, these actuators don't operate without energy loss, thus adding yet another amount to the total energy loss along the powertrain.

As with any manmade system, the powertrain needs to be controlled. Figure 3.1 shows two main modules. The first controls the engine and transmission mainly to ensure the safe and continuous supply of sufficient power to the actuators, a job that was traditionally performed by firemen, engine mechanics, and oilers. A separate module is dedicated to controlling the actuators, a task that is carried out by the engineer or operator. Special knowledge and skill was always needed to direct the actuators, which have greatly changed over time, to do the intended work effectively and safely.

William Rankine (1820–1872)

The Scottish engineer described the physical principles behind the steam engine, now called the Rankine cycle, converting heat into work. The cycle begins with water that is compressed by a pump (e.g., 50 bar) and then heated beyond its boiling point into a dry vapor. The high pressure gas is then let into a cylinder or a turbine casing where it expands (e.g., from 50 bar to 0.1 bar), pushing a piston or the blades of a rotor. The wet vapor is cooled down in a condenser, removing its high entropy and thus turning it back into a liquid.

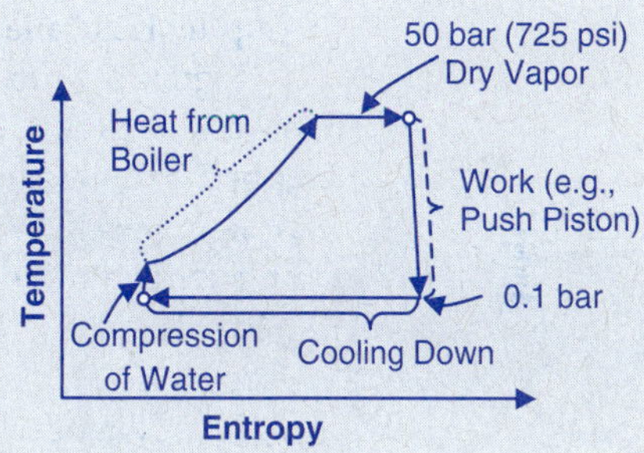

3.1.1 A HISTORICAL VIEW OF THE POWERTRAIN

As mentioned, the cost-efficient production and transmission of energy to execute heavy work with maximum effect and minimum losses have been key drivers of technology for the last 6,000 years. In this section we will look at some examples that stand for a large number of others that could have been selected.

The construction of large cathedrals, castles, and protective city walls challenged the master builders in medieval times to design new devices for lifting large objects and weights. Figure 3.2 displays one favorite crane system that was used, the mobile treadwheel, with a geared transmission to winch drums operating one or several hoist lines. It is interesting to note that instead of one block-and-tackle, the classic machine of

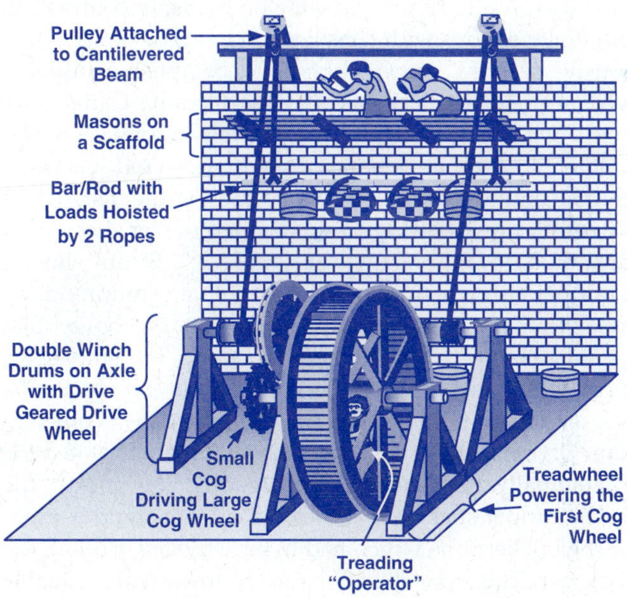

Figure 3.2 Medieval Wall Construction with Treadwheel Crane

Table 3.2 Powertrain of the Treadwheel Crane

Powertrain	Description
(1) Energy Source	Human ATP and gravity
(2) Motor	Walking human in treadwheel
(3) Transmission	Cogged and geared wheels
(4) Actuation	Winding drum of winch, ropes, pulleys, and "spreader" bar
(5) Power Plant Control	Treadwheel operator
(6) Operational Control	Walking by treadwheel operator

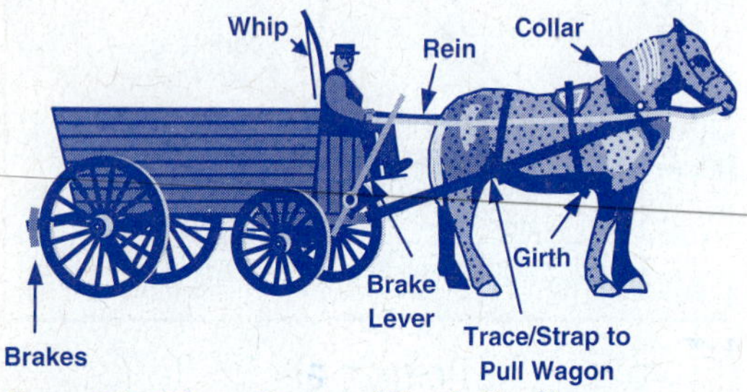

Figure 3.3 Horse-Powered Earth Hauling

Table 3.3 Powertrain of Horse-Drawn-Wagon

Powertrain	Description
(1) Energy Source	Animal ATP
(2) Engine	Skeletal muscles of horse
(3) Transmission	Horse harness
(4) Actuation	Horse pushing with shoulder and chest against collar
(5) Power Plant Control	Driver monitoring condition of horse and harness
(6) Operational Control	Driver using rein to steer, voice and whip to change speed, brakes for downhill

the wheel-with-axle has been combined with a wheel-based gear levering capability to construct a crane able to lift multiple loads suspended from a rod. Two overhead pulleys are attached to cantilevered beams supported by a wooden frame and a counterweight (not visible). Table 3.2 summarizes the main components of the powertrain for the medieval treadwheel crane for masonry construction.

Transporting heavy material for construction, including excavated soil for canals and tunnels, relied heavily on animal power, wheels, and axles. Heavy wagons, carts, scrapers, and graders were pulled by one animal or by a team of draft horses or donkeys. In other parts of the world, where the horse was less known, oxen, water buffalo, yaks, or even camels and elephants were used as energy providers and engine in one. Figure 3.3 presents a one-horse combination ready to be loaded with rock or soil. The stocky brewery horse is hitched to a two-axle wagon with a rotating front axle allowing a certain amount of steering. A key element of the harness is the collar to which the trace is attached. The padded collar does not limit the horse's movements and distributes the pulling force. Brakes are installed at the two rear wheels as the horse is unable to apply the necessary force. The driver is supplied with a levered hand, foot, or screw-brake. Table 3.3 outlines the powertrain of a horse-powered wagon.

James Watt's steam engine opened the gates for a large number of applications where heavy work had to be done. Of course, the new source of power gave rise to many innovative developments. A significant step was accomplished by mounting the steam plant onto four wheels providing the power for locomotion and the pulling of heavy loads. Unfortunately, those gargantuan road vehicles frightened and endangered animals and people to death when they steamed through the town squares. It did not take long for them to be prohibited from traveling on roads. In contrast, an immediate love story with construction ensued. Projects that previously seemed impossible to accomplish suddenly received a green light. How could the Panama Canal have been built without the large steam shovels mounted on railcars? Other, smaller versions were put on powered wheels or tracks to be employed in road construction or the installation of utilities. Figure 3.4 introduces a wheeled power shovel that was used to load the heavy-duty horse wagon shown on Figure 3.3. Steam shovels needed three to seven people to operate, depending on their size: at a minimum an operator, boiler engineer, and a fireman. Steam was generated in the boiler that surrounded the firebox, which was connected via heating pipes to the smokebox, which released the smoke and created a draft inside the firebox. The firebox had to be fed from the coal bunker that was also at the back of the shovel. The steam was piped to the engines that operated the winches controlling the ropes, the motors that rotated the engine house with the boom and finally the wheels. When loading the bucket, the operator had to crowd or thrust the dipper stick so that the bucket teeth could cut into the soil while, at the same time, the bucket rope winch had to be activated, pulling the bucket upward thus filling the dipper. Then, the shovel is rotated toward an available

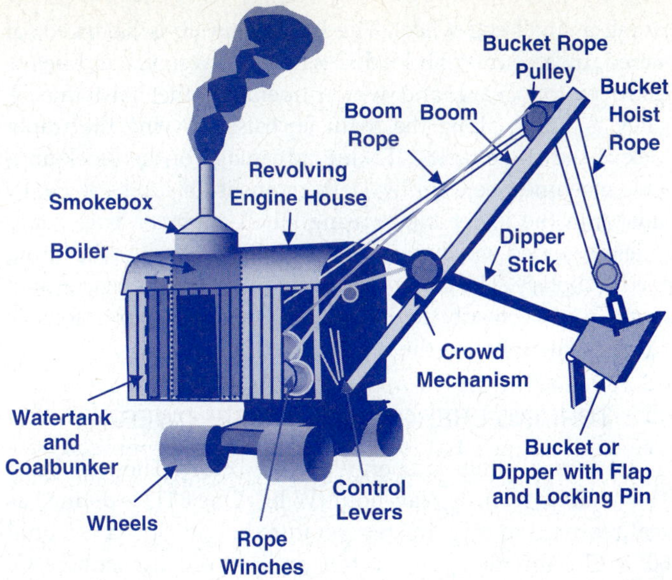

Figure 3.4 Wheeled Steam Shovel

wagon, truck, or rail car. The locking pin that kept the bucket door closed could be released remotely thus allows the load to drop out and a new cycle to start. (Table 3.4 outlines the powertrain of the steam shovel.) The small shovel shown in Figure 3.4 is built on wheels, which provide significantly more maneuverability when compared to tracked or rail-based systems. Of course, maneuverability was not a big issue in large excavations, but road construction and utility work benefited heavily from wheeled shovels.

The last example demonstrating the historic development of the powertrain for construction machinery is the tower crane. Depicted in Figure 3.5 is a common tower crane consisting of a vertical tower made of a lightweight space truss and rotating or slewing horizontal jibs mounted at the top of the tower. The 360-degree rotation is made possible by a turntable supported by heavy-duty bearings and revolved by an electric motor

Table 3.4 Powertrain for Steam Shovel

Powertrain	Description
(1) Energy Source	Chemical, burning of coal
(2) Engine	Boiler, piston, steam cylinder
(3) Transmission	Winch, ropes, block-and-tackle
(4) Actuation	Hoist line pull, boom cable, dipper stick crowding mech., rotational mech., flap lock
(5) Power Plant Control	Boiler pressure, level of water and coal, oiled joints, wear of cables, ropes and chains
(6) Operational Control	Dipper stick length, boom angle, bucket hoist cable, house rotation, bucket flap, wheels

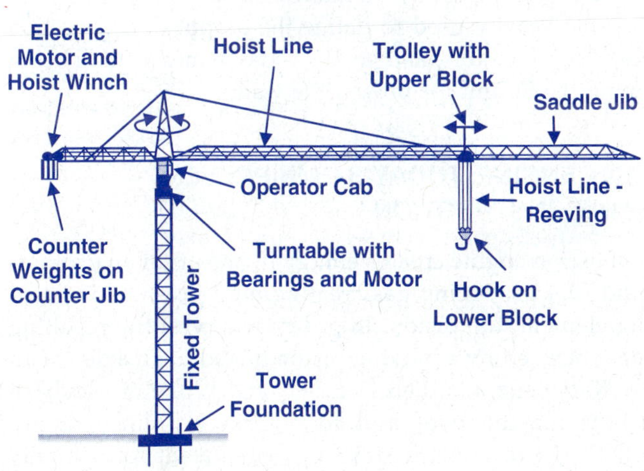

Figure 3.5 A Basic Tower Crane Configuration

Table 3.5 Powertrain of a Tower Crane

Module	Description
(1) Energy Source	AC electric current
(2) Engine	Electro motors
(3) Transmission	Gear, winch, ropes, trolley, block-and-tackle
(4) Actuation	Hoist line pull, trolley motion, slewing motion
(5) Power Plant Control	Electrical power supply, stability of crane, wear of rope lines and blocks, trolley, turntable bearings
6) Operational Control	Raising and lowering hook, slewing jib, moving trolley, speed of hoist winch and trolley

driving a small cog wheel. The large main jib is balanced, or levered, by a counter jib loaded with extra weights and heavy electro motor, gears, and winch operating the hoist line. A trolley is riding along the main jib truss carrying the upper block of the crane tackle. Taking advantage of the block-and-tackle machine (see Chapter 1), the crane is able to hook heavy loads onto the lower block using small-diameter wire ropes while the geared winches can be controlled to change the lifting speed smoothly. The crane operator can either occupy a seat in the cabin underneath the jib or use a remote control console from anywhere below the reach of the jib.

3.1.2 TURNING ENERGY INTO HORSEPOWER

The traditional units of energy are the British Thermal Units (BTU), calorie (cal), or Watt-hour (W·hr). One BTU is defined as amount of heat required to raise the temperature of one pound of liquid water from 60° to 61°Fahrenheit at a constant pressure of 1 atmosphere. Of course, energy is also an expression of its capability to do work, which has been defined as: Work = Force * Distance (newton · meter or ft-lb). In the metric system (SI) there is a simple relationship between the two in that: 1 joule (J) of energy = 1 newton·meter (N·m) of work. Not surprisingly, today the BTU is being replaced with the metric joule.

Dimensions of Energy

1 BTU = 1,056 joules (J)
1 BTU = 0.2931 watt·hr (W hr)
1 BTU = 253 calories (cal)
1 BTU = 0.000039 horsepower·hr
1 BTU = 778 ft-lb

In the 1770s James Watt and his business partner Matthew Boulton had a problem. How much should they charge the mines for the use of their steam engines powering the pumps that emptied the mines of water? As the mine owners lacked the capital to buy the engines, Watt recommended charging only for the savings they incurred. The source of energy used to power the mine pumps at that time was mules and horses that turned a vertical drive shaft. Through experiments Watt measured the work done by a mill horse turning a large wheel and found that it made 144 rounds in one hour hitched at 12 feet (3.66 meters) from the center of the center shaft. Finally, he judged that the pulling force that the horse applied was 180 lb (800 N), leading to the famous unit of 1 horsepower:

$$1\,\text{horsepower} = \frac{\text{Work of 1 horse}}{\text{min}} = \text{Force} * \text{Distance} = \frac{180\,\text{lb}_f * 144 * (2\pi * 12\,\text{ft})}{60\,\text{min}} = 32{,}572\frac{\text{ft} * \text{lb}_f}{\text{min}}$$

This number was subsequently standardized at 33,000 ft·lb_f/min. This measure became heavily used to define the output of combustion engines, turbines, electric motors, and so on. Today James Watt is being honored by naming the SI unit for power, the watt.

Dimensions of Power

1 hp = 746 watt (W)
1 hp = 33,000 ft·lb_f/min
1 hp = 550 ft lb_f/sec
1 hp = 42.44 BTU/min
1 hp = 10.68 kg-calories/min
1 hp = 75 kg-m/sec

3.1.3 THE DIESEL ENGINE: TODAY'S OMNIPRESENT "MILL HORSE" IN CONSTRUCTION

The internal combustion engine creates energy by means of an exothermic chemical reaction of a liquid fuel producing gas creating high pressure inside a solid chamber with one moving element, the piston. In order to harvest the resulting energy, one side of the chamber is blocked by a moving piston linked to an axle via an eccentric connector. Years after Karl Benz, Gottlieb Daimler, and Wilhelm Maybach received several different patents for the two- and four-stroke gasoline engines between 1878 and 1885, Rudolf Diesel received in 1893 his first patent for a highly

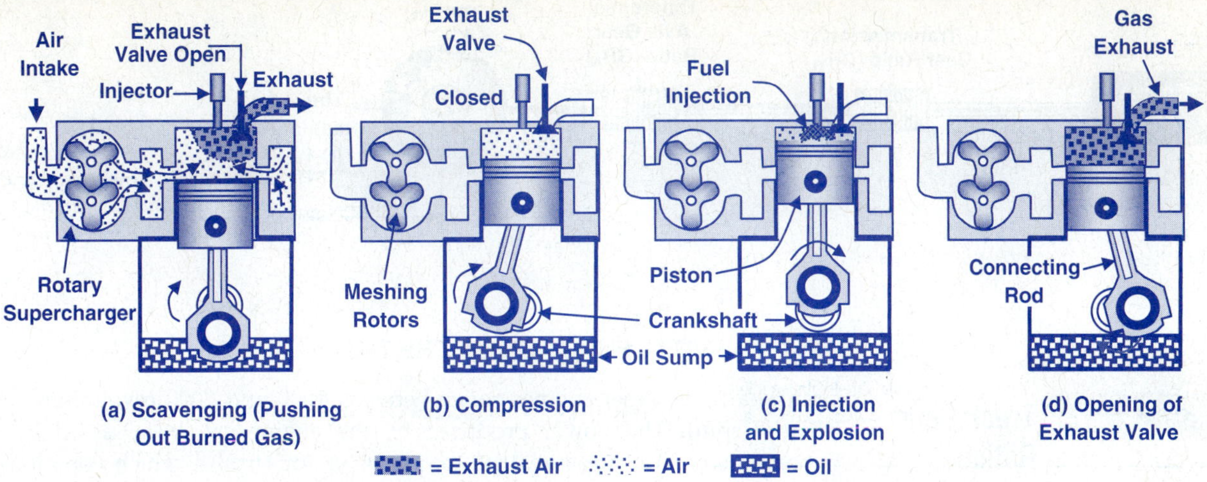

Figure 3.6 Two-Stroke-Cycle of a Supercharged Diesel Engine (Without a Spark Plug)

efficient new type of engine not needing sparkplugs for igniting an air-fuel mix. Figure 3.6 offers four views through the inside of the so-called supercharged diesel engine named for its original inventor. See Table 3.6 for an analysis of the powertrain.

As can be expected, a 2-stroke does not produce as much power as a 4-stroke diesel engine. On the other hand, it works much smoothly and is more efficient, since it does not have as many mechanical parts creating losses. The cycle begins with the piston at bottom. At this position, the air inlets around the cylinder wall, as well as the exhaust valve on the top, are open. The process of pushing the burned gases out of the combustion chamber is supported by the rotary air pumps, the so-called supercharger, that forces fresh air from the intake into the chamber (see Figure 3.6a). While the exhaust valve is being closed, the piston is rising upward, cutting off the air inlets in the cylinder wall (Figure 3.6b). As a result, the air volume residing inside the ever-decreasing chamber is being compressed to about 40 bar (about 600 psi). Rankine (see Section 3.1) showed that heat will increase the pressure in a gas. The process of the steam engine is now reversed in that the produced pressure heats up the air to 550 °C (1,022 °F), at which time the diesel fuel is injected (Figure 3.6c), creating an air-fuel mix that explodes due to the heat. The injected fuel might even be vegetable oil or biodiesel, a mix of diesel fuel and used vegetable oil (e.g., oil that was used to make french fries). The explosion is in fact a chemical reaction creating a large volume of gas, similar to dry vapor created in a boiler, forcing the piston back down (see Figure 3.6d). The rod connecting the piston to the eccentric arm of the crankshaft creates a torque onto the crankshaft that causes it to turn, similar to the axle of a winch driven by a treadwheel. At one point, the exhaust valve is again opened, and the cycle starts again.

Table 3.6 Powertrain of a Diesel Engine

Module	Description
(1) Energy Source	Diesel Fuel, Biodiesel
(2) Engine	Internal Combustion
(3) Transmission	Mechanical Gears
(4) Actuation	Crankshaft Torque
(5) Power Plant Control	Fuel Injector Opening
(6) Operational Control	Accelerator—Brakes

Rudolf Karl Diesel (1858–1913)

Already at the age of 14, Rudolf Diesel knew he wanted to become an engineer. After working on refrigeration, steam and gasoline engines, he started research using his own ideas. In 1893, Diesel received a patent for a motor that did not use a spark plug to ignite the fuel like the Otto 4-cycle gasoline engine. Nondistilled oil can be used as fuel that ignites due to its own heat created in the compression phase of the cycle. This makes it a highly efficient motor still being improved on today.

S3.1 Heavy-Duty Diesel Engine Frictional Losses, U.S Army Research

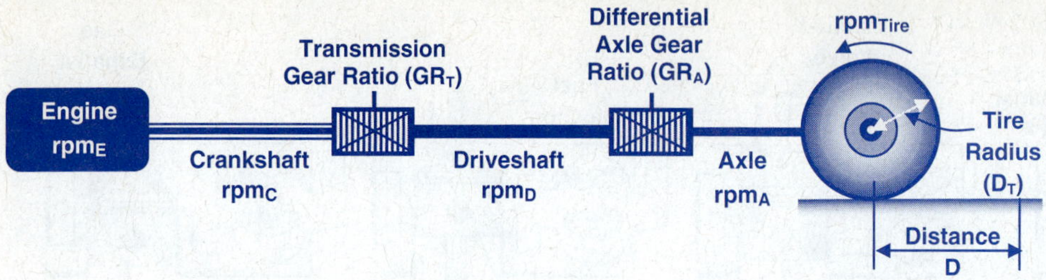

Figure 3.7 Powertrain with Gearing and Drive Mechanism

3.1.4 THE VITAL FUNCTION OF THE TRANSMISSION

Table 3.7 Truck Gear Ratios

Gear	Ratio
12	1.00
11	1.27
10	1.63
9	2.08
8	2.70
7	3.44
6	4.35
5	5.54
4	7.09
3	9.04
2	11.73
1	14.94

Crankshafts of diesel engines rotate between 1000 and 2000 revolutions per minute (rpm). The power produced by the engine can be assessed by a dynamometer that measures the torque at various rpms, which commonly peak at around 1300 rpm. However, the wheels or tracks of construction equipment turn at a much lower rpm, and in fact ramp up from 0 rpm. The element of the powertrain that makes this possible is called the gear mechanisms (gearbox) in the transmission depicted in Figure 3.7.

In fact, gear mechanisms appear at two locations along the powertrain, inside the transmission and the differential, distributing the driveshaft torque and speed to the two drive wheels while letting them rotate at different speeds. Every driver of a car with a stick shift interacts directly with the gearbox of the transmission when shifting gears up, down, or even into reverse. Lower gears turn the drive wheel slowly while high gears produce higher rpms of the wheel. On the other hand, low gears seem to provide more force to the wheel than higher gears. Let us see what happens when a truck shifts through its 12 gears, as presented in Table 3.7.

Let's first review the basic functions of engine power, work done by the crankshaft:

Power dimensions	= Work (W)/time (t) = ft-lb/min = joules/t = hp = watt
Work (W)	= Force (F) $*$ distance (s) ft-lb. (joules)
Torque (T)	= Force (F) $*$ radius (r)
Circumference	= 2π $*$ radius (r)
T(orque)	= Work (W) $*$ r/distance (s)
W(ork)	= T $*$ s/r ft-lb (joules)
Revolution of crankshaft **(rev)**	
W/rev	= F $*$ r $*$ 2π ft-lb/rev (joules/rev)
Revolutions per **minute (rpm)**	
Crankshaft **Power**	= W/t = (T$*$ s/r)/min, where s = $2\pi * r$
Power	= 2π $*$ T$*$ rpm
1 horsepower (hp)	= **33,000 ft-lbf/min (4,500 kgf-m/min)**
2π $*$ T$*$ rpm (ft-lb/min)/33,000 (ft-lb/min)	= (2π $*$ T$*$ rpm)/33,000
Imperial hp	= T (ft-lbf) $*$ rpm/5,252 = 746 watts
Metric hp	= T (kgf-m) $*$ rpm/5,224 = 1 .01 PS

At the heart of the gear box are sprockets of different diameter that can be engaged or disengaged. While one sprocket supplies the rotational power, the second sprocket takes it over via the meshing teeth on its circumference. Figure 3.8 presents graphically the effect of meshing sprockets with different radii represented by the gear ratio. If we

The gear ratio is the relationship between the radius of two gears that are meshed or two sprockets connected by a chain. Gear ratio = R/r

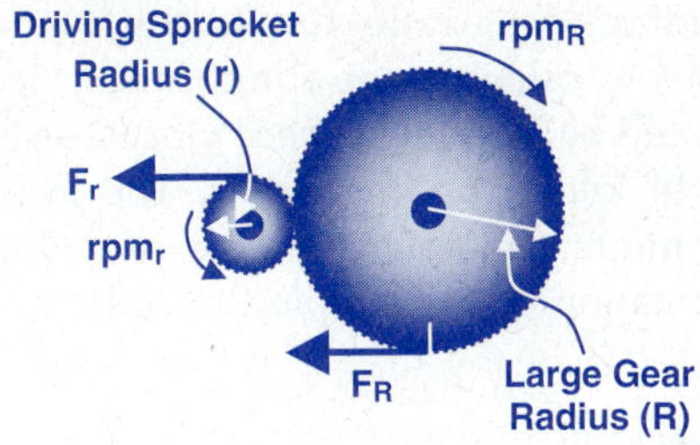

Figure 3.8 Gear Mechanism

assume that the smaller sprocket is mounted on the crankshaft coming from the engine, we can easily recognize that the larger sprocket will turn much more slowly, so that the rpm of the large R < rpm of small r. The ratio that defines this relationship is called the gear ratio.

As shown, the 12 gears all have their one ratio varying from 14.94 to 1 at the highest gear. A gear ratio indicates the difference in rpm and torque between the crankshaft (input) and the driveshaft (output) of a gear box. Let us assume that the crankshaft sprocket has a radius of r = 2.5 in (6.3 cm) and is turning at 1,400 rpm. The larger sprocket, connected to the driveshaft, has a radius of R = 6.75 in (17.1 cm). Thus the gear ratio = R/r = 6.75/2.5 = 2.7. What can we say about the rotational speed and torque that is reaching the differential at the other end of the driveshaft?

Torque of engine $\mathbf{T_E}$ (ft-lb) = (5,252 * 500) hp/1,400 rpm = 1,875 ft-lb = 2,542 joules
The force at radius r = $\mathbf{F_r} = \mathbf{T_E/r}$ = 12 in/ft * (1,875 ft-lb/2.5 in) = 4,200 lb (18.7 kN)
$\mathbf{F_R} = \mathbf{F_r}$ = 4,200 lb (18.7 kN) (teeth from the two sprockets touching each other)
Torque of driveshaft $\mathbf{T_D}$ (ft-lb) = $\mathbf{F_R * R = T_E/r * R = T_E * (R/r)}$ = 1,875 ft-lb * 2.7 = 5,062 ft-lb = 6,863 J

Let us now check out the effect of the gear ratio on the rpm. From our observation we know that a gear ratio larger than one means that the driven wheel will turn slower. Could it be that the speed difference is related to the inverse of the gear ratio?

Since the teeth of the two sprockets mesh with each other, one revolution of the smaller will turn the larger by its circumference.

Circumference of small sprocket = 2 * π * r = 6.28 * 2.5 in = 15.7 in (39.9 cm)
Circumference of large sprocket = 2 * π * R = 6.28 * 6.75 in = 42.4 in (107.7 cm)

How many revolutions of the smaller sprocket are needed for the larger sprocket to turn 1 time?

Rev of small sprocket = 42.4 in/15.7 in = 2.7

In other words, the rpm of the driveshaft is 2.7 times slower than that of the crankshaft.

In summary : $\mathbf{rpm_{Drive}} = \mathbf{rpm_{Crank}} * (1/\text{Gear Ratio})$ and $\mathbf{T_{Drive}} = \mathbf{T_{Crank}} * \text{Gear Ratio}$

The gear ratio analysis shows that torque and speed behave inversely. As a consequence, a high speed of the driveshaft will require a small-radius sprocket, which will at the same time reduce the available torque. This makes sense, as trucks need the largest torque to accelerate a heavy load from their zero speed at a red light. When traveling at maximum speed on a horizontal highway, on the other hand, only small torque is needed to overcome wind drag and losses in the powertrain.

Example Truck Tire Code: 265/70R17.5 139/136 M

265	The nominal width of the tire is approximately 265 mm at the widest point
70	Indicates the height of the sidewall of the tire to be 70% of the width
R	Specifies this is a radial tire
17.5	Tire fits 17.5 in diameter wheels
139/136	Codes that indicate the max. allowable load
M	Designates the max. speed for the tire to be 81 mph or 130 km/h

The second gear inside the differential is fixed, fitting the main use of the tractor. If heavy work needs to be done, a higher gear ratio is installed and vice versa. Let's assume our 500 hp truck tractor uses a common gear ratio for its differential $\mathbf{GR_D} = 1.85$. This allows us to calculate the rpm of the wheel axle and tire. But what would be its travel speed? Of course, to figure that out we need to know the tire radius. The insert interprets a code that one might find on a truck tire, providing us with the missing information. The example tire code indicates that

Diameter of steel wheel = 17.5 in (44.4 cm)

Tire diameter $\mathbf{D_T}$ = steel wheel + 2 * sidewall height
(sidewall height = 70% of tire width)

$$\text{Tire diameter } \mathbf{D_T} = 17.5\,\text{in} + 2 * (0.7 * 265) * (0.0393\,\text{in/mm})$$
$$= (17.5 + 14.6)\,\text{in} = 32.1\,\text{in}\ (81.5\,\text{cm})$$

Thus $\mathbf{D_T} = 2.6$ ft. (0.81 m)

$$\text{Traveling distance from one tire revolution } \mathbf{D} = \pi * 2.6\,\text{ft}$$
$$= 8.2\,\text{ft}\ (2.5\,\text{m})$$

Now we are ready to analyze the effect of the 12 gears of the 500 hp truck on the speed of the truck assuming that the example truck tires are mounted. See Table 3.8 on the effects of gear ratios.

Of course these numbers only apply when the engine operates at 500 hp, which is its maximum. Since legal limits exist to hauling speeds on highways, less hp is needed for travel at 55 or 60 mph (88–97 km/hr) in the 12th gear, and the rpm can be reduced to the most fuel-efficient revolution. In addition, the losses in the powertrain and the support of the electrical system will use up some of the power available. In other words, the truck will not be able to take advantage of all the hp to drive the wheels. In order to learn about the actual use of available engine power, field tests were made with tractor trailers weighing the allowed 80,000 lb (356 kN) traveling at 55 and 65 mph

Table 3.8 Effect of Truck Gear Ratios on Force on Road (F_R) and Speed (mph) at Full Power Use with 2.6 ft (0.81 m) Diameter Tires

Gear	Ratio	Total Gear Ratio	1800 rpm (500 hp)			1600 rpm (500 hp)			1400 rpm (500 hp)		
			T_E	mph	F_R (lbf)	T_E	mph	F_R (lbf)	T_E	mph	F_R (lbf)
1	14.94	27.64	1458.9	6.1	31017.1	1641.3	5.4	34894.2	1875.7	4.7	39879.1
2	11.73	21.70	1458.9	7.7	24352.8	1641.3	6.9	27396.9	1875.7	6.0	31310.7
3	9.04	16.72	1458.9	10.0	18768.0	1641.3	8.9	21114.1	1875.7	7.8	24130.3
4	7.09	13.12	1458.9	12.8	14719.6	1641.3	11.4	16559.6	1875.7	9.9	18925.2
5	5.54	10.25	1458.9	16.4	11501.7	1641.3	14.5	12939.4	1875.7	12.7	14787.8
6	4.35	8.05	1458.9	20.8	9031.1	1641.3	18.5	10160.0	1875.7	16.2	11611.4
7	3.44	6.36	1458.9	26.4	7141.8	1641.3	23.4	8034.6	1875.7	20.5	9182.3
8	2.70	5.00	1458.9	33.6	5605.5	1641.3	29.8	6306.2	1875.7	26.1	7207.1
9	2.08	3.85	1458.9	43.6	4318.3	1641.3	38.7	4858.1	1875.7	33.9	5552.1
10	1.63	3.02	1458.9	55.6	3384.1	1641.3	49.4	3807.1	1875.7	43.3	4350.9
11	1.27	2.35	1458.9	71.4	2636.7	1641.3	63.5	2966.2	1875.7	55.5	3390.0
12	1	1.85	1458.9	90.7	2076.1	1641.3	80.6	2335.6	1875.7	70.5	2669.3

Table 3.9 Power Needs of Tractor-Trailer

Power Need	55 mph (88 km/h)		65 mph (105 km/h)	
	hp	%	hp	%
Aerodynamic Drag	104	40	159	44
Rolling Resistance of Tires	97	38	127	35
Transmission-Driveline Losses 15% T_E	36	14	54	15
Engine and Electric Power System	20	8	20	6
Total	257	100	360	100

(88–105 km/h). Table 3.9 presents the measured power needed when traveling on a paved horizontal road.

You might ask, what can we do with the remaining hp that the engine is able to produce? Don't worry, this power comes in very handy when the truck has to climb an incline. We can predict that pulling a load uphill will require more and more hp the steeper the slope.

Worked-Out Problem 3.1 / Ensuring Sufficient Engine Power

You are working on a project to upgrade an old hydroelectric power plant in the mountains. Jack Boyd, the project manager, scheduled the delivery of a state-of-the-art turbine in one week. The shipment is planned just-in-time (JIT) because the new foundations will be ready by that time to allow for a direct installation by the overhead crane. Any delay will cause the schedule to slip. The total weight of the truck-trailer combination, called the Gross Vehicle Weight (GVW), is 110,000 lb (489 kN) larger than the max. allowable weight on highways (= 80, 000 lb), thus requiring special approvals from several states.

Jack Boyd is worried that the pulling tractor used by the trucking company, a 500 hp engine with the gearing presented in Table 3.7, might not be able to climb the 16% incline just before the summit. Although this is not Jack's direct responsibility, a delivery that is one or two days late could become his nightmare problem. He wants to know if he can sleep soundly the night before.

Background and Assumptions

The grade (incline/gradient/pitch/slope) of any physical feature such as a hill, stream, roof, railroad, or road, refers to the amount of inclination of that surface. The maximum grade of national highways is 6%, allowed in mountainous areas and hilly urban areas.

Figure 3.9 offers a sketch that will help in making sure that our mathematical calculations are based on the correct values:

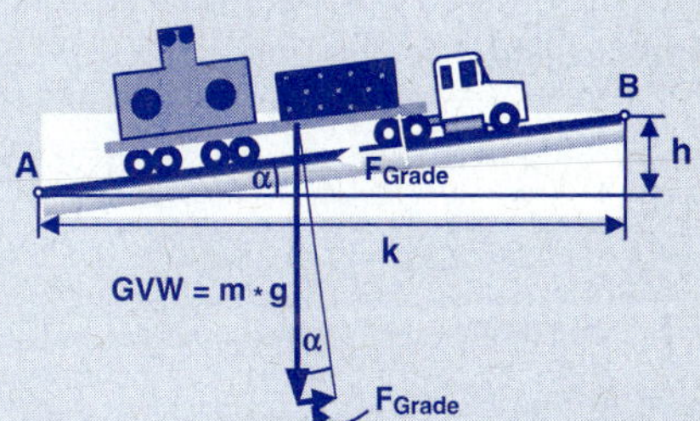

Figure 3.9 Loaded Trailer Truck at 16% Incline

$$\mathbf{GVW} = \text{Gross vehicle weight} = 110,000\,\text{lb}\ (489\,\text{kN})$$
$$\mathbf{Grade} = 16\% = \mathbf{\Delta h/k}$$
$$F_{\text{Grade}} = \mathbf{GVW} * \mathbf{\sin\alpha}$$

The truck uses common size tires of 36″ = 3 ft (0.91 m).

Calculations:

$$\mathbf{\tan\alpha} = \mathbf{\Delta h/k} = \mathbf{Grade}\ \text{and}\ \mathbf{\alpha} = \mathbf{arc\,tan\,\Delta h/k} = \mathbf{arc\,tan\,0.16} = 9.09^\circ$$
$$\text{With}\ \alpha = 9.09^\circ\ \mathbf{\sin\alpha} = 0.15799\ \text{while}\ \mathbf{\tan\alpha} = 0.16.$$
$$F_{\text{Grade}} = \mathbf{GVW} * \mathbf{\sin\alpha} = 110,000\,\mathbf{lb} * 0.15799 = 17,379\,\text{lb}\ (77\,\text{kN})$$

(**Comment:** At small angles α, sin α is equal to tan α and thus to the grade. For 16% this difference = 0.00211. For convenience, the grade resistance is calculated using the grade instead of using tan and sin functions: $\mathbf{F_{Grade}} = \mathbf{GVW_* Grade}$ = 17,600 lb (78.3 kN), resulting in a difference of 221 lb (0.98 kN), which is negligible considering all the many other uncertainties.)

Needed torque for 36 in (45.7 cm) diameter tire $\mathbf{T_{Wheel}}$ = 18 in ∗ 17,600 lb = 26,400 ft-lb (35.8 kJ)

Loss in a tandem driveline and transmission = 20%

Let's assume a low rpm (since it produces more torque) = 1400 rpm

Required **Power** (without gearing) = (2 ∗ π ∗ 26,400 ft-lb/rev 1,400 rev/min) / 33,000 ft-lb/min hp ∗ 1.2 = (2 ∗ π ∗ 26,400 ∗ 1400/33,000) ∗ 1.2 = 8,440 hp

Power to overcome aerodynamic drag is minimal	=	10 hp (low speed)
Power to overcome rolling resistance	=	40 hp
Power used by auxiliary systems	=	20 hp
Total power need for other purposes	=	70 hp
Available power at the flywheel	=	500 hp − 70 hp = 430 hp

Gearing Ratio needed = 8,440 hp/430 hp = 19.6

Table 3.8 shows that the **gear 2** provides a combined gearing of 21.7 sufficient to pull the 110,000 lb up the 16% incline at a speed of 6 mph: which results in (1.5 ft/1.3 ft tires) = 7.0 mph (11.3 km/h).

Discussion of Results:

There are several assumptions that might be off by a significant margin.

We assume

- The truck is well maintained and has a standard 20% transmission loss for tandem wheels.
- A standard rolling resistance of 40 hp (not knowing road surface condition).
- GVW is 110,000 lb (we don't know how this number has been established).
- Grade is in fact 16% (we don't know how accurately this has been measured).

In order to test the effect of a worst-case scenario, we apply a negative 10% error for all main assumptions and recalculate. The new base values are now

GVW = 121,000 lb, powertrain loss = 22%, Rolling resistances = 44 hp, Grade = 17.6%

Thus: $\mathbf{F_{Grade}}$ = 121,000 lb ∗ 0.176 = 21,296 lb (93.7 kN)

Power needed = 2 ∗ π ∗ (1.5ft ∗ 21,296 lb ∗ 1400/33,000) ∗ 1.22 ft.lb./hp)
= 10,383 hp

Power available = 500 − 74 = 426 hp

Needed **Gear Ratio** = 10,383 hp/426 hp = 24.37

Table 3.8 shows a combined gear ratio of 27.64 for **gear 1,** which is larger than the necessary 24.37.

Summary: Even in a worst-case scenario, the truck will have, in the lowest first gear, sufficient horsepower to haul the new turbine to the construction site. Nevertheless, it might be advisable to inform the trucking company of our calculations and suggest possibly using a tractor with more horsepower.

3.1.5 A WORD ABOUT THE LUBRICATION SYSTEM

The attentive reader has observed an item in Figure 3.6 that we have not discussed yet, the oil sump at the bottom of the engine block. During each piston cycle, the connecting rod and the rod bearing dive into the oil bath. Naturally, this will lubricate the moving parts inside the engine such as the piston rings. How about the valve and the gears inside the transmission? Keeping all the moving parts from "welding together" is the function of the lubricating system. Since the oil sump, collecting all the oil in a pan, is at the bottom, a pump is used to suck the oil from the pan and force it through a filter and through passages in the engine to the transmission, the crankshaft bearings, and the valves. Equally important missions of the lubrication system are to minimize the friction losses, to "flush" away heat, which is created by friction and the combustion process, and to seal the tiny openings between the moving piston and cylinder wall. Air-cooled water is pumped through small pipes circling through the lube oil carrying away its heat, thus keeping the oil temperature below a dangerous level. Anybody who has experienced the rupture of a water-pump belt knows that a lack of water cooling leads very quickly to overheating and a smoking radiator.

Finally, the cleanliness of the oil is used to assess the physical health of all the parts inside, similar to a blood check. Samples are drawn and analyzed to appraise the type and amount of tiny steel particles that have accumulated in the oil. Numbers that exceed an established average for an engine indicate that the engine is under physical stress. This information, of course, is critical in deciding about preventive maintenance before the engine is damaged and expensive to repair.

3.1.6 POWERING THE AUXILIARY SYSTEMS

While diesel engines turn fuel into torque and rpm that produce work, mechanical and electrical trucks, loaders, backhoes, and so forth depend on working nonproductive systems to function. Figure 3.10 highlights the distribution of the mechanical energy produced by the crankshaft. While the variety of revolving parts along the powertrain creates losses due to friction, many of the functions needed to operate the equipment safely depend on electric power, commonly 24 volt DC. This requires the transformation of mechanical rotation into electrical power, which is the function of the alternator that charges a battery. For this purpose, the revolving crankshaft and the electric power generator are connected via a drive belt wrapped around pulleys mounted to each mechanism.

In later chapters we will learn that today's diesel engines power most equipment in construction required to move or do "ordinary" work. As we saw in the case of the truck, mechanical gearing has its limits when moving heavy loads uphill. A different set of physical principles offered a solution and began to revolutionize the construction industry in the 1950s. The physical foundation is provided by fluid mechanics, and its application in construction is called hydraulic fluid power. Let us figure out how that works.

3.2 Taking Advantage of Hydraulic Fluid Power

It is believed that the people of Mesopotamia around 2000 BC used joined and sealed clay pipes to water the hanging gardens. How was it possible to pipe water uphill with the pumps the Mesopotamians possessed? Of course, Archimedes developed his famous screw lifting water inside an inclined channel, while the Romans used a chain of water wheels to "pump" water out of mine shafts. But those are all mechanical solutions that were not found in Mesopotamia. Discoveries in the Renaissance may give us the answer.

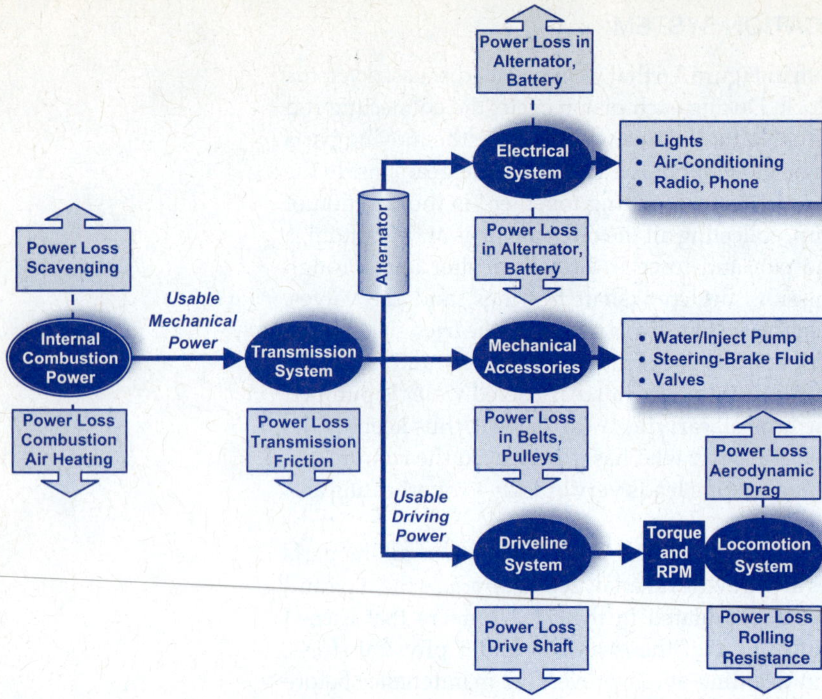

Figure 3.10 Sources of Friction Losses and Auxiliary Energy Sinks Along the Powertrain

3.2.1 FROM FLUID STATICS TO FLUID MECHANICS

The rise of systematic scientific experiments during Europe's Renaissance in the seventeenth and eighteenth centuries led to a long list of very important discoveries about behaviors of the physical world. In turn, these laid the foundations for innovative applications that were only limited by the limit of available fabrication methods and materials. One such example was the understanding of fluids, an investigation that was spurred by farmers in Tuscany, Italy. In 1643 they complained to a physics professor in Pisa, Evangelista Torricelli, about their water pumps producing less and less water at the spigot. Figure 3.11 sketches the experiments Torricelli conducted and what Blaise Pascal, living in Paris, added later on.

After filling a glass tube with mercury, which is 10 times as heavy as water, Torricelli lowered the open but capped end into a bowl of mercury. When he removed the cap (see Figure 3.11b), the mercury dropped to approximately 76 cm above the mercury in the bowl. He concluded correctly that vacuum was created in the top and that the "ocean of air" was putting pressure on the surface pushing the mercury up the tube. Because he also observed that the height of the mercury changed with weather conditions, he invented the barometer that could forecast weather based on high and low pressure, a term still used today. Figure 3.11c shows that Pascal must have read about Torricelli's work and added the next step in the discovery of fluid power. He made several bowls and tubes filled with mercury; two he left on the bottom with a friend and the others he walked onto the top of a 50 m tower in Paris. Measuring the mercury at the same time, they were able to assert that there was

Evangelista Torricelli (1608–1647)
Blaise Pascal (1623–1662)

Both were physicists and mathematicians, both worked with mercury, both were Renaissance men who believed in the power of scientific experiments, both got a dimension of pressure named after them, both died at age 39, and both were instrumental in understanding the behavior of fluids. In 1644, Torricelli famously wrote in a letter: "We live submerged at the bottom of an ocean of air."

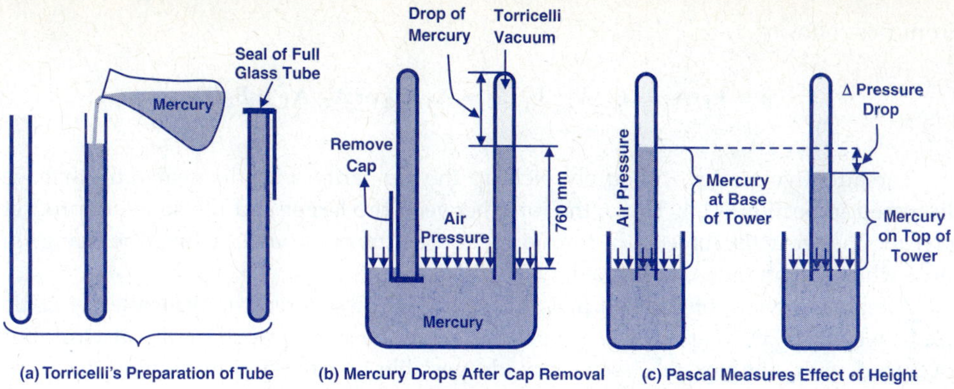

Figure 3.11 Experiments by Torricelli and Pascal to Study the Behavior of Fluids

indeed a drop of mercury that could only be caused by the 50 m difference in height. Pascal hypothesized correctly that the only way this was possible was that the force or pressure on the confined liquid acted in all directions equally, unlike the gravity force in solid objects that only worked vertically. Probably everybody has personally seen one of Pascal's next experiments, especially that with the connected glass vases shown in Figure 3.12a.

By connecting differently shaped glass vases to a common tube at the bottom, Pascal showed that the pressure depended only on the height of the liquid. Adding water to one vase made all the others lift their water surfaces, due to the increased pressure within the bottom tube, until they were all equal again. Consequently, Pascal replaced adding water, to increase pressure, with syringe plungers that fit nicely inside the glass or metal tubes and used a less viscous fluid than water. The resulting experiment is shown in Figure 3.12b. He applied a force $\mathbf{F}_1$ to the plunger on the left, which increased the pressure in the liquid by $\mathbf{p} = \mathrm{F}_1/\mathrm{A}_1 = \mathrm{F}_1/\mathrm{r}_1^2 * \pi$. As expected, the additional pressure acted on the plunger on the right, which had to be loaded up with weights in order to keep the top of the fluid equal. The outcome of this experiment is the confirmation of the hypothesis that the force $\mathbf{F}_2$ increases with the area of the plunger:

$$\mathrm{F}_2 = \mathbf{p} * \mathrm{A}_2 = (\mathrm{F}_1/(\mathrm{r}_1^2 * \pi)) * (\mathrm{r}_2^2 * \pi) = (\mathrm{F}_1/(\mathrm{r}_1^2 * \pi)) * 9\,(\mathrm{r}_1^2 * \pi) = 9\,\mathrm{F}_1$$

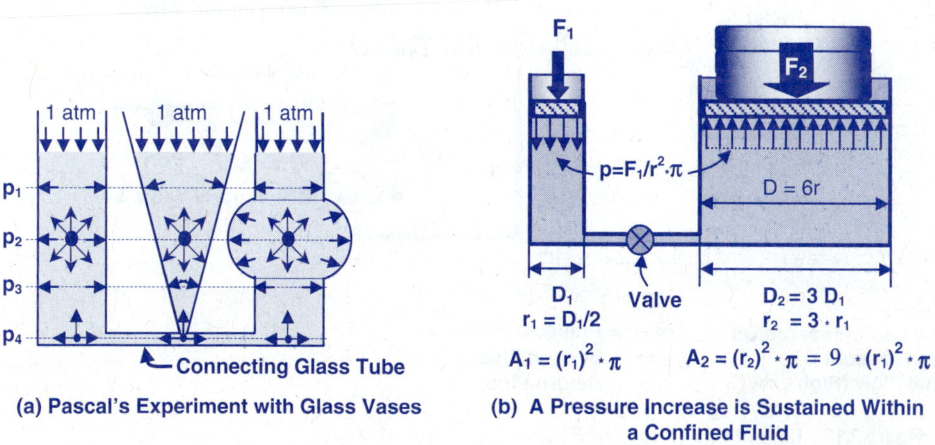

Figure 3.12 Pascal's Breakthrough Experiments with Fluids

From this follows

$$p = F_1/A_1 = F_2/A_2 : F_1/F_2 = A_1/A_2 \text{ or } A_2/A_1 = F_2/F_1$$

The attentive reader will surely pick up the similarity with the gear ratio that we discussed in Section 3.1.4. There, the ratio between the larger and the smaller sprocket gave the change in the force or the torque. Here, the ratio of the areas defines the change in force while the pressure stays fixed.

After this review of the two first pioneers who discovered the principles of static fluids, we jump 300 years forward to today where the same principles are at work but leveraged with complex mechanisms made possible by precision machining and new materials. Figure 3.12b showed the required components of a hydraulic power system: **(1)** a "force" to increase the pressure in one cylinder, **(2)** pipes and hoses to let fluid pass, **(3)** a second cylinder with a different diameter "closed" with a piston, and finally **(4)** a valve to control the flow of fluid. These four elements are exactly those that make the modern hydraulic system shown in Figure 3.13.

The first key element is a pump that pressurizes hydraulic oil from an oil reservoir similar to the force F_1 in Figure 3.12b. On the pressurized side of the pump, pressure pipes distribute the oil to various control valves that are able to direct the pressurized oil to different ports. Cylinders, closed on both ends, have two ports, port A and B, one at each side of a free-moving piston connected to a rod. In some ways this reminds one of the combustion engine without the crankshaft. The two ports allow oil to flow in or out, thus pressurizing one side while depressurizing the other side of the piston. For example, oil flowing through port A will increase the pressure in the left side of the free-moving piston. If the pressure at port B is small, the piston and rod will be pushed toward that end. However, port B needs to be open and let the oil escape; otherwise pressure will build up quickly on that side as well. This inflow and outflow of hydraulic

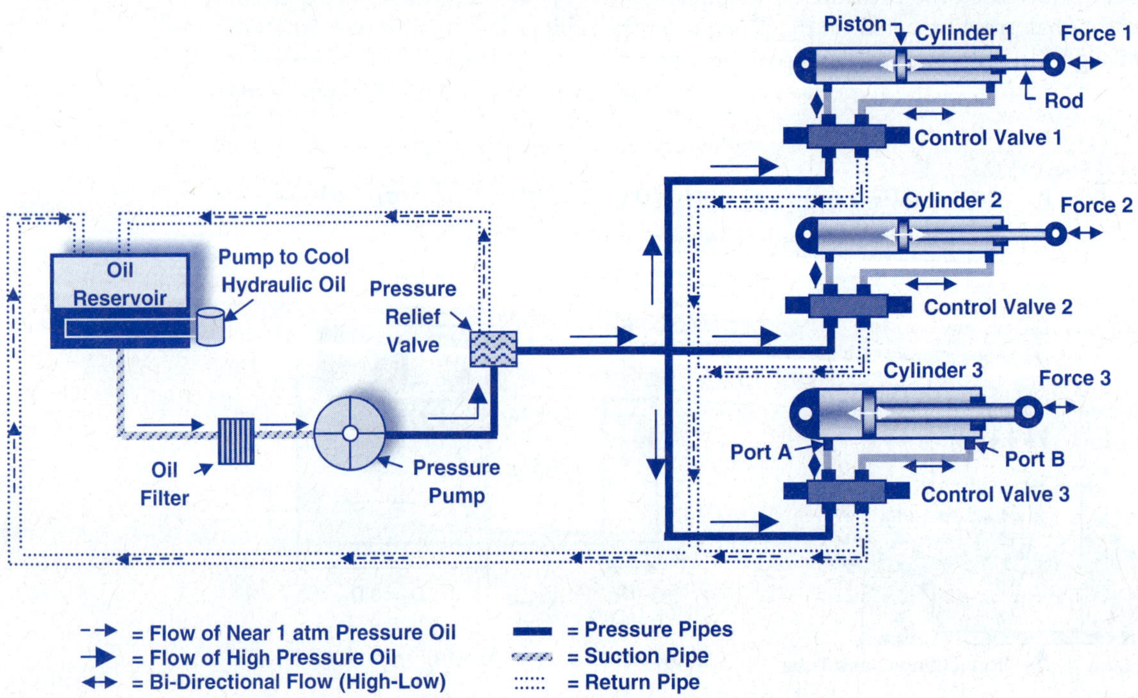

Figure 3.13 Layout of a Hydraulic System Distributing Power

oil through the two ports is "directed" by a control valve capable of opening and closing pathways between the ports A and B and the pressurized, and unpressurized, oil pipes. The escaping oil flows back to the reservoir, where cooling coils will remove the heat created during compression. In fact, this heat represents the main energy loss of a hydraulic system, a fraction of that from the mechanical powertrain. Hydraulic systems provide other advantages. Fluids are not subject to abrasive friction, as mechanical parts are, and the power distribution inside pipes can be installed very flexibly without having to worry about universal joints. In case of an overload, the pressure release valve guarantees that the system does not break at the weakest point. Besides the hydraulic cylinders, other types of actuators are available, including hydraulic motors and heavy-duty lift bags. High-pressure systems are not without danger as a hose rupture can kill people in its vicinity.

In the following, we will study some of the cornerstones of hydraulic power systems.

3.2.2 PUMPS PRODUCING POWER

There exists a wide variety of pump designs, each having strengths and weaknesses that make them most efficient for specific applications. For hydraulic equipment, we can use constant displacement pumps, which do not produce extremely high pressure, as they would require high-pressure pipes and connectors, and are mostly used in fixed applications. However, two popular types are the gear and the vane type pumps.

As the name suggests, gear pumps use two meshing gear sprockets to trap the fluid between them, compress it in the process, and push it forward toward the valves. The most common pumps are the internal and the external gear pump, the latter of which is a simple pump and is presented in Figure 3.14.

The pump is operated by an engine that turns the drive gear, positioned on the right side. The idling gear on the left is being moved forward by the teeth of the drive gear. The meshing gears are built to tight tolerances between the teeth and between the teeth and the housing of the pump.

As the drive gear turns clockwise, the idling gear is pushed to turn counterclockwise, both accelerating the incoming oil in opposite directions. The specially shaped teeth also create a suction effect on their back sides, thus pulling in more oil from the reservoir. As the gears keep turning, the top of the teeth are closing the gap to the inside of the casing until they "touch" each other, thus trapping and compressing the oil in the space between two teeth and the housing wall. This also seals off the pressure side from the suction side of the pump since the meshing teeth in the middle block the oil from passing through as well. Thus,

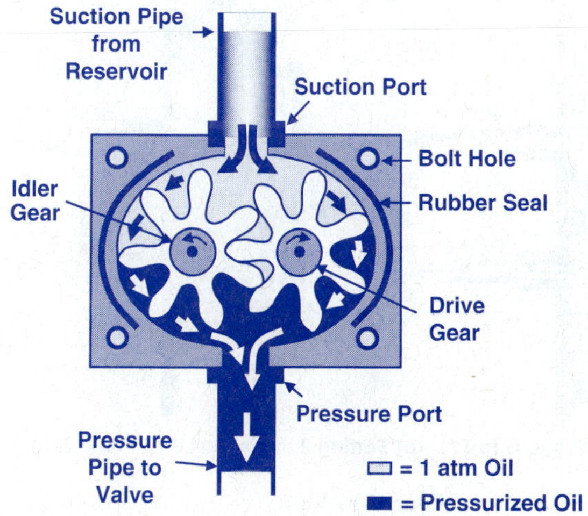

Figure 3.14 Cross-section of an External Gear Pump

inside the housing two distinct areas of pressure develop, the 1 atmosphere (atm) inflow region, shown in white, and the high-pressure area, shown in black. The amount of pressure resulting from this gear pump is simply dependent on the rpm of the drive shaft. Higher velocity will push through more hydraulic fluid, which will increase the pressure. Of course, higher speed also means more friction and higher temperatures due to the compression cycle. Both increase the risk of damaging the pump, a risk that grows exponentially after a certain rpm, creating an upper limit at which a pump should be operated.

3.2.3 HYDRAULIC VALVES CONTROL FLUID FLOW

Two different kinds of valves play important roles in ensuring safe operation as well as in the directing fluid power. Figure 3.13 introduces the first, the pressure relief valve, positioned between the pump and the actuator. It is designed to open if the pressure in the line exceeds a maximum level. As shown, in such cases the released oil is recycled back to the holding tank.

Figure 3.13 also illustrates that the cylinder rod is the element that applies the fluid power to the surrounding. We recognize that the rod can assume three states: (1) fixed position, (2) moving toward port B, and (3) moving toward port A. These states are enabled by the hydraulic valves inside a block that directs pressurized oil to either port A or B while at the same time allowing the oil to run off through the opposite port. The next, labeled the control valve, can be thought of as a gear of a transmission. It lets one move forward at different speeds or change gear into reverse. However, this is not a simple matter, since there are many different types used for a wide variety of applications not covered here. We are most interested in so-called directional control valves that direct the fluid to different ports or better "pressure chambers" where it is expected to do work. There are various valves that do that, but we will look at sliding spool valves as they are widely used on construction equipment. Figure 3.15 presents its basic mechanism with sliding pistons or spools at its center. O-rings inserted into the outer housing provide the necessary sealing that keeps high-pressure oil from leaking between the housing and the spool moving up and down. Figure 3.15 illustrates a four-way valve where a spool with three pistons is sliding up and down and blocking or opening the high pressure inlet, outlet, Port A and Port B.

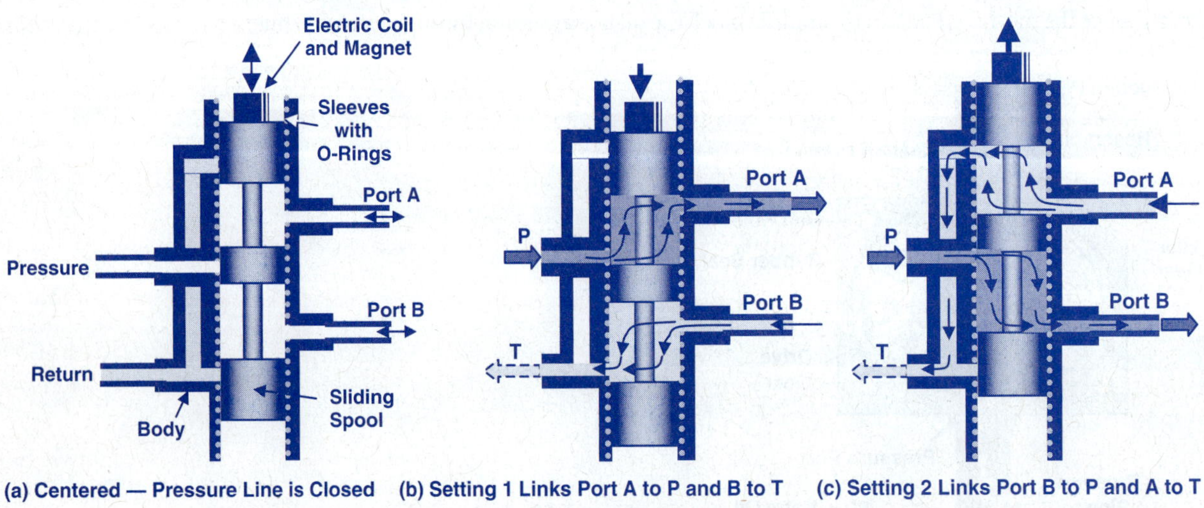

Figure 3.15 Operation of the Closed-Center Directional Four-Way Valve

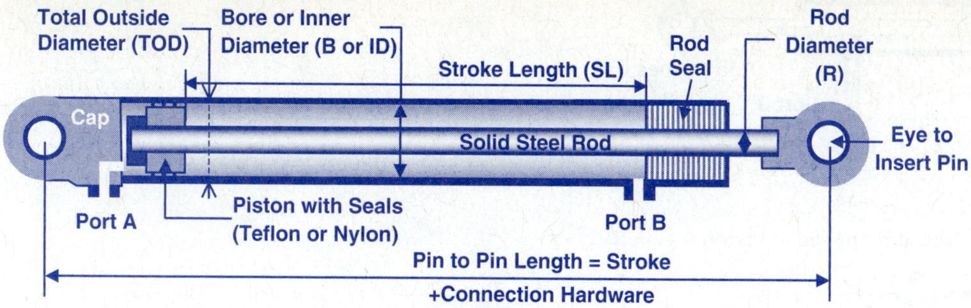

Figure 3.16 Elements of a Hydraulic Cylinder in Its Retracted State

The four-way valve is used extensively for construction equipment because of its versatility to power actuators in dual directions. Here, the pressure inlet (P) is connected to the pressure pump and the return outlet (T) to the reservoir. Finally, Figures 3.15a–c show the spool position, together with the flow of the fluid, that creates the three states of a cylinder or a motor: fixed, moving in one direction, and moving in the opposite direction. Let's look more closely at how the fluid reaches these states.

3.2.4 THE HYDRAULIC POWER "MUSCLE"

Devices that perform like muscles and move physical objects like levers are called actuators. One of the most used hydraulic actuators found in construction is the cylinder, which is basically a hollow pipe with a plunger inside and closed on both ends. Figure 3.16 depicts the main components.

There are three main values that define this linear actuator. The first is the **bore size** or **the inner diameter (B)** of the cylinder, which dictates the surface area **A** of the piston and with it the maximum force **F** that the cylinder rod can exert since $\mathbf{F} = \mathbf{A} * \mathbf{P}$. The pressure (P), of course, depends on the selected pump, setting of the relief valve, and most importantly the design strength of cylinders and hydraulic lines. In construction, a wide variety of pressure levels can be found that reach from 2,000 psi to 6,000 psi (138 bar to 414 bar, 141 atm to 423 atm, 13,800 kPa to 41,400 kPa), with the upper level constantly increasing due to stronger pumps, seals, connectors, and hoses developed by the equipment manufacturers. On the other hand, larger pressures are not always desirable as the resulting action forces in the mechanical components such as pins, joints, rod, and so on create more wear and may result in damage by operators who do not know how to use the tremendous forces carefully enough.

The second key dimension is the **stroke length (SL),** which dictates the total distance that the piston and rod is able to extend. Of course, the longer the stroke length the longer the rod has to be, which increases the risk of buckling. This leads to the third dimension, the **rod diameter (R)**. The solid rod is made of high-strength steel with a diameter that still fits into the bore. However, a large-diameter rod connected to the piston will drastically

Table 3.10 A Matrix of Pressure Conversions

	Pascal (Pa)	Bar (bar)	Atmosphere (atm)	Torr (Torr)	Pound-Force per in^2 (psi)
1 Pa	1 N/m^2	10^{-5}	9.87×10^{-6}	7.5×10^{-3}	145×10^{-6}
1 bar	100,000	10^6 dyn/cm^2	0.987	750.06	14.504
1 psi	6.89×10^3	68.9×10^{-3}	68×10^{-3}	51.715	1 lbf/in^2
	1 kgf/cm^2 = 14.696 psi				

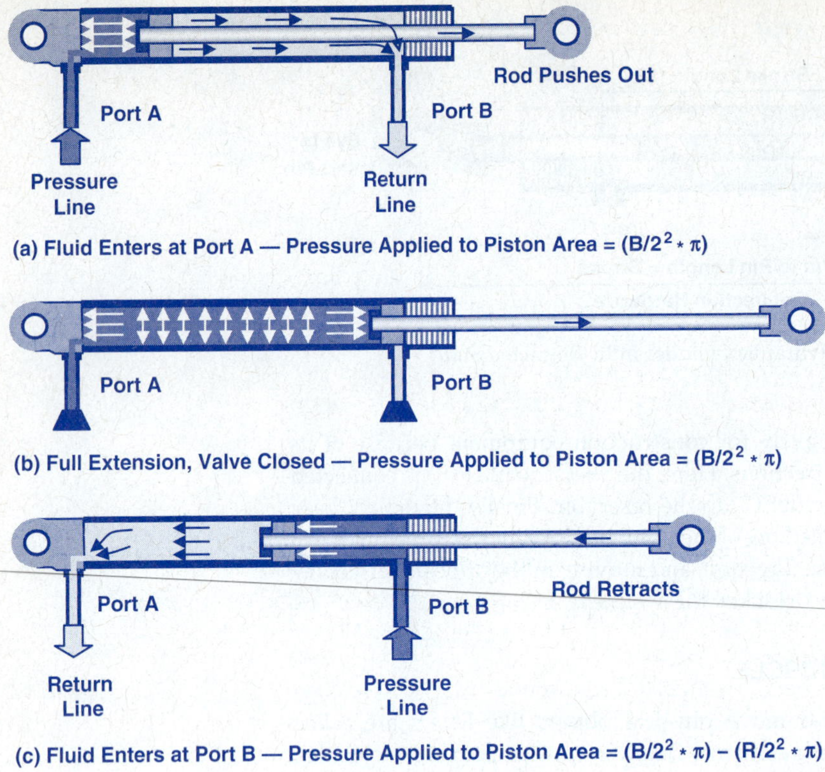

Figure 3.17 Extension and Retraction of Actuator

reduce the area for the hydraulic oil to apply pressure to the back side of the piston. This requires a close look at the application, since a large-diameter rod will allow more force when extending while reducing the achievable force when retracting. In any event, for longer **SLs**, the buckling condition becomes the deciding criterion for the rod diameter **R**. Figure 3.17 presents three states during a cylinder operation cycle.

In the first phase, the valve is set to allow pressurized fluid to leave through port A flowing into port A of the cylinder. As Figure 3.17a shows, the increased pressure pushes the piston toward the other end of the cylinder and, at the same time, forces the oil on the other side of the piston through port B of the cylinder. Logically, the return line is connected to port B of the valve with its spool in setting 1, permitting the hydraulic oil to flow into the main return line all the way back to the reservoir. Don't assume that this happens at 1 atm. The valve, in fact, serves as an orifice, which creates a counterpressure inside the cylinder to insure that the piston does not suddenly "slip away." In fact, the small back pressure is critical in situations where abrupt actions can occur in the opposite direction (e.g., slipping).

In the next phase of the process the piston hits the rod seal, representing the end of the stroke. No further travel is possible; thus the valve can be put back to the center position, in the process trapping the fluid in that state. As a result, the piston stays at that location as long as the valve setting is not changed, without the need of further power. This situation reflects the braking effect on a mechanical system keeping everything at a fixed state.

Figure 3.17c depicts the return of the piston and rod. One would be mistaken thinking that this will not require work. In fact, most applications in construction depend on the cylinder to apply a force when retracting (e.g., vertical drills). However, unlike in the first phase, the area for the pressurized fluid to act is reduced and with it the amount of force that can be created. What's responsible for the difference is the cross-section of the rod that is connected to the piston, thus preventing the fluid from acting on its surface. Hence, the larger the rod diameter **R** the less force **F** can be produced to pull the rod back into the cylinder. As mentioned above, dimensioning the rod diameter has two main effects, risk of buckling and maximum force for retracting.

S3.2 Basic Hydraulic Systems and Components (US Army)

3.2.5 PROPORTIONAL ELECTRO-HYDRAULIC CONTROL

This section introduces the state of the art in using hydraulic power. The first term, *proportional*, signifies a very important change from older valves, which could only be on or off, open or closed. As a result, switching from an off to an on would allow the oil to flow instantaneously using the entire inner cross-section of the line. Operating large construction equipment actuated by large cylinders with on-off valves, of course, can lead to drastic and undesired results. The sliding spool valve introduced in Figure 3.15 offers an opportunity to open the pressure port in a controlled fashion by moving the spool slowly over the circular or square opening of the pressure port. This technology would slowly increase the volume of fluid that could "squeeze" through the enlarging orifice. In fact, this is the basic concept of a proportional valve.

The second term in the title is *electro-hydraulic*. It indicates a transition from electronic to a hydraulic system of fluid power. Where can this switch take place? It cannot be a big surprise that the answer is the directional valve. The trick is to control the exact location of the sliding spool, using electric and magnetic "forces."

The enabling technologies are small electro-motors that respond to variable voltage that is sent to them. Negative voltage causes the motor to turn to the left, while positive voltage makes them to turn left. Furthermore, the voltage magnitude dictated the number of revolutions that a motor turns. This basic concept was modified to fit the hydraulic valves that had previously been operated by a manual lever sliding the spool against the force of a zeroing spring.

Figure 3.18 a depicts a cross-section of the ubiquitous joystick with an ergonomic grip. This particular system outputs a voltage between +5 and −5 while others reach +10 to −10. Regardless, when the potentiometer measures 0 volt, the stick is in neutral position. As soon as it is moved sideways the volt output value changes according to its angle until it reaches the max. In fact, the stick is able to follow a cone-shaped envelope, presented in a bird's-eye view in Figure 3.18b.

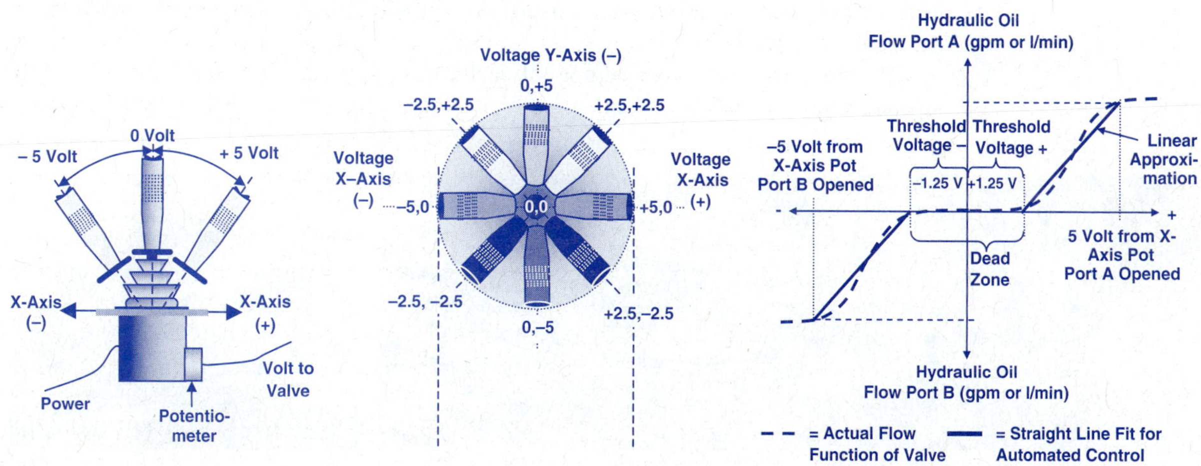

(a) X-Axis Joystick Movement

(b) Joystick Motions and Voltage Outputs on 2-Axis

(c) Proportional Control of Fluid Based on Variable Voltage Input

Figure 3.18 Joystick Control of Electro-Hydraulic Valves

The graphical basis is the Cartesian two-axis framework with the Y-axis perpendicular to the X-axis. However, the actual motions are measured via two separate potentiometers, mounted horizontally and perpendicular to each other. Hence, a joystick that is moved forward and backwards creates outputs from +5 to −5 volts. Presented in Figure 3.18b are outputs for both potentiometers at key positions where the first number is dedicated to motion along the X-axis and the second the Y-axis.

Figure 3.18c shows how the electronics of the joystick control hydraulic power by initiating movement of the spool in the four-way valve. Again, the X-axis is used to plot the angle of the stick represented as voltage that the potentiometer sends to the electronics mounted to the valve. With the help of a magnet, voltage is turned into a sliding movement of the spool, its direction and distance geared to the input voltage. In other words, as long as the input is 0 volt or below a certain threshold (e.g., +/−1.25 volt), the spool stays centered keeping the pressure and the return ports closed. As shown in the figure, the span between the upper and lower thresholds is called the dead-zone within which nothing happens. As soon as the input from the potentiometer exceeds the threshold value, the magnet is activated and pushes the spool off its center position, thus opening the two ports in one of the two settings. The so-called servo-valves, and to a certain extent the proportional valves, are able to move the spool so precisely that the flow of the created opening can be predicted for each location. The result is an S-type flow function presented in Figure 3.17c, one for each port. Within the dead-zone the spool is sliding toward the port-opening. Only at a threshold distance does a tiny area of the cross-section open, letting a small volume of oil pass through. The larger the opening gets, the larger the volume that passes through, an increase that depends on the flow mechanics through a changing shape until slowly approaching a maximum when the area is a full circle. While not very important for hand-operated hydraulic systems, such as backhoes, the linear approximation for the S-curve function, as shown, is important for automated or robotic systems that will calculate the proper change of voltage settings for the valve based on a needed flow or volume.

3.2.6 HYDRAULIC HOSE LOGISTICS

The development of hoses that were sturdy enough to sustain high pressure and abuse on a construction site added a significant new piece to the success story of hydraulic power in the field. Why were these hoses so significant? Because construction equipment is "active" in that it operates by lifting objects, digging trenches, drilling into the ground, and so forth. Rigid pipes with joints are not durable. Figure 3.19 presents several issues related to the use of hydraulic hoses.

The first picture shows that not only cylinders are used when high torque is needed, as here, to winch in a heavy pneumatic or water hose. The two hoses for pressure and

(a) Hoses Fitted to a Motor **(b) Hoses to a Moving Hydraulic Drill Protected by a Reel** **(c) A Bank of Hoses for Easy Maintenance**

Figure 3.19 Hydraulic Hoses Found on Construction Equipment

return are simple to install. To accomplish the same with rigid pipes would require more effort to create the necessary stability. Figure 3.19b presents an example of how to take care of hoses in order to keep them from being accidentally damaged. First, the hoses lead straight into a fitting that is attached to an L-connector, which leads the fluid in a right angle into the hydraulic motor. All hoses are cut to a precise length so that all of them are tight, especially around the reel, consisting of a large enough core and side walls keeping the hoses from falling out. When the drill starts to move forward, to the right of the picture, the reel will stay in place, guiding the hoses pulled by the connectors safely around the core. The last picture presents not only three large hydraulic cylinders pinned to rotating joints but also two visible rods, and many hydraulic hoses bundled together. The fixed jointing assembly on the top is especially interesting. Hoses coming from the pump to the left are being fitted to the left of the connection plate. To the right, short sections are supported by a chain with collar to keep them from hanging from the fittings at the jointing plate. At the bottom right corner one recognizes the next set of connections, one for each hose. What might be the reason for that? Consider that the picture was taken at the base of a long arm that extends toward the right with several hydraulic actuators connected to it. Consequently all the hoses supplying power to them will be exposed to the harsh environment and might be damaged easily. In case of a severe "injury" to the hose no "remedy" exists. The only way to addressing a "scratch" is to replace the hose with a new one. Now, the fittings that can be disconnected with a wrench become handy. Instead of having to replace the hose all the way to the control valve, only a "short" section would have to be replaced, thus reducing repair cost.

VG 3.2 Hydraulic Hoses

3.3 Protecting the Health of People Working with Equipment and Tools

The U.S. General Accounting Office wrote back in 1997: "Ergonomic programs can substantially reduce workers' compensation costs, with savings of up to 60%–80% over a 4- to 5-year period." One of the prime beneficiaries of ergonomics is the construction industry with its physically demanding work that, as statistics show, has a high rate of injuries, such as strains, sprains, and work-related musculoskeletal disorders (MSD). They became so prevalent in the 1990s that they became the most common injury responsible for almost half of the safety-related costs in construction. Every accident or injury not only causes pain and costs in emergency and medical treatment but also in disruption of work for investigations, paperwork, repair, and so on. Not surprisingly, one of the negative impacts of MSD was lower productivity.

Ergonomics

From the Greek words *ergon* + *nomos* = *work (by) natural laws.*

The objective of ergonomics is to match or "fit" workplace conditions and job demands to the capabilities of the human body. Effective ergonomic work results in the avoidance of illness and minimal risks for injuries and accidents. It results in most efficient work, few errors, high quality, and high workforce morale.

3.3.1 THE FRAGILE HUMAN SPINE

Lower back pain (LBP) has reached 75% of people in the developed countries. According to the U.S. Department of Labor's statistics for 2007, 30% of all nonfatal occupational injuries and illnesses in the construction industry were related to the back and shoulders. It is felt that over half of these injuries result from lifting excessive weight or lifting incorrectly. A long list of publications by the National Institute for Occupational Safety and Health (NIOSH) and the Occupational Health and Safety Council of Ontario (OHSCO), for example, not only stresses the drastic personal impact of injuries to the trunk but also how through effective body posture, use of correct lifting techniques, use of lifting equipment, and the limitation of weight LBP and MSD can be controlled. Still, the

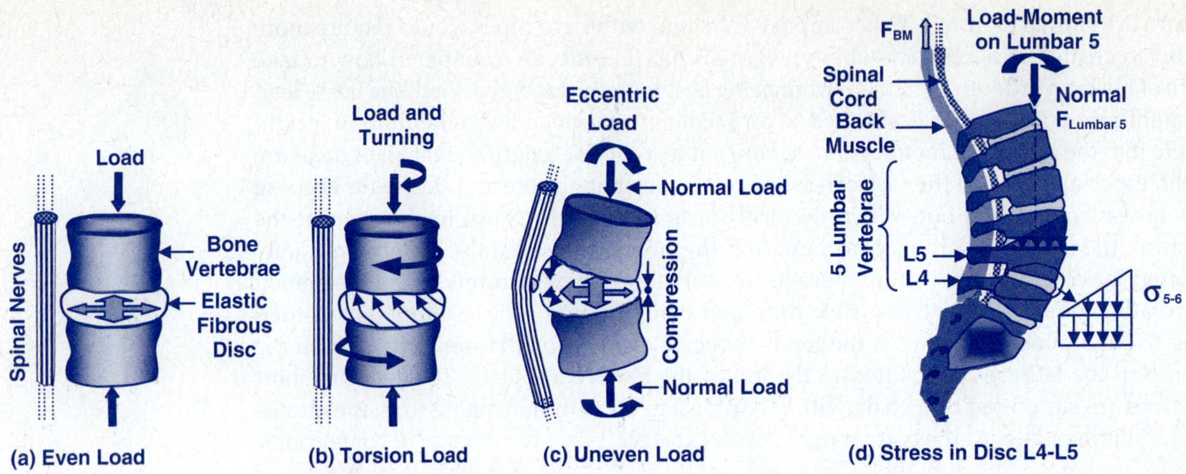

Figure 3.20 Spinal Biomechanics of Various Loading Conditions

continuing high incidence rate of this extremely painful and costly type of injury requires that one focal point of every ergonomic effort should be a review of lifting needs and procedures.

S3.3 Safety and Health Requirements –US Army
S3.4 Safety Program Management-Army
S3.5 Equipment Safety-Army Corps

The human spine is a "heavy-duty" system allowing us to walk and carry loads while remaining flexible enough to allow the upper body to bend forward and rotate. What makes this possible is the architecture of the spine made of bony vertebrae and elastic fibrous disks sandwiched in between. Nestled in close is the cord of the spinal nerves that runs from the brain to the legs and a set of strong muscles holding and "powering" the vertebrae. Figures 3.20a–c depict the deformation that a disc experiences during loading, twisting, and bending. In fact, Figure 3.20c shows a herniated disc where the disc has been overloaded, possibly a shock from loading, causing a small rupture at the tension site where a small part of the fibrous material separates itself and pushes against the spinal cord close by. As the cord has no place to go, held in place by bones and muscles, the small disc element may just be in an unfortunate location where it is painfully rubbed. Anybody who has gone through such a condition can talk about the resulting excruciating pain, the product of an inflammation of the nerves resulting from the constant rubbing. If the body is still young enough, the pain may subside after months of treatment but will commonly come back as the discs become older and less flexible.

Of course, there are many causes that can lead to back injuries that usually build up over time by wearing down a disk. Some of the more important ones are lifting of excessively heavy objects or continuous work in a bent-over posture as in nailing a floor or rebar tying. In the same category belongs receiving repeated "blows" to the spine as in driving across uneven surfaces at high speed without damping in the seat. Finally, instantaneous damage of a disk can occur during a fall or jump (from a piece of equipment) without proper flexing in the knees.

S3.6 Code of Practice MANUAL HANDLING
S3.7 Ergonomic Best Practices, Masonry
S3.8 Hand and Power Tools, OSHA

3.3.2 HAND-ARM AND WHOLE-BODY VIBRATIONS

Operators of heavy construction equipment such as bulldozers, backhoes, scrapers, haulers, loaders, skid steer vehicles, and excavators experience high-frequency vibrations combined with large shocks transmitted into the whole human body through direct contact, an effect that is called Whole-Body Vibration (WBV). NIOSHTIC-2 No. 20034733 presents the results of comparing the exposure to WBV in older and newer haulage trucks, which showed an improved protection in the newer trucks. By contrast, laborers working with large power tools such as demolition hammers, drills, hammer drills, angle grinders, chain saws, and handheld circular saws that require physical control are exposed to vibrations of their hands and arms (HAV). The effect of a vibration is exacerbated if a tight grip is required combined with low air temperature. Employees who are regularly exposed to this kind of vibration may suffer from hand-arm vibration syndrome, which causes neurological and motor disorders in the hands and fingers, as well as circulatory disorders in the fingers and disorders of the musculoskeletal system. Vascular disorders may occur in the form of white finger, which is caused by inadequate circulation. This phenomenon is generally more prevalent in colder seasons. Depending upon the duration and intensity of the exposure to vibration, it may affect only the fingertips or the entire fingers. Neurological disorders are experienced as tingling and numbness in the fingers, which become increasingly severe with increasing exposure. Examples of musculoskeletal disorders related to hand-arm vibration include wrist tendinitis and tenosynovitis.

Hand-Arm Vibration

A building worker who has worked on construction sites since the age of 14, using drilling and pneumatic hammers three or four days a week for several hours, now suffers from an occupational disease. At the age of 42, the pain in his wrist is so severe that he can no longer work. *European Agency for Safety and Health at Work*–http://osha.europa.eu.

The three main sources of whole-body vibration (WBV) on heavy equipment are

1. Low-frequency vibration caused by the tires and terrain
2. High-frequency vibration from the engine and transmission
3. Shock from running into potholes or obstacles

Measurements with vibration sensors showed, as can be expected, that different machines vibrate differently. As illustrated in Figure 3.21, mobile cranes, set up on outriggers, vibrate minimally while skid steer vehicles and scrapers develop mean accelerations that reach above 1 m/s^2 (3.3 ft/s^2). Even after exposure to only a vibration magnitude of 1 m/s^2, workers reported symptoms such as abdominal and chest pain, headaches, lower back pain, and loss of balance. However, an effect that further exacerbates the problem is the resonance between the equipment and certain body parts. Every object (or mass) has a resonant frequency, and if an object is vibrated at its resonance frequency, the maximum amplitude of the vibration will be greatly enlarged. Further complicating the matter is the fact that individual body parts and organs have their own resonant frequencies and do not vibrate as a single mass.

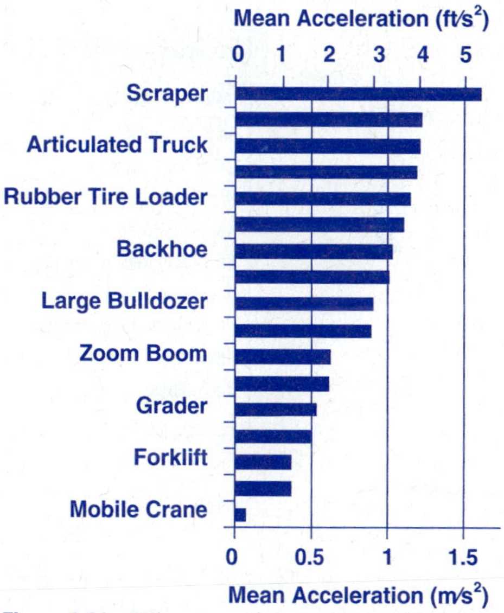

Figure 3.21 Vibration of Common Equipment
Source: ISO 2631

For example, the acceleration forces that impact the spine at 4 Hz are being amplified by up to 200%. Even worse are head, eyes, and arms vibrated at 20–30 Hz as they experience an amplification of 350%. ISO 2631, "Mechanical vibration and shock—Evaluation of human exposure to whole-body vibration" gives permissible exposure limits as a function of acceleration, frequency and length of exposure (1 min–24 hrs.) As expected, frequencies between 4 and 8 Hz have to be counterbalanced with lower accelerations since the human body is most sensitive to WBV at these frequencies. For example, a piece of equipment that vibrates at a frequency between 4 Hz and 8 Hz working should not accelerate in a vertical direction above 0.315 m/s^2 over an 8-hour day. If the operator only works only 4 hours per day the acceleration may reach 0.54 m/s^2. Figure 3.22 illustrates the effects of whole-body vibration.

S3.9 Vibration Protection-OSHA-Europe

The British Health and Safety Executive (HSE) leaflet INDG242 (rev1) addresses the dangers of vibrating hand-tools and provides guidance on maximal exposure to avoid health problems.

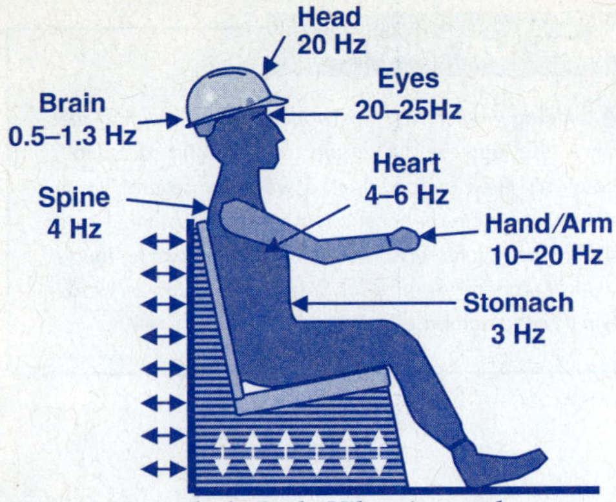

Figure 3.22 Whole Body Vibration and Dangerous Resonance

According to Table 3.11 , operating a chainsaw creates $6\,m/s^2$ vibration of the body. Using the chainsaw for one hour "earns" 60 exposure points (see Figure 3.23). With the maximum exposure action value at 100 points, an operator should stop after approximately 1.5 hours. Any longer use needs to be accompanied with mitigation measures. It is apparent that the highest vibrations by many power tools lie far beyond the hourly exposure limit (even if mitigated) of $15\,m/s^2$ and require extensive controls, such as remote control, if used for any duration.

3.3.3 FACTORS LEADING TO HEARING LOSS

An average human ear is able to detect or hear air that vibrates at frequencies that range from 20 Hz to 20,000 Hz. One way to limit or even destroy the sensitivity of the ear, or more accurately the inner ear, is by prolonged exposure to intense noise measured in decibels (dB) on a logarithmic scale. For example, an explosive noise of 150 dB may cause a permanent damage to several sensitive parts of the middle and inner ear. Longer durations at 120 dB create discomfort, slowly destroying the fine hair cells inside the cochlea.

Table 3.11 Vibration of Average Tools

Powered Tool Type	Dim	Lowest	Typical	Highest
Jack hammer	m/s^2	5	12	20
Demolition hammers	m/s^2	8	15	25
Hammer drills/combi hammers	m/s^2	6	9	25
Needle scalers	m/s^2	5	-	18
Scabblers (hammer type)	m/s^2	-	-	40
Angle grinders	m/s^2	4	-	8
Chipping hammers (metal)	m/s^2	-	18	-
Stone-working chisels	m/s^2	10	-	30
Chainsaws	m/s^2	-	6	-
Sanders (random orbital)	m/s^2	7	-	10

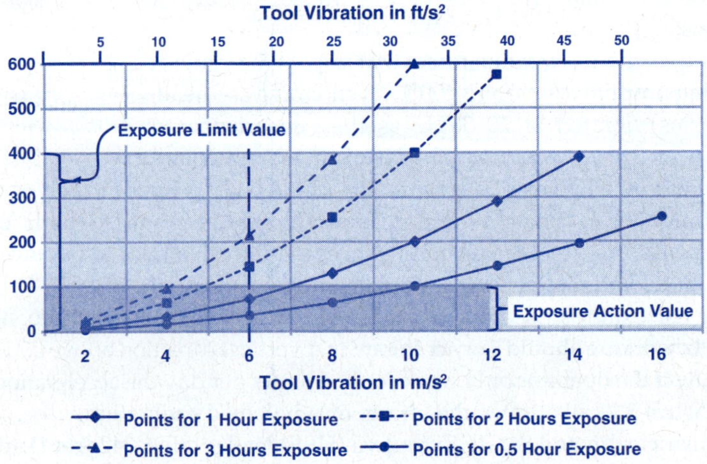

Figure 3.23 Time-Dependent Danger

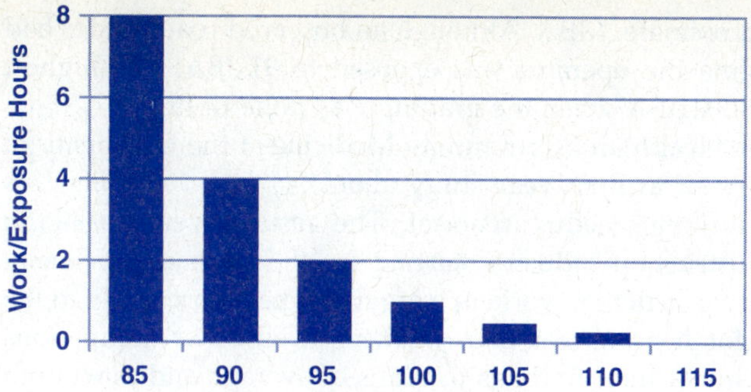

Figure 3.24 Permissible Exposure Durations for 85 PEL

According to the Federal Occupational Safety and Health Administration (OSHA 29 CFR 1910.95), the Permissible Exposure Limit (PEL) allows an 8-hour, full-shift average exposure of 85 dBA defined as an 8-hour Time-Weighed Average (TWA) of 85 dB (Action level). This number contradicts many state OSHAs and even the recommendation by NIOSH, which has set the threshold level at 85 dBA. Figure 3.24 indicates that employees who are exposed to more than 85 dB will only be able to work a full eight-hour day if they use hearing protection, such as earplugs. If this is not done, the time a person is permitted to work has to be reduced according to the noise level. Engineering or administrative noise controls are required when exposure exceeds 90 dB. Engineering controls include redesigning the space to reduce machinery noise, replacing machinery with quieter equipment, enclosing the noise source, or enclosing the noise receiver. Administrative controls include mandating the length of time an employee can be exposed to a particular noise source. Since the TWA is calculated by summing exposures multiplied by the durations, even changing work assignments can be considered. On the other hand, an exposed person might be reassigned during the day from high to low exposure areas. For every 5 dBA increase in exposure level, the allowable time is cut in half. To find if a set of multiple exposures at different dBA levels is permissible, one calculates the noise dose (D) representing the total exposure to any sound above 80 dBA during an 8-hour day as a percentage (%). The value for D needs to be below or equal 100% to be considered permissible. The equation for evaluating the noise dose is as follows:

$$\mathbf{D} = \mathbf{100} \times (\mathbf{C_1/T_1 + C_2/T_2 + \ldots C_n/T_n})$$

where

$D =$ noise dose during an 8 hr day in %
$C_i =$ hours spent at a given noise level
$T_i =$ hours permitted at noise level.

Workers must **always** use hearing protection when levels exceed 115 dBA.

Construction is a prime area for employees to receive permanent hearing losses as much of its large equipment depends on large diesel engines for power. Furthermore, powered tools use pneumatic power to drive hammering, cutting or chiseling actions to engage rock, hard soil, concrete, or even steel.

Recently, load-warning signals from equipment that is backing up have added another intensive noise source that, unfortunately, is cut out by wearing noise-protection devices. A recent study conducted by Eillmorth Spencer and Peter Kovalchik of NIOSH, *Heavy construction equipment noise study using dosimeter and time-motion studies*, showed that bulldozer operators consistently had the highest noise exposures, ranging from a 92 dBA to 109 dBA, and truck drivers without AC 90 dBA to 92 dBA. It was found that newer equipment comes with mufflers and air-conditioned operator cabins, which further reduce the noise level by an approximate 3 dBA. Although an observed road grader had a cab, even with an idling engine the operator was exposed to 91 dBA. The highest sustained dose was recorded at 104 dBA when the grading was done uphill.

In 2004, the School of Public Health and Community Medicine at the University of Washington published its results of a multi-year study about *Noise and Hearing Loss in Construction* (http://depts.washington.edu/occnoise). The results were based on six years of research in the construction industry starting in 1997. It had previously found that in Washington State, construction workers were five times more likely to file workers' compensation claims for hearing loss than workers in all other occupations combined. Figure 3.25 charts some of their findings in terms of average and maximum dBAs as well as the percentage of time construction workers were wearing hearing protection when required.

S3.10 Personal Protective Equipment
S3.11 Health Hazards in Construction-OSHA

Worked-Out Problem 3.2 / Dangerous Exposure to Noise

Frank, a laborer/operator, has been working on several jobs during one day. In the morning he was cutting masonry blocks for 1/2 hour before he operated the forklift for 1 hour lifting bricks onto the scaffold. In the afternoon he helped the formwork crew cut wooden studs and plywood using the circular saw for 1 hour. He would like to know if he should have worn ear protection.

Assumptions:

The noise level around Frank all day is high due to other operations that are going on. We assume that those are below 85 dBA and permissible. For the calculations we shall use average dBA values.

Exposures—(1–3)
.5 hr at 100 dBA (chip)
1 hr at 97 dBA (forklift)
1 hr at 94 dBA (circular)

Calculations

$D = 100 * (C_1/T_1 + \cdot\cdot)$

For PEL 90 dBA:

T_1 for 100 dBA = 2 hr
T_2 for 97 dBA = 2 hr
T_3 for 94 dBA = 4 hr

Noise Dose D_{90} =
$100 * (.5/2 + 1/2 + 1/4) =$
D_{90} = **100% (OK)**

For PEL 85 dBA:

T_1 for 100 dBA = 1 hr, T_2 for 97 dBA = 1 hr, T_3 for 94 dBA = 2 hr
Noise Dose $D_{85} = 100 * (.5/1 + 1/1 + 1/2) =$ **200%**

Discussion of Results

The PEL calculations reveal that if Frank works in an area where the rules of the Federal OSHA apply, he was okay. Anywhere else, in the United States or Europe, where PEL = 97 dBA, he was required to protect his hearing.

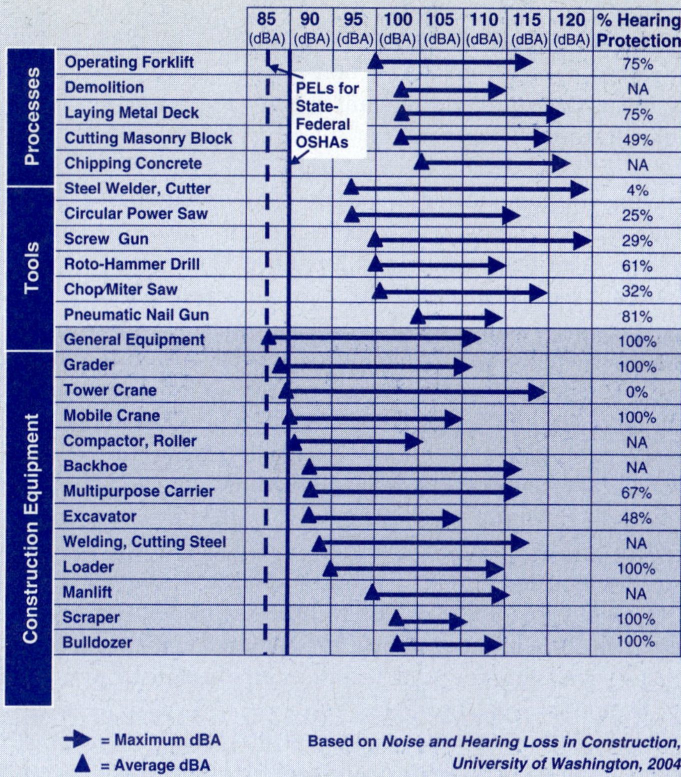

Figure 3.25 Noise Levels of Various Construction Tools

In any case, when he was working with the masonry saw, Frank was exposed to max dBA values above 115%. This means, he needed to wear protection during that operation. It is recommended that Frank immediately join a hearing-protection training program and take an audiometric test. This will enable him to assess the hazards he is in.

3.3.4 HEAT AND COLD STRESS IMPACT PRODUCTIVITY

One unique situation facing construction is its exposure to the elements (OSHA TED 01-00-015: 4). While rain and snow normally cause the operation to shut down, high or low temperature alone is not considered a reason to stop. However, the human brain tries to keep a constant temperature inside the body at about 98.6 °F (37 °C). As anybody who has worked in hot weather knows, heat will cause the body to sweat in order to cool the skin, leading to water and salt depletion. When the temperature drops, the body will increase heat creation through the metabolic process while extremities will start to get numb. The latter results in a drastic loss of muscle strength without much warning to the person who will, unaware, overestimate his capabilities. The following will review more systematically the effect of heat, cold, and humidity on productivity in construction.

S3.11 Working Outdoors-OSHA

Heat stress. The key factor related to heat stress it the *net heat load* on the body resulting from the heat created by the metabolic process and the heat exchange with the environment.

Thus the net heat load (NHL) can be calculated as

$$\mathrm{NHL} = \mathrm{M} + / - \mathrm{C} + / - \mathrm{R} - \mathrm{E}$$

Centigrade to Fahrenheit

°C = (F-32°) * 5/9

°F = 9/5C − 32°F

where

M = heat gain from Metabolism
C = heat gained or lost from Convection
R = heat gained or lost due to Radiant energy
E = heat lost through Evaporation of sweat

Evaporation has a cooling effect on the body, since the transformation from liquid to gas requires some small amount of energy, which comes from the heat. The amount of evaporation is closely related to air temperature and humidity. Humidity is a ratio that compares the present water vapor in air with the maximum possible at a specific temperature. Of course, when the humidity is low, more water evaporates from the skin than when the humidity is high. This phenomenon has been captured in a heat index, presented in Figure 3.26. The horizontal dashed line is related to an air temperature of 95° F (35 °C). At a humidity index of 25%, the body sweats at an expected rate. When the humidity rises, the amount of evaporation falls, thus reducing its cooling effect. The result is a higher net heat load for the body for the same amount of metabolism. This phenomenon is translated into an apparent temperature that is higher than the air temperature and needs to be considered when making decisions about the stress load due to work.

S3.12 Heat Stress Protection Quick Card-OSHA
S3.13 Heat Stress-OSHA Fact Sheet

A healthy body temperature, fueled by metabolism, is considered to be 98.6 °F (37 °C). If it passes above 102.0 °F (35 °C), further exposure to heat has to be brief and above 105.0 °F (40.5 °C), the body will stop functioning. Of course, these numbers are averages as other factors come to play such as age, general health, and the acclimatization level of the body.

Heat stress on the body is a gradual process resulting in increasingly severe reactions of the body. First, the muscles may start cramping up, followed by general exhaustion and tiredness. The extensive sweating will induce dehydration of the body that induces thirst, weakness, and headache. If cooling-down or rehydration are insufficient, continued heat stress will lead to fainting, a possible heat stroke, and even death.

Table 3.12 offers a general guide on how work load and apparent temperatures are related to the amount of exposure time, in % of work during an hour, for a healthy

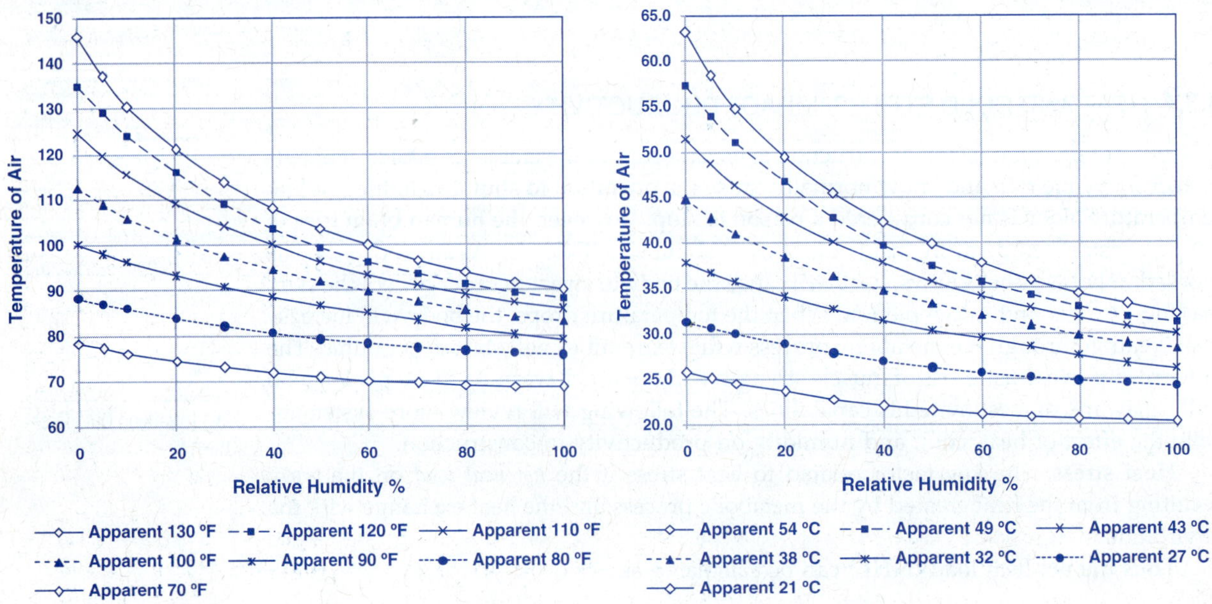

Figure 3.26 Heat Index Curves

body. Light work can be done continuously at a maximum temperature of 86 °F (30 °C), but heavy work only if it is below 77 °F (25 °C). Again, we are talking about the apparent temperature that considers the humidity index. As the table shows, the allowable temperature increases when rest periods are introduced.

Table 3.12 Permissible Work-Rest Cycles for Work Loads and Temperatures

Work %	Rest %	Work Load		
		Light	Moderate	Heavy
100	0	86°F – 30°C	80°F – 26.7°C	77°F – 25°C
75	25	87°F – 30.5°C	82°F – 27.8°C	78°F – 25.5°C
50	50	89°F – 31.6°C	85°F – 29.4°C	82°F – 27.8°C
25	75	90°F – 32.2°C	88°F – 31.1°C	86°F – 30°C

Wind Chill. The skin loses most of the heat through evaporation and convection. Both are drastically accelerated by cold wind blowing over it. Enhanced by the turbulence between a cold wind and a warmer body, the wind is able to "harvest" more energy from the warmer body and thus cool its surface and the layers below. The rate of heat loss depends on the wind speed above that surface: the faster the wind speed, the more readily the surface cools. For inanimate objects, the wind chill reduces any warmer area to the ambient temperature more quickly.

The common method to account for the windchill is to adjust the air temperature with a factor that related to the wind speed. Figure 3.27 presents a graphical means to convert air temperature to a windchill equivalent temperature value. A wind speed of 10 mph (16.1 km/h) is represented in the two graphs in Figure 3.27 intersecting curves that correspond to an air temperature of 10 °F (−12 °C). The circled points indicate that at wind speeds of 10 mph or 16.1 km/h respectively and a air temperature of 10 °F or −12 °C, the apparent temperature will be −4 °F (−20 °C).

Similar to hot and humid temperatures, cold temperatures impact the efficiency of the people working on construction sites. Different from summer heat, equipment is able to help the body by the thermal energy the engine is generating, thus reducing the impact of the cold air. The area for manual tasks is bracketed by two S-curves for lowest and highest impacts. At 0 °F the efficiency level of a human working on a manual task is

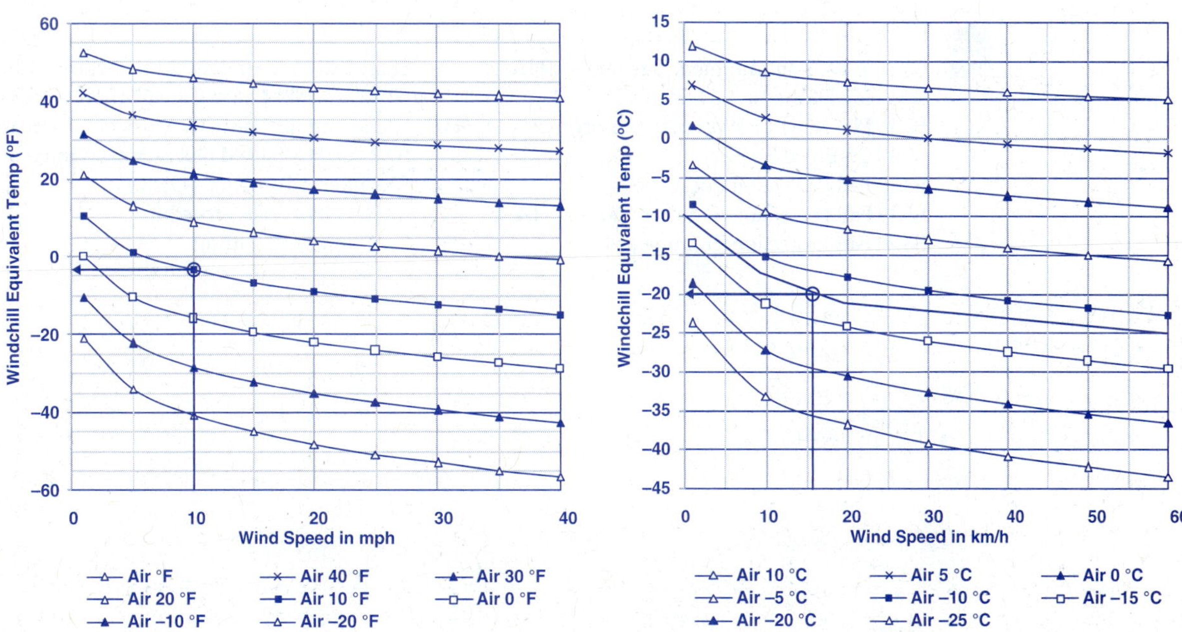

Figure 3.27 Windchill-Corrected Ambient Temperature

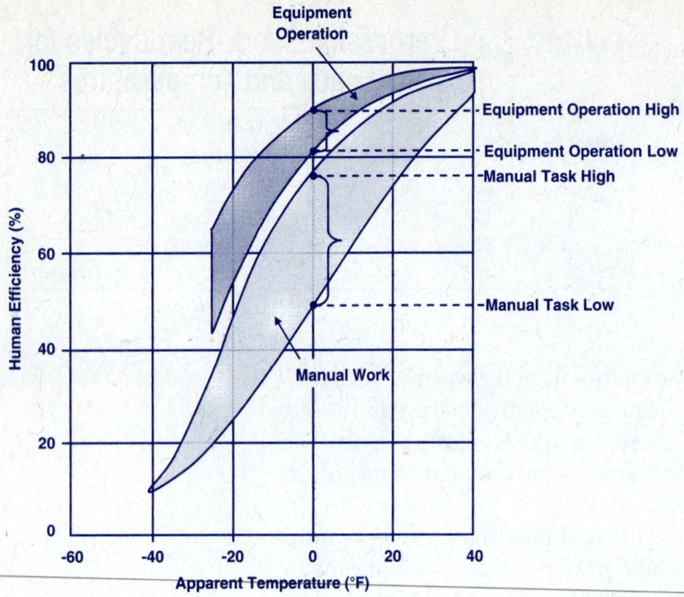

Figure 3.28 Low Temperature Affecting Human Efficiency

between 50% to 70%. In other words, work that lasts 2 hours will take up to 4 hours instead. The effect of the cold is lessened for an equipment operator to "only" 82% to 90% efficiency.

It is apparent that equipment cannot be operated at temperatures below −25 °F, while humans are able to work until −40 °F, albeit only at a 10% efficiency level.

3.3.5 SUPPLEMENTAL MATERIAL ON SAFETY AND HEALTH ON THE BOOKSITE

S3.14 What's an Imminent Danger—OSHA
S3.15 Job Hazard Analysis—OSHA
S3.16 Planning for Workplace Emergencies—OSHA
S3.17 Confined Spaces—OSHA
S3.18 Fall Protection Pocket book
S3.19 Fall Protection Requirement—OSHA
S3.20 Fall Protection Training ppt
S3.21 Fall Protection in Construction—OSHA
S3.22 Scaffold Use in Construction—OSHA
S3.23 Scaffold Use in Construction—OSHA
S3.24 OSHA'S Field Operations Manual
S3.25 Recordkeeping—OSHA
S3.26 OSHA Visit Procedure
S3.27 Injury Reporting—OSHA

3.4 Chapter Three Review

EQUIPUZZLE

The circled letters name the main source of waste in a mechanical system:

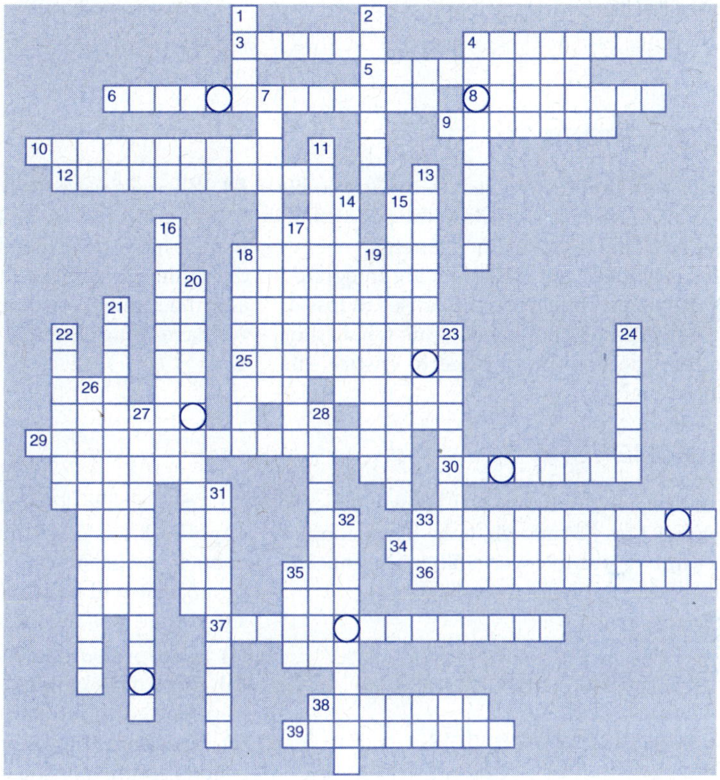

Across

3. The span between the beginning and the end of a piston "run"

4. The space at the end of the heating pipes to drop off small particles before they get into the outside of a steam engine

5. About one day

6. An injury with a debilitating effect because torn tissues rub against the spinal cord (two words)

8. A circular metal piece providing momentum to a crankshaft

9. Substance used by Pascal to measure the effect of height on fluid pressure

10. A measure for work that was arbitrarily based on the performance of a mammal

12. Acceptable based on an 8-hour Time-Weighed Average of 35 dBA

25. Defines the relationship between the number of teeth on two meshed parts (two words)

29. An expert professional (two words)

30. Describes a work room with small entry and exit

33. A small device inside an electro-hydraulic valve that opens and closed ports when activated by an electric coil (two words)

34. Hollow pipes closed at both ends and longer than the rod each one needs

36. Distributes torque and speed to two axles while letting them rotate at different speeds around turns

37. A "global" coupling between two non-aligned shafts (two words)

38. This cord safes lives when properly secured at its two ends

39. Steel rope used by a tower crane to lift heavy objects (two words)

Down

1. The abbreviation of a governmental agency that has as its mission to propose and implement laws created to avoid injuries on construction sites

2. Degree of sound intensity

4. Warning about a temporary hazard (two words)

7. Attached to the intake side of diesel engines pressurizing the air that boosts its power output

11. Keeper of the fluid needed by the pressure pump

13. Scientific field applying physics to optimize the performance of the human body

14. If not in place on a construction site, workers can get seriously hurt from falling tools or material

15. Term that defines a hydraulic valve that allows an operator to work large equipment extremely smoothly

16. A steel rod with eccentric elements needed by a combustion engine

17. Frequency of a vibrating object at which the maximum amplitude will be magnified.

18. When a human is in this condition s/he has a higher risk of making errors and having an accident

19. An undesirable by-product of powered machinery having health effects over time on the operator of tools or equipment

20. Electronic device used to measure the position of a joystick in degrees

21. Motor without spark plugs invented by a French-German (two words)
22. When not properly kept some muscles will soon cramp up
23. Last name of the scientist who first identified that the atmosphere is not weightless
24. A source of potential injury to a person
26. Mechanism consisting of meshed gears enabling the gearing up of a truck
27. Deals with forces and torques in human joints and tendons
28. A crane powered by the operators
31. A subject in engineering that takes its scientific "clues" from fluid mechanics
32. A train that does not need rails to provide torque to the rear-axle of a truck
35. Abbreviation of a disorder that destroys veins, nerves and muscles of a body part essential for working in construction

JOURNALING QUESTIONS

(1) The treadwheel used into the nineteenth century to lift heavy loads allows us to observe all parts of the powertrain operate. Discuss how the four main elements, energy source, engine, transmission, and actuation, are integrated into one piece of machinery. What are the causes of loss of energy along the powertrain? Can you imagine what kinds of control procedures need to be in place to ensure safe functioning of the crane?

(2) When we were working on Problem 3.1 to address Jack Boyd's concerns, we had to make many assumptions. What would have been needed to turn the assumptions into deterministic values? For example, the aerodynamic drag and the rolling resistance are big power-users. How could we have found the real values?

TRADITIONAL HOMEWORK

(1) The pump piston of a hydraulic car jack shown in Figure 3.29 has a diameter of 1 cm (0.4 in), while the lift piston has one of 4 cm (1.6 in). A pump lever with a mechanical advantage of 6 is used to apply force to the pump.

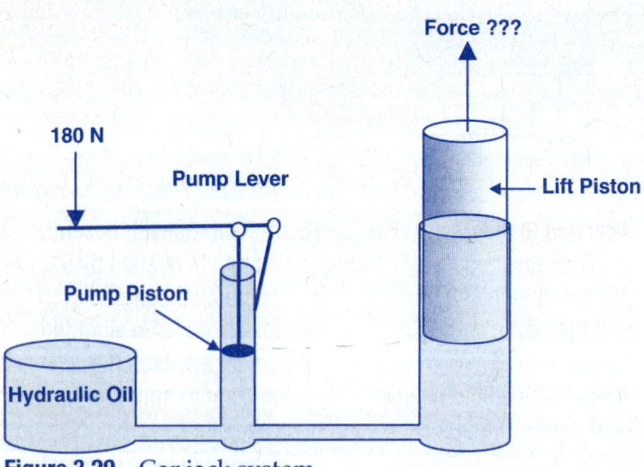

Figure 3.29 Car jack system

(a) How much mass can the jack lift if a force of 180 N is applied to the lever and efficiency of the system is 80%?
(b) Sketch a simple valve system that allows the oil from the tank to flow into the jack system when needed, to close when necessary, and to flow back when the lift piston is lowered (keep in mind that the oil from the lift has to go somewhere as the pump is closed!).

(2) A construction equipment company plans to build a new machine. They would like a preliminary idea about the hp requirement for an engine that is needed to power the necessary hydraulic system with cylinders having 5-in (12.7 cm) diameter bore, 3-in (7.6 cm) rod, and a 3-ft (91.4 cm) stroke length. The cylinders should be able to apply 50,000 lb (22.7 t) and operate at 2.2 in/sec (5.6 cm/sec) to push and 2.8 in/sec (7.1 cm/sec) to retract. The hydraulic pump that will be used produces 2.21 in^3 (36.2 cm^3) pressured fluid per revolution.
(a) What pressure is needed to operate the cylinders?
(b) How many gallons/min (gpm) or liter/min are needed to push and to retract?
(c) What are the pump revolutions-per-minute (rpm) needed to push at the desired speed?

Previous tests with the pressure pump produced following function:

$$\text{Power (hp)} = \text{gpm} * \text{rpm}/1430 \ (\text{l/min} * \text{rpm}/378)$$

(d) What is the horsepower need when only one 3-ft (91.4 cm) cylinder is used at a time?
(e) How much horsepower is needed to operate 4 cylinders at a time?
(f) What is the time difference between (d) and (e) when 8 cylinders need to be extended?
(g) What size diesel engine would you recommend and why?

(3) A delivery of a scraper to an earthwork job has to be planned. The scraper weighs 90,000 lb (400 kN), and the tractor trailer empty weighs 30,000 lb (133 kN). The final section of the delivery passes through a dense forest on a rough road with a rolling resistance of $F_{RR} = (0.15 * GVW)$ (0.1 is the rolling resistance coefficient for a rough road). There are some inclines that seem to be 18%.

Will the truck that was used in the solved problem 3.1 be able to deliver the scraper?

OPEN-ENDED PROBLEMS

(1) The problem with using the acceptable noise level function is the fact that a laborer does not necessarily know in the morning if he or she will reach the danger dose by evening. What are the long-term effects of noise overdoses (long and short)? What would you recommend that a construction manager could and should do to protect his or her construction crew from hearing problems in the future? What policies could be implemented to entice the workforce to actually follow your recommendations?

(2) The designs of operator cabs on construction equipment have dramatically improved over the last 10 years. Based on the information about a health and environmental impact, discuss what the relevant results of these improvements should be for: (a) The construction company and (b) the operator and his/her family.

BIBLIOGRAPHY

Abdelhamid, T. S., and J. G. Everett. "Physiological Demands During Construction Work." *J. Constr. Engrg. Mgmt* 128, no. 5 (2002).

Barker, T. C. "The International History of Motor Transport." *J. Contemp. Hist.* 20, no. 1 (1985).

Beavers, J. E., J. R. Moore, R. Rinehart, and Schriver, W. R. "Crane-Related Fatalities in the Construction Industry." *J. Constr. Engrg. Mgmt.* 132, no. 9 (2006).

Comfort, A."An Introduction to Heavy-Duty Diesel Engine Frictional Losses and Lubricant Properties Affecting Fuel Economy—Part I." SAE International 2003-01-3225 (2003).

Carroll-Burke, P. "Tools, Instruments and Engines: Getting a Handle on the Specificity of Engine Science." *Soc. Stud. Science.* 31, no. 4 (2001).

Danfoss, *Facts Worth Knowing in Hydraulics.* Nordborg, Denmark: Danfoss, 1998.

Davis, Gary, Ergonomic Best Practices/Acceptable Practices in the Masonry, Stonework, Tile Setting Industries. Wash. State Dept. Labor Ind. Policy and Tech. Serv., 2001.

Equipment Echoes, Magazine of the Hist. Constr. Equip. Ass. (HCEA), Grand Rapids, Ohio.

Goodyear, Factors Affecting Truck Fuel Economy, http://www.goodyear.com/truck/.

OEM Solutions Group, *Heavy Duty Bolted Cylinders.* Caterpillar Inc., Peoria Ill, 2006.

OEM Solutions Group, *Position Sensing Cylinders.* Caterpillar Inc., Peoria Ill, 2006.

OR-OSHA, *Fall Protection Requirements for Construction.* Occ. Safe. Health Admin., Region VII.

OR-OSHA, *Adopted Changes to the Vehicle Rules in General Industry and Construction.* Occ. Safe. Health Admin., Region VII, Order 6-2007.

OSHA Fact Sheet. *When a Worker Experiences a Job-Related Injury or Illness.* Occ. Safety Health Admin., U.S. Department of Labor, www.osha.gov.

OSHA. *Forms for Recording Work-Related Injuries and Illnesses.* Occ. Safety Health Admin., U.S. Department of Labor, www.osha.gov.

OSHA 3138. *Permit-Required Confined Spaces.* Occ. Safety Health Admin. 3138-01R 2004, U.S. Department of Labor, 2004, www.osha.gov.

OSHA 3146. *Fall Protection in Construction.* Occ. Safety Health Admin., U.S. Dept. Labor, (revised 1998).

OSHA 3150. *A Guide to Scaffold Use in the Construction Industry.* Small Bus. Safety Mgmt. Series. Occ. Safety Health Admin., U.S. Dept. Labor, (revised 2002)

OSHA 3080. *Hand and Power Tools,* Occ. Safe. Health Admin., U.S. Dept. Labor (revised 2002)

OSHA 3071. *Job Hazard Analysis.* Occ. Safe. Health Admin., U.S. Dept. Labor (revised 2002).

OSHA, *What to Expect During OSHA's Visit.* Occ. Safe. Health Admin., U.S. Dept. Labor (1997).

Sauer-Danfoss, *Directional Control Valves, Tech. Info.520L0564,* Revision B, Ames, IA, 2006.

U.S. Army Aviation Logistics School. *Basic Hydraulics Systems and Components.* Sub-Course Number AL 0926, U.S. Army (1994).

U.S. Army Corps of Engineers. *Safety and Health Requirements.* Dept. Army, EM 385-1-1. Washington, DC: U.S. Government Printing Office, 2008.

Vickers. *Hydraulic Hints &Trouble Shooting Guide.* General Product Support, Vickers.

Volvo Trucks. *Operator's Manual, Maintenance and Engine.* Greensboro, NC: Volvo Trucks North America, Inc., 2001.

WorkCover NSW. *Hazard Profile: Identification Tool for Bricklaying.* Gosford NSW, Australia: WorkCover Publications. www.workcover.nsw.gov.au.

WorkSafe. "Manual Handling." *West. Australia's Com. Occup. Safety Health.*www.worksafe.wa.gov.au.

chapter four

The Omnipresent Backhoes

When I say artist I mean the man who is building things—creating, molding the earth—whether it be the plains of the west—or the iron ore of Penn. It's all a big game of construction—some with a brush—some with a shovel—some choose a pen.

Jackson Pollock

Enough shovels of earth—a mountain. Enough pails of water—a river.

Chinese Proverb

No construction equipment is as prevalent across the globe as the hydraulic backhoe excavator. Not only is it capable of performing many different jobs, but it also has many different names such as the *hoe* or *back shovel*. Its success was long in coming, and its final breakthrough was made possible by several great innovations in the United States and abroad. This chapter will cover a variety of important topics related to this unique success story of construction technology; see Table 4.1.

Table 4.1 Main Topics Covered in This Chapter

Construction			Planning & Control		
Equipment	**Processes**	**Enablers & Tools**	**Managerial-Engineering Factors**	**Rules & Standards**	**Modern Tools**
✓ Steam Power Shovel ✓ Backhoe-Loader ✓ Tracked Backhoe ✓ Wheeled Backhoe ✓ Trench Compactor ✓ Backhoe "Crane" ✓ Asphalt Cutter ✓ 3-Link Arm ✓ Gooseneck Boom	✓ Cutting ✓ Lifting ✓ Tipping ✓ Curling Bucket ✓ Digging Path ✓ Shoring – Sloping – Benching ✓ Shielding – Trenchbox ✓ Loading ✓ Trenching ✓ Pipe-Laying ✓ Dewatering ✓ Backfilling ✓ Motion Control ✓ Haunching	✓ Hydraulics ✓ Joystick ✓ Buried Utilities ✓ Telescopic Stick ✓ Tool Exchanger ✓ Hydraulic Hammer ✓ Earth Auger ✓ Buckets ✓ Add-Ons ✓ Drill Rig ✓ Grapples ✓ Hydraulic Hoses ✓ Diesel Engine ✓ Crawler Tracks ✓ Outriggers	✓ Operator Skill ✓ Operator Training ✓ Soil Swell ✓ Bank Volume ✓ Shrinkage ✓ Soil Density ✓ Load Factor ✓ Call-Before-You-Dig ✓ Subsurface Utility Engineering ✓ Excavation Cycle Productivity ✓ Bucket Fill Factor ✓ Color Coded Utilities ✓ As-Builts ✓ Struck Capacity ✓ Heaped Capacity ✓ Load Capacity ✓ Friction Force	✓ SAE J296 ✓ OSHA 29 CFR, Part 1926, Subpart P ✓ OSHA 29 CFR 1926.800 ✓ One-Call Center (811) ✓ NIOSHTIC-2 No. 20024092 ✓ EM 385-1-1 Section 25 ✓ Operator Comfort ✓ Whole-Body-Vibration Damping ✓ OSHA 29 CFR 1926.651 ✓ OSHA 29 CFR 1926.550 (a)(15)	✓ Remote Control ✓ Ergonomic Controls ✓ AC Power Sensor ✓ Stakeless Excavation ✓ Line-and-Grade Laser ✓ GPS-Total Station ✓ Utility Locators ✓ Ground Penetrating Radar ✓ Computer Integrated Backhoe ✓ Visual Display Terminal

4.1 The Family Tree of Today's Backhoes

15th Century	**Medieval Scoops**	While earthmoving tools were used since the beginning of mankind, Giovanni Fontana is credited with the first design of a digging scoop in 1420. This was even before Leonardo da Vinci, in 1500, invented at least five different excavators to dredge rivers and canals in Italy.
19th Century	**The Magnificent Steam Power Shovels**	The first significant backhoe excavator was presented to the French Academy for assessment by Monsieur Macary in 1744. At that time, all the excavators dug toward the front and were thus called *front-end* or *face shovels*. They were capable of digging into the face of soil, slewing/swinging to the side and emptying the bucket into a cart pulled by horses. Over 100 of those behemoths moved 177 million cubic meters of earth to finish the Panama Canal starting in 1906. However, shovels were also needed to dig down into trenches. The first steam–driven backhoe shovel was built by Vulcan of Ohio for the American James Kewley in 1896. Although it had large steel wheels to make it mobile, it still bore a strong resemblance to a "reversed" face shovel. 
20th Century	**Losing Weight with Diesel Engines**	It was not until the 1930s that the first backhoe excavators became widely used thanks to the "slimming down" made possible by diesel engines powering the cable winches, leading to smaller equipment that could be easily transported from site to site. Two such early examples are the *"Trench Hoe"* by Austin with a 0.3 cubic meter bucket or the German mini-excavator Dolberg D 200 from 1950 that still used cables to move the boom, stick, and bucket. In addition the bucket could not be independently operated.
1957	**The Birth of the Backhoe-Loader**	The "downsizing" efforts by Case and others to create a lighter and more flexible machine led them to adapt the farm tractor that eventually became an essential addition to the equipment supply of every contractor small and large. The famous low-cost Case 320 backhoe-loader from 1957 had a hydraulically powered loader bucket in front, a hydraulic backhoe attachment in the rear, mechanical stabilizers, and was able to travel by itself from job-site to job-site. Case was the first to offer a piece of equipment where all the parts were designed to operate and "survive" as a whole.
1950s	**Hydraulics Spurs Global Innovations**	Three different kinds of innovations, however, finally secured the success of this machine: *(1) hydraulic power, (2) tracked undercarriage,* and *(3) wheel-base.* The first hydraulic excavators were built in France by Georges Bataille, founder of Poclain, mounting it on a truck chassis, while the first tracked machines were built in Italy in 1954s by Laltesi. In 1956, the German company Demag presented the first fully-hydraulic backhoe excavator B504.
1960s	**The Auto-Excavator**	Using wheels for locomotion instead of tracks added mobility to travel. A famous early example from the same time-period is the multipurpose Fuchs 301 *"Universal-Autoexcavator"* giving wheeled "trench-hoes" new impetus, similar to what Case's 320 did for backhoe-loader. 

4.2 From the Accident File

Forensic engineering is done every time a fatal accident occurs. In fact, for hundreds of years accidents have been the major source for engineers to improve their knowledge of building methods, to update construction standards, and as a result reduce the likelihood of such accidents occurring again. Thus, knowing what went wrong is an excellent source of understanding and a deterrent to repeating the same mistakes, as it constantly reminds us: *Could this happen to me?*

1. **Caught between a backhoe tail and concrete wall:** The contractor was operating a backhoe when an employee attempted to walk between the swinging upperstructure of the backhoe and a concrete wall. The employee was hit by the tail and crushed against the wall.
2. **Caught in a collapsing trench:** An employee was working in a trench 4 ft (1.2 m) wide and 7 ft (2.1 m) deep. About 30 ft (9.1 m) away, a backhoe was straddling the trench when the backhoe operator noticed a large chunk of dirt falling from the side wall behind the worker in the trench. He called out a warning, but before the worker could climb out, 6 to 8 ft (1.8–2.4 m) of the trench wall had collapsed on him and covered his body up to his neck. He suffocated before the backhoe operator could dig him out. There were no exit ladders. No sloping or shoring had been used.
3. **Turning over an excavator and surviving:** The operator was unhurt after his hydraulic excavator rolled over on its side as he was tearing down the remnants of a three-story house. He had maneuvered the backhoe onto the top of the pile of debris to reach the back of the house. A crane got the excavator partly back up before it was decided to take the arm off because there was concern that it could fall over and hit the house next door as well as all the overhead power lines.
4. **Turning over an excavator:** An equipment operator died from injuries he sustained after the backhoe rolled over the side of an embankment. The victim was performing road maintenance work using a backhoe to move dirt from an upper to a lower side. After the backhoe went over the embankment, it traveled 30 feet before flipping onto its right side. The victim was not wearing a hard hat or a seat belt, which allowed him to be thrown head-first into the roll-over protective structure. A site safety inspection had been performed by management the day before the job was to have begun, but this information had not yet been discussed with employees. The Sheriff-Coroner's Autopsy Report stated the cause of death to be craniocerebral trauma, with acute ethanol intoxication.
5. **Pulling up an invisible buried gas line:** While digging a trench behind a building, a backhoe operator damaged a $^3/_4$ in (1.9 cm) steel natural-gas service line and a 1 in (2.5 cm) water service line. This resulted in two leaks in the natural gas service line, which was operated at 35 psig. One leak occurred where the backhoe bucket contacted and pulled the natural gas service line. The other was a physical separation of the gas service line at an underground joint near the meter, which was close to the building. Gas migrated into the building at 46 Florida Avenue, where it ignited. An explosion followed, destroying three buildings. Other buildings within a two-block area of the explosion sustained significant damage. Three fatalities, five serious injuries, and one minor injury resulted.

S4.1 Backhoe Accident with Gas Line-NTSB

4.3 Modeling Forces and Motion at Work

As we learned in Chapter 3, what allows construction equipment to perform work is its capability to apply large forces at the end of a powertrain or a hydraulic cylinder.

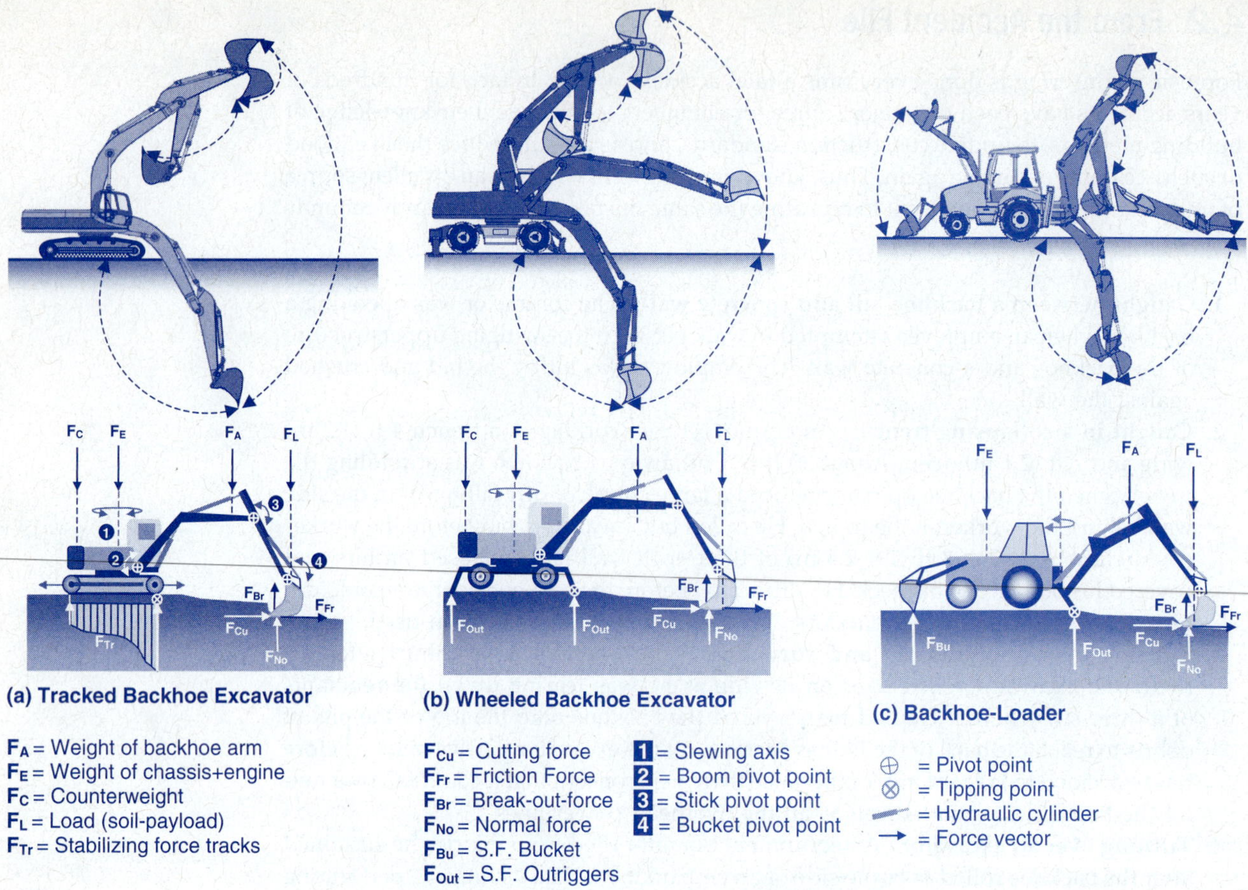

Figure 4.1 Comparison of the Motion Envelopes and Force Patterns for Three Backhoe Designs

The backhoe excavator is a perfect example of a piece of equipment that switched from one method to the other. Before we discuss the effective use of this versatile piece of machinery, we should take a closer look at how it accomplishes the basic work that it is designed to do. First one has to realize the common features of the three main backhoe designs and motion envelopes: (1) tracked, (2) wheeled, and (3) attached to loader tractor.

S4.2 Backhoe Excavator CAT 345C
S4.3 Backhoe Excavator Wheeled John Deere
S4.4 Backhoe Loader CASE
S4.5 Backhoe Excavator Compact Volvo

Digging dense soil, breaking up asphalt, or lifting concrete pipes requires large forces created at the end of a large articulated arm. As shown in Figure 4.1, each of the three designs employs an arm made of three linked segments, the boom or gooseneck boom (since the boom is slightly angled), the stick, and the bucket. Each element pivots around a pin that is attached to the end of another and is powered by its own hydraulic cylinder. For example, the stick is pinned to the end of the boom moved by the stick cylinder attached to its shorter end. By elongating this cylinder, the stick will rotate around the stick pivot point and with it the bucket attached to its long end. Of course, the operator has the capability to move all three segments, the boom, the stick, and the bucket, at the same time.

Two limitations for the backhoe to execute an action that the operator wants it to do:

(1) Lack of sufficient pressure to overcome the required force in one of the three cylinders
(2) Loss of stability due to tipping

While the first limitation case was discussed in Chapter 3, this chapter presents the second, common problem faced by equipment operators using long arms, referred to as tipping. This has been the cause for many deadly accidents. Figure 4.1 shows that each equipment has its own tipping point where the tip of the track or the front outriggers rests on solid ground. Tipping can occur when the destabilizing forces in front of the tipping point outstrips the stabilizing moment from the counterweight F_C and the weight of the chassis, engines, etc.

When the wheeled backhoe and backhoe-loader drive around with retracted outriggers, the backhoe operator needs to make sure that the arm is folded in, because the weight of the stick and bucket can make it tip, especially on sloped traveling surfaces. As accident reports #3 and #4 earlier demonstrate, rolling over sideways on an uneven surface is yet another cause for deadly accidents. Conversely, a backhoe loader may be stable driving uphill with a full bucket of dirt (in the front), but when the load is dumped into a truck, the loss of soil weight may be enough to cause the machine to tip backward.

VG4.1 Excavator Tracks

Figure 4.1 shows in addition to the three horizontal pivot points the vertical slewing axis allowing the upperstructure or the arm to rotate horizontally. While the slewing angle on the backhoe loader is limited, the two backhoes can rotate 360°.

4.4 The Capabilities of the Universal Workhorse

It is quite possible that backhoe excavators or loader-backhoes are the first truly global construction equipment. While mining prefers excavator shovels, construction worldwide depends on backhoes to dig the pits for buildings and bridge foundations. Heavy users of backhoes are utility contractors involved in the laying of pipes, be it concrete, cast-iron, or plastic. They value their many capabilities, which include (1) versatility to exchange the bucket for many different tools, (2) long reach, (3) mobility on site and on road (wheeled), (4) hydraulic power to power auxiliary tools, (5) variety of sizes reaching from tiny "tools" that fit through the front door to giants that crush trucks with their buckets. In the last 10 years, manufacturers have added new and ruggedized electronics not only to assist the operator to be more productive and safer but also to protect the equipment from "internal" damage.

S4.6 Excavation Army Corps of Engineers

Header Problem 4.1 / Installing 6 ft (1.8 m) Diameter Pipes at 16 ft (4.8 m)

Trench Inc. just signed a new contract to install 5,280 ft (1.6 km) of 72-inch-diameter (1.83 m) spiral-steel pipe in 40-foot (20 m) lengths. Each pipe is cement lined, tape wrapped, and mortared. One pipe element weighs 32,000 lbs (142 kN). Figure 4.2 presents the plan view of a section with three crossings underneath the Union Pacific Railroad (UPRR).

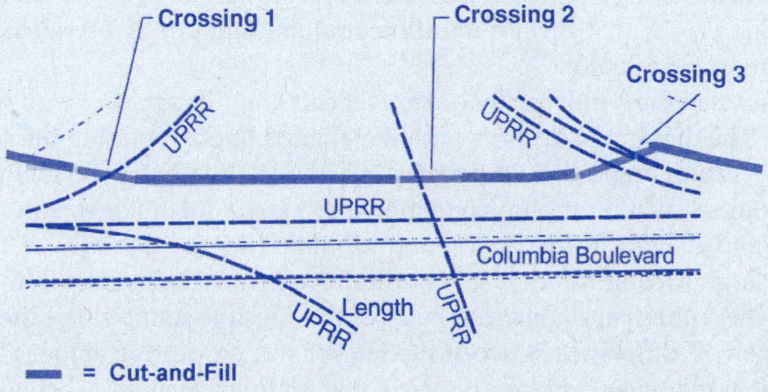

Figure 4.2 Railroad Crossings-Section of Pipe Project

Since railroad tracks can't be cut in order to excavate to lay the pipes they have to be bored and jacked—work that will be done by a specialty subcontractor. In the stretch between crossings the pipeline trench needs to be cut vertically, because of the closeness of the railroad tracks. The maximum depth of the pipe invert (bottom) is 16 ft (4.9 m), and the minimum width of the trench right above the crown of the pipe is 8 ft (2.4 m). Because a river is nearby, the soil is granular in nature, varying between coarse-grained gravel (GP-GM) to fine-grained sandy silts (SM-ML) with an average moist unit weight of 2.12 mg/m^3. The geological investigation also showed the presence of boulders in the range from 12 inches (0.3 m) to 48 inches (1.2 m).

Question 1: *What specific equipment is needed to install the required pipes and backfill the trench between the crossings?*
Question 2: *What are the preparatory measures that you must take before starting a new section?*
Question 3: *Develop a detailed work plan for the crew placing and jointing the pipe segments?*

Comment: The text that follows will enable you to answer these three questions.

4.4.1 JOYSTICKS THAT MOVE MOUNTAINS

Digging into the soil requires a force capable of pushing the cutting edge of bucket into and through the ground. On its path, the bucket will experience many forms of resisting forces that have to be overcome: (a) cohesion between the soil particles, (b) adhesion of soil to the bucket, (c) friction between bucket and the bank material at its bottom and sides, and (d) objects buried in the soil that have to be cut and broken up. Taken all together, the task of smoothly filling a bucket with soil material becomes very tricky. Why is this so difficult to do when it looks so simple? The problem is that the backhoe arm is made of a three-link mechanism, as shown in Figure 4.3, making it extremely difficult to follow a straight line. Each element rotates independently, but the location of its pivot point/pin changes with the section it is connected to. For example, pin 2 is the base for the rotation of segment pin 2—pin 3 but moves along a circle at the end of segment pin 1–pin 2. Thus, the position of the tip is dictated by the three angles in the pins 1, 2, and 3. The backhoe arm behaves like a 3-link mechanism where pin 1 is connected to the superstructure. Each rotational actuation is made possible by rod motion of a hydraulic cylinder. In Chapter 3 we learned that all modern hydraulic systems use electro-hydraulic valves that respond to positive and negative DC voltages. Figure 4.4 illustrates how the left and the right joysticks relate to the four main actuators that power a backhoe.

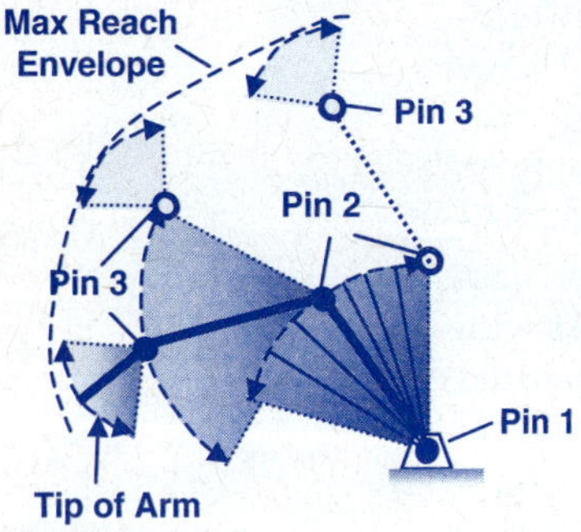

Figure 4.3 Geometry of a 3-Link Arm

The four main joystick axes control the three hydraulic cylinders of the arm, all of different sizes, and one motor that slews the upper structure 360°. During digging, each cylinder is active to achieve a desired path. As was discussed in Chapter 3, it's in fact the two-directional hydraulic valves that dictate the direction and amount of pressurized hydraulic fluid powering each actuator.

VG4.2 Backhoe Excavator at Work

Let's watch the motions of the 3-link backhoe arm as it cuts and fills a bucket with soil presented in Figure 4.5. The digging path starts with the aligned bucket entering the soil at a desired point and cutting angle. While penetrating, the stick is being rotated by extending the stick cylinder while, at the same time, bringing down the boom by retracting the boom cylinder. During this action and all along the cut the back of the bucket will experience large friction forces if it is not removed from the bank soil in its back. The 3-D model of the cutting mechanism shows that the effective countermeasure is to create an approximately 5° difference between the bucket and the cutting angle. This small gap is very critical in reducing the largest source of wasteful friction force between the bottom of the tool and the bank material (2) since it eliminates the possibly large

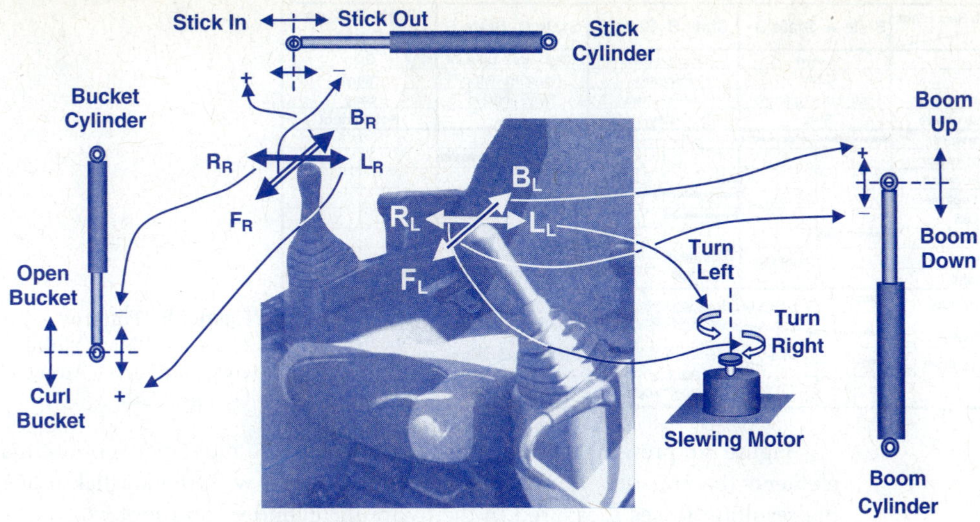

Figure 4.4 Schematic of the Hydraulic Backhoe 3-link Arm Control System

reaction forces at the bottom plate (6). The operator has no means to influence the other friction forces on the sides (5) and along the inside of the bucket (4). Unavoidable resistance forces that have to be overcome by hydraulic power originate at the cutting edges of the bucket bottom and the sides as they slice through the soil and/or push stones that lay in its path (1) and (5). As depicted in Figure 4.5, during cutting, all three cylinders are pushing forward at different speeds creating a straight path. During the extract-lift action, the bucket requires curling to retain the soil and a "booming-up."

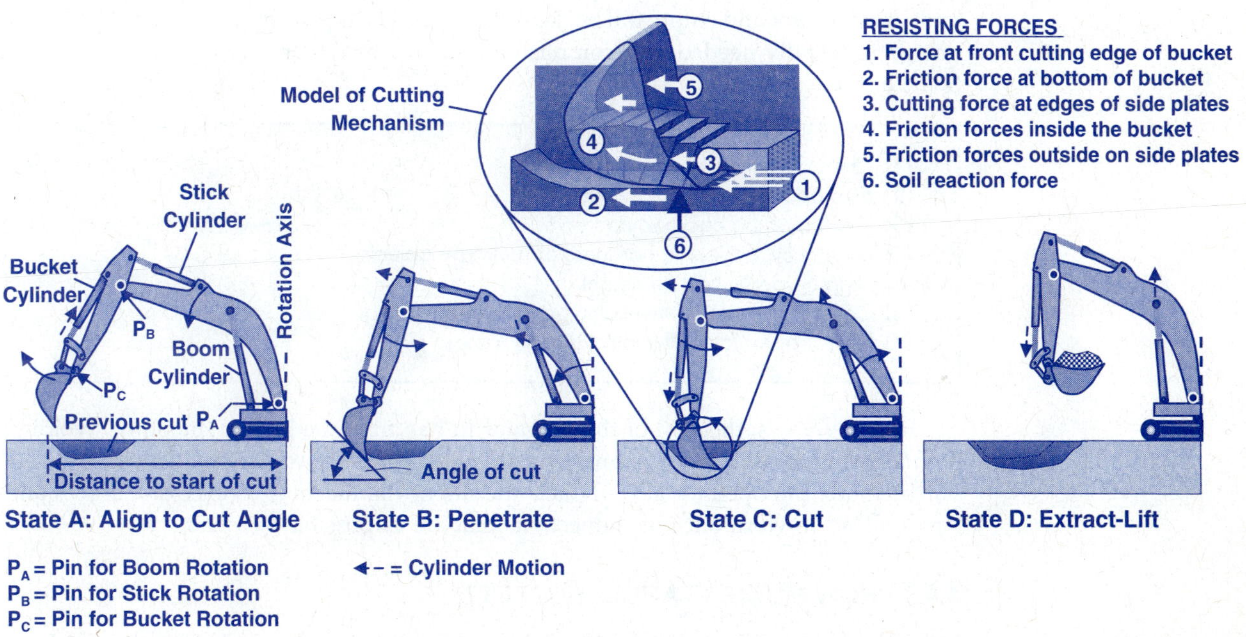

Figure 4.5 Actions of the Backhoe Arm and Cylinders Along the Digging Path

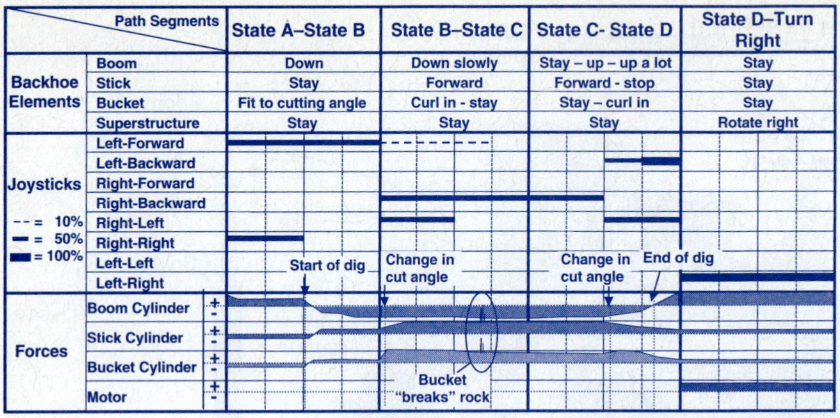

Figure 4.6 Patterns of Joystick Moves, Backhoe Motions, and Cylinder Forces Along a Basic Digging Path

Figure 4.6 presents a schematic chart aligning the movements of the main backhoe elements (boom, stick, buckets, and superstructures) with the joystick movements, and the resulting forces measured in the hydraulic cylinders and motors.

We learned in Chapter 3 that a diesel engine runs the hydraulic pump pressurizing the fluid that "powers" the cylinders and motors. Figure 4.6 displays the forces needed by each of the three arm cylinders and the motor along the digging path in a uniform soil media with a lone small rock that has to broken up. When the force line passes through the + zone, the pressure at the rod-less side of the cylinder needs to match the force and on the rod-side when the force line is in the − zone. For example, in state A (see Figure 4.5), the boom cylinder has to exert an upward force in order to keep the arm up (force line in + zone) while during the cutting phase, state C, the same cylinder has to push the boom downward (force line in − zone) in order keep the bucket on its path cutting through the soil. At the same time, the stick cylinder has to move the rod forward (force line in + zone) to rotate the stick pushing the bucket along.

In summary, the force diagram at the bottom of Figure 4.6 highlights that hydraulic power is continuously changing "sides" to (a) fill the bucket by cutting and pushing soil into the bucket, (b) break or move a buried obstacle, (c) lift full bucket out of soil media, and (d) maneuver and empty full bucket. Of course, it is impossible to calculate how much energy is actually needed to accomplish all this because there are too many variables. Some of them include

(1) Characteristics of the soil
(2) Size and shape of the bucket
(3) Weight of the backhoe arm
(4) Frequency, size, and hardness of rock-type objects
(5) Digging depth and lift height
(6) Condition of pin connections (grease and packing)
(7) Skill of operator in avoiding friction forces

In many ways, the skill of the operator in handling the bucket smoothly, avoiding both frictions and large dynamic forces, will dictate how much diesel is burned. Furthermore, the operator also impacts the life of the machine as excessive forces and stresses lead to worn-out pin connections that are impossible to repair.

4.4.2 "HOW MUCH" IS A FULL BUCKET?

While it is impossible predict the forces that have to be overcome during digging, we have a better chance to estimate the work needed to raise a full bucket to a required height.

Table 4.2 Comparison of Three Medium-Size Backhoe Buckets

	Width		Capacity		Weight		Teeth	Digging Force	
	in	mm	yd^3	m^3	lb	kg	Nr.	lbf	kN
Hitachi Zaxis 225 147 hp	28	720	0.67	0.51	1,150	530	3	34,000	151
	41	1,030	1.05	0.80	1,480	670	5	34,000	151
	45	1,150	1.19	0.91	1,590	720	5	34,000	151
	57	1,450	1.57	1.20	1,520	690	6	34,000	151
	41	***1,030***	***1.05***	***0.80***	***1,700***	***770***	***5***	***34,000***	***151***
	45	***1,150***	***1.19***	***0.91***	***1,830***	***830***	***5***	***34,000***	***151***
John Deere 200D 159 hp	30	760	0.79	0.60	1,432	650	4	28,904	128.6
	36	915	1.00	0.76	1,621	736	5	28,904	128.6
	42	1,065	1.22	0.93	1,790	813	5	28,904	128.6
	48	1,220	1.43	1.09	1,976	897	6	28,904	128.6
	42	***1,065***	***1.09***	***0.83***	***1,731***	***786***	***5***	***29,099***	***129.4***
	48	**1,220**	**1.29**	**0.99**	**1,921**	**872**	**6**	**29,099**	**129.4**
Caterpillar 320D 148 hp	24	610	0.72	0.55	1,387	629	3	28,079	125
	30	762	0.98	0.75	1,583	718	4	28,079	125
	42	1,067	1.53	1.17	1,878	852	5	28,079	125
	54	1,372	2.05	1.57	2,205	1000	6	28,079	125
	42	***1,067***	***1.31***	***1.00***	***1,909***	***866***	***5***	***32,417***	***144***
	48	***1,219***	***1.56***	***1.19***	***2,108***	***956***	***6***	***32,417***	***144***

42 = Heavy Duty Bucket

Table 4.2 presents several important values that are specified by the equipment manufacturer, such as the bucket capacity measured according to standard J296 from the Society of Automotive Engineers (SAE), its width, its number of teeth, and so on. Figure 4.7 depicts how the SAE standardized the method fabricators have to apply when they "size" the capacity of their buckets.

Buckets, made of hardened steel, are also classified according to their intended use as (a) general purpose, (b) heavy duty, and (c) rock excavation. The comparison of three medium-size backhoe excavators with around 150 hp shows that fabricators offer a wide range of bucket widths and capacities that can be used by one particular excavator. The heavy-duty buckets are generally a little bit smaller and heavier but are able to apply more digging force.

The SAE-rated capacity is the sum of the struck capacity and the material that is heaped above the strike-off plane at a 1:1 angle of repose. As the purpose of these stipulated conditions is to create a uniform measure to compare any kind of bucket, these values do not recognize the fact that the excavated materials vary widely and will not fit the 1:1 angle of repose requirement. Let us take a closer look at the reality of soil behavior.

SAE Standard vs. Actual Volume: When one watches excavators working in different soil or rock conditions one immediately observes that the actual volume in the bucket varies depending on the type of material. Figure 4.8 distinguishes among three volume capacities: (A) larger than the SAE rating, (B) SAE heaped volume, and (C) smaller than SAE rating.

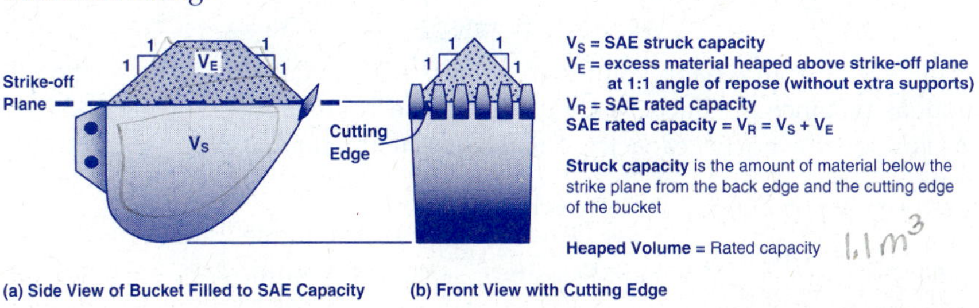

Figure 4.7 Dimensions for Defining Bucket Capacity Using SAE J296

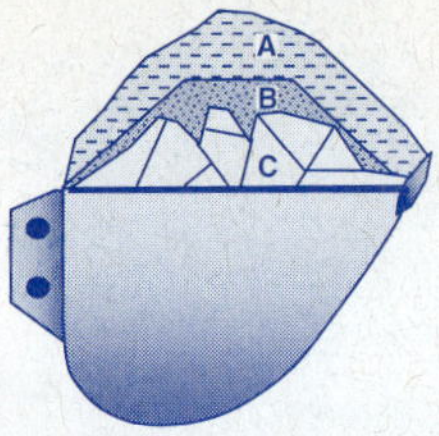

Figure 4.8 Three Conditions of Actual Bucket Volumes

The so-called fill factors are quantitative relationships between the actual volume in a bucket and the SAE-rated capacity of a bucket and are given in Table 4.3. One realizes that the value can be bigger or smaller than 100% depending on the characteristics of the soil/rock. It is easy to see that the fill factors have a wide spread, indicating that the actual soil can have a drastic effect on production.

For example, broken rock is full of irregularly shaped edges and corners that do not adjust like soft clay. As a consequence, blasted rock leaves a lot of voids between the individual pieces, thus drastically reducing the actual volume of solid rock that fits into the bucket.

Table 4.3 Adjustment Factors for Bucket Fills

Material Description	Fill Factor (%)	Condition
Moist Loam or Sandy Clay	100–110	A-B
Sand and Gravel	95–110	A-B-C
Hard, Tough Clay	80–90	C
Rock—Well Blasted	75–95	C
Rock—Poorly Blasted	40–70	C

VG4.3 Excavation & Compaction

Disturbing Bank Condition Swells the Volume: Soil material that "sits" in its natural state is referred to as being "in situ" or in its bank condition. The natural state of soil may have evolved over millions of years after it was deposited by ancient rivers, oceans, or winds. Of particular importance here is the soil density measured as the weight of a bank yd^3 (m^3) impacted by the compacting weight of the soil on top. Figure 4.9 shows the situation of an excavator digging a trench in densely compacted clayey soil. While the trench walls in their "bank condition" create vertical walls, the figure shows the piles of the loosened material dumped on the side of the trench with a fairly "flat" angle of repose.

A physical property that indicates how tight the elements of a material are packed together is called density.

$$\text{Density} = \frac{\text{Mass (lb or kg)}}{\text{Volume (yd}^3\text{ or m}^3\text{)}}$$

Based on the function, the density rises with increased mass or with lower volume. Figure 4.9c depicts how the loosening effect of excavation increases the volume of a soil in its natural bank condition. Since the mass stays fixed, its density decreases. Some of the highest density "earth" materials are metamorphic rocks (e.g., granite) that were millions of years ago exposed to high pressure and extreme heat cementing the particles together. In addition, metamorphic rock contains a lot of heavy quartz and feldspar, further pushing the density up. On the other hand, sand, gravel, common earth, and clay are the product of a sedimentary process where rivers thousands of years ago deposited them on the bottom, one layer on top of another.

The percentage of loosening that a material experiences is important in figuring out the time it takes to excavate a bank volume or to calculate the necessary space to stockpile excavation material on the side of a trench or how many truck loads have to be hauled away. Figure 4.9c presents the basic equation to calculate the loose volume of a material as a function of a material's bank volume by applying a coefficient called Swell (%).

$$\text{Swell}\,(\%) = \left(\frac{\text{Bank Density}}{\text{Loose Density}} - 1\right) * 100 = \left(\frac{\text{Loose Volume}}{\text{Bank Volume}} - 1\right) * 100$$

As the soil swells from a bank volume to a loose volume during the excavation, it decreases in density. Since the bucket only holds a loose volume, the equivalent denser bank volume is below the bucket capacity. The factor that accounts for this effect is called the load factor.

$$\text{Load Factor} = \left(\frac{\text{Loose Density}}{\text{Bank Density}}\right) = \left(\frac{\text{Bank Volume}}{\text{Loose Volume}}\right)$$

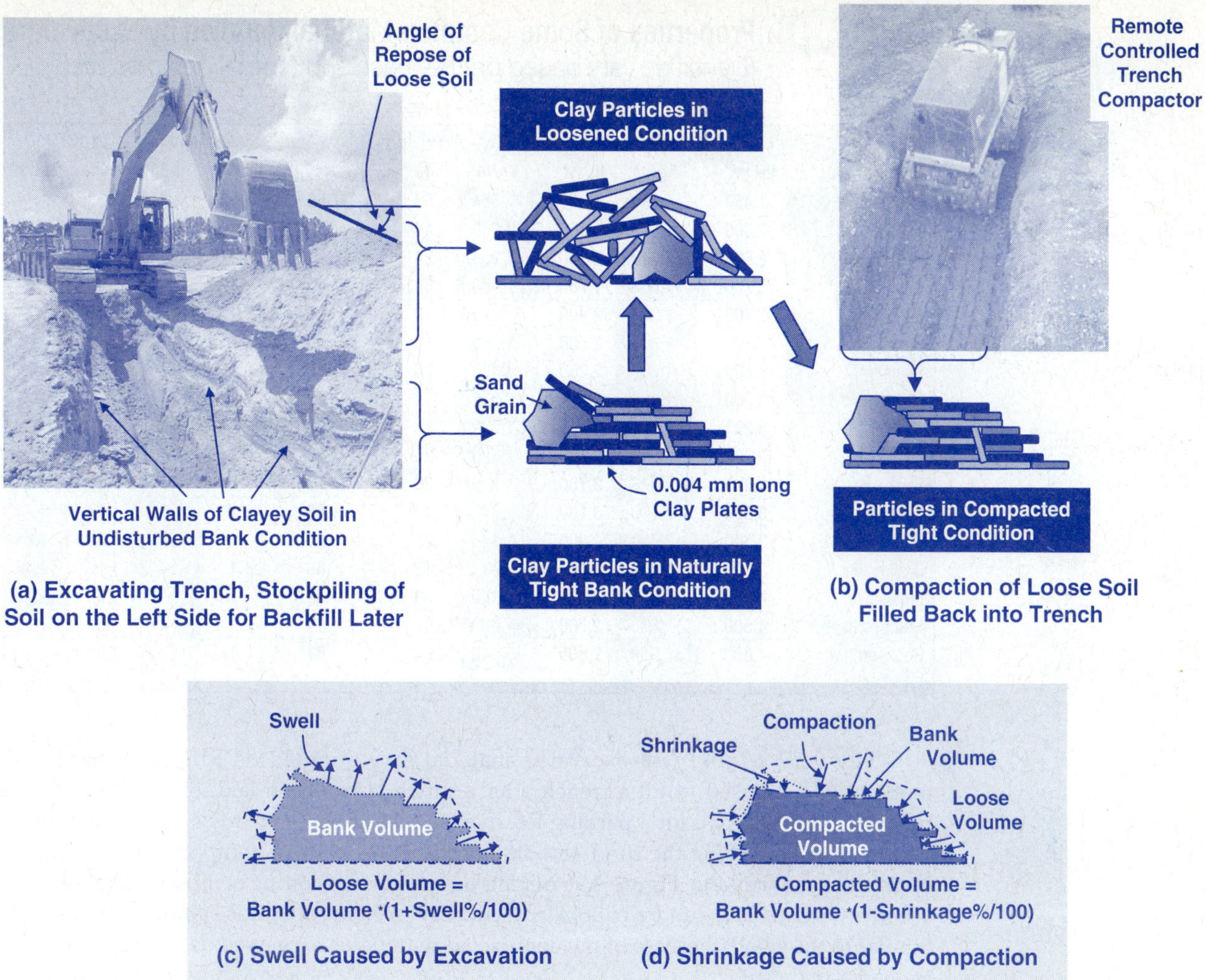

Figure 4.9 Soil Volume Changes During Excavating and Compaction of Clayey Soil

Since the bank volume is always smaller than the loose volume, the load factor is always < 1.0, which indicates how much bank material is excavated during each digging cycle with a given bucket capacity. By using the inverse of the load factor, we can assess how much bigger the loose volume of the stockpile will be than the excavated bank volume. Table 4.4 presents the load factors for a list of different materials. The values in the last column give a simple comparative ratio representing the loosening of one Bank-yd^3 (BCY or BCM) into Loose-yd^3 (LCY or LCM).

Now we have enough information to figure out what weight an excavator arm has to lift as a function of the material. Let us compare the weights of a 48-inch wide John Deere bucket filled with either dry sand (loose 2,400 lb/yd^3 or 1,420 kg/m^3) or wet gravel (loose 3,400 lb/yd^3/2,020 kg/m^3). Table 4.2 shows the bucket's volume as 1.43 yd^3 (1.09 m^3) and weighing 1,976 lb (897 kg). For dry sand, we can use an average bucket fill factor of 0.95 (small angle of repose) while the gravel heaps nicely, resulting in a fill factor of 1.10.

Bucket weight with dry sand: 1,976 lb + (1.43 yd^3 $*$ 0.95 $*$ 2,400 lb/yd^3 (loose)) = 5,236 lb (2,377 kg)

Bucket weight with wet gravel: 1,976 lb + (1.43 yd^3 $*$ 1.10 $*$ 3,400 lb/yd^3 (loose)) = 7,324 lb (3,325 kg)

Table 4.4 Properties of Some Common Material Handled by Excavators

(Densities vary based on the mass and type of particles the soil is made of)

Material Type	Bank Density lb/yd^3	Bank Density kg/m^3	Loose Density lb/yd^3	Loose Density kg/m^3	Swell %	Load Factor	Bank Vol. BCY	Bank Vol. BCM	Loose Vol. LCY	Loose Vol. LCM
Cinders	1,450	860	950	564	53	.65	1	1	1.53	1.99
Topsoil	2,300	1,370	1,600	950	44	.69	1	1	1.44	1.88
Earth	2,600	1,540	2,100	1,250	24	.81	1	1	1.24	1.61
Coal, blasted	2,700	1,600	1,780	1,058	52	.66	1	1	1.52	1.68
Sand, dry	2,700	1,600	2,400	1,420	12	.89	1	1	1.12	1.47
Clay, dry	2,800	1,660	2,500	1,480	12	.89	1	1	1.12	1.46
Shale	2,800	1,660	2,250	1,330	24	.80	1	1	1.24	1.62
Sand, moist	3,200	1,900	2,850	1,690	12	.89	1	1	1.12	1.47
Gravel, dry	3,200	1,900	2,900	1,720	10	.90	1	1	1.10	1.44
Earth, dry	3,200	1,900	2,550	1,520	25	.79	1	1	1.25	1.64
Earth, wet	3,400	2,020	2,700	1,600	26	.79	1	1	1.26	1.65
Sand, wet	3,500	2,080	3,100	1,840	13	.89	1	1	1.13	1.48
Clay, wet	3,500	2,080	2,800	1,640	25	.80	1	1	1.25	1.64
Gravel, wet	3,800	2,260	3,400	2,020	12	.89	1	1	1.12	1.47
Limestone	4,400	2,610	2,600	1,540	69	.59	1	1	1.69	2.21
Stone, crushed	4,500	2,670	2,700	1,600	67	.60	1	1	1.67	2.19
Rock, granite, blasted	4,600	2,730	2,800	1,640	64	.61	1	1	1.64	2.15

Only very seldom will excavated material stay dumped on a heap. Instead, the material will be used to fill a trench after a pipeline has been laid, or used as backfill behind a wall, or as fill for a parking lot. In each case, it will have to be compacted into a denser state by forcing the sand and clay particles back into a compact state, a process referred to as shrinking. Figure 4.9b depicts a two-drum vibrating compactor that fits into a narrow trench and can be controlled remotely from an operator standing on the top. Similar to the swell, the state of the compacted material is expressed as a percent relative to the bank density.

$$\text{Shrinkage }(\%) = \left(1 - \frac{\text{Bank Density}}{\text{Compacted Density}}\right) * 100 = \left(1 - \frac{\text{Compacted Volume}}{\text{Bank Volume}}\right) * 100$$

To end up with a shrinkage (%) that is not negative, the compacted density has to be larger than the bank density. That is exactly what happens if the contractor compacts it properly: the soil particles are packed even tighter than in its natural state. To demonstrate the changes in volume from bank to compacted, Table 4.5 provides volume factors for two

Table 4.5 Effect of Swell and Shrinkage for Two Common Materials

Material Type	Angle of Repose degrees	Swell %	Shrinkage %	Base Condition	Bank Vol. BCY/BCM	Loose Vol. LCY/LCM	Compacted Vol. LCY/LCM
Clay, wet	45	25	10	Bank = 100%	1.00	1.25	0.90
				Loose = 100%	0.80	1.00	0.72
				Compacted = 100%	1.10	1.37	1.00
Gravel, wet	35	12	5	Bank = 1	1.00	1.12	0.95
				Loose = 1	0.89	1.00	0.85
				Compacted = 1	1.05	1.18	1.00

common materials, clay and gravel. One readily recognizes that the CCY/CCM (Compacted Cubic Yard/Compacted Cubic Meter) is always smaller than the BCY. Of course, the LCY (LCM) is always larger than the BCY (BCM).

After gaining a good understanding of the properties critical to the assessment of how much a bucket weighs, we are ready to return to figure out the energy or the power that a backhoe needs to excavate a given soil.

Worked-Out Problem 4.1 / Do We Have Enough Backfill?

Earl Boles is the construction manager for AK Construction Inc., a specialty contractor for pipe-laying. He wants us to figure out the size of the pile that is created by the sandy clay material to be excavated for a new job for which he plans to use the benching method for trenching.

Earl provides us with a sketch of a cross-section that shows the pipe inside the trench with the pile of loose sandy clay on the side, and the key soil properties that he expects.

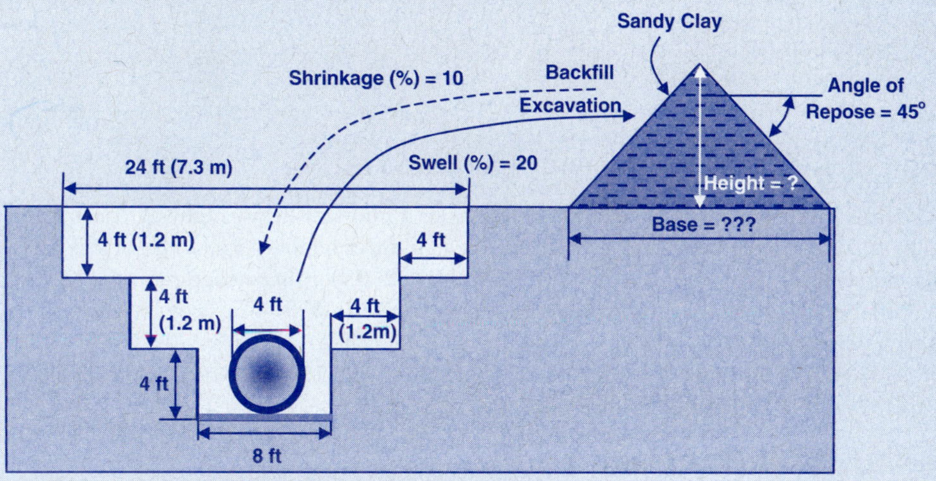

Figure 4.10 Cross-Section with Soil Properties of Pipe-Laying Job

Calculations:

Volume to be excavated per ft of trench $= ((4\,\text{ft} * 8\,\text{ft}) + (4\,\text{ft} * 16\,\text{ft}) + (4\,\text{ft} * 24\,\text{ft})) * 1\,\text{ft} = 192\,\text{ft}^3/\text{ft}$

Volume per m of trench $= ((1.2\,\text{m} * 2.4\,\text{m}) + (1.2\,\text{m} * 4.8\,\text{m}) + (1.2\,\text{m} * 7.3\,\text{m})) * 1\,\text{m} = 17.4\,\text{m}^3/\text{m}$

BCY/yd of trench $= 192\,\text{ft}^3 * 0.03704 = 7.11\,\text{yd}^3/\text{yd}$ BCM/m trench $= 17.4\,\text{m}^3/\text{m}$

LCY/yd of trench $= \text{BCY} * (1 + \text{Swell}\%/100) = 7.11\,\text{ft}^3 * (1 + 20/100) = 8.53\,\text{yd}^3/\text{yd}$

LCM/m trench $= \text{BCM} * (1 + \text{Swell}\%/100) = 17.4\,\text{m}^3 * 1.2 = 20.88\,\text{m}^3/\text{m}$

Volume of stockpile/unit length of trench $= {}^1/_2\,(\text{Height} * \text{Base}) = 8.53\,\text{yd}^3/\text{yd} = 230.3\,\text{ft}^3/\text{ft}$

Base (ft) $= 2 * (230.3\,\text{ft}^3/\text{Height (ft)})$

Base(m) $= 2 * (20.88\,\text{m}^3/\text{Height (m)})$

tan (angle of repose) $= \textit{Height}/({}^1/_2\,\textit{Base}) = \tan(45^\circ) = 1.0$

Height (ft) $= \tan(45^\circ) * ({}^1/_2\,\text{Base (ft)})$

$\text{Height}^2 = 230.3\,\text{ft}^2$; Height $= 15.2\,\text{ft}\ (4.6\,\text{m})$

Base (ft) $= 2 * ((230.3\,\text{ft}^3/\text{ft})/15.2\,\text{ft}) = 30.3\,\text{ft}$

Base (m) $= 2 * ((20.88\,\text{m}^3/\text{m})/4.6\,\text{m}) = 9.1\,\text{m}$

Volume of 1 ft pipe (ft^3) $= 1\,\text{ft} * 3.141 * r^2 = 3.141 * 4\,\text{ft}^3 = 12.56\,\text{ft}^3$

Volume of 1 m pipe $= 1\,\text{m} * 3.141 * (1.22\,\text{m}/2)^2 = 1.17\,\text{m}^3$

Volume needing compacted backfill/ft $= 192\,\text{ft}^3 - 12.56\,\text{ft}^3 = 179.44\,\text{ft}^3 = 6.65\,\text{yd}^3/\text{ft}$

Volume needing compacted backfill/m = $17.4\,m^3 - 1.17\,m^3 = 16.23\,m^3/m$
CCY after stockpile is compacted = BCY − (Shrinkage% ∗ BCY)/100
CCY/ft of trench = $7.11\,yd^3 - (10 * 7.11\,yd^3/100) = 6.4\,yd^3/ft$
CCM/m of trench = $(17.4\,m^3 - 10 * 17.4\,m^3/100) = 15.66\,m^3/m$

CCM shortage per 1 ft of trench = $6.65\,yd^3 - 6.4\,yd^3 = 0.25\,yd^3 = 6.67\,ft^3$
CCM shortage per 1 m of trench = $16.23\,m^3 - 15.66\,m^3 = 0.57\,m^3$

Discussion of Results:

The stockpile will be approximately 15 ft high and 30 ft wide. If the clay dries, the angle of repose will get smaller and with it the height. This should not happen if the backfill progresses as planned about 1 week after it is excavated.

If the assumption of the 10% shrinkage caused by compacting is accurate, there will be a need to truck soil from somewhere or, alternatively, accept that the top of the backfilled trench will be lowered by $6.65\,ft^3/24\,ft^2$ (area of trench on top) = 0.28 ft.
Drop inside the trench area = $0.57\,m^3/7.3\,m^2 = 0.08\,m$.
The CCY would be sufficient if the shrinkage % = $100 * (192\,ft^3 - 179.44\,ft^3)/192\,ft^3 = 6.5\%$ instead of 10%.

Worked-Out Problem 4.2 / How Much Energy Can We Save?

You have just started as an assistant engineer in trenching operations of a mid-size heavy-highway construction company. You are asked to calculate the possible savings in diesel fuel if a dozer removes 5 ft (1.5 m) of soil before starting a 1,000 ft (304 m) long, 6 ft (1.82 m) wide trench in wet earth mixed with clay to an average depth of 18 ft. You are told to consider the Hitachi backhoe Zaxis 225 USR with a engine power of 110 kW or 147 hp (see specifications in Table 4.1 and the downloaded brochure). The company already owns a 45-inch (1.14 m) heavy-duty bucket.

S4.7 Zaxis Backhoe 225 USR

Layout of Options: The easiest way to study the effect of the two options is to sketch them on a piece of paper (Figure 4.11).

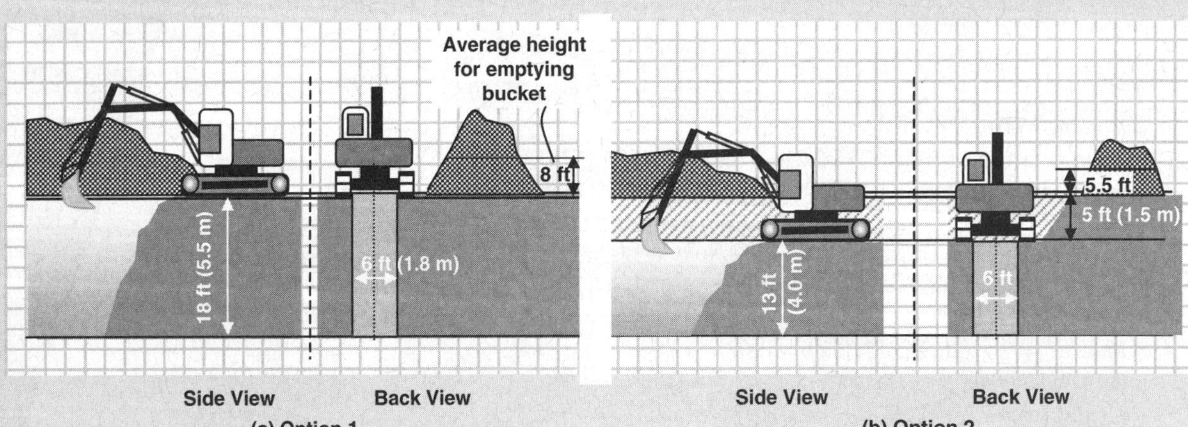

Figure 4.11 Schematic Comparison of the Two Options (Not to Scale)

Calculate the Weight of a Full Bucket:

The weight of the full bucket depends on the size that we select. From the specifications we learn that the largest bucket that could be used in "soft" material has a width of 57″ (1.45 m) and a heaped SAE volume of $1.57\,yd^3$ ($1.2\,m^3$).

However, the company already owns a 45″-wide heavy-duty bucket, which is the next size down. Let's avoid having to buy a new one and choose the 830-lb (376 kg) version, 1.19 yd^3 (0.91 m^3) and 5 teeth.

The loose soil density lies between wet earth and wet clay. According to Table 4.4, LCY should be 2,750 lb/yd^3 (1,250 kg/m^3) Let's choose the higher value, 2,800 lb/yd^3 (1,270 kg/m^3) to be conservative with our calculations.

Table 4.3 suggests a bucket fill factor between 100 and 110. We might just pick an average value.

Weight of bucket	= 830 lb (376 kg)
Bucket fill factor	= 1.05
Weight of soil = $1.05 * 1.19\,yd^3 * 2{,}800\,lb/yd^3$ = 3,499 lb ($1.05 * 0.91\,m^3 * 1{,}640\,kg/m^3$ = 1,578 kg)	
Total weight	= 4,328 lb (1,954 kg)

Calculate Average Energy to Lift the Bucket:

The appropriate equation for lifting a mass is

$$\text{Potential energy (PE)} = m * g * h = \text{Mass} * 9.81\,m/s^2 * \text{Height}$$

According to the specifications, option 1 requires a 9′7″ (2.9 m) long arm to reach the 18′-deep trench while option 2 will be fine with the smaller 7′3″ ft (2.2 m) arm. The average difference in height between the digging and the emptying of the bucket over the spoil pile is

Option 1: $18\,ft/2 + 8\,ft = 17\,ft$ (5 m)

Option 2: $13\,ft/2 + 5.5\,ft = 12\,ft$ (3.6 m)

Average energy needed: $(1\,J = 1\,kg\,(m/s)^2, 1\,Wh = 3{,}600\,J)$

Option 1: $17\,ft * 4{,}328\,lb * 9.81\,m/s^2 = (17\,ft * 0.305\,m/ft)\,m * 1{,}967\,kg * 9.81\,m/s^2 =$
$= 100{,}050\,J * 0.0002778 = 27.8\,Wh$

Option 2: $12\,ft * 4{,}328\,lb * 9.81\,m/s^2 = (12 * 0.305)\,m * 1{,}967\,kg * 9.81\,m/s^2$
$= 70{,}624\,J * 0.0002778 = 19.6\,Wh$

Calculate Average Bucket Cycles per Meter of Trench:

Option 1: Volume for 1 ft of trench = $18\,ft * 6\,ft * 1\,ft = 108\,ft^3$ (10.05 m^3)
Total number of buckets = $(108\,ft^3 * 0.037)/1.19\,yd^3$ = 3.36 full buckets/ft of trench

Option 2: Volume for 1 meter of trench = $12\,ft * 6\,ft * 1\,ft = 72\,ft^3$ (6.7 m^3)
Total number of buckets = $(72\,ft^3 * 0.037)/1.19\,yd^3$ = 2.24 full buckets/ft of trench

Total Energy Savings (Only Considering the Bucket Lifts):

Option 1: Total energy consumption for 1,000 ft long trench = 1,000 ft * 3.36 lifts/ft * 27.8 Wh/lift = 93.41 kWh (or 93.41 * 3414 = 318,894 BTU)

Option 2: Total energy consumption = 1,000 ft * 2.24 lifts/ft * 19.6 Wh/lift = 43.90 kWh (or 43.90 * 3414 = 149,874 BTU)

Total energy savings = 93.41 − 43.90 kWh = 49.51 kWh or 169,027 BTU.

Calculating Total Savings in Diesel fuel (Petro-diesel = 130,500 Btu/gallon):

Result of Calculations: Savings in gallons of diesel = 1.3 gal (4.9 litre)

Discussion of the Result:

Using the given assumptions, the energy savings are, relatively speaking, minimal. Even if we consider doubling the weight to be lifted and consider the arm and the frictions in the joints, it is assumed that a maximum of

5 gallons (18.9 liters) could be saved. On the other hand, removing the top 5 ft (1.5 m) to a width of 11 ft (for the backhoe to fit) with sloped side will take a considerable amount of time. Still, there are non-energy-related issues that could tip the balance toward choosing the second option. For example, the effort to prevent the trench walls from collapsing or to satisfy the legal requirement to ensure the safety of the laborers who have to enter the 18-ft (5.5 m) deep trench compared to the 13-ft (3.9 m) trench could be significantly higher and more expensive. In order to make an assessment about this issue we have to look at trench safety.

4.5 Principles of Safe Trenching

S4.8 Excavation-Safety-OSHA Workshop
S4.9 Excavation and Trenching-OSHA
S4.10 Excavation Trenching Safety for Project Managers
S4.11 Trenching Accident Prevention

A trench is defined as a narrow underground manmade cut that is deeper than its width and not wider than 15 feet (4.5 meters). As the accident statistics by the Occupational Safety and Health Administration (OSHA) reveal, the fatality rate for excavation work is 110% higher than the rate for general construction, which is at 12.9 fatalities per 100,000 employees. Besides falling objects, poisonous gas, or damaged utilities, cave-ins pose the greatest risk and are much more likely than other excavation-related accidents to result in worker fatalities.

According to OSHA, a cave-in means (see also Figure 4.12)

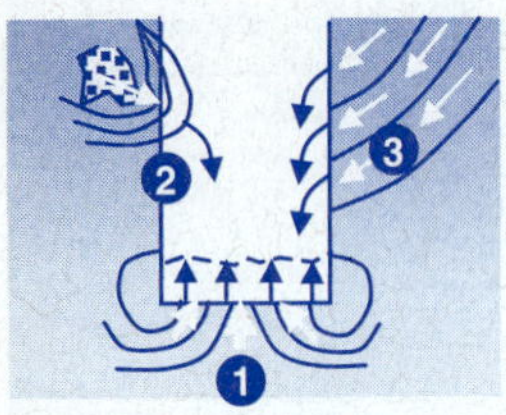

Figure 4.12 Three Cave-In Scenarios

> . . . the separation of a mass of soil or rock material from the side of an excavation (2 & 3), or the collapse of the trench floor (1), and its sudden movement into the excavation, either by falling 2 or sliding 3, in sufficient quantity so that it could entrap, bury, or otherwise injure and immobilize a person.

4.5.1 SAFETY REGULATIONS

What makes trench walls so dangerous is the fact that they can collapse without warning and can change their stability overnight. Federal regulations (OSHA 29 CFR, Part 1926, Subpart P, July 1997) require contractors to follow certain procedures and practices when working in trenches. To prevent cave-ins of 5- to 20-ft-deep trenches (1.5 m–6 m), OSHA requires that one of three protective measures be applied: ***(a) sloping and benching, (b) shielding or shoring, and (c) sheeting.*** Responsibility for determining which system is best for any given job falls to project planners and the competent person on-site. A competent person is one who can not only identify existing and predictable hazards or dangerous working conditions but also has the authority to take prompt corrective measures. For excavations deeper than 20 ft, the protective system must be designed by a licensed professional engineer. For trenches less than 5-ft deep (1.5 m), OSHA does not require a protective system, unless the competent person sees signs of a potential cave-in.

VG4.4 Working in Safe Trench in Safety

OSHA categorizes soil and rock deposits into four types:

STABLE ROCK is natural solid mineral matter that can be excavated with vertical sides and remain intact while exposed. It is usually identified by a rock name such as granite or sandstone.

TYPE A SOILS are cohesive soils with a high compressive strength. Examples of Type A soils are clay, silty clay, sandy clay, clay loam, and, in some cases, silty clay loam and sandy clay loam. No soil can be of Type A if (1) it is fissured, (2) was subject to vibration of any type, (3) has previously been disturbed, (4) is part of a sloped, layered system where the layers dip into the excavation, or 5) has seeping water.

TYPE B SOILS are cohesive soils with unconfined compressive strength that is lower than for type A. Type B soils are angular gravel; silt; silt loam; previously disturbed soils unless otherwise classified as Type C; soils that meet the unconfined compressive strength or cementation requirements of Type A soils but are fissured or subject to vibration; dry unstable rock; and layered systems sloping into the trench at a slope less than 4H:1V (only if the material would be classified as a Type B soil).

TYPE C SOILS are cohesive soils with a low unconfined compressive strength. C soils include granular soils such as gravel, sand, and loamy sand, submerged soil, soil from which water is freely seeping, and submerged rock that is not stable. Also included in this classification is material in a sloped, layered system where the layers dip into the excavation or have a slope of four horizontal to one vertical (4H:1V) or greater.

4.5.2 PROTECTIVE MEASURES

Sloping and Benching: These two approaches to protect against trench wall collapses rely on removing the heavy soil that could cave in (see Figure 4.13). Sloping involves the cutting the vertical wall to a maximum allowable slope that depends on the type of soil, the environmental conditions of exposure, and the weight that will be positioned on the top, called the surcharge load. The soil in an area that has been excavated/filled before is likely to be classified as Type C soil. This type demands that the walls of the trench must be sloped back on each side of the excavation 1.5 ft horizontally for every 1.0 ft vertically (1.5:1). The type of soil also determines the horizontal-to-vertical ratio of benched sides. Generally, the bottom vertical height of the trench must be no more than 4-ft (1.2 m) for the first bench. Subsequent benches may be as high as 4-ft vertical in Type A soil and no higher than 4-ft in Type B soil to a total trench depth of 20 ft.

In many cases, however, the lack of available space such as the close proximity to utilities and roads prevents the creation of a slope or bench.

Shoring: This method is geared toward stabilizing the vertical wall by applying a "preventive" force to the soil. The two core components are the horizontal shores, or struts, and a temporary vertical wall system. The shores are made of timber, steel, or hydraulic brace

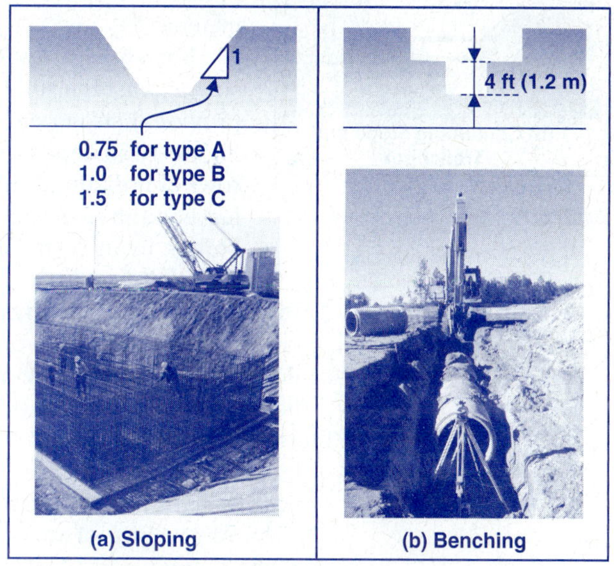

Figure 4.13 Basic Measures to Avoid Cave-Ins

Figure 4.14 Shoring Against Cave-Ins

(a) Stacked Trench Boxes and Metal Sheets

(b) Combining Slope and Trenchbox

Figure 4.15 Trenchbox Uses

spreaders applying the force to inhibit any horizontal movement of soil, underground utilities, roadways, and building foundations situated right next to the open trench. Figure 4.14a presents two flexible shoring systems deployed in a trench where several previously installed pipes and other utilities cross. Standard-size boards are set and pushed against the trench walls with lightweight, reusable hydraulic cylinders pressurized with the help of a small hand pump (see Chapter 3). Figure 4.14b offers a view into a larger opening in the ground to construct the foundation of a bridge. The larger depth and distance between the vertical openings asks for a stronger shoring and wall systems. As shown, instead of small boards, large-profile and interconnected sheetpiles have been driven into the soil by pile-driving equipment all around the perimeter of the opening. Instead of hydraulic shores, two layers of horizontal heavy steel beams brace against the movement of the sheetpiles. The horizontal beams that are placed parallel along the wall are called wales, while the beams that are support the wales are referred to as struts. While the struts commonly run perpendicular to the wall, the design shown with bracings only in the four corners keeps a large opening in the center free creating an obstacle-free space for the supply of material.

Shielding with Trenchbox: Unlike shoring, shields do not apply pressure against the soil to prevent it from collapsing; rather, their objective is to provide a strong and safe "shield" for the laborers in case of a cave-in. As Figure 4.15a depicts, a trenchbox consists of two vertical steel plates and two rows of horizontal struts that keep the walls of the box apart. The struts leave just enough space between the box and the trench walls that it will slide to the bottom by its own weight. Should the soil wall collapse, the box will prevent material from reaching the bottom, thus protecting the laborers inside. Naturally, nothing will prevent the trench without the box from caving in and possibly cutting off the working men inside, keeping them from escaping. They could even die should a water pipe break at the same time and flood the confined space. It is therefore only natural that OSHA requires a ladder as a means of egress. Figure 4.15a illustrates the proper installation of a ladder inside two trenchboxes that are stacked in order to reach the required depth while Figure 4.15b shows that trench boxes are sometimes used in combination with the sloping and benching method.

The importance of this danger zone is addressed in numerous pamphlets and workshops organized by OSHA and the Contractor Association. Despite all the efforts, the unpredictability of soil and neglect of basic precautions ensures that deadly accidents continue. Figure 4.16 presents brief safety tips that could serve as daily reminders for everybody working in and around trenches.

S4.12 SCDOT Subsurface Utility Engineering
S4.13 Common-Ground-Report
S4.14 Backhoe Accident with Gas Line-NTSB

4.5.3 CALL (DIAL) BEFORE YOU DIG

The reader may already be aware that the underground, invisible to the human eye, is slowly becoming a busy space crisscrossed by many different utilities and structures. Unfortunately, no accurate "maps" of the main utilities have been established, and the design drawings were not always followed. Thus, old drawings can't be trusted, as the contractor may have chosen a different path or the line may have had to be rerouted at a later date.

OSHA regulations require a contractor who finds underground installations to protect, properly support, or remove them. Unfortunately, by the time a gas line is exposed it may already been damaged, letting gas escape. Thus, knowing ahead of time the exact location where utilities are buried is key in protecting them.

Figure 4.16 OSHA Safety Tips for Saving Lives Know No Language Barrier

The congressional Transportation Equity Act for the 21st Century, TEA 21, found that "unintentional damage to underground facilities during excavation is a significant cause of disruptions in telecommunications, water supply, electric power, and other vital public services, such as hospital and air traffic control operations, and is a leading cause of natural gas and hazardous liquid pipeline accidents."

One does not have to look far to discover that the congressional findings are very well documented (see Accident #5). Figure 4.17 depicts sections of a new trench that passes through an existing utility "maze."

VG4.3 Buried Utilities

The increased number of accidents has led to the establishment of one-call systems in every state. Funded by members consisting of public utilities and other underground facility owner-operators, these systems provide an excavator with a single toll-free number to call in order to alert the affected member(s) about his

(a) Backhoe on Top of Gas, Drainage Pipes and Manhole **(b) Abandoned Drainage Together with Water Pipes**

Figure 4.17 Gas, Water, and Drainage Pipes Create a Jungle of Dangerous Obstacles

Figure 4.18 Reminders with One-Call Phone Number

plans to excavate in a specific area. The one-call system serves as a message-handling service, taking information about each planned excavation and distributing this information to its member utilities. It is then the responsibility of each utility to mark the location of its underground facility or facilities at the excavation site. Excavators are required by law in most states to notify the one-call organization at least two working days before starting a project (Figure 4.18). Upon notifying, excavators will be given a confirmation number, often referred to as a "dig-ticket," which excavators should keep as a reference for each excavation. The contractor needs to give detailed information while describing the proposed excavation site to the call center operator or filling out a form online or faxing it in. While street address or block and lot numbers, or both, are important for excavation notifications within a city, excavators must be aware that address information alone may not be sufficient information, especially when the proposed excavation is to take place in a new development. Landmarks, such as rivers, buildings, and structures, major streets and intersections, established boundaries or other geographic information, including GPS latitude and longitude coordinates, are useful references to help determine the location of the proposed excavation site (the request form to locate utilities in Southern California can be found at www.wiley.com/college/bernold).

The Federal Communications Commission eliminated the confusion of multiple "Call Before You Dig" numbers across the country by dedicating 811 for this purpose. 811 callers will be automatically rerouted to one of the local call centers.

Connecticut was the first state to pass a Call Before You Dig Law in 1978. It led to a 60% decline in accidental cuts of utility lines even though the length of new underground utilities being installed increased. Other states made similar observations after they implemented centers, some of which they operate themselves, while others are administered by private companies. How these systems work and what kinds of information they require vary from state to state. While the 48-hour rules do not include Saturday, Sunday, and holidays, the website or operators are available 24/7 to take requests. These operators have available a computerized mapping system that shows 90% of all utilities in the vicinity of the planned excavation. With this information, the operator is able to inform each one about the locating request. The Illinois One-Call System logged almost 1 million calls in 2001. And because one call from an excavator can result in communication with multiple utility companies, these calls resulted in more than 6.4 million notification messages being transmitted to member utilities in 2001.

Equipped with the ticket number, a professional utility locator working for or hired by the utility company will visit the site to locate and mark the existing lines, as depicted in Figure 4.19, using a variety of techniques.

S4.15 Underground Marking Standards Virginia
S4.16 Uniform Color Code Markings

Unfortunately, the technology available and the complexity of the underground cannot guarantee that the markings are perfectly accurate. The difficulty is acknowledged by creating a 3-ft (0.9 m) target zone, 18 inches in either side of the marking, where the actual location of the utility might be. Tree roots that reach far and have entangled lines or the cross-over of signals into abandoned lines are just a few examples of why we still experience accidents today. In addition, the limitations of the present systems do not allow us to predict the depth of the lines accurately. While One-Call Systems are the cornerstone for preventing utility damage today, other methods are being successfully tested as well. One such approach has been given the name Subsurface Utility Engineering (SUE), a proactive approach to damage prevention. Its core messages are (1) let's verify the exact location of underground utilities, (2) use that map to plan the route of new utilities, and (3) update the maps with accurate as-builts as we go.

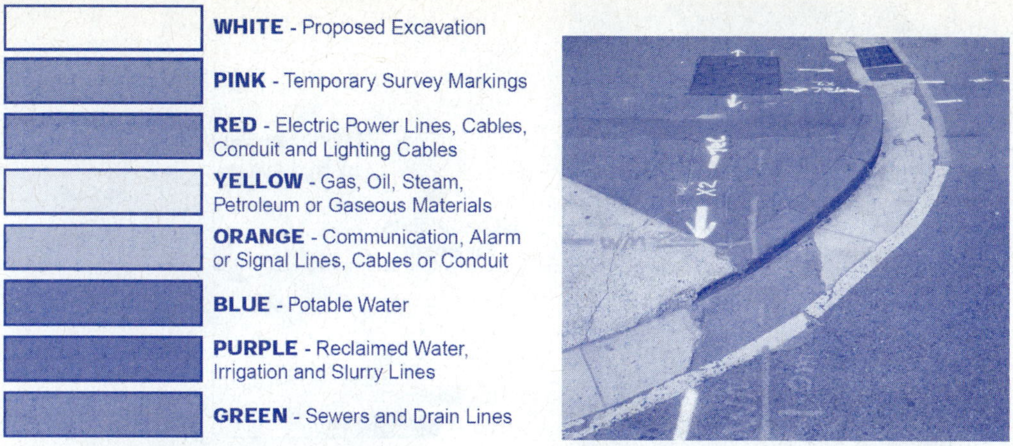

(a) Color Codes Tell What Is Buried **(b) Colorful Street Corner**

Figure 4.19 Finding and Marking Buried Utilities Before Excavation

One favorite tool for verification is visual validation of the existence of a utility by exposing each buried utility with a vacuum excavation consisting of a pressure air-lance and a vacuum hose to remove the loosened soil from the "pothole." The visual contact with the buried utility allows the identification of its type, size, and depth. After the survey crew records the exposed utility, the removed soil is placed back in the hole. The vacuum truck also has equipment to open and subsequently close an opening in asphalt, and it can use water rather than air to loosen clayey soils.

4.6 The Backhoe as a Mobile Crane

Figure 4.20a depicts how backhoe arms can also serve as a crane boom. To make the rigging of a load easier, buckets come with a hook at the backside providing a firm means to attach a chain or sling. As we learned earlier, the risk of tipping limits how much backhoes are able to lift. Let's have a closer look at the limiting factors. There are actually two independent factors that constrain the lifting capacity: (1) maximum hydraulic force in the boom and stick cylinders, and (2) tipping over of the entire backhoe. Tipping occurs when the load moment at the front tire, outrigger, or tracks exceeds the stabilizing moment resulting from the engine in the back and weight of the undercarriage (see Figure 4.20b). The published lift capacities of a machine include a factor of safety for unforeseen circumstances. In particular, the maximum capacity can either be (a) 75% of tipping load, or (b) 87% of the hydraulic force capacity, assuming that the arm itself is properly designed. One might ask: Why is the safety factor for the hydraulic limitation so much smaller? The answer is simple. Imagine what happens if an operator tries to lift a load with a weight that is higher than the forces the hydraulic cylinders can produce. In Chapter 3 we learned that the amount of force a cylinder can produce is a function of area of the cylinder (minus the rod if applicable) multiplied with the hydraulic pressure. Thus, if you try to lift a load for which the hydraulic force is insufficient the backhoe boom would simply not move. The exact opposite happens in a situation where the operator moves a load outward into the area where the lifting capacity is controlled by the tipping moment. Quite suddenly, the entire

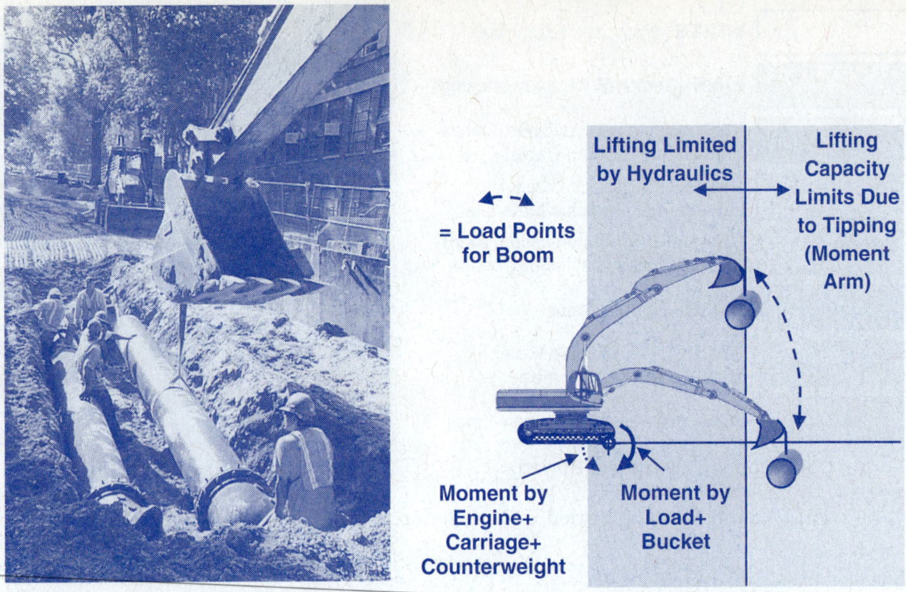

(a) Backhoe Lifts Heavy Pipe Section

(b) Areas of Lift Limitations

Figure 4.20 Lifting Situations with Bucket Attached and Ranges of Capacity Limits

piece of equipment will start to tip forward until the load hits the ground or another obstacle, at which time the backhoe, relieved from the extra load, may whip back and endanger everything around. If there is nothing holding the load, for example the empty trench, the backhoe will keep turning and eventually crash. To avoid such a condition, the maximum allowed load is 87% of what the cylinders could lift.

Here we will study the basic differences among the many types and manufacturers. For this purpose, the specifications for a variety of makes and setups were put on graph paper as shown in Figures 4.22 and 4.23. Besides comparing backhoe-loaders' standard sticks—some call them dippers—the extended telescopic sticks were also included. The first lift capacity, usually the larger one, is valid for working over the front (OF), a situation depicted in Figure 4.21a. The second is the maximum capacity if the load is picked up with the arm perpendicular to the driving direction, referred to as over the side (OS), shown in Figure 4.21b. Figure 4.21c reminds us that the calculation of the maximum load capacity assumes that a bucket is attached. Removing the bucket will allow the operator to increase the given limiting capacity by the weight of the bucket. However, removing it might be very cumbersome unless the arm is equipped with a quick tool-exchange mechanism at the tip of the stick as shown in Figure 4.21c. Finally, the backhoe loader has its arm pivot at the hitch point stabilized by two hydraulically operated outriggers as presented in Figure 4.21d.

It is apparent that the Hitachi Zaxis 225 backhoe excavator and the Volvo BL70 backhoe loader used in this comparison differ in more than one characteristic. The Zaxis operates on tracks while loader travels on wheels but uses outriggers when digging. The excavator is able to rotate 360° while the backhoe loader has a swing range of about 140°. Most important is the difference in equipment weight. The Hitachi weighs 48,510 lbf (216 kN), while the Volvo BL70 is less than half or "only" 18,976 lbf (84 kN). Of course, having more mass means increased counterweights and, at the same time, higher tipping load. The right graph in Figure 4.22 displays clearly that the Zaxis is able to stretch its arm all the way to 31 ft (9.4 m)

(a) Backhoe Loading Over the Front (OF)

(b) Working Over the Side (OS)

(c) Bucket Release

(d) Backhoe-Loader at Work

Figure 4.21 Various Backhoe Conditions Considered in Specification Charts

compared to only 16 ft (4.8 m) by the Volvo. An optional telescopic stick available for backhoe loaders would be able to reduce the gap between the two to 11 ft (3.3 m) but, as expected, will reduce its maximum lifting capacity, a fact that shows up drastically in the left graph. Because the telescopic arm is heavier than the BL 70's standard stick (SS), it lifts about 400 lbf (1.80 kN) less weight when totally retracted (TSR). This disparity doubles and triples when the telescope is extended (TSE). The curves representing the Volvo backhoe loader demonstrate the limitations of the boom cylinder. While using the boom to lift a load provides a long reach, its limited force drastically restricts the load to a maximum of 3,300 lbf (14.7 kN). The last two curves present a special lifting case, referred to as Stick Lifting (SL). In order to increase the hoisting capacity, the boom cylinder is "parked" and its valve closed. Since the stick cylinder does not have to move the entire arm, it is able to pick up more mass, boosting the lifting capacity at 4.5 ft (1.4 m) above ground 3,300 lbf (14.7 kN) in normal SS operation to 7,300 lbf (332.5 kN) when in SL mode.

S4.17 Volvo BL70

The load capacity graphs for the Zaxis show the drastic discrepancies between working OS and OF. For example when digging at a depth of 2.5 ft (0.75 m) below the surface (−2.5 ft), the safe load working OS, indicated by ❷, is approximately 6,500 lbf (29 kN). When operating OF, however, the "translation" of the tipping point to the end of

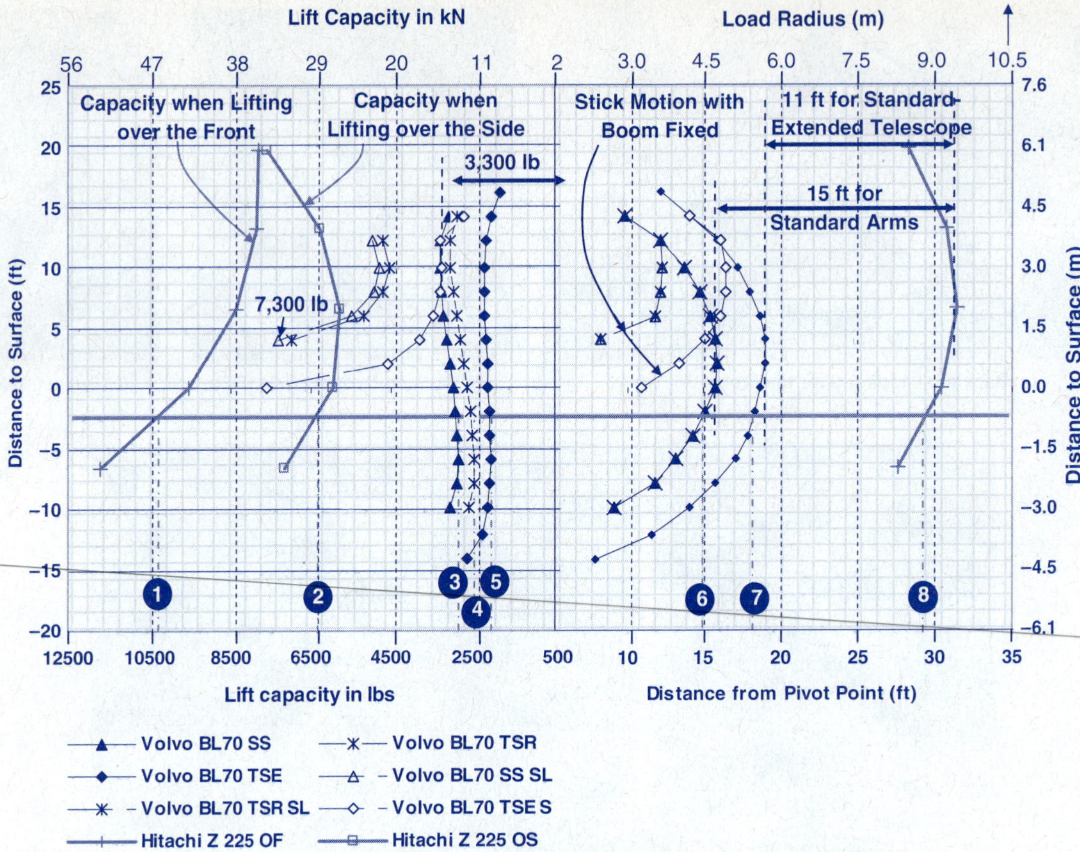

Figure 4.22 Contrasting Reach and Lift Capacities of Hitachi Backhoe Excavator and Volvo Backhoe Loader

the tracks boosts maximum load to point ❶ or 10,500 lbf (47 kN). (The labels 1–8 will be used later in the Chapter Review section.)

When purchasing a new piece of equipment, a contractor has to weigh price against performance. For a backhoe excavator, maximum reach and lifting capacity are certainly values that she will have to consider heavily. The following section uses three different makes to demonstrate how the choices may differ concerning those two factors. The comparison includes three backhoe excavators: (1) tracked Hitachi Zaxis 225, (2) tracked CAT 325C, and (3) wheeled Volvo EW 180B. For specifications, download the brochures for the three pieces of equipment from the book website.

S4.18 Backhoe CAT-325-C
S4.19 Backhoe Volvo EW180B

The weights of the machines are as follows:

Hitachi Zaxis 225	48,510 lbf (215.7 kN)
CAT 325C	78,044 lbf (347 kN)
Volvo EW 180B	43,776 lbf (194.7 kN)

Clearly, the CAT 325C is the heaviest of the three, which is converted into the largest load radius ❿ but also the highest and lowest reaches. While the EW 180B is light, not having the heavy tracks, it has, with 13,500 lbf (60 kN) ❶, a considerably higher lift capacity at a load point height of 7.5 ft (2.3 m), working over the front and with the outriggers down. Working OS, this value goes down to 10,900 lbf (48.5 kN) ❷. The significant effect of the outriggers on the lifting capacity becomes evident when inspecting the maximum lift capacity at the same height for working OF ❸. As shown in Figure 4.23, the drop from maximum lift capacity with ❶ to the one without outriggers ❸ is 2,600 lbf (11.5 kN), or almost 20%.

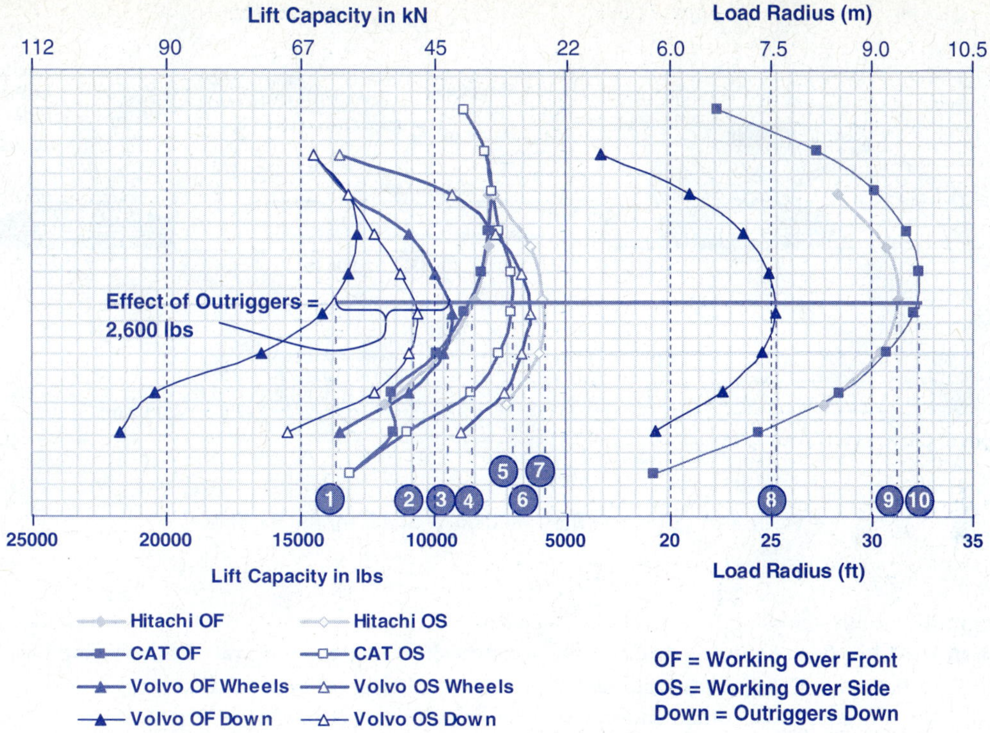

Figure 4.23 Comparing Load Capacity and Reach of Three Backhoe Excavators

The Hitachi and the CAT show similar values with the exception of the maximum lifting capacity for loading over the side, where the Zaxis 225 can lift 1,200 lbf (5.3 kN) or 17% less at the 7.5 ft (2.3 m) load point height. One reason for this discrepancy lies, not unexpectedly, in the fact that the CAT is 0.54 m (1.8′) wider than the Zaxis 225 (3.34 m (10′11″) vs. 2.8 m (9′2″).

Solving Header Problem 1 / Installing 6-Ft (1.8 m) Diameter Pipes at −16 Ft (−4.8 m)

Question 1: What Specific Equipment Is Needed to Install the Pipes at the Required Depth?

Key issues related to question 1:

a. Minimum reach of backhoe "crane" to lay a 40 ft (12.2 m) long pipe segments OF is approximately 36 ft (11 m), ($^1/_2$ * 40 ft + 6 ft to tipping point + 10 ft (3 m) vertical rotation axis).
b. Minimum lift capacity at about 37 ft (11.3 m) reach is 32,000 lbf (142 kN).
c. The length of the pipe requires the use of a spreader bar with two slings holding the pipe.
d. Digging depth is 16 ft (4.8 m).
e. Trench width is 8 ft (2.4 m) (6 ft pipe + 2 ft on each side) to allow enough space for shoring and compaction equipment.
f. Vertical walls need to be supported by shoring, trench-box, or H-piles with lag shoring made of steel or lumber (see Chapter 8).
g. Using sheet piling is difficult due to the presence of boulders, which could damage the piles.
h. The length of the pipe requires a special design for a trench-box or shoring, since 40 ft (12.2 m) have to stay unobstructed from cross-bracings.

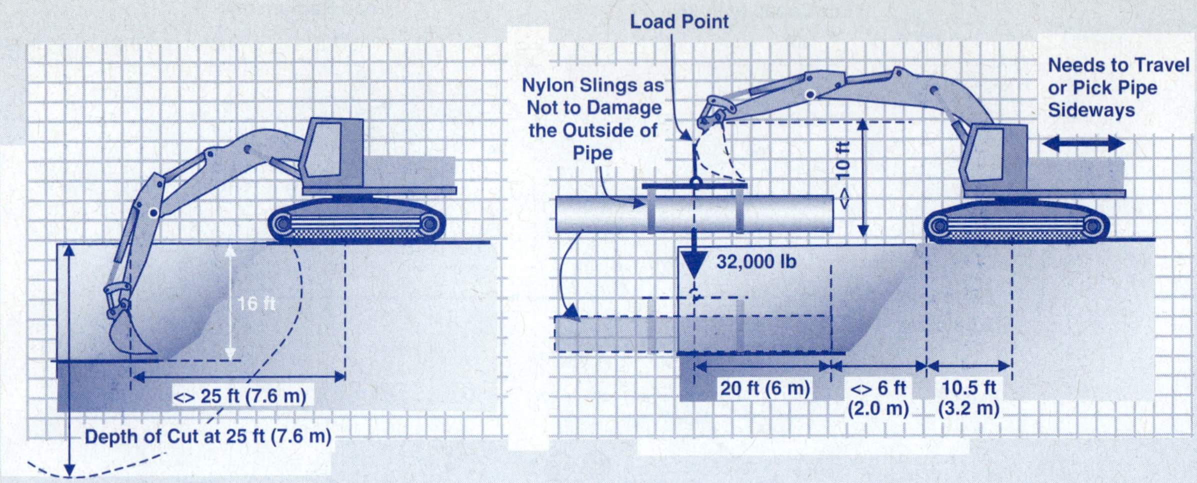

Figure 4.24 Sketch of the Two Critical States for the Backhoe

i. Underground utilities crossing the trench area must be anticipated.
j. Compaction on the side of the pipe needs to be accomplished after the removal of the shoring protection.
k. Trench, Inc., uses only Caterpillar equipment.
l. A beam laser is needed to keep line-and-grade of the trench floor and the pipe crown.
m. Dewatering requires the installation of sump-pumps ahead of trenching to lower the ground water level.

Addressing Issues a–d, k: Selection of Backhoe Excavator: The CAT 325C presented in Figure 4.24 has a maximum reach of 32 ft (9.7 m) at 10 ft (3 m) above ground and a capacity of about 7,000 lbf (31 kN). Thus, it cannot be used on this job. Caterpillar's entire product line (see booksite) shows that the CAT 345C, 365C, and 385 C L all have sufficient reach, but only the CAT 385C L has the necessary lift capacity to carry and lower the 40 ft pipes at the required distance. Table 4.6 presents a table with maximum loads capacities for 5 different load point heights. As the data shows, the given limitations for up to 35 ft OF are due not to tipping but to the maximum hydraulic pressure. Removing the 6,795 lbf (30 kN) general purpose bucket adds many more positions that can be reached, as indicated by the blue areas.

S4.20 Backhoe CAT-385C-L

At 35 ft (10.7 m) the maximum load at 10 feet is 33,400 lbf (148.6 kN) and 29,550 lbf (131 kN) at 40 ft. An interpolation between these two points results in a load at 36 ft reach of 33,400 lbf − (33,400−29,550)/5) lbf = 32,630 lbf (145 kN). This setup represents a worst-case scenario, since any backhoe operator in the real world would extend the arm to the longest reach only after the pipe is already down in the trench. Hence, the load point of 10 ft could be reduced to 0 feet, resulting in a load capacity of 37,100 lbf − ((37,100−30,150)/5) lbf = 35,710 lbf (158.8 kN) and providing a nice 11% margin of error (= 3,710 lbf) (16.5 kN).

Table 4.6 Load Capacity Matrix for the CAT 385C L with General Purpose 27 ft 7 in (8.4 m) Boom and 18 ft 1 in (5.5 m) Stick

Height of Load Point	20 ft (6 m)		25 ft (7.6 m)		30 ft (9.1 m)		35 ft (10.7 m)		40 ft (12.2 m)	
	OF (lbf)	OS (lbf)	OF (lbf)	OS (lbf)	OF (lbf)	OS (lbf)	OF (lbf)	OS (lbf)	OF (lbf)	OS (lbf)
15 ft (4.6 m)			40,750	40,750	35,000	35,000	31,000	27,100	26,900	20,650
10 ft (3 m)	59,750	59,750	46,650	46,650	38,700	34,150	33,400	25,900	29,550	19,500
5 ft (1.5 m)	67,900	62,000	51,800	43,550	42,050	32,300	35,550	24,700	30,800	19,150
0 ft (0 m)	72,750	58,400	55,400	41,150	44,500	30,750	37,100	23,650	30,150	18,500
−5 ft (−1.5 m)	74,350	56,300	57,100	39,550	45,750	29,600	37,050	22,900	29,600	18,050

■ = Lifting limitations due to hydraulic capacity.

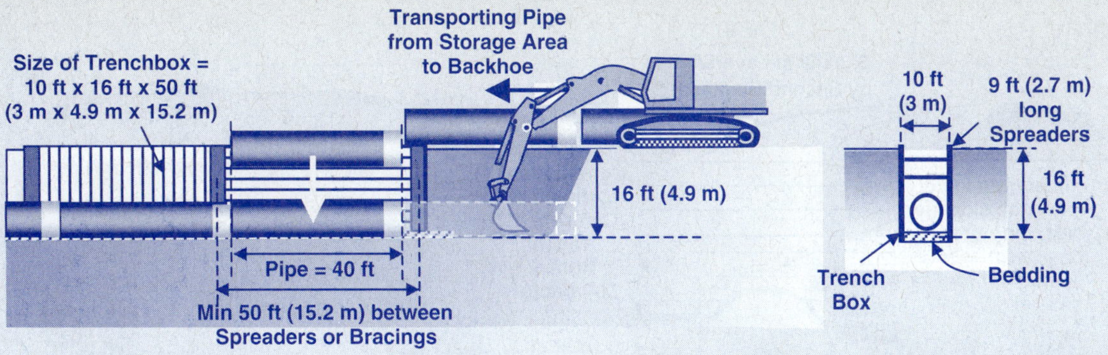

Figure 4.25 Sketch of the Trenching Operation and Pipe Installation Using a Trench-Box

In summary, it is proposed to use a CAT 385C L with general-purpose 27 ft 7 in (8.4 m) boom and 18 ft 1 in (5.5 m) stick.

Addressing Issues e-i: Design of Shoring System: The 40 ft (12.1 m) pipe length creates a challenge for designing a safe trench protection structure. The choices are shoring, trench shields, or trench boxes. Since sheetpiles are not feasible and the installation of vertical soldier piles with lagging (see Chapter 8) is too costly, we should try to find a trench box solution that is backfilled at the side of the train tracks (to better support the track foundation). (See Figure 4.25.)

As we learned above, trench boxes are dragged along the trench by the backhoe after the completion of a trench section. As a reminder, the main function of a shield is to allow the trench wall to be vertical in order to reduce the amount of excavation, backfill, and the right-of-way that has to be dug up. Such heavy-duty steel shields are much less efficient when a lot of underground utilities have to be expected, since they need to be fit around the crossing cables/pipes or they must be cut and restored later (see Figures 4.14, 4.17). The size of the box should be matched with the configurations of the trench and the trenching equipment. If the shield is too heavy, the backhoe or crane may not be able to move it efficiently. If the shield is too small, it may require many extra moves and not provide sufficient space to allow the crews to bed, set, or weld the pipes. The shield should be at least 4 ft longer than a pipe segment in order to protect its open end. If remote-controlled compactors are used, eliminating the need for people to enter the trench, the shield can be removed right after the pipe is joined, thus opening the entire space between the pipe and the trench-wall for compaction.

Common trench-boxes to be rented reach only 30 ft (9.1 m) in length. However, Trench, Inc., has worked many times in the past with the company StrongShores, Inc., which specializes in the manufacturing of unique shoring and shielding solutions for such cases (Figure 4.26). Here is their design:

To accommodate the large 6-ft (1.8 m) diameter pipe, the box incorporates five independent spreaders made of 10 ft (3 m) long, 10 in (0.25 m) schedule-120 steel pipes reinforced with end-collars. The two 10-ft (3m) high shields have a length of 45-ft (13.6 m) and its massive walls are 12-in (0.3 m) thick.

To reduce the weight and the forces on the spreader, it is recommended to slope or bench the 6 ft on the top (reducing the load from the sides). This also allows the box height to be kept at 10 ft (3 m), making it possible for the backhoe to drag along the trench bottom.

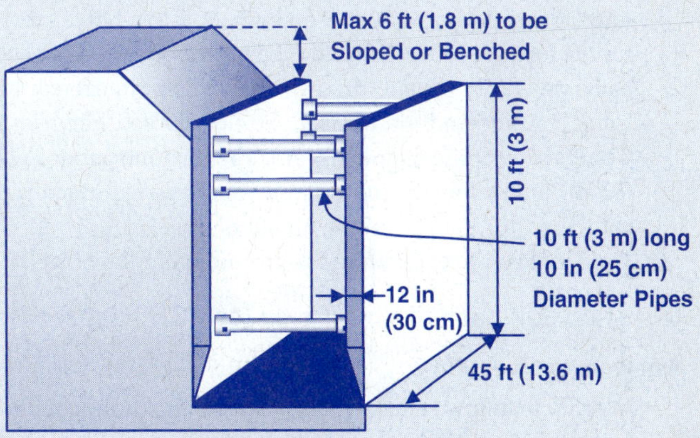

Figure 4.26 Installed Strongshores Trench-box

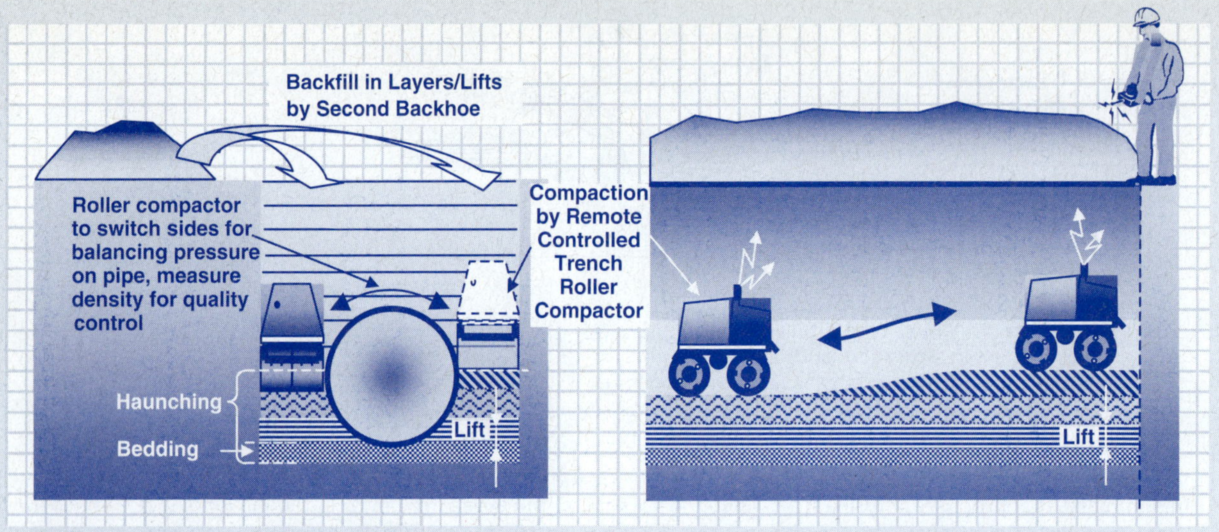

(a) Cross-section of Trench Compaction in Lifts

(b) Operation of Compactor Safely

Figure 4.27 Sketch of Layered Densification with Trench Compactor

Addressing Issue j: Selection of Backfill Compaction System: The spaces on both sides of the pipe have to be properly compacted, as this area, called the haunching area, will be critical in supporting the pipe during its life (Figure 4.27). As the trench has an overall width of 10 ft (3 m) and the pipe uses 6 ft (1.8 m), the open space on either side of the pipe measures a minimum of 2 ft (0.6 m). Several manufacturers offer remote controlled trench compactors that will fit into this narrow area (see booksite).

S4.21 Pipe Construction
S4.22 Excavation-Pipelaying-Compaction Inspection

If the project has a time constraint, a second excavator backhoe or a loader-backhoe will be needed to place backfill on both sides and then on top of the pipe. In order to achieve the required density, the compaction has to be done in layers from the bottom to the top. The region below the center of the pipe, the haunching, is most critical during compaction, as it constitutes the main support for the pipe.

Answers to Question 1:

- Both the excavation and the pipe-laying can be accomplished using the CAT 385C L. A second smaller CAT backhoe or backhoe-loader could be used to backfill the open trench as quickly as possible while the installation continues uninterrupted.
- A safe method for protecting the workers in the trench is the use of a specially fabricated trench shield that is 45 ft (13.7 m) long and 10 ft (3 m) high. In areas that exceed 10 ft of depth, benching, sloping, or a stacked segment for the top section can be used. The trench box needs to be backfilled right away on the side of the rail tracks. The backhoe will be able to drag the box along the trench bottom as the trench excavation progresses.
- Issue l: A line-and-grade laser should be used between manholes.
- Compaction of the haunching area can be done using a walk-behind trench roller compactor with a width of 2 ft (0.6 m) and remotely controllable. Several manufacturers offer such machines and one can be rented if the company does not own one already.
- Issue m: Prior to excavation the ground water level has to be lowered with a dewatering system (see Chapter 8)

Question 2: What Are Your Preparatory Measures Before Starting a New Section?

Answers to Question 2:

- In order to follow state law, a minimum of 24 hours before pushing the first bucket into the ground a request for locating all utilities in the vicinity of the trench section has to be made. This can be made either online, over the phone, or by fax. For a sample request form, see the booksite.

- Request the shut-down of overhead power lines if they are too close to the trench area.
- Verify the start date with the dewatering subcontractor so that the water level is lowered on time.
- Ensure that a competent person will be on the job site when work starts.
- Review OSHA 29 CFR, Part 1926, Subpart P (Trenching).

Question 3: What Is Your Detailed Work Plan for the Crew Placing and Jointing the Pipe Segments?

Answers to Question 3:

CAUTION: The Lining of the Pipes Needs to Be Handled with Care

1. Prepare joints of new pipe segment (cleaning, etc.) according to manufacturer's instructions.
2. Mark the position of the slings on the new pipe piece to ensure center of gravity rule.
3. Feed nylon slings under pipe and connect their end-hooks to the spreader-bar.
4. Attach a guide rope or tag line to the pipe/spreader bar.
5. Direct backhoe operator so that the load-lifting point (over front and at end of stick or bucket hook) is perfectly vertical above the pipe's center of gravity (no swinging is allowed).
6. Travel backhoe with pipe lifted over the front at a minimum 3 ft above surface to the trench box. Avoid any swinging of the load.
7. Nobody is allowed inside trench box while pipe is suspended.
8. Guide backhoe operator during lowering of pipe between the walls and spreaders of the trench-box (use tag line if necessary).
9. Jointing crew is allowed to enter in order to align and connect the new pipe to the existing pipe (using laser beam) according to manufacturer's specifications and standards
10. Carefully remove nylon slings without damaging the pipe surface.

4.7 Planning for a Productive Backhoe Operation

The efficient use of large and costly equipment is of critical importance for the contractor to achieve a profit that guarantees the survival of the business. To plan for high productivity, the construction manager has to understand how the many characteristics of a new job, the conditions of the soil, the weaknesses and strengths of various pieces of equipment, as well as the skills of his workforce can be aligned into an efficient production system. In order to identify the most important relationships that impact productivity of excavation, we need to understand its three key components: (a) repetitive digging cycle, (b) operational tasks, and (c) physical motions or subtasks. After understanding the mechanisms of the production system, we will be ready to "dig" deeper and identify those relationships that are most important in dictating the productivity of the backhoe excavation.

4.7.1 OBSERVATIONS ALONG THE EXCAVATION CYCLE

Observing a backhoe excavator or a backhoe loader digging a trench, one quickly recognizes a pattern of steps that are repeated over and over again (Figure 4.28). From what we have learned so far about the use of backhoe excavators for trenching we know that each of the 5 tasks (a)–(e) can be further dissected into subtasks and motion. For example, task (b), Fill Bucket, consists of

Subtask (b-1): Penetrate the soil surface with *suitable* bucket angle,
Subtask (b-2): Cut the soil with *appropriate* bucket motion, and
Subtask (b-3): Rotate bucket and exit *exactly* when bucket full.

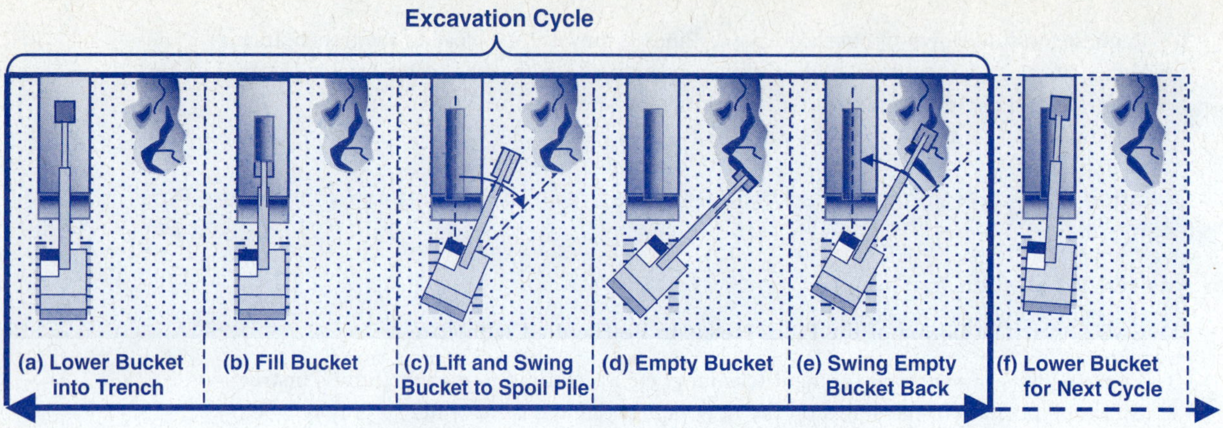

Figure 4.28 Tasks of the Backhoe-Excavation Cycle

Each of the tasks and sub-tasks requires time, which, executed in sequence, not parallel, adds up to the total cycle time:

$$\begin{aligned}\text{Total cycle time } (D_T) = &\text{ Duration for lowering bucket } (D_{low})\\ &+\text{Duration for filling bucket } (D_{fill})\\ &+\text{Duration for lifting and swinging to spoil pile } (D_{li\text{-}sw})\\ &+\text{Duration for emptying bucket } (D_{empt})\\ &+\text{Duration for swinging back } (D_{swba})\end{aligned}$$

Based on the definition of equipment productivity or production rate (= Output/Input in terms of hours), we can establish the basic function for quantifying efficiency of operation using cycle time:

$$\begin{aligned}\text{Theoretical backhoe productivity} &= \text{Cycles per hour} * \text{Average volume of material in bucket}\\ &= \underbrace{(60\text{ Minutes}/D_T)}\ * \underbrace{\text{Heaped bucket volume} * \text{Bucket fill factor}}\\ \text{Adjusted backhoe productivity} &= \text{Cycles/Hour} * \text{Operating factor} * \text{Volume per cycle}\end{aligned}$$

Naturally, the durations of the individual tasks are more or less strongly related to other factors that tend to change with the conditions of the job. In particular, the soil density relates to how easily and how quickly the bucket is able to cut, which in turn dictates the time needed to fill the bucket. On the other hand, soil density also impacts what bucket size should be selected, while the power and size of the equipment dictate the forces that can be generated to push the bucket. The way these many factors relate to each other is best explained using an example.

S4.23 Backhoe John Deere 200CLC

Worked-Out Problem 4.3 / Effect of Bucket Size on Productivity

We have been asked to propose the size of bucket that should be used with a John Deere 200 C LC Backhoe excavator to dig a trench in wet clay. According to John Deere's specifications (see booksite), you can use either the 1.5 yd^3 (1.2 m^3) general-purpose bucket or the 1.38 yd^3 (1.3 m^3) heavy-duty bucket (SAE heaped). While the

Table 4.7 Durations for Operating a John Deere 200 CLC Backhoe Excavator by Joe Banks

Task Sequence / Material Characteristic	Lowering Bucket		Dig-to-Fill		Lift and Swing (45°)		Empty Bucket		Swing Back (45°)		Total Cycle Time	
	General Bucket (sec)	Heavy Bucket (sec)	General Bucket (sec)	Heavy Bucket (sec)	General Bucket (sec)	Heavy Bucket (sec)	General Bucket (sec)	Heavy Bucket (sec)	General Bucket (sec)	Heavy Bucket (sec)	General Bucket (sec)	Heavy Bucket (sec)
Soft Clay	4	4	16	14	8	8	3	3	5	5	36	34
Tough Clay	4	4	19	15	8	8	3	3	5	5	39	35
Very Dense Clay	4	4	22	16	8	8	3	3	5	5	42	36

general-purpose bucket is lighter, it will get damaged more easily when the ground is harder than expected. In order to decide, we should compare the production rate of each option when used in wet clay. On a previous job, when Joe Banks was the backhoe operator, the average durations in Table 4.7 were collected.

From the data set in Table 4.7 we recognize that the only time difference between the two buckets, general and heavy, is during Dig-to-Fill. From the specifications, we know that the general bucket is larger in volume and, therefore, takes slightly longer to fill. On the other hand, it is more handicapped by tough and dense clay than the heavy duty bucket, resulting in slower cutting as larger friction resistances have to be overcome. One can clearly identify how these two factors, size and soil density, impact the total cycle time. The next step is to calculate productions rates, cyd/hr, for these situations using the function (see Table 4.8).

Table 4.8 Cycles and Hourly Production for Different Clays

	Total Cycles per Hour		Total yd^3/hr		Total m^3/hr	
	General Bucket	Heavy Bucket	General Bucket	Heavy Bucket	General Bucket	Heavy Bucket
Soft Clay	100	106	150	146.3	114.7	112
Tough Clay	92	103	138	142.1	105.5	109
Very Hard Clay	85	100	127.5	138	97.5	105.5

Output (excavated volume)/**Input** (time in hours)

Volume of *general bucket* with a bucket fill factor of 1.0 $= 1.5\,yd^3$ $(1.2\,m^3)$
Volume of *heavy bucket* with a bucket fill factor of 1.0 $= 1.38\,yd^3$ $(1.3\,m^3)$
Number of excavation cycles per hour with general bucket $= 3,600\,sec/36\,sec = 100$

Similarly:

We know from Chapter 2 that these numbers are purely theoretical, since construction work always gets interrupted. Although Joe is a consistent operator, he is productive for about 50 min per hour, which computes into an operating factor of $(50\,min/60\,min) = 0.833$.

Result of Bucket Selection Problem: If the wet clay to be expected on the new job can be characterized as soft, then the larger general bucket should be selected, resulting in a production rate of: $150\,yd^3 * 0.833 = 125\,yd^3/hr\,(95.3\,m^3/hr)$ for an operator with Joe's skill.

For tough and very hard clay, the heavy-duty bucket is not only desirable, because it is sturdier for the harder material, but, despite its smaller size, it allows higher overall production rates of $118.4\,yd^3/hr$ and $115.0\,yd^3/hr$ respectively $(90.2\,m^3/hr, 87.7\,m^3/hr)$.

4.7.2 TIME STUDIES OF BACKHOE OPERATIONS

Have you wondered how engineers establish the time it takes Joe Banks to accomplish the different motions and tasks in operating a John Deere 200 CLC backhoe? The father of the method that serves this purpose was Frederick W. Taylor (1856–1915), the father of scientific management. Taylor rose from common laborer to chief engineer in six years, and completed a home study course to earn a degree in mechanical engineering in 1883. Taylor began a scientific study of what workers ought to be able to produce by using time studies of elementary movements and/or tasks such those listed in Table 4.7 or Figure 4.29. Let's find out what the process or input/output (I/O) model in Figure 4.29 represents. The sole modeling elements are boxes, representing tasks, circles representing modeling resources that are in a holding position, and arrows indicating the flow of resources. We recognize that there exist two main cycles, task sequences for the operator and the backhoe: excavation of trench, lowering of pipe section, jointing new pipe, and initial backfilling followed by moving the backhoe backward to excavate next trench section. No moving of a trench box or installation of bracings is included in the process. This means that the excavation includes sloping or benching to stabilize the trench walls, a method that was used for Solved Problem 4.1 and is shown in Figure 4.10.

The only difference between the backhoe and the operator cycle is task E, taking a short break. Since the operator is commonly also the foreman on the project, such breaks are needed to discuss an issue, to give short instructions, and so on. Theoretically, the time the backhoe takes to finish its cycle ($T_{BC} = T_E + T_T + T_J + T_I$) should be used to calculate the productivity of the crew. However, task **A** requires not only the backhoe but also operator, who might be held up in **E** with a probability of x%.

We return later in the book to the subject of modeling construction operations, but right now let's focus on the question of how to establish the different task durations using Taylor's time study method. In 1911, Taylor published *Principles of Scientific Management*, a book in which he described the time studies for following a specified process. The planning of a time study includes 3 steps:

Step 1: Divide cyclic operation into smaller tasks or subtasks being executed repeatedly

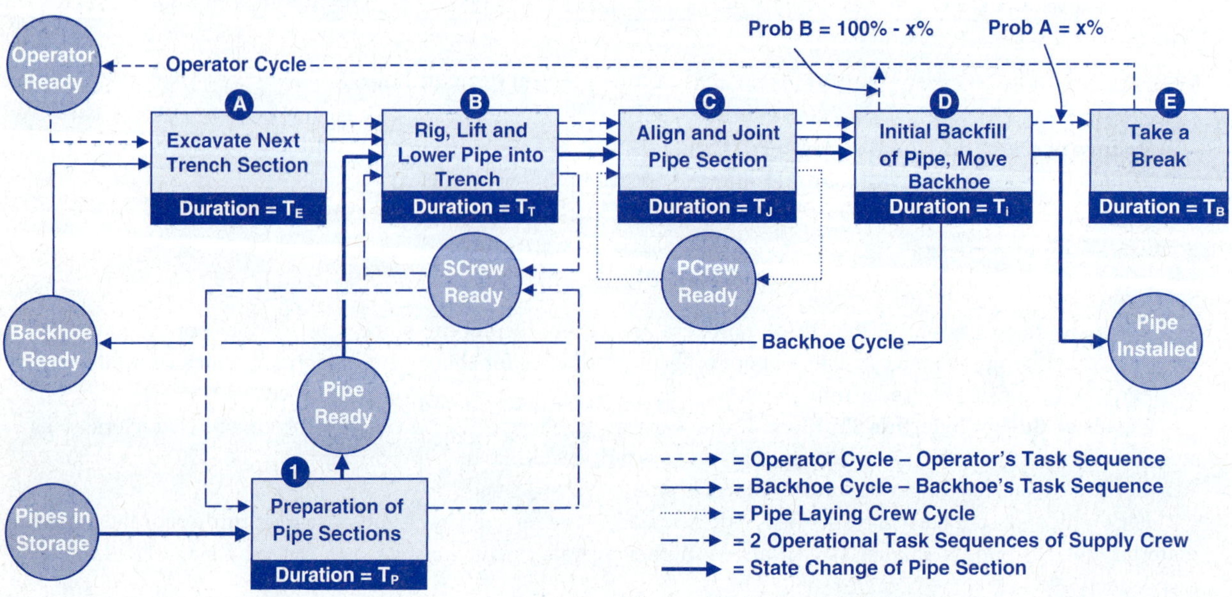

Figure 4.29 I/O Model of Trenching and Pipelaying Operation

Table 4.9 Time Study of Trenching and Pipelaying Operation

Steps 1 & 3

Project: Parkway Connector, Gaithersburg	Date: May 2, 2011
Activity: Concrete Pipes between station 10.5.38 and 10.5.39	Temperature: 68 °F
Process: Trenching and pipelaying using benching method	Rel. Humidity: 51%
Foreman: Bert Trench	
Day of the Week: Monday	Observer: Benny

	A (Excavate)		B (Transport)		C (Jointing)		D (Backfill)		Pipe Section Installed	E (Break)		Backhoe Cycle Time	Operator Cycle Time
	Start	T_E	Start	T_T	Start	T_J	Start	T_B		Start	T_B		
Cycle	Hr:Min:Sec	Min.	Hr:Min:Sec	Min.	Hr:Min:Sec	Min.	Hr:Min:Sec	Min.	Hr:Min:Sec	Hr:Min:Sec	Min.	Min.	Min.
1	7:05:15	6.25	7:11:30	2.58	7:14:05	1.08	7:15:10	0.33	7:15:30	7:15:30	0.58	10.25	10.83
2	7:16:05	4.55	7:20:38	2.78	7:23:25	2.33	7:25:45	0.50	7:26:15	7:26:15	0.25	10.17	10.42
3	7:26:30	5.58	7:32:05	2.75	7:34:50	2.17	7:37:00	0.37	7:37:22	7:37:22	0.72	10.87	11.58
4	7:38:05	2.17	7:40:15	3.55	7:43:48	2.70	7:46:30	0.42	7:46:55	7:47:05	0.25	8.83	9.08
5	7:47:10	4.08	7:51:15	2.33	7:53:35	2.42	7:56:00	0.42	7:56:25	7:56:25	0.33	9.25	9.58
6	7:56:45	5.00	8:01:45	1.50	8:03:15	1.75	8:05:00	0.33	8:05:20	0:00:00	0.00	8.58	8.58
7	8:05:20	3.50	8:08:50	2.33	8:11:10	1.83	8:13:00	0.42	8:13:25	0:00:00	0.00	8.08	8.08
8	8:13:25	5.33	8:18:45	2.25	8:21:00	1.75	8:22:45	0.38	8:23:08	0:00:00	0.00	9.72	9.72
Total Time		36.47		20.08		16.03		3.17			2.13	75.75	77.88
No. Observations		8		8		8		8			8	8	8
Average Time		4.56		2.51		2.00		0.40			0.27	9.47	9.74
Standard Dev.		0.65		0.236		0.47		0.04			0.41	0.38	0.79
Min Time		2.17		1.50		1.08		0.33			0.00	8.08	8.08
Max Time		6.25		3.55		2.70		0.50			0.72	10.87	11.58

(Margin labels: Steps 2 — cycle rows 1–8; Steps 4 — start times; Steps 5 — summary rows.)

Step 2: Decide how many times to measure the execution of the tasks of the cycles

Step 3: Prepare a form that prompts you to record all the pertinent information (e.g., date, temperature, humidity)

Table 4.9 presents the outcome of a time study that measured the installation of pipe sections. The top section of the time study addresses steps 1 and 3 of Taylor's method while the left column is linked to step 2. Of course, the tasks and subtasks A–E can also be found in the process model shown in Figure 4.29. For the sole purpose of convenience, three additional columns have been added: (a) Time when one new pipe section is finished, (b) backhoe cycle time, and (c) operator cycle time. Now we are ready to continue Taylor's recommended steps for observing and measuring task start and end times either on the construction site or through viewing video recordings followed by computing durations:

Step 4: Use a stopwatch to time and record start and/or end times of tasks repeatedly for as many times as was decided in step 2 (= number of cycles)

Step 5: Calculate individual task durations using the recorded data. Compute average times and standard deviations for each task and entire cycles.

$$\text{Average observed time} = \frac{\text{(Sum of the times recorded to perform acertain task)}}{\text{Number of observations}}$$

Table 4.9 presents the result of step 4 in the columns labeled Start. Since each task A–D is immediately followed by the next, the start of the new task also represents the end of the previous. If this is not the case, as in the operator taking a break (task E) only sometimes, a start time of a task is simply ignored. Step 5 first results in the computing of the Ts for each task (e.g., T_E). They are determined by subtracting the start-time of a task (e.g., A) from the start-time of the following task; in our case it's task B. For the first cycle, this results in $T_E = 6.25$ min, 4.55 min for the second cycle, and so on. After 8 cycles, we compute an average time of 4.56 min, with the first cycle clearly being an extreme case (Max). Finally, we are able to establish the average times for all task and of the selected cycles. We also observe that the operator took a break only 5 out of 8 times, representing a probability of 62%, using an average of 0.27 min.

4.7.3 SKILLS OF A MASTER BACKHOE OPERATOR

S4.24 Competencies of Heavy Equipment Operators

Earlier in this chapter we learned that four joystick movements control the main actuators of a backhoe excavator: (1) boom cylinder, (2) stick cylinder, (3) bucket cylinder, and (4) motor to turn superstructure. In addition, we found that along the digging path cutting and friction forces are being created, most of which are a function of how well the bucket is being handled. Key to efficiency is the smoothness with which the operator is able to execute the digging operation. In effect, the smoothness is a function of how well she or he applies and, most importantly, integrates the hydraulic power of the boom, stick, bucket, and slewing motor. The best operators use all three functions at once while maintaining the best bucket angle. By the time that the bucket is heaped, the digging should lead smoothly into the lifting and swinging motion since the extra time spent will not add any more volume. Joe Banks, the expert operator from the previous example, summarizes his "work philosophy":

> When trenching, the goal is to shave off soil layers, rather than gouge out chunks. Let the bucket teeth and cutting edge work for you. The teeth do the initial cutting by breaking the soil into pieces. The crumbling spoil begins to fill the bucket, and the curling action holds the spoil in the bucket.

Filling the bucket quickly while digging inaccurately, too deep, too narrow or too wide, will required added time for backfill or additional concrete or rebar cutting when it does not fit into the trench.

> Layout and digging techniques are extremely important because you only have one chance to get it right.

The Skill of Smooth Digging: Watching how an operator moves the bucket through the soil tells one very quickly how experienced the person is in operating a particular piece of equipment. Let us study the mechanics of the digging task (see Figure 4.30) to learn to recognize skill.

It is not difficult to realize that **Operator 1** uses a very simple control approach by sequencing the actuator actions, resulting in large friction forces all along the cutting path A1–B1. As we learned earlier, friction forces produce wasted energy and are the result of pushing the bucket with excessive force against the ground. Figure 4.30b, on the other hand, displays Joe's "digging philosophy" of shaving off layers. The joystick actions bars demonstrate, however, that shaving involves a great many more control actions. Many times, two and three actuators are active at the same time, moving the cutting edge of the bucket along a straight line rather than a circle. Such coordination is made necessary by the fact that each of the three elements of the backhoe arm is pinned, like a linked chain, to its base. Thus, the skill of the operator manifests itself in how well he is able to turn three rotations into a desired path of the last piece of the chain, the bucket. Not surprisingly, the friction forces that are produced are minimal and considered a natural "expense" of digging.

The Skill of Smooth Motion Control: We learned earlier that the digging task is just one step, albeit a critical one, along an excavation cycle of a backhoe. Anybody who has ever watched a highly experienced operator going through the entire cycle will attest to the fact that it looks like somebody is moving a conductor's baton during a symphony. Everything seems easy. In an attempt to measure this phenomenon, the Construction Automation & Robotics Laboratory at NC State University conducted a set of experiments during which backhoe operators with different skill levels were asked to excavate a 1 ft deep trench inside a soil box filled with sand. Figure 4.31 offers an overview of the experimental setup that included a hydraulically powered backhoe arm controlled from an operator seat with joysticks. The arm is instrumented with force and position sensors

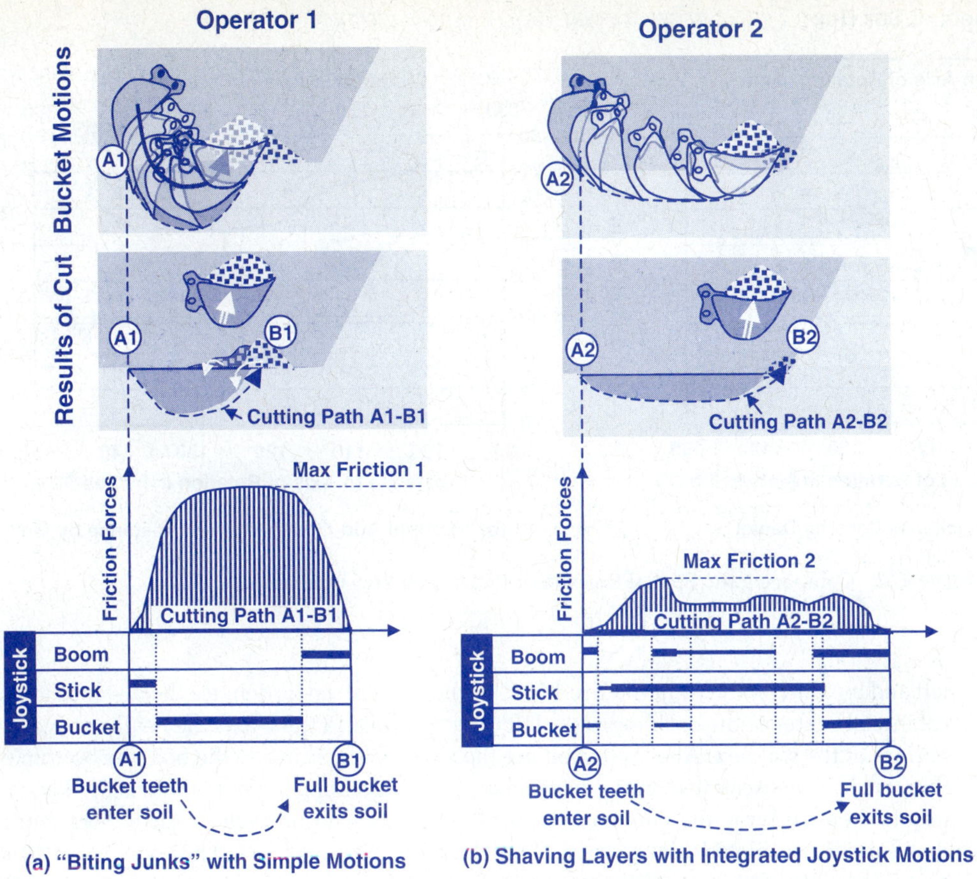

Figure 4.30 Comparing Digging Strategies and Outcomes of Operators with Different Skill Levels

that collect data about the position of boom, stick, and bucket as well as the forces exerted by the cylinders. Buried pipes or rocks are used to simulate the ground conditions. The goal of the experiment was to find out if it is possible to actually measure differences in skill levels based on the experience of an operator.

The heavy lines in the two graphs in Figure 4.30 represent the exact path of each bucket's cutting edge during the scooping and dumping cycle executed by Daniel on the

Figure 4.31 Computer Integrated Backhoe Arm

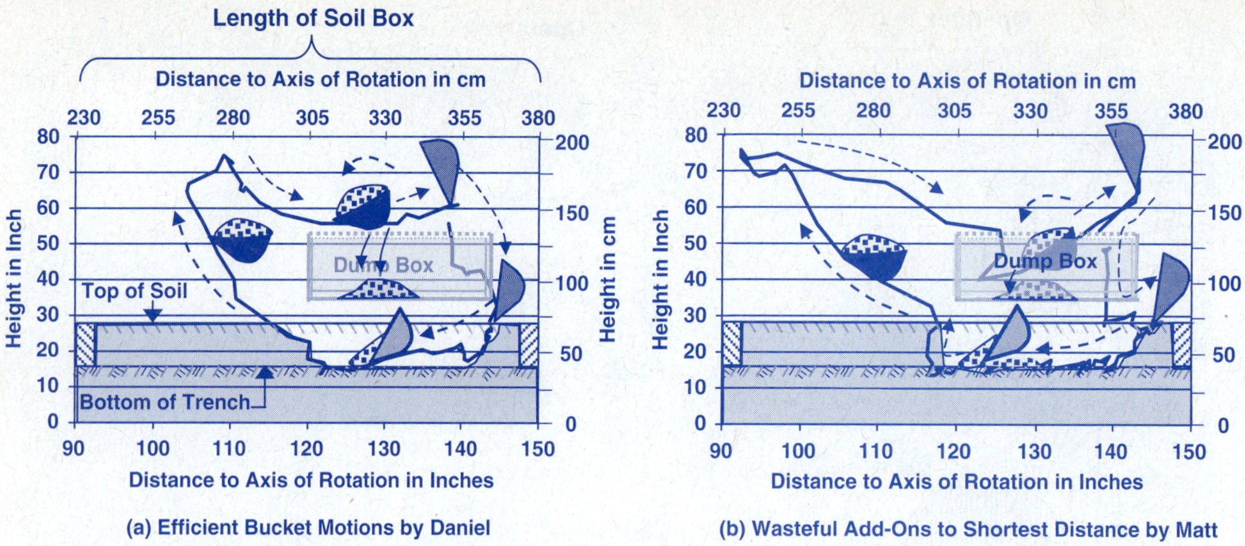

Figure 4.32 Comparing the Spatial Excavation Paths from Two Operators

left and by Matt on the right (Figure 4.32). The horizontal x-axis denotes the distance to the rotational center of the backhoe to the left of the graph and the y-axis the height above the bottom of the soil box. Also indicated are the top of the soil inside the box, the bottom of the trench to be excavated, and the location of the dump box to the rear. A fact to keep in mind when studying the path contours of the first excavation cycle of the experiment is that Daniel completed his cycle in about half the time of Matt. This is obviously no coincidence since Matt started with erratic up and forward movement, using alternatively the boom and the stick (see previous paragraph). This created a saw-tooth type motion of the bucket tip visible in Matt's plot. Daniel on the other hand cuts smoothly down to the desired 1-ft (30 cm) depth and lifts the full bucket up and over the dump box (rotation is not shown) before emptying the bucket and returning to the soil box. While both reach the same top height of 74 in (1.9 m), Matt is moving the bucket all the way back to 92 in (2.3 m) from the base while Daniel stops at 103 in (2.62 m) before switching direction toward the dump box. Matt also spends time to lower the bucket into the dump box to empty it. Another pattern that distinguishes the two is the abrupt change in direction caused by a switch from one single joystick motion to another single joystick motion. For example, Daniel's bucket leaves the trench in a smooth slope while Matt stops the horizontal cutting motion abruptly to change to a vertical exit.

In summary, the visual inspection of path patterns indicates that the two operators employed very similar excavation strategies but differed in their executions. Daniel handles the backhoe in a way that results in an overall smoother path through the soil and a shorter back-cycle to empty the bucket in the dump box. In fact, Daniel has over 10 years of experience, while Matt is a novice.

The Skill of Damage Avoidance: Trenching in urban areas always includes working around existing utilities that have been marked by locators with more or less accuracy. When lines cross through the trench space, a hunt for their location ensues involving the backhoe operator and helpers with shovels and probing rods. As the examples in Figure 4.33 demonstrate, underground obstacles have many shapes and are made of many different materials.

One of the first skills of an operator is to know when it is time to move the large bucket to the side and have the helpers do the digging so that he is able to make visual contact

(a) Working with a Large Tool Around Delicate Utilities

(b) Supports for Weak Facilities Crossing the Trench

Figure 4.33 Precision Digging and Constructing of Support Structures

with the buried facility. Even then, fragile pipes made of plastic or terracotta can shatter when the supporting soil underneath are being removed. Figure 4.33b shows such a case. Here, a bank of pipes encased in concrete has lost its foundation during trenching underneath. As shown, the backhoe operator has decided to bring in a solid steel pipe as an overhead "bridge" from which the pipe bank is being supported via strong slings, chains, and tighteners.

Finally, it is not always efficient to leave every utility intact. For example, it might be much more efficient to remove drainage pipe sections for the duration and and to fit new pipes during backfilling between the two ends. The operator has to decide on the most economical method. Overall, one of the key skills of a backhoe operator working on trenching jobs is to approach and work around buried facilities safely and cost effectively.

4.7.4 PREDICTING PRODUCTIVITY USING THE CAUSE-AND-EFFECT MODEL

We have learned that productivity improvements are critical for a company to stay competitive and profitable. In the previous sections, we found that a complex list of sometimes interconnected factors impact the production rate of an excavator operator. By observation, one realizes that the effect as well as the intensity of a factor on productivity is hard to quantify in general terms. In other words, it is impossible to create mathematical functions, such as $y = f(x)$, that allow us to precisely calculate productivity using cycle time and soil volume. In this final section we will create a framework that allows us to at least enumerate the most critical factors that need to be considered in establishing a base function of following form:

$$\textbf{Excavation productivity job}(\mathbf{i}) = \text{Standard Volume/hr} * \underbrace{\sum_{j=0}^{j=n} (a_{ij} x_j)}_{\text{Adjustment Factors}}$$

$$\textbf{Excavation productivity job(i)}=(1/\text{Average volume of loose material in bucket})*$$

$$(\text{std. dur. to fill bucket at } (D_{fill}) * \sum_{j=0}^{j=k}(a_{ij}x_j)$$

$$+\text{std. dur. to lift and swing to spoil } (D_{lift\text{-}sw}) * \sum_{j=0}^{j=l}(a_{ij}x_j)$$

$$+\text{std. dur. to empty bucket } (D_{empt}) * \sum_{j=0}^{j=m}(a_{ij}x_j)$$

$$+\text{std. dur. to swing back-lower bucket } (D_{swba\text{-}low}) * \sum_{j=0}^{j=n}(a_{ij}x_j)$$

where

$i =$ a particular excavation job
$j =$ factor identification
$x_j =$ factors that impact the standard duration (D)
$a_{ij} =$ coefficient establishing the effect of the factor x_j on the standard duration (D)

From Section 4.7.3 we know that not every section on the cycle is impacted by the same factor. For example, the denseness of the soil has little to do with the duration of swinging and emptying the bucket. At the same time, the impact of a high-powered piece of equipment affects each time segment with different intensity. Nevertheless, with the help of a factor analysis one is able to structure the many factors logically in order create a comprehensive framework to analyze productivity of excavation. Table 4.10 offers a matrix that builds on the idea that the factors that affect productivity can be distilled from established cause-and-affect relationships.

Table 4.10 Factor Analysis Using Cause-and-Effect Relationships on Backhoe Excavation

Cause	Effect	Main Factor	Affected Durations	Strength of Relation	Comments
1. Soil densities are related to the required breaking force to cut	Force requirements and time for filling bucket increase with density	Density of soil material	Filling bucket D_{fill}	Depends on engine + operator	Strong engines allow abusive force
2. Location of soil deposit affects the angle of swing	The larger the angle between digging and soil deposit, the more time is needed	Angle of swing/rotation	Swing full D_{sw} Swing back D_{swba}	Strong	Speed of rotation linked to rpm of superstr.
3. Depth of excavation	Deeper cuts will add time for lowering-lifting	Depth of cut	Lowering D_{low} Lifting D_{lift}	Strong	
4. Obstacles in the path of backhoe movement on side or above	Obstacles can restrict the free movement of the arm or superstructure	Obstacle interferences	Lifting D_{lift} Swing full D_{sw} Swing back D_{swba}	Medium	Risk of accidents and damage
5. Size and type of bucket to deal with the excavation material	Inappropriate buckets will slow digging and/or limit soil volume per cycle	Bucket size and type	Filling bucket D_{fill}	Strong	
6. Lift and digging capacity of backhoe excavator	Underpowered excavator will limit digging force and slow down process	Match of dig requirement with equipment power	Filling bucket D_{fill} Lifting D_{lift}	Strong	

7. Buried utilities that cross the excavation space	Utilities have to be found, worked around, or removed if non-essential	Avoiding damage needs time for locating	Filling bucket D_{fill} since time has to be spent locating	Strong	
8. Requirement for shoring or trench-box	Digging within trench-box limits freedom of motion, requires precision	Obstacle avoidance-precision cut	Filling bucket D_{fill} Lifting D_{lift} Lowering D_{low}		
9. Experienced operators are able to control the actuators more effectively	Amount of friction force during digging, time to completely fill bucket, speed of operation	Operator skill	Lowering D_{low} Filling bucket D_{fill} Lifting D_{lift} Swing full D_{sw} Swing back D_{swba}		

Worked-Out Problem 4.4 / Estimating a Pipe-Laying Job

We are asked to estimate the time it takes to install a section of a new water line in a suburban area. The section has to be treated separately because the open-cut trench is crossed by several buried utilities as marked by the locator but without depth information.

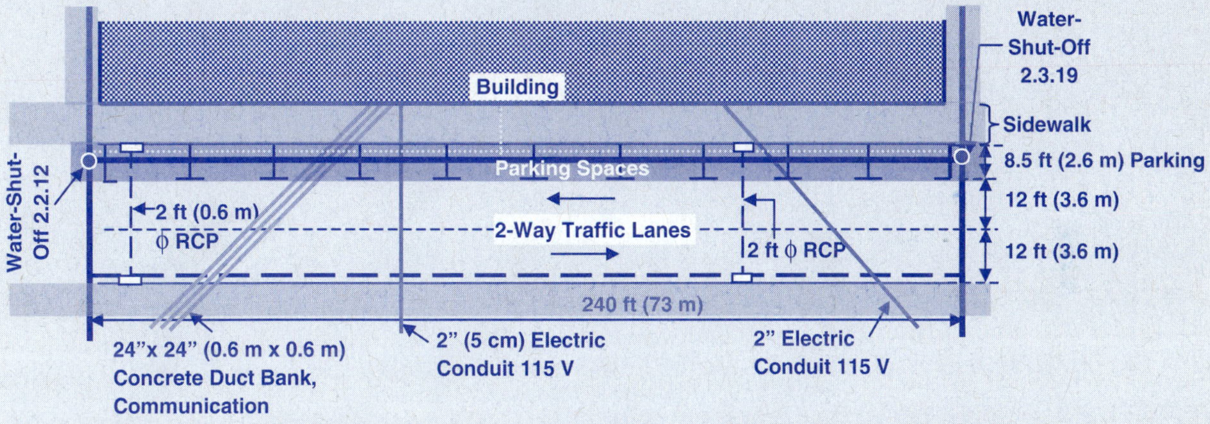

Figure 4.34 Bird's-Eye View of Markings Left by the Utility Locator

The new 10-in (0.25 m) diameter cast-iron water mains, to be installed from shut-off 2.2.12 to 2.3.19, shall be placed at a 5-ft (1.5 m) depth with a minimum trench width of 30-in (0.76 m). The 18-ft (5.5 m) long pipe pieces have compression ring gasket joints. The excavated material, consisting of previous backfill material and coarse material, has to be hauled away and stored off-site. Backfill for utility trenches shall be placed in 8-in (0.2 m) lifts or less. Both traffic lanes have to stay open, but one lane can be reduced to a minimum of 10-ft (3 m) The sidewalk also has to stay open. The entire site needs to be fenced in at all times and a max of 5-ft (1.5 m) of trench can stay open at one time.

Question: How many working days will the excavation and pipe-laying operation take if 20% of the days will be lost due to bad weather?

Collected Supplemental Information: The constraints of working within the 8.5-ft (2.6 m) parking area requires a excavator with a minimum work envelope that provides the digging force of a larger model. In addition, working within the tight area around the utilities also suggests using a smaller bucket size. The Case XR75 provides the needed capabilities, especially since its width is only 7.6 ft (2.3 m) (download brochure from website).

Since the conditions of the job require that the material be loaded on trucks, the backhoe has to turn 180° with each bucket. In order not to extend the superstructure into the traffic lane, the fence should be pushed out to reduce the lane from 12 ft to 10 ft (3.6 m to 3 m). With a tail-swing radius of 4 ft (1.2 m) and a minimum swing radius of 5.8 ft, it is able to work well within the alloted 10 ft (assuming the fence needs 6 inches). We can use the 30-in (0.76 m) wide bucket with a heaped capacity of 0.38 yd^3 (0.29 m^3). The bucket fill factor for the previously compacted material should be 0.85. The operating factor in a downtown area with lots of traffic and pedestrians is 0.75. The laying of one pipe segment takes theoretically 45 minutes on average.

Here are several pictures from a previous job in a side street in the same neighborhood, which provide additional insights on what to expect.

The cut asphalt is supported by a coarse gravel material and possibly building debris resting on clay-type natural soil material (see Figure 4.35b). There will be some "dead" conduits crossing, which have to be verified with the utility when they are dug up. It is hoped that the one known power conduit will not be exactly at the level of the new pipe. From experience we can expect that the concrete duct is at a depth of 1.5 ft (0.46 m) reaching to 3.5 ft (1.1 m). Thus we should be able to push the pipe underneath similar to Figure 4.34d. We can also assume that the 2-ft (0.6 m) drainage pipes lie below 5 ft (1.5 m) and should not interfere. The gas pipe close to the sidewalk is a constant hazard, since the previous contractor may not have been able to follow the design at all times. This is especially true in the two areas with the two stormwater inlets.

(a) Cutting the Asphalt **(b) Dead and Live Conduits**

(c) Lowering New Pipe **(d) Passing Bellow Concrete Duct**

Figure 4.35 Trenching and Pipe-Laying

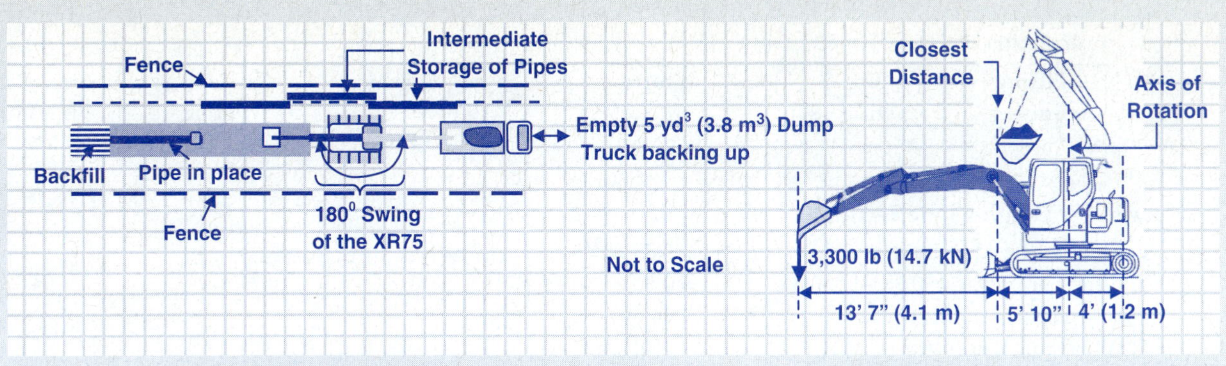

(a) Site-Layout of the Operation

(b) Case XR75 with 6'11" Stick/Arm

Figure 4.36 Sketches of the Main Components of the Operation

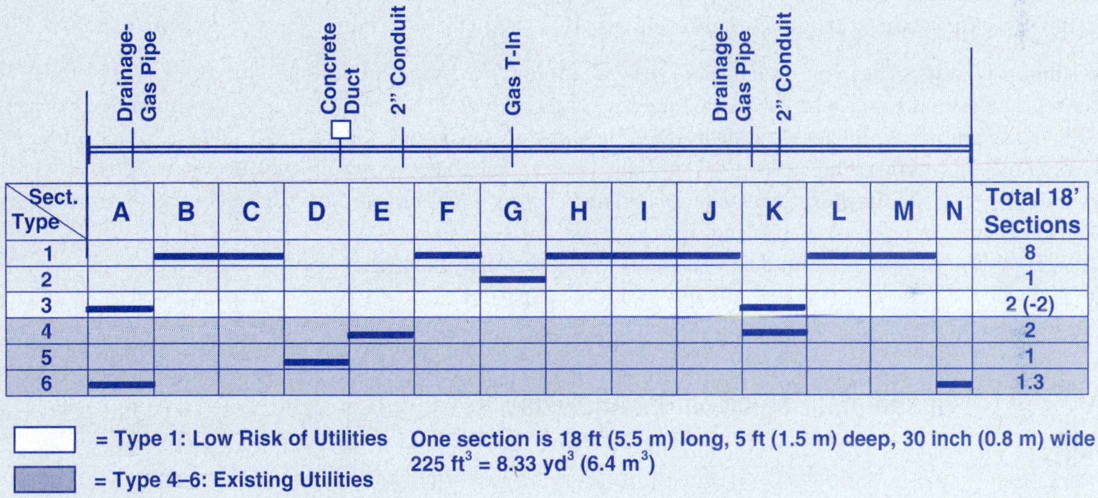

Figure 4.37 Level of Risk for Each Trench Section

There are two operators available, Joe Banks and Phil Duncan. Joe is a highly experienced operator, while Phil has never operated the Case XR 75 and is considered a below-average operator. Figure 4.36 presents a top-view of the trenching operation and a side-view of the small backhoe. Figure 4.37 highlights the level of difficulties for the different sections along the centerline if the trench is differentiated into six risk types.

Design of Trench and Pipe-Laying Operation: Table 4.11 summarizes the impact factors for each of the six section types.

Table 4.11 Impact Factors per Section Type

Type \ Factor	Swing Space	Locating Utilities	Cross Concrete Duct	Pipe Tie In	Operator Skill	Below Ave. Operator
1	0.8	1	1	1	1.2	0.9
2	0.8	0.9	1	1	1.2	0.9
3	0.8	0.95	1	1	1.2	0.8
4	0.8	0.8	1	1	1.4	0.8
5	0.8	0.8	0.4	1	1.4	0.7
6	0.8	0.8	1	0.5	1.3	0.8

Calculation of Cycle Time:

Filling Bucket	Lift & Swing 180°	Empty Bucket	Swing Back & Lower	Total Cycle Time
6 sec	4 rpm – 11 sec	2 sec	6 rpm – 8 sec	27 sec

Duration of Digging One Section

Volume to dig (densified)	$= 8.33\,yd^3 (6.4\,m^3)$
Volume per bucket $= 0.38\,yd^3 * 0.85$	$= 0.32\,yd^3 (0.24\,m^3)$
Total cycles for soil material $= 8.33/.32$	= 24 cycles
Total asphalt to remove	$= 45\,ft^2 (4.2\,m^2)$
Asphalt per cycle	$= 9\,ft^2 (0.83\,m^2)$
Cycles to remove asphalt	= 5 cycles
Total cycles per section	= 29 cycles (27 seconds/cycle)

Average time per section with operating factor of 0.75 = 17.4 minutes

Estimated average time to complete entire length = ((240 ft/18 ft) * ((45 min/0.75) + 17.4 min))/60 min = 18 hrs

Discussion of Results: The extra time needed to work around the buried utilities still has to be considered. Table 4.12 presents an estimate that is based on the time needed in each of the 14 sections (A–N). As shown, Sections A, E, K, and N take significantly longer than the average time. As a consequence, Joe would be able to trench and lay the pipes in 27.6 hours, while Phil needs 30 hrs. Still, we assume that everything will be as expected, including the weather. Further, the time for site installations/fencing, shake-out of pipes along the sidewalk, the cutting of the asphalt, the backfill operation, resurfacing, etc. is not yet considered.

Although this example is based on a very short trench distance, the impact of operational factors are apparent. While some of the minor issues are still not included (e.g., depth of excavation), the differences in operator efficiencies are significant although it only affects the trenching and not the pipe-laying.

Table 4.12 Result of a Spread-Sheet Analysis

SECTION	A	B	C	D	E	F	G	H	I	J	K	L	M	N	
Filling Bucket (sec)	60.0	6.0	6.0	12.0	30.0	6.0	6.7	6.0	6.0	6.0	30.0	6.0	6.0	60.0	
Lift&Swing (sec)	14.7	14.7	14.7	14.7	14.7	14.7	14.7	14.7	14.7	14.7	14.7	14.7	14.7	14.7	
Empty Bucket (sec)	2.0	2.0	2.0	2.0	2.0	2.0	2.0	2.0	2.0	2.0	2.0	2.0	2.0	2.0	
S. Back&Low. (sec)	10.7	10.7	10.7	10.7	10.7	10.7	10.7	10.7	10.7	10.7	10.7	10.7	10.7	10.7	
Total Trenching (sec)	**87.3**	**33.3**	**33.3**	**98.3**	**57.3**	**33.3**	**34.0**	**33.3**	**33.3**	**33.3**	**57.3**	**33.3**	**33.3**	**87.3**	
Operator Skill	**1.1**	**1.1**	**1.1**	**1.2**	**1.2**	**1.1**	**1.1**	**1.1**	**1.1**	**1.1**	**1.1**	**1.1**	**1.1**	**1.3**	
Adjust.Trench.(sec)	**79.4**	**30.3**	**30.3**	**81.9**	**47.8**	**30.3**	**30.9**	**30.3**	**30.3**	**30.3**	**52.1**	**30.3**	**30.3**	**67.2**	
# Cycles	**29**	**29**	**29**	**29**	**29**	**29**	**29**	**29**	**29**	**29**	**29**	**29**	**29**	**15**	
Time/Section (min)	**38.4**	**14.6**	**14.6**	**39.6**	**23.1**	**14.6**	**14.9**	**14.6**	**14.6**	**14.6**	**25.2**	**14.6**	**14.6**	**16.8**	
Pipe Instal (min)	180.0	45.0	45.0	112.5	45.0	45.0	45.0	45.0	45.0	45.0	45.0	45.0	45.0	180.0	
Operating Factor	0.75	0.75	0.75	0.75	0.75	0.75	0.75	0.75	0.75	0.75	0.75	0.75	0.75	0.75	
To Time Joe (hrs)	291.165	79.5	79.5	202.8	90.8	79.5	79.9	79.5	79.5	79.5	93.6	79.5	79.5	262.4	27.6
To Time Phil (hrs)	310.4	83.9	83.9	240.5	106.2	83.9	84.3	83.9	83.9	83.9	106.2	83.9	83.9	276.4	30.0

4.8 Smart Add-Ons to Boost Backhoe Safety and Productivity

Not that long ago, running a piece of powered construction equipment needed at least two people, a fireman and a machine operator. Starting the engine of the wheeled steam shovel shown in Figure 3.4 meant shoveling coal from the on-board bunker and filling the boiler with water until the required steam pressure was established (e.g., 180 psi/1.2 MPa). The operator had to be in excellent physical shape, since he had to pull, push, or step on long levers that engaged the friction drums that ran the various cables. As mentioned in

Chapter 3, hydraulic power revolutionized the operation of large construction equipment, especially the backhoe excavator. Because the cost of the ever-larger machines became prohibitive, manufacturers first started to offer more versatility through many attachments before adopting mechatronics for their products. Mechatronics, integrating electronics with mechanical systems, is the beginning of a revolution that will eventually make construction equipment the smartest machines in the world. This section will introduce only some of the devices that have made it onto construction sites.

4.8.1 ATTACHMENTS THAT INCREASE VERSATILITY

While the backhoe excavator in itself can be used for various purposes, its inherent stability, reach, and power make it an excellent platform to attach totally self-contained machines and tools that require substantial force to operate.

Hydraulic Breaker/Hammer: Probably the best known attachment to the backhoe stick is the breaker that is used to fracture rocks or concrete so that it can be loaded with the bucket. Using a backhoe arm to operate the breaker exposes the pins of the cylinders, stick, and boom to large dynamic forces caused by the hammer hitting the pin. Since these pins were designed to support the gravity force of the bucket with load and the maximum digging forces, adding dynamic forces can deform the joints. These damages will not only create a "wobbly" arm but drastically increase joint friction. Unfortunately, deformed joints cannot be fixed economically.

Earth Auger: Adding an extra hydraulic line to the arm of backhoe loaders provided the opportunity to power the motor of earth augers mounted in place of the bucket. In place of two people holding an auger, such applications are very popular where the soil consists of sand-clay type material. The exchangeable bits make it a useful tool for soil sampling, posthole making, or tree planting. Despite their simple operation, they can be very destructive when they "meet" a buried utility. As with any operation that requires digging into the subsurface, it is the law to contact One-Call 24 hours before the start of the work.

Drill/Piling Rigs: Advertisers of mountable drills ask: "Why buy a large dedicated drill rig when you can buy an attachment to a backhoe excavator?" Especially if a contractor has an old backhoe that still works, mounting a self-contained drill rig can make economic sense. The sized backhoe provides the hydraulic muscle and the reach to position the rig. A variety of hardware options are available that include pile-driving, percussion rock drills, and tie-back attachments that drill into berms and side walls. The earlier types of attachments focused on making the end-effector more dexterous by modifying or adding to the bucket. Over time a large family of grapples developed.

Simple Thumb: Like the thumb on a hand, a thumb in combination with the bucket allows the excavator to grasp objects, making it perfect for pick-and-place operations or to pick up large objects that do not fit into the bucket. Examples are placing riprap along rivers and loading logs or debris. Thumbs are also often used to pick up pieces of asphalt or concrete pavement, or in separating recyclables. The thumb can be either stiff or powered by a hydraulic cylinder.

Multi-Tine Heavy-Duty Grapples: Especially in demolition work, simple thumbs-and-buckets do cannot handle the many shapes of steel materials. By making the tines interlock, the grapple is able to close all the way, something that is not possible with a bucket. Skilled operators are able to separate the varied materials into different piles. Often, this grapple is found in combination with another heavy-duty attachment, the mobile shear.

Shear Grapples: In this application, the bucket is replaced by one, two, or three jaws or blades powered by the bucket cylinder. This tool can be found in recycling plants as well on sites to demolish bridge decking, parking structures, concrete foundation slab, floor slab, beams, and footings. They are able to cut and break large pieces to separate steel and concrete and make them loadable to be sold directly to concrete recyclers and steel mills. Shears also are used to cut beams and pipes.

A very useful feature for a job that requires the use of multiple gears is a mechanism that allows the bucket to be rapidly replaced with a different tool.

Couplers for Easy Swapping: There are different kinds of coupling mechanisms, but the objective is the same, to allow the operator to change work tools easily. Pin connectors require one pin that couples the pieces tightly. Others have hydraulic mechanisms that lock wedges into place. Pin grabbers, such as the one shown, allow the operator to grab and hold onto two pins on the work tool. Manufacturers of backhoes and tools have started to offer near standardized interfaces so that a contractor has many more choices in terms of tool usage.

As changing tools with standard coupling interfaces was made easier, a large set of specialty buckets started to appear.

S4.25 Backhoe Attachments John Deere

4.8.2 ERGONOMIC SUPPORTS FOR THE OPERATOR

Both physical and cognitive ergonomics are to be considered when searching for better ways for an operator to run a backhoe. The first field concerns itself with the human body as a component of the machine sitting inside the control cabin and interacting with it physically. The second field focuses on the sensing and cognitive processes a person employs when learning and monitoring the movements of the backhoe. Both fields try to maximize operator safety and productivity by reducing operator fatigue and the complexity of the information feedback. In recent years, both areas have made great inroads in the way the operator and the machine work together as a team. What enabled the gains in the ease with which these large pieces or equipment can be controlled is mechatronics. For example, while the early hydraulic valves required a lot of muscular effort to open and close, today's electronic joysticks send analogue current to electromagnetic hydraulic valves, requiring virtually no effort to lift tons.

Features to Reduce Operator Fatigue and Stress

The physical and mental fitness of the operator is key to safety and productivity. An operator who is fatigued, confused, or uninformed about the environment is more prone to make mistakes, with possibly deadly consequences. That's why equipment manufacturers include more and more operator-friendly features and easy-to-service components on their excavator models. The following reviews a list of ergonomic issues and how they are being addressed in backhoe excavators today.

S4.28 Whole Body Vibration Study

Operator Comfort—Heat/Cold/Humidity: Both hot and cold conditions can lead to sudden fatigue, confusion, fainting, and even a stroke. So it is not a surprise to find that air-conditioned cabs have become standard on most larger excavators. In many cases, this air-conditioning is comparable to that found on top-end automobiles with such features as set-and-forget temperature controls. Some have pressurized cabs to keep out external air and dust for more efficient cooling and cleaner air filters.

Operator Comfort—Whole-Body-Vibration: The lack of shock absorbers between the cabin and the undercarriage means that the vibration and dynamic impacts created by mounted hammers/breakers, drillers, pile drivers, and shocks from the soil-tool interaction will continually affect the body of the operator. As a result, lower-back pains are traditional complaints of excavator and loader operators. New air-suspended seating systems have dramatically reduced that occupational hazard, and cabs that are mounted on silicone-filled shock absorbers reduce the vibrations reverberating through the frame. Other opportunities are cushioned boom cylinders that dampen shocks coming from the backhoe arm.

S4.28 Whole Body Vibration Study

Operator Comfort—Healthy Body Posture: Awkward postures while operating construction equipment are due to improper cab design and work procedures. The checklist for evaluating cab design of construction equipment (NIOSHTIC-2 No. 20024092 and on booksite) stresses the need for good posture when it includes questions related to (1) adjustability and stability of the seat, (2) back and lumbar support, (3) adjustability of armrests, and (4) reach of levers and pedals.

Some of the features that are becoming standard are options to steer the excavator either by hand or by foot pedals, by joystick or by levers. The entire control console can be moved forward and backward for the best foot position or reach. The position of the joysticks can be changed, as well as the many parts that make a seat. Even gender differences are being addressed as more female operators enter the workforce.

Operator Comfort—Engine Noise: The engine of an average backhoe creates 84–93 dB (decibel). OSHA recommends wearing ear protection if the noise reaches an average of 85 dB during an extended period of time. Of course, during the operation of hammers or pile-drivers, the noise level increases. Techniques to mitigate noise level inside the operator cabin are decibel-deadening, quieting noisy engines with measures inside the engine, adding better noise barriers around, and using tighter door seals.

Operator Comfort—Visibility: A better insulated and protected operator depends more and more on visual clues from both inside and outside the cab. Visibility of the work area is being addressed in various ways: (1) decreasing the size of cab posts and adding more non-reflective glass, (2) rearview video to present the back side of the backhoe, (3) adjustable seats, and (4) adjusting the location of the cab.

Operator Comfort—Visual Display Terminals (VDT): Providing the operator with more data and information about the condition of the equipment and providing simpler tools for interacting with the machine demand improved means of communication. Taking clues from the computer game industry, many manufacturers have started to install ruggedized VDTs to present data in graphical form to the operator.

Operator Comfort—Motion Control: Mechatronics created the most drastic visible change in the operator cabin. Depicted in Figure 4.38 is a common operator seat with arm rest and joysticks. Also visible in the front is the safety lock (arm pointing down) that has to be in horizontal for the controls to engage.

The fine-tuning of hand motion was paralleled by the introduction of electro-proportional valves. Different from on-off valves, which are either totally open or totally closed, electro-proportional valves open proportionally to the amount of an electronic input such as volts. Matching the two means that the farther one moves the joystick to raise the boom or curl the bucket, the wider the hydraulic valve opens and the faster the corresponding cylinder moves. By contrast, moving the joystick just a small amount lets only very little oil flow through the hoses, allowing very slow movement of the boom or stick. Furthermore, by allowing the operator to switch from one to another control scheme (e.g., arm to track), basically all actuators can now be operated from the two joysticks. Almost endless are the possibilities of augmenting the operator controls by automatically adapting the control signals coming from the joysticks. Fuel efficiency can be improved by keeping the controls within a set range, predetermined to be most economical. Integrating pressure sensors with the operator controls offers opportunities to let the operator "feel" the weight that the backhoe is carrying. Finally, novice operators about to make dangerous maneuvers could be stopped by blocking the electronics from sending signals to the valves.

Figure 4.38 Ergonomically Designed Operator Controls with Air-Conditioning and Damped Seats

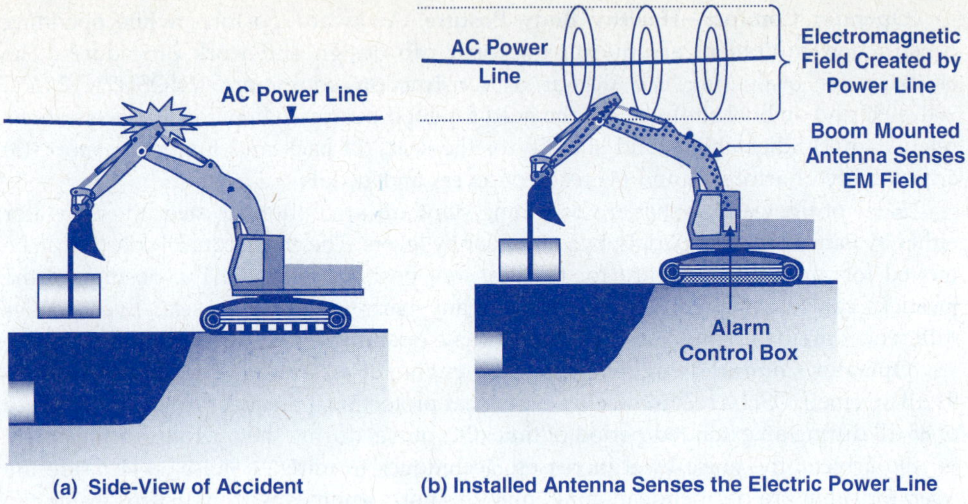

Figure 4.39 Strategy to Prevent Electrocution

4.8.3 KEEPING OPERATORS SAFE

As the accident files reveal, in addition to persons around the large equipment, operators are also in constant danger of accidentally getting hurt. Two very different approaches to improve the safety and health of operators are quickly discussed.

Preventing Electrocution: Figure 4.39 depicts the situation that led to Accident 6. While all eyes were focused on getting the pipe safely lowered into the trench, the stick touched the power line directly above the trench. The accident occurred despite several preventive steps that had been implemented and despite the fact that the crew had laid many hundreds of feet without a problem.

While the boom and hoist-lines of cranes have a higher probability of reaching power lines, this accident shows that construction equipment, even ladders, that extend more than 20 ft (6.1 m) into the air are at risk. SIGALARM provides a flexible attachment that is attachable to the arm of the excavator. Serving the function of an antenna, it detects the electrostatic field of an AC power line within a safety zone, as shown in Figure 4.39b.

Remote Control: This safety-related approach was also made possible by the introduction of electronic controls. Since the requirement of having an operator move mechanical levers to control the valves, the person could be separated from the equipment, only connected via an umbilical cord consisting of electric wires leading to the hydraulic valves.

Today the cumbersome wires have been replaced with wireless interfaces. The operator can be positioned in a safe place with good visibility. By adding video-audio feeds, the operator could in fact be sitting in a different room that would protect him even from the noise of potential explosions.

4.8.4 SMART DEVICES THAT SHOW OPERATOR POSITION

The term *stakeless excavation* has been coined to describe a situation where excavator operators will not need presurveyed stakes that tell them how deep they have to dig in a given location. Knowing where the excavator is in space provides many interesting opportunities.

Trenching in Line and on Grade: Traditionally, surveyors were employed to work with the trenching crew to ensure that the pipes would be installed with the proper grade. For two decades, that labor-intensive approach has been replaced with the help of a laser that produces a laser beam that hits the center of a target if the pipe is where it should be.

Most recently, the beam laser has been supplanted by a swirling laser, creating a laser plane above the ground and a laser receiver/target that is mounted on the backhoe stick. The receiver shows if the arm is above or below a target, using visual signals.

CAD-Integrated Excavating in 3-D: Most excavation outside a trench requires the operator to create surface contours that are not in a plane. In other words, the depth and the angle of the cut depend on the position of the excavator. The opening of the Global Positioning System (GPS) for civilian use by the U.S. Department of Defense around the world established a new platform for construction equipment to triangulate their position with the help of signals from line-of-sight satellites. With the use of two GPS stations, accuracy within the inch/centimeter range is achievable. When the use of GPS is not possible, total stations are used instead. Since total stations (see Chapter 5) are able to follow a moving line-of-sight target they can be trained to follow a prism mounted on and excavator and measure its position at all times. All the different positioning systems utilize the principle schematically illustrated in Figure 4.40. The main problem with developing spatially integrated systems using line-of-sight is the fact that the points of interest have to "submerge" into the underground, only re-emerging for emptying the bucket. One way to solve this problem is to equip the backhoe arm with angle encoders which feed their data, angles α, β, and δ to a processor that is capable of calculating the position of A, B, C, the cutting edge, and the bucket angle relative to any point on the upper-structure. The two antennas or prisms mounted on the back of the backhoe give it access to the world coordinate system x, y, and z. Since either the GPS or two total stations provide the capability to measure these two points in real time while the angle encoder tells the location of the cutting edge relative to these two points, it is possible to compute its x, y, and z coordinates. As research has shown, there is only a small step to integrating this information with the CAD design of the dig on a monitor in the cabin of the operator (see booksite). If the exact location of underground utilities is known, this information could automatically alert the operator or even prevent him from getting too close with the bucket.

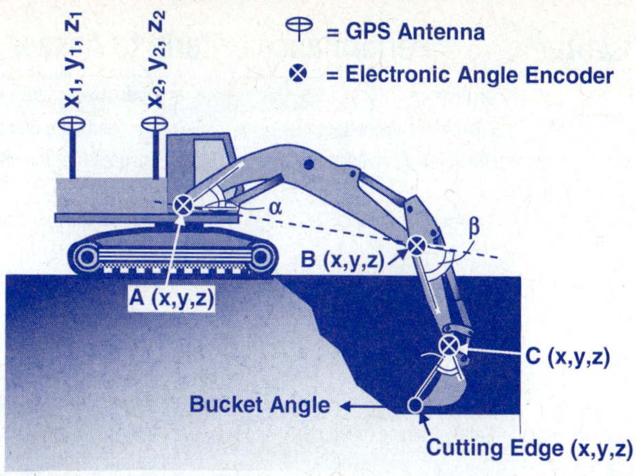

Figure 4.40 Schematic of a Spatially Integrated Backhoe Excavator and Real-Time CAD Representation

Tracking Equipment Location

Backhoe loaders are the most commonly stolen construction equipment. Satellite tracking and telemetry have changed all this. Called geofencing, this GPS-based technology allows a contractor to define a virtual fence around an operating zone within which the equipment is expected to stay. As soon as equipment crosses the "fence," the systems sends automatic messages to the owner, police, and so on, who can find its location even while in transit. Similarly, automatic machine function monitoring allows the generation of performance reports. For example, it could inform automatically about how many feet of pipe were installed so far. All this information is based on the hourly utilization of specific components of the equipment or its changes in location.

4.9 What Certifies a Skilled Backhoe Operator?

Due to their job on the controls of a central piece of equipment on a construction site, as well as their up-close position to oversee the activities in front and soon around

Table 4.13 Performance Criteria to Assess a Backhoe Operator

		Criteria	Operator Activities Linked to Safe Productive Operation
Job Planning	1.	Equipment selection	Verifies that a backhoe has the capacity to perform the necessary digging and lifting tasks.
	2.	OSHA rules in workzone	Ensures that the site layout provides sufficient access and protections according to OSHA 29 CFR 1926.800 (Excavation).
	3.	Safety equipment	Checks the availability of appropriate safety equipment (e.g., ladders, barricades, trench-boxes).
	4.	Line of authority	Clarifies the name of the *competent person* and person who is authorized to start the work.
	5.	Notification of One-Call	Follows OSHA 29 CFR 1926.651.
Preparation	1.	Walk-through	Reviews utility markings with available as-built drawings and layout of new job. Establishes action plan to protect against overhead power lines (OSHA 29 CFR 1926.550(a)(15)). Finalizes plan of work with authorized/competent person.
	2.	Safety training of laborers	Interviews workers about their knowledge of safety procedures and attitudes. Initiates additional training if necessary.
	3.	Inspection of equipment	Inspects the delivered equipment including maintenance record.
	4.	Workzone safety installations	Inspects barricades around the worksite and the backhoe (back-swing).
	5.	Inspection of add-on devices	Checks that all the devices are working properly.
Operation	1.	Enforcement of safety standards	Insists on the use of safe procedures from the beginning.
	2.	Setup of equipment	Travels at appropriate speed, and levels the machine, activates outriggers, and ensures that attachments are secured.
	3.	Equipment use	Avoids abusing the equipment and stays within load capacity.
	4.	Energy efficiency	Moves material the shortest distance without rehandling and wasteful machine idleness. Overall smooth operation.
	5.	Ergonomic and drug-free workplace	Stays clear of risky behavior caused by work induced fatigue and use of drugs.

the machine, backhoe operators play multi-faceted roles. They have to plan a job and the use of the appropriate backhoe, deal with One-Call, run and maintain the equipment optimally, and most importantly ensure the safety of the operation. Table 4.13 organizes some of the main activities that a master operator performs as part of his or her job.

S4.30 Trenching Excavation Training
S4.31 Equipment Operator Competency Tests
S4.32 Competencies for Heavy Equipment Operations

4.9.1 APPROACHES TO ASSESSMENT

Federal safety rules and regulations presently don't require any license or certification for operators of backhoes. Although some unions and larger employers do offer in-house training and certification for backhoe operators, there is no prerequisite requiring such documentation.

The Army Corps of Engineers has regulations that are valid for its own people as well as for contractors. The main article related to excavators is published with the Publication Number: EM 385-1-1 Section 25 pg. 489.

S4.33 Hand Signals Backhoe

Something that every operator should know, of course, is the universal hand signals (Figure 4.41).

An abbreviated test for backhoe operator could look as follows:

1. OSHA cites failure of the operator to provide swing radius barricades. True__, False__
2. Hazards associated with the use of hydraulic breakers include ___ and ___.
3. The excavator operator must make sure that nobody is within the _____ of the machine.
4. Clearance between the farthest reach of the excavator and power lines should be _____ feet with an additional ________ __________ for each 1000 volts over 50,000 volts.
5. OSHA issues _____ ______ for excavators used incorrectly for lifting objects.

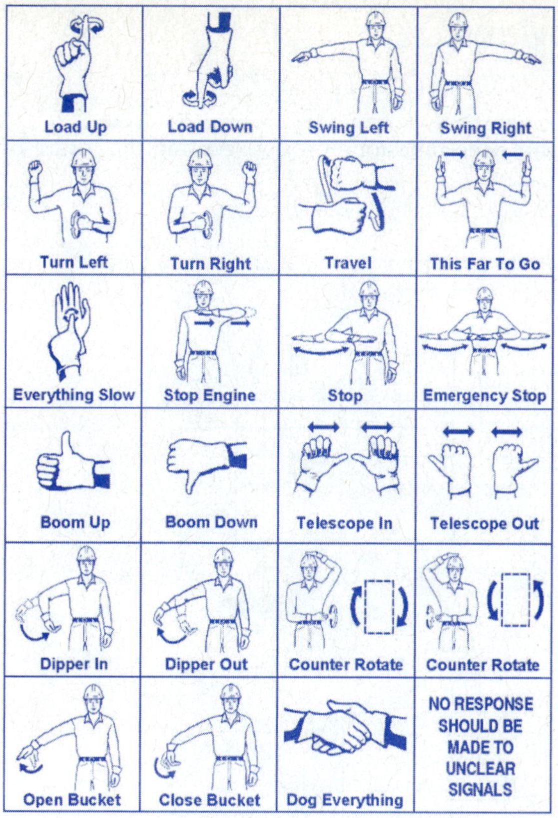

Figure 4.41 International Hand Signals to Communicate with Excavator Operators

1. T; 2. flying debris and dust; 3. swing radius; 4. 10, ½ inch; 5. crane citations

The National Center for Construction Education & Research offers courses that address a series of pertinent issues such as:

- Basic terminology and types of equipment and their uses.
- Operator responsibilities and the characteristics of a good operator.
- Purpose of signs, tags, barricades, and lockout/tagout devices.
- Guidelines for safe operation and maintenance of heavy equipment.
- Components and controls on a typical backhoe.
- Prestart inspection, startup, maintenance procedures, operational movement, and shutdown for heavy equipment.
- Accessories used on a backhoe.

Websites for the Organizations Listed Above

- UW Occupational Noise Exposure in the Construction Industry: http://staff.washington.edu/rneitzel/index.htm/
- UW Occupational Hearing Loss Study http://depts.washington.edu/occnoise/reports.html
- OSHA Construction and Hearing Conservation: http://www.osha.gov/dts/osta/otm/noise/standards.html#construction
- NIOSH Work-Related Hearing Loss: http://www.cdc.gov/niosh/topics/noise/
- Build It Smart Hearing Conservation: http://builditsmart.org/hearing.htm/
- Construction Noise Control Partnership: http://www.lhsfna.org/html/_noise_ partnership.html

Additional Reading on Noise Exposure and Hearing Conservation in Construction

- Suter AH. 2002. Construction noise: exposure, effects, and the potential for remediation; a review and analysis. *AIHA J* 63: 768-89.

4.10 Chapter Four Review

THE EQUIPUZZLE

The letters inside the circled boxes name the milestone project securing the future of excavators.

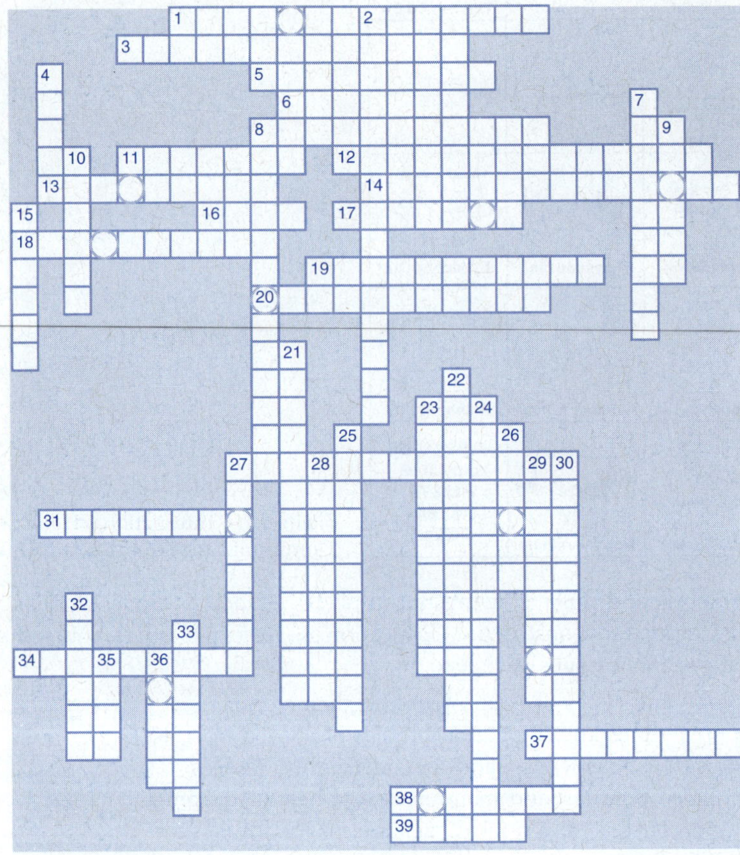

Across

1. The most accurate way of identifying the location and depth of a buried utility (two words)
3. Linear actuator that powers and rotates the stick (two words)
5. Describes a method of excavation that uses GPS and laser equipment for spatial control of the operation
6. Attachments for a backhoe excavator
8. Add-ons for backhoes
11. Accidental failure of a cross brace
12. Unique tractor on wheels that has outriggers/stabilizers, can operate as a front-end-loader, and has a digging arm with 180 degree swing
13. A safe location to attach a chain or sling to lift a pipe or manhole with backhoe
14. A type of laser that is specifically designed for laying pipes
16. The element of the backhoe arm that is attached to the backhoe swing frame
17. Action of a backhoe that minimizes the risk of a cave-in
18. Describes an undercarriage that can be widened and retracted
19. Filling and compacting around a pipe that has been laid in a trench
20. One of the first states in the U.S. that passed a "call before you dig" law
27. Other term for crowding force (two words)
31. A force that good operators avoid by shaving layers of soil when filling bucket
34. The maximum distance between the rotational axis and the rear of a backhoe (two words)
37. Status of an excavation face ready to cave in
38. Defines the "killer" incidents that can happen during trenching
39. Special add-on to a bucket that enables the operator to grasp long objects

Down

2. The office that has to be called before a dig, in most states by law, to have buried utilities located
4. Standard abbreviation for fine-grained sandy silts
7. Indicates that an excavator is powered by fluid pressure
8. A term used for the first wheeled backhoes that could travel on the road like cars (two words)
9. The status of a backhoe bucket that is related to a standardized volume
10. Special add-on to a bucket helpful when excavating hard material.
15. Color marking or the ground that indicates a buried gas-oil-petroleum pipe
21. Important device at the end of a backhoe stick if one uses many different attachments (two words)
22. Term used to describe the dangers of working inside a narrow trench with ladders for entrance and exit (two words)

23. Describes a backhoe-loader stick able to provide an extra reach

24. A general term that means "all the methods that can be used to provide safety for the laborers working in a trench" (two words)

25. Shoring system that is not continuous (two words)

26. A narrow excavation

27. Method to protect against cave-ins by creating multiple vertical walls

28. Defines a hydraulic system that is designed to power add-ons

29. Time used to load one bucket of soil into truck (two words)

30. Possible cause of death after a deadly accident of an excavator working in the vicinity of power lines

32. A trenchbox

33. Needed to safely attach a heavy pipe to the hook of a backhoe bucket

35. A layer of backfill material

36. Members of a shoring system that are placed parallel to the face

JOURNALING QUESTIONS

1. What was covered in class this week that you find most useful and why? How does it relate to what you already know about construction, physics, statics, history, and so on?

2. Sketch top and side views of the situation that led to accident #6. What do you think the judge decided? Why or why not was Concrete responsible?

3. Operator skills seem to be critical in achieving the goals of a trenching operation. Explain the different ways an operator can influence the outcome of such an operation positively or negatively.

4. The history of excavator technology is rich with great accomplishments. As an aspiring construction engineer-manager, what are the historical developments that you are most proud of?

5. Give your personal comments, evaluations, questions, insights, and poems about excavators (do you know there is a cool song about the skill of an excavator operator?).

TRADITIONAL HOMEWORK

1. Mark is operating a backhoe with the following specifications.

He is using a bucket with a heaped capacity of 1.1 m^3 (1.44 yd^3) to excavate wet clay and loading it into trucks that position themselves behind the excavator for loading. You did a time-and-motion study of Mark's present loading operation and came up with mean and standard deviations (STD) shown in Table 4.12.

Table 4.12 Result of Time-and-Motion Study in Seconds

Motions / Operator	Fill Bucket		Swing + Position		Empty Bucket		Swing Back	
	Mean	STD	Mean	STD	Mean	STD	Mean	STD
Mark	11.5	3.5	13.7	3.1	3.5	1.3	7.3	2.6

Back hoe dimensions - Standard dipper	
H - Digging depth	4386 mm
J - Loading reach at maximum height	2810 mm
K - Operating height	5495 mm
L - Reach from swing post at ground level	5810 mm
M - Reach from rear axle	1148 mm
N - Digging depth SAE 2.44 m flat bottom	4014 mm
O - Digging depth SAE 610 mm flat bottom	4359 mm
Bucket digging force	59.1 kN
Dipper digging force	38.8 kN

(a) What is the average volume of clay in the bucket?
(b) How heavy is the clay in the bucket?
(c) What is the mean cycle time?
(d) How many tons of clay will Mark excavate in an hour if he has the proper trucks in position for an average of 45 minutes per hour?
(e) What is the production in tons per hour if he has a bad hour (37.5% above mean cycle time)?
(f) What is the production in tons per hour if he has a terrible hour (42.5% above mean cycle time)?
(g) What rearrangements or modifications would you investigate to help Mark dramatically improve his production?

2. Calculating the max amount/volume of a material an excavator can have in the bucket when operating on an incline.

3. What size boom cylinder is needed to lift a pipe stuck in the mud (= double the weight) with the backhoe arm at X feet distance? Will the backhoe-loader tip over?

OPEN-ENDED PROBLEMS

1. Accident #1: What do you expect that the OSHA accident report listed as being neglected? What would you have done to avoid the accident?

2. Accident #2: What actions did the employer neglect to do or allow the workers to do that contributed to the accident and were covered by OSHA rules?

3. Accident #4: List the recommendations that you think the OSHA investigators made to avoid a similar accident.

4. Assume that you just purchased a backhoe excavator built in 1999 that is still in pretty good condition. You plan to buy several attachments that would protect the machine from being damaged by excessive and abrupt forces. In addition, you would like to give the operator a tool that would tell him when the bucket reaches the desired depth of a trench. What kind of sensors would you buy, and where would you install them on your backhoe? Describe and sketch the technology that would give the operator the tools to avoid ever over-excavating the trench.

5. The operator in Accident #4 died because several things went wrong. Elaborate on the different failures that caused the accident to happen and killed the operator (treat them as two separate issues). Design at least two separate electromechanical devices to prevent similar accidents.

6. Team up with a friend and use a watch, paper, and pencil while observing a backhoe operation (any type will do). Sketch the situation or take a picture or video. Measure durations of subtasks and breaks before computing average times of a digging cycle. Also write down the durations of

TIME-STUDY

Project:	**Date:**
Day:	**Weather:**
Crew Activity:	**Temp:**
Humidity:	
Recorder:	**Location:**

Cycle	**Start Min:Sec**	**Duration of Excavator Digging Task Sequence**								
		Extending Arm/Bucket	**Preparation to Dig**	**Filling Bucket**	**Lift and Swing**	**Waiting for . . .**	. . .	. . .	**Break**	**Cycle Time**
		Time in Sec	Time in Sec	Time in Sec	Time in Sec	Time in Sec	Time in Sec	Time in Sec	Time in Sec	Seconds
1	:									
2	:									
3	:									
4	:									
5	:									
6	:									
7	:									
8	:									
9	:									
10	:									
11	:									
12	:									
13	:									
14	:									
15	:									
16	:									
Tot Time (sec)										
Average Time (sec)										
Percent at Task										100.0%

nonproductive times such as waiting for a truck, breaks, and so on. After you have at least 16 cycles, calculate average and standard deviations for each subtask and the cycles with or without nonproductive times. What is the productivity of the operation? What do you recommend doing to improve productivity?

Appendix A: Steps to the Safe Operation of a Loader/Backhoe

The loader/backhoe is one of the most widely used pieces of construction equipment. But, as with any equipment, proper operation and maintenance are vital steps to job-site safety. Here are some basic tips for safe operation:

1. Read the operator's manual thoroughly and refer to it often. Store it where it can be found for easy reference.
2. Dress for safety. Wear a hard hat, safety glasses, safety shoes, and snugly fitting, comfortable clothing.
3. Inspect the machine each day. Perform these steps:
   - ☐ Look over the operator's area. Make sure it isn't cluttered with junk, and isn't muddy or greasy. Check the engine oil level and the hydraulic fluid level. Drain any water or sediment that has collected in the fuel system.
   - ☐ Take a walk around the loader/backhoe. Look for missing, broken, or loose parts. Make sure the tires aren't damaged or under-inflated.
   - ☐ Climb into the operator's area carefully and fasten your seat belt. The seat belt together with the rollover protection system could save your life in the event of an accident.
   - ☐ Some loaders/backhoes have left and right brake pedals. Any time you travel more than 5 mph, make sure the pedals are locked together. Using only one pedal at road speeds will cause the machine to veer sharply.
   - ☐ Always carry the loader bucket low for visibility and stability. Raise the bucket as you approach the truck you're loading.
4. When you are going to use the backhoe, follow these steps:
   - ☐ Swing the seat around so that you're facing the direction of the backhoe. Never operate the machine from behind your back!
   - ☐ Lower the stabilizer pads and level the machine.
   - ☐ If the ground is soft or muddy, place planks under stabilizer pads.
   - ☐ Before you begin digging, make sure all workers are clear of the area.
   - ☐ Check with the electric, gas and phone companies first. Know the location of utility lines before you dig!
5. To ready a unit for transport, bring the boom in and engage the locking plate. Install the swing lock pin, retract the stabilizers, and you're ready to go.
6. A safety-conscious operator is an efficient operator. By carefully studying the owner's manual and following these safety steps, you'll get the maximum benefit from your loader/backhoe.

Reprinted with permission from Nevada Milepost, Fall 1999.

Appendix B: Safety Quiz for Backhoe Operators

1. A serious danger in using a crawler excavator is having the machine fall into the hole it is digging. True__ False__
2. OSHA cites failure of the operator to provide swing radius barricades. True__ False__
3. The driver of a truck that is being loaded should get out of the truck. True__ False__

4. The tool should always be in the air when the machine is parked. True__ False__
5. Hazards associated with the use of hydraulic breakers include ______ and ___.
6. The excavator operator must make sure that nobody is within the _________ of the machine.
7. Clearance between the farthest reach of the excavator and power lines should be ______ feet with an additional ________ for each 1000 volts over 50,000 volts.
8. Before digging, you should always call the ______ service for the area.
9. Excavator operators should familiarize themselves with the local ____________ for identifying underground utilities.
10. Digging near lines requires some _______ probing.
11. Standardized hand _____ must be used between operators and other job personnel.
12. OSHA issues ______________ for excavators used incorrectly for lifting objects.
13. Any accident is ultimately blamed on the _________.
14. Safety concerns include using the rated _______ size and the condition of _______ rope and rigging.
15. Using a side sweep of the bucket for smoothing and breaking ground is a safe and effective operating technique. True_____ False______
16. A crawler excavator is inherently stable and poses little danger when digging around water holes. True______ False_____
17. A good operator can operate machines made by virtually all manufacturers. True______ False______
18. When working on side hills, it is safest to put the material (a) uphill, (b) downhill from the trench.
19. Removing material from under the upper track and depositing material under the lower track when working on a side hill is known as __.
20. When working on high slopes, the following procedures will help secure the machine from tipping over:
(a) ___________________________________,
(b)___________________________________,
(c)___________________________________.
21. Lightning is not a concern to the operator of the excavator. True____ False____
22. With most excavators, it is safe to swing the load over a truck or a person on the ground. True_____ False_____
23. Short slings on a load decrease the amount of swing, but may also decrease the capacity of the sling. True______ False_____

S4.34 Heavy Equipment Operation Quiz

chapter five

Dozers, Scrapers, and Graders

The roots of the technology discussed in this chapter began with the first people who settled down and worked the earth to grow food. Increased sophistication in agriculture led to development of plows and scoops pulled by oxen or horses, providing the power to clear and till the land. Similar tools were needed to build canals for irrigation, and later, dams, roads, and highways. As will be shown, critical innovations were the tracks and the diesel engine that enabled the economic construction of gigantic projects all over the world. Modern productivity-enhancing devices include lasers, GPS, and automatic servocontrols. Table 5.1 shows the main topics to be addressed in this chapter.

Table 5.1 Main Topics Covered in This Chapter

Construction			Planning & Control		
Equipment	**Processes**	**Enabling Tools**	**Managerial-Engineering Factors**	**Rules & Standards**	**Modern Tools**
✓ Fresno Scraper	✓ Dozing	✓ Ripper	✓ Grade Resistance	✓ Roll-Over Protection System (ROPS)	✓ 2-D Laser Plane
✓ Grader	✓ Scraping	✓ Blade—Straight	✓ Rolling Resistance	✓ SAE J1040	✓ 3-D Laser
✓ Bulldozer	✓ Hauling	- Cushion	✓ Rimpull	✓ OSHA 1926.1000	✓ GPS
✓ Scraper	✓ Ripping	- Tilt	✓ Traction–Slippage	✓ ISO 2631 (WBV)	✓ Pulsed Laser Beam
✓ Skid-Steer	✓ Cutting	- Universal	✓ Drawbar Pull		✓ Automated Blade Controller
✓ Truck	✓ Hauling	✓ Moldboard	✓ Grade		✓ Servo
	✓ Loading	✓ Scarifier	✓ Coefficient of Traction		
	✓ Grading	✓ Bowl	✓ Net Vehicle Weight		
	✓ Back-Tracking	✓ Ejector	✓ Gear Ratio–Speed		
	✓ Spreading	✓ Apron	✓ Dozer Production		
	✓ Maneuvering	✓ Diesel Engine	✓ Dozing Cycle		
	✓ Pushing	✓ Crawler Tracks	✓ Seismic Investigation		
	✓ Pitching Blade	✓ Undercarriage	✓ Wave Propagation		
	✓ Tilting Blade	✓ Geophone	✓ Cohesion–Adhesion		
	✓ Push-Loading	✓ Spinning Laser	✓ Rippability		
	✓ Slipping	✓ Laser Targets	✓ Bank Cubic Yard (BCY)		
	✓ Rolling Soil	✓ Total Station	✓ Loose Cubic Yard (LCY)		
		✓ Bogies	✓ Job Condition		
		✓ Grousers	✓ Operational Efficiency		
			✓ Material Properties		

5.1 The Family History of Dozers, Scrapers, and Graders

5000 BC	**Animal-Powered Plows**	The ancient Egyptians and Sumerians used plows and scoops powered by cattle, oxen, donkeys, and camels to work the earth. No new substantial technology was added during Roman and Medieval times related to nonmanual moving of soil.
200 AD	**Invention of Wheelbarrow**	The Chinese invented the wheelbarrow around 200 AD, which found its way to Europe around the thirteenth century. The Canal du Midi in France (17th century) and the 363-mile Erie Canal, started in 1817, were still built mainly with spades, wheelbarrows, and animal-drawn plows and straight push boards.
19th Century	**Fresno Scraper**	In the late 1900s, the need for canals in California challenged the Scottish immigrant James Porteous, from Fresno, to come up with ingenious improvements to the buck scraper. He turned the scraper blade into a C-shaped bowl and added a horizontal cutting edge and curved sled runners that facilitated the controlled deposit of the soil. Between 1884 and the advent of tractor-drawn scrapers in the 1910s, thousands of "Fresnos" were used for excavating canals and ditches, leveling land, and grading for roads and railroads. It dominated the field for 40 years, riding first on runners, then wheels.
1880	**Horse-Drawn Road Graders-Levelers**	The first road graders date from 1880s. They were, of course, horse-drawn, but they were remarkably similar to today's very sophisticated motor graders. Called road draggers, scrapers, or levelers, they were used to build, grade, and maintain dirt and gravel roads. Starting with 1910, they became self-propelled when the horses were replaced with a motor that powered the front wheels.
1904	**Invention of the Track System**	On July 23, 1904, David Roberts, working for the British firm Richard Hornsby & Sons, received his first patent: ". . . two pitched chains of links and pins with cross bars and blocks of metal and wood make contact with the ground and are passed around the front and rear sprocket wheels, one on each side of the vehicle and form a track. . . . Steering may be accomplished by varying the speed of the driving sprocket wheels on either side of the vehicle." The term *caterpillar* was coined by British soldiers who tested the first vehicles produced by Hornsby to pull artillery pieces. Around 1914, the patents were sold to Holt Manufacturing Company of Stockton, California, which in 1924 merged with L. Best Gas Traction Company to form the Caterpillar Tractor Co. In 1931, the first Diesel Sixty Tractor rolled off the assembly line.
1923	**Invention of the Bulldozer**	If you "bulldozed" someone in the 19-century, you coerced him, often with firearms. Thus, "bulldozing" came to mean using strong force to push away obstacles. This short term became convenient when during the 1920s a new device took form. The young farmer James Cummings from Wyoming thought that the backfilling of pipeline trenches could be done much more efficiently with his tractor than with mules and the Fresno. He built his first bulldozer in 1923 and promptly received a backfilling contract for a pipeline job. During the 1920s, many different inventors worked on creating more operable dozer blades.
1930s	**Motorgrader**	Finland and Sweden face long, cold winters followed by road maintenance every spring. Several companies that later folded into today's Volvo Group, Akerman and Munktells, spearheaded the design of motor graders for easy operation. Since so many different elements needed to be adjusted, hydraulic fluid power found an early application. Scarifiers (a series of short metal teeth) were commonly added behind the front wheels. They could be lowered and allowed the operator to scarify or scratch up the old uneven surface before grading it smoothly with the blade.

1930	Motor Scraper	During the 1930s, Robert Gilmore LeTourneau built the first machines that integrated the single-axle bowl scraper with a single-axle wheel tractor. With hydraulic fluid power still in its infancy, the LeTourneau scrapers used cables and pulleys to lower and lift the bowl and to power the ejector that actively pushed the soil out of the bowl. Today's wheel tractor-scrapers have not only hydraulic actuators, but also a separate engine that powers the trailing wheels, and an elevator mechanism that lifts and mixes the cut material, while the operator is provided with air conditioning inside a noise/dust-insulated cabin that is cushioned from the rough ride across uneven travel surfaces.
1940s	Skid-Steer Dozers	Beginning in 1943, LeTourneau developed large rubber-tired dozers, called Tournadozers, to speed up the push-loading of scrapers. They eventually reached a speed of 20 ml/hr(31 km/hr) and a weight of 30,000 lb (15, 400 kg). They were steered by slowing or locking the one side of the wheels, a method that is now called "skid-steer." Today they are mostly articulated and are found in mining, where they are loved for their speed and traction.
1965	Crawler-Transporters Hans and Franz	Although it does not push or pull vehicles in soft soil, the crawler-transporter belongs to the family of tracked vehicles that are connected to David Roberts and the Holt Manufacturing Company. Designed by Bucyrus International and built by the Marion Power Shovel Co. two crawlers, Hans and Franz, were delivered in 1965. They transport the rockets along the 3.5 ml (5.6 km) Crawlerway on a 5% incline from NASA's Vehicle Assembly Building along the to Launch Complex. One track consists of 57 shoes, and one shoe weighs approx. 2000 lb (900 kg). The entire crawler is 6 million lbs, or 2,700 tons.
1980s	Laser Plane-Controlled Bulldozer Blade	Bulldozers have become high-tech workhorses that are omnipresent. For one thing, the blades that are mounted on tracked tractors come in many different shapes and sizes and can be hydraulically controlled to various orientations. One-shank or multi-shank rippers have been added to the large 300 hp pieces of steel to break up rock to be pushed or loaded into scrapers or trucks. Torque converters, hydrostatic transmissions, a multitude of electronic sensors, GPS or laser systems, and wireless communication support the operator of these giants. The comfortable cabin has a built-in roll-over protection system.
1990s	High-Tech Graders	In a straight configuration, a motor grader of today operates much like a horse-drawn "Fresno" 100 years ago. On the other hand, the additional capabilities offered by articulated graders with "creeping" speed controls challenge any equipment operator. The moldboard/blade can be articulated into almost any orientation via the load-sensing hydraulic cylinders and a rugged direct-drive ring system. Most have front wheels that can be leaned, allowing compensation for the side forces caused by the angled moldboard pushing the soil. An articulated frame model has a hinge in the frame behind the cab. Rippers mounted on the back are used to loosen the surface before windrowing it with the moldboard.

5.2 From the Accident File

The following accidents are examples of what can happen. The underlying problems provide focal points for many of the discussions, and the reader will be again challenged to design solutions at the end to the chapter.

(1) Bulldozer kills operator: A bulldozer owner and a mechanic were having difficulty starting the diesel engine. They worked from the left side, the mechanic sitting and the owner standing on the track three to four feet away. As the mechanic tightened the main lead wire to the starter, the owner leaned into the cab of the bulldozer and pushed the start button. The two men did not realize the dozer was in reverse gear. When the owner pushed the button, the bulldozer jumped backward, throwing the mechanic off and throwing the owner underneath the track. The owner died at the scene from massive injuries to his pelvis and chest.

(2) Spotter run over by backing dozer: A male construction worker playing the role of a spotter was directing a truck driver to dump a load of gravel for a parking lot construction. At the same time, a bulldozer and a skid-steer loader were involved in spreading the gravel. While traveling backward toward the gravel pile, the track of the bulldozer struck the spotter on the back of his legs and rolled over his legs and torso. The operator, unaware of the run over, drove the bulldozer forward about 50 feet and stopped. The skid-steer loader, watching the accident occur, yelled warning messages but nobody heard him because of the noise. A police officer arrived approximately one minute after the 911 call and assisted in providing CPR. Emergency medical services (EMS) personnel arrived approximately two minutes after the 911 call and found, through examination and cardiac monitoring, that the victim had no signs of life.

(3) Operator crushed by overturning dozer: A male supervisor for an excavation construction company was loading a bulldozer onto flatbed trailer to return it to the company shop. Truck and trailer were parked with the right wheels on the sloped shoulder and the left wheels on the paved surface. The victim drove the bulldozer up two metal-and-wood ramps that led from the ground to a wooden "beavertail" platform, which was wet due to a light rain. While halfway on the platform, the tracks of the machine began to slide sideways toward the ditch and off the trailer. Being pinned beneath the cab structure the driver, was unable to jump off when it landed on top of him. EMS responders arrived, and elevated the bulldozer with air bags and blocks to remove the victim's body. The victim was pronounced dead at the scene.

(4) Motor grader runs over a surveyor: A surveyor and a laborer, both wearing high-visibility safety vests and hard hats, were checking the grades at a level intersection construction site. The weather conditions were clear, sunny, and warm. The road was completely closed to all traffic. Since the grader could not go over the manhole cover without damaging it, a second laborer assisted the grader operator by spotting it so that the manhole cover could be circumnavigated. The motor grader operator stopped as he approached the manhole cover, checked his side mirrors for any persons or any other machine that might be behind him, and then proceeded to back up, looking at the grade. When he was run over by the grader, the victim was bent down tying a ribbon on a stake with his back to the grader. The backup warning device was working properly.

(5) Overturning scraper kills operator: A male scraper operator working on parkway construction project was moving earth from a cut area to a fill area. As he was completing a U-turn into the cut area, the right-side tire of the scraper's tractor ran up on an elevated berm, causing the machine to slowly tip over to the left. As the machine tipped, the victim apparently abandoned the operator's cab, jumping to the ground, where he slipped and fell. As he was attempting to get up and retreat to safety, he was crushed by the top bar of the machine's rollover protective structure (ROPS). A local emergency medical service (EMS) responded in 15 minutes; however, because of the extent of the victim's injuries, no life-saving attempts were made, and the victim was pronounced dead at the scene.

5.3 Modeling Forces and Motions

Before we analyze some of the basic principles that underlie the effective utilization of these superb pieces of equipment, we should take a close look at the main forces that interact with the machine, as well as the main tool motions (Figure 5.1).

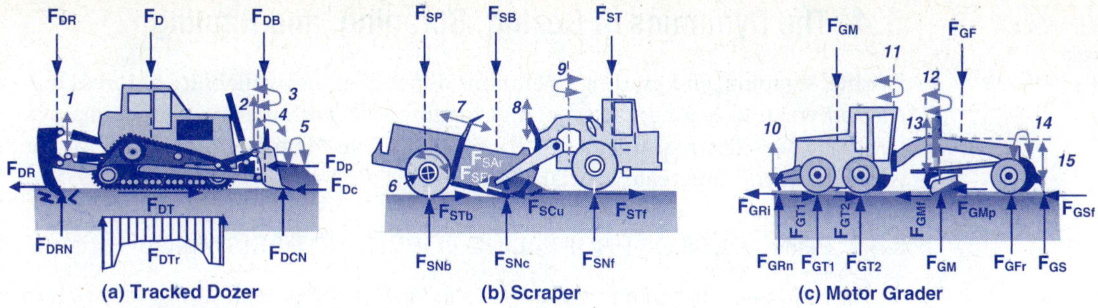

= Hydraulic cylinder

1= Ripper up/down
2= Blade up/down
3= Blade angling
4= Blade pitching
5 = Blade tilting

F_D = Weight of dozer tractor
F_{DR} = Weight of ripper attachment
F_{DB} = Weight of blade attachment
F_{DRN} = Normal reaction force to ripper
F_{DTr} = Normal reaction force to tracks
F_{DCN} = Normal reaction force to blade
F_{DR} = Forces caused by ripping
F_{DT} = Traction forces
F_{Dc} = Cutting force at blade
F_{Dp} = Force due to pushing soil

= Pan pivot point

6 = Bowl rotation
7 = Ejector
8 = Apron open/close
9 = Scraper Hitch - Articulation

F_{ST} = Weight of scraper tractor
F_{SB} = Weight of scraper bowl
F_{SP} = Weight of scraper back engine
F_{SNf} = Normal reaction to front wheels
F_{SNc} = Normal reaction force to cutting
F_{SNb} = Normal reaction force back wheels
F_{STf} = Traction force front wheels
F_{SCu} = Cutting force at cutting edge
F_{SFri} = Friction force inside bowl
F_{SA} = Adhension force at bowl
F_{STb} = Traction force back wheels
F_{SNb} = Normal reaction force back wheels

= Force vector

10 = Ripper up/down
11 = Articulation
12 = Moldboard angling
13 = Moldboard up/down
14 = Moldboard tilting
15 = Scarifier up/down

F_{GM} = Weight motor tractor
F_{GF} = Weight front frame
F_{GS} = Reaction to scarifier
F_{GFr} = Reaction force on front wheel
F_{GM} = Reaction force to moldboard
F_{GT2} = Reaction force traction wheel 2
F_{GT1} = Reaction force traction wheel 1
F_{GRn} = Normal reaction force for ripper
F_{GSf} = Horizontal force by scarifier
F_{GMp} = Push force by material on moldboard
F_{GT2} = Traction force on wheel 2
F_{GT1} = Traction force on wheel 1
F_{GRi} = Horizontal force by back ripper

Figure 5.1 Overview of the Main Motions and Forces for Three Earth-Shaping Pieces of Equipment

Header Problem 5.1 / Uphill or Downhill Loading?

VG4.1 Excavator Tracks

The French company Creuser Co. is planning the earthmoving operation for a new highway. One interesting problem that they are facing is the question how to excavate in section A-B. The subsurface in that area consists of several layers of soft soil, but harder material has to be expected. It has been found that it is more effective to use scrapers and a ripping dozer instead of backhoes and trucks. One last problem that is left to be studied is the loading procedure for the scrapers. There is a general rule that the scraper should load on the downhill slope, shown as option 1. This rule might not be optimal in this case as the scrapers would have to drive uphill with a full bowl, slowing them down severely. The alternative is to load while moving uphill after driving down empty. This decision is considered important as the distance between A and B is significant and the time spent in the cutting area critical for productivity. Creuser owns a fleet of Caterpillar standard 627G scrapers and a 340 hp Liebherr PR 754 Litronic dozer that they are planning to use.

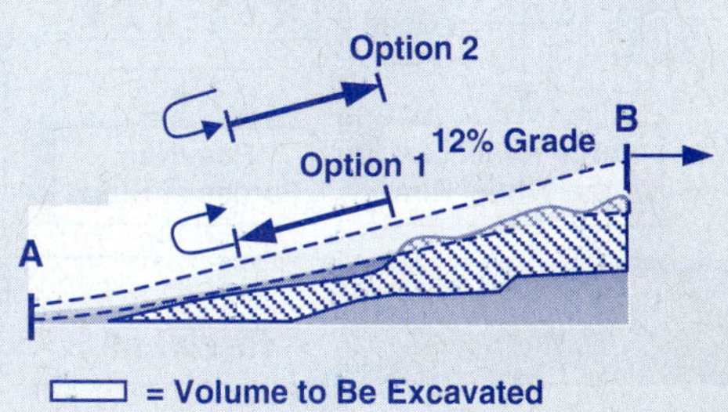

Question 1: *Is it possible for a 627G scraper to travel up the slope when full?*
Question 2: *Will the scrapers be able to load on the uphill slope on their own?*
Question 3: *What is the overall recommendation for how Creuser should proceed?*

S5.1 Dozer Liebherr PR754Litronic
S5.2 Scraper CAT 627G

5.4 The Dynamics of Dozing, Scraping, and Hauling

Dozing, scraping, and grading operations depend on the availability of forces that can be exerted when loosening, pushing, and loading soils with many different characteristics. Different densities require special efforts to remove them from their natural condition, while boulders can create special challenges.

5.4.1 PRINCIPLES OF PROPELLING FORCE AND RESISTANCE

Anyone who has driven a powered vehicle in snow or in sand dunes knows how easily one gets stuck even though plenty of engine power is available. Also, only a minimal amount of water on a hard road surface may prevent a fast car from slowing down when it applies its brakes. Similar observations can be made on an earthwork site where a clayey soil material has the same effect as packed snow. On the other hand, soft uncompacted soil acts like dry sand. The underlying physical principles that create a resisting force where the wheel or the track touches the surface that is sufficient for a powered vehicle to advance are summarized in the complex map depicted in Figure 5.2.

In order to propel a large mass of machinery, with its inertia and various friction-producing bearings, the equipment has to produce a driving force. The first indicator that tells us something about the "muscles" of a piece of earthwork equipment is the rimpull, the name given to the driving force that is available at the rim of a wheel. Indicated on the left side of Figure 5.2, the rimpull force is created by the powertrain and its range a function of the gear. We learned in Chapter 3 that the amount of torque available at the end of the powertrain depends on the horsepower that the engine is able to produce minus the losses in the transmission (*rule: the higher the gear, the faster the machine and the lower the producible torque at the wheel*). The rimpull is determined by measuring the torque **T** at the driving axle and dividing it by the radius **R** of the wheel base. As indicated, the

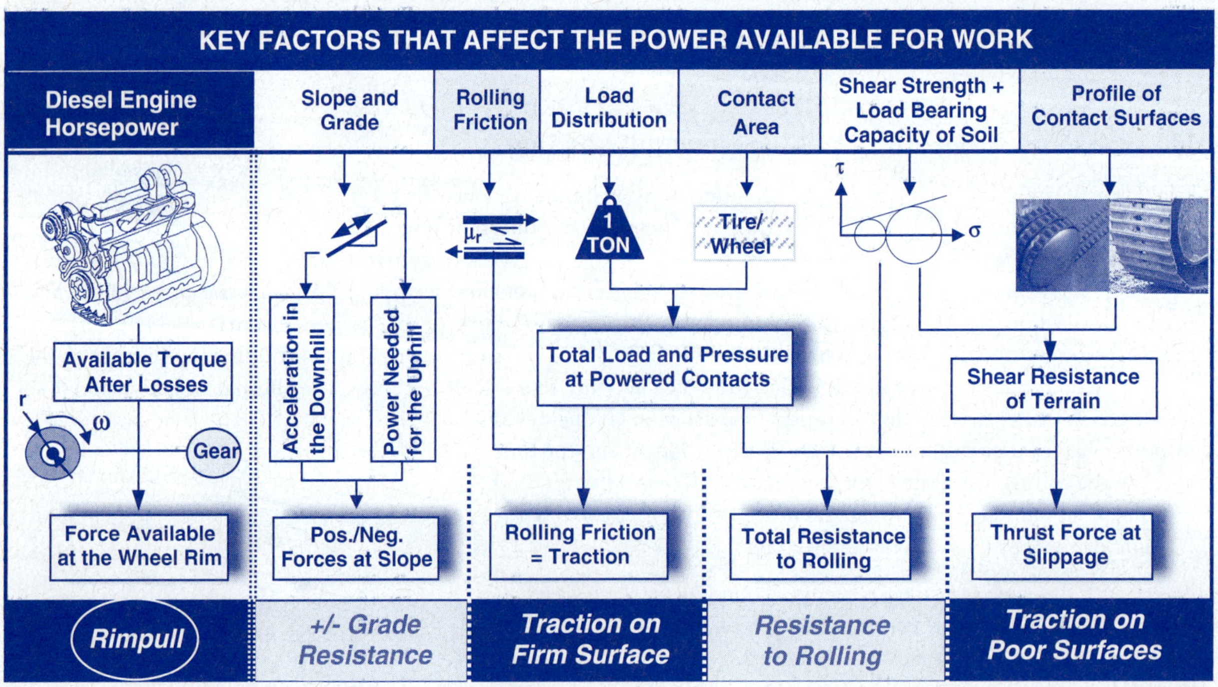

Figure 5.2 Relationships Between Power, Conditions, and Useable Forces

rimpull force has to not only accelerate the mass of the machine (**F_{drive}= m∗a**), but also overcome the frictions in the drive bearings and the resistance to rolling called *Rolling Resistance (RR)*. As Figure 5.3 suggests, a soft driving surface will require large forces to move on. Unless the engine of the equipment is too weak, there will be a force left over that can be used to pull or push extra loads. Called the *Drawbar Pull,* this remaining force is defined as

Figure 5.3 Tire Experiencing Large Rolling Resistance (RR)

$$\text{Drawbar Pull}_{\text{horiz.}} = \text{Rimpull} - F_{\text{drive}} - \text{Rolling Resistance (RR)}$$

Equipment involved in road construction is rarely driving across a horizontal surface. Instead, it has to drive uphill and downhill to pull and push soil material from one place to another. Overcoming the slope (degrees) or the grade (%) of the road requires or creates force since potential energy has to be produced when traveling uphill or is being produced when traveling downhill. Without adding power to a truck driving uphill, the creation of potential energy ($= \Delta\, m\, g\, h$) will lead to a reduction in kinetic energy ($= \Delta \tfrac{1}{2}\, m\, v^2$). Since m and g do not change, the speed of the vehicle will have to change, either negatively when the vehicle drives uphill or positively when going downhill. The adverse force that has to be overcome in the uphill grade is called grade resistance (GR) and is negative on a downhill slope. Thus, we can augment the function for the available drawbar pull for a horizontal surface as follows:

$$\text{Drawbar Pull} = \text{Rimpull} - F_{\text{drive}} - \text{Rolling Resistance (RR)} +/- \text{Grade Resistance (GR)}$$

The basic function for continuing the driving motion without acceleration is:

$$\textbf{Drawbar Pull} + \textbf{RR} +/- \textbf{GR} = \textbf{Rimpull} = f(\textbf{hp} - \textbf{Powertrain Losses})$$

In Chapter 3 we learned that the horsepower created in the engine ($1\,\text{hp} = 550\,\text{lb} * 1\,\text{ft}/1\text{sec} = 5{,}252\,\text{lb} - \text{ft} * 1\,\text{rev}/1\text{min} = 33{,}000\,\text{ft} - \text{lbs/min} = 746\,\text{N m/sec}$) is transferred, in the form of torque, to the driving axle via the many components of the powertrain. Having a large torque coming from a large engine, however, does not guarantee that a vehicle will in fact move forward. Tires may burn turning at high speed or digging a hole by ejecting the road material to the back. What is needed is a friction force sufficient to resist the rimpull at the tire. The French Charles Coulomb in the eighteenth century demonstrated that the resistance to movement by an object resting on a surface is dictated by its weight and a so-called coefficient of friction (μ), a factor that hinges on the exterior roughness of the object and the surface. However, we are not talking about a plane but a rolling wheel. Rolling friction is the "cousin" of the Coulomb's friction force and is called traction. While Coulomb was mostly interested in finding out what keeps surfaces from sliding (e.g., wall made of stone blocks), traction "connects" a wheel or a track to the driving surface by activating

Who Discovered Friction?

Leonardo da Vinci was the first to talk about a rule of friction: *the frictional force was proportional to the weight.* The mathematician Charles Augustin de Coulomb graduated from the École du Génie at Mézières in November 1761 and became a trained engineer with the rank of lieutenant in the Corps du Génie. In February 1764 he was set to Martinique in the West Indies. Coulomb was put in charge of the building of the new Fort Bourbon. On his return to France Coulomb continued his research, in particular using the shipyards in Rochefort as laboratory for his experiments. His studies led to Coulomb's major work on friction, *Théorie des machines simples,* which won him the Grand Prix from the Académie des Sciences in 1781. Coulomb was honored as the one who had: ". . . investigated both static and dynamic friction of sliding surfaces and friction in bending of cords and in rolling. From examination of many physical parameters, he developed a series of two-term equations, the first term a constant and the second term varying with time, normal force, velocity, or other parameters."

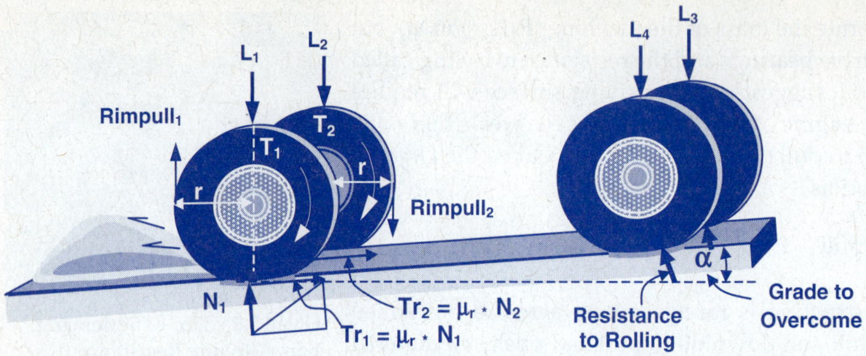

Figure 5.4 Forces Acting on the Wheels of a Two-Drive Vehicle

the traction forces between the two. As in describing the friction force, the traction function uses a coefficient to describe its more or less linear relationship with the normal force N.

$$\mathrm{Tr} = (\mu_r * N)$$

where

Tr = Traction force
μ_r = Coefficient of traction
N = Force normal to the driving surface

It is apparent that the coefficient of traction, representing the degree, between 0 and 1, at which the materials on both surfaces lock up while in contact, has the same function as the coefficient of friction.

Before continuing in our quest to understand what is needed to drive a full truck up the hill, let's summarize the key forces that enter the equation. Depicted in Figure 5.4 are two torques, T_1 and T_2, acting on two drive wheels producing two rimpull forces, $\mathrm{Rimpull}_1 + \mathrm{Rimpull}_2$. To avoid slipping, the wheels need traction with the driving surface, resulting in traction forces $\mathrm{Tr}_1 + \mathrm{Tr}_2$. The amount of traction available is dictated by the traction coefficient (μ_r), the load on a wheel (L), and the angle of grade (α). Two outside forces are acting on the wheels, rolling resistance (RR) and grade resistance (GR).

5.4.2 TERRAMECHANICS: FORCES RESISTING ROLLING

Let's compare the mechanics of rolling across a solid flat surface with rolling over a wet clay. Figure 5.5a illustrates the relationships between the main forces that come to play. We can establish the following equilibrium functions:

At wheel	x: Rimpull = Drawbar Pull
At point P	x: Rimpull = Tr(action)
	y: N(ormal) = L(oad)
Coulomb:	$\mathrm{Tr} = (\mu_r * N) = (\mu_r * L)$
Vector analysis:	$\mathrm{Tr} = \tan(\alpha) * N$

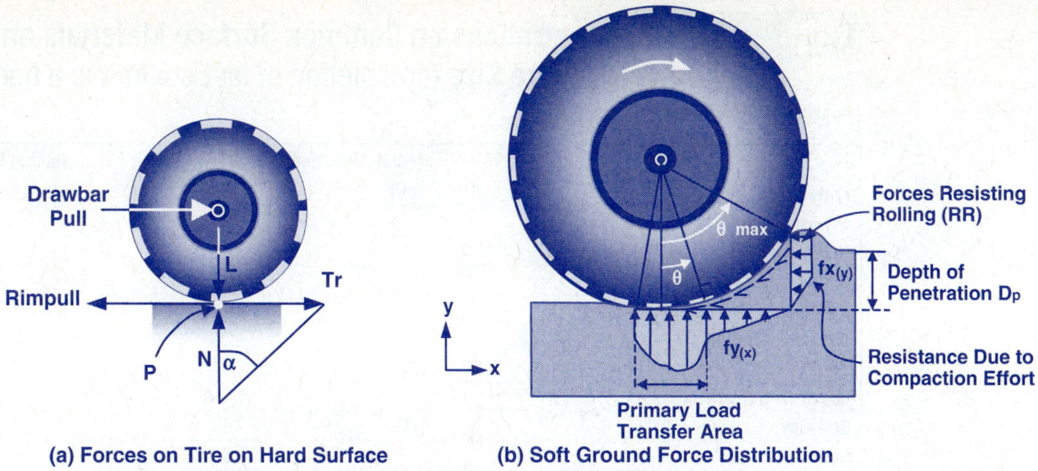

Figure 5.5 Force Distributions During Rolling Motion on Hard and Soft Ground

In order for the vehicle to roll forward, the traction force has to be equal to or larger than the rimpull, or, in other words, the drawbar pull has to be smaller than the available rimpull, which in turn has to be smaller than the traction force created at P:

$$\textbf{Drawbar pull} < \textbf{Available Rimpull} = \textbf{Rimpull} - \textbf{RR} +/- \textbf{GR} < \mu_r * \textbf{L (Traction)}$$

Driving on soft ground with a tire causes the wheel to compact the soil material and as a result, the tire starts a "dozing" action. Figure 5.5b highlights how the soft material is pushed at the surface, as happens in front of a dozerblade, while the deeper layers are compacted due to the increasing normal forces perpendicular to the curvature of the wheel:

$$N(\theta) = \sqrt{(fx_{(y)})^2 + (fy_{(x)})^2}$$

The resistance to rolling can now be defined as the sum of all horizontal components of $N(\theta)$ from 0 to depth of penetration.

$$RR = \int_0^{D_P} \mu\, fx_{(y)}$$

With the help of experimental tests it was found that the force a wheel loaded with 1 ton (t) needs to drive out of a rut with a D_P of 1 inch (2.5 cm) is

$$rr_P = 30\,\text{lb}/ton = 0.15\,\text{kN/t}$$

In an effort to standardize the computation of **RR** for pertinent soil conditions, a second set of experiments was conducted establishing ratios that link the load **L** at the contact area with the depth penetration D_P for a particular surface material. Some example ratios are listed in Table 5.2.

Table 5.2 **Tire Penetrations on Common Surface Materials and Average Tire Size (completion of last column is a homework assignment)**

Rubber Tire Driving on:	Median Tire Penetration		Rolling Resistance
	cm/t	in/ton	%
Dry-wet concrete or asphalt	0	0	
Gravel road—Well compacted, dry	0	0	
Gravel road—Loose, dry	8	3	
Dirt road—Compacted, maintained	4	1.6	
Dirt road—Rutted	6	2.4	
Soft dirt road	20	8	
Loose sand or gravel	18–35	7–14	
Soft clay	40	16	
Coal	8	3	

For a knowledgeable soils engineer, it comes as no surprise that a soft clay is at the top of the list, allowing a tire to sink 40 cm for every ton of load. On the other end of the spectrum, the rolling resistance of hard and compact surfaces is zero. However, the flexing of the treads and the sidewalls of an inflated rubber tire creates a frictional effect that has to be counted as Rolling Resistance. For the average size tire used for grader, trucks and scrapers, the following rolling resistance due to tire flexure ($\mathbf{rr_T}$) can be used:

$$\boxed{\mathbf{rr_T = 40\,lb/ton = 0.2\,kN/t}}$$

Worked-Out Problem 5.1 / Rolling Resistance

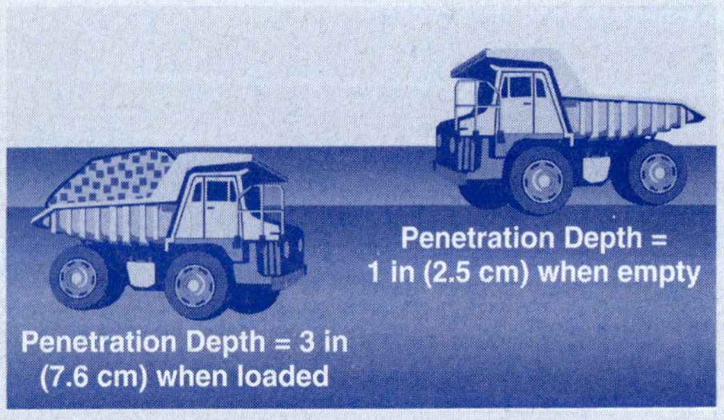

Figure 5.6 Cycling Trucks

You are wondering about the power losses that a hauling truck is incurring when cycling from loading to dumping (Figure 5.6). The truck weighs 41 tons (37 t), and the average load 27.5 tons (25 t).

Rolling resistance has two sources: Tire flexure and penetration of road surface. Thus

$$\mathbf{RR}(\text{lbs}) = (\mathbf{rr_T}+\mathbf{rr_P}) * \mathbf{L},$$

$$\text{Truck weight when full} = L_{full} = 41\text{ tons} + 27.5\text{ tons} = 68.5\text{ tons}$$

Rolling resistance when loaded on hauling road:

$$\text{RR} = (40\,\text{lb/ton} + ((30\,(\text{lb/ton})/\text{inch}) * 3\,\text{inch})) * 68.5\,\text{ton}$$
$$= 130\,\text{lb/ton} * 68.5\,\text{ton} = 8,905\,\text{lb}$$

$$\text{RR} = (0.2\,\text{kN/t} + ((0.15\,(\text{kN/t})/2.5\,\text{cm}) * 7.6\,\text{cm})) * 62\,\text{t} = 0.656\,\text{kN/t} * 62\ \text{t} = 40\,\text{kN}$$

Rolling resistance when on return road:

$$\text{RR} = (40\,\text{lb/ton} + ((30\,(\text{lb/ton})/\text{inch}) * 1\,\text{inch})) * 41\,\text{ton} = 70\,\text{lbs} * 41\,\text{ton} = 2,870\,\text{lb}$$

$$\text{RR} = (0.2\,\text{kN/t} + ((0.15\,(\text{kN/t})/2.5\,\text{cm}) * 2.5\,\text{cm})) * 37\,\text{t} = 0.35\,\text{kN/t} * 37\ \text{t} = 12.8\,\text{kN}$$

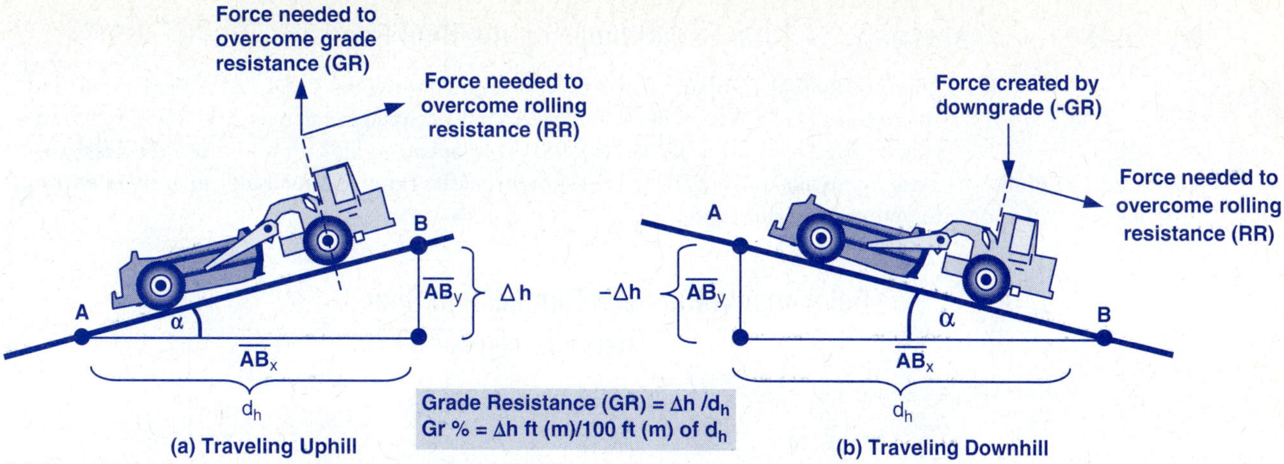

Figure 5.7 Dimensions to Characterize the Force Related to Grade

5.4.3 FORCE NEEDED TO DRIVE UPHILL

So far we have focused on a vehicle moving on a flat surface. Now, we let a scraper overcome a vertical distance Δh as in Figure 5.7a.

Grade Resistance GR

From physics we know that an object that is being lifted is gaining in potential energy. Hence, the scraper traveling uphill from A to B (see Figure 5.7a) requires a force that pushes it the vertical distance of Δh. The needed work can be expressed as ΔEnergy between A and B $= g * m * \Delta h$, representing the gravitational energy increase of the scraper. The only force that can produce the energy needed is the rimpull. How much rimpull is needed is now a function of the weight of the scraper $(m * g)$ and Δh. In an effort to combine the two forces needed to overcome both grade and rolling resistance, the % (percentile) emerged as a common denominator. Hence, the grade is expressed in $\% = \Delta\, h\, m\, (ft)$ per 100 m (ft) of horizontal distance.

Again, empirical studies were used to establish a base function relating GR% to resistance in lbs and kg. This resulting function is called the Grade Resistance Factor (GRF):

$$\mathbf{GRF = Unit\ of\ resistance\ force/GR\% = 20\,lb/ton = 0.1\,kN/t}$$

Both dimensions, ton and t, refer to the weight of the vehicle either loaded or empty. A common term used to identify the related variable is Gross Vehicle Weight (GVW), which allows us to establish a function for the Grade Resistance measured in lbs or kN:

$$\mathbf{GR\,(lbs) = GR\% * GVW(ton) * 20\,lb/ton}$$
$$\mathbf{GR\,(kN) = GR\% * GVW(t) * 0.1\,kN/t}$$

Now we are ready to calculate the average resistances that a vehicle has to overcome when hauling material or pushing soil.

Worked-Out Problem 5.2 / Total Resistances on the Haul Road

S5.3 Scraper Terex TS14G

Assume that the company that you are working for owns a Terex TS14G scraper, which has a Net Vehicle Weight of 62, 922 lb (28.5 t). Its heaped capacity is 20 yd^3 (15.3 m^3), and the average payload is 48, 000 lb (21.8 t). The operators don't want to use the first gear when hauling uphill. Your boss would like to know the steepest grade that the Terex scraper is able to overcome with a full load traveling on a temporary soft dirt road.

Table 5.3 Relationship Between Rimpull and Speed

	TEREX TS14G PARAMETERS			
	Rimpull		Speed	
Gear	lb	kN	mi/hr	km/hr
First	26,200	116.59	2	3.2
Second	19,800	88.11	4	6.4
Third	17,600	78.32	6	10
Fourth	13,200	58.74	9	15
Fifth	11,000	48.95	13	21

Assumptions

Approximate rimpull available in second gear = 19, 800 lb or 88 kN
Rimpull (lb) = RR (lb) + GR (lb) + Drawbar Pull (lb)
Drawbar pull = lb (nothing is being pulled uphill)
Surface penetration depth on soft dirt road = 8 in = 20.3 cm
RR (lb) = (40 lb/ton + ((30 (lb/ton)/in) ∗ 8 in) ∗ GVW (ton)
RR (kN) = (0.2 kN/t + ((0.15(kN/t)/2.5 cm) ∗ 20.3 cm) ∗ GVW(t)
GVW = 62, 922 lb + 48, 000 lb = 111, 000 lb = 55.4 ton
GVW = 279.9 kN + 213.5 kN = 493 kN = 50.3 t

Calculations

GR% = (Rimpull(lb) − RR(lb) − Drawbar Pull(lb))/(GVW(ton) ∗ 20 lb/ton)
RR (lb) = (40 + 240)lb/ton ∗ 55.4 ton = 15, 510 lb
RR (kN) = (0.2 + 1.2) kN/t ∗ 50.3 t = 7.04 kN
GR% = (19, 800 − 15, 510 lb)/(55.4 ton ∗ 20 lb/ton)
= 4, 290/1, 110 = 3.8%
GR% = (88.2 kN − 70.3 kN)/(50.3 t ∗ 0.098 kN/t) = 17.9/4.8 = 3.8%

What-if?: If the soft surface of the road could be compacted, less power would be needed to overcome rolling resistance, which would in turn allow a steeper grade.

Tuning: Compacted road surface reduced depth penetration by 50% to 4 in = 10 cm

RR_{comp}(lb) = (40 lb/ton + (30 (lb/ton)/in) ∗ 4 in) ∗ 55.5 ton
= (160 ∗ 55.5) lb = 8, 880 lb (39.5 kN)
GR% = (19, 800 − 8, 880) lb/(55.4 ∗ 20) lb = 10, 920/1, 110 = 9.8%

Discussion

The rolling resistance caused by a soft surface requires horsepower, thus reducing the amount of power available to overcome grade resistance. As the calculations show, reducing the depth penetration in the travel surface by 50% will allow the Terex to overcome a 9.8% instead of the 3.8% grade.

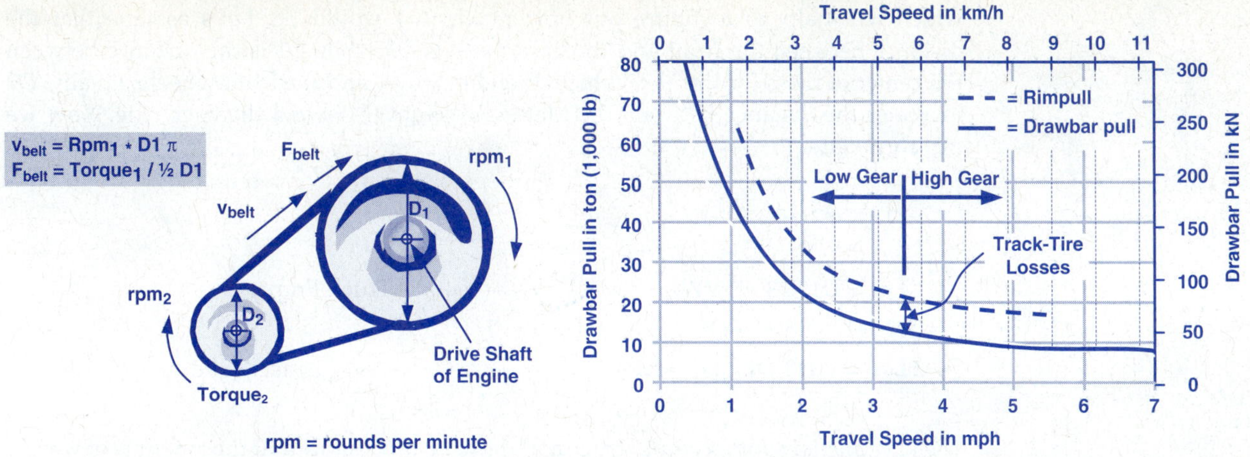

(a) Gearing Ratios and Speed **(b) Available Rimpull at Different Gears – The Rimpull Curve**

Figure 5.8 The Effect of the Gear on Speed and Rimpull

5.4.4 THE ACID TEST: DO WE HAVE SUFFICIENT TRACTION?

Each individual piece of equipment is able to produce a maximum amount of rimpull force that, as we learned above, is dictated by the hp of the engine to produce a torque at the drive wheel, the losses inside the transmission, and the radius of the tire. Figure 5.8 shows that a major factor that affects the torque produced by an engine is the gear ratio between the driving pulley or sprocket on the drive shaft of the engine and the driven pulley or sprocket.

This material is linked to Chapter 3, where, in section 3.1.4, we discussed the relationship among gear, force, and speed. Reviewing the graph in Figure 5.8b, we recognize that speed and available rimpull create a convex curve. A higher rimpull (= force) leads to a lower speed (= rpm of driving wheel). Finally, since every piece of equipment has its unique engine and powertrain, each earthmoving equipment has its own rimpull curve.

Table 5.3 Coefficients of Traction

Rubber Tire Driving on:	Median μ_r
Dry concrete	0.9
Wet asphalt	0.5
Dry asphalt	0.9
Gravel road—Well compacted, dry	0.6
Gravel road—Loose, dry	0.36
Dirt road—Compacted, maintained	0.55
Dirt road—Rutted	0.45
Soft dirt road	0.4
Loose sand or gravel	0.2
Soft clay (dry)	0.45
Coal	0.45

Do we have sufficient traction not to slip? We saw above that to answer this question we need to ascertain two factors, load and coefficient of traction.

Table 5.3 shows that the coefficients of traction vary from a 0.9 on concrete and asphalt to a 0.2 in loose sand. No wonder that vehicles get stuck in sand dunes!

Are we ready to figure out the traction force ($T = \mu_r * N$) that can be expected? Not quite! We need to know the load on the wheels that propel the mass forward. Figure 5.9 presents a two-wheel powered large truck with two normal forces underneath the driving wheels, L_{W1} and L_{W2}. L_{W3} and L_{W4} underneath the steering wheel do not count as they do not exert a rimpull force.

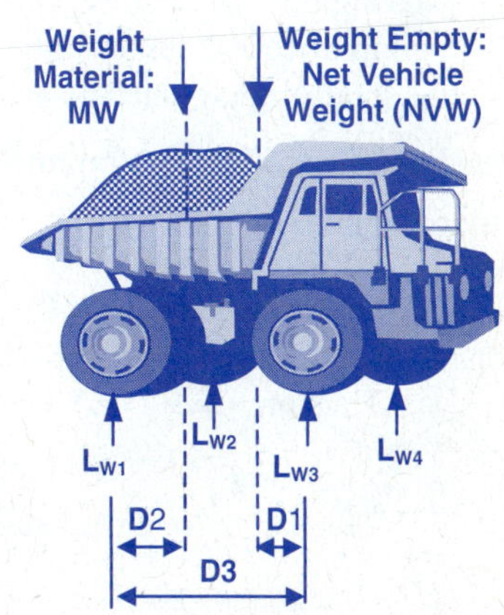

Figure 5.9 Schematic of a Loaded Truck

As a consequence, only one part of the total or gross vehicle weight (GVW) can be counted to create the force normal to the driving surface that can be multiplied with the coefficient of traction μ_r. Good thing we know Newton's principles, which will help us to figure out how much that would be. Let's assume that the distance between the front and the back axles is D3 while D1 is the distance between the center of mass for the Net Vehicle Weight (NVW) and the front wheels. Finally, D2 indicates the distance between the Material Weight (MV) and the back axle. Now we can formulate:

- $L_{W1} = L_{W2}, L_{W3} = L_{W4}$
- $D1 * NVW - D3 * (L_{W1} + L_{W2}) = 0$
- $\mathbf{2\,L_{W1} = D1/D3 * NVW}$ ⟵ Truck Empty
- $(D3\text{-}D2) * MW - D3 * (L_{W1} + L_{W2}) = 0$
- $\mathbf{2\,L_{W1} = ((D3\text{-}D2)/D3) * MW}$ ⟵ Load Alone

The key functions for an empty truck and the load alone show that the amounts of weight that will find their way to the back axle are related to the positions of the center of gravity relative to the two axles. We observe that the amount of the truck weight that helps create traction increases with D1. The same basic relationship holds true for the material. Similarly, a smaller D2 will increase the percentage of MW that is resting on the back axle, thus increasing the $2\,L_{W1}$.

$$\textbf{The total normal force on the drive wheels} = \underbrace{(D1/D3 * NVW)}_{\text{Empty truck}} + \underbrace{(((D3\text{-}D2)/D3) * MW)}_{\text{Material}}$$

It is now clear why four-wheel drives are able to keep moving in situations where other vehicles get stuck. They are able to use their total Gross Vehicle Weight (GVW = NVW + MW) to create traction.

Worked-Out Problem 5.3 / Will It Get Stuck?

After reading the material about the traction problem, you must be worried that the Terex TS14G scraper will not be able to get up the maximum grades of 3.8 and 9.8 degrees that we calculated in the previous example (see Figure 5.10). Before giving your boss the result of your analysis, we had better make sure that the scraper will in fact have sufficient traction. You recall that you used the total rimpull that was available = 19,800 lb or 88 kN.

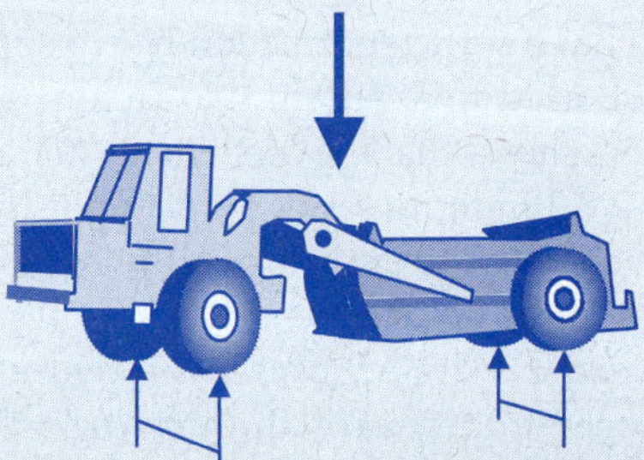

Figure 5.10 Front-Wheel-Drive Scraper

Base Functions and Parameters

Only front wheels are powered
GVW = 55.5 ton = 50 t
Soft dirt road $\mu_r = 0.4$
Compact dirt road $\mu_r = 0.5$
Traction = $\mu_r *$ Load on drive axle

Calculations

Traction on soft road surface = 0.4 * 111,000 lb * 0.509
= 22,600 lb or 100.5 kN

Traction on compacted road surface = 28,250 lb or 125.7 kN

Discussion of Results

On the soft road surface, the available traction is slightly larger than the 19,800 lb (88 kN) of rimpull that is needed to climb the 3.8% in second gear while the compaction increases the traction coefficient to make the climb possible.

Should the company decide that it is too costly to compact the road for the scraper, the only solution is to reduce the grade. What happens if we reduce the grades by 0.8%?

Rimpull (lb) = RR (lb) + GR (lb) + Drawbar Pull (lb)
RR (lb) = (40 lb/ton + (30(lb/ton)/in) * 8 in) * 55.5 ton = 15,540 lb
RR (kN) = (0.2 kN + (0.15(kN/t)/2.5 cm) * 50 t = 70 kN
GR (lb) = GR% * 55.5 ton * 20 lb/ton = (3%) * 55.5 ton * 20 lb/ton = 3,330 lb (14.8 kN)
Required rimpull = (15,540 lb + 3,330 lb) = 18,870 lb (83.9 kN)
Min. coefficient of traction = (18,870 lb/(111,000 lb * 0.509) = 0.33 (gravel road according to Table 5.3)

5.5 Power and Traction to Cut, Rip, and Doze

The three types of equipment covered in this chapter are designed to cut and break the soil and soft rock so that it can be relocated. Separating portions of the dense soil material, as done by the dozer and the scraper, requires the peak amount of force. Since this mechanism dictates the sizing and the management of the equipment, it is necessary to have a good understanding of the critical impact factors.

5.5.1 MODELING SOIL-TOOL INTERACTIONS

S5.4 The Fresno Scraper 1884

As we have learned in previous chapters, modeling the real world is an omnipresent tool used by engineers both to understand and to design solutions to a problem. Figure 5.11 models several models with different views of the two main tool-soil interactions, dozing and scraping.

Ali Touran (Touran et al., 1997) presented a simplified approach to calculating the force P needed to overcome the resistance of soil that is being "sheared" away from its

Courtesy Position Partners Pty Ltd

(a) Bulldozer with Straight Blade

(b) Horse-Drawn Fresno Scraper

Figure 5.11 Comparing Geotechnical Models for Dozing and Scraping

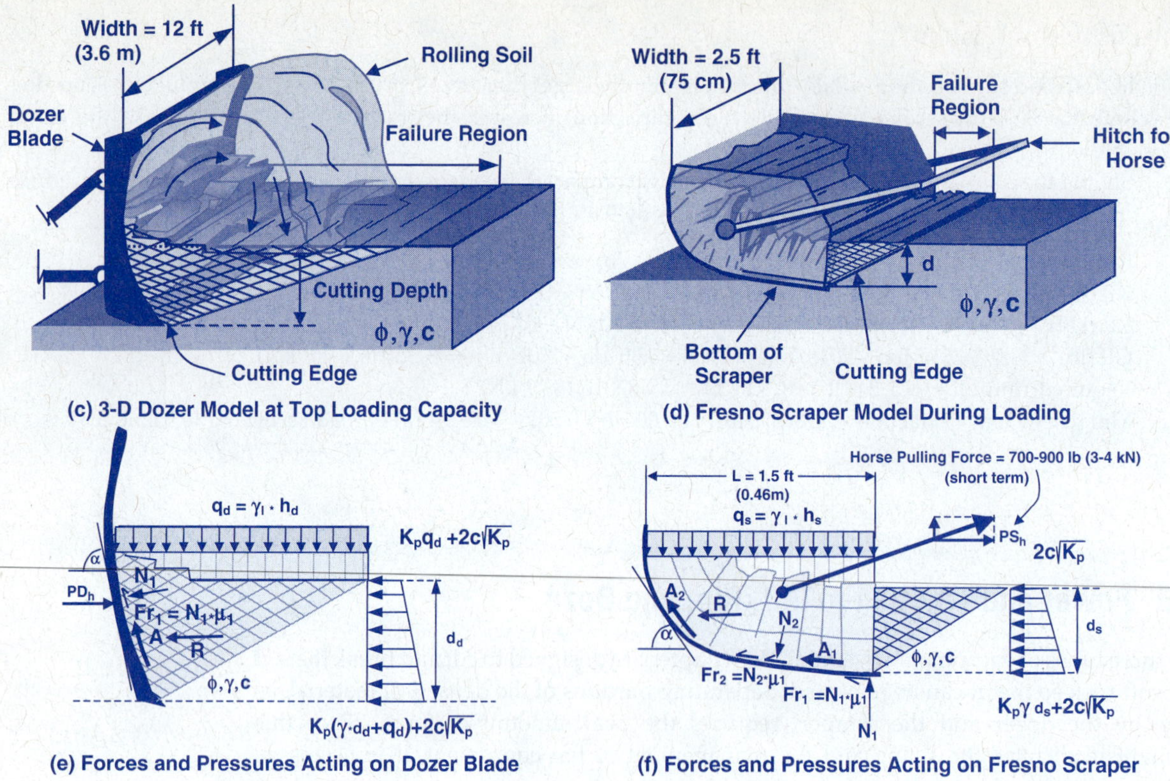

Figure 5.11 *(Continued)*

original place. The basic functions for calculating the resultant force P_h for cohesive and noncohesive soil have the following forms:

$$\textbf{Simplified } \mathbf{R} = \sqrt{(\mathbf{R_p^2} + \mathbf{A^2})} \qquad \mathbf{R_p} = \mathbf{1/2\,d} * \mathbf{K_p}(\gamma * \mathbf{d} + \mathbf{2q}) + \mathbf{2c} * \mathbf{d}\sqrt{\mathbf{K_p}}$$

$$\mathbf{A} = \mathbf{a} * \mathbf{d} * (\mathbf{cosec}\,\alpha)$$

where

R_p = Coulomb's passive resistance in kN/m or lb per ft
A = Resistance due to adhesion of soil to the steel
R = Resultant force from passive resistance and adhesion
Fr = Resistance due to friction between soil and steel
P_h = Total horizontal pull or push needed to overcome total resistance
Kp = Coefficient of passive earth pressure model, which is 1 for cohesive soil
d = Depth of cut
h = Height of loose soil overlay
q = Distributed load from the loosened soil overlay
γ = Soil's unit weight, approx. 115 lb/ft^3 (18.0 kN/m^3) for firm clay
γ_l = Unit weight of loose soil with 25% swell = 88 lb/ft^3 (13.5 kN/m^3)
c = Soil cohesion
a = Soil adhesion factor typically = 0.5 c
α = Average inclination of blade or backside of pan
μ_1 = Dynamic friction coefficient clay-steel approx. 0.20

The resultant force for a dozer blade and the Fresno scraper cutting the firm clay is:

Maximum Push for Dozer Blade:

$c = 600\,lb/ft^2 (28.7\,kPa)$

Depth of cut by dozer $d_d = 1\,ft (0.33\,m)$

Height of overlay $h_d = 3\,ft (1\,m)$

Av. inclination of dozer blade in cut area $\alpha = 80°$

$R_p = 1/2\gamma * d^2 + \gamma l * h_d d + 2c * d =$

$d(0.5\gamma * d + \gamma_{**} h_d + 2c =$

$1\,ft(0.5 * 115\,lb/ft^3 * 1\,ft) + (115\,lb/ft^3 * 3\,ft)$
$+(2 * 600\,lb/ft^2) = (57.5 + 345 + 1,200) lb/ft$
$= 1,602\,lb/ft\ (24\,kN/m)$

$A = 0.5 * 600\,lb/ft^2 * 1\,ft(1/\sin 80°) =$

$(300/ - 0.9939) lb/ft = -302\,lb/ft\ (4.5\,kN/m)$

$R = \sqrt{(1,602)^2 + (-302)^2} = 1,630\,lb/ft = P_h = 24.4\,kN/m$

Max push needed $= P_h *$ width of dozer blade $=$

$1,630\,lb/ft\ 12\,ft = \underline{19,560\,lb} = 87\,kN$

Maximum Pull for Fresno Scraper:

$c = 400\,lb/ft^2 (19.1\,kPa)$

$d_s =$ Depth of cut by Fresno $= 0.25\,ft (0.08\,m)$

$P_h = \sqrt{(R_p^2 + A^2)} +$ Friction forces from N_1 and N_2

Height of overlay $h_s = 0.75\,ft (0.25\,m)$

$L_2 =$ area of adhesion inside pan $= 0.5\,ft (.16\,m)$

$N_1 = 0.75$(Weights of soil in pan + pan)/ft $= 125\,lb/ft$

$N_2 = L_2 * (h_s\gamma + d_s\gamma_l) = 1\,ft(.25 * 115 + .75 * 88) lb/ft^2 =$
$95\,lb/ft$

Average inclination of scraper backwall $\alpha = 45°$

$R_p = 1/2\gamma * d^2 + 2c * d_s = d(0.5\gamma * d + 2c)$

$0.25(0.5 * 115\,lb/ft^3 * 0.25ft + 2 * 400\,lb/ft^2 =$

$0.25(14.4 + 800) lb/ft = 203\,lb/ft\ (3\,kN/m)$

$A_1 = 0.5c * L_2 = 100\,lb/ft\ (1.5\,kN/m)$

$A_2 = 0.125 * 400\,lb/ft(1/\sin 45°) = (50/.85) lb/ft =$
$58\,lb/ft$

$R = \sqrt{(303)^2 + (58)^2} = 308\,lb/ft\ (4.6\,kN/m)$

$Fr_s = (N_1 + N_2) * \mu_1 = 220\,lb * 0.20 = 44\,lb/ft\ (0.66\,kN/m)$

$P_h = 352\,lb/ft\ (5.3\,kN/m)$

Max. pull needed $= P_h *$ width of scraper $=$

$352\,lb/ft * 2.5\,ft = \underline{880\,lb\ (3.9\,kN)}$

The results of this analysis of the maximum forces of a dozing and an historical scraping operation indicate that the assumptions and formulas utilized seem to be accurate. While it is difficult to assess what horses in the early twentieth century were able to pull, draft horses today draw 500 lb on a continuous basis and a maximum of 2000 lb (8.9 kN). Thus, the 880 lb (3.9 kN) as a peak pull force to be reached at the end of each scraping operation should not cause a well-fed horse to get overexerted.

The calculated 19,560 lb (87 kN) for a dozer with a straight blade also lies within the power range of a bulldozer. The net horsepower of the three "small" dozers compared in Figure 5.12 ranges from 168 to 240 hp. As Table 5.4 shows, the 185 hp D6 by Caterpillar is

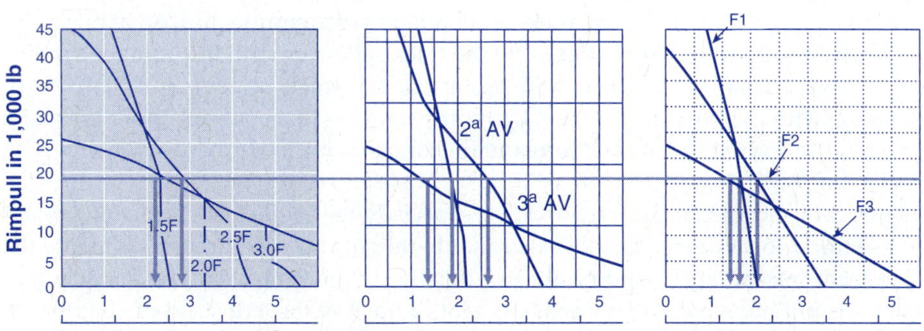

Figure 5.12 Rimpull-Graphs for Three Dozers

Table 5.5 Operating Data for Three Dozers Applying a Rimpull of 19,560 lb (87 kN)

		Caterpillar D 6	New Holland D 255	Komatsu D 61
Operating Weight	lb-kN	41,298–183.7	66,800–297.1	36,840–163.8
Net Power	hp	185	240	168
First Gear	mph/kmph	2.1/3.4	1.9/3.05	1.9/3.05
Second Gear	mph/kmph	2.9/4.7	2.7/4.3	2.2/3.5
Third Gear	mph/kmph	2.1/3.4	1.4/2.2	1.5/2.4

S5.5 Dozer CAT D6
S5.6 Dozer New Holland D255
S5.7 Dozer Komatsu D61

second in operating weight and net power but reaches, comparatively, the highest speeds. Overall, however, the three selected dozers show very similar performance characteristics. While comparing brochure data, we always need to keep in mind that they are theoretical in nature. That means that a good operator will outperform any less trained operator working a higher-rated machine.

Can we assume that a crawler tractor does not ever get stuck in soft soil or snow? Not at all! Of course, traction of a track depends on a different set of terramechanical functions. Let us find out how a crawler finds traction.

5.5.2 TRACTION WITH TRACKS

The history files earlier in the chapter told us that tracks were invented and patented in the early 1900s and first used to replace horses in pulling large cannons for the British Army. Since then, tracks have found widespread use in construction, agriculture, snow vehicles, and so on. Tracks fulfill two goals: (a) distribution of the operating weight, and (b) resistance in soft material. Figure 5.13 depicts the main components of this mechanical masterpiece.

The locomotion is provided by a track chain and a drive wheel with sprockets powered by the powertrain. In Figure 5.13, the drive wheel of the Caterpillar D11 is positioned off the ground and does not have to carry any load. On many dozer models, the sprocket drive wheel takes the place of the rear idler. The track chain has predrilled holes allowing track shoes and plates to be bolted on. Those plates come in various widths and even shapes, including relief holes (Figure 5.13b) to reduce pressure when operating in clayey material. Traction on soft material is provided by the grousers that are welded onto the track plates. Most common is the single-grouser plate, also providing space to mount rubber or steel pads in between. This can be useful if the tracks are being operated on hard surfaces, such as concrete or asphalt, where the grousers not only provide little traction but also destroy the surface. Bogies with two rollers each fulfill a double function in that they help in distributing the load evenly along the length of the track, and they also provide the agility to travel over uneven surfaces. As shown in Figure 5.13a, rocks and boulders concentrate the load at one point on the track and can cause large stresses to the track rollers around the point of contact. Three or four bogies are used to add agility to the track system in that they, in combination with elastic pads, function like sturdy springs that contract under pressure. As a result, the entire track is able to maintain contact with the ground while hugging the rocks.

Figure 5.14 shows that the traction laws, especially the coefficient of traction, do not apply to a track dozer operating in weak soil. It highlights how the grousers bite into the soil like teeth into an apple and establish a totally different mechanism. Imagine that the tractor is pulling a heavy load needing a lot of traction force. Instead of slipping on the surface, the grousers push on soil-wedges under the track plates, compacting the

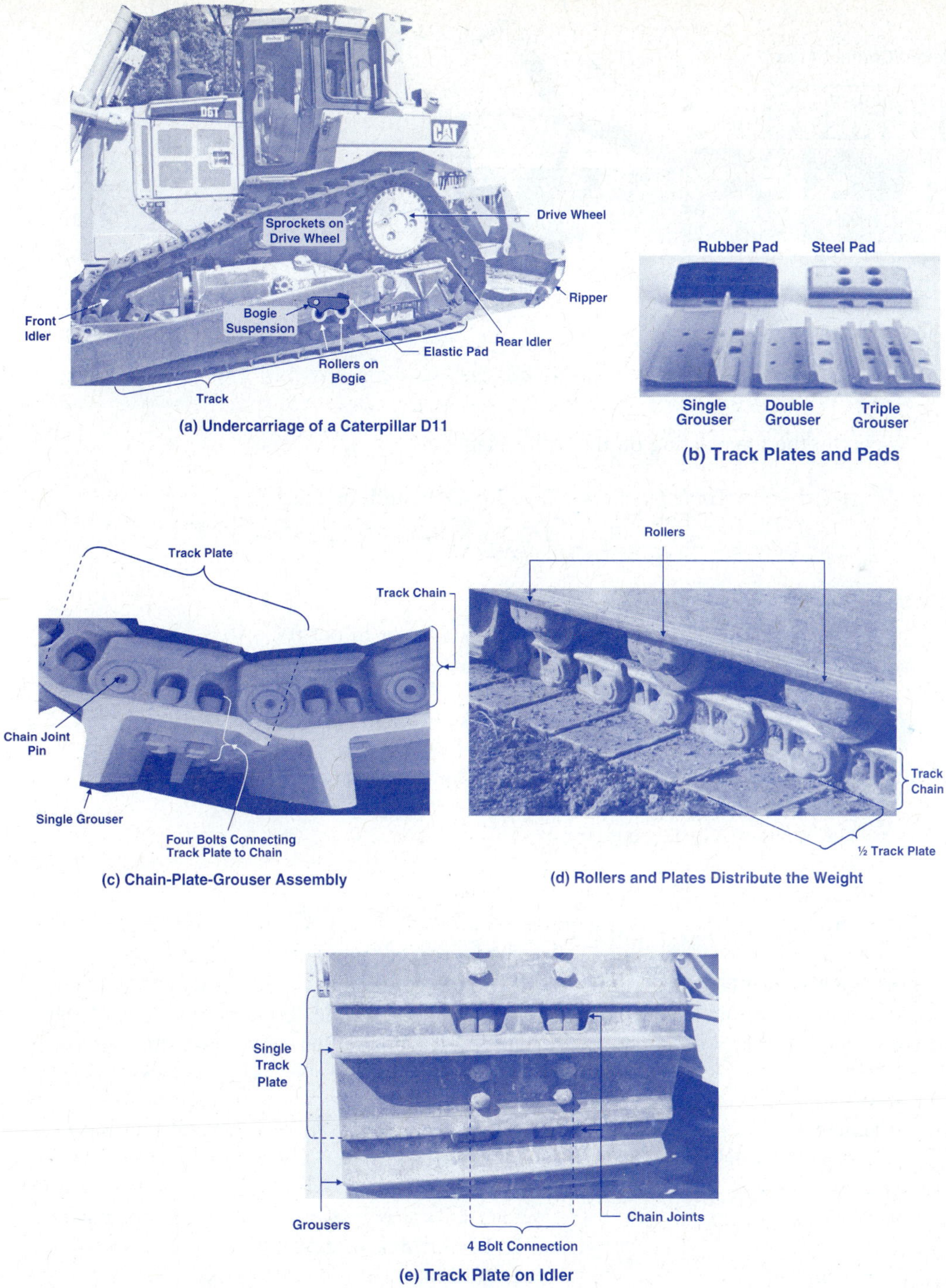

Figure 5.13 Elements of the Steel Track Mechanism for Crawlers

sheared-off sections and thus opening a slippage gap. Research has shown that the traction force increases during the initial phase of slippage before it collapses when the gap reaches about 20% of the distance between the grousers. In order to calculate the resistance that the soil creates, we can again employ the simplified function used to

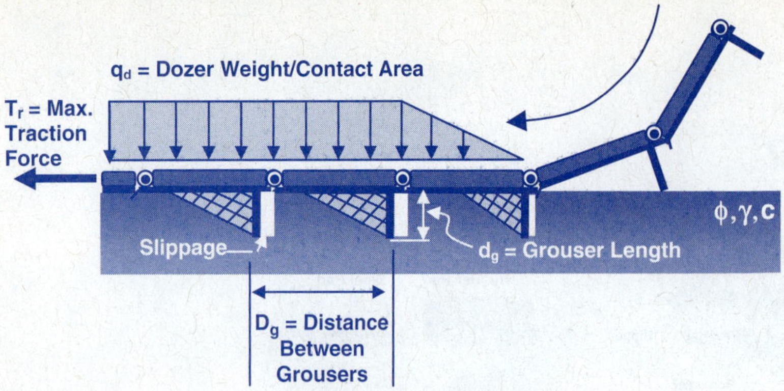

Figure 5.14 Model of Soil-Track Interaction

ascertain the force acting on the dozer blade:

$$\text{Total Traction} = 2 * (\text{Rp}/\text{D})\ \text{Length of Track}$$

$$R_p = 1/2\,d * K_p(\gamma * d + 2q) + 2c * d\sqrt{K_p}$$

where

R_p = Coulomb's passive resistance in kN/m or lb per ft
K_p = Coefficient of passive earth pressure model which is 1 for cohesive soil
D = Distance between grousers
d = Depth of soil penetration
q = Distributed load acting on soil
γ = Soil's unit weight, approx. 115 lb/ft^3 (18.0 kN/m^3) for firm clay
c = Soil cohesion

Worked-Out Problem 5.4 / Dozer Traction

While the rimpull chart for the Komatsu D 61EX-15 provides us with the maximum rimpull when driving in first gear, it does not tell us anything about the maximum traction that can be achieved. Let's try to figure out the maximum traction that can be produced on soft/wet clay with a basic track configuration (see brochure for Komatsu D 61 on booksite).

Basic Functions and Parameters

W_g = gross operating weight	= 36,840 lb (163.8 kN)
Kp = Coeff. of passive earth pressure	= 1 for cohesive soil
Number of plates/shoes on ground	= 14
d = grouser height	= 2.3 inch (5.75 cm) = 0.1917 ft
Width of track plate	= 2 ft (0.6 m)
Length of track on ground	= 8.5 ft (2.6 m)
D = distance between grousers	= 102 inch/14 = 7.3 inch (18.5 cm)
A_c = Area of ground contact	= 4,836 in^2 (3.1 m^2)
q = Distributed load	= W_g/A_c = 7.6 lb/in^2 = (52.4 kPa)
γ = Soil's unit weight	= 115 lb/ft^3 (18.0 kN/m^3)
c = Soil cohesion for wet clay	= 600 lb/ft^2 (28.7 kPa)

Calculations for 1 Track Plate

$R_p = 1/2\,d * K_p(\gamma * d + 2q) + 2c * d\sqrt{K_p} = 0.5 * d^2 * \gamma + d(q + 2c)$

$R_p = (0.5 * (0.1917\,ft)^2 * 115\,lb/ft^3) + 0.1917\,ft(144 * 7.6\,lb/ft^2 + 2 * 600\,lb/ft^2)$

$R_p = 2.11\,lb/ft + 210\,lb/ft + 230\,lb/ft = 442\,lb/ft\ (6.5\,kN/m)$

Total resistance from 1 grouser = 2 ft(width) * 442 lb/ft = 884 lb

Total resistance from 1 grouser = 0.6 m * 6.5 kN/m = 3.9 kN

Traction from one track = 14 grousers * 884 lb (3.9 kN) = 12,370 lb (55kN)

Total traction from two tracks = 2 * 12,370 lb = 24,740 lb (110 kN)

What–if? The basic function for calculating soil resistance shows that there are three main values that are important: (a) soil pressure, (b) cohesion, and (c) grouser height. While increased soil pressure can only be achieved by reducing the width of the track shoes, total resistance from 1 shoe will stay the same. Cohesion is a soil property and can't be changed easily. The only option we have to increase traction is to change the length of the grousers. What if we add 0.5 inch to the grouser height?

Alternative: New grouser height d = 2.3 in + 0.5 in = 2.8 in = 0.233 ft (7.1 cm)(= 22% increase)

$R_p = (0.5 * (0.233\,ft)^2 * 115\,lb/ft^3) + 0.233\,ft(144 * 7.6\,lb/ft^2 + 2 * 600\,lb/ft^2) =$ 538 lb/ft (7.9 kN/m)

Total traction from two tracks = 30,128 lb (134 kN)(= 22% increase)

Discussion: Assuming the base shearing model and function used for the calculation of the total resistance, we recognize that the coefficients of traction can be modified by changing the height of the grousers. In fact, the amount of total traction is linearly proportional to the grouser height, a fact that finds its equivalent, albeit not to the same extent in the depth of the treads on rubber tires. The coefficients of traction in clay-type material for the two cases are **0.67** and **0.82**, respectively. Most importantly, the result of both calculations show that the total traction forces are sufficient to push the 1 ft of soil for which a rimpull of 19,560 lb (87 kN) is needed.

Table 5.6 provides the average coefficients of traction for a variety of surfaces that have been established using empirical methods.

We notice with great satisfaction that the results of our calculations in the previous example correspond with the empirical data. While we calculated a traction coefficient of 0.67 for soft/wet clay using standard grousers, the table tells us to use an almost equivalent coefficient of 0.7. The difference of 0.03 lies well within the margin of error, particularly since the soil's cohesion can vary widely between $400 - 1{,}000\,lb/ft^2$ (19.1 kPa – 47.9 kPa).

So far we made sure that a dozer has sufficient rimpull and traction to begin work. Now we are ready to figure out how much work can be accomplished.

Table 5.6 Traction Coefficients for Vehicles with Metal Tracks

Travel Surface	Coefficient of Traction	Travel Surface	Coefficient of Traction
Concrete	0.45	Quarry pit	0.55
Dry clay, loam	0.90	Gravel road (loose)	0.50
Wet clay, loam	0.70	Loose earth	0.60
Dry sand	0.30	Ice	0.12
Wet sand	0.50	Packed snow	0.27

5.5.3 DOZER PRODUCTION

Dozers are flexible enough to be used for several tasks, such as (1) dozing soil, coal, or waste; (2) clearing trees from land; (3) ripping hard soil or soft rock; and (4) pulling/pushing scrapers. Each of these tasks requires a different set of tools to be mounted on that tractor. The following provides a short overview of the operational characteristics for each case.

Dozing with Crawlers

The main dozing tools are the different blades to be mounted on the C-frame or push arm with a hydraulic pitch control cylinder. The many blade shapes and sizes available need to be selected according to the soil characteristics and the job to be done (see www.wiley.com/college/bernold). Figure 5.15 shows several dozing blades that can commonly be found on earthmoving sites.

S5.5 Dozer CAT D6
S5.6 Dozer New Holland D255
S5.7 Dozer Komatsu D61

There are three main occasions when we would like to know the production rate of one single dozer or an entire equipment fleet: (1) to forecast the amount of time a job takes and estimate its cost, (2) to assess the efficiency of the ongoing operation, and (3) to establish historical records and study critical success factors. In this chapter we will focus on the first example, the prediction of time and cost based on the work that needs to be done and the calculated production rate of our equipment.

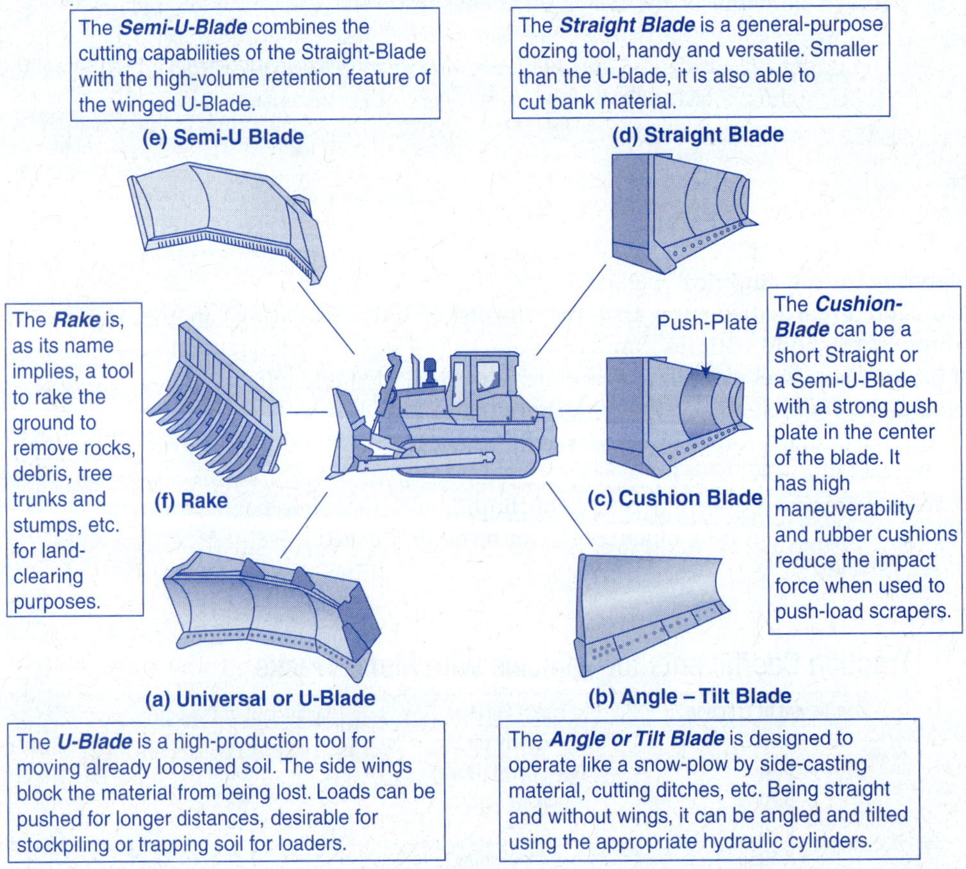

Figure 5.15 Common Dozer Blades and Their Performance Characteristics

Let's apply the basic function for computing the output rate of a dozer that we used for excavators in the last chapter, namely

Production Rate of Dozer (P_{Dozer}) = Volume of Output/Time to Produce Output (O/T)

We have already learned that the establishment of both output volume and duration of output creation requires the use of models and mathematical calculations to reach satisfactory results. The following presents a structured four-step approach to estimate volume and time in preparation to calculating the dozer production in step 5.

Estimating Volume

Step (1) Base Function:
Predicted Output = Theoretical Output ∗ (Adjustment Factors)

Step (2) Theoretical Model: (Based on SAE Practice J 1265)

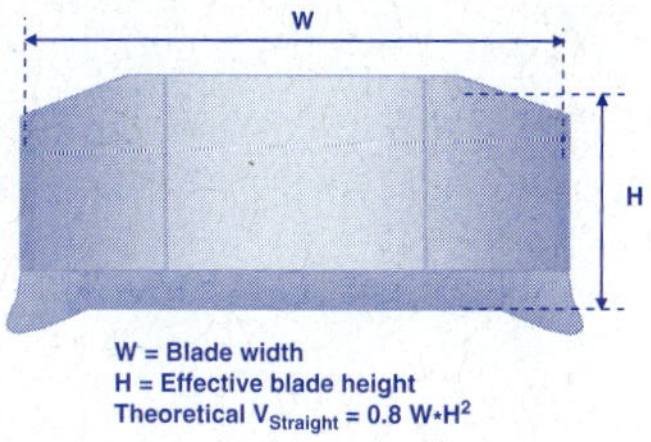

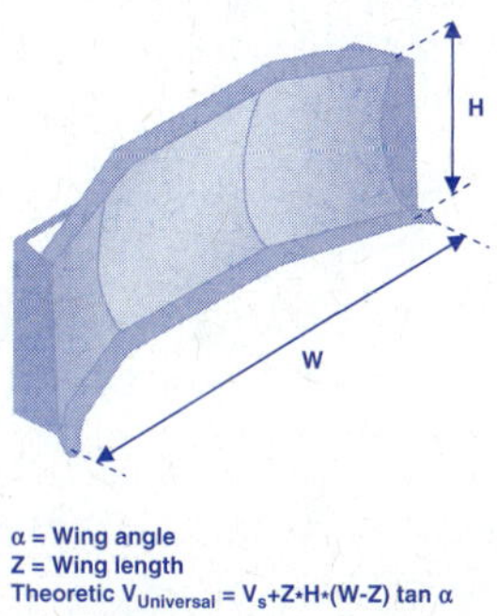

(3) Adjustment Factors:

Material Properties (Oa_M)

a. Soil density	1.0–1.2
b. Hard to cut	0.6–0.8
c. Sticky material	0.8
d. Blasted rock	0.6–0.8

Operational Efficiencies (Oa_O)

a. Slot dozing	1.2
b. Operator skill	0.8–1.1
c. Work condition	0.8–1.0

4) Output Prediction:

Volume of 1 blade = V_s or V_u ∗ $(Oa_{M\,(a-d)} * Oa_{O\,(a-c)})$

Estimating Time

Step (1) Base Function:
Predicted Time = Theoretical Time ∗ (Adjustment Factors)

(2) Theoretical I/O Model:

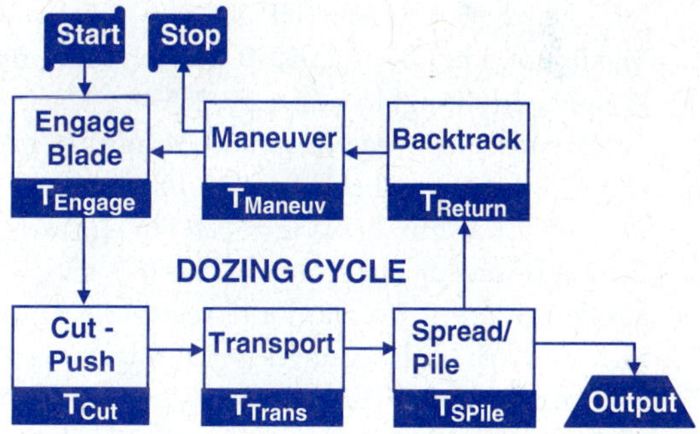

Theoretical Cycle Time = T_{Cycle} =

$$T_{Engage} + T_{Cut} + T_{Transp} + T_{SPile} + T_{Return} + T_{Maneuv}$$

(3) Adjustment Factors:

Job Condition (Ta_J)

a. Slope, % Grade	0.4–1.6
b. Material Properties	0.6–1.2
c. Visibility-Environment	0.8–1.0

Equipment Condition (Ta_E)

a. Direct Drive Transmission	0.8
b. Blade size/maneuverability	0.8–1.2

Operational Efficiencies (Ta_O)

a. Productive time	0.7–0.9
b. Operator skill	0.6–1.2
c. Complexity of operation	1.0–1.2
d. Electronic Guidance	0.6–1.0

4) Dozing Time Prediction:

Time for 1 blade = $T_{Cycle} * (Ta_{J\,(a-c)} * Ta_{E\,(a-b)} * Ta_{O(a-d)})$

(5) Estimate of Dozer Production:

P_{Dozer}(cyd/hr-m^3/hr) = O/T = Volume of 1 blade (cyd-m^3)/Time for 1 blade (hr)

Worked-Out Problem 5.5 / Estimating Dozer Production

Marc Unger, your supervisor at E-Movers, Inc., asks you to estimate the hourly production of a new dozer that the company bought, a John Deere 950J (download brochure from booksite).

The job that you should consider is a parking lot that needs to be cleared from 2 ft of topsoil followed by leveling of the clay subsurface. Let's assume that the parking lot measures 330 ft × 330 ft (100.6 m) with a 0% grade. Marc gives you two critical rules that lead to high production dozing. The operator should (1) fill load on the blade in two lengths of the tractor at the most, and (2) stop cutting once the blade is full, slide the dirt in front of the blade through the slot created by the first pass and deposit it on the pile, which is called slot dozing. Because it is a simple job, Gregg Kirlin, who just passed the training program, will be the operator. Marc also mentions while leaving that the productive time to use should be 45 min out of an hour, and that for removing the topsoil no electronic guidance would be needed.

Basic Functions and Parameters

P_{Dozer} = Volume of 1 blade (cyd)/Time for 1 blade (hr)
John Deere 950J, 247 net HP
Maximum rimpull = 82, 055 lb (365 kN) at .09 mph (1.5 km/h)
~~Weight with straight blade = 59, 255 lb~~
Track widths between 22 in and 36 in (0.56 m–0.91 m)
H = height of straight blade = 5 ft 1 in (1.55 m)
W = width of straight blade = 12ft 1 in (3.7 m)
Length of tractor incl. blade = 20 ft (6.1 m)
Maximum speed (forward and reverse) 6.8 mph (11 km/h)
Maximum rimpull required for cutting 1 ft of topsoil = 25, 000 lb (111 kN)
Maximum rimpull required for "slotting" = 10, 000 lb (44.5 kN)

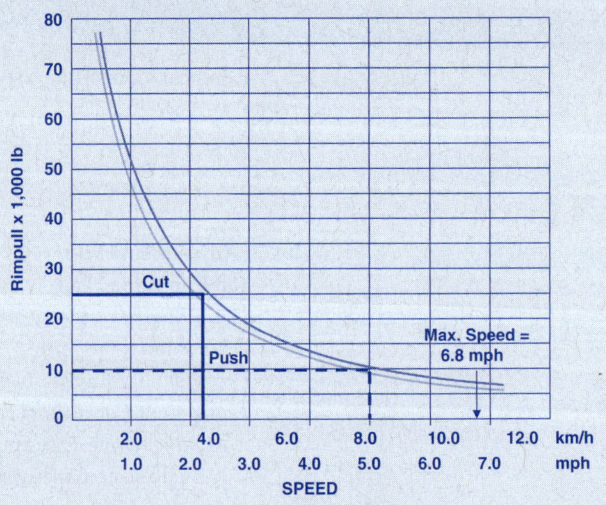

$$\text{Speed during cut} = 2.2\,\text{mph} = (5280/60)\,\text{ft} - \text{hr/mile} * \text{min} * 2.2\,\text{mph}$$
$$= 194\,\text{ft/min}\ (59\,\text{m/min})$$

Speed during push = 5.0 mph = 88 ∗ 5.0 mph = 440 ft/min (134 m/min)
Speed during backtrack = 6.8 mph = 88 ∗ 6.8 mph = 598 ft/min (182 m/min)
Productive time adjustment $Ta_O = 45/60 = 0.833$

Operator skill adjustment $Oa_O = 0.9$
Operator skill adjustment $Ta_O = 1.2$
Direct drive transmission considered in rimpull-speed chart

Calculations

Theoretical **load straight blade** $= V_s = 0.8\,W * H^2 = (0.8 * 12.1\,ft * 5.1^2\,ft^2)/27\,(ft^3/yd^3) =$ **$9.4\,yd^3$ ($7.2\,m^3$)**
Average pushing distance= $(330\,ft/2)/2 = 82\,ft$ (25 m) (soil is pushed to all sides of the 330 ft (100.5 m) square)

Cycle time: Engage engine and blade	$= 0.05$ min
+ Time for cut $= (2 * 20\,ft)/194\,ft/min$	$= 0.2$ min
+ Average time to push $= 82\,ft/440\,ft/min$	$= 0.18$ min
+ Time to pile soil $= 15\,ft/150\,ft/min$	$= 0.1$ min
+ Average time to backtrack $= 97\,ft/598\,ft/min$	$= 0.15$ min
+ Maneuver time	$= 0.1$ min
Theoretical **cycle time**	**$= 0.78$ min**

Adjustments of theoretical output $= Oa_O = 0.9$ (operator skill)
Adjustments of cycle time $= Ta_O = 0.833 * 1.2 = 0.99$ (prod. time, skill, complex.)
Estimated dozer production $= (0.9 * 9.4\,yd^3/0.78\,min * 0.99) * 60\,min = 657\,yd^3/hr$ ($503\,m^3/hr$)

What-If? It is well known that the skill of the operator has a large impact on the productivity of large equipment. If Gregg Kirlin had already had 6 years of experience, no skill adjustments would be necessary. What would happen to the production rate?

Alternative

Operator skill adjustment $Oa_O = 1.0$
Adjustments of cycle time $Ta_O = 0.833$ (productive time)
Modifies dozer production $= (9.4\,yd^3/0.78\,min * 0.833) * 60\,min$
$= 868\,yd^3/hr$ ($664\,m^3/hr$)

Discussion

The good workability of the soil, simplicity of the job, and the large blade result in a relative high production rate of $657\,yd^3/hr$ ($503\,m^3/hr$) that is limited mainly by the low skill level of Gregg Kirlin. While this job provides a good opportunity to practice on the new John Deere dozer, an average operator could boost the productivity by over 30% to $868\,yd^3/hr$ ($664\,m^3/hr$).

5.5.4 RIPPING ROCK TO AVOID EXPENSIVE BLASTING

Rather than using drilling and blasting to loosen hard soil or rock, a brute force method, called ripping, has been developed. The main purpose of ripping is to save money and time. As a general rule, the cost of excavation using ripping is one-third to one-half of the cost of having to drill and blast.

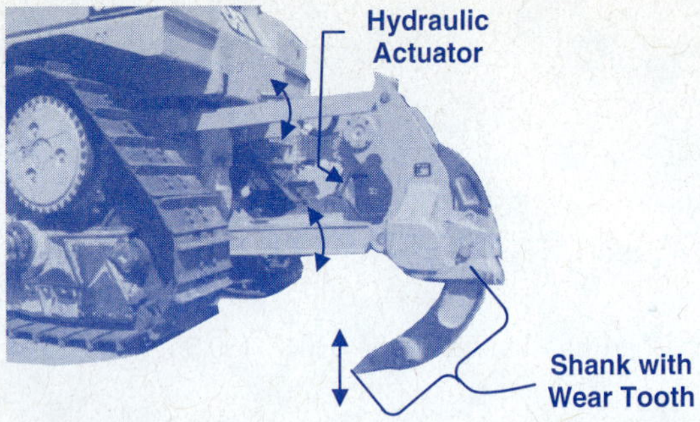

Figure 5.16 Drilling Is Replaced by Heavy-Duty Single Shank Ripping

As the weight and the power of crawler tractors has increased over time, the applicability of ripping has increased, since harder material can be broken up to greater depths. Not surprisingly, the wear and tear of this brute force method causes a significant increase in repair cost of the undercarriage. A single shank with a short tip is usually used for heavy ripping where penetration is difficult and shock is severe. Long tips are used for unconsolidated abrasive materials that are easy to rip. Two or more shanks are used for easily ripped materials (see Figure 5.16).

Planning the operation effectively will allow for a mitigation of the shock forces while helping in optimizing the fragmentation of the rock. For example, the sizes of the broken-up rocks can be controlled by modifying the direction, the depth of penetration. and the spacing of the ripping passes.

> **Refraction versus Reflection**
>
> The Dutch mathematician Willebrord Snell (1591–1626) established the law of the refraction of light:
>
> $$\frac{\sin \theta_1}{\sin \theta_2} = \frac{v_1}{v_1}$$
>
> 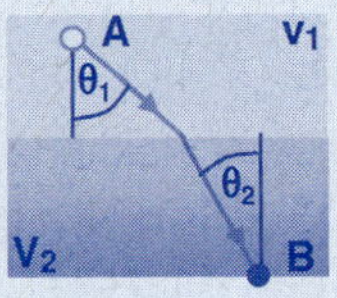
>
>
> Snell's law states that the ratio of sines of the incident (θ_1) and refraction (θ_2) angles of light or other waves passing through the boundary of two isotropic media is equal the wave velocities in the two media. However, the law only applies when the wave enters into a higher velocity/density material.
>
> 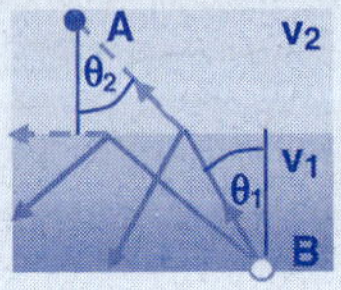
>
>
> As the wave enters a less dense material, its direction moves away from the normal to the boundary, and the refraction angle increases with the incident angle until it becomes 90° and $\sin \theta_2 = 1$. From that point on the wave is reflected internally.

Not every rock is rippable, however. The rippability of soil and rock depends on a variety of geological characteristics that are related to the hardness, stratification, and degree of fracturing.

Bad candidates are massive and homogeneous formations without weaknesses, consisting of fine-grained material held together by a solid cementing agent. How can we figure out if a cut area will require drilling or ripping when it is all covered with topsoil? The obvious answer is through a geophysical investigation. What might not be that well known is the most common method, a seismic refraction survey.

Because the speed at which shock/seismic waves travel through subsurface materials changes with material density and other factors, geotechnical engineers found that Snell's law also applies to waves passing through soil and rock. For example, the wave speed in a hard, tight rock can be up to 20,000 ft/sec (6,080 m/sec) while in loose topsoil it is only 1,000 ft/sec (304 m/sec). The geophysicists developed a method to artificially create sound-waves in a geo-material using various initiators such as a sledgehammer hitting a metallic plate planted into the material, a weight being dropped, or explosive charge. The wave that is created propagates through the subsurface at a velocity dependant on the elastic properties of the material. At a given distance from the source of the wave, a geophone is used to record the arrival of the artificial wave. The term *geophone* derives from the Greek word "geo" meaning "earth" and "phone" meaning "sound." Figure 5.17 presents the basic elements of measuring the velocity of sound waves in a subsurface material.

A sound wave is created at time = 0 at the start position, which travels in all directions, in particular toward the location of the geophone. After w seconds, the first amplitude reaches D and is captured electronically. The velocity of the sound-wave can now be easily computed by dividing distance D between the sound source and the receiver and T the elapsed time.

As each type of rock can exist in many different fractured conditions, the velocity in one and the same material may vary drastically. It is easy to understand that a lower velocity indicates an increased likelihood that a rock is rippable. However, rippability is also a function of a dozer's power and weight. Figure 5.18 highlights these two phenomena.

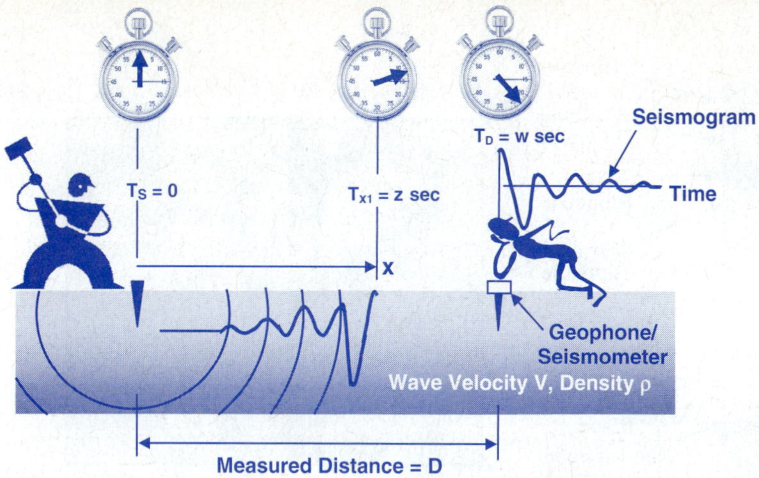

Figure 5.17 Measuring the Wave Velocity in the Ground

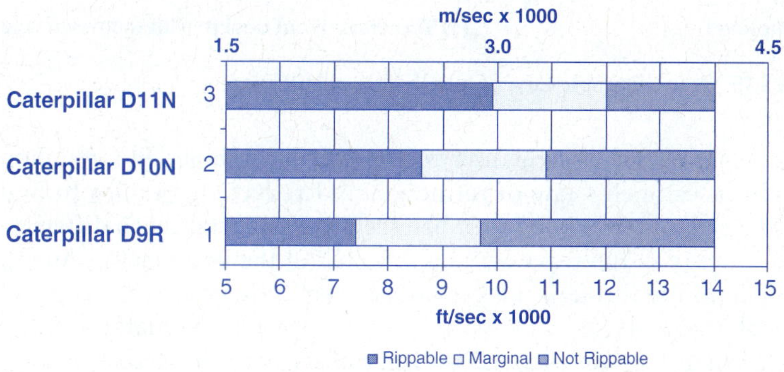

Figure 5.18 Rippability of Sandstone

As sandstone most commonly consists of layered deposits from old river basins that have been compacted over time, even relative dense material, with velocities of 10–12 ft/sec (3.05–3.6m/sec) can be ripped by large dozers. Figure 5.18 shows, however, that size matters. Not surprisingly, the 405-hp and 52-ton (47 t) D9R is not able to handle sandstone with a velocity of 10 ft/sec (3.05 m/sec) that would not create a problem for the 770 hp and 107 ton (97 t) heavy D11N. Ripping should be done in first gear because 1 to 1.5 mph (1.6 km/h–2.4 km/h) gives economical production, and because the wear to tractor undercarriage and ripper tip increases rapidly with higher speeds. When working with easily ripped materials, it is also recommended to use two or three teeth rather than increase speed.

In most circumstances is it insufficient to know the geological characteristic of the top layer, especially since we have not found its thickness. Figure 5.19 presents a case where a deep cut has to be made for a highway. While open cuts and pre-existing agent riverbanks will provide us with some information about the deeper layers, the 3-D surface between the visible points may vary widely, thus making any volume estimation a wild guess. Naturally, the financial risk in guessing wrong is significant.

The non-rippable shale in Figure 5.19 is sandwiched between two marginally rippable rock sandstones of approximately 9,050 and 10,000 ft/sec. It is apparent, however, that the thickness of the shale layer may change drastically between A and B. Knowing the contour of the top and bottom surface would allow the calculation of the exact volume of the shale that has to be drilled and blasted. Thus, the risk of underestimating the cost of excavating the cut could be eliminated. Because of the value of knowing the depth of critical rock layers, geotechnical engineers have come

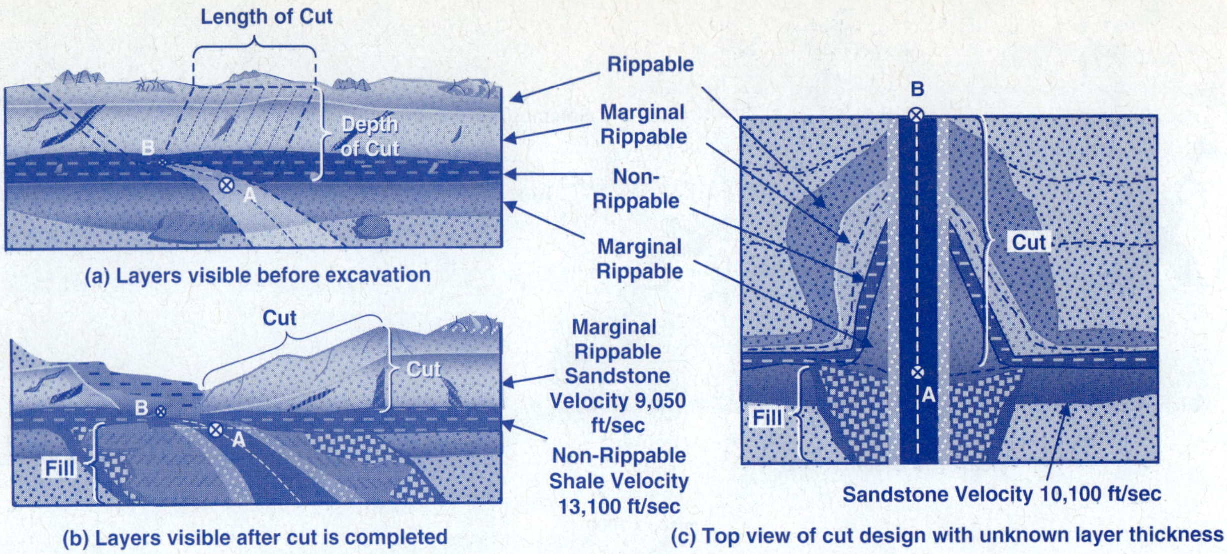

Figure 5.19 Non-Rippable Layer Only Visible at Hillside

up with a seismic soil investigation method that combines the geophone or seismometer and uses Snell's law of refraction and reflection. Similar to light, when a seismic wave encounters a boundary between two materials with different densities, some of the energy in the wave will be reflected at the boundary, while some of the energy will continue through the boundary. The acoustic waves propagate into the subsurface at a velocity dependant on the properties of the material through which they travel. When the waves reach an interface where the density changes significantly, they are either refracted or reflected.

As Figure 5.20 demonstrates, the refractions at the boundary between the first and the second layer with densities of ρ_1 and ρ_2 respectively are caused by an increase in velocity from v_1 to v_2. However, sound wave A never reaches the boundary and stays within the slowest top layer. It will reach geophones 1 and 2 before any other at T_1 and T_2, which gives us two data points to calculate an average speed $v_1 = 1/2\,(x_1/T_1 + x_2/T_2)$ or $v_1 = (x_2 - x_1)/(T_2 - T_1) = 1/\text{slope}$. The first arrival sequence changes at geophone 3 where suddenly

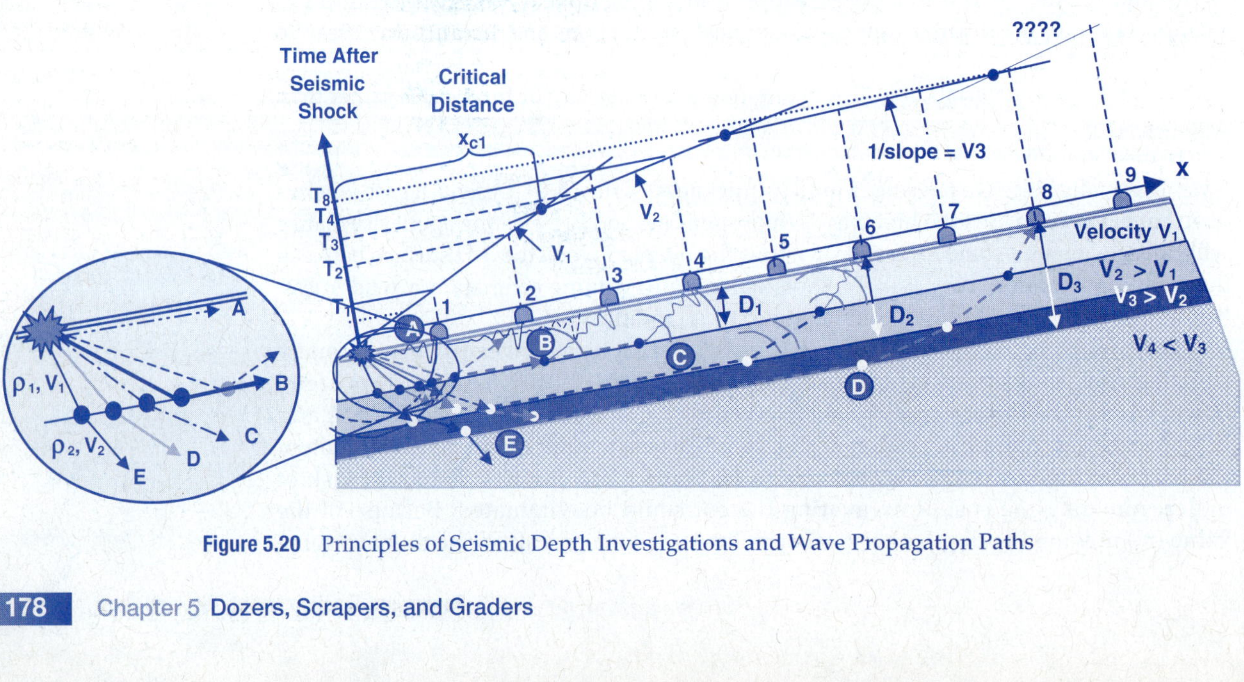

Figure 5.20 Principles of Seismic Depth Investigations and Wave Propagation Paths

wave B arrives at T_3 shortly before A. This is the result of what is called the critical refraction of wave B at the boundary. Critically refracted waves travel along the interface at the velocity of the lower layer and continually refract energy back to the surface. Receivers laid out in linear arrays on the surface are able to record the refracted and reflected waves, most importantly, the first and fastest one. Waves that have an incident angle that is larger are reflected back to the surface and are, of course, slower than the surface waves.

Waves that enter the boundary at a steeper angle, such as C, D, and E, are refracted at a smaller angle and enter the second layer to continue at a higher speed until they hit the next large change in density. As shown, wave B arrives first at geophones 3 and 4 and is second at geophone 5. Similarly to v_1 we are now able to calculate v_2 and v_3 and plot them on a time-distance graph. The intersections between the slopes indicating velocity represent the critical distances at which the wave along the next-lower faster boundary is overtaking the next-higher but slower wave. This technique works well until a deeper layer is less dense and thus slows down the wave at the boundary. For example, the fraction angle of sound wave E is getting smaller because, as we already know $v_4 < v_3$, it might not be able to refract or even reflect back to the surface. Thin layers whose influence is less than three data points cannot have velocities confidently interpreted, and typically may not be detected. Although we assumed the boundaries between layers to be planar, irregular surfaces can be readily interpreted.

In addition, the same test can be repeated parallel or perpendicular to the selected line.

The velocities can be used not only to track rippable rock but also to predict the depth of the boundaries applying the function:

$$D = \frac{X}{2}\sqrt{\frac{V_2 - V_1}{V_2 + V_1}}$$

where

$D =$ depth
$X =$ distance along x axis to the point of intersection
$V_1 =$ the velocity wave in the upper layer
$V_2 =$ the velocity in the next lower layer

Software packages are now available that analyze and present seismic velocity as a continuously varying gradient across a grid or mesh. Such packages utilize optimizing methods coupled with finite-element or finite-difference concepts to achieve interpretations. The resulting velocity gradient style of interpretation presents very different representations of the subsurface from traditional interpretations.

To estimate how much a dozer is able rip per hour it is necessary to make several rule-of-thumb decisions, sometimes with little background information. The total cycle time is the summation of rip, turn, and delay times. As we learned above, the optimal ripping velocity, v_{Ripper}, is in first gear and lies in the range of 1 to 1.5 mph (1.6 to 2.4 km/h), including slippage and stalls. Turn time is time required to turn the tractor and maneuver to begin the next pass. Turn time usually is 0.20 to 0.35 min. Thus, the hourly production of a dozer operating at 75% can be calculated as follows:

$$P_{Ripper}(BCY/hr) = 1/27 * 60\,min/hr * 0.75 * L(ft) * S_R(ft) * P_D(ft)/C_T(min)$$

where

$L =$ Length of one pass
$S_R =$ Space between two passes
$P_D =$ Depth of penetration by ripper shank
$C_T =$ Time for one pass $= (L\,(ft)/v_{Ripper} + T_{Maneuv})$ min
$BCY =$ Bank cubic yard
$LCY =$ Loose cubic yard

One might think that the depth of penetration should be the maximum possible in order to get the highest production per hour. However, if one includes the efficiency of follow-up operation, loading of the ripped material, into the overall calculation, spending a little more time on ripping may very well pay off handsomely. For example, the thickness of a layer may be less, and setting the tip to that layer can produce excellent and uniform breakage of the rock that in turn will speed up the loading either by scraper or loader. Additional benefits could be less wear on the equipment and less slippage of the tractor.

The ripper production rate is expressed in bank volume or BCY per hr (bank m^3/hr), while loading and hauling are expressed in loose cubic yards that can be calculated by considering the swell factor of the excavated material (see Chapter 4).

Worked-Out Problem 5.6 / Ripper Production

Consider a dozer with a single-shank ripper ripping rock with a swell percent of 55%. The dozer maintains an average velocity along the pass of 600 ft (182 m) of 1.3 mph (1.9 km/h) and needs 0.35 min as turn/maneuver time. The distance between two passes is 3 ft (0.9 m), and the average depth of penetration is 2 ft (0.6 m). Due to the rough work, the operator performs only 60% of the time. How much LCY or loose m^3 (LCM) has to be hauled away in order to keep up with the ripper?

$$P_{Ripper} = 1/27 * 60\,\text{min/hr} * 0.60 * L(\text{ft}) * S_R(\text{ft}) * P_D(\text{ft})/((L(\text{ft})/v_{Ripper}) + T_{Maneuv})\text{min})$$
$$P_{Ripper} = 60\,\text{min/hr} * L(\text{m}) * S_R(\text{m}) * P_D(\text{m})/((L(\text{ft})/v_{Ripper}) + T_{Maneuv})\text{min})$$
$$L = 600\,\text{ft}(183\,\text{m})$$
$$S_R = 3\,\text{ft}(0.91\,\text{m})$$
$$P_D = 2\,\text{ft}(0.61\,\text{m})$$
$$v_{Ripper} = 1.3\,\text{mph} = 5,280\,\text{ft/mile} * 1.3\,\text{mph} = 6854\,\text{ft/hr} = 114\,\text{ft/min}(34.7\,\text{m/min})$$
$$T_{Maneuv} = 0.35\,\text{min}$$
$$\text{Swell\%} = 55\%$$

Calculations

$$\text{Cycle time} = (600\,\text{ft}/(114\,\text{ft/min})) + 0.35\,\text{min} = 5.6\,\text{min}$$
$$\text{Cycle time} = (183\,\text{m}/(34.7\,\text{m/min})) + 0.35\,\text{min} = 5.6\,\text{min}$$
$$P_{Ripper} = (1/27 * 60\,\text{min/hr} * 60\% * 600\,\text{ft} * 3\,\text{ft} * 2\,\text{ft})/5.6\,\text{min} = 857\,\text{BCY/hr}$$
$$P_{Ripper} = (60\,\text{min/hr} * 60\% * 183\,\text{m} * 0.91\,\text{m} * 0.61\,\text{m})/5.6\,\text{min} = 653\,\text{BCM/hr}$$
$$\text{Haulage} = ((857\,\text{BCY/hr})/55\%) = 1,558\,\text{LCY/hr}$$
$$\text{Haulage} = ((653\,\text{BCM/hr})/0.55) = 1,187\,\text{LCM/hr}$$

Discussion

The result of the calculation shows that hauling production is almost double that of the ripping dozer. Special attention should be given to the loading process, which also has to be aligned with the 1,558 LCY/hr (1,187 LCM/hr).

Solving Header Problem 5.1 / Uphill or Downhill Loading?

Question 1: Is It Possible for a 621G or a 627G Scraper to Travel Up the Slope When Full?

Key data points to address question 1

Maximum grade to overcome = 12%

Maximum surface penetration by tires on soft dirt road = 8 in (20 cm)

Traction coefficient on soft dirt road = 0.4

Weight empty	621G 33.4 t	73,789 lb	627G 37.85 t	83,441 lb
Weight full	621G 57.4 t	126,589 lb	627G 61.8 t	136,241 lb
Weight on rear axle	621G	47% when full, 22% when empty	627G	52% when full

Figure 5.21 presents the Rimpull-Resistance curves for both scrapers.

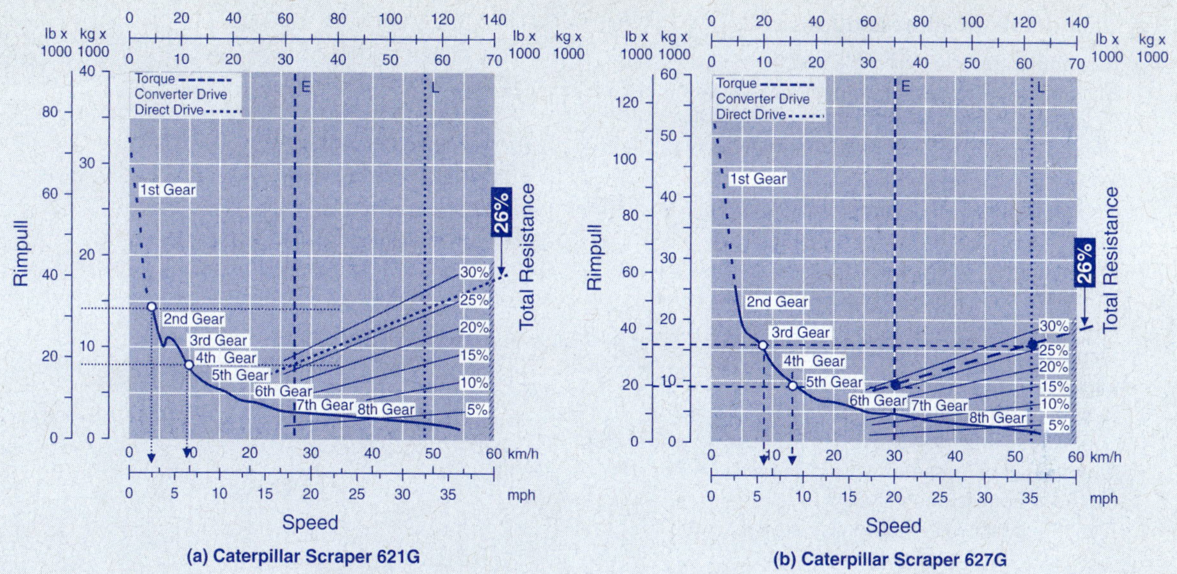

Figure 5.21 Relationships Between Total Resistance, Rimpull, and Speed

Total Resistance Driving Uphill

Grade Resistance $GR\% = 12\%$

Rolling Resistance $RR\% = (40\,\text{lb/ton} + ((30\,\text{lb/ton})/\text{in} * 8\,\text{in})/20\,\text{lb/ton/\%}) = 14\%$

Total resistance $TR\% = GR\% + RR\% = 12\% + 14\% = 26\%$

Required Rimpull

S5.8 Scraper CAT 621G
S5.9 Scraper CAT 627G

Weight of loaded 621 G = 126,589 lb (57.4 t)

Rimpull to overcome 26% resistance $= (126{,}589\,\text{lb}/2{,}000)\text{ton} * 20\,\text{lb/ton/\%}) * 26\% = 32{,}916\,\text{lb}$ (14.9 t)

Weight of loaded 627 G = 136,241 lb (61 t)

Rimpull to overcome 26% resistance $= (136{,}241\,\text{lb}/2{,}000)\text{ton} * 20\,\text{lb/ton/\%}) * 26\% = 35{,}423\,\text{lb}$ (16.1 t)

Necessary Gears According to Rimpull Tables

The 621 G has to use first gear with a speed of 2.6 mph (4.18 km/h) while the 627 G can use third gear with a speed of 5.5 mph (8.85 km/h).

Traction

Load on drive wheels of 621 G tractor when full $= (53\% * 126{,}589\,\text{lb}) = 67{,}092\,\text{lb}$ (30 t)

Max traction of 621 G on dirt road $= 67{,}092\,\text{lb} * 0.4 = 26{,}836\,\text{lb}$ (12.2 t)

Load on drive wheels of four-wheel drive 627 G when full $= 136{,}241\,\text{lb}$ (61.8 t)

Max traction of 627 G on dirt road $= 136{,}241\,\text{lb} * 0.4 = 54{,}496\,\text{lb}$ (24.7 t)

Response to Question 1: The 621 G scraper will have barely enough rimpull to drive uphill with a full bowl if the road consists, as assumed, of soft dirt. While the rimpull might be just enough to climb, the lack of traction will cause it to spin its wheels. Creating a well-maintained dirt road would reduce the TR% to 17% (12% + 5%) and the required rimpull to 21,520 lb (9.76 t) and increase the traction coefficient to 0.55 with a new traction force of 36,900 lb (16.73 t). The 627 G scraper will have sufficient rimpull and traction to drive uphill.

Question 2: Will the Scrapers Be Able to Load on the Uphill Slope on Their Own?

Key data points to address question 2

Maximum grade to overcome = 12%
Maximum surface penetration by tires on soft ground = 10 in (25 cm)
Traction coefficient on soft clay ground = 0.4

Weight empty	621G	33.47 t	73,789 lb	627G	37.8 t	83,441 lb
Weight on rear axle	621G	32%		627G	41%	
Weight full	621G	57.4 t	126,589 lb	627G	61.8 t	136,241 lb
Weight on rear axle	621G	47%		627G	52%	

Width of bowl-opening when loading = 10 ft (3.05 m)
Cohesion of clay material c = 400 lb/ft^2 (19.1 kPa)
Kp = 1 for cohesive soil
Depth of cut = 0.5 ft (0.16 m)
Height of loose overlay h_s = 2 ft (0.61 m)
Length of adhesion inside pan L = 3 ft (0.91 m)
Soil's unit weight $\gamma = 115\,\text{lb/ft}^3 (18\,\text{kN/m}^3)$ for firm clay
Unit weight of loose soil with 25% swell $\gamma_l = 88\,\text{lb/ft}^3 (13.5\,\text{kN/m}^3)$
Soil adhesion factor a = 0.5c
Dynamic friction coefficient clay-steel $\mu_1 = 0.20$

Total Resistance Loading Uphill

Rolling Resistance RR% = (40 lb/ton + ((30 lb/ton)/in * 10 in)/20 lb/ton/%) = 17%
Total resistance TR% = GR% + RR% = 12% + 17% = 29%

Required Rimpull

Empty 621 G (TR = 29%) = (73,789 lb/2,000)ton * 20 lb/ton/%) * 29% = 21,398 lb (95 kN)
Empty 627 G (TR = 29%) = (83,441 lb/2,000)ton * 20 lb/ton/%) * 29% = 24,197 lb (108 kN)
Full 621 G (TR = 29%) = (126,589 lb/2,000)ton * 20 lb/ton/%) * 29% = 36,714 lb (163 kN)
Full 627 G (TR = 29%) = (136,241 lb/2,000)ton * 20 lb/ton/%) * 29% = 39,510 lb (175.7 kN)

Traction

Maximum traction of empty 621 G in loading area = (78% * 73,789 lb) * 0.4 = 23,002 lb (102 kN)
Maximum traction of empty 627 G in loading area = 83,441 lb * 0.4 = 33,376 lb (148 kN)
Maximum traction of full 621 G = 26,836 lb and for 627 G = 54,496 lb (as on road) (242.4 kN)

Approximate Resistance to Fill Scraper When Empty

Total resistance R_T = Width * (Friction + Adhesion + Resistance to cut R_C (Coulomb simplified)/ft)

$$\mathbf{R_C = \sqrt{\left(R_p^2 + A^2\right)} \quad R_p = 1/2\,d * K_p(\gamma * d + 2q) + 2c * d\sqrt{K_p}}$$

$$\mathbf{A = a * d * (\operatorname{cosec}\alpha) + L * 0.5c}$$

At the beginning $\alpha = 0$, since soil has not reached the back of the bowl yet.
Friction causing load N_1 inside bowl = $2 * L * (d_s\gamma + h_s\gamma_l) = 6\,\text{ft}(.5 * 115 + 2 * 88)\text{lb/ft}^2 = 1,400\,\text{lb/ft}$ (20.4 kN/m)
Assumed friction causing force N_2 underneath cutting edge = 1,400 lb/ft (20.4 kN/m) (caused by hydraulic cylinder pushing edge of bowl into the soil)

$$R_P = 1/2\,d_s(d_s * \gamma + 2\,h_s * \gamma_1) + 2\,c * d_s = 0.25\,\text{ft}\,(.5\,\text{ft} * 115\,\text{lb/ft}^3 + 4\,\text{ft} * 88\,\text{lb/ft}^3) + 1\,\text{ft} * 400\,\text{lb/ft}^2 = 458\,\text{lb/ft}$$

$$A = 0.5\,c * L = 200\,\text{lb/ft}^2 * 3\,\text{ft} = 600\,\text{lb/ft}\,(8.79\,\text{kN/m})$$

$$R_C = \sqrt{(458)^2 + (600)^2} = 755\,\text{lb/ft}\,(10.8\,\text{kN/m})$$

$$Fr_s = (N_1 + N_2) * \mu = 2,800\,\text{lb} * 0.20 = 560\,\text{lb/ft}\,(8.8\,\text{kN/m})$$

$$R_T = 10\,\text{ft}(755\,\text{lb/ft} + 560\,\text{lb/ft}) = 13,150\,\text{lb}\,(58.5\,\text{kN})$$

Response to Question 2: Table 5.7 is a summary with all the pertinent results:

Table 5.7 Summary of Calculations

			621 G		627 G	
			lb	kN	lb	kN
1.	Required Rimpull for TR% Empty		21,398	95.2	24,197	107.6
2.	Required Rimpull for TR% Loaded		36,714	163.3	39,510	175.7
3.	Pull to Begin Loading		13,150	58.5	13,150	58.5
4.	Needed Rimpull to begin loading	1+3	34,548	153.7	37,347	166.1
5.	Traction Empty		23,002	**102**	**33,376**	**148.5**
6.	Traction Full		26,836	119.4	54,496	242.4
7.	Available Rimpull Empty	5-1	**1,604**	**7.1**	**9,179**	**40.8**
8.	Available Rimpull Loaded	6-1	**0**	**0**	**14,986**	**66.6**

Using the above assumptions, it becomes evident that the 621G is unable to load itself. The 627G has rimpull available to start loading to a depth of maybe 4 in (10 cm), which will cause a reduction of the friction and the cutting force. However, the continuous loading will increase in the overlay load and thus the friction inside the bowl and the resistance to cut, although the traction increases slowly as the bowl fills to 22 yd^3 (16.82 m^3).

Assuming that the scraper travels at a speed of 2.5 mph (4 km/h), this will take a travel distance of (22 yd^3 $*$ 27 ft^3/yd^3)/(10 ft $*$ ((4 in)/12 in/ft)) = 178 ft and a loading time of (198 ft/5,280 ft/ml $*$ 2.5 mph) $*$ 60 = 0.9 min. (16.82 m^3/3.05 m $*$ 0.1 m) = 55 m.

Question 3: What Is the Overall Recommendation for How Creuser Should Proceed: Option 1, Option 2, or a Third Option?

Key Data Points to Address Question 3

S5.10 Dozer Liebherr PR 754

Maximum grade to overcome = 12%

Operating weight of Liebherr PR 754 Dozer: 89,948 lb (40.8 t)

Traction coefficient for metal tracks in loose earth = 0.60

Rolling resistance = 10% (assumption)

Total Resistance Dozer Traveling Uphill

$$\text{Total Resistance (TR)} = (89,948\,\text{lb}/2,000)\text{ton} * 20\,\text{lb/ton/\%}) * (12\% + 10\%) = 19,788\,\text{lb}(88\,\text{kN})$$

Traction:

$$\text{Max traction}: 89,948\,\text{lbs} * 0.6 = 53,968\,\text{lb}\,(240\,\text{kN})$$

Available Drawbar Pull to Push Scraper:

Max traction	= 53,968 lb
Total resistance TR for dozer	= 19,788 lb
Drawbar pull available to push scraper (max traction – TR)	= 34,180 lb (152 kN)

Force needed to start loading (see above) = 13,150 lb
TR scraper empty = 21,398 lb
Total rimpull needed to start = 34,548 lb (153.7 kN)

Available scraper rimpull when empty 627 G = 9,179 lb (40.8 kN)
Pushing force needed from dozer to start loading uphill = 25,369 lb (112.8 kN)

From Figure 5.22 we learn that at a travel speed of 2.3 mph (3.7 km/h) the PR 754 will be able to produce a total drawbar pull force = 46,000 lb (= 204.6 kN), of which it needs 19,788 lb (88.0 kN) to overcome TR, leaving 26,000 lb (115.6 kN) for pushing the scraper. The maximum amount for pushing force depends on the maximum traction. For the PR 754 the maximum traction is = 53,968 lb (240 kN) at a speed of 1.95 mph (3.1 km/h).

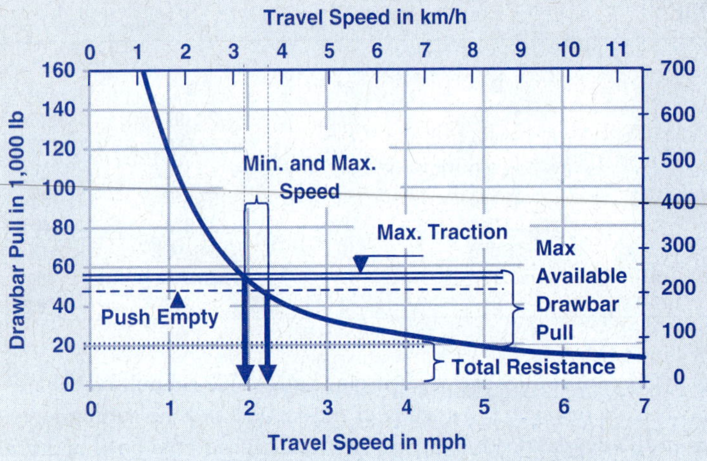

Figure 5.22 Maximum Drawbar Pull for the 627G

Response to Question 3

Option 1: With the assumed road condition, this option is not feasible for the 621 G because there is not enough traction to drive uphill on a soft dirt road. A well-maintained road, however, will allow the 621 G to travel uphill with a full load.

The 627G has no problem with this option, because the engine in the back makes it a four-wheel drive.

Loading the 621 G downhill will change 12% Grade from + to − resulting in a TR% = 17% − 12% = 5% and a TR = 3, 689 lb (16.4 kN) instead of the 21,398 lb (95.1 kN) going uphill (traction = 23,002 lb (102 kN)).

Loading the 627G downhill will change TR to 4,172 lb (18.6 kN) from 24,197 lb (107.6 kN) going uphill (traction = 33,376 lb (148 kN)).

For both scrapers, the initial force to cut the soil to a depth of 6 in, requiring an additional force of 13,150 lb (58.5 kN), will be below the limit created by traction.

The expected time for loading downhill will be 0.6 for the 627 G and 0.8 min for the 621 G.

Option 2: The 621G does not have enough traction even to begin loading uphill, while the 627G has sufficient traction and rimpull to load if the cutting edge is not lowered below 4 in. Having the Liebherr PR 754 dozer assisting will allow both scrapers to load uphill. The dozer has 34,180 lb (152 kN) available for pushing a scraper, allowing them to cut deeper than the 6 in (0.16 m) that was assumed for the calculations. Still, the trouble would be for the 621G to get out of the loading area by itself despite the large rolling resistance. Based on the calculations, it must be assumed that it would get stuck many times. Again, the 627G will be able to travel uphill after it is loaded thanks to its four-wheel drive.

Because of the rocky material that has to be expected, it is preferable to have a dozer with a ripper available on site. Also, a pushing dozer is able to reduce the loading time of a scraper by up to 50%, an issue that will be important if cost is considered. By the same token, a 627G is more expensive than the 621G because of the additional engine in the back.

In summary, with the exception of exiting the uphill loading area by the 621G, both options can work with both scrapers if the necessary conditions are created. Having the push-dozer available would be a great benefit and would provide insurance in case a scraper got stuck despite all precautions. A cost comparison of all the options should be added in order to come to a final recommendation.

5.6 Boosting Productivity with Electronics

Excavations normally require that a certain elevation and slope or grade be established. The surveyor's job is to translate the designer's plans out into the field by locating points that provide visual as well as written information concerning slopes, depth of cuts, location of centerlines, top elevations of fills, and so on. Historically, the key tools for this work were the different surveying instruments and wooden stakes with a colored flag. Each stake will have written on it some key information such as the offset distance to the beginning or end of a slope, the depth of cut or fill at the point, and so forth. The goal of construction surveys is to hammer grade stakes into the ground at locations that are offset from where the excavation will take place so that they "survive" until the completion of the project. However, under field conditions, these stakes are often lost, stolen, broken, or buried. This will require the resetting of grade stakes during operations—resulting in operational delays.

The motor grader and the dozer are the two most common pieces of equipment used for jobs that require the creation of a predefined surface contour. Commonly, the bulldozer will start by roughing out the uneven surface, and earth will be moved assisted by scrapers. The closer the excavation gets to the desired grade, the more care is needed to control the dozer and the blade. An uneven surface causes the dozer to pitch forward and backward, and with it, of course, the blade. When the undercarriage pitches forward, the blade digs deeper into soil in the front and vice versa. Thus, instead of leveling the terrain, a dozer that is not controlled expertly will actually increase the unevenness of the terrain over which it drives. A skilled operator can counter-move the blade by anticipating the pitching of the carriage. This ability to anticipate the motion of the dozer and to pitch the blade in the opposite direction takes a good deal of skill, and that skill can only be acquired through experience.

However, as almost any earthwork contractor will attest, the most complicated piece of earthmoving equipment to operate is the motor grader. Motor graders have traditionally been the domain of only the most experienced equipment operators. The complexity of the controls and the skill necessary to maintain precise grades have often led to long learning curves and years of regular use. A motor grader can have up to 14 levers in addition to the steering wheel, as shown in Figure 5.23. Before the introduction of the

VG5.1 Motor Grader

Figure 5.23 View of the Grader Control Console

The Laser Plane Technology

The term *laser* is an acronym for *L*ight *A*mplification by *S*timulated *E*mission of *R*adiation. The first working laser was demonstrated in May 1960 by Theodore Maiman at the Hughes Research Laboratories. Today, the ubiquitous laser is employed to read bar codes, to create light points, cut steel, or read compact discs. The laser range-finder measures the return-time *t* it takes a laser pulse that it sends toward a reflecting object. Since the light pulse travels at the speed of light $c = 186{,}282$ mi/s (299,792 km/s), twice the distance between the laser and the reflector the distance can be calculated as $D = (c * t)/2$. Using a rotating/spinning mirror, the distances to objects in a 360° plane can be measured. Similarly, a fast-spinning mirror is able to reflect a laser light and create a 360° laser-plane. With the help of a light-sensitive target, one is able to electronically sense a value Δ that indicates how much the center of a target is off from the plane.

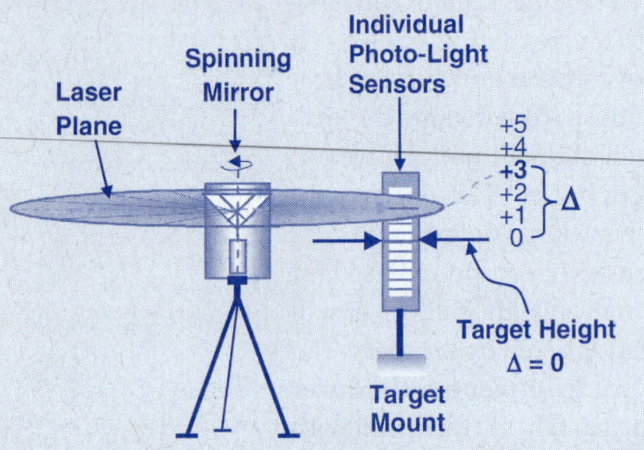

hydraulic control, an operator needed significant muscle power, leading to fatigue. Up to 8 forward and 4 reverse gears allow the operator to drive at precise speeds that match the complexity of each task. The productivity and accuracy of both the grading dozer and the motor grader depend heavily on the motor skill of the operator. However, there is still only so much that even an excellent operator can accomplish if he does not have key sensing capabilities to determine the exact elevation and slope of the blade relative to a target. One technology that has opened the door to bridge this chasm is the laser, in particular the line, the rotating or plane, and the pulsed laser. For the discussion of automation, let's consider the following road construction project for which a dozer will do the rough and the grader the fine grading.

5.6.1 2-D LASER GUIDANCE FOR THE OPERATOR

Today, electronic gauges provide critical information to any driver of a car or even pedestrians crossing a street. While pedestrians are guided by a digital time clock about the seconds until the light will turn yellow electronic displays in the car show, for example, how fast the car is traveling, the temperature in and outside, how much gas is left, or the location and direction of the car on an electronic street map. Dozer operators, however, are able to get information about the distance of the blade relative to a target elevation in 2-D or in 3-D, both made possible by lasers.

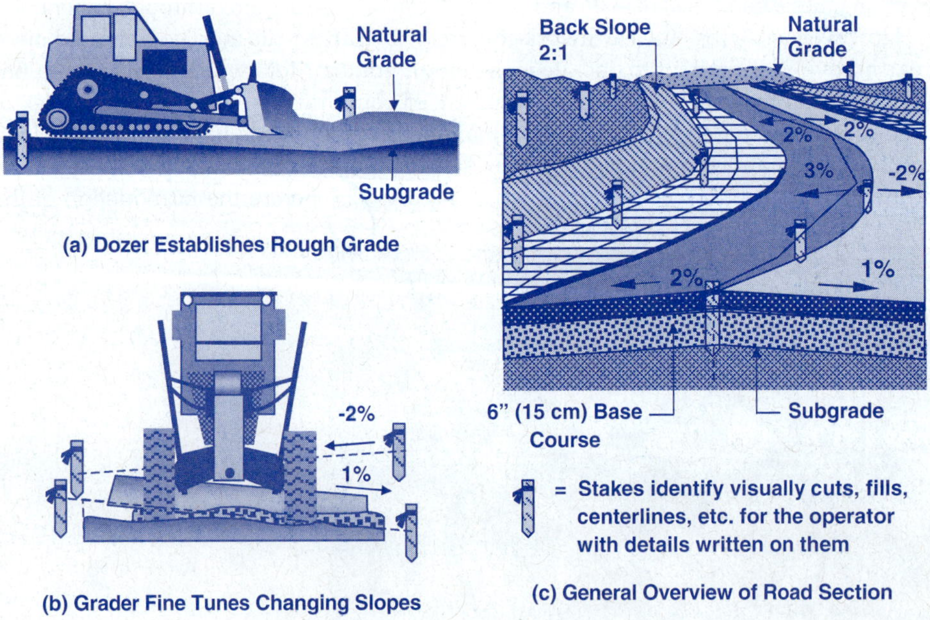

Figure 5.24 Construction Layout for Grading of Road Base to Be Done by Dozer and Grader

In 1965, Robert Studebaker invented the laser plane technology that combined early lasers with a spinning mirror. Here, laser technology is used to spread a harmless, stable plane of laser light of approximately 2,000 ft (600 m) in diameter that creates a reference for setting, adjusting, and marking elevations. By mounting a physical laser target or an electronic light sensor array on a pole connected to the dozer blade, its operator is able to gain information that will allow the precise control of the grading operation from the cab of the machine. By tilting on the mirror, the laser beam can also by aligned to sloping surfaces such as parking lots.

As Figure 5.24d indicates, curves and smoothly changing slopes are still a great challenge even with the help of a laser plane. The problem, of course, is the fact that changes in elevations are not constant (i.e., slope angle) but follow geometrical contours that are linked to design requirements. For example, the slight super-elevation of the road curve shown in Figure 5.22 requires a smooth transition from a +1% slope to a −2% and back to +2%. Today, such surfaces can be achieved, even without a guiding concrete curb, with the 3-D laser technology and the downloading of CAD data into the operator cabin.

The GPS is made up of three parts: 24 satellites orbiting the Earth in 12 hours in almost the same track; control and monitoring stations on Earth; and the GPS receivers owned by users. GPS satellites broadcast signals from space that are picked up and identified by GPS receivers. Each GPS receiver then provides the three-dimensional location (latitude, longitude, and altitude) plus the time. http://www.gps.gov/.

Due to ionospheric and tropospheric distortions of the signals, errors are introduced into the calculation of coordinates of the target receiver. Differential GPS—using the location of a known reference station to constantly recalibrate—found a way to correct most of the inaccuracies in a GPS receiver used in the vicinity of the reference station.

5.6.2 3-D GUIDANCE FOR THE OPERATOR

There are basically two approaches to provide the operator with 3-D information, the Total Station or the GPS (Global Positioning System) technologies.

A total station extended the functions of a theodolite, a precision instrument to triangulate a distant point, by adding a pulsed laser transmitter, a laser receiver, and a precision clock controlled by software. The distance to a remote beam reflector can now be obtained by measuring the time t it takes a laser pulse to reach and return from the target. Since the light pulse travels at the speed of light c = 186,282 ml/s (299,792 km/s),

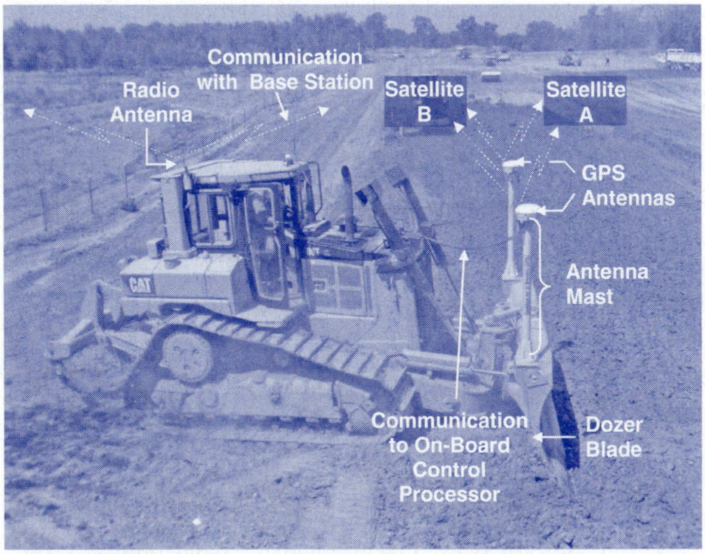

Courtesy SITECH Solutions Pty Ltd, Australia

(a) Dozer Equipped with Trimble's GPS SiteVision System

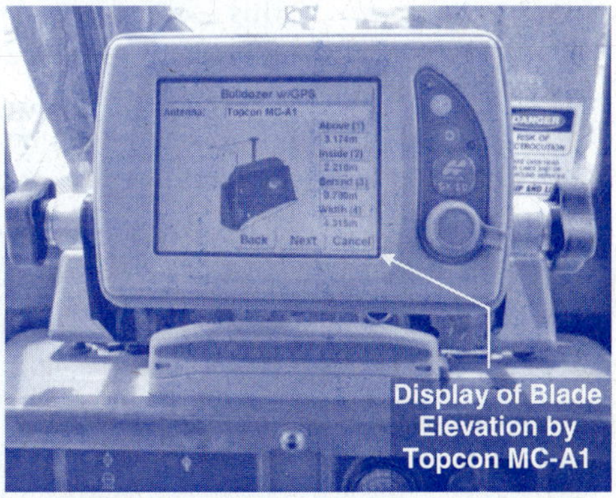

Courtesy of Dale Forsyth, Position Partners Pty Ltd, Australia

(b) Dozer Operator Display by Topcon GPS

Figure 5.25 Rotating Laser Beam Provides Invisible 2-D Reference Plane for Grading Operation

A *Total Station* is an optical instrument used in modern surveying as well as by police, crime scene investigators, private accident reconstructors and insurance companies to take measurements of scenes. It is a combination of an electronic theodolite (transit), an electronic distance measuring device (EDM), and software running on an external computer. (Wikipedia)

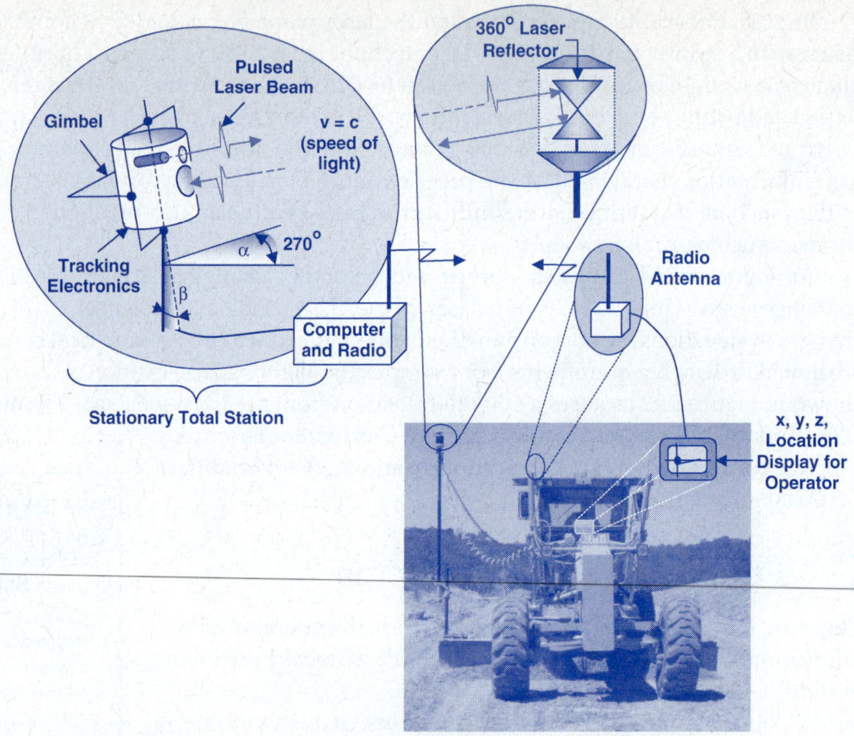

Courtesy Position Partners Pty Ltd, Australia

Figure 5.26 Concept of Real-Time Electronic 3-D Position Measurements

twice the distance between the laser and the reflector the distance can be calculated as $D = (c * t)/2$. Since the speed of the laser is so high, such calculations can be made many times a second if the controller and the clock are able to monitor accurately the time between transmission and receiving. Modern technology provides such capabilities. Still, because of the time it takes to communicate internally, to compute, and to radio results, latencies occur between the time the laser hits the reflector and the time the operator is able to read the elevation on the screen. Although minute, these latencies can add up if they are not monitored. Overall, however, this technology provides elevation accuracies in the low millimeter range (1/20 in) for distances of 200 m (600 ft). However, one total station can only work with one reflector.

The one-to-one limitation can be overcome with Differential GPS, which is not as accurate as the total station approach, but which allows several targets to get position information simultaneously. Here, the reflector on the mast is replaced by a GPS receiver feeding real-time position information to an on-board computer that simultaneously receives reference data from a target with known coordinates somewhere on the construction site. Inaccuracies due to incontrollable conditions are detected by the fixed target-station and radioed to the on-board computer to be plugged into its real-time calculations. In this scenario, multiple pieces of equipment can have multiple receivers all being fed information from the same target station. However, this solution is not cheap, as it requires the purchase of ruggedized GPS receivers and on-board computers for every machine. For

Automation

Automation (ancient Greek: = self-dictated) uses computer power and electronics to control mechanical systems and processes. These self-acting or self-regulating mechanisms are able to take over difficult, repetitive, and dangerous tasks. In certain instances electronic sensors, built into the feed-back control loop, are able to collect critical operational data in places where humans can't go, and to do so faster and more accurately.

more information about 3-D position measurement systems on the market, visit the booksite.

5.6.3 AUTOMATING THE OPERATION OF THE BLADE CYLINDERS

Despite the fact that the dozers, graders, or scrapers are full of electronic controls and that the operator receives position data electronically, operating these sophisticated piece of equipment is still done by a human. However, having the commands to the mechanical actuators and pumps all done electronically opens the door to closing some of the loops to establish fully automated feed-back controls. In other words, for suitable tasks the human can be replaced with smart and fast-reacting automatic controllers. One such opportunity is the manipulation of the blade to follow 3-D contours in real time. Closing the loop means having a computer compare the actual elevation Z_{actual} with $Z_{desired}$ at a particular (X,Y), the information radioed by a positioning system and the 3-D data provided by a computer-aided-design software (CAD). If the computer finds that the actual elevation $Z_{actual} <> Z_{desired}$ at a particular point in the x,y plane, the computer not only calculates the discrepancy $\Delta Z(x,y)$, but is also enabled to initiate corrective actions. In the example of the blade control, a computer is programmed to first decide which action should be taken, if more than one path is available, and to send commands to the cylinder(s) that move the blade. All this time, the human operator is focusing on steering the equipment safely and quickly.

Figure 5.27 highlights how the main components of an automatic blade control system are integrated. As shown, both the operator and the position controller are able to send electronic signals to the electrohydraulic valve directing the pressurized oil either to extend or retract the rod. That also means that the operator can override the automatic mechanism at any time.

Fundamental to achieving high precision is the ability to actuate the hydraulic cylinder, shown in Figure 5.27, an exact distance from the present to a desired point. This can only be achieved with feedback about the position of the rod, information that is provided by position-sensing cylinders (PSC) with the help of a linear-displacement sensor integrated with the cylinder. The cylinder feedback, labeled in Figure 5.27, sends the information concerning the end of the rod location, along the length of the stroke between A and D, to the position controller.

In order to allow the different operations to run in parallel, a servo system is commonly put between the blade control system and the hydraulic cylinder. The servo mechanism has two inputs, desired and actual position of cylinder, and one output channel, electro-hydraulic valve.

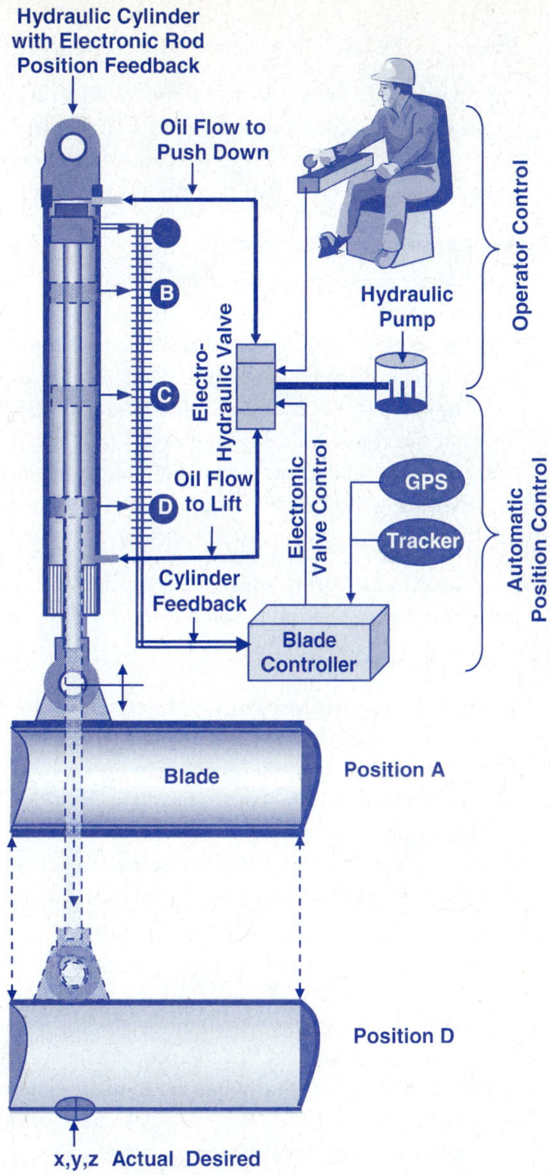

Figure 5.27 Schematic of Automatic Blade Control

The Servomechanism (Servo)

A servo is a self-contained system able to reach and stay at a desired target such as a position, temperature, pressure, or speed. A good example is the car's cruise control, which uses a closed loop feedback where discrepancies from a desired speed lead to acceleration or deceleration. Servos are commonly partially electronic in nature, using electric, hydraulic, or pneumatic motors for activation with an electronic feedback from a transducer (e.g., position-sensing cylinders). The servo compares the actual position of the activator with the desired position, which is given as an input to the servo. Any difference between the actual and wanted values (an "error signal") will be immediately corrected by sending updated signals to the motor. This corrective cycle is repeated until the error is eliminated, and begins anew with a different input or actual value.

Worked-Out Problem 5.7 / Automatic Control

The operator of a large dozer equipped with an automatic blade control system comes to you reporting that the blade on his crawler is constantly off by 5 inches (11 cm). He first thought that it was an abnormal high inaccuracy of the system, but he came to believe that the system was off since the 5-inch error seems to be consistent in contrast to the smaller random errors that occur. He wants you to check if everything is set and calibrated properly.

Assumptions and Parameters

- The site is using a subcontractor that operates and maintains a differential GPS for the dozer. It is assumed that it operates within the normal error range between +/−0.8–1.6 in (2–4 cm).
- The stroke length of the vertical blade cylinder rod is 3 ft (0.90 cm) and the linear sensor provides an analogue feedback between 0 mV and 12 mV (milli-volt) a linear indication of how much the rod is extended.
- The servo for the hydraulic blade control receives a digital input from the controller that tells it the exact amount that the rod should be extended.
- The servo sends an analogue signal to the electro-hydraulic valve. The range of the analogue signal is +/−12 mV, indicating the direction and the amount of hydraulic fluid that is allowed to pass. A negative value retracts that rod, and a positive value causes the rod to extend out further until it reaches the end of the cylinder.

The dozer is parked on the rough grade of the new highway within the reach of the GPS system. Equipped with the necessary manuals and volt testers, we climb into the operator cabin, which houses all the blade control system.

System Tests

1. The first test is to verify the correctness of the input data to the servo. From the display one can read that the GPS shows an actual blade elevation of 456 ft 6 in (139.14 m). Typing in the exact x,y-coordinates from the GPS into the CAD software provides you with the desired elevation of the rough grade, 455 ft 11 in (138.96 m). Thus the input to the servo should read +7 in (+18 cm), meaning that the cylinder rod should be extended by that amount, thus lowering the blade to the required elevation. Reading the digital display on the servo-box, we can see that this value corresponds with the input value from the blade controller.
2. The second test is the verification of the feedback from the cylinder. According to the servo, the rod is presently at position C = 17 in (42.5 cm) (see Figure 5.25). In order to verify whether it is actually at this position, we can measure the mV linear sensor on the cylinder using the voltmeter we brought along. After it is hooked up, we get a digital read-out of 3.3 mV, much less than the expected 5.66 mV. In other words, the rod is at point B instead of point C as the servo system assumes.
3. According to the operator, he observed a constant error. Thus it is unlikely that the servo program assumes a different cylinder length. It is most likely that there is a glitch in the servo control program that offsets the blade elevation by a constant 5 in. A quick check of the downloaded servo program, using the laptop, tells us that the instructions indeed included an offset value which must have "snuck" in by accident during setup. After the variable is set to 0, the servo will use the actual feedback to instruct the cylinder to extend.

5.7 Planning for Safe and Healthy Work

Dozers and scrapers have to operate in rough and dangerous physical environments. Uneven roads shock-load the speeding scraper, and ripping in rocky areas can cause the dozer to shake violently up and down. Similar to excavator operations, dust, humidity, and heat can be tiring. Let's start the review of safety mechanisms with the rollovers that can be so deadly.

5.7.1 ROLL-OVER PROTECTION SYSTEM

For dozers, scrapers, and graders, the risk of accidental overturning is significantly high, because they frequently operate on sloping uneven ground, sometimes in close proximity

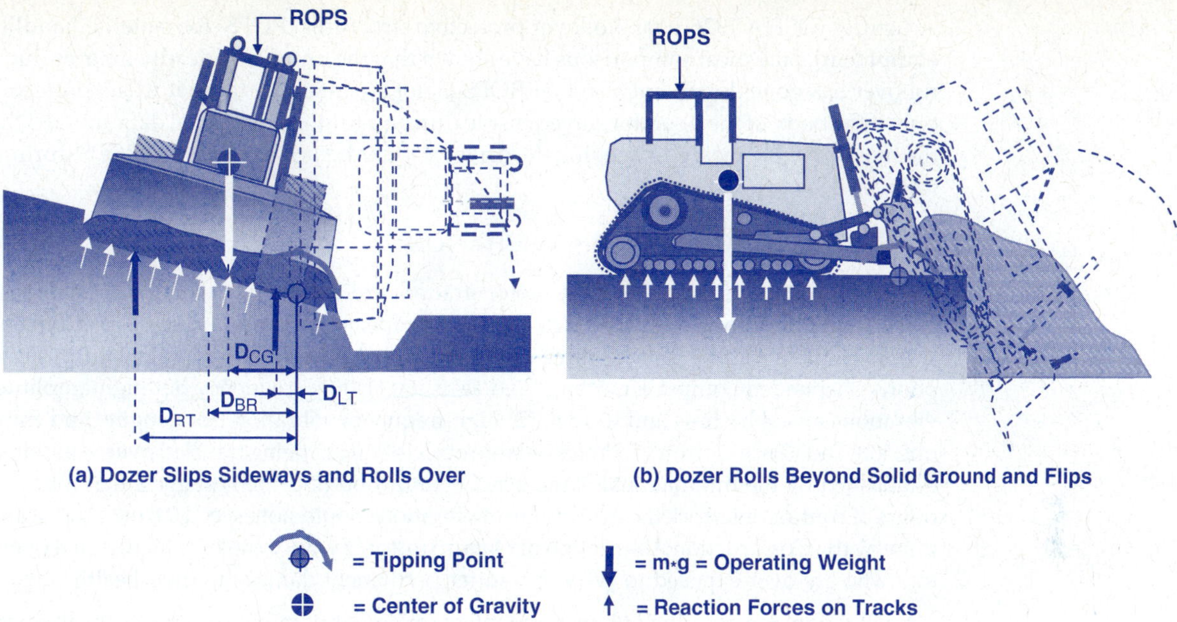

Figure 5.28 Situations with Risks for Overturning

to ditches and trenches. In the past, operators had very little in the way of body protection, as in the cases in Figure 5.28 where the heavy machine turned over.

Frequent deadly accidents involving rubber-tired and crawler tractors/loaders in agriculture, forestry, and construction led the Society of Automotive Engineers (SAE) in North America in 1974 to stipulate performance requirements for Rollover Protection Structure or ROPS (SAE J1040: Performance Criteria for Rollover Protective Structures (ROPS) for Construction, Earthmoving, Forestry and Mining Machines).

A ROPS is a strong steel frame compartment intended to protect the operator from crushing injuries if the vehicle turns over. Acknowledging that tractors have to operate in risky environments, various regulatory agencies, including the U.S. OSHA, now require and regulate both the strength as well as the energy-absorption capabilities of such

OSHA Regulations (2 excerpts)

Part Title: Safety and Health Regulations for Construction

Subpart: W
Standard Number: 1926.1000
Standard Title: Rollover protective structures (ROPS) for material handling equipment.

(a) This section applies to the following types of material handling equipment: To all rubber-tired, self-propelled scrapers, rubber-tired front-end loaders, rubber-tired dozers, wheel-type agricultural and industrial tractors, crawler tractors, crawler-type loaders, and motor graders, with or without attachments, that are used in construction work. Rollover protective structures and supporting attachment shall meet the minimum performance criteria detailed in 1926.1001 and 1926.1002, as applicable or shall be designed, fabricated, and installed in a manner which will support, based on the ultimate strength of the metal, at least two times the weight of the prime mover applied at the point of impact.

(b) Material handling machinery described in paragraph (a) of this section and manufactured on or after September 1, 1972, shall be equipped with rollover protective structures which meet the minimum performance standards prescribed in 1926.1001 and 1926.1002, as applicable.

structures (OSHA 1926.1000: Rollover protective structures (ROPS) for material handling equipment). Statistical comparisons have shown that the number of deadly injuries due to rollover has gone down, but also that ROPS is not a panacea because it was a protection only if the body of the operator stayed inside during a tumble. Empirical data showed that the most effective means of keeping the driver within the protection of the ROPS during a rollover is a seat belt.

5.7.2 MITIGATING SHOCK AND VIBRATION

Operators of dozers, scrapers, and motor graders are exposed to whole-body vibration (WBV) caused by rough vehicle rides, leading to repeated exposure to static and dynamic shocks to the entire body. The three main sources of whole-body vibration from heavy poorly-suspended equipment, cabins, and seats are (1) low-frequency but high-amplitude vibration caused by tires and terrain, (2) high-frequency vibration from engine and transmission, and (3) shock from potholes or large objects that are being traveled over. Guidelines established by the International Standards Organization (ISO 2631) suggest that for seven hours of continuous work the magnitude of vibration should not exceed 0.8 m/s2. Research done by the Construction Association of Ontario (http://www.csao.org) reveals that operators who are over-exposed to WBV can suffer permanent damage to their health, such as

- Bone damage, rubbing of bones and joints causing inflammation, especially along the backbone
- Stomach and digestive problems from shaking of organs and the abdominal cavity
- Heart problems, varicose veins, varicocele, and piles due to constant variation in blood pressure
- Disruption of the nervous system resulting in weakness, fatigue, loss of appetite, irritability, headache, insomnia, and impotence

A recent estimate shows that there are currently 540,000 operating engineers (OEs) in the United States (BLS, 2003; this estimate does not include mechanics or oilers) who are exposed to whole-body vibration. A majority of these OEs (90%) perform excavating and paving work (e.g., operating dozers, loaders, excavators), while the remaining are crane operators (10%). K.N. Kittusam and B. Buchholz (2004) reported about a study from 1987 where 352 operators of earthmoving equipment with at least three years of work experience were studied. The most frequent (and significant) health impairment among the operators was lumbar syndrome (81%) and was significantly higher than found in the control group. The authors concluded that long-term exposure to whole-body vibrations causes morphological changes in the lumbar spine

The Construction Association of Ontario sponsored a study to compare the vibrations experienced by various pieces of construction equipment (Figure 5.29). Not surprisingly, the scraper leads the list with an average of 1.6 m/s^2 (0.63 in/s^2), exactly double the level recommended by ISO 2631. Both small and large bulldozers also exceed the limit, while graders, with 5.5 m/s^2 (2.16 in/s^2), are relative safe.

There are basically three strategies to avoid the trauma of WBV: (1) risk avoidance by slowing down or driving around danger zones, (2) cabin isolation from the main chassis, and (3) seat suspension. Figure 5.30 shows how equipment manufacturers have responded to the call for more ergonomic solutions. Views into three modern equipment cabin units show that they all come with comfortable and adjustable seats that have shock absorbers. In addition, the cabins are mounted on springs that represent the second measure to prevent dangerous levels of body vibration.

Not only are the seats suspended, but they can also be adjusted to the body of each operator with flexible armrest and back supports, shock absorbers that adjust for the weight of the operator and, seats that swivel to left and right in order reduce the need for upper-body twisting.

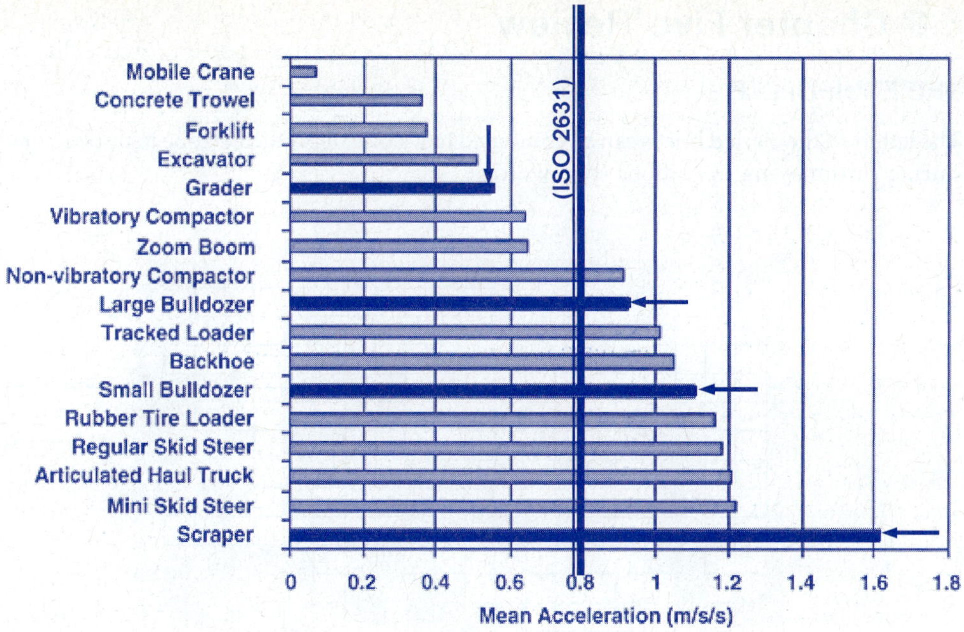

Figure 5.29 Measuring Vibration in Unprotected Cabins (CSAOhttp://www.csao.org)

While not all equipment is at the same ergonomic level, managerial techniques have to suffice. Some of the management actions that are recommended include the following:

- Include the danger of WBV in safety training
- Upgrade older equipment with ergonomic seats that mitigate WBV
- Ensure the installation of ergonomic WBV protection in new equipment
- Establish "vibration danger areas" and "vibration danger tools" where job rotation and regular work breaks can be introduced
- Label or designate all vibration hazard processes and equipment and develop safe working practices
- Carry out and record regular vibration monitoring

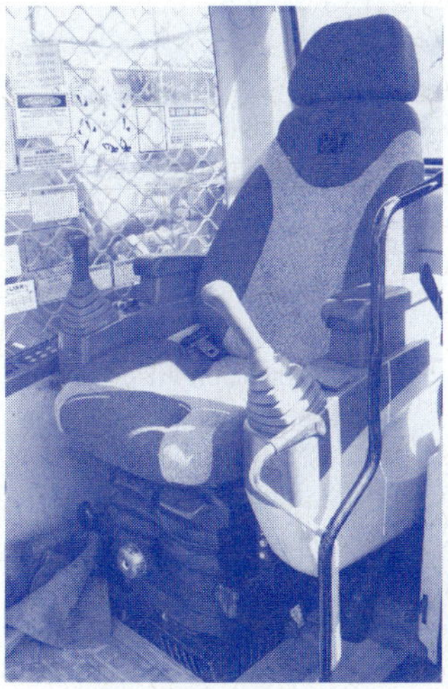

(a) Backhoe Excavator

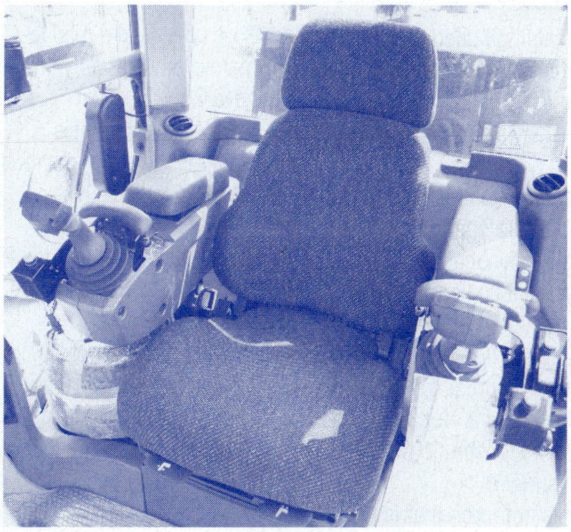

(b) Grader

(c) Skidsteer Loader

Figure 5.30 Modern Seats Providing Ergonomically Designed Work Stations

5.8 Chapter Five Review

THE EQUIPUZZLE

The letters in the circled boxes can be combined to identify a protected zone to be observed during earthmoving operations (two words).

Across

1. Last name of the person who defined the value of a horsepower
4. Greek for geophone
7. A long curved piece of straight steel mounted underneath a motor grader
8. Term coined by British soldiers who saw the first crawlers pulling their heavy material across soggy fields
10. Increases with the use of differential GPS
11. Invisible rotating light beam that can be picked up by sensors to measure their distance to a desired mark
12. Last name of a famous engineer who combined the scraper bowl with a tractor and developed skid-steer
13. Hydraulically operated attachment to a dozer to loosen hard rock/soil material
15. Hinge to attach the C-frame of a dozer
16. Leave visual indentations in the soil after a crawler drove over it
17. Friction between the treads of a wheeled vehicle and the ground
18. Short for a control system that very precisely maintains a target value
22. Time-consuming work eliminated by ripping (three words)
29. Short for a famous scraper pulled by horses, mules, and tractors
31. Short for a structure that wraps around the front of a dozer and supports the blade
32. Slang for a scraper
33. Special hitch used by scraper tractor to make tight turns

Down

2. Term used to indicate a closed-loop operation
3. Blade motion that allows a dozer to lift up heavy objects
5. The maximum force available at the hitch of a tractor (2 words)
6. Tool used by a motor grader to loosen up the road surface
9. A resistant force that is related to the condition of the ground a vehicle is traveling over (two words)
14. Slope change of a road to improve driving safety in a curve
19. A resistant force that is related to the slope of a road or ground surface (two words)
20. Special dozer blade used to push-load scrapers (two words)
21. They allow a dozer to traverse smooth surfaces without leaving heavy marks
23. Force provided by the engine measured at the rim of the tire
24. Enables the measurement of the distance between an emitting and a reflecting target using the speed of light (two words)
25. Abbreviation for a force needed to overcome gravitation
26. Mechanical device that allows a tracked vehicle to travel smoothly across obstacles on the ground
27. One source of rolling resistance to a rubber tire (2 words)
28. Hydraulic mechanism used by the scraper to push the material at of the bowl
30. Acronym for a structure that safeguards the operator of a heavy equipment from being crushed

JOURNALING QUESTIONS

1. The construction of the Erie Canal was a major achievement in its time. Discuss how the majority of the excavation was accomplished and how this would be done today. Sketch some key scenes of the two operations, today and yesterday.

2. Calculations of the traction forces for a scraper and a bulldozer are based on totally different approaches. Compare the two using appropriate models.

3. A modern motor grader is able to articulate the moldboard and the front-wheels into extraordinary configurations. Elaborate on the many ways the grader is able to operate.

4. Being able to accurately control the blade of a dozer or a motor grader in 3-D space can provide many benefits to a highway contractor. Discuss why this statement is true. How is this accomplished using modern position technologies?

BASIC CONCEPTS AND DEFINITIONS

1. ROPS are providing more protection to operators of dozers, scrapers, and motor graders. What does an equipment manufacturer have to do to certify a new ROPS? Why is it not sufficient to protect an operator from being killed?

2. Automation offers opportunities to achieve higher productivity of grading while making the work of an operator simpler. What aspects of automation are crucial for the productivity of a grading dozer or a motor grader?

3. The calculation of traction depends on a coefficient of traction. What physical phenomena are covered by this coefficient? What are its boundaries? How dependable is the value of the coefficient in representing the actual conditions?

4. Complete Table 5.2 by filling out the last column containing the rolling resistance of a rubber tire on various surfaces. *Hint:* Don't forget the fact that each tire flexes under the load.

5. To unload the scraper, the operator opens the apron and activates the ejector. Sketch a cross-section of the scraper pan and show how the operator is able to create a new layer of soil, called a lift, of 6 in (0.15 m) and not 10 or 12 in (0.25–0.3 m).

LESSONS FROM THE ACCIDENT FILE

1. Accident #1: The two people involved in this tragic accident violated basic rules of operating a bulldozer. Focusing on fixing the stubborn starter, the two working on the repair forgot the possible effect of having the engine suddenly start up with the gear engaged. Present at least two mechanical or electronic guards that would prevent such an accident.

2. Accident #2: The noisy environment of a construction site makes it hard for somebody focusing on a particular job to hear a warning sound. Come up with one managerial and one technical intervention that would have protected the spotter.

3. Accident #3: The conditions leading up to this accident contributed greatly to its occurrence. What mechanical addition would have avoided the slipping of the dozer? What could have protected the operator during the overturning of the dozer?

4. Accident #4: Similar to Accident #2, the warning signal of the motor grader was not heard. What devices could have helped the operator and the surveyor to recognize the danger?

5. Accident #5: This is an example where having a ROPS did not prevent the operator from getting crushed. What requirement and what rule did the operator of the scraper ignore?

TRADITIONAL HOMEWORK

1. You are planning to deploy a Terex TS14G Scraper to a new job to move soil with a bank density of 3,200 lb/yd^3 (1,900 kg/m^3) and a swell of 20%. You would like to ensure that there is enough rimpull to begin loading, expecting that the initial force to cut is approximately 15,000 lb (66.7 kN). Consider a negative grade of 3% and a median tire penetration of 10 in/ton (27.8 cm/t).

(a) Verify if the tractor is able to provide sufficient rimpull to begin loading.

(b) Is there enough traction?

(c) How much load is added when bowl is half-full? By how much did the traction change?

(d) The needed pulling force at half time rose to 18,000 lb (80 kN). Will the Terex be able to continue? Why did the pulling force increase?

2. Caterpillar sells an elevating scraper, the 365 hp 623G having, an operating weight empty of 81,196 lb (36.8 t) and a bowl opening of 138 in (2.5 m). Besides serving as a soil mixer, the elevator reduces the some of the cutting forces that the tractor needs to overcome.

(a) Considering the initial loading force calculations to be the same as for the Header Problem, by how many lbs does the elevator reduce the resistance to loading (1,315 lb/ft (19.1 kN/m))?

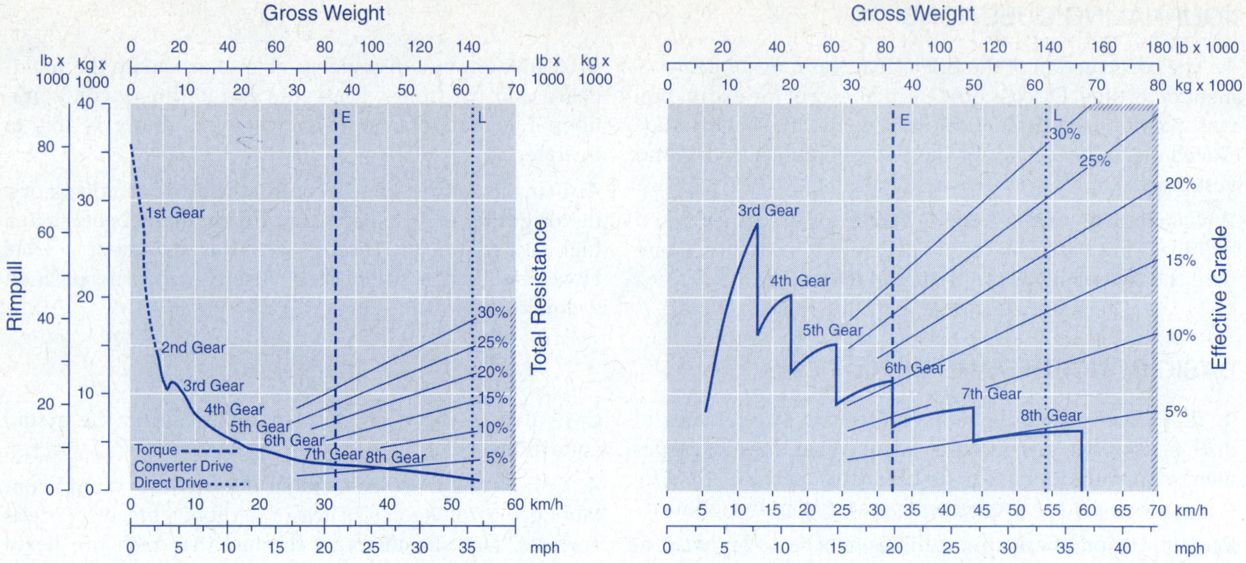

Figure 5.31 Rimpull Curves for the CAT 623G

(b) Considering the given heaped capacity of 23 yd^3 as the maximum the scraper can be loaded with gravel/sand material having a bank unit weight of 2,900 lb/yd^3 (1,720 kg/m^3) and a swell of 15%. What is its operating weight when fully loaded? How much traction can the tractor create at the same time?

(c) Nevertheless, even an elevating scraper needs rimpull to load, since it must travel in loose gravel/sand with a penetration of 12 in/ton (33.3 cm/t). Using the rimpull chart for the 623 G below and reserving 8,000 lb (35.6 kN) to overcome cutting resistance, what would be the maximum grade% that it could climb while loading?

(d) The scraper has to travel on the haul road having 8% rolling resistance. At one spot the loaded vehicle has to drive loaded downhill at a negative 18%. What is a safe speed that the retarder will allow the operator to use (Effective Grade for Retarder = Grade% − Rolling Resistance)?

3. You just bought a 140 hp Case 1650K dozer and wonder what the production will be in grading using a straight blade with the dimension of 10 ft × 4 ft (3.05 m × 1.2 m). Bob, a young operator who has finished training on a Cat, will be assigned to the Case. For the calculation assume an average push-slotting distance of 80 ft (24.4 m) (second gear) and a cut distance in first gear of twice the length of the undercarriage (12 ft (3.6 m)). Backtracking will be done 50% in second and 50% in third gear, and maneuvering + engaging time is 0.15 min. Assign an operator skill factor of 0.8 and a productive time of 50 min.

What hourly production will you expect from Bob?

OPEN-ENDED PROBLEMS

1. Scrapers can be equipped with a bail in the front and a hook in the back. This allows two scrapers to combine their rimpull and the traction forces. Explain how such a push-pull combination could be used most effectively, and how the extra cost for the attachments could be worthwhile.

2. Contractors use scrapers only up to a certain haul distance. List some alternatives and present facts and reasons why one would be preferred over the other. Can you establish a function that could be used to define the "switching" distance?

BIBLIOGRAPHY

Abdelhamid, T.S., and J.G. Everett, "Physiological Demands During Construction Work," *J. Constr. Engrg. Mgmt.* 128, no. 5 (2002).

Basarira, H., and C. Karpuz, "A Rippability Classification System for Marls in Lignite Mines," *J. Engrg. Geology* 74 (2004).

Bauer, R., W. Leung, and T. Barfoot, "Experimental and Simulation Results of Wheel-Soil Interaction for Planetary Rovers," Proc. IEEE IROSEdmonton, CAN, August 2–6, 2005.

Benoit, O., Ph. Gotteland, and A. Quibel, "Prediction of Trafficability for Tracked Vehicle on Broken Soil: Real Size Tests," *J. Terramech.* 39 (2003).

Bygden, G., L. Eliasson, and I. Waesterlund, "Rut Depth, Soil Compaction and Rolling Resistance When Using Bogie Tracks," *J. Terramech.* 40 (2004).

Elazouni, A. M., and I. M., Basha, "Evaluating the Performance of Construction Equipment Operators in Egypt," *J. Constr. Engrg. Mgmt.* 122, no. 2 (1996).

Franco, Y., D. Rubinstein, and I. Shmulevich, "Prediction of Soil-Bulldozer Blade Interaction Using Discrete Element Method," *Trans. ASABE* 50, no. 2 (2007).

ISO Standard 3471, *Earth Moving Machinery—Roll-Over Protective Structures—Laboratory Tests and Performance Requirements*, ISO, Geneva, Switzerland.

Ito, N., "Bulldozer Blade Control," *J. Terramech.* 28, no. 1 (1991).

Jonasson, S., P. Dunston, K. Ahmed, and J. Hamilton, "Factors in Productivity and Unit Cost for Advanced Machine Guidance," *J. Constr. Engrg. Mgmt.* 128, no. 5 (2002).

Karwatka, D., "R. G. LeTourneau and His Massive Earth-Moving Equipment," *Techdirections* (May, 2006).

Kittusamy, N.K., and B. Buchholz, Whole-Body Vibration and Postural Stress Among Operators of Construction Equipment: A Literature Review, NIOSH, Spokane, WA.

Kobayashi, T., H. Ochiai, R. Fukagawa, S. Aoki, and K. Tamoi, "A Proposal for Estimating Strength Parameters of Lunar Surface from Soil Cutting Resistances," *Proc. Earth & Space 2006*, League City/Houston, TX, March 5–8, 2006.

Maclaurin, B., "Comparing the NRMM (VCI), MMP and VLCI Traction Models," *J. Terramech.* 44 (2007).

Muro, T., "Tractive and Braking Performance of a Flexible Tracked Tractor Moving Up and Down Weak Sloped Terrain," *J. Terramech.* 30, no. 6 (1993).

Nardin, A., and B.A. Schrefler, "Modelling of Cutting Tool—Soil Interaction—Part II: Macromechanical model and Upscaling," *J. Comp. Mech.* 36 (2005).

Nardin, A., G. Zavarise, and B.A. Schrefler, "Modelling of Cutting Tool—Soil Interaction—Part I: Contact Behavior," *J. Comp. Mech.* 31 (2003).

North Carolina Dept. of Transportation, *Manual for Construction Layout*, NCDOT Public Information Program, Raleigh, NC, 2007.

Ok, S.C., and S.K. Sinha, "Construction Equipment Productivity Estimation Using Artificial Neural Network Model," *J. Constr. Mgmt. Eco.* 24 (2006).

Phair, M., "Information Technology," *Build. + Constr.* (February 2007).

Ren, L., Z. Han, J. Li, and J. Tong, "Effects of Non-Smooth Characteristics on Bionic Bulldozer Blades in Resistance Reduction against Soil," *J. Terramech.* 39 (2003).

Saarilahti, M., *Soil Interaction Model*, Quality of Life and Management of Living Resources Contract No. QLK5-1999-00991, Project Deliverable D, Finland, May 2002.

Scarlett, A.J., J.S. Price, and R.M. Stayner, "Whole-Body Vibration: Evaluation of Emission and Exposure Levels Arising from Agricultural Tractors," *J. Terramech.* 44 (2007).

Schreiber, M., and H.D. Kutzbach, "Comparison of Different Zero-Slip Definitions and a Proposal to Standardize Tire Traction Performance," *J. Terramech.* 44 (2007).

South Dakota Dept. of Transp., *Earthwork*, SDDOT, Bureau or Personnel, Rapid City, 2006.

Stayner, R.M., "Whole-Body Vibration and Shock: A Literature Review," Research Report, 333/2001, RMS Vibration Test Laboratory, Ludlow, UK.

Tatum, C.B., M. Vorster, and M. Klingler, "Innovations in Earthmoving Equipment: New Forms and Their Evolution," *J. Constr. Eng. Mgmt.* 132, no. 9 (2006).

Touran, A., T.C. Sheahan, and E. Ozcan, "Rational Equipment Selection Methods Based on Soil Conditions," *J. Constr. Engrg. Mgmt.* 123, no. 1 (1997).

Van Wyk, D.J., J. Spoelstra, and J.H. de Klerk, "Mathematical Modelling of the Interaction between a Tracked Vehicle and the Terrain," *J. Appl. Math. Modelling* 20 (Nov. 1996).

Wong, J.Y., and W. Huang, "'Wheels vs. Tracks'—A Fundamental Evaluation from the Traction Perspective," *J. Terramech.* 43 (2006).

chapter six

Trucks, Loaders, and the Quarry Operation

How could we get tower cranes erected without trailer rigs hauling them onto the site? They constitute the backbone of the many supply chains that begin in steel foundries, cement factories, manufacturers of mechanical systems, or rock quarries. While most trailer trucks travel the highways, many are too large to fit the lanes or too heavy for the bridges. These off-highway trucks commonly haul earth and rock within a restricted area such as a site for road construction, a mine, or a quarry. In the latter, they are served by loaders or excavators and dump the rock material into crushers. These heavy movers have to follow a myriad of federal and state laws when on roads or highways. Efforts to reduce costs have led to an increasingly sophisticated set of embedded electronics ensuring the efficient and safe operation of these expensive vehicles. In parallel, loaders have become more agile, stronger, smart, and operator-friendly. This chapter will review the historical background of these interesting pieces of equipment before we follow the supply chain of aggregates.

As in the other chapters, Table 6.1 presents the main topics that will be discussed organized in six categories.

Table 6.1 Topics Covered in This Chapter

Construction			Planning & Control		
Equipment	**Processes**	**Materials**	**Managerial-Engineering Factors**	**Rules & Standards**	**Modern Tools**
✓ Trucks ✓ Off-Road Trucks ✓ Articulated Truck ✓ Trailers ✓ Shovel Excavator ✓ Wheel Loader ✓ Feeder ✓ Jaw Crusher ✓ Conveyor ✓ Rock Screen ✓ Cone Crusher ✓ Impact Crusher ✓ Gyratory Crusher	✓ Truck Loading ✓ Securing Cargo ✓ Hauling ✓ Crushing ✓ Conveying ✓ Screening ✓ Closed Side Setting (CSS) ✓ Bill of Lading ✓ Rebar Bundling ✓ Staging Rebar	✓ Shotrock ✓ Aggregate ✓ Material Density ✓ Crushed Rock ✓ Rip-Rap ✓ Gradation #57 ✓ Gradation #67 ✓ Gradation #78	✓ Gross Vehicle Weight ✓ Cycle Time ✓ Hourly Fleet Production ✓ Grade Resistance ✓ Rolling Resistance ✓ Loading Strategies ✓ Operator Skill ✓ Balance Point Analysis ✓ Bunching ✓ Gradation Curve ✓ Conveyor Capacity ✓ Tension Force ✓ Crusher Capacity Curve ✓ Rimpull-Speed	✓ Federal Bridge Formula ✓ Truck Sizes and Weights ✓ FHWA Vehicle Classification ✓ ASTM C 33 ✓ AASHTO M 80 ✓ ASTM E 11	✓ RFID ✓ Bar Codes ✓ Telematics ✓ GPS ✓ Electronic Engine Management ✓ Geo-Fencing ✓ Trailer Tracking ✓ Traffic Guidance

6.1 From 4 to 800 HP Trucks Within One Century

Wagons pulled by horses, mules, or oxen (bullocks) were the main means for transporting freight over short or long distances before steam engines gave rise to the trains on tracks and on roads (train is from the Anglo-French word *trainer* meaning to drag or draw). While railroads (= permanent road on rails fixed to cross-ties) flourished on long-haul distances, the road-worthy but heavy steam wagons destroyed European roadbeds and were a hazard to normal traffic as well as to the investors in railroads. The legal restrictions on weight and speed hampered their growth but paved the way for wagons powered by smaller gasoline and later diesel engines.

Roman times	**Carts**	Trucks (from the Latin word *trochus*, meaning iron pulley, or the old English word *truckle*, meaning small iron wheel) or lorries got off to a slow start, since they challenged traditional ways of doing things such as the use of horses and steam engine.
1896	**The Phoenix**	In October 1896, the Daimler company produced its first "Lastkraftwagen" or LKW (powered cargo wagon), called the Phoenix. It was built with traditional wagon wheels and a 2-cylinder 4-hp engine positioned underneath the driver's seat, thus offering space for a 3-ton load (2.7 t).
1904	**Enabling Technologies**	New enabling technologies, such as ball bearings, improved spark plugs, and pneumatic tires
1910	**Main Players**	Many truck companies still operating today received their first orders, such as Mack, Scania, Magirus-Deutz, Renault, Saurer, Leyland, and Fiat.
1913	**Weight Restriction Laws**	State truck weight laws were passed to limit damage to the earth- and gravel-surfaced roads caused by the iron and solid rubber wheels of heavy trucks: in Maine (18,000 lbs GVW), Massachusetts (28,000 lbs GVW), Pennsylvania (24,000 lbs GVW), and Washington (24,000 lbs GVW).
1916	**Ready-Mixed Concrete Truck**	An important year for concrete construction was 1916, when Stephan Stepanian from Grandview, Ohio, filed a patent application for the first ready-mix truck. Because his prototype was too heavy for the roads at that time, he had to wait until 1920 when his new application was finally approved. 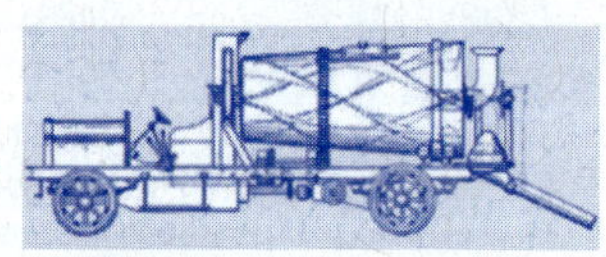 While the first trucks carried only 1 yd^3 (0.76 m^3) weighing 4,000 lb (1.8 t), today's trucks move 10 times that much while they themselves weigh about 26,000 lb (11.8 t).
1920	**Dump Truck**	Truck frames started to be equipped with wooden boxes that were hinged in the back and that could be raised in the front. The lifting was accomplished with the help of a winch-cable-pulley assembly behind the cab.
1926	**Pneumatic Tires**	Solid rubber tires for trucks were finally replaced by air-pressured tires pioneered by Goodyear and Dunlop in 1919.
1930s	**8-Wheelers**	In Great Britain, commercial vehicles were taxed based on the weight that was put on the road by one axle. This led to the first eight-wheelers in the early 1930s used specifically for heavy fuel trucks.
1930s	**Off-Road Trucks**	The first large 20 yd^3 (15.3 m^3) off-road heavy-duty rigid haulers were sold by Euclid Company of Ohio and Mack.
1930s	**Fifth Wheel**	Rigid-frame trucks had a fixed bed and also required a large turning radius. Copying from horse wagons that could be unhitched, companies like the Holland Hitch Company started to produce a similar coupling mechanism for tractors. A tractor would reverse under a trailer equipped with a kingpin that would slide into a slot in the skidplate to be locked by jaws.

1940s	Bottom Dumper	Euclid borrowed from railroad cars and used the fifth wheel coupling to design the articulated bottom dump. Extremely flexible, this combination of a tractor (wheel of track) and trailer(s) was able to haul large amounts of material and unload it easily in a windrow.
1966	Articulated All-Terrain Hauler	Volvo offered the first hauler without a separate front axle for the tractor. The four-wheel-drive articulated dump truck steered by articulating the joint between the cab and trailer, using hydraulic rams. The load capacity was 10 tons. This concept became very popular and is now being sold by all major equipment manufacturers.
2008	Ultra Class Haul Truck (Mining)	Bucyrus introduces the MT6300AC that has a diesel/AC electric powertrain driven by a 3,750 horsepower (2,800 kW) engine. It weighs 264 tons (240 t) carries a maximum load of 400 tons (363 t). Height 44 ft 4 in (13.5 m), length 51 ft 1 in (15.6 m), and width 31 ft 10 in (9.7 m).

6.2 From the Accident File

1. A grader operator was leveling dirt that trucks were dumping on the top of a four-foot high berm alongside a road. He had gotten off his grader and walked over to the driver of a two-axle tractor pulling two bottom-dump trailers. The first trailer had a single-axle with a GVW of 35,000 lb (15.8 t). The second had two axles, with a GVW of 40,000 lb (18.1 t). As the tractor stopped while making a left-side U-turn, the operator walked over and gave the driver instructions as to where to dump his load. The driver of the truck thought the victim was at a safe distance from the truck when he continued his U-turn. As the truck was completing the turn, the rear tires of the second trailer caught the victim and ran over him.
2. As two trucks were delivering asphalt to a pavement operation, one driver noticed a problem with the pneumatic braking system and stopped on the side of the road. Both drivers started to investigate the airline connections at the air tank behind the cab. The tank provided air for the brakes as well as the tailgate trip handle system locking the gate. After raising the bed of the dump truck they inspected the airlines, which revealed no problems. They proceeded to check the connections to the brakes and the tailgate system at the underside of the truck. After lowering the bed, one driver crawled under the rear and heard air escaping from the tailgate cylinder, which suddenly opened the trip mechanism and buried the victim, standing at the tailgate, up to his waist in hot asphalt. The first driver was unhurt.
3. A slip-form curber was being supplied with concrete from a ready-mix truck while the mixer of the second truck was being washed on the other side of the street. At the same time, the concrete quality inspector was taking a sample of 600 lb from the chute of the first truck directly into his wheelbarrow and was pushing it back to his truck parked on a cross-road behind the truck, where the driver was finishing the washing. The inspector crossed the road close behind the truck when the driver got back into the cabin to drive backward toward the cross-road. Not seeing anybody in his mirrors and hearing his back-up alarm, he assumed that the inspector had heard him and was at a safe distance. One has to assume that, instead, the victim tripped or fell over the wheelbarrow because he was struck and run over by the rear wheel of the ready-mix truck.
4. A truck driver was pulling a lowboy trailer with a 60,000-lb (27.2 t) wheeled front-end-loader when he suddenly had to stop, leaving skid marks. For the rigging at least four 3/8-inch (1 cm) chains rated at 6,600 pounds (3.0 t) each had been used. Due to the large braking force, the chains and rigging securing the loader broke, which caused it to roll onto the cab, crushing the driver.

5. The operator of a jaw crusher decided to clear a blockage in the feeder using a long steel bar to dislodge the rocks—without stopping the crusher. While he was working with it, the steel bar was caught by the moving jaw and forced against the plate of the fixed jaw. As a result, the upper end of the bar slammed against the operator's neck, killing him instantly.

6.3 About Trucks and Trailers

Figure 6.1 Road Signage for Trucks

According to American Trucking Association (ATA), the trucking industry in 2006 hauled 69% of the total volume of freight transported in the United States. This equates to a carrying load of 10.7 billion tons and $645.6 billion in revenue, representing 83.8% of the nation's freight bill. The Federal Highway Administration's (FHWA) Highway Statistics 2006 reports that private and commercial trucks and truck-tractors with four wheels or more amounted to 9.1 million vehicles compared to 5.2 in 2002 and 4.4 million in 1997. For that year the FHA produced a more detailed analysis that showed that 28.5%, or 1.3 million, had more than 4 axles, of which 110,000 (8.2%) were large dump trucks. An analysis of the cargo shows that 208,381 (16%) hauled construction-related materials. Applying the same percentages to 2006, we can approximate that 224,000 large dump trucks or 432,000 large trucks/trailers were working in construction. Finally, we can assume that 1.44 million trucks of all sizes (not including pickups) are involved in the construction industry, while trucking costs account for over 7 cents of every dollar of output in the construction industry (FHWA). See Figure 6.1.

In 1913 the first regulations for the operation of motorized trucks were implemented by U.S. states such as by Pennsylvania, which restricted the Gross Vehicle Weight (GVW) to 28,000 lb (125 kN or 12.7 t). The main purpose of those weight restrictions was to protect the roads from rapid deterioration due to the large forces caused by the steel or hard rubber wheels. Interestingly, the height or width was left to the trucker, who had to find a way to pass under bridges. In fact, the states were responsible for most regulations until the Federal-Aid Highway Act of 1956, which regulated the truck sizes and weights (TS&W) for trucks traveling the interstate system. The maximum width was set at 8 ft (2.4 m) and is now 8.5 ft (2.6 m). Today, interstates allow a maximum single-axle weight of 20,000 lb (9 t), tandem-axle weight limit of 34,000 lb (15.4 t) and a GVW of 80,000 lb (36.3 t). While the GVW has the absolute value, trucks also have to submit to the Federal Bridge Formula (FBF), which establishes the maximum weight based on the number of axles and the total distance between the outer tires. As the name indicates, the intention of this formula is to prevent bridges from overstressing by requiring trucks to distribute their weight onto a larger area. In other words, more axles and longer trucks are allowed higher loads.

Federal Bridge Formula

$$W = 500\left(\frac{L*N}{N-1} + 12*N + 36\right)$$

where

W = Maximum weight to the nearest 500 lbs

L = Distance in ft between the outer axle of any two or more consecutive axles

N = Number of axles being considered

The goal of the federal law is to protect the interstate highways, for which the federal government pays, while allowing industries to function efficiently. The individual states pursue their own interests when setting legal weight limits on their roads. For this reason, Congress created grandfather exemptions, which were kept in place for a certain time. Today almost all states have adopted the Interstate TS&W rules but implemented a myriad of exemptions through so-called routine or one-time permits for higher weights on the axles or the GVW. Of course, the state is collecting money for each.

6.3.1 HOW MUCH ARE YOU ALLOWED TO LOAD?

A truck, like any other piece of construction equipment, has fixed as well as variable costs related to many different factors, most of them linked to the amount of load that is being transported. It is clear that the economic goal of the truck operator is to minimize the total cost/ton. Considering the fixed costs only, one finds that loading more will lower is the cost/ton. Adding a variable cost creates a function with a slope that is negative. Figure 6.2 combines the two simplified cost functions.

Consider the hypothetical fixed, variable, and total costs of a dump truck starting at point 1 when it is driving around empty. As the truck is loaded, its fixed cost per GVW is being reduced linearly while the variable costs slowly rise due to the cost for added fuel and so on. When loaded beyond its maximum according to the FBF, at point 2, there may be a need to pay for a special permit. If that permit cost is high, the cost per GVW will step up before declining again. At point 3, the maximum GVW of 80,000 lb (36.3 t) is reached, which will require a special permit and will step up the fixed and total cost. At point 4, the maximum weight permitted in the state is reached. Comparing the curves for total costs A and B, one realizes that the slopes of the cost curves as well as the size of the permit cost dictate the load with the minimum cost. For example, for option A with the high permit cost, the minimum GVW is at point 2, while for option B it's at point 3. Since the ultimate goal of the trucker is profit, one still has to consider the revenue per ton loaded a curve that starts at zero at point 1.

6.3.2 THE MAXIMUM ALLOWABLE GROSS VEHICLE WEIGHT (GVW)

Figure 6.2 shows that not having to get special permits for loads that exceed legal limits may add cost while not adding much revenue for the extra 500 lb (0.2 t). Thus, a good strategy would be to stay below the federal limit. Let's figure out what that could be.

The first thing we see when looking at trucks a little more closely is that there are many configurations. In fact, the FHWA has defined 9 different categories from Class 5 to Class 13, as shown in Table 6.2.

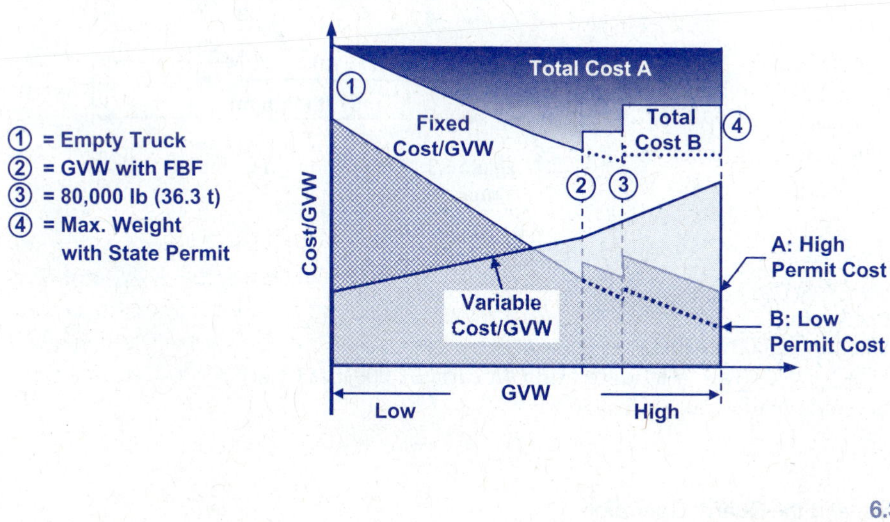

Figure 6.2 The Changing Cost per GVW for Loaded Trucks

S6.1 Truck Drivers Guidebook
S6.2 Trucks FM 5-434

Table 6.2 FHWA Commercial Vehicle Classification Schema

Vehicle Class	Schema	Description
5		Two-axle, six-tire, single-unit trucks
6		Three-axle single-unit trucks
7		Four- or more than four-axle single-unit trucks
8		Four- or less than four-axle single-trailer trucks
9		Five-axle single-trailer trucks
10		Six- or more than six-axle single-trailer trucks
11		Five- or less than five-axle multi-trailer trucks
12		Six-axle multi-trailer trucks
13		Seven- or more than seven-axle multi-trailer trucks

Worked-Out Problem 6.1 / Trucking Steel

You are working for a steel fabricator who has to ship finished steel for a new section on a construction site. The shipment consists of 31,500 lbs (14.3 t) of long main beams, 26,500 lbs (12 t) of columns, and 21,400 lbs (9.7 t) of joists. The company owns its own semi-trailers and truck tractors. You are asked to find out how many trailers are needed, without special permits for overloading, as the truck will have to drive through several states. The weight of the tractor is approximately 26,000 lbs (11.8 t) lbs and the semi-trailer 10,000 lbs (4.5 t).

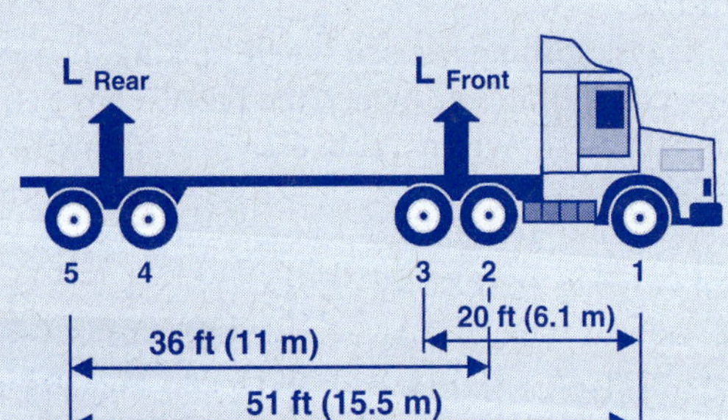

Figure 6.3 Dimensions of Company's Trailer Truck

Assumptions

(1) The steering axle carries 12,000 lbs (5.4 t).

(2) Interstate maximum for single-axle is 20,000 lbs (9.1 t).

(3) Tandem-axle weight limit is 34,000 lbs (15.4 t).

(4) Maximum GVW for truck is 80,000 lbs (36.3 t).

(5) FBF $\mathbf{W} = 500\left(\frac{L*N}{N-1}+12*N+36\right)$

Calculations

Maximum Tractor Load

Load based on axle rules : Axle 1 = 12,000 lbs (5.4 t)
Axle 2 + Axle 3 are a tandem and can carry 34,000 lbs (15.4 t)
Max axle load = 46,000 lbs (20.8 t)

$$\text{FBF}: \quad W = 500\,((3*20)/2) + 12*3 + 36) = 500*102 = 51{,}000\,\text{lbs}\ (23.1\,\text{t})$$
Taking the smaller value, we can allow $L_{Front} = 34{,}000$ lbs (15.4 t)

Maximum Trailer Gross Load

Load based on axle rules = 34,000 + 34,000 = 68,000 lbs (30.8 t)

$$\text{FBF}: \quad W = 500\,((4*36)/3 + 12*4 + 36) = 500*132 = 66{,}000\,\text{lbs}\ (29.9\,\text{t})$$
Taking the smaller value, we can allow $L_{Front} = L_{Rear} = 33{,}000$ lbs (14.9 t)

Maximum Total GVW

GVW based on the axle rule = $12{,}000 + 34{,}000 + L_{Rear} = 80{,}000$ lbs (36.3 t)

$$\text{FBF}: \quad W = 500\,((5*51)/4) + 12*5 + 36) = 500*160 = 80{,}000\,\text{lbs}\ (36.3\,\text{t})$$

Result of Calculations

The lowest allowable weight comes from the FBF for the two tandems directly below the trailer. As s result the maximum allowable GVW for the semi-trailer tractor trailer is

$$(2*33{,}000 + 12{,}000) = 78{,}000\,\text{lbs}\ (35.4\,\text{t})$$

The tanker truck industry managed to win the only exception to the use of the FBF in federal law. Two consecutive sets of tandem axles can carry a gross load of 34,000 lbs each if the distance between them is 36 ft. or more.

Discussion: Applying the axle and the FBF rules for allowable GVW calculations showed a maximum value of 78,000 lbs (35.4 t). But since the distance between the tandem axles is exactly 36 ft (11 m), we can take advantage of the only exemption, which results in an allowable GVW of 80,000 lbs (36.3 t).

The net weight of the semi-trailer truck combination amounts to 36,000 lbs (16.3 t). As a result the total additional load on one trailer can be 44,000 lbs (20 t). Thus, two semi-trailers need to be used. In order to even the load, a good distribution could be to have the first carry the large beams and the second the columns, while the joists could be split between them. The recommendation could look as follows:

$$\begin{aligned}\text{Trailer 1} &= 31{,}500\,\text{lbs} + 8{,}500\,\text{lbs} = 40{,}000\,\text{lbs}\ (18.1\,\text{t})\\ \text{Trailer 2} &= 26{,}500\,\text{lbs} + 12{,}900\,\text{lbs} = 39{,}400\,\text{lbs}\ (17.9\,\text{t})\end{aligned}$$

VG6.1 Work Trucks

6.3.3 ADDING A LIFT AXLE

It is easy to recognize that the FBF "punishes" those trucks that are short, such as dump trucks or ready-mix trucks used widely in construction. Making trucks longer and adding more axles, however, will not only make them heavier but also add to the rolling resistance and wear. Hydraulics offers a simple solution to add an extra axle that can be lifted and lowered on demand. It is estimated that 70% of all 3-axle trucks are equipped with such lift axles. Let's see what they do to the ready-mix truck shown in Figure 6.3.

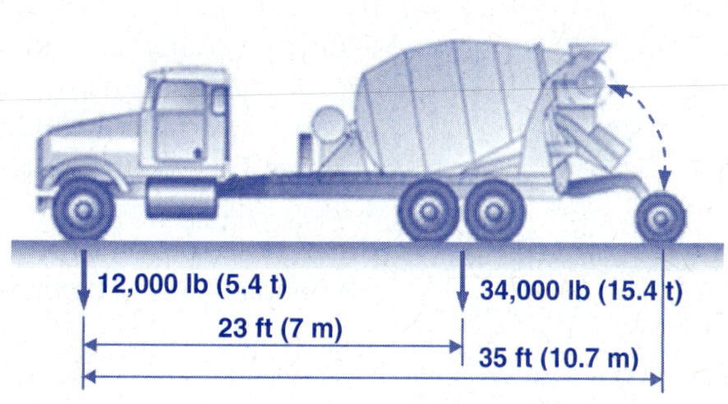

Figure 6.4 Ready-Mix Truck with Tail Lift Axle

At the tail of the truck, a mechanical arm with a fourth axle is installed. If lowered, it extends the footprint of the truck by adding approximately 12 ft (3.6 m), increasing its stability but making it more difficult to maneuver. Most importantly, it adds to the allowable GVW. Using the axle rule, the GVW increases from (12,000 lbs + 34,000 lbs) 46,000 lbs (20.8 t) to 66,000 lbs (29.9 t) by adding 20,000 lbs (9.0 t) for a single axle.

FBF with axle up: $W = 500 * ((3 * 23/2) + 12 * 3 + 36) = 53{,}200$ lbs (24.1 t)

FBF with axle down: $W = 500 * ((4 * 35/3) + 12 * 4 + 36) = 65{,}500$ lbs (29.5 t)

The quick analysis shows that without the lift axle the Maximum GVW is limited by the axle rule to 46,000 lbs (20.8 t) and increases to 65,500 lbs (29.7 t), a GVW where both rules coincide, in that the axle rule results in 66,000 lbs (29.9 t). Assuming that the chassis weighs 18,000 lbs (8.1 t) and the mixer 10,000 lbs (4.5 t), an empty weight of 28,000 lbs (12.7 t) can be calculated, leaving approximately 38,000 lbs (17.2 t) of concrete to be loaded. With these assumptions the mixer will hold 10 yd^3 (7.6 m^3) of a concrete with a unit weight of 3,800 lb/yd^3 (2.25 t/m^3).

S6.3 WorkTrucks Kenworth

6.3.4 LOADING AND SECURING TRUCKS AND TRAILERS

In more ways than one, a moving truck can become a deadly weapon where its cargo turns into flying spears or catapulted boulders. The apparent mechanism is the Newton's first law of motion. The net force on an object is equal to the mass of the object multiplied by its acceleration: $F = ma$. Thus, a truck coasting down the road at a constant speed hitting a large obstacle, like another truck or a barricade, will experience a force that causes it to decelerate. Of course, if the cargo on the truck is not secured strongly enough, it will separate and continue to travel. The force needed to keep a cargo unit attached equals the mass times the truck's deceleration.

> A truck blew a tire, lost control and hit the median. As a result, the tailgate flipped off the back of the truck and crashed into an SUV heading the opposite way, killing both people inside.

The second situation where acceleration can have strong consequences is a sharp turn. The centripetal force is being applied at the tire rim, should the road surface supply sufficient traction and keep the truck on track. However, the cargo has to be kept in place as well, since it experiences the centripetal acceleration (v^2/R). Should the rigging of the trailer or its walls not be strong enough, the cargo will accelerate outward away from the center of the turn.

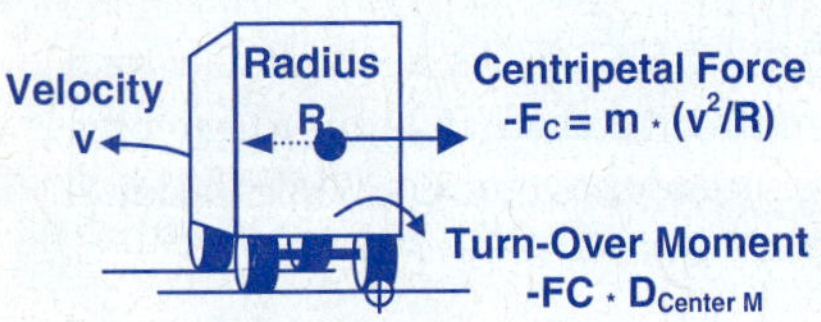

Figure 6.5 Trailer Turn Force Model

Securing Modular Cargo: The *Australian Load Restrain Guide* provides the following guideline for providing a safe rigging for a cargo unit weighing 11 ton (10 t).

As Figure 6.6 indicates, the goal is to apply force from all sides with the front, with 80% of the cargo weight, given the highest priority. Still, the two sides and the back side should be restrained with a force that is 50% of the weight.

As Figure 6.7 indicates, two main loading cases exist. Case A includes cargo that rests on the truck-trailer bed, directly or on pallets, while creating direct contact between the surfaces. Case B, however, represents loads that result in extremely small contact areas as with wheels or pipes. As shown, securing these two different cases requires drastically different strategies. Case A is able to take advantage of the friction that is created by the load and enhanced by the tie-down force. For example, if the friction coefficient between

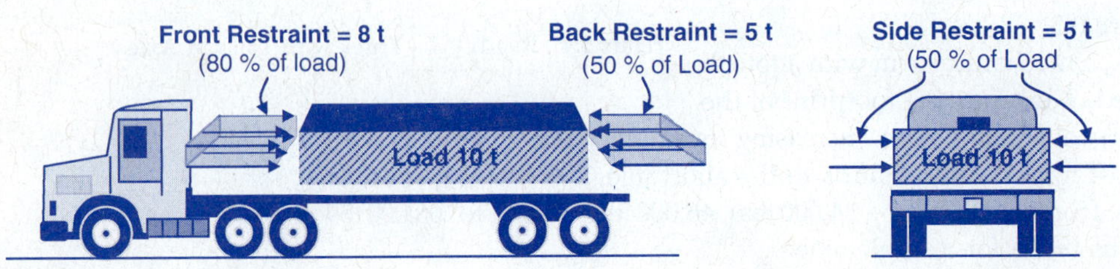

Figure 6.6 Recommendation Concerning the Application of Restraining Forces

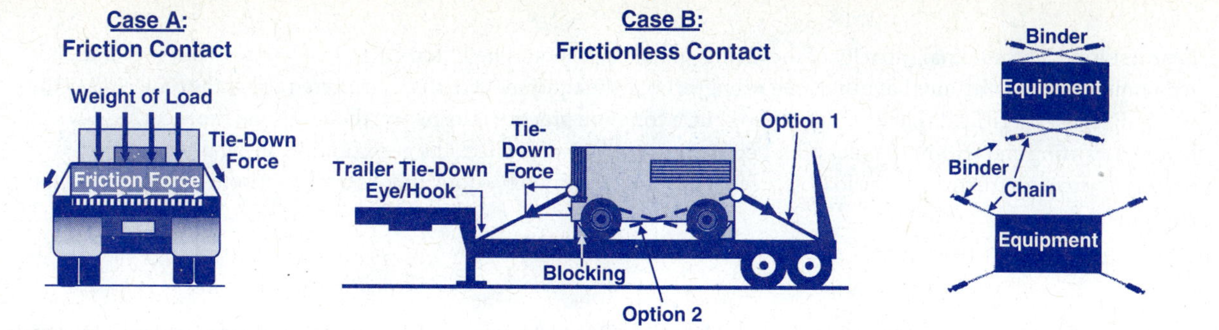

(a) Creating Friction Forces (b) Blocking and Tying (c) Symmetrical Tie-Downs (Top View)

Figure 6.7 Strategies for Securing Bulk Cargo and Equipment

the load and the pallet/truck bed is assumed to be 0.3, the load alone will create a restraining force of 30%. In order to reach the recommended 80% restraining force for a 5.5 ton (5 t) cargo, in the traveling direction an additional 50% or 2.75 ton (2.5 t) of friction force has to be created by the nylon straps or chains. With 0.3 as the friction coefficient, the tie forces in the vertical direction have to add up to 9.1 ton (8.3 t) or approximately 13.2 ton (12 t) in the combined tie-down forces.

Worked-Out Problem 6.2 / Lashing Forces to Secure Trailer Load

You are working for a construction company that owns and transports its equipment. It has just bought a mobile articulated people platform weighing 20 tons (18 t) that needs to be transported to a construction site. The company wants to know the forces to expect in the chains and binders that will be needed to secure the equipment on one of their trailers.

Assumptions:

The effects of blocking and braking the four wheels will be ignored.
The horizontal restraining force in the travel direction should be 80% of the equipment weight.
The 4 chains and binders will be arranged as shown in Figure 6.7c) in a non-crossed arrangement.
The location of the tie hooks will create a 3:1 slope of the chain in a 30 degree angle to the center axis of the trailer.
As a binder to tighten each chain, a ratchet binder turnbuckle will be used.
Safety factor for chains and binders = 5.

Calculations

Required horizontal force in traveling direction = 80% of 20 ton	= 16.0 ton (15.4 t)
Horizontal force in direction of travel per chain = 16 ton/2 chains	= 8.0 ton (7.3 t)
Horizontal component in the plane of the chain = 8 ton/cosine 30°	= 9.2 ton (8.4 t)

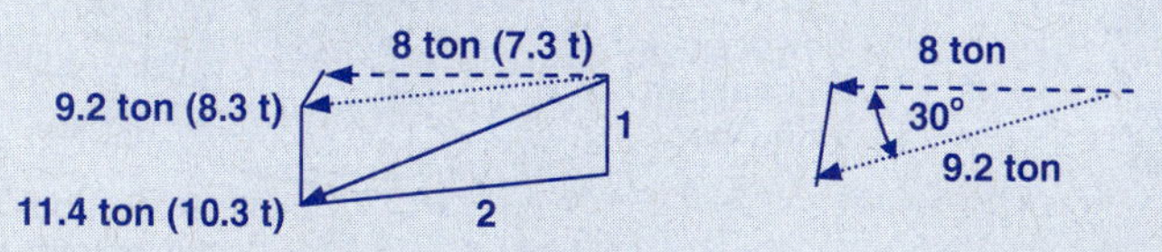

Force in chain with an incline of 3:1 = 1.162∗9.2 ton	= 10.7 ton (9.7 t)
Required loading capacity of chains and binder (safety factor) = 5∗10.7 ton	= 53.5 ton (49 t)

Discussion: The nominal capacity of the chains in the back should be 11 ton (10 t). In order to avoid mistakes, it is recommended that all chains have the same strength. Load the equipment perfectly centered above the centerline of the trailer. Because cargo loads have a tendency to shift, a common practice is to inspect the cargo and the securing devices before departing and within 25 miles after beginning a trip. Always check the cargo and securing devices as often as necessary during a trip to keep the load secured. Only use dedicated tie-down hooks or eyes at both the trailer and the equipment to be transported.

Securing Loose Material. Dump trucks haul loose material such as soil, aggregate, and dry concrete. Sometimes, they also transport viscous substances like asphalt. Anybody who has driven behind such a loaded truck that spilled or dropped small pieces of its load has experienced hazards such as broken windshields and side-slipping in curves.

In some states and on some deployment locations, it is a requirement that all loose-material loads carried in dump trucks must be covered. This includes not only construction waste material (e.g., plywood pieces) but also dry sand that can turn into sandpaper-like clouds when hit by a car. A large percentage of highway litter originates from uncovered pickups or dump trucks coming from construction sites.

One main hazard represents the tailgate that does not seal or is not locked and latched properly. However, many deadly accidents involving the truck driver happen during dumping, more exactly when the driver is opening the tailgate before raising the bed. In order to understand, let's quickly review the many different ways tailgate systems operate.

Conventional dump trucks have tailgates that can be opened by swinging motions around two pins on the bottom or the top. Generally, the top pin is only released in order to extend the bed by the height of the tailgate to transport cargo that is too long to fit inside the box. For dumping loose material, the tailgate is being opened at the bottom, as depicted in Figure 6.8a, by releasing the jaw hook arresting the bottom hinge pin. Chains on both sides of the gate prevent the gate from opening too far, and also serve as a safety should the tailgate fall off the hinges by accident.

The jaw hook can be actuated either by hand or with the help of a remotely controlled motor or cylinder situated underneath the bed of the truck, as shown in Figure 6.8b. The hand-operated lever mechanism used to be located in the center of the back, an

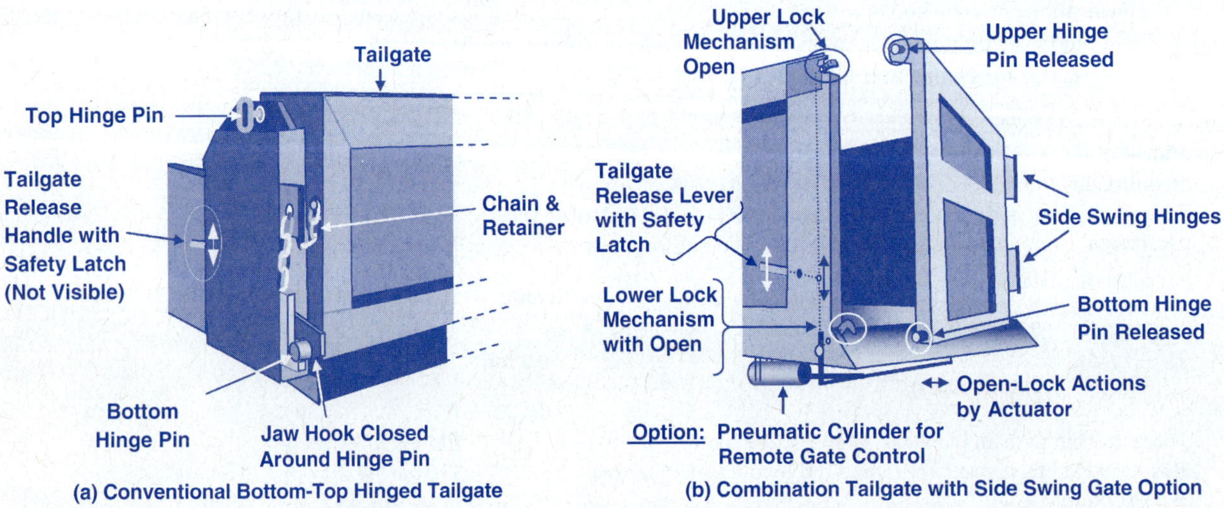

Figure 6.8 Tailgate Designs with Safe Main Release Mechanisms

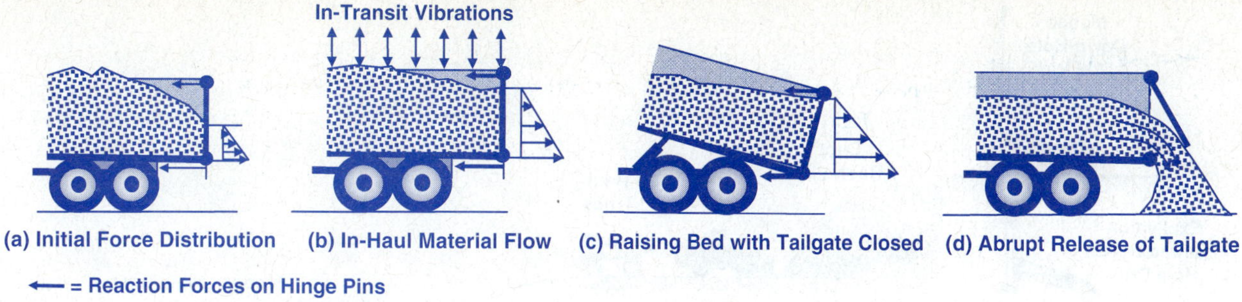

Figure 6.9 Changing Forces Caused by Shifting Loose Material Inside the Dump Box

arrangement causing many accidents, some of them deadly. Despite having main release handles located on the side or having remote controls, accidents still occur commonly when the driver is trying to fix a problem with the mechanism. Figure 6.9 demonstrates the principles of loose and heavy material that easily moves inside the box invisible to anybody outside.

Vibration is a simple way to rearrange loose material and is used to help compact gravel in road construction. During hauling of loose material across uneven surfaces, loose material inside the dump box experiences the same type of vibratory effects, here without the extra loads. Similar to liquefaction, the angle of repose is drastically reduced, albeit not totally eliminated, which results in a rearrangement of the particles as indicated in Figure 6.9b. There are many other instances that may result in a further movement of the material toward the tailgate, such as raising the truck bed to check on a problem underneath or backing down an incline. The result is the same: the heavy material pushes against the tailgate, which may cause the release mechanism to stick only to abruptly open. Should a person stand in the vicinity of the tailgate, even when the truck bed is down, large amounts of material will flow out explosively, as depicted in Figure 6.9d.

6.4 Equipment in the Quarry

According to the United States Geological Survey, the United States in 2008 produced 1.44 billion tons (1.3 billion t) of crushed stone worth $11 billion, of which 1.2 billion tons (1.1 billion t) was used as construction aggregate to produce concrete, asphalt, or road-base material.

The sources for this kind of material are metamorphic or igneous rocks that are hard and durable but that can be crushed. While such rocks are abundant and easy to mine in mountainous areas, coastal regions depend on old bedrock that rises to the surface. Because the cost of aggregates is closely linked to transportation costs, the distance from source to the location of need has to be as short as possible. As a consequence, all along the east and west coast of the United States one will find aggregate mines, commonly referred to as quarries, that are located on top of bedrock outcroppings. The main method of breaking the rock loose from its bed is to drill and blast using explosives. Naturally, over time those quarries dig deeper and deeper into the bedrock, requiring the quarry operators to install equipment that allows them to transport the mined rock to the surface. Figure 6.10 depicts a generic layout of a quarry operation that begins on top of a hard-rock bench where down-hole drills create openings in the solid rock that can be filled with explosives and charged.

The quarrying of hard rock has two distinct phases: (a) pit operation and (b) aggregate processing. Although having to work in sync, each one has a distinctive set of

VG6.2 In the Quarry

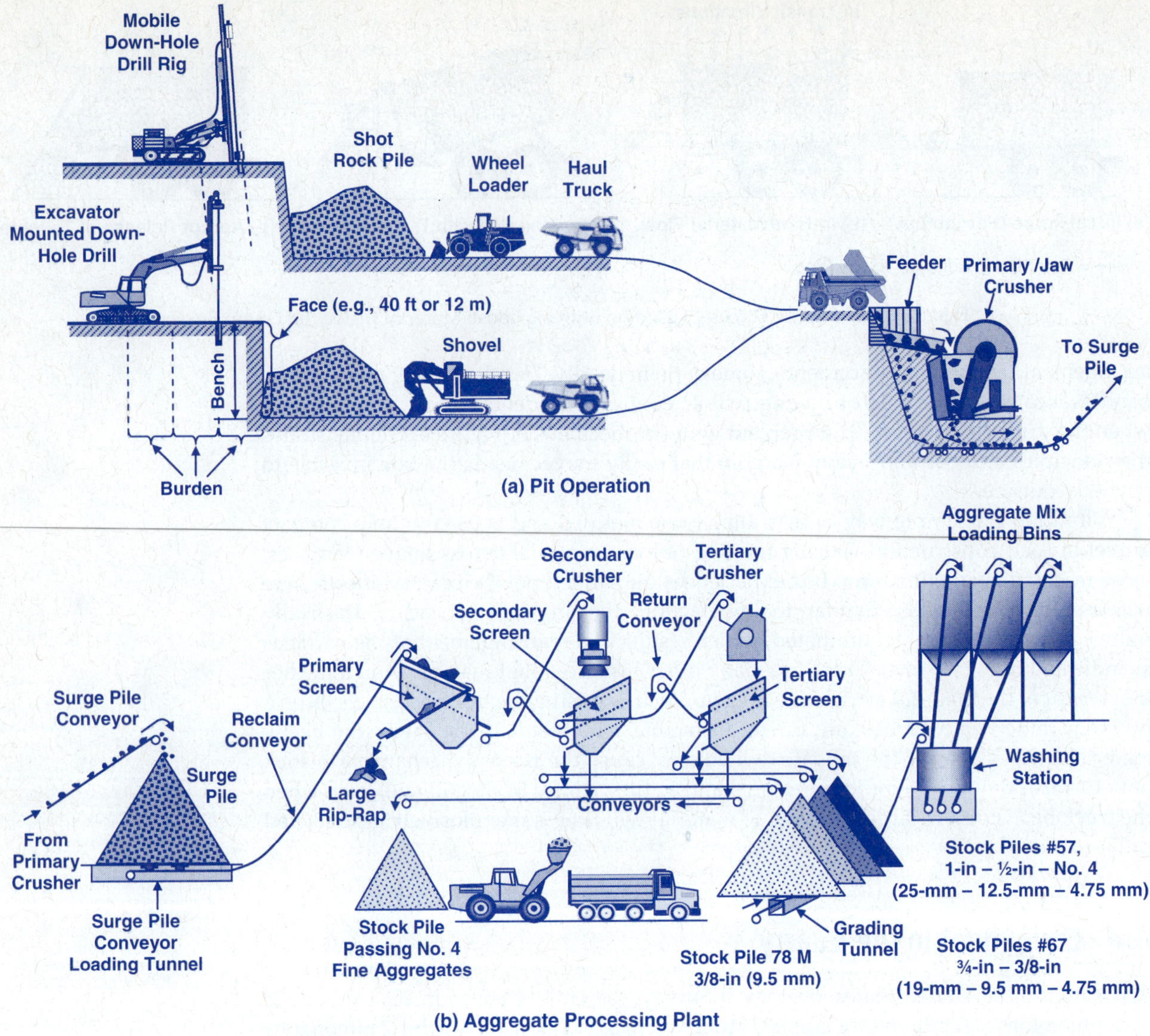

Figure 6.10 Schematic Layout of a Common Quarry Operation

equipment that needs to be designed and managed. The following will discuss each in more detail.

6.4.1 LOADING AND HAULING SHOTROCK

Figure 6.10a provides the background for this phase of aggregate production. It begins with the drills establishing several rows of holes that provide the opening for explosives to be inserted into the solid rock. As shown, the depth of the bore holes depends on the bench height that is optimal for a given type of rock (e.g., 40 ft). Similarly, the burden distance depends on the characteristics of the rock and the selected explosives. The explosives, such as ANFO (ammonium nitrate fuel oil), are delivered in trucks and fed into each borehole together with a detonator. The top of each hole is "stemmed" with quarry dust or fine aggregate to prevent the gas from the explosives from escaping into the atmosphere.

After all the detonators have been properly prepared and the quarry is cleared and the sirens are blasted to make sure that everyone nearby is warned. The detonators are subsequently connected to the electric trigger wire, and the circuit is checked.

The energy from the blast fragments and throws the rock creating a shot rock pile or shotpile in front of the bench face. The size of the rock pieces depend upon how well the blast was designed and executed. The goal is to have the rock in a size that can be easily loaded into haul trucks and fit into the primary crusher. Of course, the rock is not as consistent as one would like, resulting in boulders that are too big and need to be broken down with a separate rock hammer. The efficiencies of the tasks that follow, loading, hauling, and dumping of the shotrock, will control the productivity of the entire quarry operation. Let's look at the basic factors underlying an effective use of heavy equipment.

Truck Loading

The average blast at a mid-size quarry can range from 20,000 tons (18,200 t) to 40,000 tons (36,400 t) of rock, which requires large loaders and mining trucks. The booksite for Chapter 6 lists specification files for two common mining trucks, the CAT 775F (CAT-Truck-Mining) and the Hitachi EH1100 (Hitachi-Truck-Mining) that can be found in average-size quarries. For loaders, two efficient options are wheel loaders and shovel excavators, both effective loaders from a plane surface. The booksite for Chapter 6 contains the specifications for two of each type:

Wheel Loaders:	Liebherr L 586 2plus2 (Liebherr-Wheel-Loader)
	Komatsu WA900-3 (Komatsu-Wheel-LoaderWA900 or WA600)
Shovel Excavator:	Hitachi EX1200 (Hitachi-Shovel-Mining)
	Bucyrus RH90C (Bucyrus-Shovel-Mining)

S6.4 Loader Liebherr L 586
S6.5 Loader Komatsu WA900-3
S6.6 Shovel Hitachi EX1200
S6.7 Shovel Bucyrus RH90c

Of course, the two loading technologies cycle through a series of shovel loading and dumping motions. Figure 6.11 compares a loading cycle of the two machines from a bird's-eye and side view perspective.

The rule of thumb is that a hauling truck should be filled with four to five buckets. For example, a truck having a volume of 40 yd^3 (30.6 m^3) should be loaded with a bucket that is able to hold between 8.25 yd^3 (6.3 m^3) and 10.5 yd^3 (8 m^3). However, as we learned in Chapter 4, the theoretical capacity of a bucket is only a baseline and not the actual amount that is being dumped each time the excavator or loader empties its bucket. Not only does the density of the loaded rock come into play, but equally important is the skill of the operator and the relationship between the height of the stockpile and the reach of the excavator or loader.

The key of an optimal bucket-filling motion, as shown in Figure 6.11c–e, is to end up with a 100% full bucket when it reaches the top of the face. Figure 6.11d portrays the motion for filling the bucket to a heaped level. While the bucket moves upward, it is continuously being curled (rotated) so that the opening points upward, allowing a heap of rock above the bucket rim. We have already learned in Chapter 4 that this final capacity is called the heaped volume, an amount that depends on the condition of the material and that is expressed with a fill factor as a percentage of the theoretical bucket capacity. According to Table 4.3, the bucket fill factor for well blasted rock, or shotrock, is between 75% and 90%.

If the bucket is not full, less than the desired volume is delivered to the truck and if filled too early, the overflow material will fall out of the bucket or from the top of the face to the floor, as shown in Figure 6.11c. The latter case requires extra time for cleaning up the floor at the beginning of the next cycle, leading to a filling motion that starts with material already in the bucket. Inadvertently, this leads to the bucket being filled prematurely, resulting in even more material at the top of the face to fall. It is therefore recommended to work with the operator to find the balance point in terms of bench height that allows a

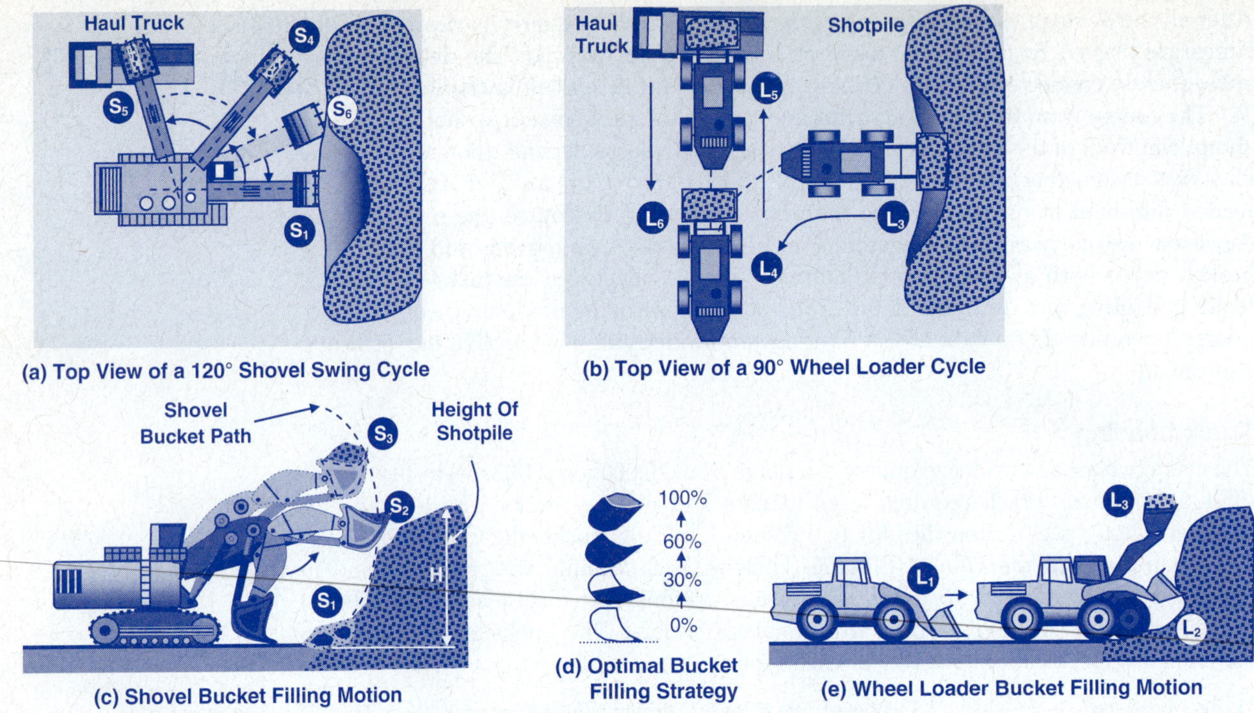

Figure 6.11 Loading Processes for Shovel Excavators and Wheel Loaders

completion of the upward path with an optimal face height. A rule of thumb is that it should be around 70% of the bucket path. The second issue is the amount of spillage that may occur during swinging and driving, as the soil and rock may not be cohesive enough to sustain the vibration and may fall off. This might suggest a reduction of the heap size on the bucket.

The swinging and the traveling motions of the shovel and the wheel loader represent nonproductive times that should be held at a minimum. As a general rule, the closer the haul truck parks relative to the place where the bucket is filled, the shorter the nonproductive time will be. Figures 6.11a and b depict two common scenarios for loading trucks. For example, the shovel excavator may have to rotate 120° to dump the material into the truck bed box. Is this really the best location for parking the truck for loading? Figure 6.11c also presents the repetitive phases of a shovel excavator cycle, which dictate the production rate of the equipment. Figure 6.12 presents the function to calculate the volume loaded per hour.

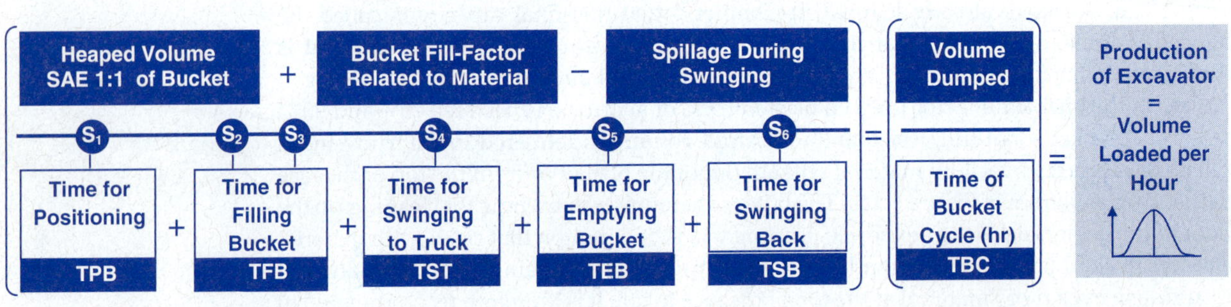

Figure 6.12 Calculation of the Production Rate of an Excavator (Shovel or Backhoe)

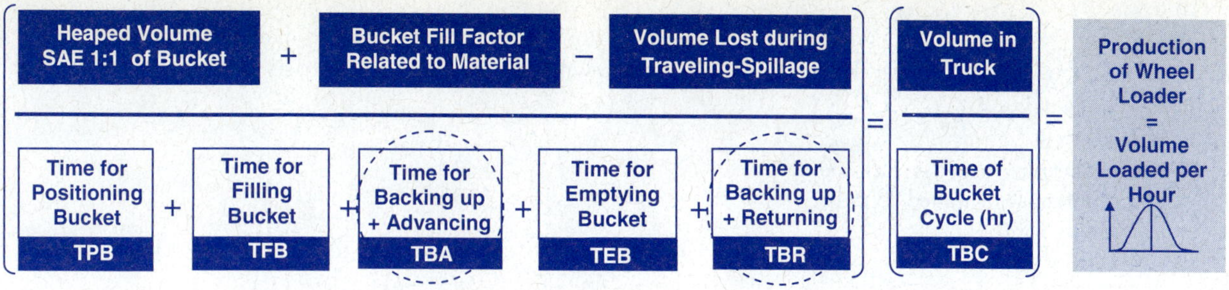

Figure 6.13 Calculation of Cycle Time and Production for Wheel Loader

The cycle is lined up in six steps S_1–S_6 but five time increments TPB–TSB. The discrepancy in the number originates in the two options for completing the filling motion that start the swing. Theoretically, the bucket is full at S_6, but the operator might not be aware or not be attentive enough and lift the bucket unnecessarily higher, up to S_3. The time of the bucket cycle as well as the volume will vary around a mean value and will shift up or down depending on (1) the operator's skill, (2) the condition of the rock, (3) the power exerted by arm, and (4) the speed with which the excavator is able to rotate around its center axis.

A wheel loader will require a slightly different layout as the equipment is mobile and has to maneuver, as indicated in Figure 6.11b. Naturally, the wheel loader operator will vary the position where he or she digs into the soil, making it more difficult to minimize its travel distance. Good communication between operator and truck drivers to coordinate parking positions is very important. As always, establishing a routine that is simple to understand and follow is needed for safe operation. Like the shovel excavator, wheel loader operators repeat several loading cycles in order to fill a truck.

The two dashed circles in Figure 6.13 highlight the only difference between the shovel excavator and loader cycle. Instead of just swinging the bucket, the loader must maneuver in reverse and in forward to approach the truck and to return to the digging position.

Worked-Out Problem 6.3 / Shotrock Loading

You are working for HardQuarry, Inc., which mines aggregates for the construction industry. The management intends to beef up the capacity of the entire plant after it invested in expanding the crushing capacity last year. Now they ask for a study to optimize the loading and hauling of shotrock inside the pit to feed the primary crusher. There, the average bench height is 40 ft (12.2 m). At present they use two Caterpillar off-highway trucks 775F, holding 70 tons (63.5 t) with a heap capacity of 55.6 yd^3 (42.5 m^3), to transport the material. The longest mean cycle time of a truck to dump a load is 18 minutes, with a standard deviation of 1.5 minutes. The shortest, when the loading is closest to the primary crusher, is 15 minutes with the same standard deviation.

Management wants to compare two options for buying a new piece of loading equipment: (1) the Komatsu wheel loader WA900, or (2) the Bucyrus shovel excavator RH90C. The question is which of the two options would fit the CAT truck better.

Background Information and Assumptions

Table 4.3 shows rock having a bank density of 4,600 lb/yd^3 (2,730 kg/m^3) and a load factor of 61%.

We assume that both pieces of equipment will be run by qualified operators able to meet the production factors shown below on a consistent basis.

S6.8 Truck CAT 775F

CAT truck 775F

Loading height (rim of the truck body)	= 13 ft (4 m)
Maximum capacity heaped 2:1 (SAE)	= 55.6 yd^3 (42.5 m^3)
Inside body width	= 13 ft (4 m)
Gross operating weight	= 242,000 lb (110 t)
Weight empty	= 99,615 lb (45.2 t)
Max load capacity	= 142,385 lb (64.4 t)

Weight of rock in truck (lb) = (55.6 yd^3 ∗ 4,600 lb/yd^3) ∗ 0.61 = 156,013 lb

Weight of rock in truck (t) = 42.5 m^3 ∗ 2,730 kg/m^3 ∗ 0.61 = 70.7 t

Loading the truck to a heaped 2:1 volume makes it too heavy. We need to reduce the amount so that the weight meets maximum capacity.

Maximum volume to be loaded = 142,385 lb/(4,600 lb/yd^3 ∗ 0.61) = 50.7 yd^3

Maximum volume to be loaded = 64,400 kg/(2,730 kg/m^3 ∗ 0.61) = 38.7 m^3

Komatsu WA900

Engine power net	= 638 kW (856 HP)
Heaped capacity	= 17.0 yd^3 (13.0 m^3) (WA900 brochure, p. 11)
Bucket fill factor	= 90%
Actual capacity	= 15.3 yd^3 (11.7 m^3)
Dumping clearance, max. height at 45° dump angle	= 15.3 ft (4.6 m) (sufficient for the 775F)
Bucket width	= 15.9 ft (4.8 m)
Speed in first gear forward	= 4.3 mph (7 km/h) (WA900 brochure, p. 10)
Speed in first gear reverse	= 4.4 mph (7.1 km/h) (WA900 brochure, p. 10)
Average travel distance to dump (reverse and forward)	= 75 ft (23 m)
Time to position the bucket for dig	= 1 sec (mostly while advancing)
Time to fill bucket	= 14.0 sec
Time for dumping bucket	= 5.0 sec
Accelerations, breaking, direction change	= 3.0 sec
Spillage during travel	= 6%

Bucyrus RH90C

Engine power net	= 858 kW (1,150 HP)
Heaped bucket capacity SAE 1:1	= 15.7 yd^3 (12.0 m^3)
Bucket fill factor	= 90%
Actual volume in bucket	= 14.1 yd^3 (10.8 m^3)
Max. dumping height	= 33 ft (10.1 m) (sufficient for the 775F)
Bucket width	= 6.25 ft (1.9 m)
Max. swing speed	= 4.6 rpm
Time for 120° swing (0.3 r ∗ 60 sec/min)/ 4.6 rpm	= 4.0 sec
Speed-up and breaking of swing	= 2.0 sec
Time to fill shovel bucket	= 17.0 sec
Time to position the bucket for the next dig	= 3.0 sec
Time to dump full bucket	= 5.0 sec
Spillage during swing	= 4%

Calculations

Loading 775 F by Komatsu WA900 wheel loader

Maximum volume to be loaded in one truck	$= 50.7\,yd^3$	$38.7\,m^3$
Volume in one loader bucket	$= 15.3\,yd^3$	$11.7\,m^3$
Spillage during travel 6%	$= 0.9\,yd^3$	$0.7\,m^3$
Volume of rock loaded during each cycle	$= 14.4\,yd^3$	$11.0\,m^3$

Number of loader cycles required to fill truck $= 50.7\,yd^3/14.4\,yd^3 = 3.6$ cycles $= 4$ cycles

(Remark: In case the weight/volume of the rock is lower, 55.6 yd^3 could be loaded. The number of cycles would be 3.86, still requiring 4 cycles)

Cycle time for 1 load = TBC = TPB + TFB + TBA + TEB + TBR (see Figure 6.11)

$$TBC = 1\,sec + 14\,sec + ((75\,ft/4.35\,mph) + 3\,sec) + 5\,sec + ((75\,ft/4.35\,mph) + 3\,sec) =$$
$$26\,sec + 2 * ((75\,ft * 3{,}600\,sec/hr)/(((4.4 + 4.3)/2)\,mph * 5{,}280\,ft/mile) = 49.5\,sec$$
$$TBC = 26\,sec + 2 * ((23\,m * 3{,}600\,sec/hr)/((7 + 7.1)/2)\,km/h * 1{,}000\,m/km) = 49.5\,sec$$

Time for loading one truck $= 4 * 49.5\,sec = 198\,sec = 3.3\,min.$

Loading the 775 F with Bucyrus shovel excavator:

Volume to be loaded onto the truck	$= 50.7\,yd^3$	$38.7\,m^3$
Actual volume in bucket	$= 14.1\,yd^3$	$10.8\,m^3$
Spillage during swing 4%	$= 0.56\,yd^3$	$0.4\,m^3$
Actual amount of rock dumped per cycle	$= 13.5\,yd^3$	$10.4\,m^3$

Total number of cycles needed to load $= 50.7\,yd^3/13.5\,yd^3 = 3.75$ cycles $= 4$ cycles

(Remark: In case the weight/volume of the rock is lower, 55.6 yd^3 could be loaded. The number of cycles would be 4.11, slightly more than 4 cycles. If we could bring the bucket fill factor from 90% to 95%, the amount of rock dumped per cycle would be closer to 14.8 yd^3 and the necessary loading cycles 3.9.)

Cycle time for 1 load = TBC = TPB + TFB + TST + TEB + TSB (see Figure 6.10)

$$TBC = 3\,sec + 17\,sec + 4\,sec + 2\,sec + 5\,sec + 4\,sec + 2\,sec = 37\,sec$$

Time to load one truck $= 4$ cycles $* 37.0\,sec = 148\,sec = 2.5\,min$

Discussion of Results

The assumptions of the rock density (4600 lb/yd^3 and of the bucket fill factor of 90% and a load factor of 61% led to an overloading of the truck as its weight limitation at the maximum capacity would be exceeded.

Both pieces of equipment are able to reach the necessary height to dump their buckets into the body of the truck, which has a rim height of 13 ft. The Komatsu wheel loader has the smaller reach, with 15.3 ft, but will be able to do the job.

Both are able to load a truck within 4 loading cycles. If the shotrock has a lower loose density and the 55.6 yd^3 truck capacity can be used, then the shovel will need 5 cycles to totally fill the truck while the loader has enough extra capacity to finish in 4 cycles. However, if we increase the bucket fill factor for the shovel from 90% to 95%, which might be possible, the shovel will also finish the truck in 4 cycles.

The main difference between the two is in the cycle time due to the necessary travel time of the loader for the assumed 75 ft (23 m) distance (23.5 seconds compared to the 8 seconds of swing time).

While the shovel will load (4 cycles) in 2.5 minutes, the wheel loader will need 3.3 minutes or 33% longer than the shovel.

Naturally, the shovel needs to relocate from time to time to a new loading position, which has to be considered in the overall evaluation.

Although the shovel seems to be the more productive, it is too early to decide which piece of equipment to choose. For one thing, we need to consider the truck hauling time, since the two pieces of equipment will depend on each other to supply the rock to the primary crusher. In fact, even the production capacity of the first crusher might limit the loading.

In Chapter 5 we learned about the effect of grade and rolling resistance on the speed of scrapers. Of course, the concept of resistance to scraper tires also applies to trucks, including the off-highway CAT truck 775F (see supplemental material S6.8 CAT 775F).

Let us review how to calculate the travel time by going through a short exercise. Assume that the 775F has to travel uphill in a loaded condition to the primary crusher at a steady 9% on a road with a rolling resistance of 5%. The distance between the loader and the dumping place is 2800 ft (850 m). We can assume that, according to Table 5.4, the coefficient of traction is approximately 0.5 (for a compacted gravel road that is covered with small amounts of spillage).

Hauling Time

The total resistance for uphill travel is $9\% + 5\% = 14\%$.

The loaded weight of the truck is 242,000 lb (110 t). On the graph on page 25 of S6.8, Truck CAT 775F, we recognize that the rimpull for this condition (loaded and 14% total resistance) is 34,000 lb or 15.4 t. As shown, the fifth gear will provide the necessary torque to the wheel to turn. From page 19 of S6.8, we learn that when loaded 67% of the trucks weight, or 162,000 lb (73 t), are carried by the powered axle. Thus, the available traction is $0.5 * 162,000\,\text{lb} = 81,000\,\text{lb}$ (36 t), which is sufficient for the 34,000 lb (15.4 t) rimpull. Projecting vertically down from the intersection of the horizontal line related to the load and the total resistance and the fifth gear, we can deduce that the traveling speed would be 17.5 mph or 28 km/h.

On the other hand, driving downhill empty would result in a different speed. First, the total resistance would change from the 14% to the −9% grade, resistance +5%, rolling resistance to a −4% total resistance. According to its retarding performance on page 22 IPR. of S6.8, we learn that the maximum speed downhill is 42 mph or 67 km/hr. This speed, however, is too high for the condition of a haul road in a quarry and needs to be reduced to 30 mph (48.2 km/h):

$$\text{Traveling time full uphill (mph)} = 60\text{ min/hr} * (2800\text{ ft} * 1\text{ mile}/5280\text{ ft})/17.5\text{ mph} = 1.8\text{ min}$$

$$\text{Traveling time full uphill (km/h)} = 60\text{ min/hr} * (850\text{ m}/28,000\text{ m/h}) = 1.8\text{ min}$$

$$\text{Traveling empty downhill (mph)} = 60\text{ min/hr} * (2800\text{ ft } 1\text{ mile}/5280\text{ ft})/30\text{ mph} = 1.1\text{ min}$$

$$\text{Traveling empty downhill (km/h)} = 60\text{ min/hr} * (850\text{ m}/48,200\text{ m/h}) = 1.1\text{ min}$$

Total traveling time = 1.8 min + 1.1 min = 2.9 min.

If we assume that the time to maneuver and dump at the primary crusher = 2.5 minutes, we can compute that the time for a full truck to return to the loader = 5.4 minutes.

Design of Most Efficient Loading Scheme for Shovel Excavator. Figures 6.11a and b presented one loading situation for each machine. The question is, are they the best possible choice for achieving the highest production possible? From calculating the loading cycle time, we know that really only the swing or travel time can be varied. Consequently, the key to minimizing the time needed to load a truck is its position relative to that of the loading machine when filling its bucket. Let us look a little more closely how truck positions affect loading time. Figure 6.14 depicts several possible locations for a truck loaded by a shovel excavator.

The layouts show three swing angles reaching from 45° to 180° as the largest. Besides the angle to the digging axis, there is a second characteristic of the truck position, indicated by L_{180} to L_{45}, namely the angle between the truck centerline and the tangent of the bucket rotation labeled A, B, and C. At issue is the efficiency of emptying the bucket

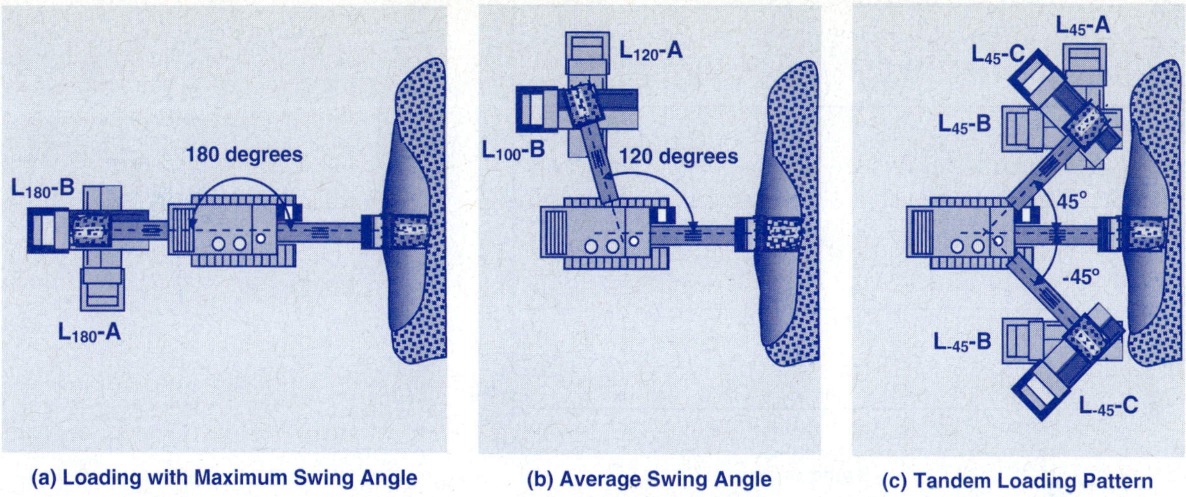

Figure 6.14 Loading Area Layouts for a Shovel Excavator and Truck

into the truck, which is hampered if the operator has to stop and aim for the truck center. The most effective line-up is a truck that has its centerline parallel to the tangent, marked as C, at a distance that is most convenient for the operator.

Figure 6.14c points out another important feature of an effective layout, namely the staging of the trucks on both sides of the excavator. By having two loading locations, a second truck is able to position itself while the first is still getting loaded, thus eliminating the need for the excavator to wait for a new truck to back up into place. This layout is referred to as a tandem or double-sided loading pattern.

In order to get some quantitative means to measure the consequence of changing swing angles, three different shovel operators, Jim, Mark, and Len, were observed while loading trucks positioned at three different swing angles. Table 6.3 presents the statistical results of the observations.

The effect of the larger swing angle is also evident, in that the mean durations increase with degrees of swing. The data in the table reflect the fact that Jim is the most experienced operator on the particular excavator, as he ends up with a consistently lower mean duration but most importantly, with much smaller standard deviations than Mark and Len. Let us find out what the differences amount to in percent for a 120° loading scenario.

Back-cycle time at 120°	*= 13.6 + 12.8 + 4.0 = 30.4 sec*	***: 100%***		
Jim: Back-cycle time	= 11.5 + 10.9 + 3.6 = 26.0 sec	: 85%	StDev:	2.8 sec
Mark: Back-cycle time	= 13.8 + 12.9 + 4.0 = 30.7 sec	: 101%	StDev:	6.6 sec
Len: Back-cycle time	= 15.4 + 14.6 + 4.4 = 34.4 sec	: 113%	StDev:	7.1 sec

Table 6.3 Performance Statistics for Three Shovel Operators

Operator	45 Degrees		120 Degrees		180 Degrees		
	Swing Full (sec)	Swing Back (sec)	Swing Full (sec)	Swing Back (sec)	Swing Full (sec)	Swing Back (sec)	Empty Bucket (sec)
Jim-Mean Duration	5.5	5.3	11.5	10.9	14.8	14.4	3.6
Jim-Standard Deviation	0.5	0.6	1.1	1.2	1.3	1.8	0.5
Mark-Mean Duration	6.4	5.9	13.8	12.9	18.3	17.6	4.0
Mark-Standard Deviation	2.1	1.5	2.7	2.8	3.7	3.2	1.1
Len-Mean Duration	7.2	6.9	15.4	14.6	20.3	19.5	4.4
Len-Standard Deviation	2.0	1.6	3.1	2.9	3.9	3.2	1.1
Average Duration	6.4	6.0	13.6	12.8	17.8	17.2	4.0
Ave. Standard Deviation	1.5	1.2	2.3	2.3	3.0	2.7	0.9

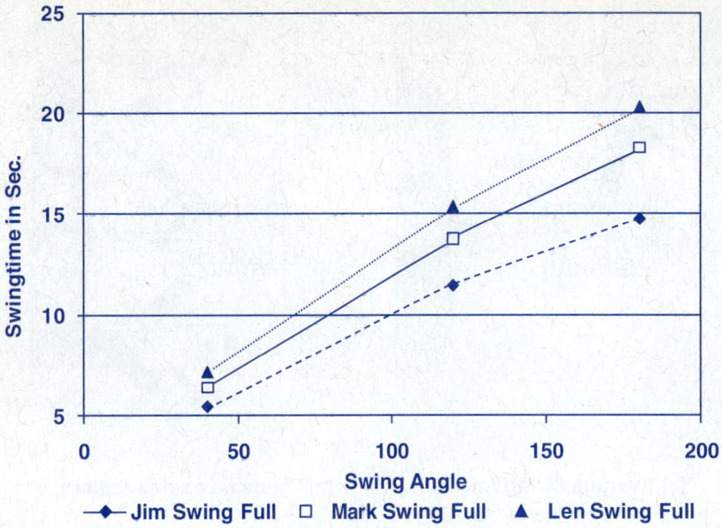

Figure 6.15 Swing Duration as a Function of Swing Angle and Operator Skill

As expected, Jim beats the average duration for swinging and emptying the bucket by 15%, while Len takes 13% above the average. Another indicator of Jim's excellent skill is the significantly lower standard deviation, pointing to a high consistency of his control movements. Such steadiness can only be acquired through long years of practice.

While this is but a small sample, it demonstrates how skill levels of operators have a considerable effect on the production rate of large equipment. In addition, skill not only results in shorter cycle times but also in more consistency.

The data also show the effect of the swing angle on the swing time, including the acceleration and deceleration before dumping the load. Figure 6.15 presents a graph that compares the average durations for swinging the shovel bucket to the three observed angles.

The graph shows nicely the differences between the three operators in swing time. It also shows impressively the declining effect of the time for acceleration and breaking at the beginning and end of each swing. As more time is spent at high rpm, the percentage spent at the slower rpm is decreasing.

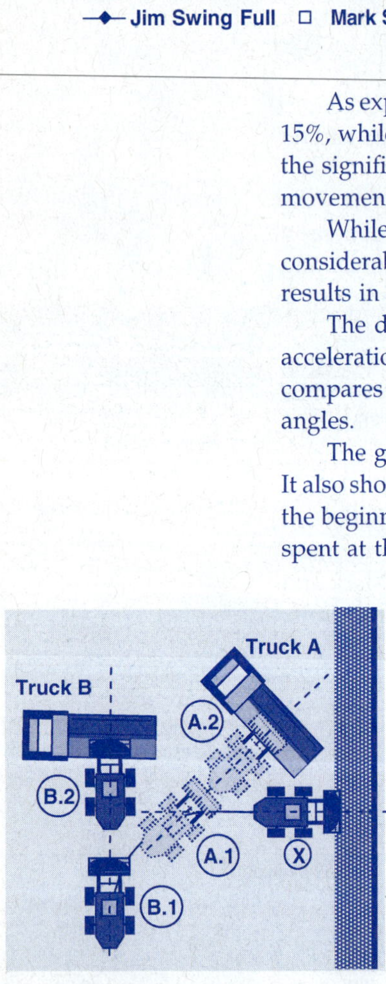

Figure 6.16 Wheel Loader Strategies

Design of Most Efficient Loading Scheme for Wheel Loader

A wheel loader will require a slightly different layout, as the equipment needs space to maneuver. The layout in Figure 6.11b required the operator to make a 90 degree turn in reverse to line up before advancing toward the waiting truck. Again, is it really necessary to make a full 90 degree turn? Figure 6.16 offers a 45 degree alternative labeled as position Truck A. Naturally, the wheel loader operator will vary the rock-loading position, marked as X, making it difficult for the truck driver to park at the best location for the 4 loading cycles.

Good communication between operator and truck drivers to coordinate parking positions is very important. As always, establishing a routine that is simple to understand and follow is needed for a safe operation. The most dangerous phases are the backing-up sections from X to A1 or B1, due to the limited area of vision especially relative to truck B.

Exactly like the shovel excavator, wheel loader operators repeat several loading cycles in order to fill a truck. In the Worked-Out Problem 6.3 we assumed that the distance that the loader has to travel for each dump, reverse from X to B.1, and forward from B.1 to B.2 is 75 ft or 23 m. We need to affirm that the operator will use only first gear while driving with the full bucket, which is 4.3 mph (7 km/h) forward and 4.4 mph (7.1 km/h) in reverse. In addition, accelerations at X and B.1, as well as the decelerations at B.1 and B.2, will add further time. The position of truck A reduces the 75-ft (23 m) distance that the loader has to travel to approximately 2/3 the distance or 50 ft (15 m), of which 50% is in reverse and 50% in forward gear. Alternatively, we could safely assume that the operator will use second gear for returning to load the next bucket (7.6 mph or 12.2 km/hr in forward and 7.7 mph or 12.4 km/hr in reverse). Thus, the average speed with a full bucket is 4.35 mph, and with empty bucket, 7.65 mph.

Now we are able to compute the travel times for the loader when the truck is in position A, including the 3 seconds for accelerating and breaking:

$$\text{Time to unload bucket} = \begin{aligned}&3\,\text{sec} + ((50\,\text{ft} * 3600\,\text{sec/hr})/(4.35\,\text{mph} * 5{,}280\,\text{ft/mile})\\&+3\,\text{sec} + ((50\,\text{ft} * 3600\,\text{sec/hr})/(7.65\,\text{mph} * 5{,}280\,\text{ft/mile})\\&= 6\,\text{sec} + 7.8\,\text{sec} + 4.4\,\text{sec} = 18.2\,\text{sec}\end{aligned}$$

By adding the 1 second for positioning, 14 seconds for loading and 5 seconds for unloading, we get 38.2 seconds compared the 49 seconds used in the worked-out example 6.3. It comes without a surprise that truck position A reduces cycle time significantly. This improvement comes with a caveat, since it will require excellent communications between the truck and loader operator.

Worked-Out Problem 6.4 / Fine-Tuning the Rock Hauling Operation

Vicky Oakley, the operations manager of HardQuarry, Inc., reviewed the discussion of Problem 6.3. She totally agrees with the final comment about having to consider the entire operation instead of the loading task by itself. She also thinks that some of the parameters that were used are not optimal. In particular, she challenges the cycle times for both the shovel and the loader. She thinks that by positioning the trucks to be loaded in a better position, the cycle times for loading could be significantly improved. Finally, she points out that the comparative analysis has to pay close attention to the strategy for switching from one full truck to the next empty truck. Much time could be lost if the empty truck was not in place by the time the shove or loader was ready to dump the next load.

Vicky would like to know what the best hourly production in tons/hr would be for the two loading machines with the CAT 775F trucks. Since the loading location changes based on the location of the shot, she would like a comparison of hauling distances between 2,800 ft (850 m) and 5,600 ft (1,700 m) with an average incline of 10%. Vicky also insists that the time to maneuver and dump at the primary crusher, presently at an average of about 2.5 minutes, is too high. She attributes the excessive time to the fact that the truck drivers get no guiding communication when the crusher is down: should they wait, or dump on the buffer pile? She says she is working on a system that lets the drivers know what to do before they arrive. For this reason, she wants us to use 2 minutes for maneuvering and dumping.

Remark: The buffer pile is used when there an interruption at the crusher, which we will discuss later. Instead of leaving the trucks waiting to dump they will empty the body onto a buffer pile that will be worked off at a later time.

Assumptions

The traction is sufficient for the necessary rimpull to travel uphill.
Time to maneuver and dump at the primary crusher = 2.5 min.
Maximum downhill speed for the road condition in the quarry = 25 mph (40.2 km/hr).
Operating factor = 0.85.
Operator skill = Average.
The body of the CAT 775F will be filled to its maximum of 50 ton (45 t).

Loading Strategy: Based on the effect of the swing angle on loading time, we select the 45 degree tandem or double-sided loading that should be achievable in most loading situations for both the shovel and the wheel loader.

This layout will minimize the loading cycle time and eliminate the need for the loading equipment to stay idle while the next truck gets in position to be loaded.

Loading Time: Both loaders will need 4 cycles to load (as calculated in Worked-Out Problem 6.3).

We assume that the shovel operator will be able to achieve a bucket fill factor of 95%, while the wheel loader has sufficient capacity with 90% and more spillage, as can be expected.

Loading 775 F by Komatsu WA900 wheel loader: = 4 cycles ∗ 37 sec = 2.5 min

Loading the 775 F with Bucyrus shovel excavator:

Cycle time for 1 load = TBC = TPB + TFB + TST + TEB + TSB

Time for 45° swing = (0.125 r ∗ 60 sec/min)/4.6 rpm = 1.6 sec

TBC = 3 sec + 17 sec + 1.6 sec + 1.5 sec + 5 sec + 1.6 sec + 1.5 sec = 31.2 sec

(It is assumed that that the time for braking and acceleration will fall slightly due to the shorter swing time.)

Loading time = 4 cycles ∗ 31.2 sec = 2.08 min (2.1 min)

Truck Travel Times

Loading Location 1: 2,800 ft (854 m) at 10% incline and 5% rolling resistance

(Please review S6.8 CAT truck 775F pp. 22–25)

Driving uphill the total resistance is 10% + 5% = 15%

According to the chart on pg. 25 in S6.8 the loaded truck requires 37,000 lb (16.8 t) of rimpull, and the speed would be 6.5 mph (10.4 km/hr)

Duration for uphill travel = (60 min/hr ∗ 2,800 ft ∗ (1 mile/5,280 ft))/ 6.5 mph = 4.9 min

Duration for uphill travel = (60 min/hr ∗ 850 m)/10,400 m/hr = 4.9 min

Driving downhill the effective grade is 10% − 5% = 5%

According to the 3,000 ft chart on pg. 24 in S6.8, the empty truck could have a maximum speed of 42 mph. However, according to the assumptions, the maximum speed for the road condition is 25 mph, or 40.2 km/hr.

Duration for downhill travel = (60 min/hr ∗ 2,800 ft ∗ (1 mile/5,280 ft))/ 25 mph = 1.3 min

Duration for downhill travel = (60 min/hr ∗ 850 m)/40,200 m/hr = 1.3 min

<u>Total travel time for location1 = 4.9 min + 1.3 min = 6.2 min</u>

Loading Location 2: 5,600 ft (1,700 m) at 10% incline and 5% rolling resistance

Uphill speed = 6.5 mph (10.4 km/hr)

Duration for uphill travel = (60 min/hr ∗ 5,600 ft ∗ (1 mile/5,280 ft))/6.5 mph = 9.8 min

Downhill speed = 25 mph (40.2 km/hr)

Duration for downhill travel = (60 min/hr ∗ 5,600 ft ∗ (1 mile/5,280 ft))/25 mph = 2.6 min

Duration for downhill travel = (60 min/hr ∗ 1,708 m)/40,200 m/hr = 2.6 min

<u>Total travel time for location 2 = 9.8 min + 2.6 min = 12.4 min</u>

Truck Cycle Times

CAT 775 F with Komatsu WA900 wheel loader:

Loading time	= 2.5 min		
Unloading time	= 2.0 min	= 4.5 min	
Total cycle time for 2,800 ft (850 m) (6.2 min + 4.5 min)			= 10.7 min
Total cycle time for 5,600 ft (1,700 m) (12.4 min + 4.5 min)			= 16.9 min

CAT 775 F with Bucyrus RH90C shovel excavator:

Loading time	= 2.1 min		
Unloading time	= 2.0 min	= 4.1 min	
Total cycle time for 2,800 ft (850 m) (6.2 min + 4.1 min))			= 10.3 min
Total cycle time for 5,600 ft (1,700 m) (12.4 min + 4.1 min))			= 16.5 min

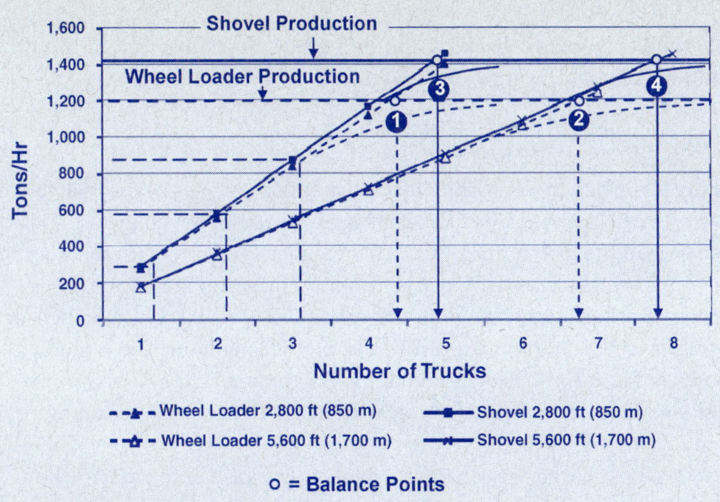

Figure 6.17 Balance Point Analyses for Loaders and Trucks

Balance Point Analysis

Figure 6.17 presents the balance points for the two loaders and the two truck travel distances:

The maximum loading productions in tons per hour (ton/hr) are a function of how many trucks can be loaded per hour, and their maximum capacity = 50 ton (45 t).
Wheel loader production = ((60 min/hr)/2.5 min) ∗ 50 tons = 1,200 ton/hr (1,088 t/hr)
Shovel production = ((60 min/hr)/2.1 min) ∗ 50 tons = 1,428 ton/hr (1,295 t/hr)

The theoretical balance point defines the number of trucks that are needed to keep the servers, here the loaders, busy at all times. Figure 6.17 shows that four balance points need to be computed.

Balance point ❶: Hourly production of one truck = ((60 min/hr)/10.7 min) ∗ 50 ton = 280.4 ton/hr (254.4 t/hr)

Trucks needed = (1,200 ton/hr)/(280.4 ton/hr) = 4.3 trucks

Balance point ❷: Trucks needed = (1,200 ton/hr)/((60/16.9) ∗ 50 ton/hr) = 6.8 trucks
Balance point ❸: Trucks needed = (1,428 ton/hr)/((60/10.3) ∗ 50 ton/hr) = 4.9 trucks
Balance point ❹: Trucks needed = (1,428 ton/hr)/((60/16.5) ∗ 50 ton/hr) = 7.8 trucks

Finally, we need to include the operating factor since the operators will have to take short breaks now and then. In addition, the travel and loading time are only averages or means with standard deviations. In a smooth-running operation, where everything works like clockwork, this can have immediate consequences. For example, if one truck driver or loader operator takes longer than the average cycle, loading or hauling, then all the trucks that follow

Table 6.4 Maximal Fleet Production

	Number of Trucks	Theoretical Fleet Production (ton/hr)		Production Adjusted for Bunching (ton/hr)			Operating Factor	Actual Production (ton/hr)	
		WA900	RH90C	Bunching Factor	WA900	RH90C		WA900	RH90C
Short Haul	4	1,121	1,165	4%	1,076	1,118	0.85	915	950
	5	1,200	1,428	10%	1,080	1,285	0.85	918	1,092
	6	1,200	1,428	5%	1,140	1,356	0.85	969	1,153
Long Haul	6	1,065	1,091	4%	1,022	1,047	0.85	869	890
	7	1,200	1,272	10%	1,080	1,208	0.85	918	1,027
	8	1,200	1,428	5%	1,140	1,356	0.85	969	1,153

tightly behind each other will be pushed back by the difference. Speeding up, on the other hand, will have no positive impact, as it only results in a waiting time behind the truck in front.

Discussion of Results

The redesign of the loading layout led to significant improvements. However, the Komatsu wheel loader WA900 still has a slightly longer loading cycle. Although its bucket is bigger than the shovel bucket, it is not enough to reduce the number of cycles needed to load the 50-ton truck. Both of them need 4 cycles, with the Bucyrus having to make sure that the fill factor stays at 95%; otherwise the truck will not be full after 4 cycles.

Because of its shorter loading cycle, the Bucyrus shovel RH90C is able to reach a slightly higher actual production with 5 or 6 trucks on the short haul and 8 trucks on the long haul. The bunching effect is largest around the balance point, as shown in Figure 6.17. For this situation it is assumed to be a maximum of 10%, as the drivers are most probably equally skilled. For this reason, the production is still increasing when more trucks are being added beyond the balance point.

In response to Vicky Oakley's charge, we can offer Table 6.4. The answer to what the hourly production would be is: It depends! The loader fleet varies between 920 ton/hr and 970 ton/hr, and the shovel fleet between 1,027 ton/hr (932 t/hr) and 1,153 ton/hr (1,045 t/hr), depending on the number of trucks that are used. Since adding trucks beyond the balance point adds cost while leading only to marginal improvements (e.g., less than 5%), it is recommended to study how big the benefit of the addition production would be. For example, using 6 trucks instead of 5 on the short haul distance leads to 51 tons and 61 tons more per hour for the WA900 and the RH90C respectively. The question is: Are the extra tons worth the cost of a truck with operator?

The basis for finding an answer to the question raised in Worked-Out Problem 6.4—is it worth hauling an extra 50 tons (45 t)?—is a good understanding of how the aggregated plant works and, of course, what its capacity is. The following section will introduce the main equipment and methods used in producing crushed aggregates.

6.4.2 ABOUT CRUSHERS, CONVEYORS, AND SCREENERS

Figure 6.10b shows that the processing begins at the primary crusher at the end of the hauling cycle. Here the trucks back up and empty the body into the feeder bin that separates and sends the smaller fragments around the crusher. Thus, only the larger rock pieces up to 4 ft × 4 ft × 5 ft (1.3 m × 1.3 m × 1.5 m) are crushed to fragments that are between 8 in and 12 in (20–30.5 cm) in diameter. Because of the large size and the weight of the rock that thunders into the feeder, it has to sustain large forces. Not surprisingly, feeders are made of massive steel elements and are appropriately referred to as grizzly feeders.

Breaking those large pieces of rock not only requires heavy-duty machines but also depends on lots of energy to drive the breakers. Figure 6.18 presents a common distribution of energy usage in a quarry plant.

Brute force is applied to break the stone into smaller and smaller pieces. Appropriately, the machines that do the work are called crushers. For the reduction of rock sizes quarries use three main types of crushers in a row: Primary, Secondary, and Tertiary. Belt conveyors move material from the discharge end of one to the intake of the next. Let us study the inner workings of these special machines.

S6.9 Jaw Crusher Terex
S6.10 Jaw Crusher Lippman

The Jaw Crusher

This is one of the two typical primary crushers. Figure 6.19 depicts the key components of the machine in action. A jaw crusher consists of a vertical fixed jaw and one inclined moving jaw. The rocks fall from the feeder into the opening between the jaws that are lined with strong tungsten plates. The elliptical motion of the top end of the moving jaw is created by the eccentric shaft mounted on a large powered flywheel. This end of the jaw is called the pitman.

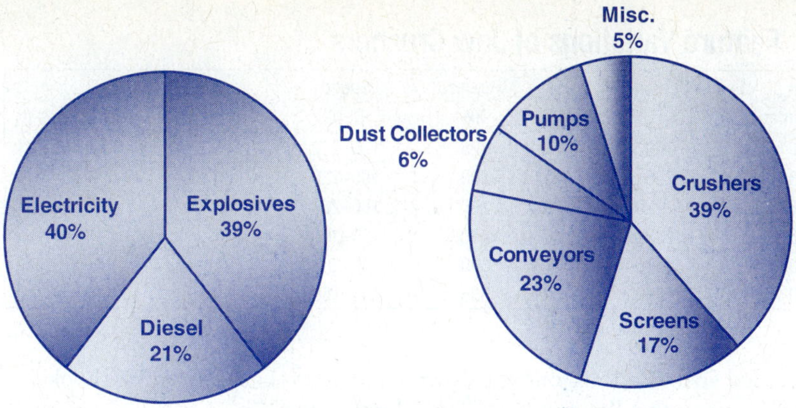

Figure 6.18 A Quarry's Energy Sources and Needs

The bottom end of the moving jaw is kept in place by the combination of a tension rod and the toggle. As shown in Figure 6.19, the toggle plate controlled by a hydraulic piston are used to adjust the jaw opening, which is referred to as the closed-side setting (CSS). Of course, this opening is responsible not only for the sizes of the produced material, but also the volume of rock that can be crushed within an hour. Still, the characteristics and shape of the fed rock have to be brought into the equation. Consequently, fabricators of jack crushers can provide only guiding performance data for their machines, each having to be adjusted to the local rock conditions. Table 6.5 provides examples of key crusher characteristics. As shown, the size of the crusher is related to the feed opening with the first number indicating the width and the second its depth. Of course, a larger opening allows larger rocks to enter the crusher, which in turn requires more power to crush and a heavier frame, liner plates, and bearings to sustain the forces that develop in the crushing process.

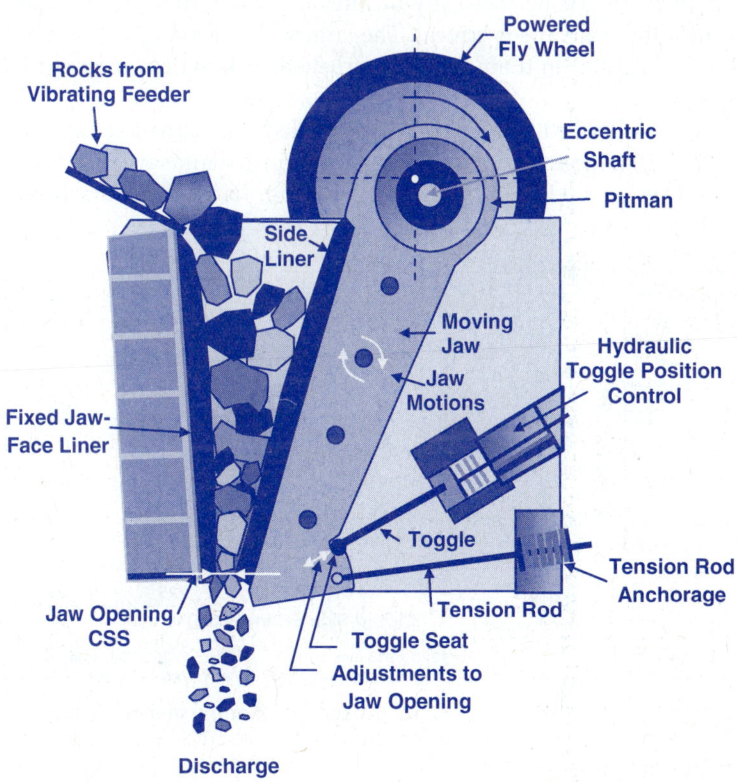

Figure 6.19 Cross-Sectional View of a Jaw Crusher in Operation

Table 6.5 Feature Variations of Jaw Crushers

Crusher	Feed Opening		Power	Speed	Weight	
	inch	mm	HP	RPM	Tons	Metric Tons (t)
Small A	30 × 36	750 × 910	125	190–220	16	14
Small B	30 × 62	750 × 1600	200	190–220	35	31
Medium A	48 × 36	1220 × 910	200	180–200	42	37
Medium B	48 × 42	1220 × 1065	250	180–200	46	41
Large A	50 × 62	1275 × 1600	300	180–200	95	85
Large B	60 × 50	1500 × 1275	335	145–180	85	76

For more detailed specifications, one can download two brochures from the booksite, one from Lippmann, Inc., and the one from Terex-Jaques, Inc.

As mentioned earlier, each crusher can adjust the jaw opening for different CSSs or closed-side settings by resetting the toggle plate and the tension rod. Figure 6.20 shows the relationships between size of the jaw openings and production in short or metric tons/hour.

One needs to keep in mind that hourly production is normalized to a dry, free-flowing rock with a density of 100 lb/ft^3 or 1,600 kg/m^3. Before the final selection of a new crusher, the manufacturer commonly will conduct tests with the local rock to ensure the best sizing for its desired use.

The comparison of the six example crusher sizes shows that the smaller ones provide a smaller jaw opening and with it, will provide smaller size rock as output. Of course, the smaller the opening, the smaller the output (e.g., 3-in with 150 tons/hr; 75-mm with 136 t/hr) while larger CSSs allow more throughput (e.g., 10-in with 450 tons/hr; 250-mm with 408 t/hr). Larger crushers are clearly designed to handle larger feed rock, as indicated by the larger feeder opening, to produce larger-sized output with a higher production rate. As the primary crusher basically dictates the production of the entire quarry, the output/hour of this machine is critical. The largest rock leaving the primary crusher is typically 8 inch (20.3 cm) in diameter. This crushed rock is then moved to a stockpile via a conveyor belt.

Anybody who has closely looked at a conveyor belt filled with crushed rock has noticed that the sizes of the rock pieces are not uniform, as in "8-inch diameter." Hard rock will break in a range of sizes limited only by the jaw opening. In fact, manufacturers

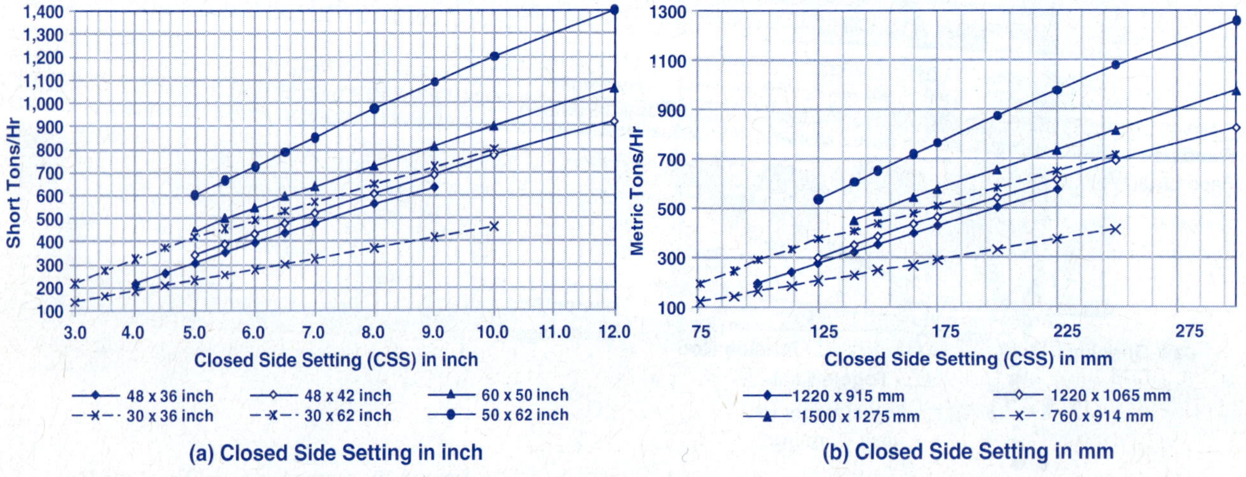

Figure 6.20 Jaw Crusher Performances as a Function of Size and CSS

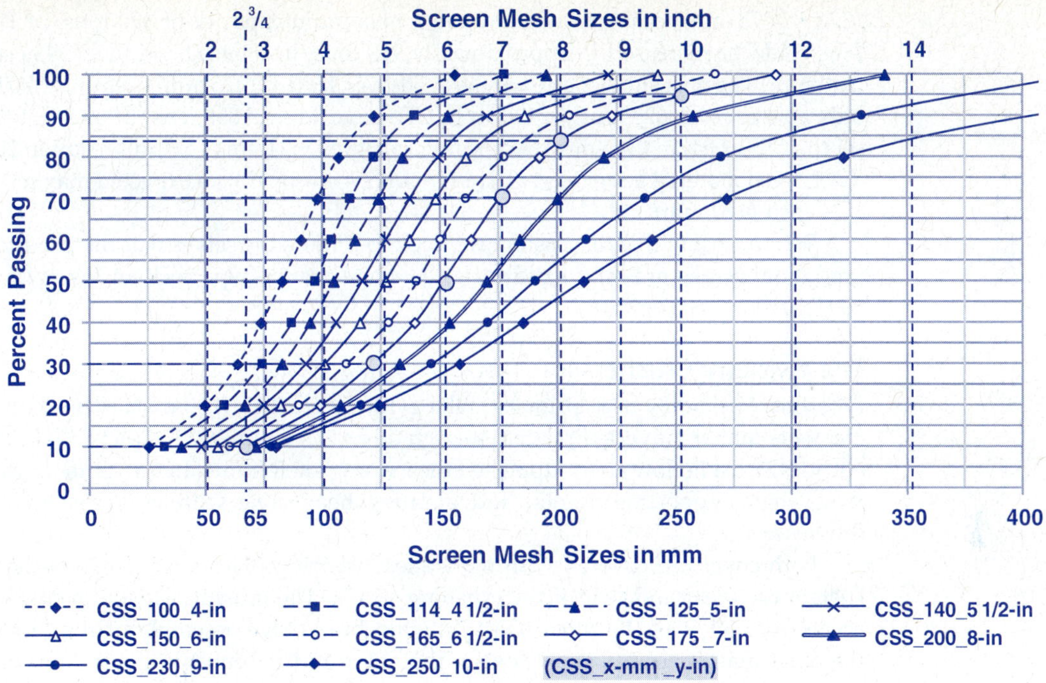

Figure 6.21 Gradation Curves for a Large-Jaw Crusher

provide data describing the size distribution for a middle-of-the-road rock that has gone through the various CSS settings as gradation tables or plots. Figure 6.21 presents two sample graphs, one in inches and on in millimeters, with curves that relate each CSS with a rock size distribution shown as percent that passes through an assortment of screen sizes. Let's have a closer look to understand how they work.

As we will see later, the smallest CSS of 4 inches or 100 mm is still rather large. On the other hand, a big CSS setting has to be expected for a jaw crusher as the primary having to break big shotrock clumps. In addition, this first breaker will normally be followed by secondary and even tertiary crushers, each one designed to deal with smaller sizes more efficiently. Let's consider a situation where the crusher is set to a CSS of 7 inches or 175-mm marked with gray circles at some intersections with percent passing. For example, the dashed horizontal 10% line meets the curve at 2¾ inches or 70 mm. This means that only 10% of the crushed material will pass a screen with a mesh size of 2¾ in or 65 mm. On the other hand, 84% will pass a sieve size of 8 inches (200-mm) and 70% a sieve size of 7 inches (175 mm). From this we can compute that 10% of the crushed rock will be retained by the 7-in sieve. Starting at the top of the gradation curve, we can see that 95% will pass through a 10-in (250-mm) mesh. This in turn means that 5% of the rock is larger and will stay on the screen. Since only 90% will pass the 9-in mesh, a second 5% will not pass and will stay on the screen with the maximum size approximately 11¾ in (300 mm). Now we can infer that 5% is larger than mesh 10 in and 5% is between mesh 9 and 10 in. Finally, approximately 7% is too large to pass 8-in mesh and will be retained.

It's important to notice that despite setting the jaw opening to 7 in (175 mm), 30% of the discharged material is larger than this setting. This, of course, is caused by elongated rock pieces that slip through the opening. This fact needs to be considered when trying to achieve a certain size distribution for the secondary crushers.

Earlier we saw that the CSS sizing is related to the production rate of a crusher. In the case of a CSS of 7 in (175 mm), Figure 6.20 indicates the largest (50 × 62-in or

1220 × 1575 mm)) crushers would have a maximum capacity of 850 tons/hr (771 t/hr). This could be increased to approximately 980 tons/hr (889 t/hr) if the CSS is set to 8-in (200-mm) The next smaller crusher (60 × 50-in or 1500 × 1275 mm) uses only 300 HP but is only able to crush 625 tons/hr (567 t/hr). To have it reach 850 tons/hr one would have to set the CSS at 9½ in (240-mm). According to Figure 6.21, this CSS will result in 20% of the discharged material being larger than 11½ in (290 mm) with an expected maximum size of 18½ inch (470 mm).

Before closing this discussion, we need to remind ourselves that the production and gradation graphs are only valid for an average granite-type rock and is normalized to 100 lb/ft^3 or 1,600 kg/m^3.

The Gyratory Crusher

A second commonly used primary is the gyratory crusher. It consists of two cone-shaped elements that are inverted relative to each other. As Figure 6.22 depicts, the feeder bowl empties into a hollow cone supported by a strong shell frame. In the center of the sits the main shaft supporting a solid and gyrating head cone with its larger end pointing downward.

S6.11 Gyratory Nordberg
S6.12 Gyratory Sandvik

Both cones are covered with manganese wearing plates and a mantel that have to sustain the wear and tear of the crushing forces. The crushing action is created by the eccentric motion of the inner shaft initiated by a drive assembly at the lower end of the shaft that closes the gap between the outer and the inner cone. As you observe in Figure 6.22, the upper end of the shaft is held in place by a spider assembly and sleeve that allow it to be pushed vertically to change the CSS at the bottom. Presently the CSS is on the right bottom end of the outer cone as the main shaft is being pushed to the right side by the eccentric bushing at the bottom end. The size of the gap can be adjusted by hydraulically

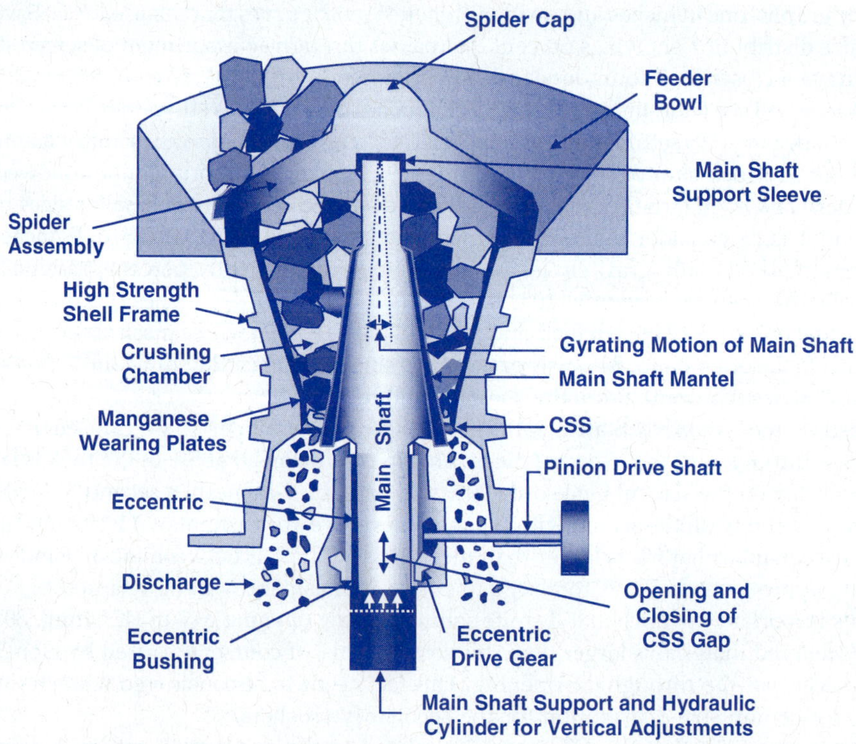

Figure 6.22 Cross-Section of a Gyratory Crusher in Operation

lifting or lowering the main shaft. As the bushing is rotated around by the drive gear, the gap between the outer and inner cone is opened and closed in a circling motion. Consequently, unlike a jaw crusher, the crusher never has a back cycle time when no crushing is taking place. The gyratory crusher is continuously fracturing rock. Hence, it is not surprising to find that gyratory primary crushers have larger production rates than jaw crushers that have to spend unproductive time to let the rock slide downward. The booksite contains a brochure of large gyratory primaries with capacities that range from 1,800 tons/hr (1,635 t/hr) to up to 8,385 tons/hr (7,605 t/hr) with feed openings from 42 × 65 inches to 60 × 110 inch. In comparison, the largest jaw crusher in Figure 6.18 only reached 1,400 tons/hr (1,270 t/hr). As can be expected, their power needs are also much higher, from 500 to 1,400 HP, when compared to the 335 HP of the largest jaw crushers. At the same time, the gyratory operates at 85 to 100 rpm, while the jaw crusher can reach 180 rpm. Again, the slower rotation does not include dead motion. As with the jaw crusher, the liners plates wear off. As a result, the gap for the CSS will have to be adjusted; otherwise the particle size of the output will change over time. Commonly, the quality control of the quarry will sample the output of each crusher over time to assess if the gradation of the crushed material has moved out of a required range. Should the rock pieces get bigger over time, the input to the secondary crushers will change with it and thus create an imbalance in the entire resulting operation.

Before moving to the secondary crushers, we need to learn about two fascinating technologies that are needed to deliver the crushed rock, the belt-conveyor and the vibrating screens.

Designing the Surge Pile Conveyor

Belt conveyors are the preferred method for high-capacity transportation of bulk material, such as rock or coal. Probably the most interesting element is the continuous belt that has to be flexible enough to make tight turns around the two main pulleys, one on each end, strong enough to sustain the tension forces that have to be induced into the belt to keep it moving, and durable in order to sustain the wear and tear from the material being loaded onto the belt. The two main questions that design engineers have to address are related to the power that is needed to move the belt and the kind of belt that will survive under the particular situation. The following will address some of the key aspects that need to be considered. For a more detailed review please visit the book website containing S6.15, a conveyor handbook.

S6.13 Conveyor Safety
S6.14 Conveyor Sandvik
S6.15 Conveyor Handbook

The commonly used inclination α for crushed rock is 12 degrees. Starting with this value, one can compute the lift height and the length of the conveyor. While they can reach over 3,300 ft (1,005 m), lengths between 1,300 ft (396 m) and 1,970 ft (600 m) are more common. The loading station of the conveyors is commonly placed directly under the discharge of the crusher. The loading chute should be specially designed so that heavy and sharp rock pieces can't fall from a large height onto the belt. As Figure 6.23 demonstrates, steps can be built into the chute that break the fall of the rocks and redirect them so that they "roll" onto the belt in the direction of travel. This will not only minimize the energy needed to accelerate the feed but also minimize the wear and tear on the belt itself. If something should fall, an impact table is positioned underneath to protect the belt from being punctured or overstretched.

The conveyor system shown is powered at the top by a drive pulley but shows a wide variety of pulley applications—but all redirect the belt. In order to minimize the bending stresses in the belt, which commonly consists of multiply plies that are protected by a rubber cover, the pulleys have minimal radius for each belt type.

In order to avoid the sagging of the belt between the supporting idlers, a certain amount of tension needs to be induced into the belt. Figure 6.23 depicts one of the two

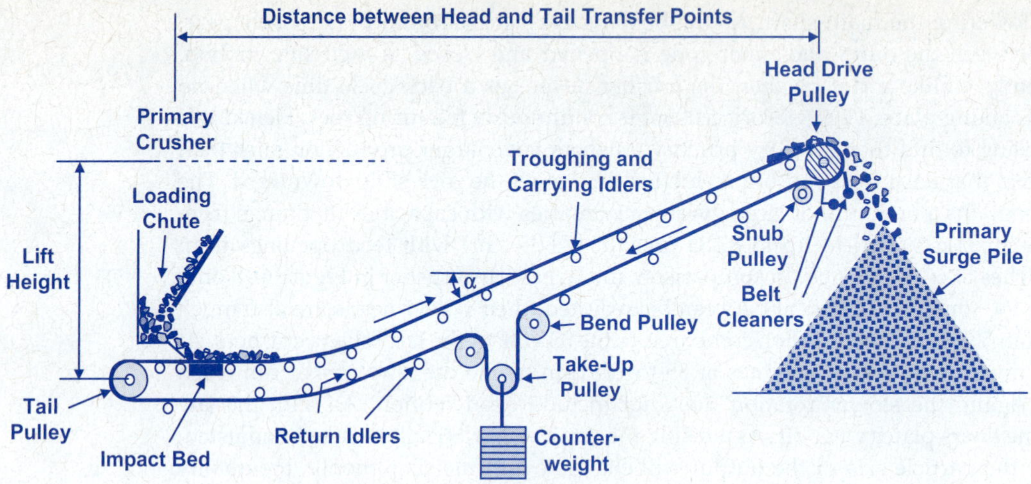

Figure 6.23 Common Conveyor Configuration for Feeding the Surge Pile

common methods, the gravitational take-up system. The free-moving weight suspended from the take-up pulley ensures a consistent tension independent of whether the belt is loaded or empty.

The main factor that drives the selection of the belt is the hourly capacity of the belt, which depends of course on the speed and the volume that can be put on the belt. Figure 6.24 displays the main factors underlying belt design.

The most common conveyor designs today are based on three equally sized idler rollers, two of which are angled upwards, most commonly at an angle of 30°. The belt is placed on top with a space left empty at each edge in order to avoid spillage of the material during transport. A sufficient edge distance is about 7–8% of the belt width. Finally, the surcharge angle, as depicted in Figure 6.24, depends upon the material's characteristics, especially the angle of repose. Since the material will shake and vibrate during transfer on the belt, the surcharge angle will be smaller than the angle of repose measured when the material is static. Table 6.6 presents a sample of common bulk materials in construction with the recommended angle of surcharge and maximum conveyor slopes.

The small sample of material shows angles of surcharge that lie between 20 and 30 degrees, with the larger angles linked to wet or damp material. Thus, it is not surprising that the most commonly used angle for design is 20 degrees, a conservative assumption that can only be verified when the actual material is loaded and moved on the belt. The Dunlop Conveyor Belt Design Manual offers a series of tables to quickly gauge the capacity of running belts with different widths and trough angles. Figure 6.25 presents

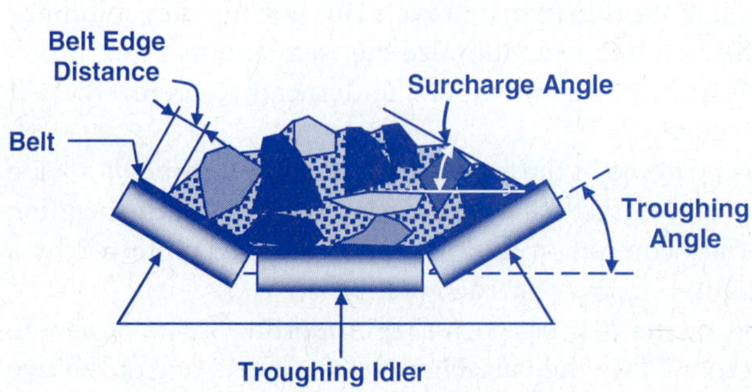

Figure 6.24 Cross-Section of a Full Troughing Belt

Table 6.6 Maximum Surcharge and Conveyor Angles for Various Materials

Material Characteristics	Bulk Density lbs/yd^3	Bulk Density t/m^3	Angle of Surcharge Degrees	Conveyor Slope Degrees
Clay, dry, fine	2,000	1.13	20	22
Clay, wet, lumpy	2,200	1.27	25	20
Stone, crushed	2,700	1.60	20	22
Gravel, bank run, wet	2,600	1.52	25	20
Gravel, dry	2,200	1.31	20	18
Limestone, crushed	2,400	1.40	20	18
Sand, bank, damp	2,400	1.40	30	22
Sandstone, broken	2,500	1.44	20	18

the cross-sectional areas for six belt widths between 29.5 inches and 59.0 inches (0.75 m and 1.50 m) and trough angles from 20° to 45°.

In order to compute the hourly capacity measured in tons/hr we only need to decide on the speed of the conveyor and consider the density of the bulk material. The Dunlop Conveyor Belt Design Manual includes a series of tables for the various belt widths, trough angles and speeds using a normalized material density of 2,240 lbs/yd^3 (1.3 t/m^3). Figure 6.26 depicts the results for a belt width of 47.2-in (1.2-m).

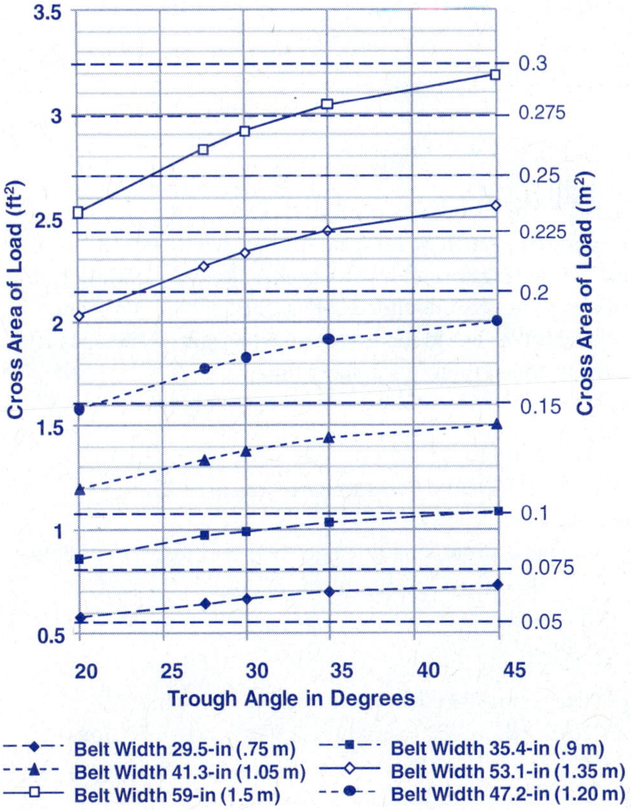

Figure 6.25 Cross-Sectional Load Areas of Troughed Belt with 20° Surcharge Angle

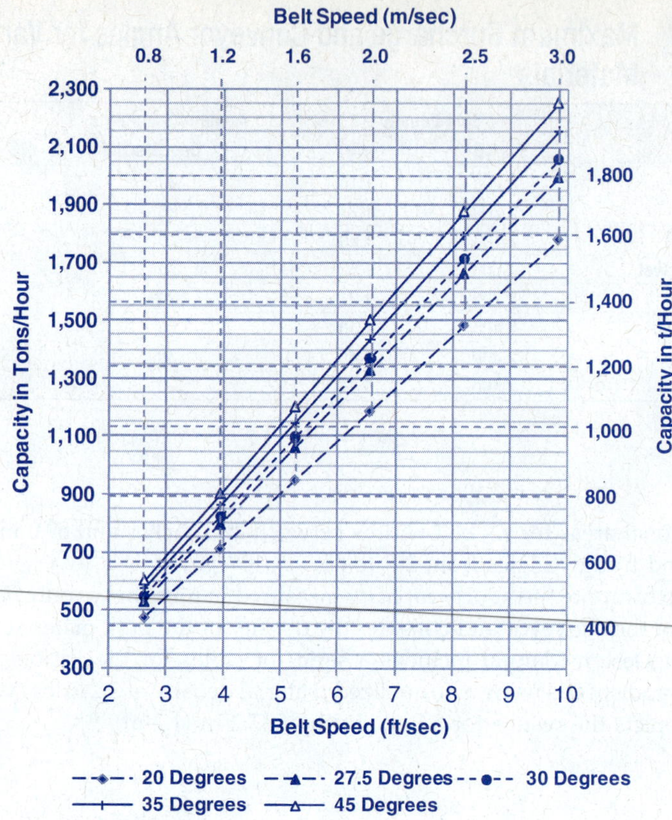

Figure 6.26 Hourly Output of 47.2-in (1.2 m) Troughed Belt Conveyor

Worked-Out Problem 6.5 / Belt Conveyor Capacity

The company that we are working for is planning to build a new conveyor for a new primary crusher that will produce 800 tons/hr (714 t/hr) of crushed rock with a 9-in CSS (see Figure 6.18). The distance between loading and the head will be about 2,460 ft (750 m), and the slope will be 12%, as recommended for crushed rock. Let's assume that we are asked to calculate what conveyor size we would need. We should include a safety margin of 5% to avoid a loading breakdown in situations where the crusher might produce more for a short time.

Assumptions

For crushed rock, Table 6.6 shows a density of 2,700 lb/yd^3 (1.6 t/m^3), and the surcharge angle of 20°.

We will consider the most common troughing angle of 30°.

For the first design, we should use a low-to-average belt speed promising average wear and tear. A possible speed is 4 ft/sec (1.2 m/sec).

Calculations

The hourly design capacity for the 2,700 lb/yd^3 = 1.05 ∗ 800 tons/hr (714 t/hr) = 840 tons/hr (762 t/hr).

The capacity normalized to a standard rock with 2,240 lb/yd^3 is (840 tons/hr)/(2,700/2,240) = 700 tons/hr (635 t/hr).

We consult the graph in Figure 6.24 that represents the values for a width of 1.2 m starting with speed of 4 ft/sec. For a troughing angle of 30°, we get a capacity of 800 normalized tons/hr (725 t/hr), which is sufficient for the 700 tons/hr that have to be transported up the incline.

Alternative Solution 1: The speed of the belt could be reduced to (1.0 m/sec) 3.3 ft/sec, which would be sufficient to deliver the desired amount.

Alternative Solution 2: The next-smaller belt is 41.3 in (1.05 m). The graph in Figure 6.2 shows the effect of a smaller belt on the cross-sectional load area. The intersection of the vertical line representing a 30° trough angle and the 47.2-in (1.2-m)wide belt shows that the load cross-area is about 1.82 ft^2 (0.17 m^2). By moving downward to the intersection with the next-smaller belt size, 41.3 inches, we see that the area is now reduced to 1.38 ft^2 (0.129 m^2), a 24% reduction.

By considering the same design speed of 4 ft/sec (1.2 m/sec) and the same 30° troughing angle, but a 24% smaller cross area for the load, we can establish a linear equation for the new capacity with the 800 normalized tons/hr for the large belt as its base.

Capacity for the 42.3-in (1.1–m) belt with the same = 800 ton/hr ∗ 0.76 = 608 tons/hr (551 t/hr), which is not sufficient. Since a 35° trough angle only adds 2% we need to consider an increase in belt speed. The belt speed that would produce (700 tons/hr)/0.76 = 921 tons/hr (835 t/hr). The equivalent on the 47.3-in belt is 4.5 ft/sec (1.4 m/sec) (Figure 6.24). The actual capacity considering the actual density is 700 tons/hr ∗ 1.2 = 840 ton/hr (762 t/hr) of crushed rock.

Discussion of Results

The initial solution is a design that is able to transport 14% more than what is required. The first alternative uses the initial solution but with a speed of about 3.4 ft/sec (1.0 m/sec), which is 14% lower than the desired design speed of 4 ft/sec (1.2 m/sec). The second alternative offers a smaller and thus cheaper system with a belt width of 41.3 in (1.05 m), with a slightly higher belt speed of 4.5 ft/sec (1.4 m/sec) still in the range considered an average speed.

The next issue to be solved is the horsepower of the engine and the tension needed in the belt. As we will see, the tension is not only needed to keep the belt from sagging too much between adjacent idlers, but it is also very critical in providing the necessary friction/traction at the drive pulley.

Figure 6.23 shows a surge pile conveyor with the drive pulley on its upper end, which is one of many positions to locate the motor. Since this is the only point where the belt is being powered, the drive pulley has to provide sufficient force to overcome all the resistances to the movement of the belt, of which there are many.

The force model at the drive pulley indicates that the key tension force is T_e, called the effective tension force. It is the force that has to be matched by the cumulative friction forces between the belt and the drive pulley within the warp angle, the angle of contact between the two. Figure 6.27b shows the convex curve of the radial forces between a and b providing the necessary normal forces that, when multiplied by the existing coefficient of friction between the belt and the pulley, allow us to calculate the total available friction force.

The tension force T_e consists of four major components that need to be summed:

1. Force T_B to move the empty belt
2. Force T_L to move the load horizontally
3. Force T_S to lower or lift the load along a slope
4. Force T_A to overcome the resistance of accessories (i.e., scrapers, cleaners)

Thus the function for the effective tension force can formulated as

$$T_e = T_B + T_H + T_S + T_A$$

and the necessary power to move the belts at a speed s (ft/sec):

$$\text{Drive-Power} = T_e\,(\text{N}) * s\,(\text{m/sec}) = \text{kW}$$

Calculation of T_B (Moving Empty Belt): The approach to compute T_B is to first establish an equivalent mass per a unit length (e.g., ft) and multiply the mass by the total conveyor length L and the friction coefficient f_B for an

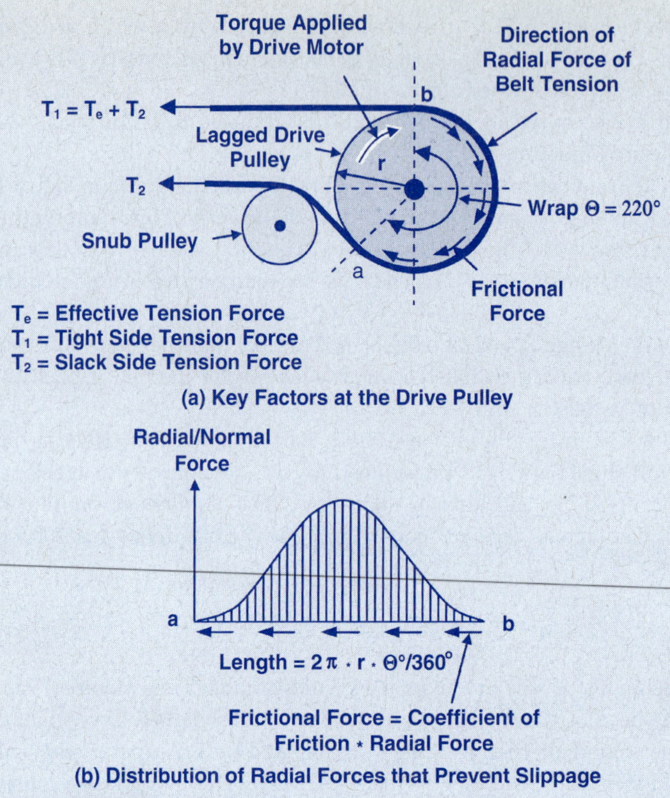

Figure 6.27 Forces at the Drive Pulley Responsible for Keeping the Belt in Motion

empty belt. The mass, called the mass of moving parts m_p, is computed by adding to the mass of the belt the masses of all rotating parts such as the troughing and return idlers and all the pulleys. At this stage of design, however, what is still missing are the idlers and pulleys; engineers, however, work with empirical values that come from experience and confirm the assumptions with detailed calculations at the end.

Hence, we can establish the following function for the tension force needed for moving the empty belt:

$$T_B\,(N) = g * m_p * f_B * L$$

Here are two set of data from Dunlop's Belt Design Manual

Not only is crushed rock heavy when compared to other material, it is also abrasive and requires a heavy-duty belt with sufficient rubber coverage to ensure a long life.

The conveyor that is the basis for Worked-Out Problem 6.4 is 2,460 ft (750 m) long with a width of either 47.2 in (1.2 m) or 41.3 in (1.05 m). From Table 6.7 and 6.8 we get the following information:

Mass of moving parts for empty conveyor:	41.3-in: $m_P = 55.1$ lbs/ft (82 kg/m)
Coefficient of friction for a conveyor > 500 m	47.2-in: $m_P = 63.8$ lbs/ft (95 kg/m)
	$f_B = 0.02$

The total force needed to drive the empty belt can now be calculated as:

$$T_B\,(lb) = 55.1\ \text{lbs/ft} * 0.02 * 2{,}460\ \text{ft} = 2{,}710\ \text{lbs}$$
$$T_B\,(t) = *82\ \text{kgf/m} * 0.02 * 750\ \text{m} = 1.23\ \text{t}$$

Calculation of T_L (Moving Load Horizontally): Similar to T_B, we calculate the weight of the material on the belt for a unit of length and multiply it by the total length and an appropriate coefficient of friction.

Table 6.7 Empirical Mass Values for Moving Parts

Belt Width		Mass of Moving Parts m_p							
		Light Duty Idlers Light Belt		Medium Duty Idlers Moderate Belt		Heavy Duty Idlers Heavy Belt		Extra Heavy Duty Idlers Steel Cord Belt	
mm	inch	kg/m	lbs/ft	kg/m	lbs/ft	kg/m	lbs/ft	kg/m	lbs/ft
450	17.7	23.0	15.5	25.0	16.8	33.0	22.2		
600	23.6	29.0	19.5	36.0	24.2	45.0	30.2	49.0	32.9
750	29.5	37.0	24.9	46.0	30.9	57.0	38.3	63.0	42.3
900	35.4	45.0	30.2	55.0	37.0	70.0	47.0	79.0	53.1
1050	41.3	52.0	34.9	64.0	43.0	82.0	55.1	94.0	63.2
1200	47.2	63.0	42.3	71.0	47.7	95.0	63.8	110.0	73.9
1350	53.1	70.0	47.0	82.0	55.1	107.0	71.9	127.0	85.3
1500	59.1			91.0	61.2	121.0	81.3	143.0	96.1
1650	65.0			100.0	67.2	132.0	88.7	160.0	107.5
1800	70.9					144.0	96.8	178.0	119.6
2100	82.7					168.0	112.9	205.0	137.8
2200	86.6					177.0	118.9	219.0	147.2

Dunlop Belt Design Manual

Table 6.8 Empirical Friction Coefficient Values

Symbol	Friction Coefficient for	Value of the Friction Factor			
		Normal Operating Conditions. Under 250 m (820 ft)	Normal Operating Conditions. Over 250 m (820 ft)	Very Well Aligned Structure with No Tilted Idlers etc. Over 500 m (1,640 ft)	Regenerative Conveyor (Decline)
f_B	Empty belt	0.022	0.020	0.020	0.018
f_L	Loaded belt	0.027	0.022	0.020	0.018
f_S	Skirtboards	0.650	0.650	0.650	0.650
f_C	Scrapers	0.600	0.600	0.600	0.600

Dunlop Belt Design Manual

Weight of crushed rock (Table 6.6) = 2,700 lbs/yd^3 (1.6 t/m^3)
The cross-sectional areas for 41.3-in: 1.38 ft^2 (0.129 m^2)
47.2-in: 1.82 ft^2 (0.17 m^2)
Hourly capacity at a belt speed of 4.5 ft/sec (1.4 m/sec) is 840 tons/hr (725 t/hr)

The weight of 1 meter of crushed rock = Unit Load = 0.129 m^2 ∗ 1,600 kg/m^3 = 206.4 kg/m
The weight of 1 foot of crusher rock = Unit Load = 1.38 ft^2 ∗ (2,700/27) lbs/ft^3 = 138 lbs/ft

From Table 6.7 we get the coefficient for a loaded belt, which is assumed to be the same as for an empty belt, 0.002. This only holds for a well-maintained system where friction increases linearly with the weight.

$$T_L = 206.4\,kg/m * 0.02 * 750\,m = 3.1\,t\ (6{,}834\,lbs)$$

Calculation of T_S (Moving Load Up/Down a Slope): The origins of the two forces so far have been friction and the acceleration of mass in a horizontal plane. T_S, however, only has to be considered when the conveyor is on an incline, as in a quarry plant. Power is needed to provide the potential energy to elevate the rock from the loading point, inside the quarry, and the top of the surge pile. Naturally, the amount needed is a function of the slope and the length of the conveyor.

The conveyor in the worked-out sample problem is 2,460 ft (750 m) long and has slope $\alpha = 12\%$.

We can calculate the lift height of the conveyor H = 0.12 ∗ 2,460 ft (750 m) = 295 ft (90 m).

$$T_S = \text{Unit Load Mass} * H = 206.4\,kg/m * 90\,m = 18.5\,t\ (40{,}800\,lbs)$$

Calculation of T_a (Friction by Accessories): Figure 6.23 depicts a conveyor with two scrapers at the head. They consist of angled steel blades forced against the entire width of the belt, scraping off the material that still sticks to it.

Table 6.8 shows the coefficient of friction as $f_C = 0.6$. We may assume the force that pushes the two blades as 100 lbs (45.3 kg).

$$T_A = 2 * (\text{Normal Force} * f_C) = 2 * (*45.3\,\text{kg} * 0.6) = 54.3\,\text{kg}\ (119\,\text{lbs})$$

Finally, we are ready to compute the total effective tension force:

T_B	=	1.23 t	2,710 lbs
$+T_L$	=	3.10 t	6,834 lbs
$+T_S$	=	18.50 t	40,800 lbs
$+T_A$	=	0.05 t	119 lbs
T_e		22.88 t	50,463 lbs

The required power to create the needed force at a belt speed S = 4.5 ft/sec (1.4 m/sec) can now be computed as:

$$\text{Power (kW)} = T_e\,(\text{kN}) * S\,(\text{m/sec}) = 22.88\,\text{t} * 9.81\,\text{kN/t} * 1.4\,\text{m/sec} = 315\,\text{kW}\ (422\,\text{hp})$$

Figure 6.27 explains that the drive motor required sufficient frictional force to create the forces to overcome the resisting tension forces T_e. Figure 6.27b indicates that friction is linked to the uneven radial force created by the belt between the two contact points, a and b. Furthermore, we realize through vector analysis that the size of the radial force is a function of the tension forces in the belt.

There are two basic types of drive pulley surfaces used: (a) bare and (b) lagged. The bare pulleys offer a bare steel surface to create friction at the contact with the belt, while the lagged drive pulleys sport a surface that is modified to increase friction with the belt. Such pulleys are covered with a layer of special rubber and a friction-enhancing surface profile. As a result, the lagged drive pulleys provide a higher coefficient of friction and thus require a lower radial force to transmit the driving force compared to the bare pulleys.

The snub pulley, shown in Figures 6.23 and 6.27, plays an important role in controlling the contact area between the belt and the drive pulley. Without any extra help, the wrap or angle of contact is limited to 180°, as the incoming and the outgoing belt planes are parallel to each other. Since the total friction force is the integral of the normal forces multiplied by the coefficient of friction, increasing the contact area has a direct impact on the total friction force. As shown, by installing a snub pulley at a desirable angle and distance from the drive pulley, the wrap angle can be controlled.

Again, to simplify the design process and to codify the experience from working systems, tables have been created. The data in Table 6.9 combine the effect of the pulley surface friction and the wrap angle (contact area) into a factor k that relates T_e to the base tension that is needed to create the necessary friction force. The underlying function is as follows: Minimum tension force in belt $= k * T_e$,

From Figure 6.25 we infer that the minimum tension force is the slack-side tension force T_2.

Hence, we can establish the following functions: $T_2 = k * T_e$ and $T_1 = T_e\,(1 + k)$,

Table 6.9 contains two terms that have not been specifically addressed, namely screw and automatic gravity take-up. Both systems are used to introduce the minimum tension needed for powering the belt, the screw take-up limited to short belts. Here, the tail pulley is mounted on a sliding frame that can be adjusted forward and backward with the help of a long screw, the take-up screw. The automatic system, depicted in Figure 6.23, uses a heavy mass suspended from a free pulley, the take-up pulley, to create tension forces into the return side of the belt, which is half its weight. By nature, these forces are constant and should be T_2 (possibly minus the weight of the return belt at the position on the return). When temperature changes cause an elongation or shortening of the belts, the counterweight simply slides up or down.

Now we are ready to calculate the slack side tension T_2 for $T_e = 225$ kN (50,584 lbs) and $T2 = k * T_e$.

Because of the length of our conveyor, we need a gravity take-up system providing a constant tension. Figure 6.23 suggests that we use a snub pulley, which offers higher wrap angles and thus a longer area of friction. The effect on k is drastic. For example, for a bare pulley with a gravity take-up, changing the wrap angle from 180° to 200° changes k from k = 0.84 to k = 0.72 resulting in a 14% reduction of needed slack side tension.

Table 6.9 k-Values for a Variety of Conveyor Settings

Angle of Belt Wrap at Drive	Type of Drive	Screw Take-Up		Automatic Gravity Take-Up	
		Bare Pulley	Lagged Pulley	Bare Pulley	Lagged Pulley
150	Plain	1.5	1.0	1.08	0.670
160	Plain	1.4	0.9	0.99	0.600
170	Plain	1.3	0.9	0.91	0.550
180	Plain	1.2	0.8	0.84	0.500
190	Snubbed	1.1	0.7	0.77	0.450
200	Snubbed	1.0	0.7	0.72	0.420
210	Snubbed	1.0	0.7	0.67	0.380
220	Snubbed	0.9	0.6	0.62	0.350
230	Snubbed	0.9	0.6	0.58	0.320
240	Snubbed	0.8	0.6	0.54	0.300

Dunlop Belt Design Manual

The drive pulley shown in Figure 6.27 recommends a lagged drive and a wrap angle of 220°. For this design, Table 6.9 suggests k = 0.35, which means that the slack-side tension has to be 35% of the effective tension:

$$T_2 = k * T_e = 0.35 * 225\,\text{kN} = 79\,\text{kN}\ (17{,}760\,\text{lbs})$$
$$T_1 = T_e * (1 + k) = 225\,\text{kN} * 1.35 = 304\,\text{kN}\ (68{,}345\,\text{lbs})$$

T_1 is the maximum tension force that the belt experiences. This allows us to finally pick a belt that is sufficiently strong to take the abuse from the heavy materials and tension forces at the drive pulley. In a first step we establish the maximum unit-operating tension T for the two belts we have considered, 41.3-in (1.05 m) and 43.7-in (1.20 m); this will help us to identify the belt class.

For the smaller belt = 304 kN/41.3-in = 7.36 kN/in (289 kN/m)
For the larger belt = 304 kN/43.7-in = 6.95 kN/in (253 kN/m)

The Dunlop Belt Design Manual shows in Table 12 that we need a steel-reinforced belting for either case, recommending for both the use of the belt class ST 2000 that should have 5 plies and would weigh 3.13 lb/ft^2 (15.3 kg/m^2). Three different minimum pulley sizes are recommended: drive pulley 31.5 in (0.8 m), tail pulley 24.8 in (0.63 m), and bend pulley 19.7 in (0.5 m).

Furthermore the number of plies for both belts should be 5 due to the high tension and the large rocks that they have to carry. Finally, the carrying idlers should be spaced 1 every 3.2 feet (1 m) and the return idlers the common 9.8 feet (3 m)

After we have designed a conveyor system that will deposit the crushed rock on the surge pile, we could repeat the exercise for the much shorter reclaim conveyor, which is loaded inside a tunnel underneath the surge pile and transfers the rock onto the first screening system, the primary screen. But instead we should take a closer look at this simple but very important equipment in the plant.

Designing the Screens

S6.16 Vibrating Screen Tabor

The main customers of rock quarry aggregates are asphalt and concrete plants, both having to follow rules and standards set by the American Association of State Highway Transportation Officials (AASHTO), the American Society for Testing and Materials (ASTM), and the American Concrete Institute (ACI), to name the most important ones. There are four basic categories that have been defined to separate the products according to sieve size:

(a) Coarse Aggregate: All the material retained on and above the No. 4 (4.75 mm) sieve.

(b) Fine Aggregate: All the material passing the No. 4 (4.75 mm) sieve.

(c) Mineral Dust: That portion of the fine aggregate passing the No. 200 (75 μm) sieve.

(d) Mineral Filler: A finely divided mineral product, at least 70 percent of which will pass the No. 200 (75 μm) sieve.

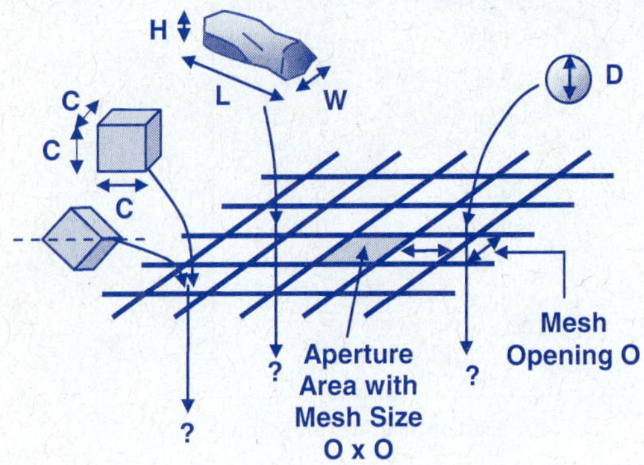

Figure 6.28 Rock Features Critical to Passing a Sieve

Figure 6.28 presents the physical situation of rock particles falling on a wire mesh defining a sieve with square openings of size O. As shown, crushed rock differs not only in size but also in shape such as cubical, spherical, elongated, and flaky or any combination.

One condition for particles to pass is that C of a near-cubical element needs to be smaller than the mesh opening (C < O). However, should the rock piece drop into the aperture with one corner pointing vertically down, it might get stuck in the sieve rather than staying on the top. This will not happen with a sphere when the diameter is smaller than the opening (D < O). In case of an elongated rock, if two dimensions are smaller than the mesh opening (H < O and W < O) it will able to pass, although one side is large. This condition leads to the situation that one can always find rock that is larger than the mesh size in the passing material.

By definition, coarse and fine aggregates are divided by a sieve No. 4. The definition of what this means can be found in the ASTM standard E 11, partially shown in Table 6.10.

How will this help us in setting the primary screen? The answer is critical, because it is responsible for the production rate of the entire plant. A key rule is not to crush rock that is already at a size that we can sell. The first opportunity to implement the cost saving guideline is the primary screen. What are the sizes that are desired? Again, we have to consider what the quarry customers' needs are. We already understand that they need to make concrete and asphalt! For the quarry operation, this means to produce and stockpile those sizes that fit the requirements of the customers of the concrete and the asphalt plants. In every state and county, the Department of Transportation (DOT) is one of those major clients, and their requirements concerning aggregate gradations are available on the Internet. While they might differ among regions, due to the characteristics of local rock, one will find five aggregate gradations that dominate. Table 6.11 defines those five according to ASTM and AASHTO.

Table 6.10 Standard Sieve Sizes According to ASTM E 11

Standard Sieve Designation (ASTM E 11)		Nominal Sieve Opening	
Standard	Alternate	mm	in
Coarse Sieves			
50.0 mm	2 in	50.0	2
37.5 mm	1-1/2 in	37.5	1.5
25.0 mm	1 in	25.0	1
19.0 mm	3/4 in	19.0	0.75
12.5 mm	1/2 in	12.5	0.5
9.5 mm	3/8 in	9.5	0.375
Fine Sieves			
4.75 mm	No. 4	4.75	0.1870
2.36 mm	No. 8	2.36	0.0937
1.18 mm	No. 16	1.18	0.0469
600 μm	No. 30	0.60	0.0234
300 μm	No. 50	0.30	0.0117
150 μm	No. 100	0.15	0.0059

Table 6.11 Grading Requirements for Coarse Aggregates (ASTM C 33 and AASHTO M 80)

#	Nominal Size, Sieves with Square Openings	Amounts Finer than Each Laboratory Sieve, Mass Percent Passing							
		50 mm (2 in)	37.5 mm (1 ½ in)	25.4 mm (1 in)	19 mm (¾ in)	12.5 mm (½ in)	9.5 mm (⅜ in)	4.75 mm (No. 4)	2.36 mm (No. 8)
467	1½ in - No. 4 (37.5–4.75 mm)	100	95–100	—	35–70	—	10–30	0–5	—
57	1 in - No. 4 (25.4–4.75 mm)	—	100	95–100	—	25–60	—	0–10	0–5
67	¾ in - No. 4 (19.0–4.75 mm)	—	—	100	90–100	—	25–55	0–10	0–5
7	½ in - No. 4 (12.5–4.75 mm)	—	—	—	100	90–100	40–70	0–15	0–5
78	½ in - No. 8 (12.5–2.36 mm)	—	—	—	100	90–100	40–75	5–25	0–10

A gradation is in fact a "recipe" for a final product that defines, in percentage of mass, how much of it needs to be contained or pass a certain sieve size as defined in Table 6.10. We find out the gradation by passing an aggregate mix through a series of sieves stacked with progressively smaller openings from top to bottom and then weighing the material retained on each sieve. For example, if the concrete plant requests # 57 and # 67 a sieve analysis should produce the following result.

Because #78, consisting mainly of ⅜-in and No. 4, is needed for the production of asphalt, we are able to rank the stone sizes needed most:

1. ⅜-in (9.5 mm)
2. ½-in (12.5 mm)
3. No. 4 (4.75 mm)
4. ¾-in (19.0 mm)
5. 1-in (25.4 mm)

Before we move on to look at the sieves that are being used in a plant, we need to understand that the ASTM standards are designed to be used in a materials laboratory with small amounts of stone with closely controlled sieves. Of course, efficiency requirements make it impossible to use small sieving equipment to reach the capacities that are needed. The heuristic rule designers of quarry operations use a sieve size that is one number smaller than the nominal size that the laboratory will use to inspect the final output gradation.

Figure 6.29 presents the cross-section of an inclined vibrating screen able to scalp off oversize rock, meaning rock too big for the secondary crusher, plus four different aggregate sizes. This is accomplished with the use of three decks, each with different mesh openings, as shown in Figure 6.28. The screens are stacked in an incline on top of each other inside a rectangular screening box ("multi-deck") vibrated by a rotating eccentric transverse shaft, driven by an electric motor by v-belts, to move the stone over the screens and down the small incline. The screen box is connected to its support frame via springs or resilient rubber dampers.

Table 6.12 Percentage of Rock Size Needs Based on Concrete Aggregate "Recipes"

Gradation 67		Gradation 57	
¾-in	0–10%	1-in	0–5%
⅜-in	45–75%	½-in	40–75%
No. 4	15–55%	No. 4	25–60%
No. 8	0–5%	No. 8	0–5%
No. 16–100	0–5%	No. 16–100	0–5%

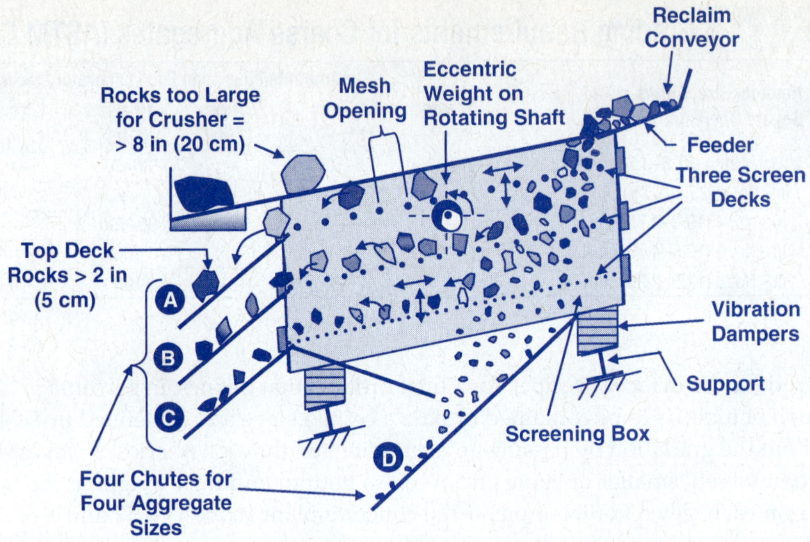

Figure 6.29 Cross-Section of Three-Deck Inclined Vibrating Screen

The unsorted material is fed by the reclaiming conveyor on the right end, where the stones have to be able to pass metal bars attached to the top, called the grizzly. The few pieces of rock there that are oversized and would block the secondary crushers are removed. Commonly they are used as riprap to protect shorelines, streambeds, and bridge abutments against erosion. The first or top deck has the largest mesh openings but retains those stones that are too big to be used as a final product Ⓐ. In other words, the stones that do not pass the top deck need to be crushed further by the secondary crusher. The same need for further crushing may hold true for the second deck retaining the next-smaller size Ⓑ. On the other hand, one may decide to pick a screen size that collects the largest-size aggregate that can be sold. This decision has to be made based on an overall design of the entire plant that begins with the setting of the primary crusher, which dictates the gradation of the crushed rock arriving at the first vibrating screen. Most probably, size Ⓒ will retain stone sizes that can be used directly as a product, such as No. 4, since this size will be needed heavily for concrete and asphalt. Since further crushing of the larger stones will create additional small-size aggregates, it would make sense to collect No. 4 and move it to a stockpile while expecting that after further crushing more No. 4 size stones will be produced and added to the stockpile. The third screen deck that retains Ⓒ lets pass rock pieces that are smaller than the selected mesh size, such as No. 4 with the nominal size of 0.187 in or 4.75 mm, which does include No. 8. On the other hand, it will contain quite a lot of sizes that are not acceptable, such as those smaller than No. 100. Of course, the secondary and tertiary crushers will produce more of these small sizes. As seen in Figure 6.21, screening is carried out after each crushing process. Thus, for the purpose of reaching the highest production possible, it might be desirable to remove this size and put it on a stockpile. In fact, the stones that pass No. 4 constitute a mix of fine aggregates that is referred to as crush and run. While not a highly prized product, it is sold as a base material for gravel walks and driveways or as a cover of clayey soil before pouring concrete for driveways, and so on.

Numerous factors affect the volume that screens can handle before blocking up. For example, the selection of mesh openings has to be geared to the shape, weight, and gradation of stones, the degree of inclination, the hertz (Hz) of the vibration, the size, and the position of screens within a deck. Finally, only long years of experience and tests with a particular rock will ensure that the desired capacity of a screening box meets the capacity of the conveyor feeding the crushed rock.

Secondary Cone Crushers

Because of their flexibility and consistency, cone crushers are extensively utilized as secondary crushers. Figure 6.30 introduces the cone-crusher belonging to the gyratory crusher "family." As the name implies, the key element is the cone, in fact two cones, one convex, and one concave. As shown, the fixed outer cone is supported by the main frame, while the solid inner cone is attached to the gyrating main shaft.

S6.17 Cone Crushers Nordberg
S6.18 Cone Crushers Sandvik

Again, the rock is fed in from the top. The feed bowl serves as an in-process storage container to make sure that there are always stones at the bottom of the bowl and in the crushing chamber. This will ensure that the cone operates in a balanced state and at top capacity. The top frame holds the head and the top bearing for the main shaft. Of course, strong plates and covers are used to protect them against the impact of the rock pieces falling into the feed bowl. Figure 6.30 shows also two hydraulic clamps to keep the upper and middle frame in place but also to stop them when work needs to be done inside the crushing chamber. Figure 6.31 displays in more detail the crushing operation by the two cones.

The two situations depicted in Figure 6.31 represent cross-sectional views of the two most extreme positions of a cone as it is being gyrated (not rotated) 360°. While the main shaft is much smaller than the one from the gyratory crusher, its principal function is the same. The top of the main shaft is held in place by a bearing allowing it to gyrate, the bottom is moved in a circle around a vertical axis by an eccentric bushing. As the cone is fixed to the main shaft, it follows its motion, and with it the mantle moving toward and away from the hard-steel plates of the outer cone. As a rock enters from the feeder into the crushing chamber at the top at the largest opening Ⓐ, it becomes wedged between the swiveling mantle and the outer cone liner. At the cone position the smallest gap between

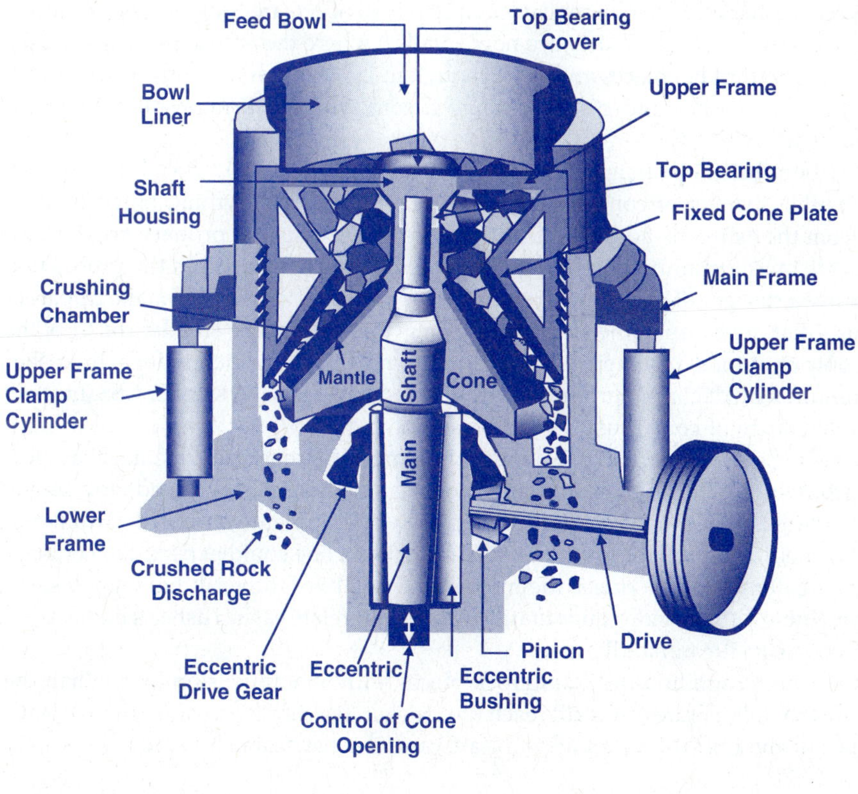

Figure 6.30 Cross-Section of a Cone Crusher in Operation

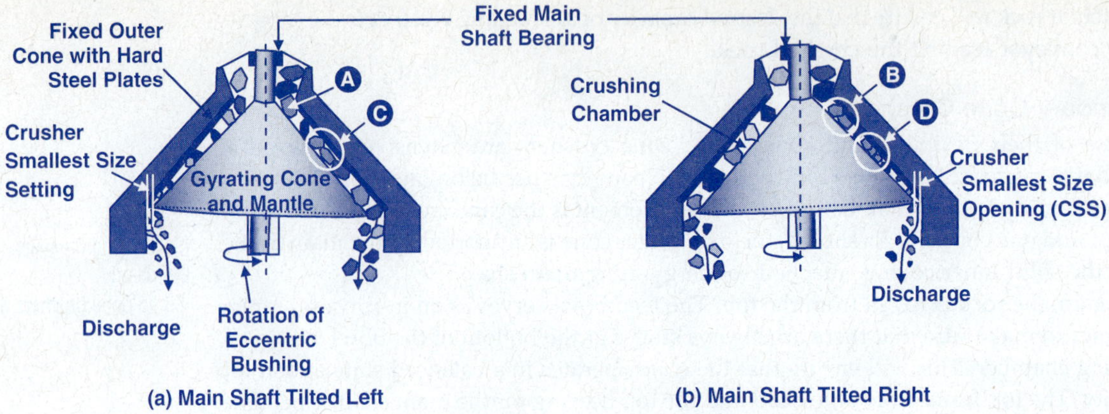

Figure 6.31 The Mechanics of Rock Breaking by a Cone Crusher

the two cones is at the left side on the bottom edge constituting its CSS. Now the eccentric bushing keeps swiveling the cone 180° to the position shown in Figure 6.31b. Since the rock from Ⓐ is wedged and stuck but the opening is closing just a little bit on the top as well, the crusher forces will be strong enough to break the stone down into two smaller ones at Ⓑ. At this maximum right-tilted position, the smallest opening has moved to the right bottom edge while the largest opening is 180° off, on the left side. Naturally, the cone will now swivel back toward a left-tilted shaft position, which will open the right crushing chamber a little bit, thus causing the two stones from Ⓑ to slide lower into the chamber getting wedged again at Ⓒ. As the inner cone keeps swiveling, it will now narrow the space at Ⓒ and break the stones further into progressively smaller pieces in Ⓓ. This process will continue until the pieces reach the lower edge, where they will leave the crushing chamber right after the cone has created the smallest gap and crushed the stone in the vicinity of the bowl liner or concave. Large pieces of ore are broken once, and then fall to a lower position (because they are now smaller) where they are broken again. This process continues until the pieces are small enough to fall through the narrow opening at the bottom of the crusher and discharge onto a conveyor belt bringing it to the next vibrating screen.

When we studied the primary crusher, we learned that commonly the maximum rock size that fits the opening of cone crusher is 8 in or 203 mm. Everything bigger will be removed from the "stream" and used as riprap. As in the case of a primary crusher, we need to decide on the minimum opening or the CSS. Figure 6.32 displays the production curves for five crushers of different strengths. The graphical depictions make apparent relationships that are comparable to those of primary crushers: The smaller the CSS, the lower the output in tons/hr. There is an interesting new phenomenon that each crusher has an optimum with declining production on either side, larger and smaller CSS. In other words, the openings of cone crushers can be set too big.

Each crusher, such as the SB220, is shown with three different-size crushing chambers to choose from, indicated by the three extensions C, M, and F. The C stands for coarse, meaning the largest rock that that particular piece of equipment can handle. Similarly, M stands for medium and F for finest or smallest rock. As one could expect, each crusher type with the extension C produces the most in the family of three, all having the same drive motor. Table 6.13 provides quantitative values that relate each crusher's name to the maximum rock size it can handle.

Based on the recognition that some rock pieces with one dimension larger than the sieve size, depicted in Figure 6.28, will reach the crusher, we should consider the SB350-C, the EB540-C, or the EB620-C as an appropriate cone crusher. Table 6.13 lists all of them

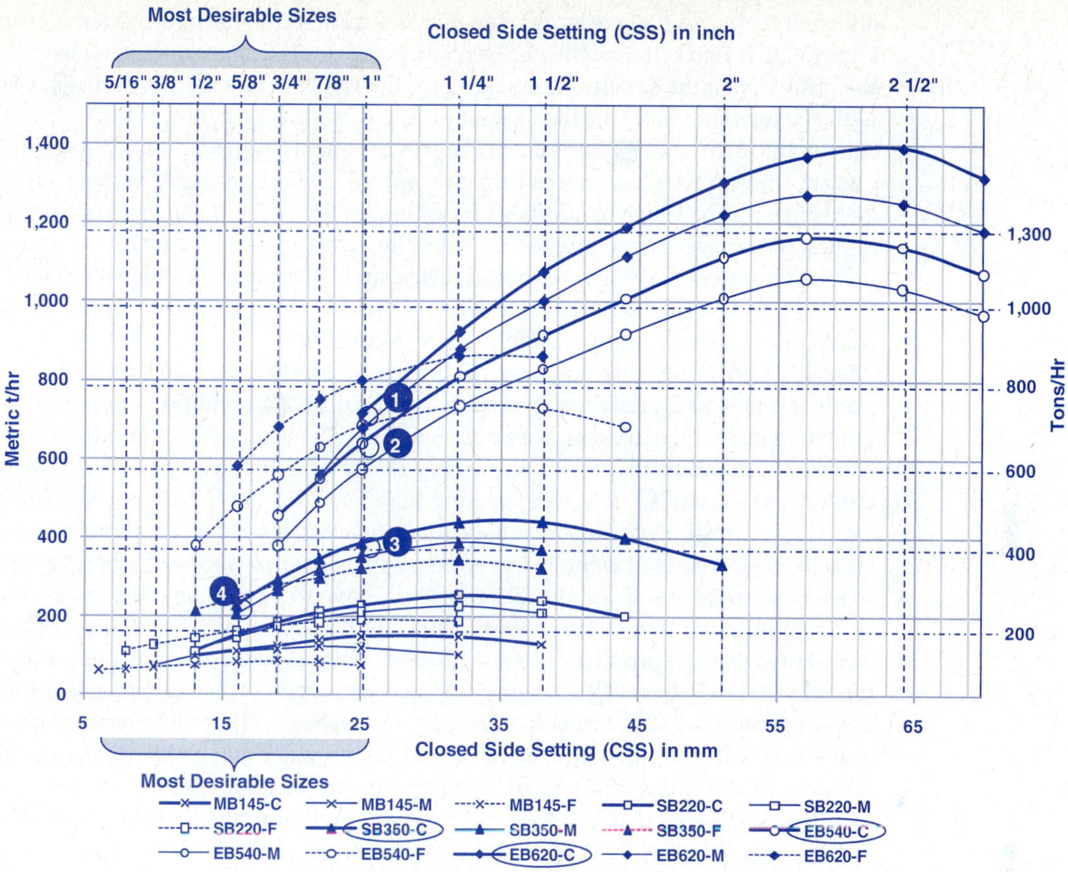

Figure 6.32 Production Curves for a Sample of Cone Crushers

Table 6.13 Crusher "Families" and Their Maximum Rock Sizes

Crusher "Family"	Motor-Size		Crushing Chamber Size	Max Rock Size	
	kW	HP		mm	inch
MB110	110	148	MB110-C	90	3.54
			MB110-M	65	2.56
			MB110-F	38	1.50
MB145	145	194	MB145-C	145	5.71
			MB145-M	90	3.54
			MB145-F	50	1.97
SB220	220	295	SB220-C	175	6.89
			SB220-M	110	4.33
			SB220-F	70	2.76
SB350	350	469	SB350-C	215	8.46
			SB350-M	135	5.31
			SB350-F	85	3.35
EB540	540	724	EB540-C	240	9.45
			EB540-M	155	6.10
			EB540-F	90	3.54
EB620	620	831	EB620-C	330	12.99
			EB620-M	195	7.68
			EB620-F	120	4.72

allowing rock sizes that are larger than 8-in (200 mm). With this information we can look at Figure 6.32 to find out how their CSSs relate to the desired aggregate sizes (see Table 6.12). We observe that the smallest CSS setting for the EB620-C is 1-in (25.4 mm), ❶, but having with 720 tons/hr (640 t/hr) the highest output of all three options. EB540-C will produce 700 ton/hr (630 t/hr), ❷, but allows the crusher openings to be changed to ¾-in (19 mm). Clearly, the SB350-C produces at point ❸ the lowest amount, namely 405 tons/hr (360 t/hr); on the other hand, its CSS can be set to 5/8-in. If our goal is to achieve the highest output we are encouraged to choose the strongest crusher. If our goal is to produce the sizes we need the most, we should pick a smaller crusher, allowing us to set a smaller CSS leading to a higher percentage passing the desired screen sizes. Thus, we have two competing solutions that we should try to reconcile.

Let's verify the effect of the CSS on the amount of desired sizes that we would produce using the gradation curves in Figure 6.33. Ⓐ identifies the point where the gradation curve of a crusher with a CSS of 1-in (25 mm) crosses the mesh size of 1½-in (38 mm). It indicates that 8% of discharged stones will not pass the 1½-in (38 mm) screen while, marked with Ⓑ, 30% will be larger than 1-in (25.4 mm), and only 33% smaller than ½-in (12.7 mm) Ⓒ. Unfortunately, only about 40% will lead to a useful product, with only about 8% to be retained in the 3/8-in screen, which amounts to 58 tons/hr (51 t/hr). On the other hand, 60% or 432 ton/hr (386 t/hr) will have to be sent to a tertiary crusher. Let's compare this with the output of the SB350-C with a setting of 5/8-in. Marker ❹ in Figure 6.32 shows that the production of the SB350-C with a CSS of 5/8-in is approximately 230 ton/hr (205 t/hr). Label Ⓔ in Figure 6.31 marks the percentage of the output that passes the 3/8-in sieve as 52% and 67% passing ½-in mesh size. These two passing percentages also tell us that 15% is retained on the 3/8-in screen, since that much did not pass it. Thus, the amount of 3/8-in is 15% of 230 ton/hr or 34.5 ton/hr (31 t/hr). Of course, this is smaller than the 58 tons/hr that EB620-C can produce. Similarly, the SB350-C generates 13% of 5/8-in at Ⓖ, 7% of ¾-in at Ⓖ and 6% of 1-in at Ⓖ. Overall, with a CSS of 5/8-in 95% of the SB350-C's discharge can be used as it passes the 1-in mesh screen. In other words, 95% of 230 ton/hr (20 t/hr) can be retained and only 5% has to be sent for recrushing.

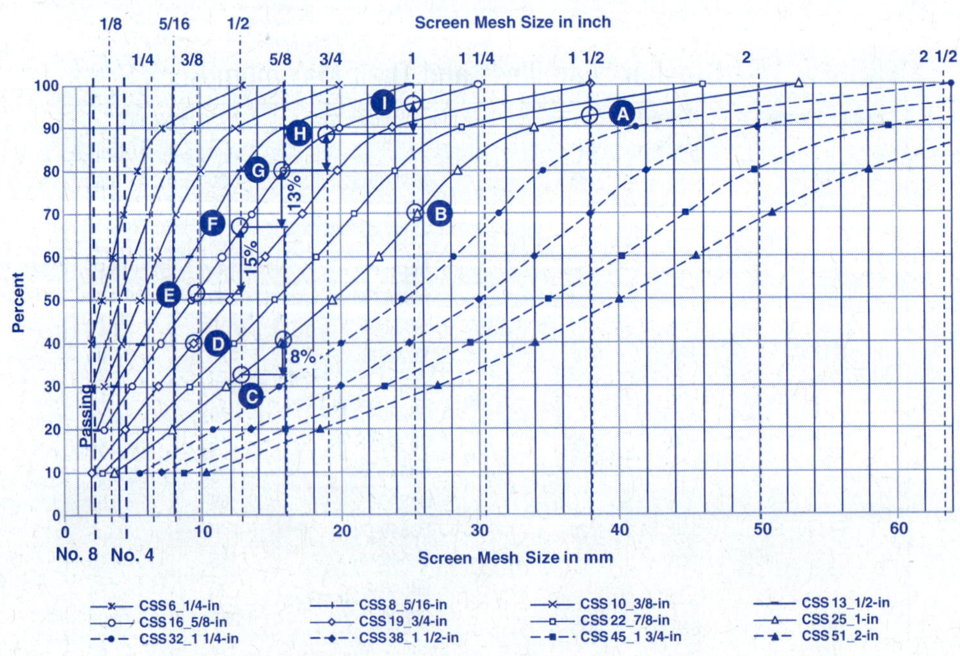

Figure 6.33 Gradation Curves for Cone Crushers

Unfortunately, this crusher is able only to process 230 ton/hr (209 t/hr), which is much smaller than the 800 ton/hr (726 t/hr) that the primary jaw crusher or the dump trucks can haul to the grizzly feeder. There would be a need for 3 or 4 SB350-Cs to match one jaw crusher alone.

Tertiary Impact Crushers

Impact crushers utilize the dynamic forces of two objects colliding at high speed rather than the static forces applied to break rocks. Inside the crushing chamber, steel hammers rotating at 300 to 1500 rpm hit pieces of feed rock and smash the initially broken up stones against the breaker plates covering the walls of the chamber. Manufacturers offer two different types for slightly different purposes: (1) Horizontal Shaft Impact (HSI), and (2) Vertical Shaft Impact (VSI) crushers. HSI crushers are generally used for coarser and VSI crushers for finer material. Figure 6.34 presents an HSI, which is more common in quarry operations.

The rock feed enters the chamber on an inclined plate at the inlet in the direction of the hammers moving at a speed that can be varied between 100 ft/sec and 200 ft/sec (30 and 60 m/sec). Higher speeds tend to make a higher percentage of fine material, while lower speeds result in a more uniformly graded material. Of course, the velocity of the hammer is a function of the radius of the rotor, holding the blow bars, and the chosen rpm. Rotor dimensions vary from 16-in (406 mm) diameter and 24-in (610 mm) width to 6.5 ft (2.0 m) diameter to 7.5 ft (2.3 m) in width.

The high-speed hammers (see arrows indicating the direction) strike the material at the inlet to the chamber and throw the broken-up pieces against the first breaker plate on the top of the chamber. As Figure 6.34 depicts, hydraulic actuators are available to change the angle of the two breaker plates via a setting rod. Of particular importance is the gap between the tip of the first breaker plate and the tip of the blow bar. The arrows demonstrate that the rock pieces ricochet from the breaker plates back into the chamber, colliding with rocks traveling in the opposite direction and resulting in rock-on-rock breakage. Pushed by the blow bars, they all have to pass through gap 1, which opens and

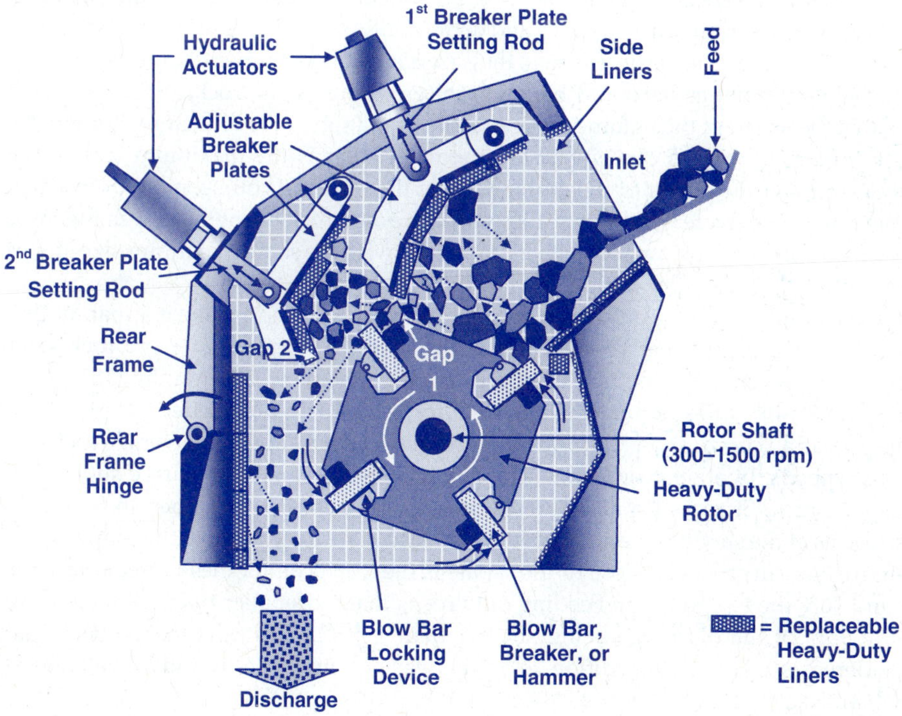

Figure 6.34 Cross-Sectional View of an Impact Crusher

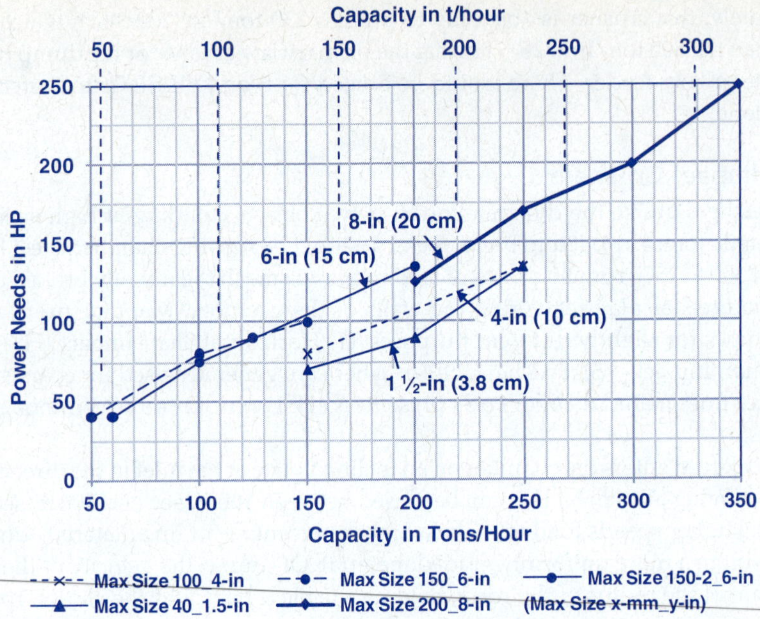

Figure 6.35 Capacities of Impact Crushers as a Function of Power and Feed Size

closes slightly with the rotation of the square rotor. In the smaller second section of the crusher chamber, the second breaker plate is set much tighter, thus increasing the rock-on-rock fracturing. Finally, passing gap 2, the rock pieces hit the last, vertical breaker plates in the discharge section.

Access to the rotor and hammers, for preventive maintenance, is made easy by hinging the rear frame in the back and two hydraulic cylinders on the sides of the crusher (not shown in Figure 6.34). As a result, the entire top frame, including the two breaker plates, opens like the shell of a clam allowing personnel easy entrance into the chamber.

HSI crushers come in different sizes and configurations to fit various uses. While they are most widely used as tertiary crushers, they are also built as primaries and secondaries for material that is not as hard and big as granite coming as shotrock. Above we have learned that the speed of the hammers impacts the gradation of the discharge. But what is critically important for the capacity of an impact crusher is the maximum size of the incoming rock and the power of the motor driving the rotor. Of course, more power will allow the rotor to be made wider, thus allowing more feed rock to enter through the inlet. Figure 6.35 graphs several impact crusher "families" with the same maximum size stones reaching from 1½ in (40 mm) to 8 in (200 mm).

The five graphs, plotting only the upper limits of each crusher, make it apparent that capacity has an almost linear relationship with power. As expected, larger rock sizes require more power to break. For example, for a capacity of 200 ton/hr (181 t/hr), breaking the 1½-in stones requires 82 hp while for 8-in 125 hp are needed.

The gradation curves shown in Figure 6.36 represent the results from average rock and at average speeds. In other words, one is able to change the gradation curve slightly by changing the rpm. Increased rpm will result in a higher percentage of fines and, thus, in steeper slopes at the smaller sizes. The smallest-size feed rock included in the graph is 1⅛–in (30-mm). Not surprisingly, close to 100% passes the ¾-in (19 mm) mesh screen, 48% the No. 4, and 30% the No. 8 screen. Feeding only rocks that are smaller than 1½ in (40-mm) results in 95% of the discharge being smaller than 1-in (25-mm), the largest rock that we considered as desirable. In addition, 11% is between $^1/_2$-in and $^3/_8$ in, and 7% between 1-in and $^3/_4$-in.

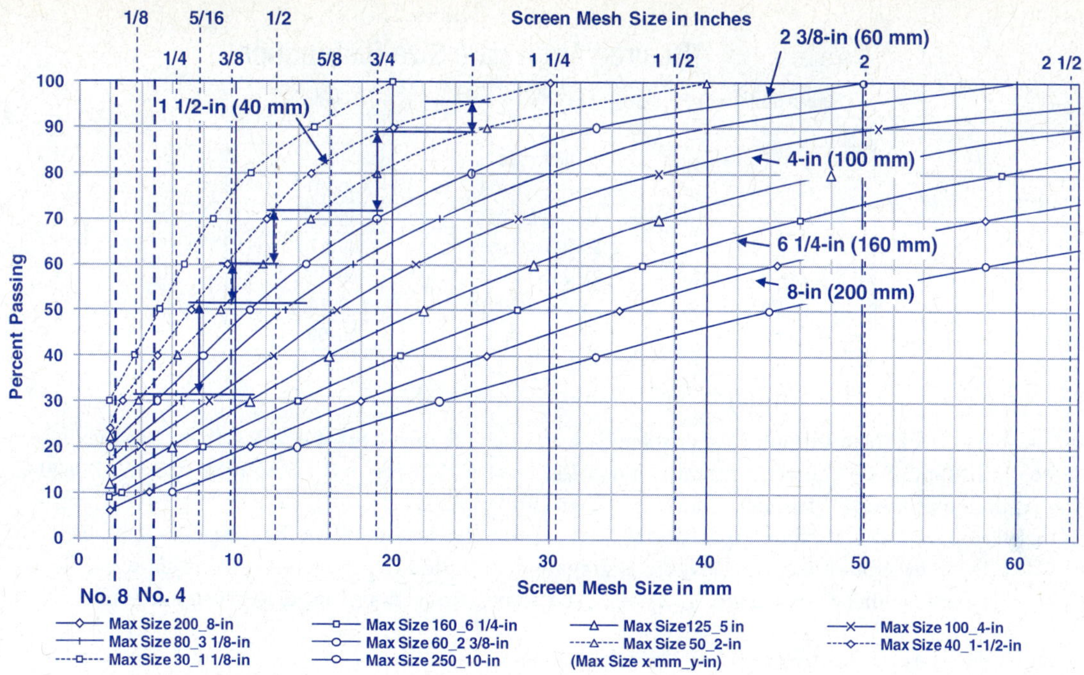

Figure 6.36 Gradation Curves for Impact Crusher with Average Speed and Rock

6.4.3 PLANNING THE PLANT OPERATION

We started the study of the quarry production with an overview in Figure 6.10. In between, we studied four different crusher types, conveyors, and screen operations. We also learned that the most efficient DSS setting for the primary jaw crusher is 8-in (200 mm), which would allow the processing of between 700-ton/hr and 1,000 ton/hr (625 – 893 t/hr) depending on the size. We found the proper conveyor size to move 800 ton/hr of crushed rock to the surge pile up at the plant level where it is stored before it is picked up by the reclaiming conveyor. This conveyor feeds into the primary screen, consisting of three inclined decks with a grizzly on top to remove oversize rock to be used as riprap. The selection of the mesh sizes for the primary screen is a major decision point as it directs the flow of up to four sizes and the output capacity of the additional crushers that will be necessary. For this reason, we need to consider the entire plant operation when making any kind of decision on equipment size and setting. We have three overarching goals when designing the plant operation:

1. Produce aggregate in the needed sizes
2. Produce aggregate in the needed quantities
3. Meet the capacity of the primary crusher with the fewest movements

Worked-Out Problem 6.6 / Design of Crushing Plant

The Blue-Mountain rock quarry is planning to recalibrate its output in order to better serve the needs of concrete and asphalt plants in the region. It owns a jaw-crusher that produces 800 ton/hr crushed rock at a DSS of 8-in. The setup of its equipment is drawn in Figure 6.10. They presently run the cone crusher EB540-C (see Table 6.12) as a secondary and a 250 ton/hr (227 t/hr) impact crusher with maximum 4-in feed material as a tertiary. It has been found that the

Table 6.14 Desired Aggregate Size Distribution

Screen Size (inch)		Ton/hr
10	11%	90
1 ½	89% pass	0
1	6–11% retained	40–80
¾	6–11% retained	40–80
½	20–24% retained	140–170
⅜	21–25% retained	150–180
No. 4	15–20% retained	110–140
No. 4	7–13% pass	50–90

EB540-C is able to work the maximum 10-in rock sufficiently despite the fact that it is designed for a maximum rock size of 9.45-in (240-mm). Of course, it requires much less electricity compared to the next size, the EB620-C.

The gradation of the jaw-crusher material is shown in Figure 6.21.

Of the 800 ton/hr (726 t/hr) they expect about 11% as riprap that is larger than the 10-in (250-mm). From the 712 ton/hr (646 t/hr), they would like to receive an aggregate distribution that matches the one shown in Table 6.14.

Fine aggregates smaller than the No. 100 sieve need to be removed and used as crush-and-run.

Background and Assumptions

From the gradation curve in Figure 6.21, we learn that no gradation information exists below the 10% of the crushed rock that is smaller than 2.8 in or 72 mm. Figure 6.37 presents the complete gradation curve with an assumed segment that closes the 10% gap.

The largest useful size is 1 in (25-mm), which is only 1% of the 712 ton/hr (645 t/hr) (difference between 1 ½ and 1 in passing), while 2% will pass this screen. In other words, only 3% of the entire intake will be screened off. Considering this minimal amount of useful aggregate coming from the primary, it seems to be nonproductive to separate any it. In other words, it would make sense to send the entire 712 ton/hr to the crushers.

Presently the impact crusher is set to accept a maximum rock size of 4 in (100 mm). Figure 6.38 shows the gradations of some alternative sizes together with the relevant screen mesh sizes. As shown, 19% will not pass mesh size 1 ½ in.

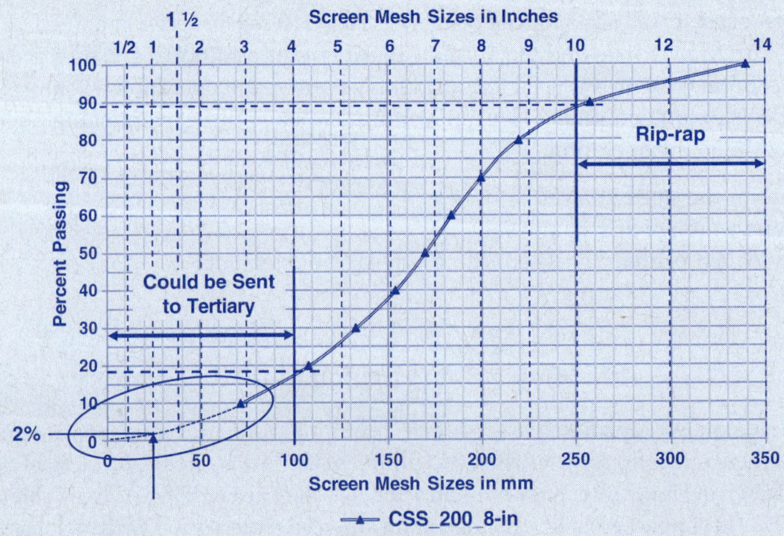

Figure 6.37 Complete Jaw-Crusher Gradation for CSS 8-in (200-mm)

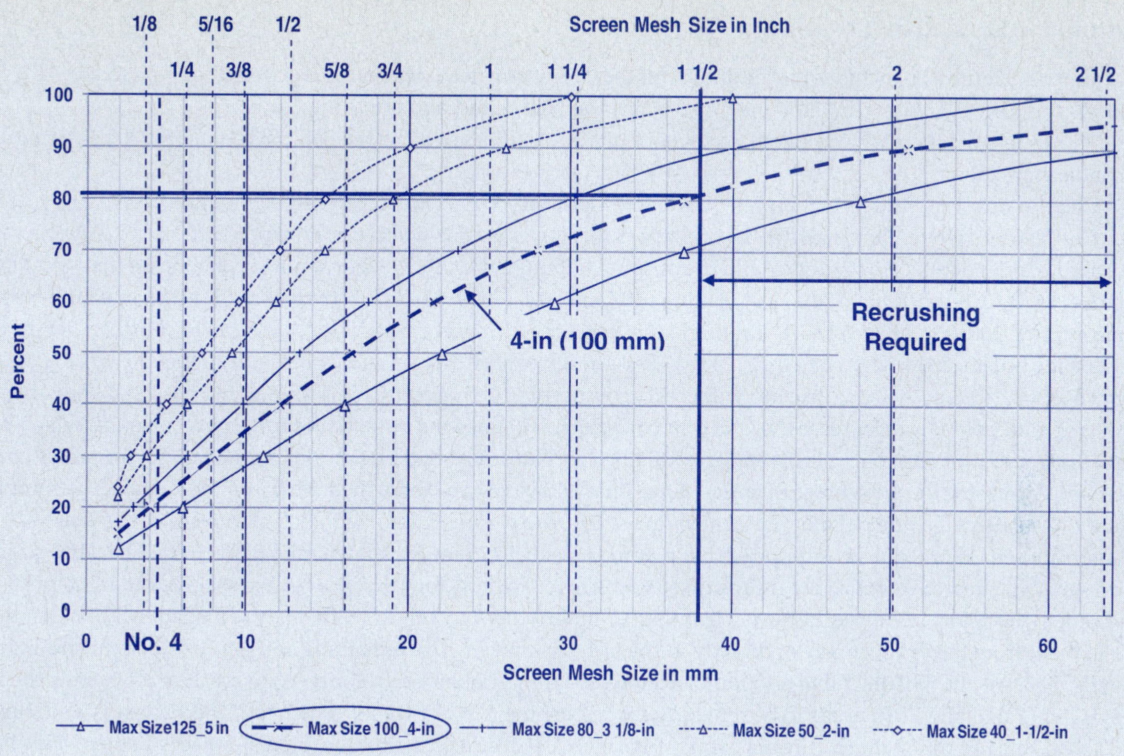

Figure 6.38 Gradations for the Impact Crusher with Various Feed Sizes

Setting Options for Secondary Crusher:

Figure 6.32 presented the output for the EB540-C. With a setting of 1 ¼ in it can handle slightly more than 800 tons/hr (725 t/hr), and with a setting of 1 in (25 mm) it handles 640 ton/hr (580 t/hr).

Setting of Primary Screen

As before, rocks larger than 10 in (250-mm) or 11% of the total jaw crusher output will be removed via a grizzly on top of the primary screen. Because the tertiary presently accepts a maximum of 4-in (100 mm), we might start with the same configuration. Based on Figure 6.37, approximately 18% of the total 800 tons/hour could be sent directly to the tertiary impact crusher. This would reduce the amount that flows to the secondary to

$$(100\% - (11\% + 18\%)) * 800\,\text{tons/hr} = 568\,\text{tons/hr}\,(515\,\text{t/hr})$$

which is smaller than the 640 ton/hr that can be handled by the EB540-C with a CSS of 1 in.

Setting Options for the Tertiary

The present 250-ton/hr impact crusher accepts a maximum of 4-in rock. According to the gradation curves in Figure 6.38, this maximum rock size results in 81% of the incoming material being able to pass 1½-in (40 mm). In other words, 19% have to be sent back into the inlet to be recrushed. This would mean that of the 144 ton/hr (18% of 800 ton/hr) 25.4 ton/hr would have to be recrushed. Thus, the total inflow at the tertiary crusher caused by screening the 4-in at the primary screen will be 170 tons/hr (154 t/hr). This amount is acceptable, as it is smaller than the 250 ton/hr (226 t/hr) capacity of the present impact crusher. Yet, it does not include what will come off the secondary screen that should be crushed by the tertiary. On the other hand, we might set the secondary screen so that only 80 ton/hr (72.5 t/hr) will be sent to the tertiary.

Setting the Secondary Screen

The graph in Figure 6.33 includes the gradation for the CSS of 1-in. It shows that 93% passes the 1½-in screen, 70% the 1-in, 49 the ¾-in, 32% the ½-in, 25% the 3/8-in, 12% the No. 4, and about 8% the No. 8.

From this gradation we can calculate the amount of useful sizes from the 568 tons/hr (515 t/hr) entering the secondary crusher.

Clearly, much too much 1-in and ¾-in are being produced, and much too little ½-in and 3/8-in.

One solution would be to narrow the CSS to 7/8 in. However, this will reduce the production to 590 ton/hr. On the other hand, not much is changing with the output. Although the CSS 7/8-in opening will increase the percentage retained in the 3/8-in screen to 10%, its tons/hr is still only 0.1 ∗ 590 tons/hr = 59 tons/hr (53 t/hr), still far from the minimum of 200 ton/hr (181 t/hr), although the tertiary will surely be a large percentage. Further, while the percentage of 1-in size will decline from 23% to 13% the amount of 3/4 -in will increase from the 21% to 28%. All in all, closing the CSS at the cone crusher will create counterproductive results but a small increase in 3/8-in output.

A second option is to increase the CSS to 1¼ in, which will raise the potential output to 820 tons/hr, which will leave 110 ton/hr for recrushing (820 ton/hr – 710 ton/hr). Naturally, this larger opening will move the gradation to the right, increasing the percentage of larger stones that will have to be recrushed. To better understand the complex situation, we use a flow model diagram that shows where the various sizes of crushed stone move. The exact values of the different flows can be supported by a spreadsheet program as shown in Figure 6.39. This first design represents an attempt to utilize the setups discussed above. First, 88 ton/hr (80 t/hr) are scalped off at the primary and sent to the riprap pile since they are larger than 10 in. Another 144 ton/hr (130 t/hr) of crushed rock, smaller than 4-in, is sent from the primary screen directly to the tertiary crusher. The remaining 568 ton/hr (515 t/hr) flows into the EB540-C set at CSS 1-in. Table 6.15 indicates the resulting graduation from the cone crusher. Of course, the 7% that is larger than the 1½ in needs to be sent back for recrushing. Should we keep the rest? The answer is also shown in Table 6.15. There is too much 1-in and even ¾-in to meet the desired distribution. Here it is necessary to decide if all or part of the flow should be sent for recrushing or if some should be sent to the tertiary crusher. Naturally, of crucial importance is the maximum capacity of the secondary crusher, 640 ton/hr (580 t/hr). Because of this limitation, the maximum amount of crushed stone sent back onto the reclaiming conveyor already supplying 568 ton/hr is 72 ton/hr (640 ton/hr −568 ton/hr) (65 t/hr).

Table 6.18 makes clear that the amount of stones to be recrushed is, with 40 + 131 + 119 ton/hr, much larger than the 72 ton/hr of empty capacity. Consequently, we decide to send the entire 131 ton/hr of the 1-in to the tertiary crusher while the 40 ton/hr of 1½-in can be recrushed by the EB540-C. Of course, the ¾-in size, with its 119 ton/hr, is also out of bounds. As a way out, it is proposed to split the total amount, using a splitter, to an approximately 1:3 split, which would result in 32 ton/hr being recrushed and 89 ton/hr sent to the tertiary. Alas, the recrushed 72 ton/hr, all smaller than 3 in, will have to be added to the discharge amounts. Since the recrushed material is much smaller than the reclaimed rocks from the surge pile, their gradation should be kept separate. For this reason, the underlying spreadsheet uses the following spread for the recrushed rocks: ¾-in 27%, ½-in 27%, 3/8-in 17%, No. 4 17%, and No. 4 pass 13%. As a result, the 119 ton/hr crushed from the surge pile will be augmented by the 27% of the

Table 6.15 Actual-Desired Output Comparison for Cone Crusher EB540-C with CSS 1-in

Screen Size (inch)	CSS 1-inch Gradation Curve 100% = 568 ton/hr	Actual ton/hr	CSS 1 1/4-inch Gradation Curve 100% = 820 ton/hr	Desired Sizing 100% = 800 ton/hr	Desired ton/hr
1½	7% retained	40	13% retained	89% pass	710
1	23% retained	130 *(too much)*	35% retained	5–10% retained	40–80
½	21% retained	119 *(too much)*	14% retained	5–10% retained	40–80
½	17% retained	97 (too little)	14% retained	20–25% retained	160–200
⅜	7% retained	39 (too little)	6% retained	25–30% retained	200–240
No. 4	13% retained	74 (too little)	10% retained	15–20% retained	120–160
No. 4	12% pass	68 OK	8% pass	5–10% pass	40–80

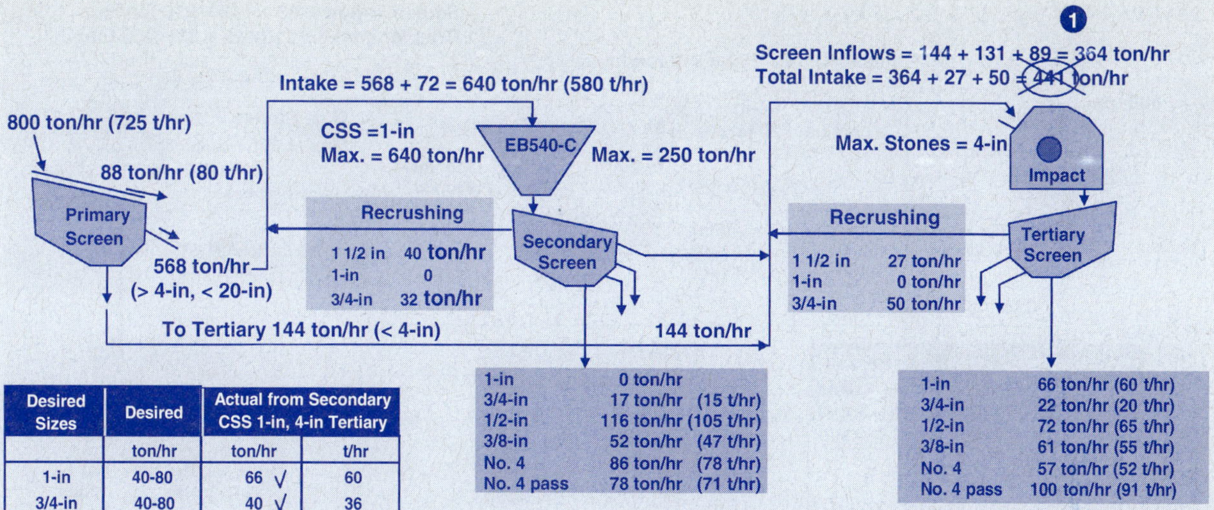

Desired Sizes	Desired	Actual from Secondary CSS 1-in, 4-in Tertiary	
	ton/hr	ton/hr	t/hr
1-in	40-80	66 √	60
3/4-in	40-80	40 √	36
1/2-in	140-170	187 √	170
3/8-in	150-180 ↔	112 (2)	102
No. 4	110-140	143 √	130
No. 4 pass	50-90 ↔	178 (3)	162
Total		726	659
Secondary		348	316
Tertiary		~~379~~ (1)	344
Useful		726	659

Figure 6.39 Flow Diagram of Plant with Secondary CSS at 1-in and Tertiary at 4-in

72 ton/hr of recrushed material. Thus the total amount of $\frac{3}{4}$-in size rock coming from the secondary at all times is 119 ton/hr + 19 ton/hr = 138 ton/hr. In order to meet the balance, 17 ton/hr is split off and sent to the $\frac{3}{4}$-in stockpile. Of course, this amount can be varied by sending more or less to the tertiary.

The first problem arises at the feeder to the impact crusher with a maximum intake capacity of 250 ton/hr. First, 144 ton/hr of rocks smaller than 4 inches will arrive via a conveyor from the primary screen. Second, 131 ton/hr of 1-in and 89 ton/hr of $\frac{3}{4}$-in rock are coming from the secondary screen. As noted in the flow diagram shown in Figure 6.39, this total of 364 ton/hr is already above the maximum capacity although it does not even include the recrushed material. In order to get a feeling for the size distribution, the spreadsheet was used to meet the desired amounts as closely as possible. As the summary sheet in Figure 6.39 shows, it's easy to meet the goals of the larger sizes, however, not enough 3/8-in aggregate (2) and too much fine sand (3) was produced. As one would expect, the hammers of the fast-spinning rotors of the impact crusher create with 100 ton/hr passing No. 4 more fine "dust" than the larger capacity statically crushing cone-breaker (78 ton/hr).

After this first iteration, we can make some observations that will help in modifying the present setup.

1. The capacity of the cone crusher has to be drastically improved, as the capacity of the tertiary cannot be changed without getting a new and larger system.
2. Changing the CSS of the cone crusher from the present 1 in to $1\frac{1}{4}$ in to $1\frac{1}{2}$ in will increase its capacity to 820 ton/hr and 920 ton/hr, respectively (see Figure 6.30).
3. A CSS of $1\frac{1}{4}$ inches will produce 6% of 3/8-in size and the $1\frac{1}{2}$ in 5% of 3/8 in.
4. A larger-capacity cone crusher will allow a larger amount to be recrushed. Recrushing will further reduce the sizes and add to the 3/8-in amounts in a higher percentage.
5. 6% of 712 ton/hr = 43 ton/hr, 5% of 712 ton/hr = 35.6 ton/hr.
6. Both CSS $1\frac{1}{4}$ in and $1\frac{1}{2}$ in for the cone crusher will not produce any additional material that passes No. 4.
7. Changing the tertiary from a 4-in to a 2-in intake will increase the amount of 3/8-in to from 7% to 11%. However, it will also increase the material passing No. 4 from 22% to 32%, thus producing more, not less, of the fine sizes.

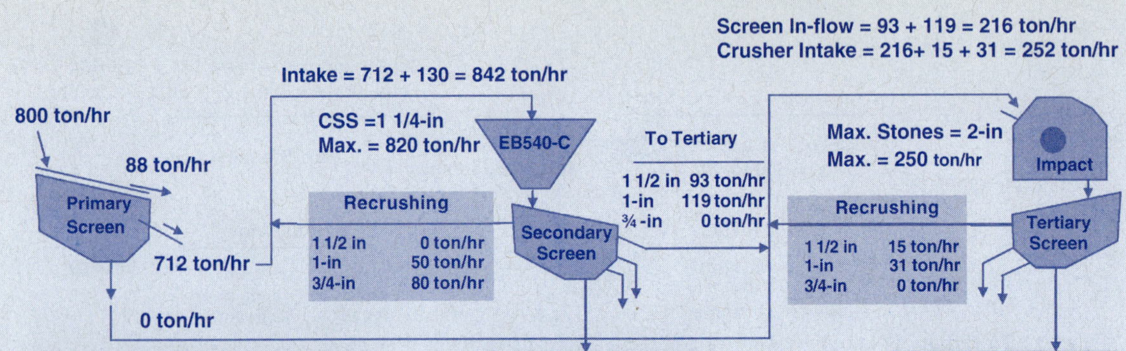

Desired Sizes	Desired	Secondary CSS 1 1/4-in, 2-in Tertiary
	ton/hr	Actual ton/hr
1-in	40-80	80 √
3/4-in	40-80	76 √
1/2-in	140-170	181 √
3/8-in	150-180 ↔	99 ❶
No. 4	110-140	150 √
No. 4 pass	50-90 ↔	130 ❷
Total	Intake	715
Output Sec	ton/hr	500
Output Ter	ton/hr	214
Tot Output	ton/hr	714

Size	%	ton/hr
1 1/2-in	13	93
1-in	35	249
3/4-in	14	131
1/2-in	14	139
3/8-in	6	70
No. 4	10	103
No. 4 pass	8	57

Size	%	ton/hr
1 1/2-in	7	15
1-in	13	31
3/4-in	10	25
1/2-in	15	41
3/8-in	10	28
No. 4	18	47
No. 4 pass	28	73

Figure 6.40 Flow Diagram of Plant with Secondary CSS at $1\frac{1}{4}$-in and Tertiary at 2-in

8. By reducing the speed of the impact rotor while feeding a maximum of 2-in rocks from an average to a low speed, it is possible to attain a gradation curve that results from rock feeds with a maximum size of $2\frac{3}{8}$-in or 60 mm (see Figure 6.36).

Based on these observations we can reset the crusher parameters and run the spreadsheet a second time. Figure 6.40 presents the results.

As anticipated, changing the CSS for the secondary to a larger $1\frac{1}{4}$-in drastically changed the flow pattern and, most importantly, allows the reduction of the intake at the tertiary to 252 ton/hr, which meets the capacity limit. On the other hand, the secondary is fed 5% over capacity that may be possible because only rocks of sizes 1 in and $\frac{3}{4}$ in are recrushed. Field tests have to determine whether the cone crusher is able to handle the resulting 842 ton/hr. If not, more crushed 1-in, in addition to the 93 ton/hr $1\frac{1}{2}$-in stones, can be sent directly to the impact crusher geared to handle 2-in rock. Possibly, the tertiary is able to handle the slight overload easily because of the sizes smaller then 2-in being fed into the crushing chamber.

What is disappointing is the fact that the desired $\frac{3}{8}$-in size is not being met ❶. On the other hand, this must be expected, in that the mesh size of this sieve is only $\frac{1}{8}$-in smaller than that of the $\frac{1}{2}$-in a very small span to collect the wanted quantity. There is nothing significant we can do about the gradations that would add a sufficiently large amount of $\frac{3}{8}$-in. However, there must be a solution to this imbalance, as other quarries face the same dilemma. Let us go back to the origin of the desired rock size quantities—the grading requirements for concrete and asphalt in Table 6.11. Based on ASTM C 33, for the concrete gradation # 67 the $\frac{1}{2}$-in size is not being measured separately. This means that 25–55% of $\frac{3}{8}$-in aggregate include all the sizes that pass $\frac{3}{4}$-in but not $\frac{3}{8}$-in which of course, includes $\frac{1}{2}$-in. The same holds true for # 467. Similarly, gradation # 57, $\frac{1}{2}$-in can also include $\frac{3}{4}$-in.

We can conclude that ASTM C 33 provides significant flexibility concerning $\frac{3}{8}$-in stones. For all intents and purposes, we can add the desired quantities and outputs of $\frac{1}{2}$-in and $3\frac{3}{8}$-in and in addition, a small amount of $\frac{3}{4}$-in. Thus we can compute

Desired $\frac{1}{2}$-in $+\frac{3}{8}$-in $= (140-170 \text{ ton/hr}) + (150-180 \text{ ton/hr}) = 290-350$ ton/hr

Actual $\frac{1}{2}$-in $+\frac{3}{8}$-in $= 181$ ton/hr $+ 99$ ton/hr $= 280$ ton/hr and

Desired $\frac{3}{4}$-in = 40–80 ton/hr
Actual $\frac{3}{4}$-in = 76 ton/hr

Discussion of Results

With great relief, we observe that the combined quantity of 3/8-in and ½-in is, with 280 ton/hr (254 t/hr), only 3.5% off the lowest target of 290 ton/hr (263 t/hr). An additional 36 ton/hr (33 t/hr) could be used from the $\frac{3}{4}$-in, enough to meet both requirements.

This leaves only the problem of producing too much that passes No. 4 sieve or too much of the fine aggregate ❷. Although the extra material can be sold as crush-and-run, the unit price would be low. On the other hand, it might be possible to sell more of the fine aggregate for concrete for mixes that require a very dense mix. Finally, it might be possible that by opening the cone crusher to a CSS of $1\frac{3}{8}$-in or $1\frac{1}{2}$-in, less fine aggregate could be produced.

6.5 Off-Road Trucks: Earth Haulers

What makes a truck an off-road truck? We already know about the size and weight limitations on roads and highways; a quick look at the physical characteristics of samples might provide us with an answer. We already met the CAT 775F working hauling shotrock in the quarry that weighs 110 ton (100 t). Figure 6.41 presents two, the articulated truck D350 from Deere and the rigid-frame earth hauler EH1100-3 from Hitachi (for more information visit the booksite). As shown in Figure 6.41a, the 350D has a maximum GVW of (46,000 + 46,000 + 40,000) 132,000 lb (59.0 t) while the EH1100-3 weighs 244,000 lb (111 t).

Let's check the GVW limitations. Based on the axle rule, the 350D is allowed a GVW of (12,000 + 20,000 + 20,000) = 52,000 lb (23.6 t).

Conversely, the maximum GVW of the EH1100-3 is (12,000 + 20,000) = 32,000 lb (14.5 t).

Based on the FBF rule, the max GVW of the 350D is: $W = 500 * ((3 * 21/2) + 12 * 3 + 36) = 51{,}6501$ (23.4 t).

For the EH1100-3: $W = 500 * (2 * 14 + 12 * 2 + 36) = 44{,}000$ lb (20 t).

Checking the allowable GVWs we realize very quickly that the two trucks will not be allowed onto roads and highways. In addition, the restrictions for width of 8.5 ft (2.6 m)

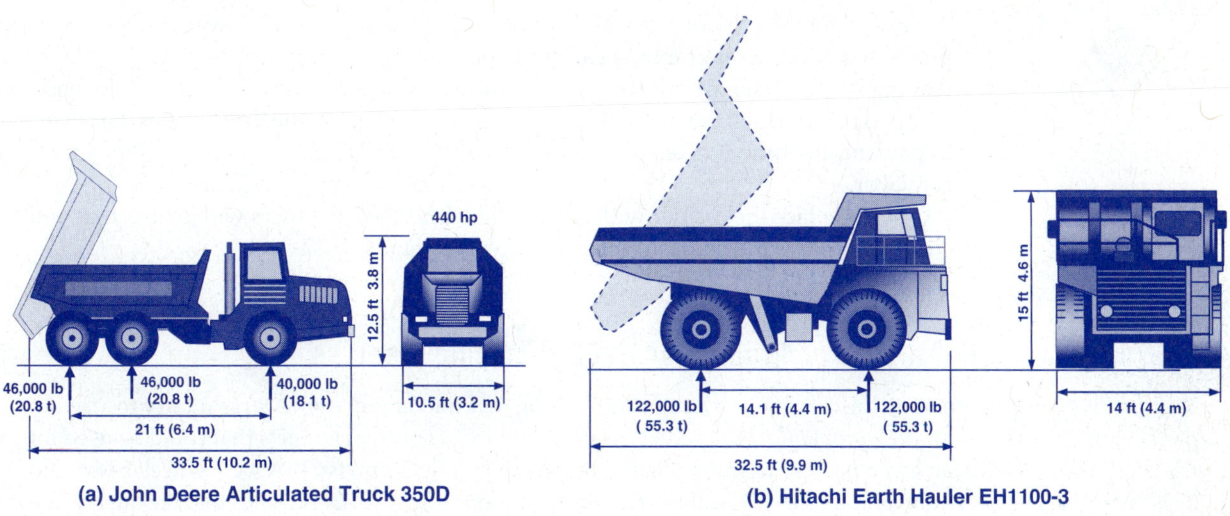

Figure 6.41 Examples of Off-Road Trucks

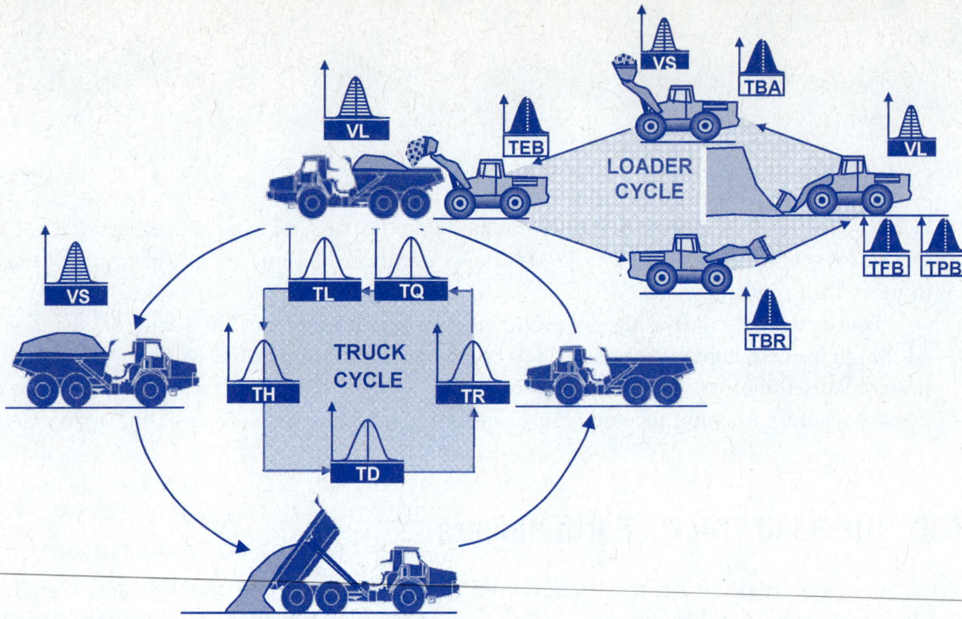

Figure 6.42 Modeling the Earthmoving Operation Using Wheel Loader and Truck

and height of 14 ft (4.3 m) are not met by either of them. As a result, large haulers are only visible on highways on top of trailers decorated with large signs: OVERSIZED LOAD.

Trucks and haulers depend on a second resource for loading. The production rate of an earth-moving truck depends on a variety of factors, some more important than others. Beginning with the definition of productivity of one unit, volume of soil transported per hour from point A to point B, we can quickly identify two key measures: (a) volume of soil hauled per trip, and (b) number of hauling trips per hour. While the volume depends a lot on how much the excavator was able to load, the second measure is a function of the cycle time for a truck.

6.5.1 OPTIMIZING THE EQUIPMENT FLEET

An excavator or a loader resembles a cashier at a supermarket who fills a cart with purchased goods after grabbing each unit and entering the price by hand or via a barcode reading system. The overall throughput, measured as groceries per minute, depends on the speed with which the cashier is able to process each item, the time it takes the customer to pay and the time it takes the next customer to put the goods on the cashier's table or conveyor.

Let's replace the cashier with an excavator and the customers with trucks that, unlike customers, stay in a constant loop, as shown in Figure 6.42.

Worked-Out Problem 6.7 / Production of Earthmoving Fleet

You are involved in estimating the cost of excavating and trucking rock and gravel from a large quarry to an earth dam being built. One of the key problems is to predict the productivity of equipment fleets that could be used for the project. You have been given the task to study what a combination of the Komatsu WA500-6 wheel loader and 350D articulated trucks from John Deere will be able to move in one hour. Figure 6.43 indicates that the quarry is at a slightly higher altitude and can be reached on 1.5-mile (2.4 km) road that is fairly well maintained, resulting in a RR of 3%. Of course, your boss also wants to know how many trucks will be needed.

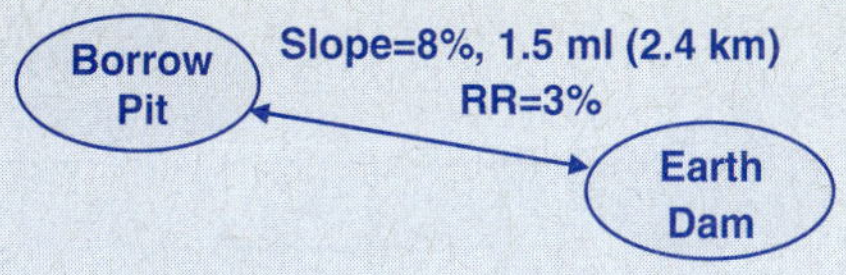

Figure 6.43 The Haul Road

Assumptions

It is assumed that the traveling time within the quarry area is on average 1 minute and 1.3 minutes at the dam. Of course, since the loading and dump locations change as the work progresses, these durations will change. The specifications for the two types of equipment are available on the booksite. For the loading operation, it is assumed that a tandem loading strategy will be applied where the loader will never have to wait for a truck to be available. The skill of the loader operator is presumed to be average to good, resulting in the following mean durations:

$$TPB = 13\,\text{seconds}, TFB = 31\,\text{seconds}, TBA = 72\,\text{seconds}, TEB = 7\,\text{seconds}, TBR = 57\,\text{seconds}$$

See Figure 6.13 should the meaning of the abbreviations be unknown. The heaped volume of the excavating bucket used by the WA500-6 loader is 6.8 yd^3 (5.2 m^3), and the struck volume is 5.5 yd^3 (4.2 m^3).

The heaped 2:1 volume of the 350D = 26.3 yd^3 (20.1 m^3), and the struck volume is 19.9 yd^3 (15.2 m^3).

Truck speed for a full truck on a 8% decline and an RR of 3% results in a Total Resistance TR = −5%. From the Retardation chart in Figure 6.44, considering the fully loaded GVW, we get an average speed of 35.7 ml/hr (57 km/hr).

For the empty 350D traveling up the 8% incline and experiencing the 3% RR, the chart in Figure 6.44 tells us that the average speed will be 14 ml/hr (22.5 km/hr).

Calculations Step 1: Number of Loader Cycles

Expected average dumper volume = 25 yd^3 (19 m^3)
Average loader bucket volume VL-VS (spillage) = 6.2 yd^3
Number of loading cycles = 25 yd^3/6.2 yd^3 = 4 cycles (= 24.8 yd^3)

As these calculations use mean values, the expectation of achieving 24.8 yd^3 (19 m^3) is reasonable.

Calculations Step 2: Truck Loading Duration

Mean duration of one loader cycle = 13 sec + 21 sec + 52 sec + 7 sec + 42 sec = 135 sec
Mean duration for loading one truck = 4 ∗ 135 sec = 540 sec = 9 min

Calculations Step 3: Truck Cycle Time

Mean loading time = 9.0 min
Travel time inside quarry area = 1.4 min
Travel fully loaded to the dam = (60) ∗ 1.5 ml/(35.7 ml/hr) = 2.6 min
Travel inside dam area = 2.3 min
Emptying dumper (including maneuvering) = 2.2 min
Travel inside dam area = 2.3 min
Travel with NVW to quarry TR = 11% = (60) ∗ 1.5 ml/(14 ml/hr) = 6.6 min
Travel time inside quarry area = 1.4 min
Total cycle time without queuing at the loader = 27.8 min

Calculation Step 4: Hourly Production of One Truck

Average volume in one truck with 5% spillage 95% ∗ 24.8 yd^3 = 23.5 yd^3 (18 m^3)
Total time for one truck cycle = 27.8 min
Hourly production = 60 min/hr ∗ (23.5 yd^3/27.8 min) = 50.7 yd^3/hr (38.8 m^3/hr)

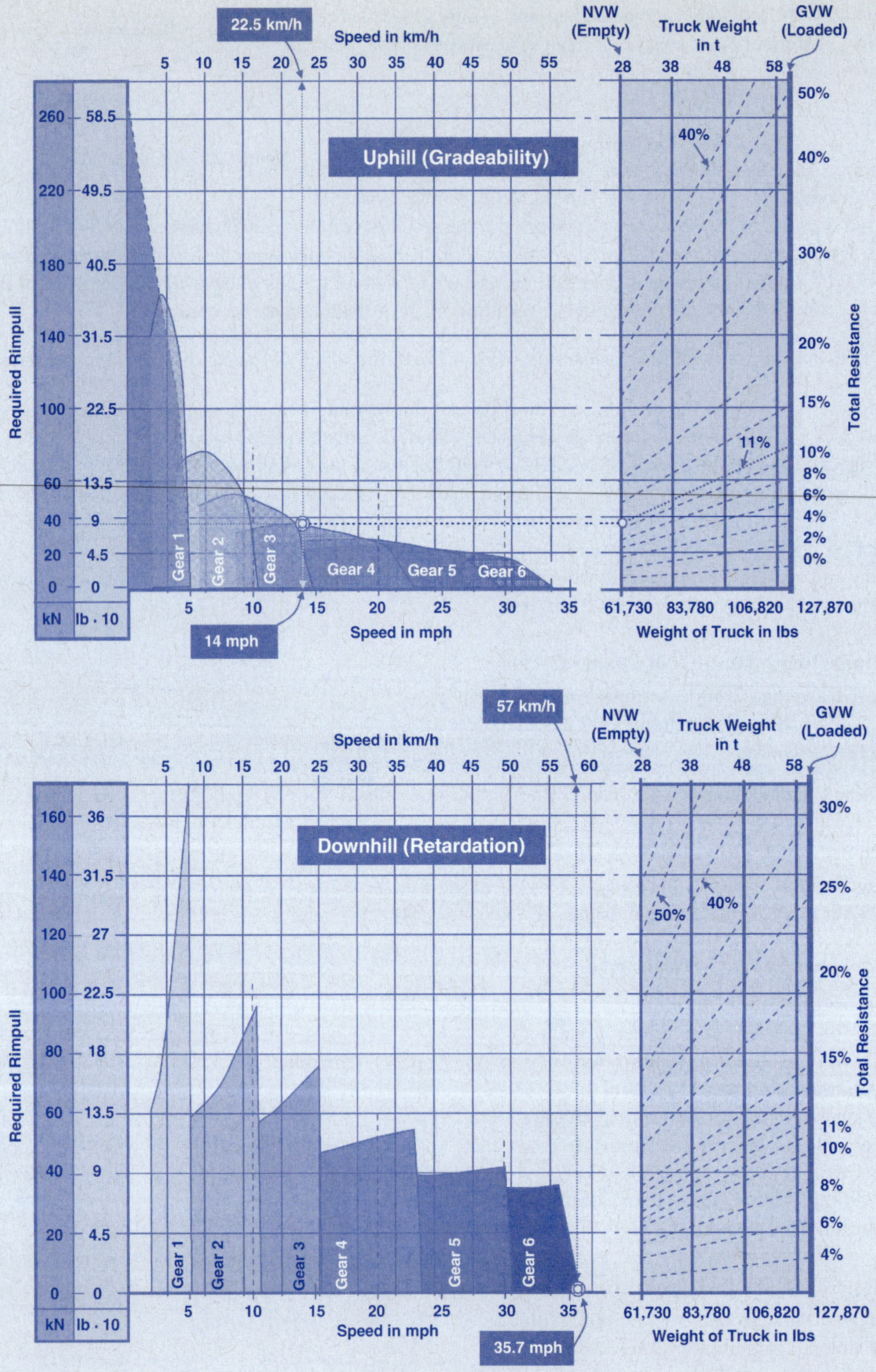

Figure 6.44 Rimpull Curves for the 350D

Calculations Step 5: Hourly Production of Loader

Average volume in one truck = 6.2 yd^3 (4.7 m^3)
Time for one loader cycle = 135 sec
Maximum production/hr = 60 min/hr * (6.2 yd^3/(135 sec/60) min) = 165.3 yd^3/hr (126.4 m^3/hr)

Calculations 6: Number of Trucks Needed

A balance point analysis is used to find the number of trucks that should be used and what the resulting production would be. The bottleneck production rate is set by the wheel loader at 165.3 yd^3/hr (126.4 m^3/hr). The loader is considered the bottleneck since this equipment is not able to load any greater amounts unless technical changes are made such as a larger engine, bucket, or even a different loading method. The balance point graph (see Figure 6.45) provides the data for the final analysis. The horizontal line marks the possible production rate of the loader that is independent from the number of trucks available. When we assume that only one 350D truck would be hauling, then the two pieces of equipment would be able to load, transport, and unload 50.7 yd^3/hr (38.8 m^3/hr) at the dam. This would also mean that the loader would be waiting at the quarry for the lone 350D to come back. Obviously, this is not very efficient. The production doubles to 101.4 yd^3/hr (77.6 m^3/hr) if a second truck is added but is still far short of the 165.3 yd^3/hr that could be loaded. Adding a third truck improves the rate to 145.1 yd^3/hr (111 m^3/hr) and not to 152.1 yd^3/hr (116.4 m^3/hr), as should be expected. The difference of 7 yd^3/hr is caused by the bunching effect, which is related to the randomized effect of operating trucks within a loop.

Discussion of Calculations

The balance point is at 3.26 trucks, requiring a decision whether to use 3 or 4. The bunching effect is small at 3 trucks, but there is an additional loss of production due to lack of 0.26 trucks, forcing the loader to stay idle and wait for the next to arrive.

On the other hand, 4 trucks will not be able to haul 202.8 yd^3/hr (155 m^3/hr), as the loader is only able to load 165.3 yd^3/hr (126 m^3/hr) with the bunching effect, reducing it to 157.1 yd^3/hr (120 m^3/hr).

Adding the fourth truck will increase the hourly production by 12 yd^3/hr (+10%) from the 145.1 yd^3/hr (111m^3/hr).

A balance point .75 units below the next higher integer causes a drastic loss of volume that could be hauled. Although the operational costs do not have to be considered at this time, it is clear that solutions should be found to improve the loading capacity of the wheel loader.

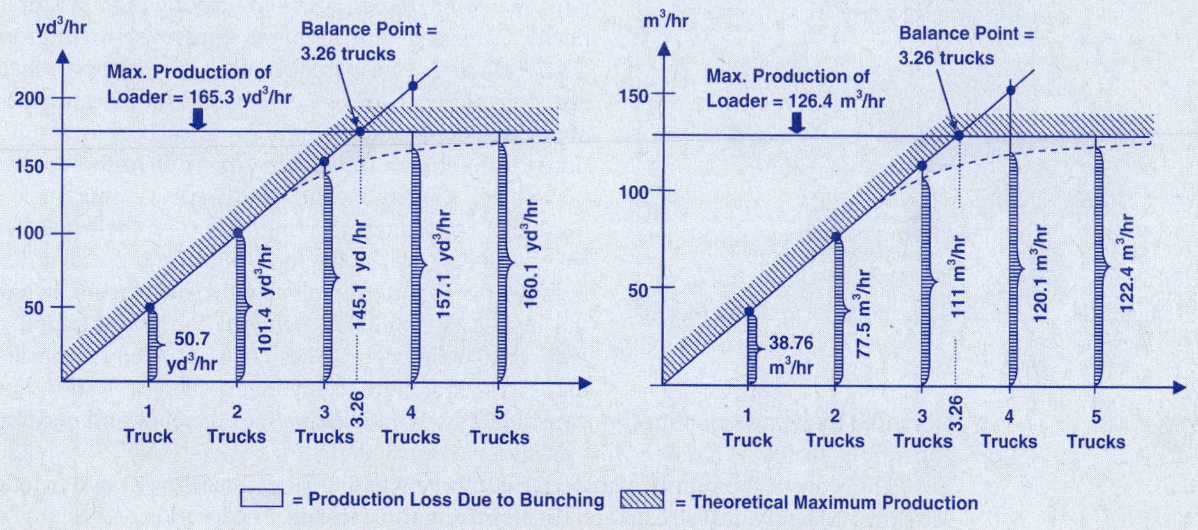

Figure 6.45 Balance Point Analysis of Earth Hauling Operation

Bunching – Up in the Queue

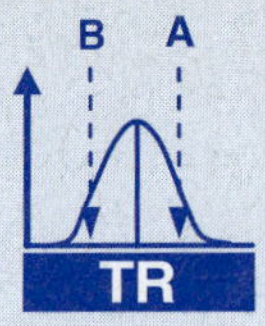

The source of bunching at a bottleneck, such as cycling trucks at the loading point, is the randomness of driving times. If driver B is following driver A on the back-cycle return, he will have to wait in the queue (TQ) until A is loaded, although he was faster.

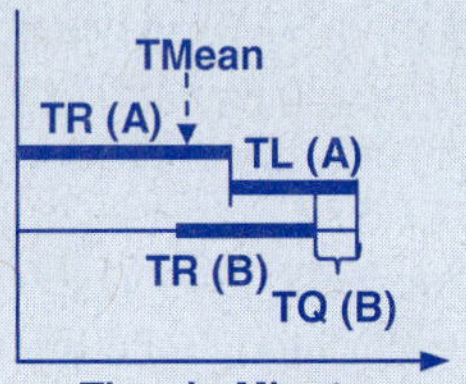

TR(A) > TMean and TR(B) < TMean.

TQ(B) =(TR(A) – TMean)+ (TMean – TR(B))

Option 1: When we compared the three operators Jim, Mark, and Len, we learned that a good operator is able to beat the average productivity of a wheel loader. It might be possible to hire an operator who could increase the loaded productivity by 10%, bringing the balance point to $181.8/50.7 = 3.6$.

Option 2: Replace loader with a truck with slightly stronger engine and a larger bucket such as the Komatsu WA600-3 with a heaped bucket capacity of 8.0 yd^3 (6.1 m^3) instead of the 6.8 yd^3. This change would increase the loading capacity by 17% and the theoretical production rate from 165.3 yd^3/hr to 194.0 yd^3/hr (148 m^3/hr), with a new balance point of 3.83 trucks and an approximate fleet production, considering an 8% bunching loss, of 179 yd^3/hr (137 m^3/hr).

Final Recommendation

With a switch to a slightly larger loader, such as the WA600-3, 22 yd^3/hr (17 m^3/hr) could be added to the projected 157.1 yd^3/hr (120 m^3/hr). This would better balance the fleet and possibly reduce the cost per yd^3. Thus, a change in the selected equipment should be considered. Should the operating cost of the WA600 be too high, a highly skilled operator should be hired.

How Does RFID Work?

At the core is a physical tag consisting of a digital memory chip, a transceiver, and an RF antenna. Active tags have a battery to transmit data to a reader over a long distance (100 ft) and have up to 1 MB of read/write memory, while passive tags only transmit when activated by a reader's magnetic field. Most passive tags hold fixed "license plate" data, while active tags collect, store, process, and communicate 2-way with an approved reader.

The remote reader decodes/encodes the data passing through to/from a host computer.

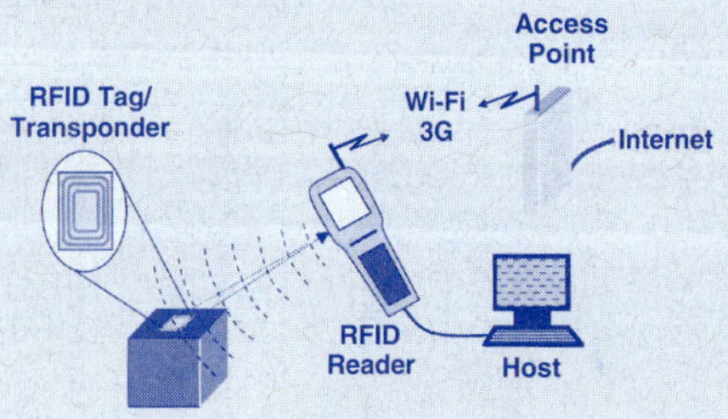

6.6 Electronics and Satellites Boost Truck Performance and Security

6.6.1 RFIDS STREAMLINE THE SUPPLY CHAIN

The quintessential goal of trucks in construction is to ship material, components, tools, and equipment. As the amounts, the sources, and the speed of these movements have increased, the difficulty in tracking them has drastically increased. Furthermore, pressures to improve efficiency and transparency along the supply chain are demanding new ways of collecting information about the location and condition of each unit. A traditional technology used to identify an item to be transported is a written tag that is affixed to a unit, as in a bundle of rebar, and traveling with it on the truck. The success of barcodes in the food, retail, and postal industries has not found its equivalent in construction, where the necessary standards between the many suppliers and construction companies have not been established industry-wide. New technologies such as RFID, short for radio frequency identification, are already revolutionizing the efficiency and meaning of supply chains. What does this mean for construction?

Think about the supply of material, equipment and tools as consisting of two parallel streams: (a) a **physical stream** and (b) an **information stream**. The location and quality of the units that you see moved by trucks and cranes are only known to you and those watching it. Even you might not know its final location or if it is the right kind, and so on. All that should be included in the information stream that has its origin in a design and a request order and stops when the unit becomes part of a bigger unit, is returned back or is dismantled at the end of its life. Consider the traditional method for fabricating, delivering, and placing reinforcement bars for an elevated concrete slab (see Figure 6.47). This starts with the design and detailing of rebar (1) on a computer.

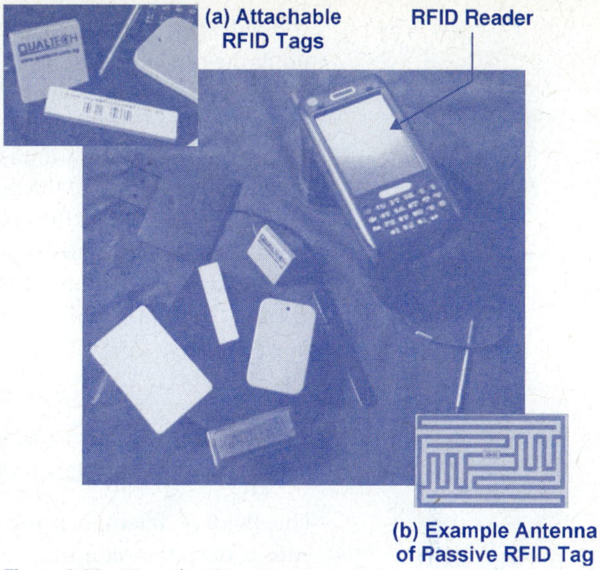

Figure 6.46 Tags for Construction

Based on the list of needed rebar pieces, organized by shape and length, the fabrication shop cuts and bends (2) the raw material (3) to be assembled into bundles according to shapes (4) to be laid out on the shop floor. As soon as a shipment of a job is complete, the bundles are loaded on trailers that meet the weight limitations for which bills of ladings or material list are created (5). When needed on the construction sites, tractor will pull the trailers to the site (6) to be unloaded by crane onto the empty formwork (7). Subsequently, a rebar-laying crew will bust the bundles (8) and place the bars according to the rebar design (1). Figure 6.47 presents the two streams mentioned above. The "source" of the physical reinforcing bars is the deformed steel in coils or straight lengths (2) and ends with the rebar being placed and tied (8) to be cast in concrete to become a concrete slab. The information stream starts earlier in time during the design process (1) and lasts beyond the encasing of the rebar with concrete. On the other hand, the stream consists of many substreams that spring up: (a) blueprints, (b) electronic data for shear line/rebar bender, (c) printed bundle tags, (d) loading list/bill of lading, and finally (e) an (electronic) bill for the shipped rebar (9).

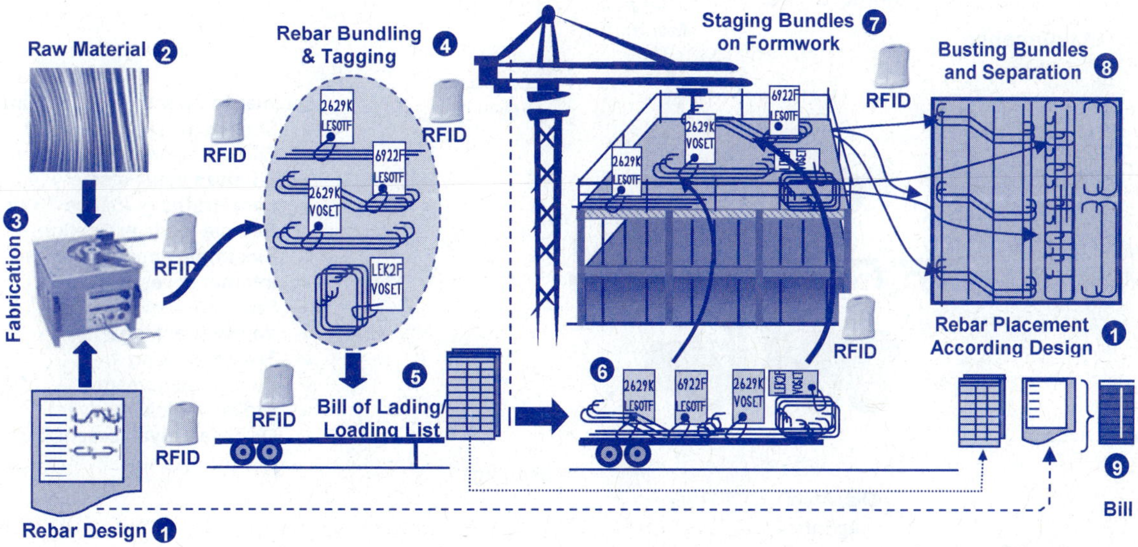

Figure 6.47 Traditional Rebar Supply Chain Using Printed Tags with RFID

RFIDs connect all the subchannels into one electronic data stream accessible anytime along the physical supply chain using authorized readers. RFID tags are created during the design (1) by storing all the data necessary for cutting and bending (3), bundling (4), trailer-loading (5), shipping (6), staging (7), placing (8), and billing (9). RFID readers can be used when information is needed, for example, for the loading of a trailer. A search algorithm identifies all the bundles that needed to be loaded and activates a light on each tag for a crane or forklift operator to see. The sequence of loading is also prescribed to meet weight and load-balancing requirements. If the tags have been properly programmed, they contain data about how to stage and the most efficient sequence for placing the bars. Finally, the reusable tags are read in as the basis for creating the bills electronically.

6.6.2 TELEMATICS CONNECTS AND PROTECTS CONSTRUCTION EQUIPMENT

The field of telematics represents a conglomerate of other areas, mainly telemetry and information-communications technologies (ICT). Telemetry (derived from Greek terms tele = remote and metron = measure) had its birth during the Space Race in the 1960s and 1970s when measurement had to be taken on space vehicles and transmitted using radio signals to Earth for monitoring. The emergence of the satellite networks opened up vast avenues for expanding telemetry from space to Earth-based applications such as the Global Positioning System (GPS). Two other large domains that are using digital communication have since arisen and have linked up with the original telemetry to make the still-growing field of telematics. These are the Internet and the wireless (Wi-Fi) networks covering entire nations. The integration of those three areas has opened many new opportunities for users of trucks and will continue to do so in the future. Figure 6.48 gives a graphical overview of the vast information network that has been created to assist the driver, protect the equipment, and send emergency aid.

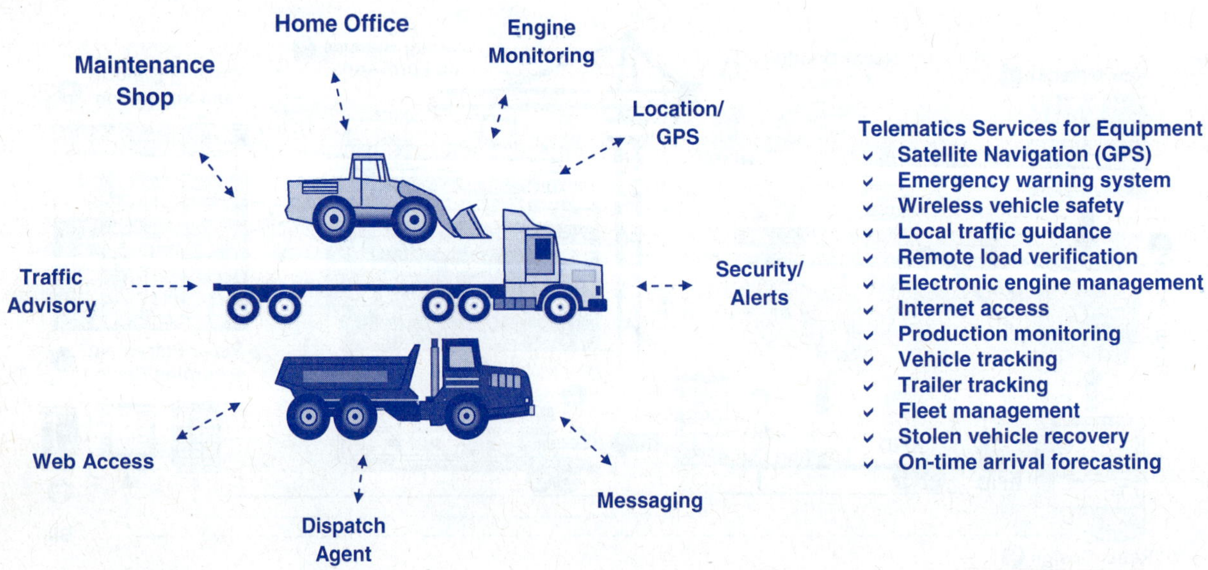

Figure 6.48 Telematics Providing Support for Construction Trucks

Telematics offers useful services not only for long-haul semi-trailer trucks, but for any expensive production vehicle. The list of telematic services shown in Figure 6.48 which is prone to grow longer over time, covers ubiquitous issues such as stolen vehicle recovery and engine management. The latter example is an especially useful service to keep expensive off-highway trucks from breaking down unexpectedly. On-board sensors collect data about the performance of critical components to be sent automatically to the maintenance shop for analysis. During off-time, the truck can be maintained when needed, and replacement parts can be ordered and installed before they stop working. In the same vein, the efficiency of earth haulers, dump trucks, and concrete trucks can be easily verified and improved through managerial steps. For example, the tracking component may show that a concrete truck has to wait a long time at a particular construction site before it is unloaded. This can be easily addressed by sending fewer trucks or rerouting trucks. In fact, wireless cameras installed on the truck may provide the dispatcher visual images about the situation as well as pictures of the traffic on the road.

A steel company that delivers to construction sites using semi-trailers can appreciate the trailer tracking component that is independent from the trailer tractor. Although much cheaper than the tractor, the theft of a trailer together with the cargo can create significant disruptions and cost to reship the engineered components. Trailers also might get dropped off at the wrong place or simply forgotten, thus reducing the shipping capacity of the company.

6.7 Chapter Six Review

THE EQUIPUZZLE

The letters in the identified fields can be arranged to form the name of a Greek person who invented a mechanism that is installed in many trucks used in construction.

Across

7. The facts about something. It is helpful to know a lot of this before you turn in your report

10. "Opposite" to an articulated truck

11. Acronym for bridge gross weight formula

12. Media used by RFID tags to communicate with a remote reader

13. Alternate term for a un-powered tail axle

15. Descriptive term indicating that a truck is not designed to drive on the highway

18. One application of telematics that helps avoid truck breakdowns

23. All-wheel drive dump truck without a traditional steer axle

24. Front axle of a truck

25. Strong tool used to provide tension in chains when tying-down cargo

26. Name of first truck powered by a internal explosion engine

Down

1. Important factor impacting the cycle time of an excavator bucket

2. The cause that leads to bunching at truck loading (. . . of travel time)

3. Functional descriptor of a conventional swing tailgate for end dump trucks

4. Latin origin of the word truck, meaning iron pulley

5. Soil/gravel/sand that is lost by the truck on the road

6. 99,000 foot-pounds of work in one minute

8. Mechanical coupler between a tractor and a semitrailer

9. Term describing a system of integrated communication networks

14. Term describing a special truck axle with a weight limit on the Interstate system of 34,000 lb

16. Alternate term for a trailer with a clam shell type bottom dump gate

17. Theoretical point that defines the optimal number of trucks for an earthmoving operation

19. Unpowered axle in front of drive axle

20. The cheaper RFID tags are . . . .

21. A "Jake Brake"

22. Open flat-bed trailer used to transport heavy loads

JOURNALING QUESTIONS

1. Telematics offers a long list of application areas that save money. Pick one from the list and describe how it would work and what the benefits would be.

2. Active RFIDs offer many interesting opportunities for cutting time to search for materials or to protect tools from being stolen or misplaced. Describe an opportunity that you can think of.

3. Assume that one driver out of five off-road trucks hauling gravel for a road construction is always slower than the average of the four others. Describe how this can reduce the entire fleet production. What are some possible remedies that you would propose to a contractor to boost the productivity?

4. When loading a tracked excavator to be transported on the highway, one has to think about many laws/regulations and safety measures to protect the equipment during loading and unloading as well as in transport. Draw up a checklist of all the issues that need to be covered.

5. Planning the settings for the crushers and screens of a quarry is not as straightforward as calculating the volume of an earth dam. What did you find particularly interesting when you followed Worked-Out Problem 6.5?

TRADITIONAL HOMEWORK

1. A straight-frame flatbed semi-trailer has to be loaded with a shipment of fabricated steel for a new building. The steering axle is allowed to carry a maximum of 14,000 lbs (63.5 t). The trailer alone weighs 12,000 lbs (54.4 t) while the tractor weighs 19,000 lbs (86.2 t).

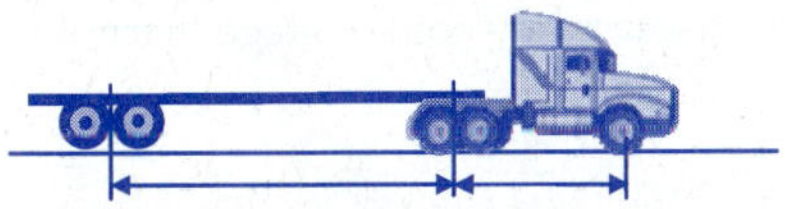

Figure 6.49 Semi-Trailer Dimensions

(a) What are the maximum allowable loads on the different axles?
(b) How heavy can the steel elements be that you can load onto the trailer?
(c) What are basic rules to be used for tying the steel elements to the trailer?

2. There are two types of load limit calculations for trucks. Describe the engineering concepts that provide the basis for each limitation.

3. You are asked to calculate the hourly production of a shovel-truck combination used to create a dam against ocean hurricane waves. The contractor decided to use the Komatsu rigid off-highway HD325-7 truck (rimpull curves in Figure 6.50) with 498 net-hp, and heaped capacity is 24 m^3 (31.4 yd^3). As the shovel excavator Liebherr R 974 C with a heaped bucket capacity of 5.2 m^3.

The material is rocky material, which will slow down the filling of the bucket an average of 10%. The average times for loading trucks with a 90% swing angle is shown in Table 6.16.

Table 6.16 Averages for Shovel Cycle

Swing Full	Swing Back	Empty Bucket	Positioning Bucket	Filling Bucket	Spillage
18.3 sec	17.4 sec	4.9 sec	6.8 sec	13.6 sec	5%

Since this job is excellent for younger operators to improve their skills, you expect that at least at the beginning you will have to expect a 20% overall slower operation than what the average should be. One can expect that this will improve over time, as the project will last at least one entire year.

The HD325-7 truck experiences a 3% RR and travels first 9% downhill from the quarry for 0.5 mile (0.8 km) to be followed by 0.5 mile horizontal section that increases over time to a total of 3 miles (4.8 km). The expected spillage will be at 5%, and the time for maneuvering and dumping is 3.2 min.

(a) Calculate the hourly production of the excavator shovel for three different operator skill levels 80%, 100%, 120% while keeping the spillage and the impact of the rock unchanged.
(b) What is the cycle time of the Komatsu at three different horizontal distances of 0.5, 1.5, and 3 miles (4.8 km)?
(c) Calculate the loading times for the three operator skills.
(d) Draw the balance-points for all the possible combinations on a piece of graphing paper.
(e) What actual production will you expect if you assume a maximum of 15% bunching and the balance point?
(f) Propose a strategy for the fleet (e.g., number of trucks) during the one-year project if you assume that it will progress linearly over the year. Formulate the best arguments that you can use when presenting your results to the project manager, who is eager to keep the fleet running at an optimum.

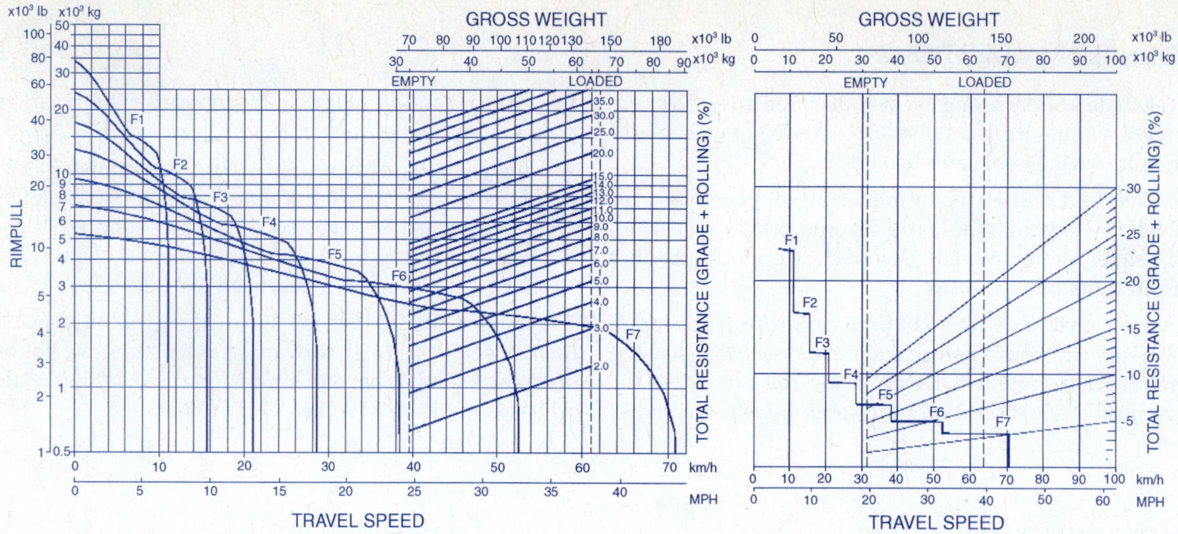

Figure 6.50 Rimpull Curves for Komatsu Off-Highway HD325-7

4. In the final discussion of Worked-Out Problem 6.5, it was suggested that the amount of fine aggregate might be reduced by opening the CSS of the cone crusher 1½ in. Run the calculations for this opening to find out if this prediction is true. You might want to test the spreadsheet available on the booksite, www.wiley.com/college/bernold.

LESSONS FROM THE ACCIDENT FILE

1. Case A involved a tractor pulling one semi-trailer and a second "pup" hitched to the first trailer. This combination is also called a Rocky Mountain Double. Assume that the two 8-ft-wide trailers were side dumps similar to the SSD from Trail King Industries and that the turn radius of the tractor was 33 ft (10 m).

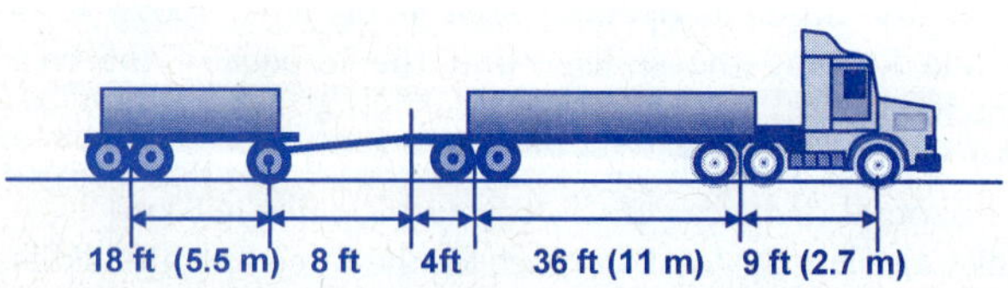

Figure 6.51 Dimension of the Rocky Mountain Double

(a) Draw on a piece of graph paper the off-tracking situation when the entire combination made the 180 degree turn (check the road design literature if you are not familiar with off-tracking problems).

(b) How can it be possible that the victim did not realize what was happening? What could have been done to avoid the accident?

(c) Is there a technology that would warn the driver and stop the truck in a situation like this?

2. Draw the situation of accident Case 2. Sketch the truck bed, the tailgate, the asphalt, and the two drivers right before the abrupt opening of the tailgate. Draw using an accurate scale. Explain what happened at the air-cylinder that allowed the gate to open. What simple mechanism would have prevented the tailgate from opening?

3. Let's assume that all four chains rated at 6,600 lb (3 t) used to tie down the wheeled loader in Case 4 were tight. Since the nominal strength should be about 20% of the actual, we could assume that they should have been able to withstand a force of 130,000 lb (59 t). Explain what rules might have been broken, and present two possibilities of what might actually have gone wrong.

4. The incident where the jaw-crusher operator was killed by a steel bar does seem strange. Please sketch the situations prior and during the instant where the operator was killed. In other words, how did it happen?

OPEN-ENDED PROBLEMS

1. Telematics may speed up the weighing of a truck hauling steel from Alabama to Missouri. Installing an active RFID with a communication distance of 200 ft (60 m) would even eliminate the need for a truck driver to be weighed again and again. Describe how such an integrated system could work and what other benefits beyond the time at the scales could be possible.

2. Assume that the manager of a large precast concrete operation with approximately 30 concrete ready-mix trucks in a large urban area is considering the installation of the most modern telematics system in order to optimize fleet utilization and delivery time.

(a) Write down a list of specific and quantitative goals that the manager might establish to define the meaning of optimal fleet utilization and optimal delivery time (e.g., preventive maintenance with zero breakdowns).

(b) What information would be needed in real time to reach the goals?

(c) Design a telematics system for the concrete plant and trucks to automatically collect the needed data. Mark the location and type of electronic sensors including the transceivers and power sources. Distinguish between satellite and Internet communication systems.

chapter seven

Cranes—Gentle Giants in Construction

Cranes have one of the richest and longest histories of any machine used today and can look back at numerous fantastic achievements. No obelisk could have been raised without them, no cathedral, no Empire State Building. Even bridges or hydroelectric dams depend on those heavy lifters capable of spanning across entire valleys. Nevertheless, the basic components are simple, consisting of a rope, a force able to pull the rope, a mast, and finally the magnificent pulleys. Amazing mechanisms such as gantries, telescoping, lattice or articulating booms are able to move a load horizontally. This chapter will introduce you to the history of the technology, reviewing the mechanisms of its key components while keeping in mind that lack of maintenance and operational planning can quickly lead to disasters.

As you have seen in the previous chapter, the effective and safe operation of construction equipment depends on applying the physical principles of mechanics properly. Thus you again need to be able to use moment and force vector calculations. We will use spreadsheets and apply some basic rules of statistics, while you will be asked to connect to the booksite and search the Internet to find additional information.

Table 7.1 Topics Covered in This Chapter

Construction			Planning & Control		
Equipment and Major Mechanisms	**Processes**	**Enabling Components**	**Managerial-, Engineering Factors**	**Rules, Laws, Standards**	**Control Elements**
✓ Telescoping Boom	✓ Choker Hitching	✓ Boom Line	✓ Mechanical Advantage	✓ OSHA 29 CFR	✓ Inclinometer
✓ Outrigger	✓ Basket Hitching	✓ Stiff Legs	✓ Friction	✓ Subpart N	✓ Pressure Sensor
✓ Tower Crane	✓ Hoisting	✓ Saddle Jib	✓ Person in Charge	✓ 1926.550(g)(3)(ii)	✓ Extensometer
✓ Articulated Boom Crane	✓ Slewing	✓ Slewing Ring	✓ Load/Lifting Capacity	✓ ANSI/ASME B30.5 1926	✓ Interference Protector
✓ Wire Rope	✓ Luffing	✓ Load Block	✓ Safety Factor	✓ EM 385-1-1	✓ Tilt Sensor
✓ Block-and-Tackle	✓ Shuttle Trolley	✓ Sheave	✓ Tipping Load	✓ ANSI/ASME B30.9	✓ Overload Protection
✓ Capstan	✓ Reeving	✓ Boom	✓ Lift Plan	✓ OSHA 1926.1419	✓ Anti-Two-Block
✓ Bottom Slewing Crane	✓ Two-Blocking	✓ Pulley Block	✓ Side Loading		✓ Spotter
✓ Top Slewing Crane	✓ Hand Signaling	✓ Outriggers	✓ Crane Cycle		
✓ Anti-Two-Block	✓ Load Tilting	✓ Winch	✓ Crane Productivity		
✓ Self-Erecting Crane	✓ Two-Crane Lifting	✓ Slings and Chains	✓ Part Lines		
✓ Luffing Jib	✓ Side-Loading	✓ Snap Hook	✓ Operator Skill		
✓ Hook Block	✓ Tipping	✓ Load Line	✓ Load Capacity Chart		
✓ Hitch, Hitch Point	✓ Telescoping	✓ Counter-Jib	✓ Probability		
✓ Reeve		✓ Jacking Frame	✓ Center of Gravity		
✓ Gantry Crane		✓ Trolley (Jib)	✓ Stabilizing Moment		
✓ Sheave		✓ Pad or Float	✓ Shock Loading		
✓ Lattice Boom		✓ Rope	✓ Safe Working Load		
			✓ Cycle Time		

7.1 An Impressive Family Tree

3000 BC—AD	**From the Shaduf to the A-Frame**	Mast, hoist drives, ropes, winches, block and tackle were heavily used by the Egyptians, ancient Greeks, and Romans. The simple shaduf lifts water buckets for irrigation in Egypt with the help of a lever/beam and counterweight. Two long wooden beams are made into a leaning A held back by guy ropes. A cross bar serves as winch that controls a rope that is held by a pulley connected to the top. 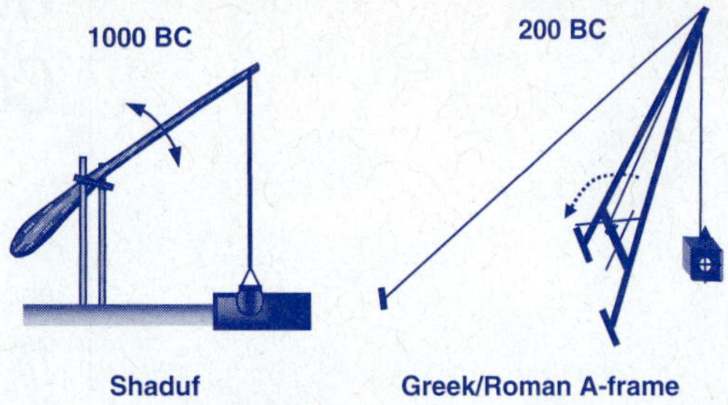
Europe in the Middle Ages	**Treadwheel with Gears**	In the Middle Ages many Roman technologies were reinvented because they had been forgotten in the meantime—for example, the treadwheel, inside which one or several people turn the gears that operate winches and ropes to unload ships or lift mortar, stone, or timber for the construction of city walls or the European cathedrals. A famous crane designer was Leonardo da Vinci (1452–1519).
1900s	**Steam-Powered Cranes**	The second wave of success of the crane began during the early 1900s with the emergence of a new source of power. The steam engine replaced the large treadwheel, thus speeding up the entire hoisting process. Its compact size made it possible to mount it on rail carts or powered rollers, together with a luffing boom and a rotating platform.
1920	**Truck Crane**	Truck-mounted slewing and luffing jibs, operated by cable winches and powered by gasoline engines, showed up and were heavily used by utility departments.
1930s	**Cableway Cranes**	Large dam construction sites between steep mounted sides gave birth to cableway cranes. The latest of them have a 40-ton (36.3 t) lift capacity spanning 4,200 ft (1,281 m). 
1940	**Crawler Crane**	The crawler crane was really a modified tracked shovel excavator. Equipped with long lattice jibs, a large counterweight, and a block and tackle, they were readied for heavy lifting and on-site mobility.
1960	**Tower Cranes**	Top slewing climbing saddle-jib tower cranes made their first appearance. The mast is able to gain in height either by jacking itself up inside the growing building or by inserting additional mast segments. 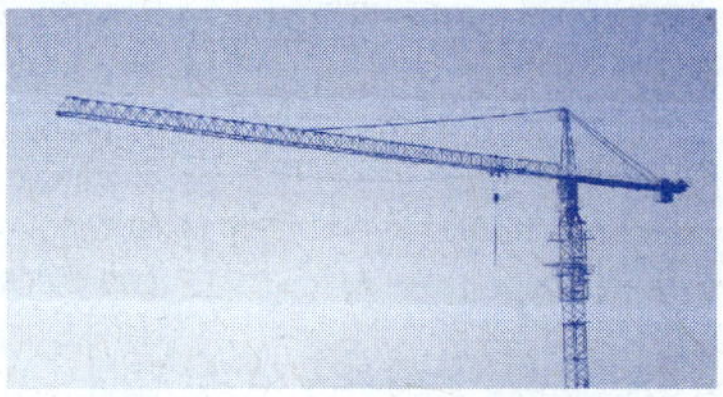
1965	**Hydraulic Telescoping Truck Crane**	The first telescopic boom carrier cranes appeared around 1965, combining mobility with large load capacities. The hydraulic boom mounted on a carrier is able to travel on the highway and is quickly installed.
1980	**Space Shuttle Crane**	The space shuttle needed an articulated crane that could be remotely controlled and be mounted inside the shuttle bay. Today's crane mounted on the International Space Station added many more capabilities. 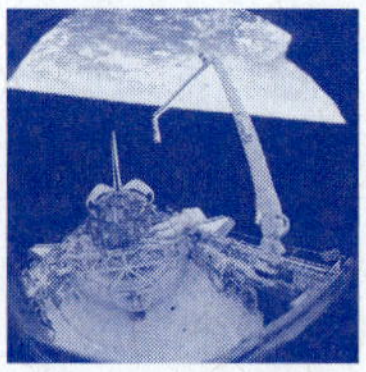

7.2 From the Accident File

1. **Tipping Over:** A 39-year-old crane operator was killed when the crawler crane he was operating on a barge tipped over. At the time of the incident the victim was lifting a diesel pile-driver weighing approximately 15,500 lb (7 t). The crane was not positively secured to the barge at the time of the incident.
2. **Electrocution:** The victim and one welder were assisting the crane operator in unloading a truck. The crane operator lifted a ladder off the truck and swung it over the top of a tank. The boom was extended approximately 45 ft (13.7 m) and the jib was not attached. The victim held onto the ladder and gave the crane operator a hand signal to continue swinging and lowering the ladder behind a tank. The next thing the crane operator knew was that a portion of the crane cable near the end of the boom was on fire and he swung the boom in the opposite direction, away from the powerlines. The crane operator then left the crane and found the victim lying on the ground.
3. **Crushed by Load:** On the morning of the accident, the foreman and his crew of six employees were in the process of unloading the day's first truckload of pipes. Two slings were used to rig a load of six pipes to the hoist line. One eye on the end of each sling was attached to the hook. The eye on the opposite end of each sling encircled the hoist cable above the hook. This configuration did not allow the two slings to close down around the load, but created a loose cradle to support the load when lifted by the crane. When a load was picked up from the truck, it did not look level, and the foreman told the operator to stop hoisting and to boom up a bit. After this maneuver the foreman approached the load, at which time the pipes started to slip and twist, forcing some loose pipes to roll off the truck. As the foreman tried to push the load away from the truck he tripped and fell. When he stood back up, he was crushed between the swinging load and the truck.
4. **Two-Blocking:** The feeder and hopper assembly had to be lifted off a primary crusher, using a 75-ton (68 t) mobile crane. When the accident occurred, the crew had extended the crane's outriggers and roped off the working radius of the crane. The crane operator had raised the crane's boom to approximately 71° and telescoped the boom to about 50 ft (15.2 m). Suddenly the hoist line snapped, and the headache ball with hook fell, striking the foreman on the ground, killing him instantly.
5. **Scissor Lift Tipping:** The incident occurred when an electrician was contracted to install 36 high-intensity lights around the perimeter of a two-story building. A battery-powered, hydraulic scissor lift was used to reach the soffit. The soft ground had to be covered with a large sheet of plywood so that the wheels of the lift would not sink in. The supervisor raised the lift to test it before turning it over to the victim with his job instructions. At about 4:30 P.M. a co-worker of the victim's was sawing wood on the ground below the lift when he heard the victim cry out and saw the lift fall over. The lift's work platform crashed to the ground at the edge of a bay, throwing the victim headfirst into the shallow water.

7.3 Forces and Moments Affecting Crane Tipping

Two of the five accidents summarized in Section 7.2 are related to the tipping of a crane, a situation where the entire crane, including the truck carrier, turns over. In other words, while booms or ropes may break and result in a catastrophic accident, a tipped-over crane may be still intact when put back up. Let's review quickly what basic conditions may result in tipping (Figure 7.1).

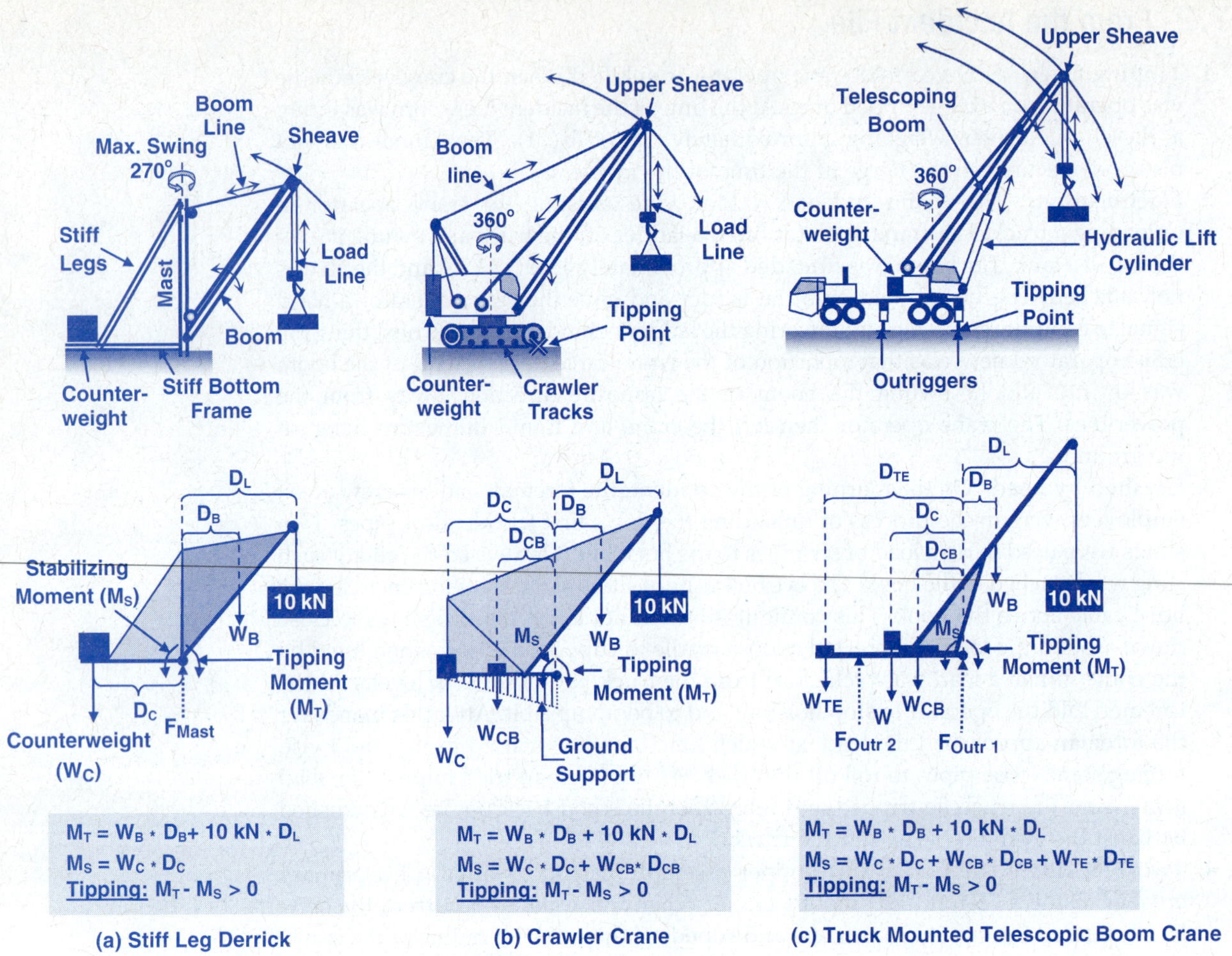

Figure 7.1 Modeling Tipping Conditions and Base Moment for Five Cranes

> ### Of Triangles and Trusses
>
> 
>
> In civil engineering, the *triangle* has long gained a seat of honor at the table of fundamental building concepts. Its simple geometry conceals a feature that is key to building large structures: Triangles keep their planar shapes as long as we maintain the lengths of the edges.
>
> 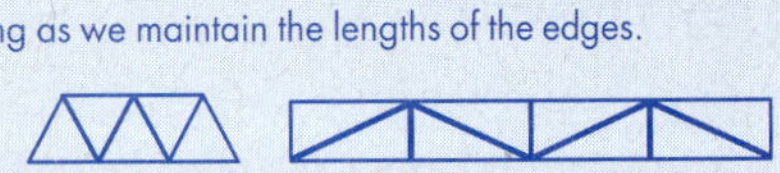
>
> A *truss* is an assembly of triangles that are pinned together at their corners. By pinning together planar triangles in three dimensions, space trusses can be built. The elements that constitute a truss are called members. In a planar truss we talk about horizontal top and bottom chords, diagonals, and posts. Each member is in compression, tension, or neutral.

The five selected cranes represent different basic types that can be found in construction. However, they all have one feature in common: all have a tipping point located where the crane support element closest to the load makes contact with solid ground. The force diagrams show another key feature that keeps a crane up: the triangle with real or virtual straight members. In fact, the triangle is the simplest geometric figure that will remain stable as long as the sides stay fixed (see insert). By connecting several plane triangles at the corners, a truss can be built that is able to carry large loads within the plane. The shaded areas indicate the actual or virtual planar truss, which helps one to see the basic structural concept that supports the booms and jibs.

When will a crane tip over? Anybody who has ridden a seesaw knows that the heavier person sitting on one end is able to keep the person on the other end high up in the air. Two actions will change the balance and tip the board. First, the heavier person moves closer to the pivoting center when

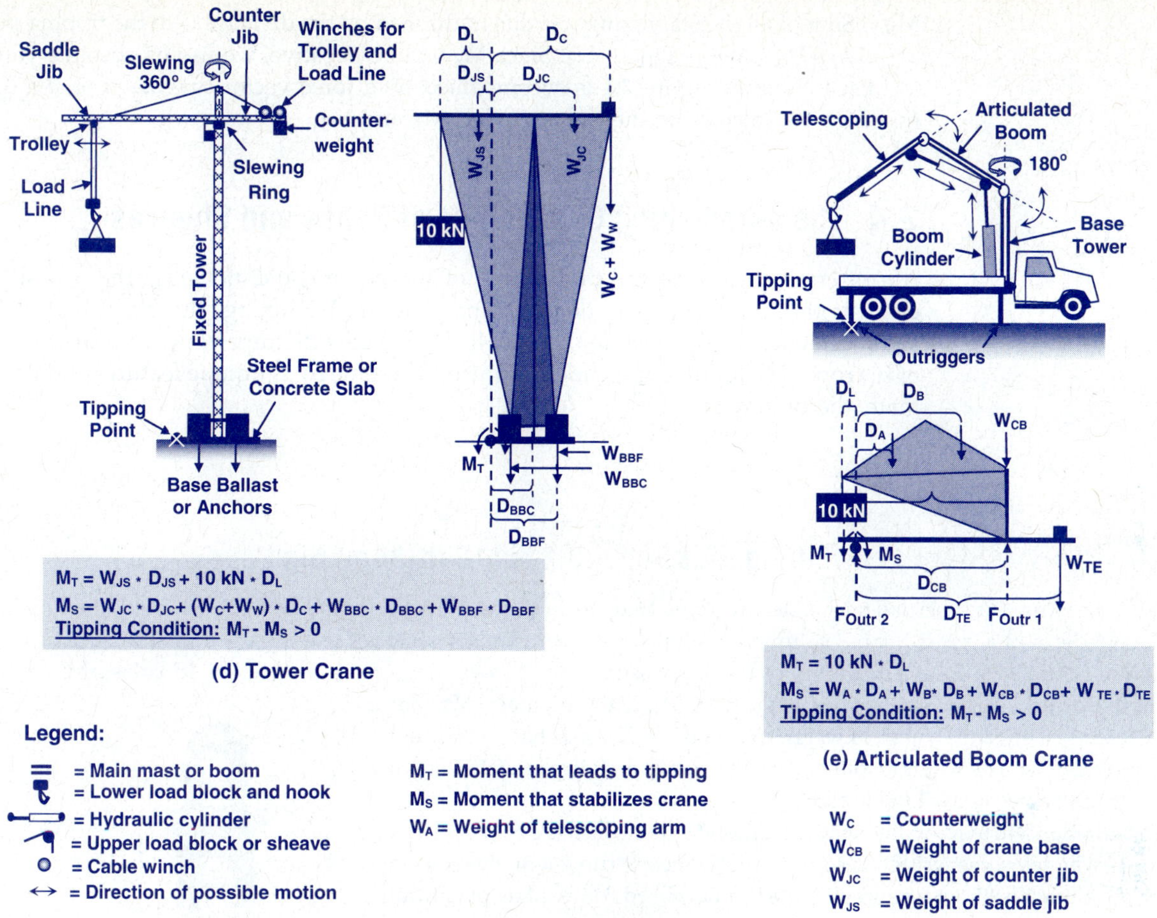

Figure 7.1 (Continued)

suddenly the two move into a state of equilibrium and the board will tip to a horizontal level. Second, a second person might join the lighter one. If their combined weight is higher, the board will tip rapidly to their side. From this example, we learn that both the load's distance to the tipping point, as well as its mass, influences the equilibrium. In fact we talk about moment or bending moments created at both sides of the pivot. The simple definition of a moment or torque is

$$\mathbf{M_A = Force * Distance\ to\ A}$$

Returning to the seesaw, we recognize that the board is stable or in equilibrium when the bending moments on each side are equal:

$$\mathbf{M_{A\ Positive} = M_{A\ Negative}}$$

Equating $\mathbf{M_{A\ Positive}}$ with the tipping moment $\mathbf{M_T}$ and $\mathbf{M_{A\ Negative}}$ with the stabilizing Moment $\mathbf{M_S}$ with $\mathbf{M_{A\ Negative}}$, we find the condition to keep the crane from toppling over: The stabilizing moment M_S has to be larger than M_T or $|M_S| > M_T$.

$\mathbf{M_T}$ = Sum of all weights on the side of the boom multiplied by the horizontal distance D to the tipping point $= (W_B * D_B + W_{Load} * D_{Load} + \ldots)$

$\mathbf{M_S}$ = Sum of all the stabilizing weights multiplied by the distance D to the tipping point $= (-W_C * D_C - W_{TE} * D_{TE} + \ldots)$. Since M_S will be negative, we use its absolute value.

Each crane in Figure 7.1 is accompanied by a force vector diagram and the basic functions for calculating the two moments M_T and M_S.

7.4 The Astonishing Capabilities of Ropes and Sheaves

An observer reviewing the family tree of the crane could not miss the one simple element that has been part of the crane since its beginning, the rope. In fact, the advancements of crane technology are closely linked with improving the characteristics of the rope. The following section will introduce some of its unique features and its soul mate, the pulley.

Header Problem 7.1 / Lifting 53.3 Ton (48.2 t) 112 ft (34 m) High

After the Roman emperor Trajan's defeat of the Dacians in the year 106, bringing home much gold, the Roman Senate decided to erect a large column to the memory of his major victories (see Figure 7.2). The structure is about 125 ft (38 m) high including its large pedestal. The shaft is made from a series of 18 colossal Carrara marble drums, each weighing about 40 ton (36.3 t), with a diameter of about 13 ft (4 m). The capital block of weighs 53.3 tons (48.2 t) and had to be lifted onto the 112 ft (34 m) high column. Inside, a spiral staircase of 185 stairs provides access to a viewing platform at the top.

Marcus Vitruvius Pollio and Heron of Alexandria were Roman engineers who left some written descriptions about construction methods used at that time. We know for sure that the Romans used ropes made of natural fibers with blocks and tackle, a mechanism invented by the Greek engineer Archimedes who lived 287–212 BC. The blocks consisted of free-spinning pulleys/wheels and a center pin around which the wheels were able to turn. The rope was reeved through two blocks and connected to the load on one side and to a winch or a capstan at the other. The commonly used A-frame or derrick crane arrangements, however, did not provide enough lifting height.

Question 1

How did the Romans create a hoisting structure without the boom/mast cranes? Sketch a possible configuration.

Question 2

How many sisal ropes, blocks, and tackles did they need to lift the capital block? Assume that they used a rope size that could be pulled by hand for an extended time and wrapped around a winch.

Question 3

How long might it have taken them to lift the stone 115 ft (35 m)? Assume that it took 30 seconds to turn the winch or capstan one revolution.

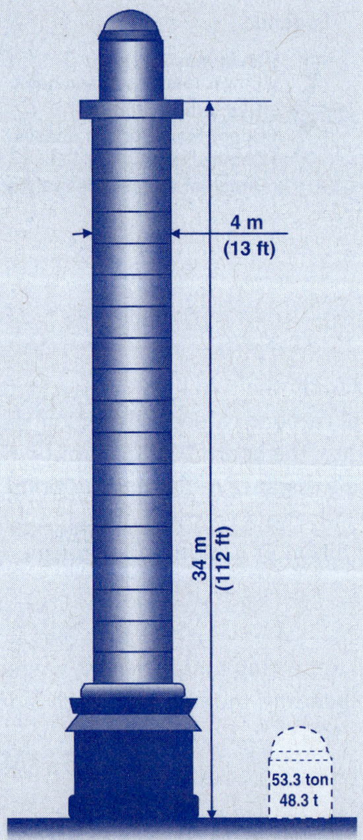

Figure 7.2 Trajan's Column

7.4.1 GAINING THE MECHANICAL ADVANTAGE

The mechanical advantage that Archimedes was able to utilize in his invention was later defined by Newton in his First Law that says that the forces on a motionless object add up to zero. Here is how Archimedes' important invention worked.

According to Newton's First Law,

$F_L = F_A + F_B$ and $F_A = F_B$
$F_A = F_B = 1/2\,F_L$ or $\mathbf{F_L = 2 * F_A}$

As Figure 7.3 shows, if John pulls with a force F_A, the load he is able to lift is two times larger assuming that the friction between center pin and pulley and the weight of the rope is negligible. Thus, the **mechanical advantage** (MA) = 2.

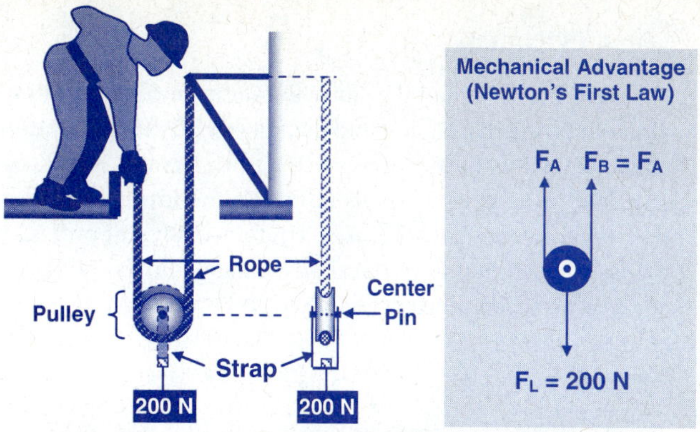

Figure 7.3 John Takes Advantage of the Rope and Pulley System

Worked-Out Problem 7.1 / Gaining a Mechanical Advantage

John's foreman, Betsy, did not like the way John set up the rigging operation because of safety concerns (fall through opening). She asked us to recommend a hoisting mechanism so that he can stand back from the opening while pulling the rope. We are allowed to use a second pulley.

Options: In Figure 7.4, John presents two different solutions, A and B, with two options for attaching the end of the rope, called the dead end.

Calculate the MA and the amount of force that John has to exert.

Answer for design option A: Nothing has changed when the ropes connecting to the free pulley are replaced by forces necessary to hold the load of 200 N in place. The upper fixed pulley does not add any mechanical advantage; thus John will have to pull with 100 N.

Answer for design option B: Again, the rope linking the free with the fixed pulley carries 100 N. The fact that John is positioned below does not change the force. It is still 100 N.

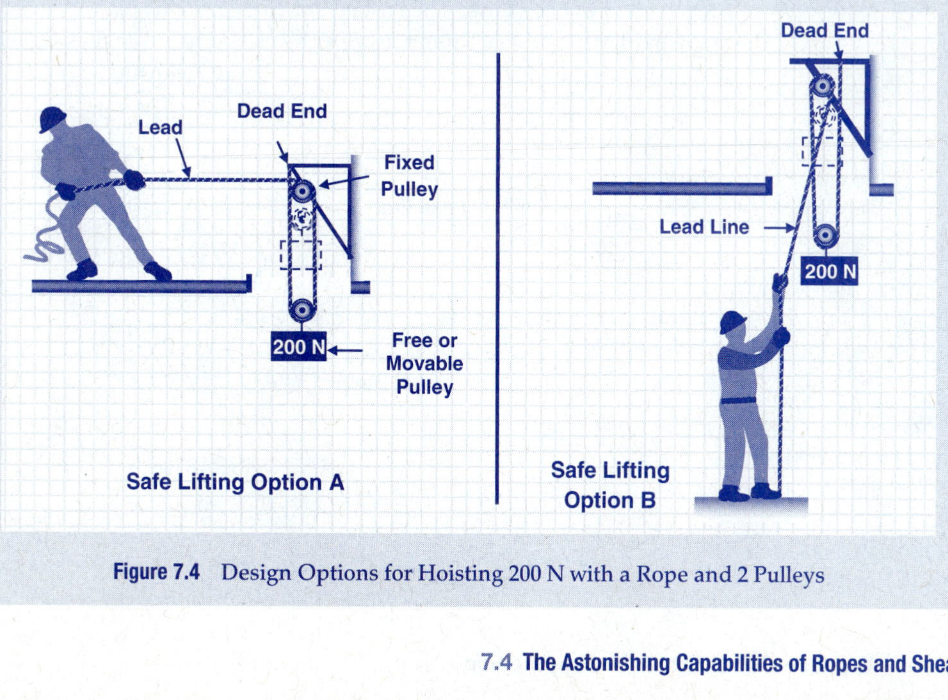

Figure 7.4 Design Options for Hoisting 200 N with a Rope and 2 Pulleys

Final Evaluation

Design options A and B require the same amount of pulling force and work. Both setups for option B will not allow John to swing the load over to the floor or push a cart underneath when reaching the top. In fact, option B setup 2 will not even allow the load to reach the upper level. Overall, option A is preferable, since John does not risk being hit by the load if he slips when standing on the lower level.

Archimedes saw that if he combined several pulleys into a block, he could reeve the rope back and forth and thus increase the mechanical advantage. On the other hand, each pulley added friction forces, which had to be overcome when a heavy load had to be lifted.

(a) Boom head with two separate upper blocks for heavy and light lifting (shown with one rope)

(b) Heavy-duty hook blocks with multiple sheaves and reeving

Figure 7.5 Block-and-Tackle with Hooks

Exercise

Assume that you are operating an electric **winch** (drum diameter = 1 ft/0.3 m) with a reeved block system that consists of an upper block, similar to Figure 7.5a, with three **sheaves** and a lower block, Figure 7.5b shows a quartet, with two sheaves and a hook as portrayed in Figure 7.6. The friction force of each pulley can be assumed to be 10% of the force that opposes movement. The load that has to be hoisted is 6,000 lb (2.7 t). The weight of the rope and the slings can be neglected while each block weighs 100 lb (.04 t).

Calculate the minimum force that a crane winch has to be able to apply in order to hoist the load. How many winch rotations are necessary to lift the load 66 ft(20.1 m)?

Solution

Step 1: Establish the mechanical advantages
Step 2: Calculate loads in each line (when stationary)
Step 3: Assess friction forces in each sheave
Step 4: Calculate total load in lead line (when hoisting)
Step 5: Compute total length of lead line to be winched
Step 6: Number of turns based on the circumference of drum

Step 1: Model of Hoisting System:

Step 2: Load in Each Line: The dashed line in the reeving model symbolizes the cut that would sever the load from the upper block and has to be replaced with forces at A–E. Applying Newton's First Law, we can write following equation:

$$F_A + F_B + F_C + F_D + D_E = 6{,}100\text{ lb } (6{,}000\text{ lb} + 1\text{ weight of block})\ (2.7\text{ t})$$

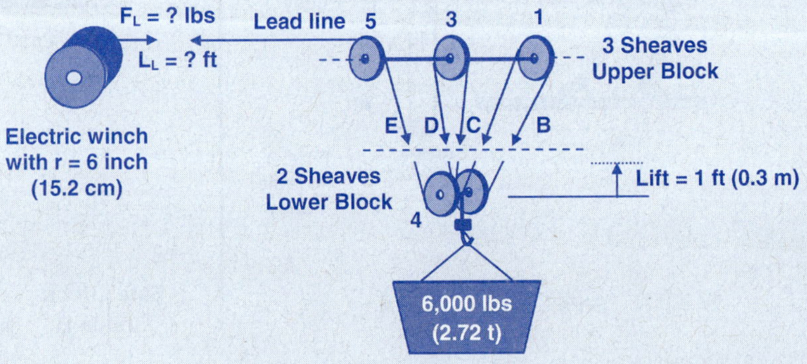

Figure 7.6 Reeving Design

Furthermore,

$F_A = F_B = F_C = F_D = D_E$
Thus : $F_A = 6{,}100\,\text{lb}/5 = 1{,}200\,\text{lb}\ (0.54\,\text{t})$

Step 3: Friction: The amount of friction depends mostly on the type and condition of bearings between the center pin and the pulley. In the present situation, we assume a friction factor of 10% or 10% of the resisting force.

Step 4: Adding of friction forces: In order to lift the load, line A has to be shortened, which can only happen if line B is increasing its force at pulley 1, which will create a torque that will turn it to the left. In other words, F_A is resisting the motion and is thus the basis for the friction force calculation at pulley 1. Based on the same logic for the other pulleys, Table 7.2 can be created:

Table 7.2 Compounding Friction Forces

Line Section	Resisting Force lb	Resisting Force t	Friction %	Pulling Force lb	Pulling Force t
A	1,200	0.544	10	1,320	0.599
B	1,320	0.598	10	1,452	0.659
C	1,452	0.658	10	1,597	0.724
D	1,597	0.724	10	1,756	0.797
E	1,756	0.796	10	*1,932*	0.876

As we observe, the compounding effect of friction, if fixed, leads to this function:

Lead line force = stationary force $* (1 + \text{friction}\,\%)^{(\text{number of lines})}$
$1{,}932\,\text{lb}(0.876\,\text{t}) = 1{,}200\,\text{lb} * (1 + 0.1)^5$

Step 5: Lead Line Movement: In order for the load to rise 1 ft, rope A must be shortened by 1 ft. In fact, every rope A–E connecting the lower with the upper block must be shortened by 1 ft. As a consequence, pulley 1 will have to turn, creating 1 ft of slack in B. As in the calculation of friction forces, the effect is compounding and leads to Table 7.3.

The total travel of the lead line has to not only shorten rope section E, but also remove all the slack that is "handed" over from the other rope sections through the pulley movements. The total lead line travel is thus compounded:

LL = 66 ft × (1 original + slacks) = 66 ft × number of lines between blocks = 66 ft × 5 = 330 ft (20.1 m × 5 = 100.5 m).
The circumference of the winch drum = 2 ∗ 0.5 ft × 3.14 = 3.14 ft (0.96 m)
Necessary revolutions of winch drum = 330 ft/3.14 ft = 105 rev.

Table 7.3 Compounding Rope Shortening

Line Section	Shortening of Section ft (m)	A ft	A m	B ft	B m	C ft	C m	D ft	D m
A	66 (20.1)	→	→	→	→	→	→	→	
B	66 (20.1)	66	20.1	→	→	→	→	→	
C	66 (20.1)	66	20.1	66	20.1	→	→	→	
D	66 (20.1)	66	20.1	66	20.1	66	20.1	→	→
E	66 (20.1)	66	20.1	66	20.1	66	20.1	66	20.1

Total at lead line: 330 ft 100.5 m

7.4.2 THE INNER LIFE OF WIRE ROPES

In the nineteenth century, American rope making involved several competing rope factories serving a rapidly growing demand from the gold mines in the West, the canal boats being towed on railcars over steep inclines, bridge construction, and the many streetcars adopted by major U.S. cities (San Francisco, for instance). Two designs are still used to identify ropes today: **Seale**, invented by Thomas Seale, and the **Warrington (inventor unknown),** an identification used by **John Roebling,** who later designed and built the Brooklyn bridge in New York.

Wire ropes require six key characteristics:

1. Resistance to sudden breaks
2. Resistance to bending fatigue
3. Resistance to abrasion
4. Resistance to crushing in pulley grooves
5. Resistance to rotation during lift
6. Resistance to corrosion

Each application (e.g., suspension for a bridge or hoisting loads) leads to unique designs with characteristics that best meet the need of the application, but they all use the same building blocks: (a) flexible core made of natural or manmade fiber (polypropylene or nylon), (b) individual steel wires, and (c) strands made of twisted wires. The number of wires, possibly of different sizes, and the number of strands identify a wire rope. Figure 7.7 presents three examples of common designs, each with special properties.

While the ordinary rope contains wires of the same size, for example 19 in a strand, both the Seale and the Warrington designs use two and three sizes, respectively. The Seale places larger-diameter wires, having better abrasion resistance, at the outside while the smaller-size wires, providing reduced resistance to bending, are laid around the center of a strand. In a second process, 6 or 8 strands are twisted around the core, made either of fiber or of an independent set of small wires. The importance of the core lies not only in its elasticity but also in its ability to support the outer strands holding the strands in place. The so-called independent wire rope core not only adds to the strength of the rope but also protects from crushing, which could severely damage the rope.

"Healthy" ropes and rope attachments are of course critical to safe lifting. Only frequent inspections and proper maintenance of the ropes will guarantee hazardless work

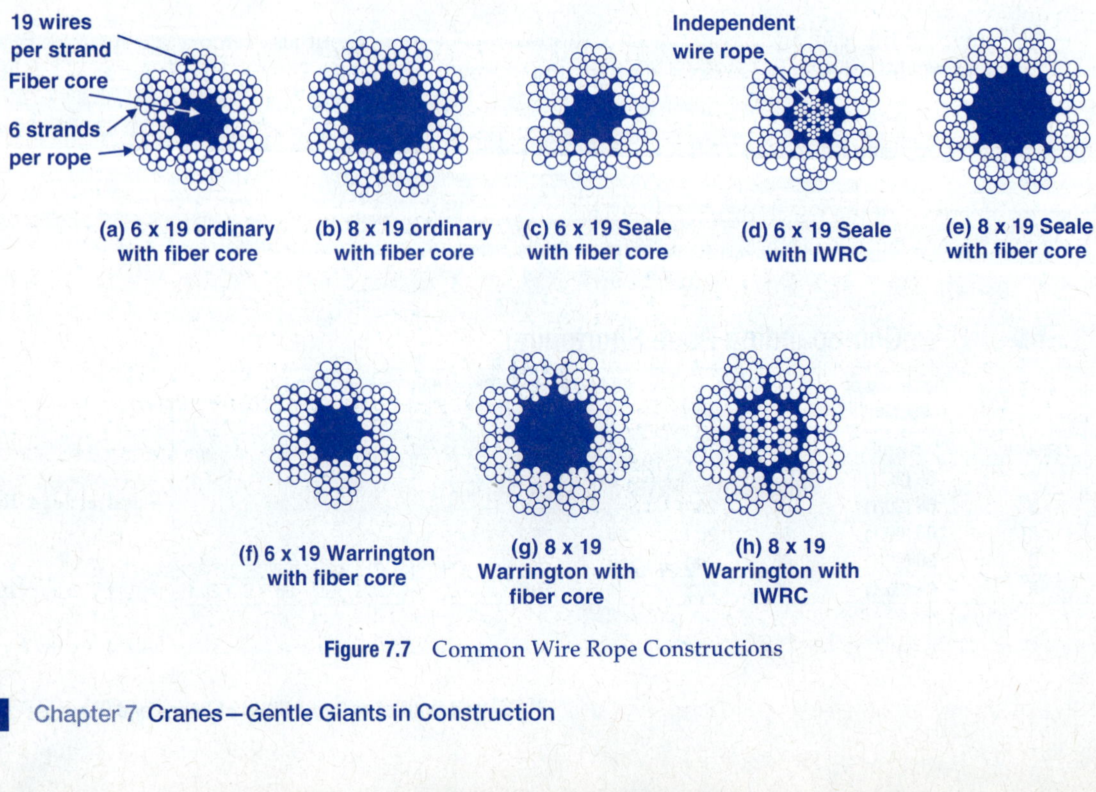

Figure 7.7 Common Wire Rope Constructions

(a) Multi-strand wire rope come of the drum
(b) Spliced eye with thimble and pressed metal sleeves
(c) Spliced eye with thimble and 3 U-bolt clips
(d) Dangerous wire rope with frayed/ broken wires and a kink

Figure 7.8 Examples from Visual Inspections of Wire Ropes

with an extremely strong man-made system. Sufficient maintenance involves regular lubrication to reduce friction between the rope's components as well as the friction between rope and sheaves or drums. Figure 7.8 shows trustworthy and questionable situations.

Because of the importance of inspection and the serious consequences if neglected, many agencies and associations have created inspection procedures in addition to the ones recommended by the equipment manufacturers. One such example is Manual No. 385-1-1 from the U.S. Army Corps of Engineers. It can be downloaded from the book's website:

DEPARTMENT OF THE ARMY EM 385-1-1
U.S. Army Corps of Engineers
CESO-ZA Washington, D.C. 20314-1000
For sale by the U.S. Government Printing Office
Superintendent of Documents, Mail Stop: SSOP, Washington, DC 20402-9328
ISBN 0-16-048877-x
Manual No. 385-1-1 3 November 2003 Safety
SAFETY AND HEALTH REQUIREMENTS

1. Purpose. This manual prescribes the safety and health requirements for all Corps of Engineers activities and operations.

Appendix F: RIGGING INSPECTION AND REMOVAL CRITERIA

Rigging shall be inspected by a competent person and replaced in accordance with ANSI/ASME B30.9 and the manufacturer's recommendations. Rigging degradation not only indicates that the rigging is becoming unsafe and requires replacement; it also often indicates problem(s) with the rigging setup, use, or maintenance. Evidence of failure is cause for replacement of the rigging

1. WIRE ROPE.
2. FIBER ROPE.
3. CHAIN.
4. METAL MESH SLINGS.
5. SYNTHETIC WEBBING SLINGS.
6. ATTACHMENTS.

7.4.3 THE INCONSPICUOUS SLINGS "AT THE END OF THE ROPE"

A variety of hook designs are being used. Figure 7.9 presents a single and a double hook as examples. **Slings** connect a load to the hook. Anybody who has been involved in moving heavy loads with cranes as well as reviewing the long list of crane accidents will attest that

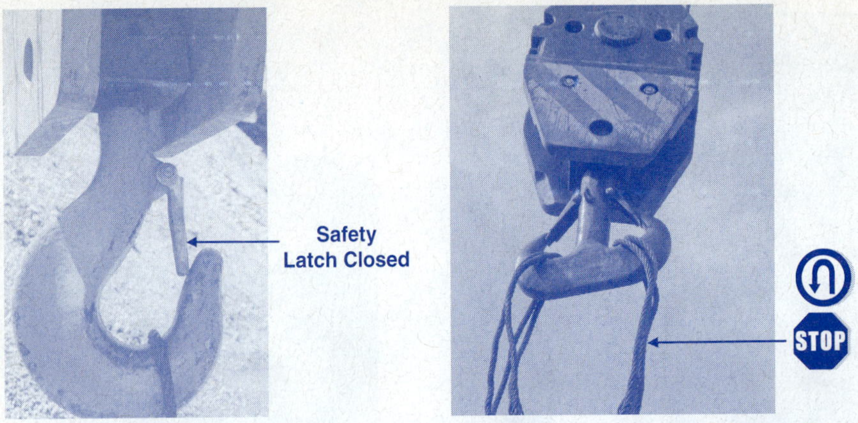

(a) Single Hook (b) Swiveling Double Hook on Multi-Sheave Block

Figure 7.9 Snap Hooks with Closed Safety Latches

rigging loads should be left to the real experts. I, for one, always take a step backward, having personally had the sobering experience of being pinned between a heavy concrete formwork and a concrete wall trying to stop the formwork from swinging.

The primary goal of a safe load is to be able to lift it without any side movement or rotation from its location. This has two basic requirements: (1) the crane hook must be vertically above the object's center of gravity, and (2) the element must be equally distributed between two or more slings causing them to stretch evenly. Figure 7.10 highlights basic principles for hitching a load to the hook.

The different shapes, densities, and weights of loads require different hitching methods and sling types. Figure 7.10a illustrates the importance of the sling angle on the resulting tension stresses that can exceed the weight of the object to be lifted. For example, if the sling angle is 30°, each of the two slings will experience a tension force that is equivalent to the load. Figure 7.10b presents a chocker hitch. As indicated,

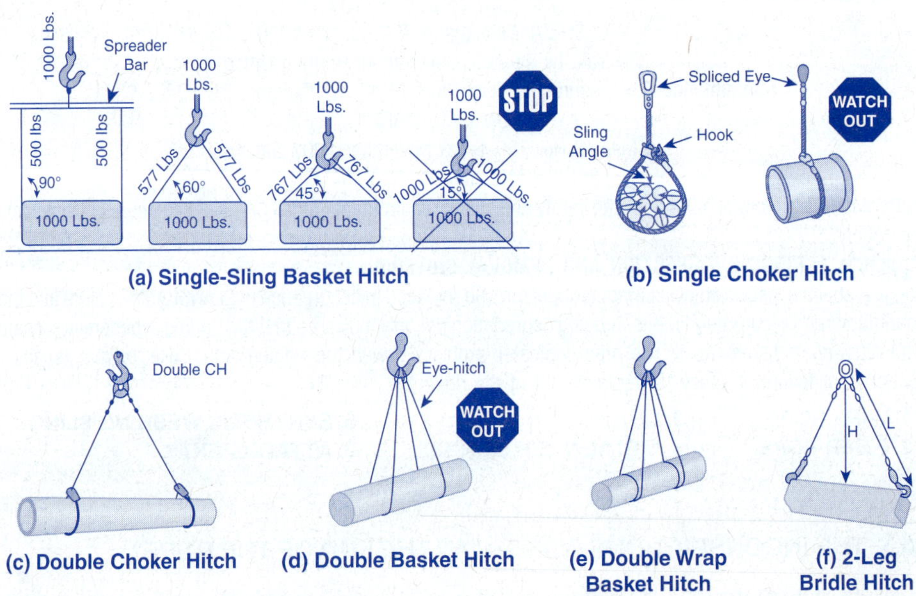

Figure 7.10 Common Hitching Configurations and Some Associated Problems (for a more exhaustive listing, visit the booksite about wire ropes and slings)

the sling chokes the load like a noose and will crush it if it does not have the necessary rigidity. For example, a bundle of long steel rebar will easily resist the choking force while neon-lamps might get crushed. The rule of thumb for single-sling chokes with a sling angle larger than 45° is that the actual Safe Working Load (SWL) for the hitch is 75% of its nominal SWL. Attention! Smooth surfaces such as on heavy pipes may not have sufficient friction and let a pipe slide off. Figure 7.10d illustrates how a double basket hitch, if balanced properly, allows a long, smooth object to be lifted. Finally, Figure 7.10f presents another common hitching configurations for long objects that have sufficient stiffness not to break between the two hooks, called the bridle hitch. Here the actual SWL is calculated by dividing the height between the load and the hook by the length of the hitch and multiplying it by two times the nominal SWL of the hitch. When more than 2 legs are needed, it is strongly recommended to use the same actual SWL value as for a 2-leg, since in reality two of the hitches will carry most to all of the load.

Figures 7.8c and 7.9b and now 7.11 all emphasize a very common treatment of wire ropes ending in permanent damage. The commonly used term is for the damage is kink, which is defined as an "imperfection of something that is likely to cause a problem." Figure 7.8c shows perfectly how the rope is plastically deformed and some of the wires have started to break. If the rope it forced into a tight loop, as shown in Figures 7.11a and b, some of the wires and strands will move from the elastic zone of steel into the plastic zone and stay elongated compared to other wires and strands in the same cross-section. As a result, the shorter wires are forced to carry all the load, ending up with higher stresses relative to the rest of the rope. It becomes the weakest point and will break here first. Even "restraightening" will not be able to remove the damage.

Figure 7.11 presents a safe solution to lifting heavy loads with a wire rope, the metal thimble. It fulfills two functions that are very important for the safety of the rope: (a) preventing rope from over-bending or kinking while giving it a safe shape, and (b) protecting against abrasion and pinching at the inside of the loop. The data from the U.S. Navy Specifications FF-T-276B on Thimbles and Ropes stipulate the minimum thimble dimension for various rope diameters. One can easily see that the minimum inner diameter D is between 2.5 to 3 times the diameter of the rope. For pulleys on cranes where the lines have to constantly bend and rebend, this ratio is between 30 and 50, depending on the construction of the rope.

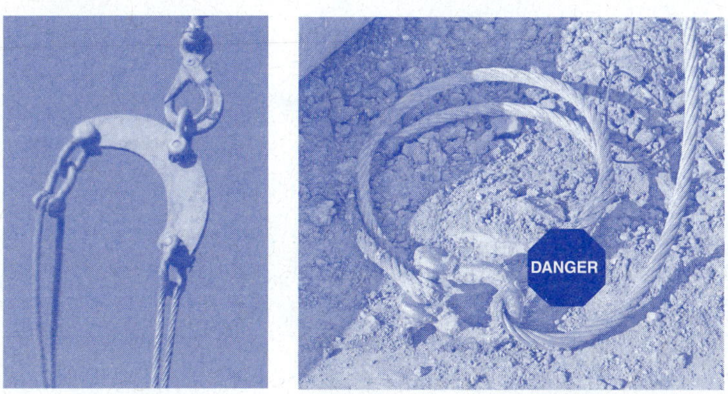

(a) Damaging Loop

(b) Shackles Supported by Tightly Looped Ropes

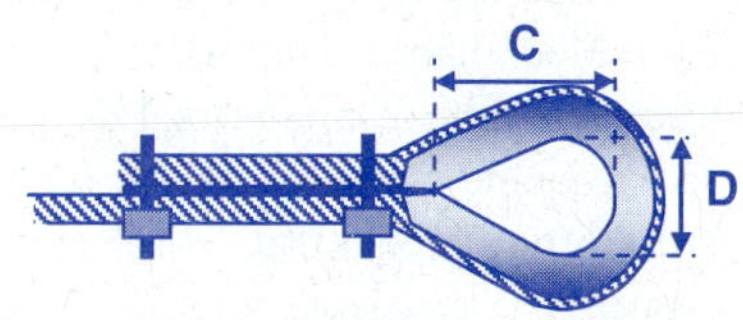

Rope Diameter		C		D	
in	cm	in	cm	in	cm
¼	0.6	1 ⅝	4.6	⅞	2.2
½	1.3	2 ¾	7.0	1 ½	3.8
¾	1.9	3 ¾	9.5	2	5.1
1.5	3.8	6 ¼	15.9	3 ½	8.9
2 ¼	5.7	14	35.6	7	17.8

(According to Navy FF-T-276B Specifications)

(c) Thimble Dimensions for Safe Rope Eye

Figure 7.11 Preventing the Dangerous Kinking of Rope Loops

Solving the Header Problem 7.1 / Lifting 53.3 Ton (48.2 t) 112 ft (34 m) High

Question 1

How did the Romans create a hoisting structure without boom/mast cranes? Sketch a possible configuration.

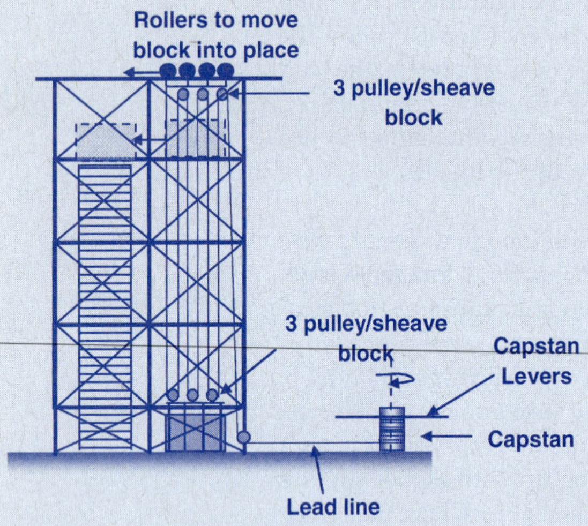

Figure 7.12 Temporary Hoisting Structure

Answer to Question 1: We can assume that the Romans used the same technique as the Greeks. They normally framed a wooden truss structure around the object to be built (e.g., column) and high enough to fit the last stone. Figure 7.12 offers a sketch of how the Romans may have created this feat.

Question 2

How many sisal ropes, blocks, and tackles did they need to lift the capital block weighing 53.3 tons (48.3 t)? Assume that they used a rope size that could be pulled by hand for an extended time and wrapped around a winch.

Additional Information Collected to Question 2: Tug-of-war might be a good indicator of the rope size that they used, and the strength of today's sisal could be corrected by a factor of 1.5 for lower quality. Safety factors for ropes today are between 5 and 12, depending on the application. The friction of the wooden pulleys could be assumed as 20%. Capstans are winches turned 90 degrees sideways. They consist of a wooden drum that is mounted on a vertical iron axle. Levers, known as handspikes, were inserted through holes at the top of the drum and used to turn the capstan. A rope wrapped several turns around the drum could be pulled. A holding ratchet provided the people or animals on the handspikes an opportunity to rest. It is more likely that the Romans used a capstan for this job rather than winches, since several "pushers" could operate at the same time, circling the capstan. We can assume that a capstan had an approximate diameter of 1.64 ft (50 cm).

Our Assumptions:

- **a.** The largest rope used for tug-of-war has a circumference of 5.5 in (14 cm) with r = 0.88 in (22.3 mm).
- **b.** Sisal rope has an average strength of 100 MPa = 100 N/mm^2 = = 14,503 psi.
- **c.** Each block had 3 pulleys.
- **d.** Capstan had a diameter of 1.64 ft (50 cm), and it took 30 seconds for one revolution.
- **e.** The frequent breaks during operation can be considered with a operating factor of 0.7.
- **f.** Total lifting height is 115 ft(35 m).

Sketch of Answer to Question 1 (not to scale)

Formulation of Equations

F_{SRM} = Max. force of modern sisal rope = Cross-sectional area $*$ unit strength

$F_{SRM} = (0.88\,\text{in}^2 * 3.14 * 14{,}503\,\text{psi} = 35{,}300\,\text{lb}\ (17.6\,\text{ton})$

$F_{SRM} = (22.3\,\text{mm}^2 * 3.14) * 100\,\text{N/mm}^2 = 157\,\text{kN}\ (16\,\text{t})$

F_{SRR} = Max. force of Roman sisal rope

$F_{SRR} = 157\,\text{kN}/1.5$ (low quality) $= 105\,\text{kN}\ (23{,}600\,\text{lb})$

F_{SRS} = Safe load of Roman sisal rope (with 10 as safety factor)

$F_{SRS} = 105\,\text{kN}/10 = 10.5\,\text{kN}\ (1.18\,\text{ton or } 1.07\,\text{t})$

MA = Mechanical advantage of the 2 blocks

$MA = 2 * 3$ pulleys $= 6$

Total $F_{Friction}$ = Friction forces compounded over 6 pulleys with friction factor of 20%

Total $F_{Friction} = (53.3\,\text{ton}/6) * (1 + 0.2)^6 = 8.88\,\text{ton} * 2.986 = 26.52\,\text{ton}(24\,\text{t})$

T-Ropes = Total number of ropes needed to lift the capital block using 2 * 3 blocks

T-Ropes $= ((53.3\,\text{ton}/6) + 26.52\,\text{ton})/1.18\,\text{ton} = 35.38\,\text{ton}/1.18\,\text{ton} = 30$ ropes

Answer to Question 2

This means that 30 lead lines, each with a 1.18 ton (1.07 t) pull, will each have to be connected to a separate capstan.

Question 3

How long might it have taken them to lift the stone 115 ft (35 m)? Assume that it took 30 seconds to turn the capstan one time.

D_L = Distance that lead line has to travel with 6 pulleys and a height of 115 ft (35 m)

$D_L = 115\,\text{ft} * 6 = 690\,\text{ft}\,(210\,\text{m})$

H_D = Hoist duration using capstan with $d = 1.64$ ft (50 cm) turned 2 revolution/min and an operating factor of 0.7 (the people turning the capstan had to rest 30% of the total time).

$H_D = (690\,\text{ft}/(1.64\,\text{ft} * 3.14)/\text{revolution}) * 0.5\,\text{min/revolution} * 1/0.7 = 96\,\text{min}$

$H_D = (210\,\text{m}/(0.5\,\text{m} * 3.14)/\text{revolution}) * 0.5\,\text{min/revolution} * 1/0.7 = 96\,\text{min}$

Answer to Question 3

It took 30 capstans to hoist the marble and 96 minutes of hoisting time that includes 29 minutes of breaks.

7.5 Keeping Mobile Cranes Erect

Most of today's cranes are mobile, meaning that they are able to move around with their own power. They are organized into three basic subtypes: (a) crawlers, (b) truck-mounted, and (c) wheeled. The next section will discuss their capabilities in more detail.

7.5.1 THE CRAWLER CRANES

Spurred by the development of more rugged track technology for World War II fighting vehicles and dozers used for the rapid construction of airfields, crane manufacturers in the 1940s adopted the concept because of the stability and the large footprints offered by the large steel plates. They equipped the tracked base with a heavy-duty vertical turntable on which they mounted the upper part, including the operator cab, counterweights, winches and, of course, the boom (see Figure 7.13). When wide steel treads are used, the crawler crane has a low ground bearing pressure of only 5 psi (34.5 kPa), making it possible to travel over soft terrain such as those found on construction sites.

7.5.2 THE WHEELED CRANES

Wheeled mobile cranes consist of two basic types: (a) all-terrain crane and (b) truck crane, shown in Figures 7.14 and 7.15. Both run on tires for fast travel, which creates a problem during operation.

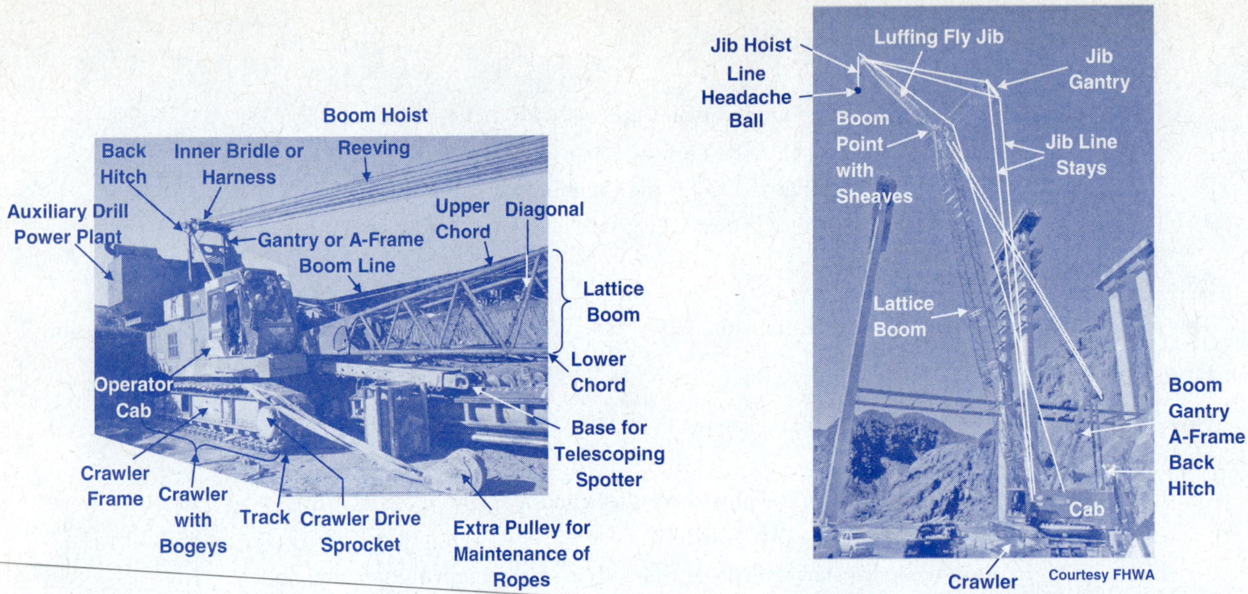

(a) Crawler Crane with Drill Rig Attachments During Maintenance (b) Erected Crawler Crane with Luffing Jib

Figure 7.13 Key Components of a Crawler Crane

As the load on the tires increases during a lift, the air in the tires will compress, resulting in a tilting of the crane chassis in the direction of the load. Naturally, the maximum safe load of a crane given by the manufacturer is only valid for a perfectly horizontal crane base. For this reason, wheeled cranes always come with at least four outriggers, two in the front and two in the back. Displayed in Figure 7.14a is a

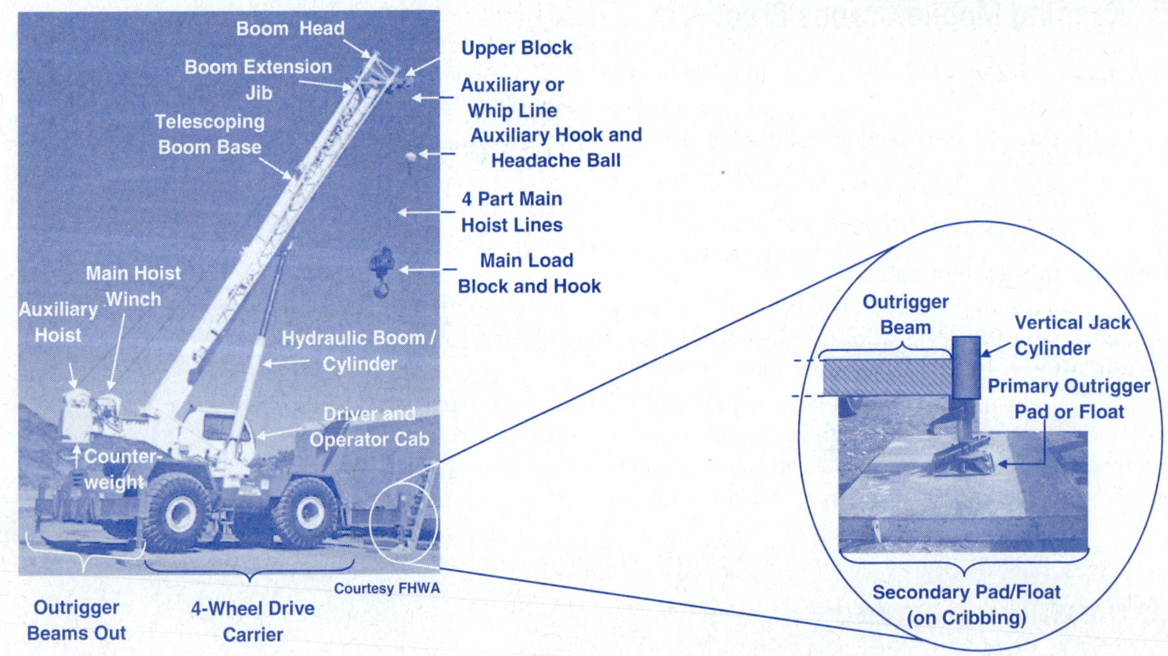

(a) Crane on Outrigger with Wheels Off Ground (b) Outrigger Resting on Large Secondary Float

Figure 7.14 Rough-Terrain Telescope Boom Crane

rough-terrain crane with four fully deployed outriggers with wheels clear of the ground. Figure 7.14b shows that the primary pad or float rests on a secondary much larger pad to distribute the weight over a greater area. All four outrigger beams must be equally extended to the appropriate vertical stripe. The leveling of the crane is accomplished by activating the hydraulic lift jacks at the tips of the outrigger beams and by using an electronic indicator or bubble level.

Depending on the size of a crane, the telescopic boom consists of 2 or more telescopic sections with a fixed base section pinned to the turret drive and lifted by the hydraulic boom lift cylinder, as depicted in Figures 7.14 and 7.15. The first section is powered by a hydraulic cylinder mount in its center. With the help of pulleys and cables, the subsequent sections are "telescoped" at the same time, all powered by the one cylinder. To extend the boom, flying jibs can be mounted to the boom head as Figure 7.15 demonstrates. When not needed, the jib can be unbolted from the head on one side of the boom and swung around to rest on the other side of the boom, as shown in Figure 7.14a. Of course, this convenience sacrifices the possibility of luffing the jib when in place.

Finally, Figure 7.16 provides two picture of what happens when cranes on outriggers are overloaded. The all-terrain crane in Figure 7.16a shows two properly deployed outriggers, but the jack rods had collapsed. Thus, the tipping over was really a result of a structural failure. On the other hand, the truck crane in Figure 7.16b tipped with the outriggers still fully deployed.

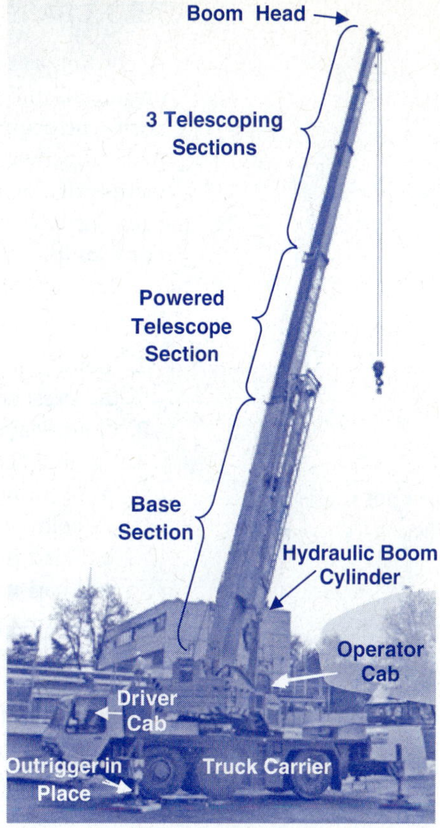

Figure 7.15 Fully Erected Truck Crane

(a) All Terrain Crane with Structural Failures of Two Jack Rods

(b) Two Tipped-Over Truck Cranes

Figure 7.16 Results of Overloading Cranes

Header Problem 7.2 / The Crashing Cranes

We are assigned to do a forensic analysis of what caused the two truck cranes in Figure 7.16b to turn over simultaneously, both being telescoping cranes.

The tower structure to be lifted and moved weighed approximately 18,000 lbs (8.16 t) and was 80 ft (24.4 m) high, well within the capacity of two cranes.

7.5.3 CRANE LIFTS THAT REQUIRE DETAILED PLANNING

According to the ASCE's Manual on Crane Safety on Construction Sites, lifts should be organized into three groups: (1) critical, (2) production, and (3) general or ordinary. A critical lift requires a specific lift plan, as it covers situations where multiple cranes have to work together or the load is close to the allowable capacity, is difficult due to the complexity of the situation, involves a toxic material, and so. Various governmental agencies have established their own definition of what a critical lift plan has to include. For example, the DOE's standard on Hoisting and Rigging states:

The person-in-charge (PIC) shall ensure that a pre-job lift plan is prepared that defines the operation and shall include the following:

1. Identification of the items to be moved, the weight, dimensions, center of gravity, and the presence of hazardous or toxic materials.
2. Identification of cranes to be used by type and rated capacity.
3. Rigging sketches that include (as applicable):
   a. Identification and rated capacity of slings, lifting bars, rigging accessories, and below-the-block lifting devices.
   b. Load-indicating devices.
   c. Load vectors.
   d. Lifting points.
   e. Sling angles.
   f. Boom and swing angles.
   g. Methods of attachments.
   h. Crane orientations.
   i. Other factors affecting equipment capacity.
4. Operation procedures and special instructions to operators, including rigging precautions and safety measures to be followed as applicable.

S7.1 DOE Hoisting and Rigging

Other recommended or required measures are:

1. Develop a "what-if a failure happens" plan that seeks to minimize any possible impact.
2. Get a weather report, especially a wind forecast.
3. Pre-approve the location of the crane through an inspection of the ground.
4. Verify the selected communication channels are appropriate and functional.
5. Call a meeting of all the involved personnel prior to the lift to discuss in detail the lift plan and any questions and reservations that anybody may have.

7.5.4 THE "SECRETS" ABOUT LIFTING CAPACITY AND LOAD RATING

The accidents demonstrate impressively that cranes have limits that are not always clearly understood even by a crane operator or the qualified person. That is why one of the first items on the lift plan is to clearly establish the rated capacity of the equipment being used. The definition of a crane/load capacity states: the lifting capacity [must be] established by the certified agent for various angles and positions.

The matrix in Table 7.4 lists allowable loads as a function of their distances from the centerline of the crane rotation in feet, shown in the column labeled Radius, and the length of the telescopic boom in feet. The two distinctive areas correlate with two different crane failure modes. The maximum loads in the darker area are associated with small radii where tipping is not a problem. In other words, the load limits in the dark area are not due to the potential of tipping but are a function of the breaking limit of a key structural element of the crane. Such a case is shown in Figure 7.16a, where the cylinder rod of the outrigger had collapsed before the crane tipped over. The load limits in the white boxes,

Table 7.4 Maximum Boom Load Capacities in Lb (t) for a Mobile Crane with 8, 500 Lb (3.8 t) of Counterweight and Fully Extended Outriggers

Radius (ft)	Boom Length (ft)										
	38	50	60	70	76.5	90	100	110	120	130	140
10	154,600	152,100	138,900	81,700							
20	65,100	63,500	62,200	60,900	58,500	50,600	46,000	40,800			
30	41,900	41,700	41,600	40,600	40,100	39,600	38,200	37,300	33,800	29,800	22,500
40		29,100	28,400	27,900	28,100	28,000	28,000	27,500	27,100	26,900	20,000
50			21,100	20,600	20,400	20,200	20,100	19,900	19,600	19,100	16,100
60				15,400	15,000	14,600	14,500	14,300	14,000	13,600	12,700
70					10,800	11,000	10,900	10,800	10,400	10,100	9,700
80						8,400	8,300	8,200	7,900	7,500	7,200
90							6,300	6,300	6,000	5,600	5,300
100								4,800	4,500	4,100	3,800
110									3,300	3,000	2,700
120										2,000	1,700
130											900

Radius (m)	Boom Length (m)										
	11.58	15.24	18.29	21.34	23.32	27.43	30.48	33.53	36.58	39.62	42.67
3.0	70.13	68.99	63.00	37.06							
6.1	29.53	28.80	28.21	27.62	26.54	22.95	20.87	18.51			
9.1	19.01	18.91	18.87	18.42	18.19	17.96	17.33	16.92	15.33	13.52	10.21
12.2		13.20	12.88	12.66	12.75	12.70	12.70	12.47	12.29	12.20	9.07
15.2			9.57	9.34	9.25	9.16	9.12	9.03	8.89	8.66	7.30
18.3				6.99	6.80	6.62	6.58	6.49	6.35	6.17	5.76
21.3					4.90	4.99	4.94	4.90	4.72	4.58	4.40
24.4						3.81	3.76	3.72	3.58	3.40	3.27
27.4							2.86	2.86	2.72	2.54	2.40
30.5								2.18	2.04	1.86	1.72
33.5									1.50	1.36	1.22
36.6										0.91	0.77
39.6											0.41

(shaded) = Load Limits Due to Possible Structural Failure
(unshaded) = Load Limits Due to Crane Tipping

on the other hand, are clearly linked to the larger radii. These load limits are a function of tipping moments that exceed the stabilizing moments.

Figure 7.17 presents the 2-D path of the boom from highest to lowest point for each of the boom extensions.

The horizontal pin of the boom rotation is 10 ft(3 m) off the ground approximately 7 ft(2.1 m) behind the centerline of crane rotation. As shown, the boom angle α covers approximately 62 degrees between 18 to 80 degrees. Each crane will have its own spatial limitations, mainly due to the mechanisms for lifting the boom such as the hydraulic cylinder in the case of a telescopic boom crane.

Further inspection of Table 7.4 reveals an interesting phenomenon in that the load data for the same radius are similar with a slight tendency to decrease for larger lengths.

Figure 7.18 depicts that each load capacity curve shows two distinct segments. The load limits for larger radii create a polynomial function. For example, the loads related to the larger radii for the 100 ft(30.4 m) boom follow the function: $\mathbf{y} = -\mathbf{0.385x^3} + 78.496\mathbf{x^2} - \mathbf{5{,}621}.7\mathbf{x} + \mathbf{152{,}420}$ with an excellent fit (R2 = 0.9992). In other words, between

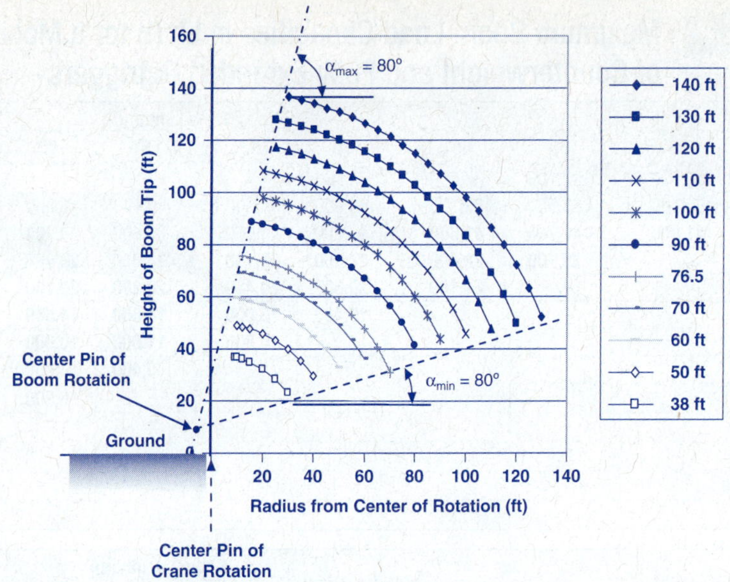

Figure 7.17 Boom Reach for Mobile Crane Represented in Table 7.4

the radius = 50 and 90 ft one is able to calculate the load capacity = y using this function. However, the maximum loads below 50 ft (15.2 m) do not fit. Why not?

The reader will quickly infer that this must be related to the switch of possible failure mode from tipping to a structural break. Indeed, Table 7.4 shows the border between the two failure modes exactly between 40 ft (12.2 m) and 50 ft (15.2 m) radius.

Why is the load capacity based on tipping not linear? Finding the answer to this question will get us closer to understanding one of the single most important points that dictates load capacity of a well-maintained crane, the tipping point or tipping fulcrum.

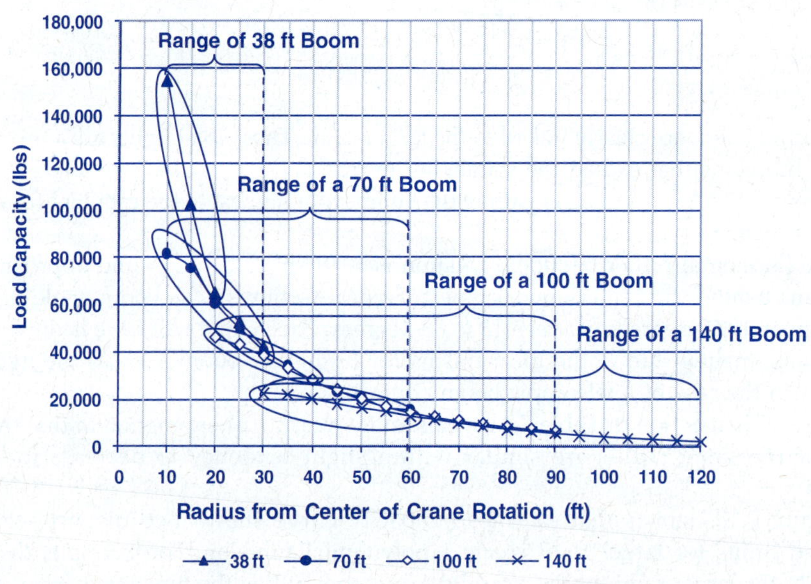

Figure 7.18 Load Capacity Curves for Four Boom Lengths

7.5.5 SAFETY FACTORS TAKE IMPERFECTION INTO ACCOUNT

It is standard engineering procedure to introduce a design safety factor to consider the randomness of the real world, which does not adhere perfectly to the assumptions that lie behind the "perfect" calculations. In cases such as cranes, where accidents can have serious effects, operational safety factors to reduce the allowable maximum load are used.

It has been shown that cranes fail because of two causes: (1) collapse of a critical structural element of the crane (e.g., outrigger), and (2) tipping over.

In the case of the mobile crane shown in Table 7.4, the manufacturer warns the operator only about tipping with a warning below the capacity chart:

> Rated lifting capacities shown on fully extended *outriggers do not exceed 85% of the tipping loads* and *on tires do not exceed 75% of the tipping loads.*

As the definitions of operational safety indicates, the 85% or 75% with which the tipping load is reduced does **NOT** consider the abnormal conditions that could exist on the jobsite. It only considers imperfections in the material, welds, and so on used to make the crane.

Thus, the actual load leading to tipping is larger than the load capacity provided by the manufacturer. It is apparent that a tipping of the crane due to a load represents a case where the forces acting on the structure are so unbalanced that they make the system flip. Assuming that the outriggers are strong enough, tipping will result in the boom and load crashing to the ground while the carrier is twisted straight up, as shown in Figure 7.16b. The condition where this occurs is discussed in section 7.3 and can be depicted with following equation: $\Gamma M_T = \Gamma M_S$, where M_T are the moments adding to tipping and the M_S are moments helping stabilize the crane.

Worked-Out Problem 7.2 / Planning the Crane Setup

We have recently been hired by the craning and hoisting company BridleHitch, Inc. headquartered in Philadelphia. They have been asked to provide a bid for erecting a tower crane, drawn in Figure 7.19, with the assistance of a mobile crane. Trek Tucker is in charge of estimating and asks you to investigate if (a) one of the economical cranes that they own, with the capacity chart shown in Table 7.4, would be large enough to handle the job, and (b) if it would be possible to erect the entire tower crane from one truck crane setup location. Naturally, not having to reposition the crane would save time and money. We are given the dimensions of the tower crane, including the weights of the main component that have to be lifted.

Goal: The problem allows us, as newcomers to craning, to get used to the capacity charts as well as the spatial limitations of cranes.

Approach to Solving Problem

Step 1: Find the most challenging lift(s) in terms of weight and height, and consult the capacity chart to see if the crane would be able to handle them

Step 2: Position the crane in a location where it is able to handle the two most critical lifts without repositioning. Check if the other lifts could be done from the same location

Step 1: Table 7.5 shows that the **heaviest** piece to be lifted is the slewing platform with the attached cab weighing 16,500 lbs (7.5 t) followed by the jib with 15,048 lbs (6.82 t). The **highest** lift is needed for the tower head as its top reaches 132.2 ft (40.1 m). Figure 7.20 presents the situation of attaching the tower head. Although the weight of the tower head is "only" 5,632 lbs (2.5 t), it could be the critical position as the height is close to the limit of the mobile crane.

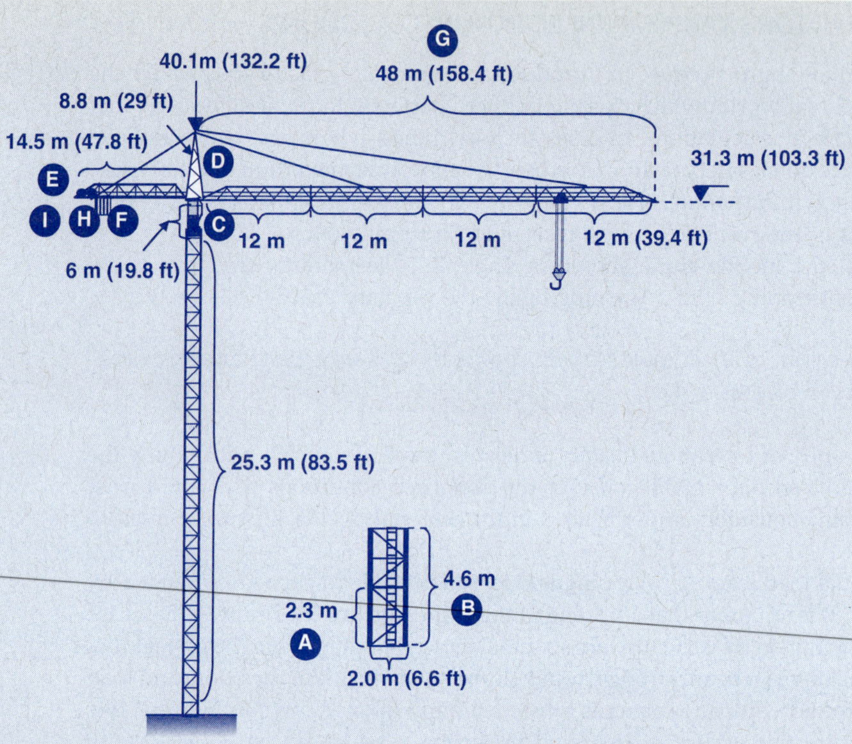

Figure 7.19 Tower Crane After Erection

Figure 7.20 also shows that the height of the boom still has to consider the reeving and the size of the lower block with hook. For the mobile crane at hand, we need to assign 4 ft for this combination. Consequently, the desired boom height is 136.2 ft (41.5 m).

Table 7.6 highlights that only one configuration will be able to satisfy the height requirements of the tower head, namely a boom length of 140 ft (42.7 m) with a radius of 30 ft (9.1 m). A very short bridle hitch should be used or a heavy-duty shackle. Table 7.4 reassures us that its height is not a problem, as the crane capacity is 22,500 lb (10.2 t).

Now we need to verify that we are able to lift the heaviest piece, the 16,500-lb slewing platform, which needs to be lifted to a height of 103.3 ft (31.5 m). Again, we should add 4 ft for the hook and lower block, resulting in a tip

Table 7.5 Specifications of Tower Crane to Be Erected

	Element	Max. Weight		Length		Max. Lift Height		Hook Height		Mobile Crane	
		lb	t	ft	m	ft	m	ft	m	Radius	Boom
A	Tower Section	2,420	1.098	7.6	2.3	83.5	25.3	90.1	27.3	?	?
B	Long Tower Section	4,312	1.956	15.2	4.6	83.5	25.3	90.1	27.3	?	?
C	Slewing Platform with Cab	16,500	7.484	19.8	6.0	103.3	31.3	110.0	33.3	?	?
D	Tower Head	5,632	2.555	29.0	8.8	132.1	40.1	133.1	40.3	?	?
E	Counter Jib	9,790	4.441	47.8	14.5	103.3	31.3	113.2	34.3	?	?
F	Erection Counterweight	7,920	3.592	-	-	103.3	31.3	106.6	32.3	?	?
G	Jib	15,048	6.826	158.4	48.0	103.3	31.3	116.5	35.3	?	?
H	Hoist Winch Unit	7,040	3.193	45 hp		103.3	31.3	106.6	32.3	?	?
I	Final Counterweights	5,500	2.495	-	-	103.3	31.3	106.6	32.3	?	?

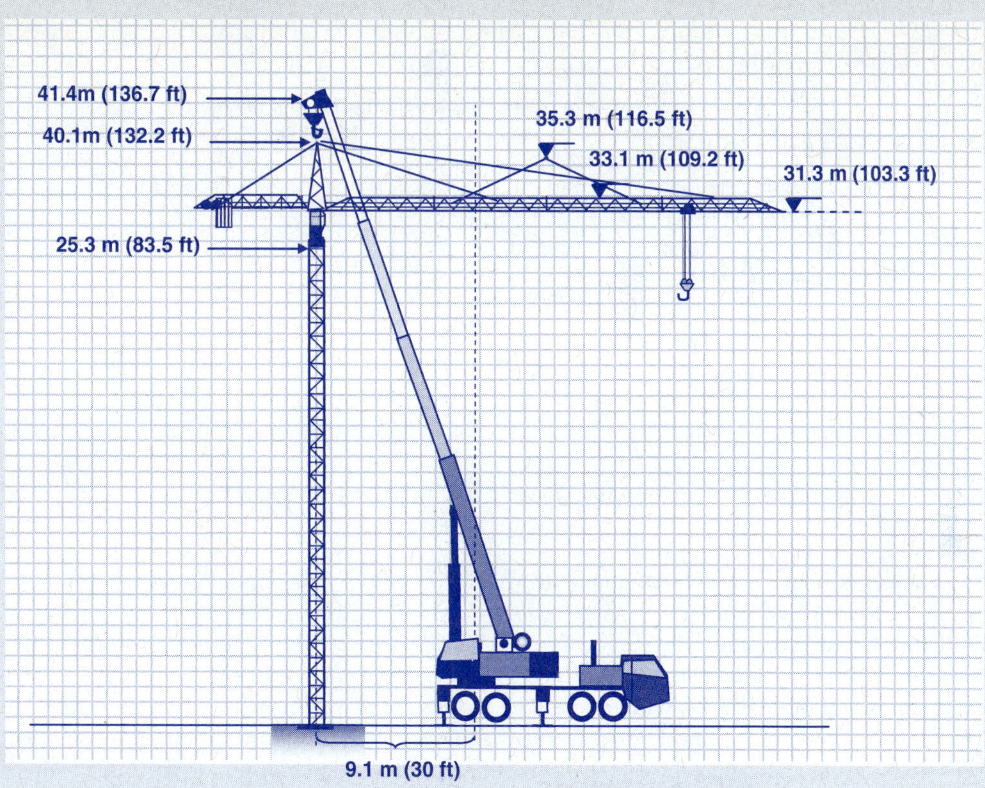

Figure 7.20 Schematic of the Lift to Attach Tower Head at 132.2 ft (not to scale)

height of 107.3 ft (32.7 m). Checking Table 7.4, we are assured that this will be not a problem, since boom lengths of 110 ft (33.5 m) and higher, combined with a radius of 50 ft (15.2 m) or less, will provide the necessary capacity. Since no other element is heavier or has to be lifted higher, we are able to answer the first question posed by Trek Tucker. Yes, we will be able to erect the tower crane with the economical truck crane.

Step 2: Lifting all the elements from one position is a challenge since the height of the tower head forces us to position the crane 30 ft (9.1 m) from the center of the tower. Let's review the sketch of the footprint of the situation in Figure 7.21.

The second most critical element is the jib weighing 15,048 lb (6.82 t). Its center of gravity is 25 m (82.5 ft) from the center of the tower. By positioning the crane between these two points but 30 ft from the tower center point, it might be possible to reach the center of the jib. From Table 7.6 we learn that the largest radii that allow us to reach 116.5 ft + 4 ft (hook and block) are 70 ft (21.3 m) for a boom length of 140 ft (42.7 m) and 50 ft (15.2 m) for a boom length of 130 ft (39.6 m). Table 7.4 tells us that for both boom lengths, the maximum radius is around 50 ft (a little more for the shorter boom).

Table 7.6 Truck Crane Boom Tip Heights for Maximum Extensions

			Radius (ft)					
Boom Length			**30**	**40**	**50**	**60**	**70**	**80**
140 ft	Boom Tip Height	ft	136.7	134.2	130.8	126.5	121.2	114.9
130 ft		ft	126.5	123.7	120.0	115.3	109.5	102.5
120 ft		ft	116.2	113.1	109.1	103.9	97.5	89.4

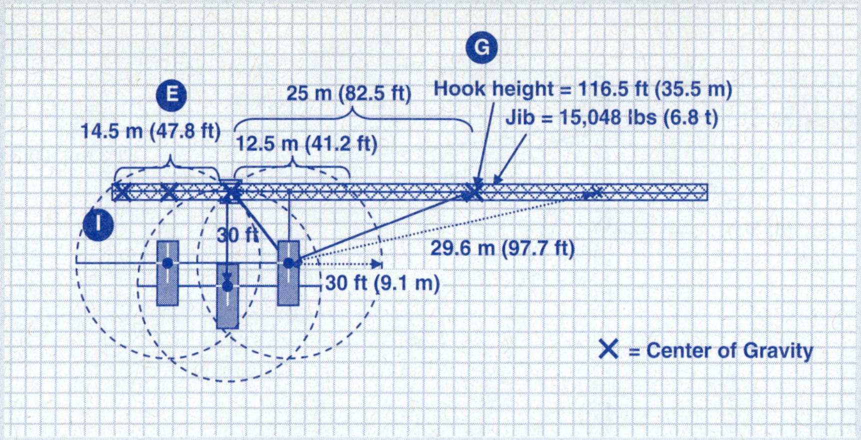

Figure 7.21 Footprint of Tower Crane Erection with Truck Crane

As the distance between the tower center and the jib center is only 82.5 ft (25 m), it will be impossible to lift both elements from the same position.

An alternative to lifting the entire jib in one piece is to install it in two halves. Figure 7.21 indicates the centers of gravity for this option. While the half close to the tower will be no problem, we do need to verify that the installation of the second half, weighing 7, 524 lbs, will be safe. The distance of its center of gravity is 36 m from the crane or 37 m (122 ft) from the center of the tower. On the other hand, the minimum radius to install the first half of the jib needs to be 30 ft (9.1 m) as well, which limits the distance of the crane position from the jib. Still, the tip of the crane boom needs to reach a height of 120.5 ft, which is only possible with radii of 70 ft (21 m) for a boom length of 140 ft and 50 ft (15.2 m) for a boom length of 130 ft (40 m). According to Table 7.4, both setups have sufficient capacity to lift $^1/_2$ of the jib. Nevertheless, based on geometry, the distance between the center of the crane rotation and the center of gravity of the second half of the jib is 97.7 ft (29.6 m). Of course, this is far above the maximum radius of 70 ft due to the height requirements.

Result: The answer to question 2 is No. It is not possible to erect the entire crane from one position. It is possible to erect the tower and the counter jib from one position, but the crane needs to be repositioned to lift the jib in one piece. The 5, 500 lb (2.5 t) counter weights can be installed from the latter position if it is properly selected.

7.6 Reaching High and Out with a Jib

Using straight boom designs provides an extremely safe and reliable lifting structure. However, its physical limitations are constricting once a crane has been erected. Today most cranes provide an add-on for cases where one does not use the available load capacity but would like to lift or place a load higher up, further out, or over an existing high obstacle. Let's review the mechanics of the luffing jib used by the 80-t on (72.6 t) crawler crane CC-280-1 from Terex-Demag (Figure 7.22).

S7.2 Crawler Crane CC-280-1
VG7.1 Cranes

The GVW of the crawler without the 57,860 lb (26.2 t) counterweight is 91,960 lb (41.7 t) (carrier boom crane weight was 95,000 lb (43.1 t) and its counterweight 8,500 lb). Its maximum radius with the main boom is 138 ft, and its maximum boom length is 147 ft (44.8 m). Compared to the load capacity of 8,300 lb (3.8 t) at a radius of 80 ft (24.4 m) and a boom of 100 ft (30.5 m), the CC-280 is 12,100 lb (5.5 t). We can safely assume that the larger capacity was gained by replacing the relative heavy telescopic boom with the light lattice boom and by providing a much heavier counterweight.

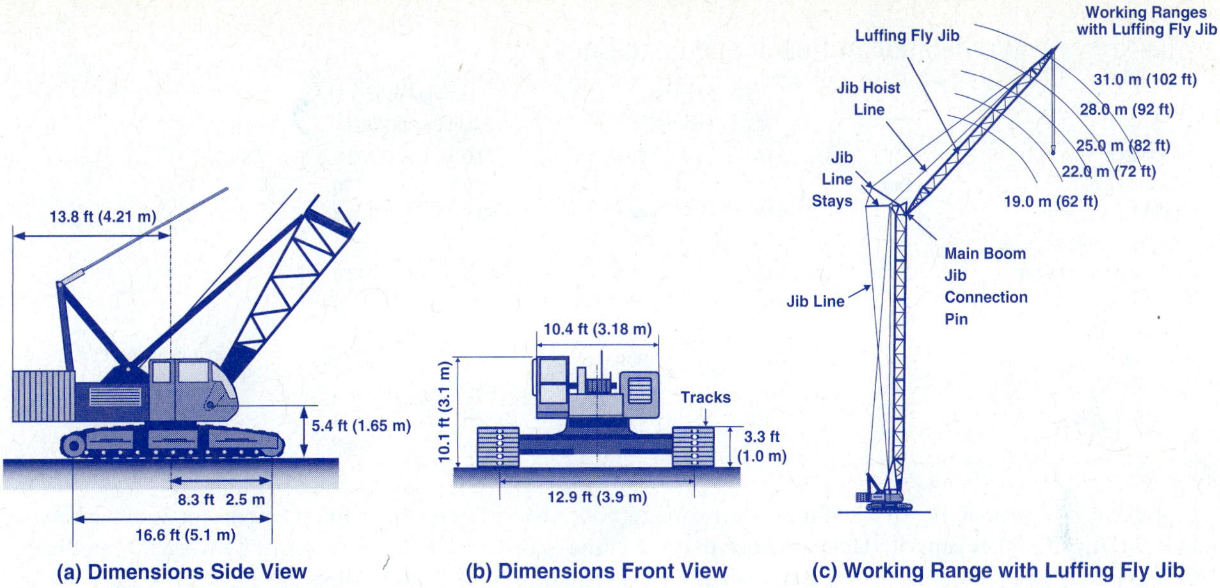

Figure 7.22 Base Dimensions of the CC-280

Worked-Out Problem 7.3 / Lift Design

Consider the following layout for a lift that involves putting a 5,500 lb (2.5 t) heavy elevator hoisting mechanism onto the roof of a new building that is 118 ft (36 m) high. We also would like to know how long it will take to lift the load.

Lift Plan

a. *Operational plan:* One possible scenario for this critical lift is shown in Figure 7.23. The crane will be readied at 41.5 ft (12.6 m) from the face of the building and at the shortest horizontal distance of 91.5 ft (27.9 m) from the target point C. The load will be picked directly from the truck at point A at an approximate distance of 52 ft (15.8 m) from the rotational axis of the crane. The jib is set to less than 10° for the pickup. Between A and B, the hoist line will be shortened until the load is as high as possible or at least 10 ft higher than the roof of the building (= 118 ft). After the load safely passes the roof edge at B, the crane keeps rotating until the boom is lined up with C. At this time, the main boom can be lowered to a maximum of 70 degrees from the horizontal or 20 degrees from the vertical. Now the jib block should be able to reach point C.

b. *Crane Configuration:* From the crane specifications on the booksite we learn that there are two fly jib options: (1) fixed at 10° or 30° and (2) luffing from 90° to 60° (70°) relative to the main boom. Figure 7.23 allows us to define the extreme radii that the crane has to be able to reach.

Point A: $R_{min} = 52\,\text{ft}\ (15.8\,\text{m})$

Point C: $R_{max} = (50\,\text{ft} + 45\,\text{ft} + 3\,\text{ft}) = 98\,\text{ft}\ (29.9\,\text{m})$

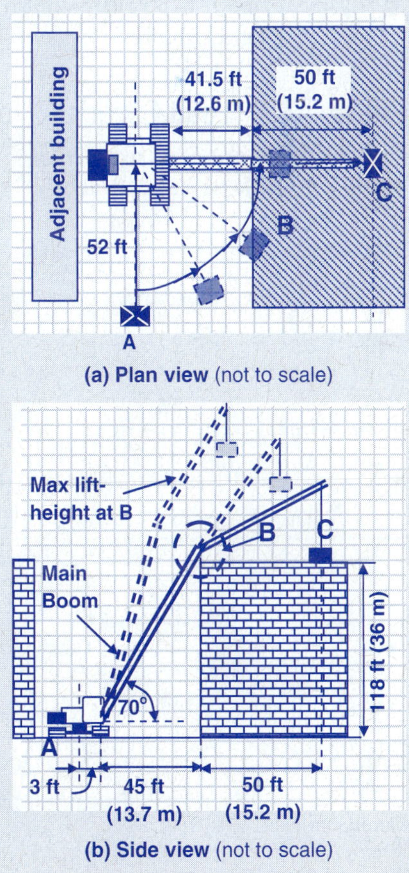

Figure 7.23 Sketch of Lift plan

Table 7.7 Verification of Reach and Load Capacity

		Fixed Fly Jib		Luffing Fly Jib	
		Main Boom 148 ft		Main Boom 128 ft	
Jib length	ft	39.4	59.0	72.0	102.0
Boom angle	Deg.	70	70	70	70
Height at boom tip	ft	144.5	144.5	125.4	125.4
Height at building edge	ft	129.0	129.0	-	-
Total Reach of boom	ft	50.6	50.6	43.8	43.8
Jib angle	Deg.	10	10	70	70
Reach of jib	ft	19.7	29.5	46.3	65.6
Max reach (boom + jib)	ft	70.3	80.1	90.1	109.3
	m	21.4	24.4	27.5	33.3
Load Capacity	lbs	21,320	14,760	10,496	10,496
	t	9.70	6.70	4.80	4.80

←Required radius = 98 ft or 29.9 m

←Required capacity = 5,500 lb or 2.5 t

Table 7.7 shows that the fixed and the luffing configurations have different maximum lengths for both the boom and the fly jib. Either capacity table assumes that the crane is able to luff its main boom between 90° and 60°. However, the side view of the operational plan illustrates that the building height limits the boom angle to 70°. As a consequence, it is necessary to verify that the minimum and maximum radii can be reached with one of the jib combinations available.

The result of the calculations allows us to make two observations:

1. Only one setup will be able to reach 98 ft (29.9 m), namely the longest luffing fly jib. Although the main boom is 20 feet shorter, the longer jib provides the necessary reach. The load at A can be picked up with a luffing angle of 88°.

Table 7.8 Load Capacities for CC-280 with Fixed and Luffing Jibs

Radius	LIFTING CAPACITIES FIXED FLY JIB Fly jib - Hilfsausleger - Fléchette					
	12 m		15 m		18 m	
	10°	30°	10°	30°	10°	30°
m	t	t	t	t	t	t
14	6,5	-	-	-	-	-
16	6,5	-	5,5	-	4,5	-
18	6,5	5,0	5,5	-	4,5	-
20	6,5	5,0	5,5	4,0	4,5	-
22	5,9	5,0	5,5	4,0	4,5	3,5
24	5,1	5,0	5,1	4,0	4,5	3,5
26	4,5	4,5	4,5	4,0	4,5	3,5
28	4,0	4,0	4,0	4,0	4,0	3,5
30	3,6	3,6	3,6	3,6	3,6	3,5
32	3,2	3,2	3,2	3,2	3,2	3,2
34	2,8	2,8	2,8	2,8	2,8	2,8
36	2,4	2,4	2,4	2,4	2,4	2,4
38	2,1	2,1	2,1	2,1	2,1	2,1
40	1,9	1,9	1,9	1,9	1,9	1,9
42	1,6	1,6	1,6	1,6	1,6	1,6
44	1,4	1,4	1,4	1,4	1,4	1,4
46	-	-	-	-	-	-

Main boom max = 148 ft (45 m)
Jib min = 29.5 ft (12 m)
Jib max = 59 ft (18 m)
Jib at either 10° or 30°

Radius Ausladung Portée	LIFTING CAPACITIES LUFFING FLY JIB LW,								
	25 m			28 m			31 m		
	90° 88°	88° 80°	80° 70°	90° 88°	88° 80°	80° 70°	90° 88°	88° 80°	80° 70°
m	t	t	t	t	t	t	t	t	t
6	-	-	-	-	-	-	-	-	-
7	12,5	-	-	-	-	-	-	-	-
8	12,5	-	-	10,9	-	-	9,4	-	-
9	12,3	-	-	10,8	-	-	9,3	-	-
10	12,1	12,1	-	10,7	10,7	-	9,1	-	-
12	11,4	11,4	-	10,4	10,4	-	8,7	8,7	-
14	10,6	10,6	-	9,9	9,9	-	8,3	8,3	-
16	9,5	9,5	-	9,0	9,0	-	7,8	7,8	-
18	8,1	8,1	8,1	7,9	7,9	-	7,2	7,2	-
20	6,9	6,9	6,9	6,8	6,8	6,8	6,6	6,6	6,6
22	5,8	5,8	5,8	5,9	5,9	5,9	5,9	5,9	5,9
24	5,0	5,0	5,0	5,1	5,1	5,1	5,2	5,2	5,2
26	4,3	4,3	4,3	4,4	4,4	4,4	4,4	4,4	4,4
28	-	3,7	3,7	3,8	3,8	3,8	3,8	3,8	3,8
30	-	3,2	3,2	3,2	3,2	3,2	3,2	3,2	3,2
32	-	2,9	2,9	-	2,8	2,8	2,7	2,7	2,7
34	-	-	2,7	-	2,6	2,6	-	2,5	2,5

Main boom max 128 ft (39 m)
Min Jib = 38 ft (25 m)
Max Jib = 102 ft (31 m)

2. The pickup point A is so close to the crane (52 ft or 15.8 m) that the fixed fly jib has to be set at 10°. This fly jib angle prohibits it from reaching point C by approximately 18 ft (5.5 m).

As there is no apparent reason why the pickup point is so close, we might investigate the effect of a change. Table 7.8 shows that at a radius of 22 m (72 ft) even the longest fixed-fly jib can be set to 30°, while any of the luffing jib angles will work. The reach for the 18 m (59 ft) fly jib set to 30° with the main boom at 70° is 30 m (96 ft). The missing 2 ft could be achieved by reducing the crane angle to 69°, which will still provide a height of the boom at the building's edge of 37.5 m (123 ft) (1.5 m above the roof) but a total reach of 30 m (99 ft). The lifting capacity of this configuration would be 3.5 t or 11,480 lbs. The rigging details cannot be addressed because we lack the necessary information about the load, such as its center of gravity. It is customary, however, that the manufacturer delivers the machine with proper hooks for slings, particularly since they have to handle it first.

Discussion of Results

The fixed as well as the luffing fly jib will be adequate, although the pickup location would have to be changed for the fixed jib. The fixed jib with 11,480 lb (vs. 10,496 lb) has a slightly higher lifting capacity. Both options provide a minimal operational safety factor of 1.9.

Overall, it would be preferable to select the luffing jib option, since it offers more flexibility and room for errors in both reach and weight. While it probably would cause no problem in changing the load-pickup point, the requirement to lower the main boom to 69° or less is a safety risk, even though preventive measures could be implemented. In summary, it is recommended to use the 31-m (102-ft) long luffing fly jib on the 39-m (128-ft) main boom.

Solving Header Problem 7.2 / What Can We Learn from a Crane Accident?

The tipped truck crane shown in Figure 7.16b was not the only truck crane that crashed at that accident. There were two trucks involved, both of them employed to remove an old steel structure from its footings, to be dismantled on the ground. Here are two views of the situation right before the events started.

Layout of the Rigging Operation

Operational plan: Based on the photograph, the following plan view of the equipment setup can be sketched out (not to scale).

Factual Observations

S7.3 Link Belt HTC-8690

The top and the side views shown in Figures 7.24 and 7.25, not to scale, disclose several facts:

1. Both cranes are on outriggers. Crane 1 is boomed out to the side and crane 2 to the back of the truck carrier.
2. A boom radius of 70 ft (21.2 m) and tip height of 105 ft (31.8 m) has been chosen. The corresponding boom length is 120 ft (36.4 m).
3. The lift capacity of the 90-ton Link-Belt HTC-8690 at these boom configurations with the standard 8,500 lb (3.9 t) counterweight on fully extended outriggers is 10,400 lbs (4.7 t) (see brochure on booksite).
4. The tower is 70 ft (21.2 m) high and weighs approximately 18,000 lbs (8.2 t) (nobody really knows). In addition, the center of gravity is estimated to be right at the 5th platform.
5. The one-leg bridle hitches were hooked to two tower legs approximately at the height of the center of gravity. The roof of the building on top of the tower caused the sling to bend.
6. The main axis of the trucks is not in line with the trajectory of the motion needed to lay the tower on its side. This complicates the control of the crane: all three, the boom, the turret, and the winch, have to be changed at the same time during the operation.

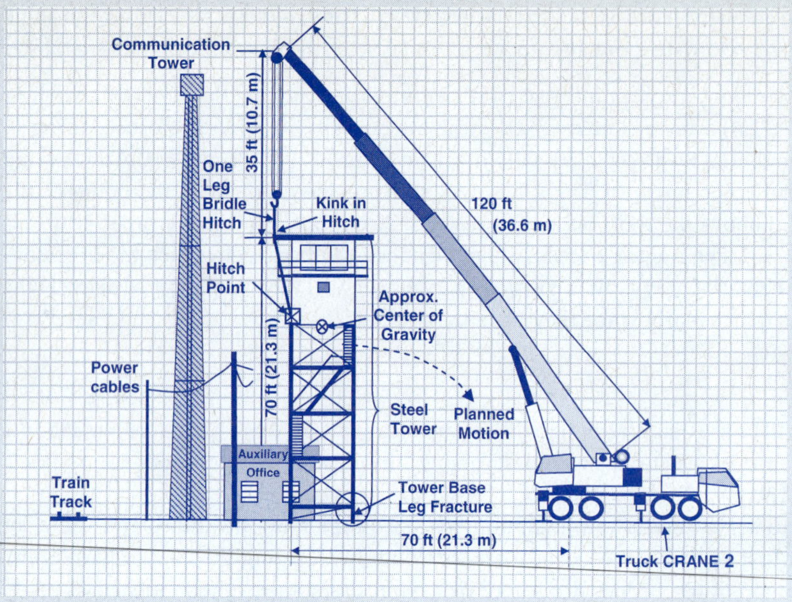

Figure 7.24 Side View of Pre-Crash Situation

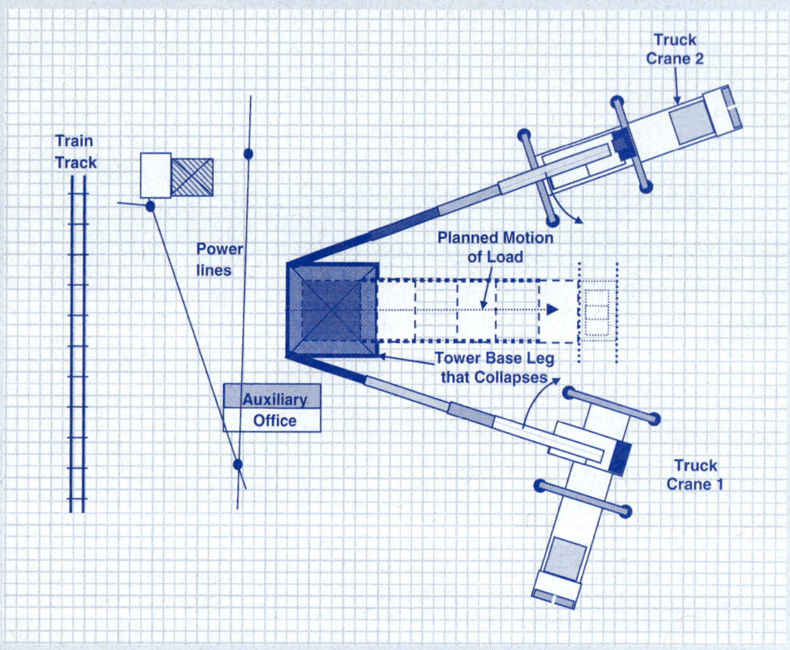

Figure 7.25 Top View of Pre-Crash Situation

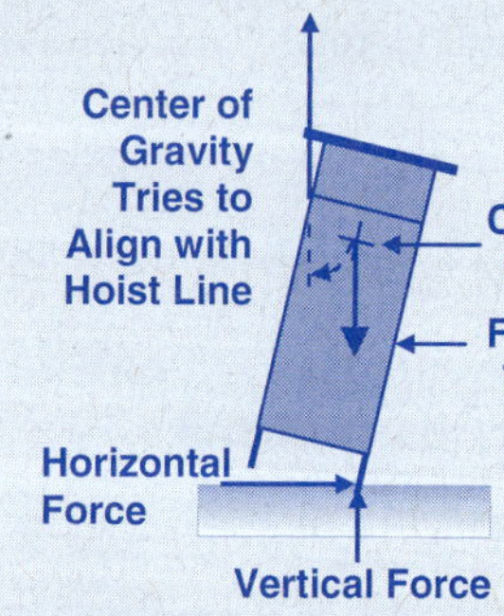

Figure 7.26 Forces Acting on the Front Legs

7. The four vertical legs of the tower were originally designed to sustain compression loads and were stabilized with simple cross-bracings attached to their main joints.
8. As must have been expected (based on the design), the tower started to lean forward as soon as the hoist lines lifted two legs off their footings. However, as Figure 7.26 indicates, the center of gravity tilts forward while the front legs keep their footings on the ground. As shown, this creates two kinds of reaction forces, vertical and horizontal, on the legs designed only to sustain forces in the direction of their central axis.
9. The left leg on the front collapses, causing the top of the tower to lean toward the lost support. Of course, the loss of the leg changes the equilibrium that was balanced by the two hitches and the two legs. Figure 7.27 shows that

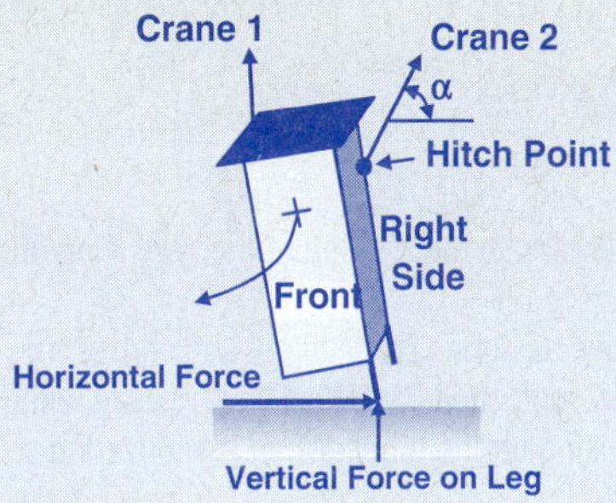

Figure 7.27 Collapsed Leg Rotates Structure

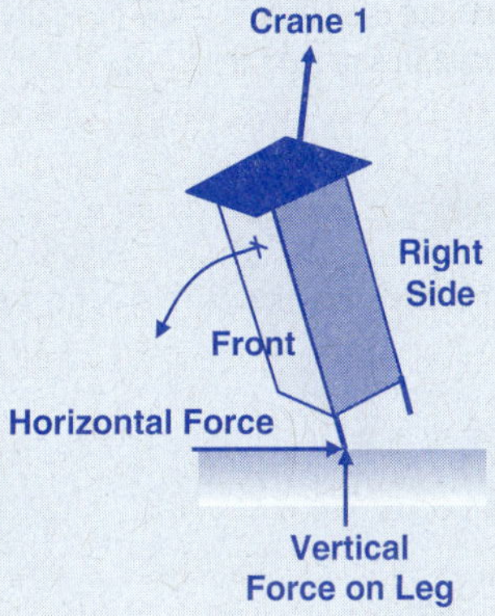

Figure 7.28 Crane 1 Hoist Boom Overstressed

the falling tower moves right away towards the left and front side of the tower. In addition, the one leg that is not broken experiences additional horizontal forces for which it was not designed. However, the most disastrous result is the angle α that results from the jerking of the tower to the left. The critical problem arises from the fact that cranes are designed to lift loads vertically, meaning $\alpha = 0$. Side-loading the boom at the tip, as occurs here, can add torque to the boom, bringing the crane quickly to a critical state. In fact, the rapid acceleration of the falling steel structure sufficed to tip crane 2.

10. With the lift capacity of crane 2 gone, the hoist line of crane 1 suddenly has to take over the forces of the accelerating tower structure rotating around the one front leg still intact as shown in Figure 7.28.
11. This additional load is too high and causes crane 1 to tip over as well.

Cause-and-Effect (Fishbone Diagram): A convenient approach to figure out how this accident could have been avoided, or in other words, how the structure could have removed safely from its foundation, is the use of the cause-and-effect diagram. Figure 7.29 offers a partial example that includes two relationship levels.

The fishbone diagram shows four main branches that contributed to the accident, but one is clearly dominant. Tilting a load while depending on the structural support of the load itself is a high-risk operation. The reason, of course, is the fact that in such cases a load will most probably be exposed to forces for which it was not designed. If it is nevertheless necessary to go ahead with such a tilt, one could reinforce the weak elements, such as the base legs of the tower. In this case, however, as soon as the tilting started, the front legs experienced horizontal forces for which they were not designed, leading to collapse and subsequent disaster. Tilting the load with eccentric lifting, causing a kinked hitch, is yet another major lapse that contributed to the accident. Finally, the positioning of the crane did not contribute directly to the collapse but did not allow possibly avoiding the final consequence.

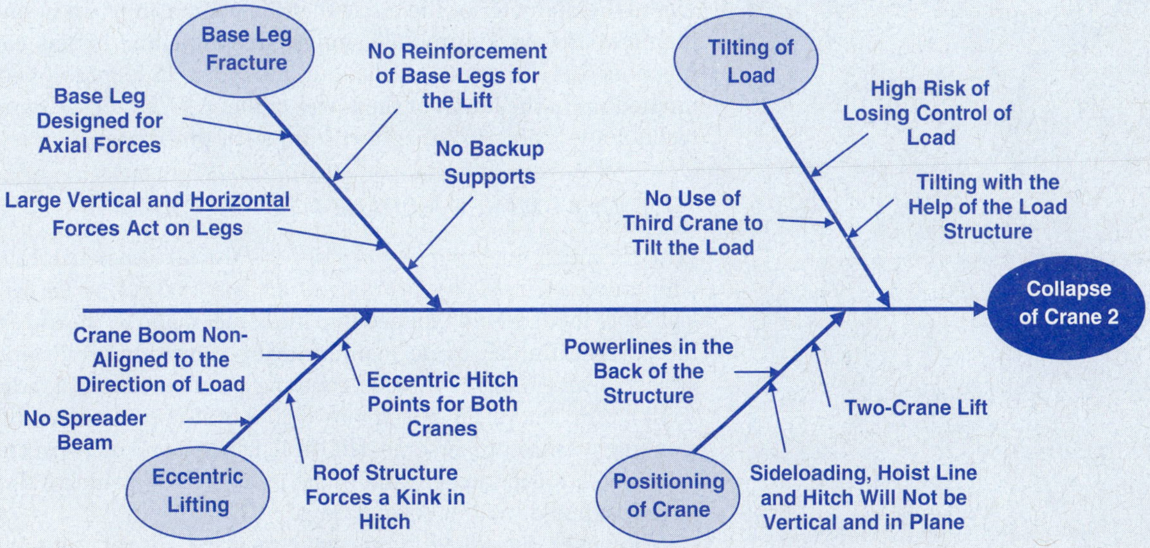

Figure 7.29 Cause-and-Effect Diagram for Tipping of Crane 2

Lessons Learned

The rigging experts planning the lift accepted not one but two extremely high risks:

(1) A **two-crane lift,** and (2) **lifting and tilting an off-centric load.**

A **two-crane lift** can only be done with experienced crane operators and a signal person to coordinate the lift. Before the beginning of the lift, a dry run needs to be executed.

Not only is there a risk of miscommunication, but the shifting of the load's center of gravity, as during the tilting of the load, or the change in crane radius will change not only how much each crane is carrying but also the max load capacity. When lifting a load with two cranes, both hoist lines have to remain vertical at all times to prevent one crane from pulling on the other. When maneuvering a load, it is easy for one crane to pull on the other without the operator's realizing it. To prevent pulling from occurring, a monitor for each crane needs to be used to keep the hoist lines vertical.

Tilting of an off-centric complex load unleashes unknown forces caused by the center of gravity trying to align itself vertically underneath a sole hitch point or somewhere underneath the axis between two hitch points. By having one segment of the load still touching the ground, there is a great risk that the segment will experience a force for which it was not designed, and will therefore break. As was the case in the above accident, the structure to be handled was fairly complex, and built a long time ago.

Actions to Avoid Such Accidents

a. **Reinforcement of front legs that would sustain the horizontal (shear) forces**
b. **Use of a spreader beam, supported by both cranes, with hitch points slightly above the structure's center of gravity**
c. **Use of a third crane to tilt the structure when off the ground**
d. **Lifting and removing the structure in sections.**

7.7 The Omnipresent Tower Crane

The night-and-day presence of the tower crane has made it a symbol for construction, much like the hard hat. Among many unique features, its small footprint is extremely valuable in areas where space for storage and work is a premium. On the other hand, not being able to move the fixed towers can be detrimental in cases where the diameter of the circle that the boom can reach is smaller than the site. As Figure 7.30 shows, slewing crane booms can be positioned a different levels, but the hoist line might become tangled up in the boom of the lower crane. As we will see, modern technology is able to safeguard from such problems.

Figure 7.30 Three Tower Cranes in Close Proximity Need Careful Position Planning

7.7.1 WHAT MAKES A CRANE A TOWER CRANE?

Without much guessing, we can say it's the (a) *slender lattice tower* supporting a (b) *slewing jib arrangement* with a (c) *cab for the operator high above* the "scenery" tugged into the corner made by the tower and the jib. The Danish crane manufacturing company Kroll built the largest tower crane, the K 10000, covering a circle area with a radius of 285 ft (86 m) and a lift of 132 tons (120 t) at its tip. Apparently, the climbing was so cumbersome that Kroll installed and delivered tower cabins with toilet, kitchen, and living room so that operators did not have to come down for several days.

However, the tower cranes have so many special features that their common denominators end here. Let's try to organize this large "family."

7.7.2 THE SELF-ERECTING BOTTOM SLEWERS

The bottom-slewing tower cranes can be considered cousins of the crawler and carrier crane using a fly-jib. All three have a main boom or mast mounted on a sturdy turntable together with counterweight, winches, gears, and motors. The platform itself rotates on vertical bearings stabilized by a base held down by additional ballast, anchors and/or outriggers. As Figure 7.22 shows, the crane base has been put on rails, truck carriers, or crawlers.

In the early 1940s, manufacturers of cranes realized that having all the motors and winches at the foot of a lightweight mast with top jib created the opportunity to add a feature that was extremely important for small, cash-strapped contractors in Europe. Right after World War II, the need for construction was big, and thanks to the U.S. Marshall Plan, money was made available. However, contractors did not have the capital for large equipment and site installations. Thus, the innovation of having a crane erect itself without the help of a second crane was both strange and economical. After the initial skepticism was diffused, the self-erecting crane that can be towed to the construction site has seen a steady growth. Affixed to a base and powered by electricity, it is able to unfold itself, ready for the bolts and locking mechanisms. Figure 7.31 presents the key stages of the basic concept, which that has not changed much over the years.

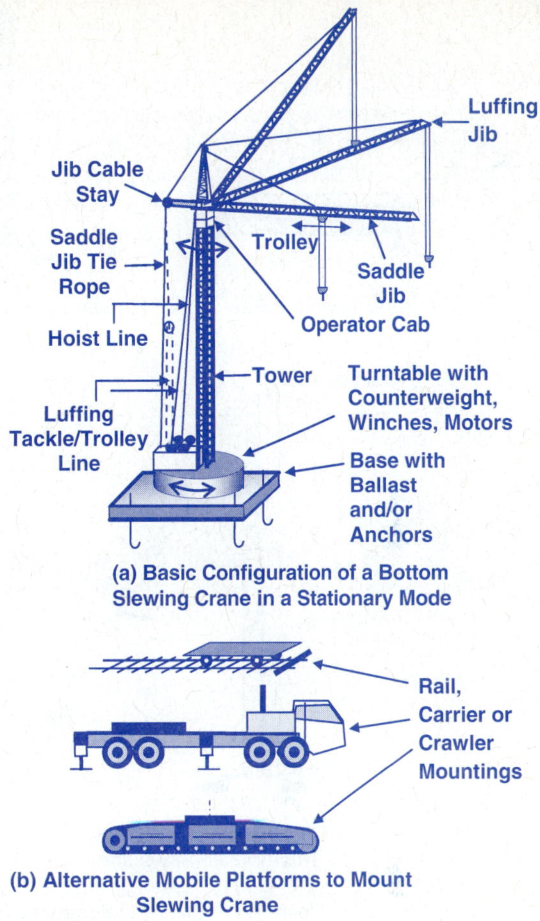

Figure 7.31 Bottom-Slewing Crane in Various Configurations

What make a bottom-slewing crane easily recognizable are the vertical cables along the back of the tower as, the motors and winches are located on the turntable at the foot of the mast. Besides the line that hoists a load, a luffing tackle services the jib, and another line actuates the trolley. The stabilizing moment against crane tipping is provided by the counterweight on the turntable and a base loaded down with more ballast or anchored. Figure 7.31b illustrates that the turntable has also been affixed to mobile bases such as a cart on rails or the ubiquitous truck carriers and crawlers. It is quite obvious that the latter two are able to take advantage of their capabilities to speedily self-erect, as discussed in the next section. If configured with a saddle jib, a trolley provides the means to change the distance between the hoist line and the tower mast. Depicted in Figure 7.32, having a luffing jib makes the trolley unnecessary, since the head of the jib changes the horizontal as well as the vertical coordinates of the sheave on the tip. Unlike in the mobile cranes, the operator is able to climb the tower to reach a lone cabin equipped with all the controls needed including radio communication (for crane control see Section 7.9). Since they are so remote and constrained when the line of sight is obstructed, modern cranes feature wireless remote control interfaces that can be carried around.

Figure 7.32 Top-Slewing Tower Crane with Luffing Jib

Self-erecting cranes have the ability to raise themselves from a trailer configuration into a tower-crane without additional assistance. As there are many "roads leading to Rome," there are many approaches to accomplishing this extraordinary feat. Figure 7.33 highlights the key steps of one of them—all starting with the compact crane components being towed onto the site (Figure 7.33a). After positioning the base onto the prepared footing (step 2) and hooking up the power, the bottom tower segment is turned upright (step 3).

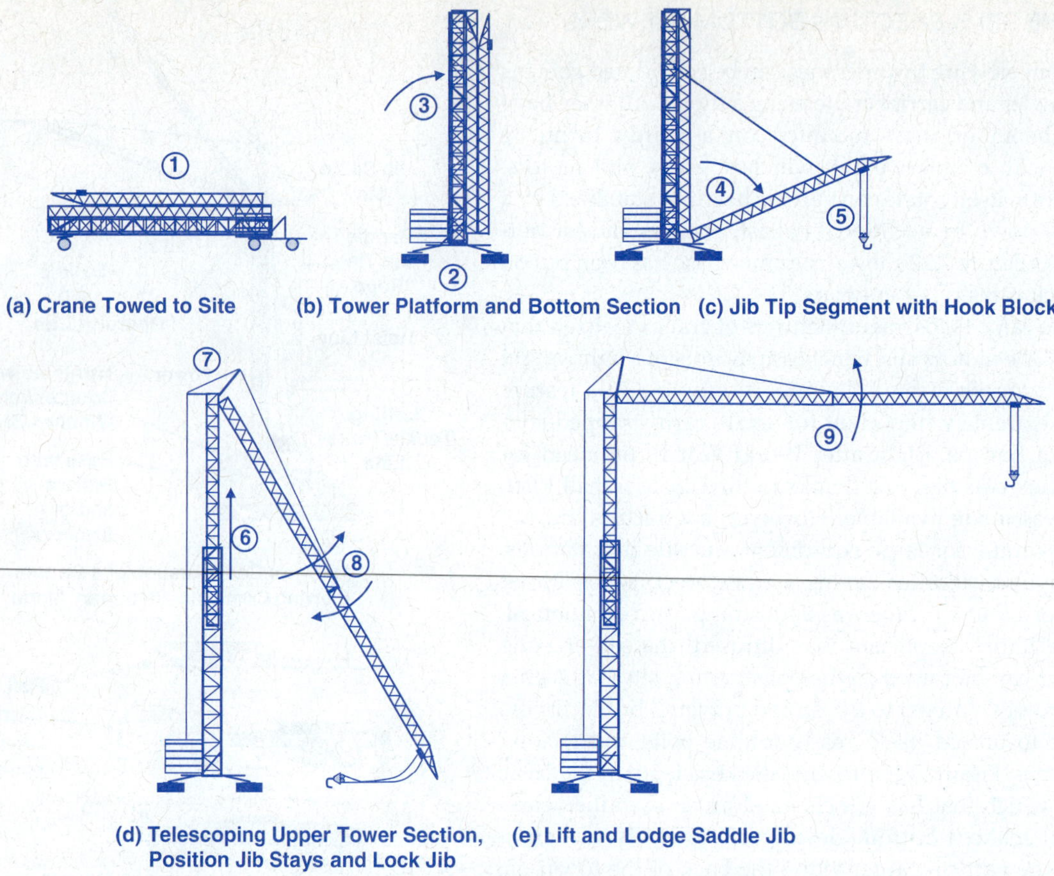

Figure 7.33 Example Sequence of a Self-Erecting Bottom-Slewing Saddle Crane

Figure 7.33b shows that with this action the other crane pieces are also put into a vertical orientation. After the outer jib section, hinged at the bottom, is lowered to the ground (step 4) and the hook block is attached (step 5), the upper tower segment is pushed upward in a telescopic mode (step 6). Now the cable stays/gantries can be swiveled into place (step 7), which provides the necessary leverages to pull the inner jib section, hinged at the head of the tower, outward to be aligned and locked with the outer section (step 8). Last, the jib line can be used to pull and rotate the saddle jib into its horizontal position (step 9).

While the crawler lattice crane was most popular in the United States and the luffing bottom slewers could be found on almost every construction site in Europe, the need for higher and higher buildings in tighter city areas spurred another innovative development: the tower crane with a stable mast.

7.7.3 THE CLIMBING TOP SLEWERS

Having a non-slewing tower provided a series of advantages that balanced out the need for a second crane to erect. In particular, those benefits were able to help meet the constraints of downtown construction of high buildings. They included:

1. Positioning of tower inside the building footprint where there was a continuing opening, such as an elevator shaft
2. Tying the tower to the rising building to add stability even at great heights.

3. Pushing the crane base up inside the concrete elevator shaft to minimize the need for tower section and the time for dismantling at the end.
4. Putting motors and winches on a counter-jib together with the counterweight to shorten the lines and increase the accuracy of synchronizing trolley and load lines.
5. Replacing the space-consuming and slow rail system with long saddle-jibs with fast-moving trolleys.

Figure 7.34 presents the many different developments that were made possible by keeping the tower fixed and putting the slewing ring on the top rather than at the bottom.

Very quickly, the fixed tower cranes became very popular, which allowed European manufacturers in the 1960s to expand their reach. Companies like Schwing, Peiner, Potain, and Liebherr continually pushed the lifting capacities and the reach of the jib to 3,100 lb

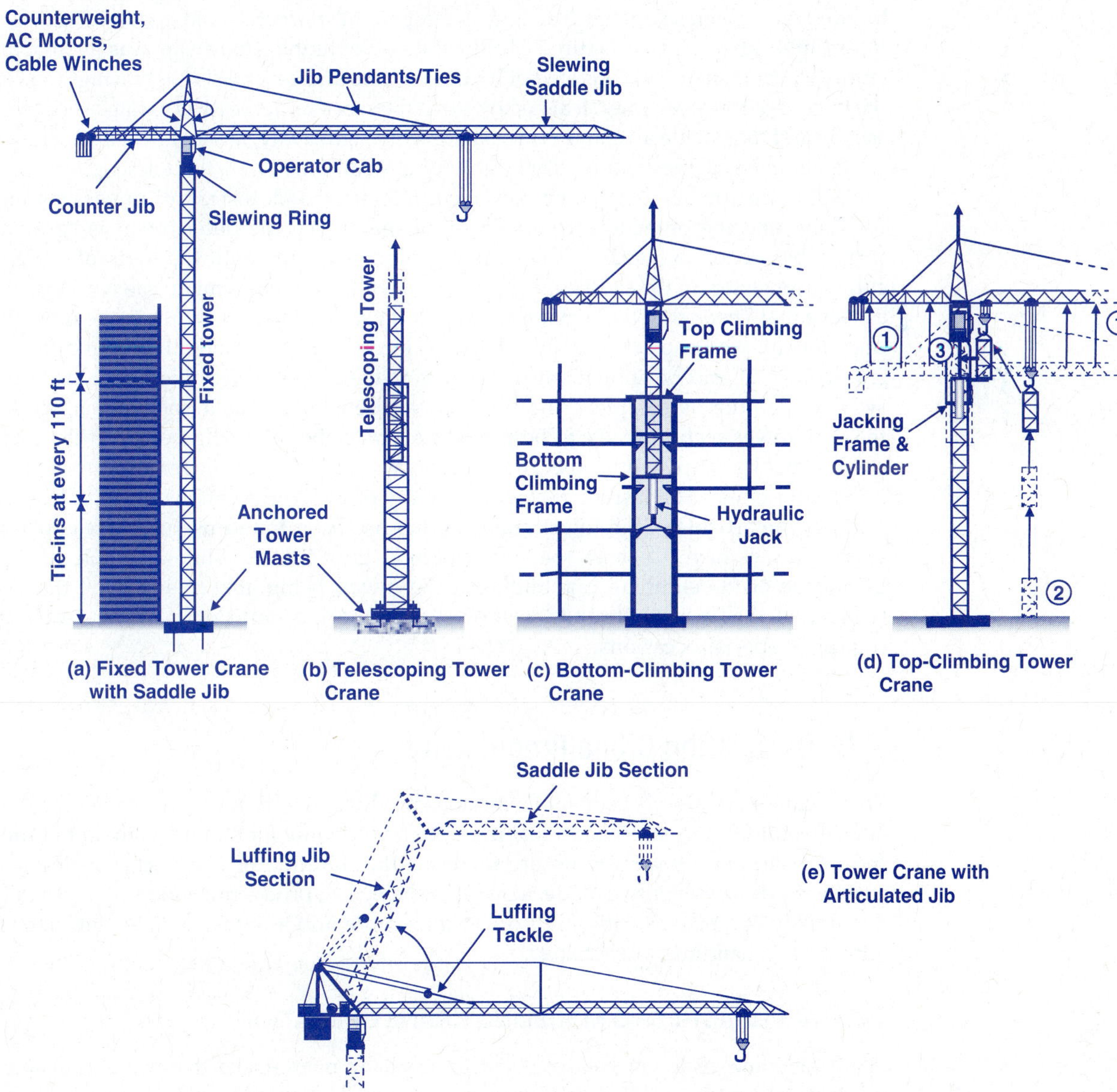

Figure 7.34 Technical "Outgrowths" of the Top-Slewing Tower Crane Concept

(13.7 kN) at 100 ft (30 m). Today, common fixed-tower cranes, such as the one shown in Figure 7.25a, are built to lift 6,000–7,000 lb (26.7 kN–31.1 kN) at a radius of 240–60 ft (73.2–79.3 m). While they can be freestanding with a hook-height of over 200 ft (61 m), tie-ins into the structure of the growing building allow the crane to become taller while always maintaining a maximum unsupported mast height. As shown, the slewing ring is located directly underneath the operator's cabin, allowing him to slew with the job. Other manufacturers put the ring on the very top of the mast but attach the cabin to the jib itself. Jib and jib support designs vary from crane to crane, some having one pendant and some two for either the main or the counter jib. Of course, the need to bump the saddle jib up as the building adds more levels required a new mechanism. Again, several competing concepts evolved and can be observed today. The design shown in Figure 7.34b applies the same telescoping approach as some of the self-erecting bottom slewers shown earlier. Very early in the development of the tower crane, manufacturers realized that they could save money for the contractor by "crawling" upward inside the building without having to add new tower segments that had to be bought or rented and required time for disassembly at the end. As Figure 7.34c illustrates, the climbing hardware was put into the sturdy elevator shaft and consisted of hydraulic jacks and several sets of bracings to keep the tower vertical at all times. It also shows that the space underneath is vacated section by section and thus made available for finishing work right away, another benefit of a tower crane climbing at the bottom. There is only a small step from the telescoping tower concept to the top-climbing crane portrayed in Figure 7.34d. Instead of pre-assembling the inner and the outer tower completely at the outset, the outer tower is "grown" section by section as needed. The climbing process begins with the hydraulic lifting cylinder pushing up the saddle jib, guided by a temporary climbing frame working like a sleeve that keeps the saddle connected with the tower (step 1). At the same time, the operator can pick one of the standard section elements, to be shared between different cranes (step 2), and position it onto a cantilevered arm on the climbing frame. After the jacking operation is complete, the cylinders are retracted, which opens the space for the new tower section ready to be pulled into place before being bolted to the main crane structure (step 3). If need be, the temporary frame can now be removed to be used on another crane. The last example of the many other configurations that can be seen today is the articulated jib tower crane, which has been found useful in areas where several cranes have to share the same space (Figure 7.34e). While also able to take advantage of the climbing capabilities of the tower, being able to articulate the two sections of the jib provides added flexibility to work around obstacles or reach up higher on special occasions.

7.8 Calculating Crane Productivity

Tower cranes are mainly used for continuous production within the area covered by the jib. Different from mobile cranes that most often work only for a limited amount of time from one location, tower cranes are there for the duration of a project, hoisting and moving loads in a repetitive mode. Referred as being in a production mode, lifts generally do not require a plan as critical lifts do. What is more important is the cycle time of a lift and the safe handling of each load.

7.8.1 APPROACHES TO MODELING CRANE OPERATION

A lift or crane cycle consists of a list of small or micro-tasks that are executed by production resources. Table 7.9 contains 26 small steps related to a crane cycle, which starts with a laborer identifying the approximate center of gravity of an object

Table 7.9 Three Task Taxonomies Describing a Craning Cycle

Micro-Tasks	Act	Combination Tasks	Process Tasks
1. Load container, etc., used in supply	L1	Preparing lift	A Preparing lift
2. Identify center of gravity of a new object	L1		
3. Hitch object to hook-block with slings	L1		
4. Tighten slings	L1, 0		
5. Move boom to create vertical load line	0	Maneuver out of pickup area	B Hoisting load to destination
6. Raise load off ground and stabilize	0		
7. Maneuver load out of pickup area	0		
8. Hoisting	0	Lifting to destination area	
9. Travel on tracks	0		
10. Slewing	0		
11. Luffing jib	0		
12. Shuttle trolley	0		
13. Slow down into destination area	0	Maneuver to destination point	
14. Wait for signal/space	0		C Wait
15. Lower load and align to position	0		D Maneuver, Unloading/unhitching
16. Holding load in place (steel)	L2, 0	Unloading	
17. Unload batch or unhitching slings	L2		
18. Maneuver hook out of area	0	Maneuver out	
19. Hoisting	0	Returning to pickup area	E Return to pickup area
20. Slewing	0		
21. Luffing jib	0		
22. Travel on tracks	0		
23. Shuttle trolley	0		
24. Slow down into pickup area	0	Maneuvering to pickup point	
25. Maneuver in pickup area	0		
26. Set down load or slings	0		
26. Unhitch load or slings	L1		

in order to hitch it appropriately to the crane hook. As we learned earlier, a sling has not only to fit the shape and weight of a load but provide a safe means for the crane operator to keep the center of gravity vertically beneath the hook-block. What follows are 25 additional small tasks that are acted on by the operator and one or two laborers, either at the beginning or end of a lift. As micro-task 13 indicates, a crane operation is sometimes linked to another process in such a way that its progress can delay the unloading or unhooking of a load.

Several associated micro-tasks are joined together into a separate category called combination tasks. A third category, labeled process tasks, groups micro-tasks on a still higher level for the purpose of modeling an entire construction process, as shown in the next section. The guiding principle for defining tasks is its intended use. Figure 7.35 presents how the smallest tasks are useful in modeling the various motion-control strategies of a crane operator. Three strategies to control the hoisting of the load are modeled using the micro-tasks as labeled according to Table 7.9. The first strategy, sequencing, relies on doing only one micro-task at a time, since a novice operator would probably start learning the functionality of each, because most of them are associated with a separate actuator. For example, Micro-Task 8, hoisting, requires the control of the hoist line winch while 10, slewing, is coupled with the slewing motor. Each actuator has its own speed, delays, and so on, which a new operator needs to get familiar with in order to lift the load safely. The partial multi-tasking might be the strategy of somebody who has worked with the crane for some time but who is extremely careful when maneuvering within the loading or the destination area. Such a strategy would also be chosen by an experienced operator when those two areas are so congested that collisions may

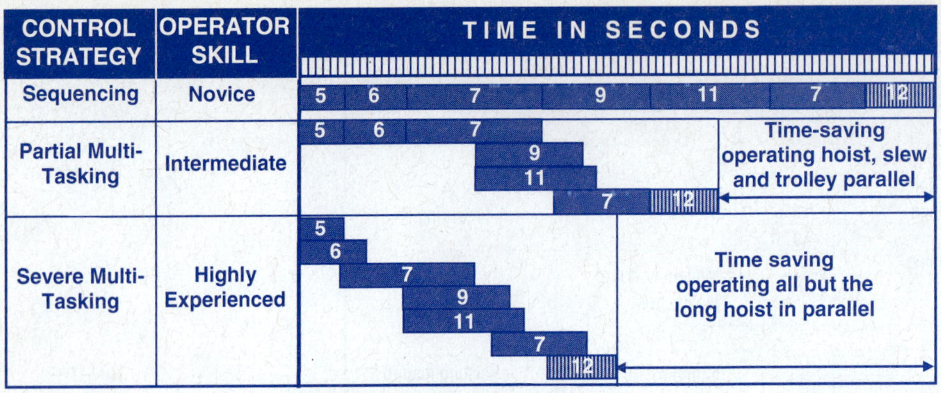

Figure 7.35 Modeling Control Strategies Using Micro-Tasks, as Seen on a Bar Chart

easily occur. In between, however, hoisting, slewing, and moving the trolley are done mostly in parallel, thus speeding up the time the entire operation takes. Lastly, an experienced operator who is able to work the crane in an uncongested area would most probably run several actuators in parallel, in a severe multi-tasking mode. As a result, more time can be saved, as shown in the chart.

One can easily see that hoisting can actually be modeled and studied in even greater detail when one considers the fact that winch motors commonly have 4 gears. In order to achieve smooth takeoff and setdown while traveling at high speed in between, an appropriate gearing up and gearing down will be crucial. Furthermore, the gearing may have to be adapted according to the weight of the load. Training a new operator the time-sequencing of gearing could conveniently be accomplished using an approach similar to Figure 7.35.

Modeling crane operation at different levels also highlights what affect the total duration of a crane cycle. Table 7.10 lists some of those factors and explains how they impact the production time.

S7.4 Pecco SK 180

Table 7.10 Operational Factors of Craning and Their Relationships

	Operational Factors	Effected Craning Maneuvers	Affected Process Tasks
1.	Mechanical advantage of lines	Duration of lifting and lowering	B: Hoisting load to destination
2.	Hoist winch speed		
3.	Lift height		
4.	Slewing motor speed	Time for slewing toward destination and pick location	B: Hoisting load to destination
5.	Slewing angle		
6.	Luffing winch speed	Luffing time to change pick-and-place radii	B: Hoisting load to destination
7.	Change in radius, no. pick-and-place		
8.	Trolley speed	Trolley travel to change pick-place radii	
9.	Congestion in pick-/unload areas	Time to maneuver within pick-destination space	A: Preparing lift D: Maneuver in unloading area
10.	Dependency on other tasks/labor	Waiting time in loading-unloading areas	C: Waiting for available space, guidance, or approval in destination area
11.	Simplicity of hitching/CoG load	Time for hitching and load alignment	A: Preparing lift of a new object
12.	Speed of loading and unloading	Time for loading and unloading when hooked (bucket)	A: Time needed to re-fill a bucket or hook a load D: Time needed to empty container or unhook load
13.	Operator skill	Minimized times with multi-tasking, efficient and smooth motion paths without corrections	B: Hoisting load to destination D: Maneuver, unloading/unhitching E: Return to pickup area

Worked-Out Problem 7.4 / Assessing Crane Productivity

Assume that you are placing concrete on the fourth floor of a multi-story building using a crane and bucket method. The tower crane that has been rented is a Pecco SK 180 with a 185-ft (56.4 m) saddle jib. The concrete is delivered by concrete trucks from a nearby batch plant and filled into a 1 yd^3 (0.76 m^3) bucket with clamshell gate weighing 750 lb (3.3 kN). The height of the 4th floor is 36 ft above the point where the buckets are filled.

The two views of the operation presented in Figure 7.36 provide an overview of the concrete operation at a stage where concrete pouring onto the formwork is progressing along the far edge, starting at one corner of the building. Two points on the concrete cycle, labeled Fill and Empty, mark the two key locations where one would be able to measure the hourly production of the operation. By counting the number of 1 yd^3 buckets filled, or alternatively emptied, in one hour, one is able to calculate how many yd^3 (m^3) the concrete crew was placing. The goal of this exercise, however, is to predict what the hourly production will be for the given situation.

Search for Missing Information: It is apparent that the performance of the crane will be crucial in predicting how many times the bucket can be cycled back and forth in an hour. From the specifications on the booksite we are able to create the graphs shown in Figure 7.37. However, before we are able to use the order-to-use load capacity and hoist speed function we need to establish the maximum load that has to be hoisted, the filled bucket weighing 750 lb empty:

Empty bucket	= 750 lb	= 3.3 kN
1 yd^3 of concrete weighing 150 lb/ft^3 (27 ft^3/yd^3)	= 4,050 lb (0.76 m^3 * 23.5 kN/m^3)	= 18.0 kN
Total weight of filled bucket	**= 4,800 lb**	**= 21.3 kN**

From the crane specifications we further learn that the average trolley speed is 120 ft/min (36.6 m/min) and the slewing/swing takes 1 minute for 360°.

Assumptions

a. The crane operator is good–excellent and does not require any correction factors.

b. Because the batch plant is close, there will always be a concrete truck ready to unload.

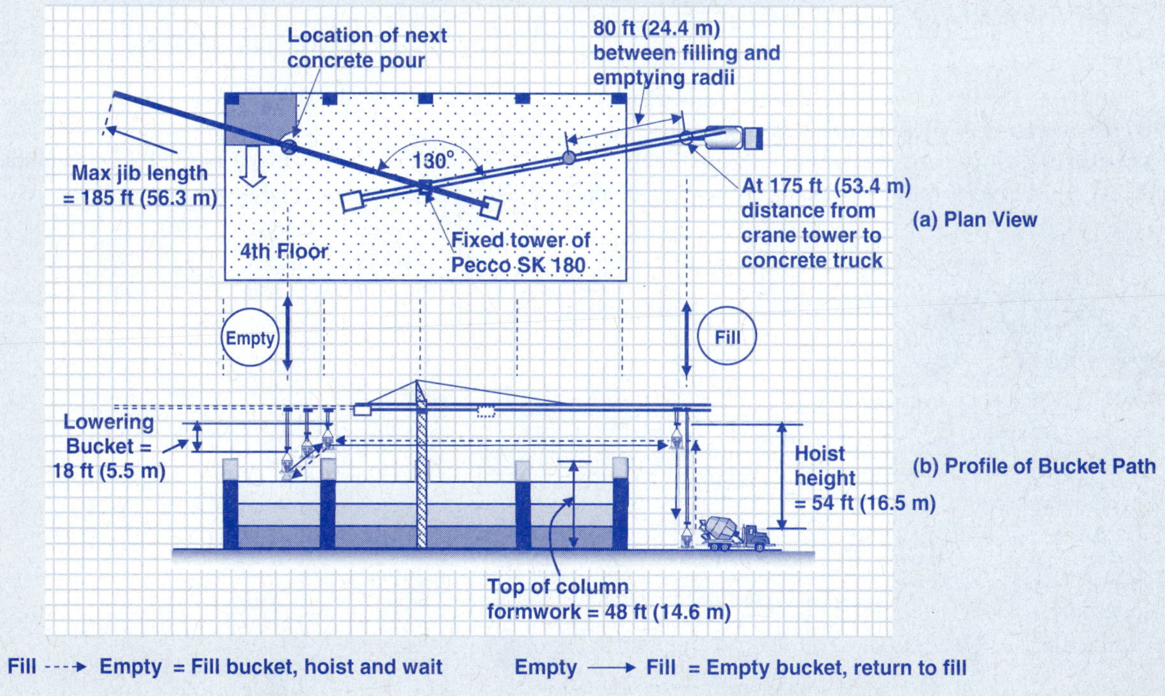

Figure 7.36 Situation of Concrete Placement Operation on Fourth Floor Using Crane and Bucket (not to scale)

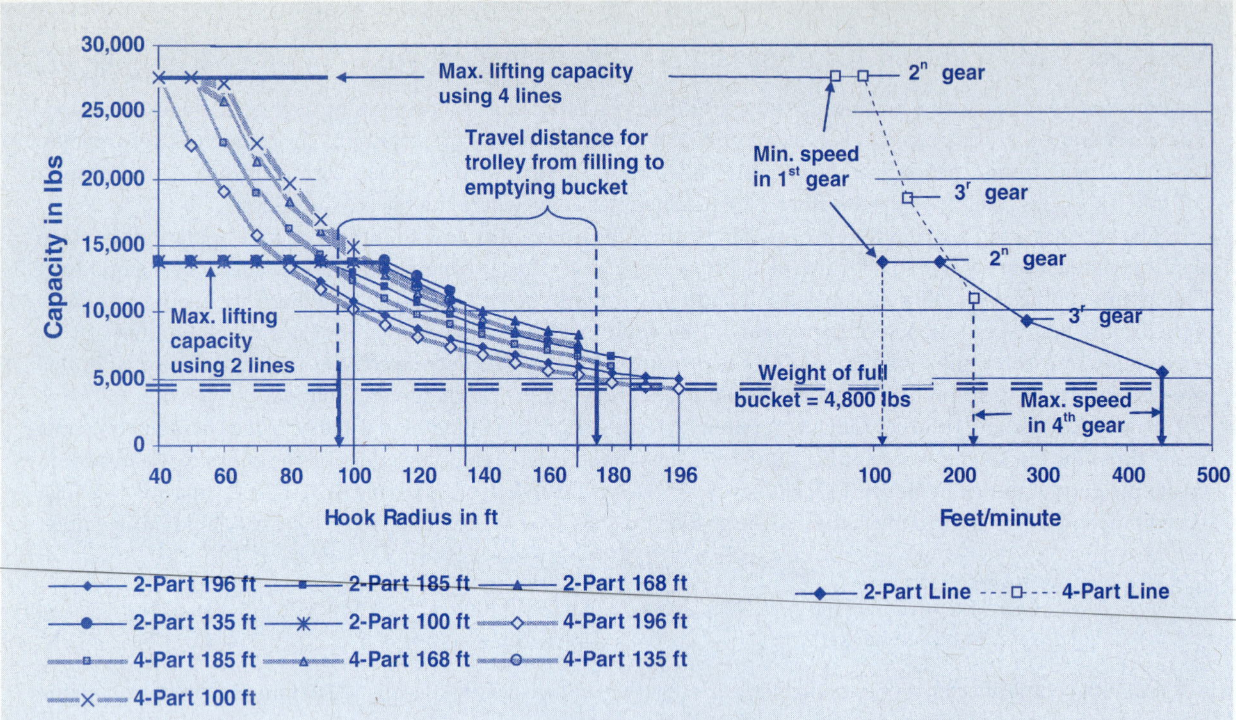

Figure 7.37 Hoist Capacity and Speed as a Function of Rope Parts, Jib Lengths, and Hook Radius

c. The concrete crew is experienced and works with well-maintained tools and equipment, so the risk of major delays due to equipment failure is extremely small.

d. Small personal times for water breaks, restroom, rests, etc. account to 10 minutes for every hour (this is also called a 50-min hour).

e. Spreading, vibrating, and finishing of the concrete are synchronized with the supply.

Modeling the Operation: The Input-Output process model is used to describe the concrete supply by bucket and crane. Figure 7.38 presents the 5 tasks, identified earlier, connected by arrows that indicate the cyclic "flow" of the bucket and crane between the Fill and the Empty areas. Filled or empty circles affixed to the bucket symbol identify if the bucket is full and empty or being filled or emptied. The cycle time of supplying one bucket of concrete

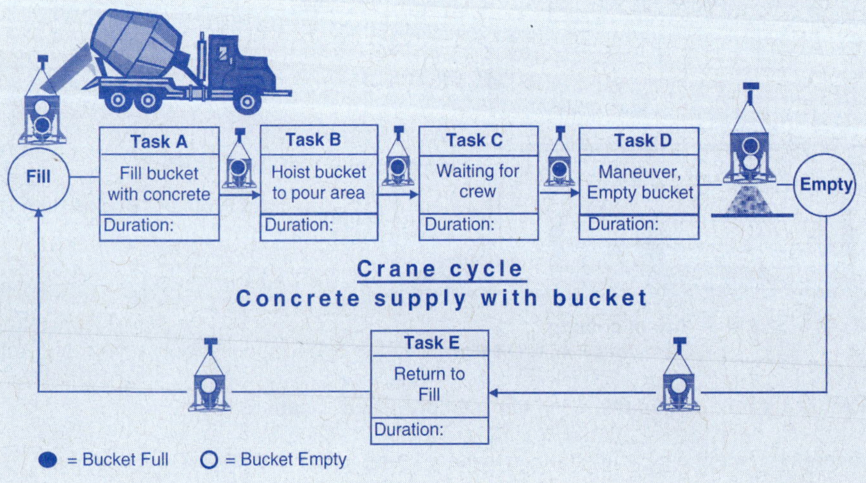

Figure 7.38 Process Model of the Cyclic Operation of Concrete Supply by Bucket

is the elapsed time between two consecutive bucket fillings. In other words, it is the time it takes the crane to sequence from Task A–Task E back to Task A or the summation of all task durations $D_A - D_E$. Therefore, hourly production of the concrete operation can be defined as

$$\text{Production} = \text{Concrete supply/hr} = (50/60) * \underbrace{(60\,\text{min}\,/\overbrace{(D_A + D_B + D_C + D_D + D_E)}^{\text{Total cycle time}})}_{\text{Number of cycles/hour}} * \text{ bucket capacity}$$

for D_A, D_B, D_C, D_D, D_E in minutes

Naturally, the duration of each task depends on the nature of the operation, the many physical conditions, and the constraints created by non-controllable factors. Table 7.11 provides a simple structure for considering the most important items and shows how to include their effects in establishing the cycle time and, with it, the production of the operation.

Calculation of Production: According to Table 7.9, the total time for the assumed conditions and operator efficiencies, which will be different for every construction site, is 285 seconds or 4.75 minutes. Using the above equation for hourly production, the 10-minute breaks per hour, and the 1 yd^3 bucket size, we calculate

$$\begin{aligned}\text{Expected concrete supply for job} &= (50/60) * (60\,\text{min}/4.75\,\text{min}) * 1\,\text{yd}^3 \\ &= 10.5\,\text{yd}^3/\text{hr}\ (8\,\text{m}^3/\text{hr})\end{aligned}$$

Table 7.11 Structured Approach to Calculating Cycle Times for Concrete Supply

Task	Impact Factors	Qualitative & Quantitative Unit Measures	Factor for Skill/Attentiveness Adjustment	Values for Concrete Operation	Fixed Dur. (sec)	Var. Dur. (sec)	Total Dur. (sec)	Task Dur. (sec)
	Bucket size, concrete	Cu-ft of concrete/ min.	% for dry/slow concrete	18 cu-ft/min * 100%	90		90	
A	type	from truck, delay due to		27 cu-ft bucket fixed	0			
	Preparedness to	driver						
	release concrete			0%				$D_A = 90$
	Obstacles, space to	Time for take-off prep.	% for operator skill	15 * 100%	15	-	15	
	man., slewing	1 min./360°,	-	130° one way	-	22	22	
	degrees, hoist height,	gear 1 110 ft/min.	-	gear 1,10 ft up	6	-	6	
B	luffing-trolley travel,	gear 4 440 ft/min,	-	gear 4,44 ft up	6	-	6	
	hoist down, multi-	200 ft/min trolley speed,	-	80 ft between radii	-	24	24	
	tasking	gear 2 180 ft/min,	-	gear 2 18 ft down	6	-	6	
		motions done parallel	reduced tot. time	trolley–slew in sequ.	−24	-	−24	$D_B = 55$
	Appropriate crew size	Crew is available for new	Added sec. per cycle for delays	The present crew is	0		0	
C	Equipment failures	concrete pour		matched, equip. in	00		00	
	Supervision/spotting	Crane bucket is guided		good condition				$D_C = 0$
	position bucket to	Positioning time,	% for unskilled operator	15 sec. to maneuver	15		15	
D	empty bucket size,	cu-ft of concrete/ min. flow	% for slow concrete	60 sec. to empty-	60		60	
	ease of operating	from bucket,	% for poor maintenance	move bucket				
	bucket gate, concrete	delay due to gate problem						$D_D = 75$
	flow,							
	Obstacles, people in	Time for takeoff	% for operator skill	10 * 100%	10		10	
	area hoist height,	gear 2 180 ft/min	-	gear 2 18 ft up	6		6	
	slewing degrees,	1 min./360°,	-	130° one way	-	22	22	
E	trolley travel, lower	200 ft/min trolley tr.	-	80 ft between radii	-	24	24	
	bucket, position	gear 4 440 ft/min	-	gear 4,44 ft down	6		6	
	bucket, multitasking	gear 1 110 ft/min	% for operator skill	gear 1,10 ft down	6		6	
		maneuvering bucket	reduced tot. time	15 * 100%	15		15	
		Amt. parallel motions		trolley–slew in sequ.	−24		−24	$D_E = 65$
							Total Cycle Time =	**285**

Discussion of Results

Here again, the result of our calculations is only as good as our assumptions. Will the operator or the truck driver really perform perfectly and reverse the concrete drum when needed? Or will there be breakdown of equipment, and if so, what is the effect? It is certain, however, there will never be a cycle that takes exactly 4.75 min. or that $D_B = 55$ seconds. The reason for this, as was discussed earlier, is the randomness of life or of human-controlled activities such as controlling a crane or driving a truck. For this reason, we should assess the effect of the randomness. An appropriate tool for studying its effect on this simple operation is a spreadsheet that provides some basic statistical functions such as random numbers and standard deviations. Figure 7.39 presents both the inputs and the outputs of two crane cycles which could be easily expanded to 10 or 100 cycles.

Figure 7.39a shows three columns that describe the normally distributed duration of each task as its mean, standard deviation, and the absolute minimum time one is able to complete the task, also referred to as the crash time. All three values have to be generated beforehand, employing sound statistical observations. At this point it is sufficient to understand that the mean is the arithmetic average of a series of observations (e.g., crane cycle times measured on site). The standard deviation, also calculated from the collected data, represents a measure of how widely the duration are dispersed from the average value (the mean). Although mean and standard deviations are based on real-world data, employing the NORMINV function can produce a time that is, although statistically correct, far below a crash duration. In order to eliminate such occurrences, an IF statement is used, as shown in Figure 7.39b, that replaces impossibly low values with the minimum duration listed in column E.

The results of the two cycles with randomly picked durations from the normal distributed time functions show, not surprisingly, two more cycle times, one with 265 seconds below and one with 309 seconds above the 285 seconds calculated using the mean durations.

Overall it can be said that, based on the assumptions, the average hourly supply of concrete can be expected to be 10.5 yd^3/hr (8 m^3/hr), but it could go as low as 9.8 yd^3/hr (7.5 m^3/hr)(cycle time = 305 *sec*) or as high as 11.3 yd^3/hr (8.6 m^3/hr) (cycle time = 265 sec).

		Column C	Column D	Column E	CYCLE 1			CYCLE 2		
Row Number	Task ID	Mean Duration (Sec)	Standard Deviation (Sec)	Minimum Duration (Sec)	Random Duration (Sec)	Expected Duration (Sec)	Task Finished (Sec)	Random Duration (Sec)	Expected Duration (Sec)	Task Finished (Sec)
6	A	90	15	85	92	92	92	95	95	404
7	B	55	25	50	64	64	156	41	50	454
8	C	0	45	0	3	3	159	-45	0	519
9	D	75	45	65	16	65	224	59	65	999
10	E	65	25	55	84	84	309	10	55	574
					Cycle Time 1 (sec) =		309	Cycle Time 2 (sec) =		265

(a) Spreadsheet Output for 2 Cycles

=NORMINV(RAND(),C6,D6)	=IF(G6<E6,E6,G6)	=H6
=NORMINV(RAND(),C7,D7)	=IF(G7<E7,E7,G7)	=I6+H7
=NORMINV(RAND(),C8,D8)	=IF(G8<E8,E8,G8)	=I7+H8
=NORMINV(RAND(),C9,D9)	=IF(G9<E9,E9,G9)	=I8+H9
=NORMINV(RAND(),C10,D10)	=IF(G10<E10,E10,G10)	=I9+H10
	Cycle Time	=I10

(b) Spreadsheet Window Showing Underlying Formulas

NORMINV(probability,mean,standard_dev) (EXCEL Spreadsheet Help)

Returns the inverse of the normal cumulative distribution for the specified mean and standard deviation
Probability is a probability corresponding to the normal distribution. **RAND()** returns an evenly distributed random number greater than or equal to 0 and less than 1. A new random number is returned every time the worksheet is calculated.
Mean is the arithmetic mean of the distribution.
Standard_dev is the standard deviation of the distribution. The standard deviation is a measure of how widely values are dispersed from the average value (the mean).

Figure 7.39 Spreadsheet Simulation of Concrete Supply with Normal Distributed Task Durations

7.9 Electronic Devices to Boost Safety and Productivity

The advances in electronic gadgets have also made big changes in the cranes of today. Instead of large levers that controlled the cable drums of cranes in the last century, the emergence of electric and hydraulic articulators paved the way for the use of wired or wireless control networks using miniature switches and relays. In fact, the first tools were simple mechanical devices such as a metallic arrow hanging from the side of the boom indicating the boom angle. While still in use, their simplicity did not meet the needs of complex cranes and they are now being slowly augmented with electronic, microprocessor-based systems. For example, load measuring devices not only sense the load on the hoist line but also the boom angle, the length of the boom, or the position of the trolley. From these values, an integrated microprocessor calculates the load-moment (load $*$ distance between load and tipping axis) and compares it against the electronic load chart provided by the manufacturer. Even video cameras are not uncommon now, as they can provide views to the operator that lets him work safer and more productively.

The spectrum of electronic devices can be organized in three application areas: (a) accident prevention, (b) crane "health" monitoring, and (c) operator control augmentation. Most of them, however, fulfill multiple functions. As we have seen, the causes of crane accidents are varied and are addressed by extensive OSHA rules, operator training/certifications, and lift planning efforts. However, as the statistics show, deadly accidents are hard to "kill." One of those recurring accidents is referred to as two-blocking.

7.9.1 THE ANTI–TWO BLOCK SYSTEM

The inherent cause for two-blocking is the existence of two blocks because of the mechanical advantage that can be gained from multiple lines. Figure 7.40 highlights the situation that leads to two-blocking.

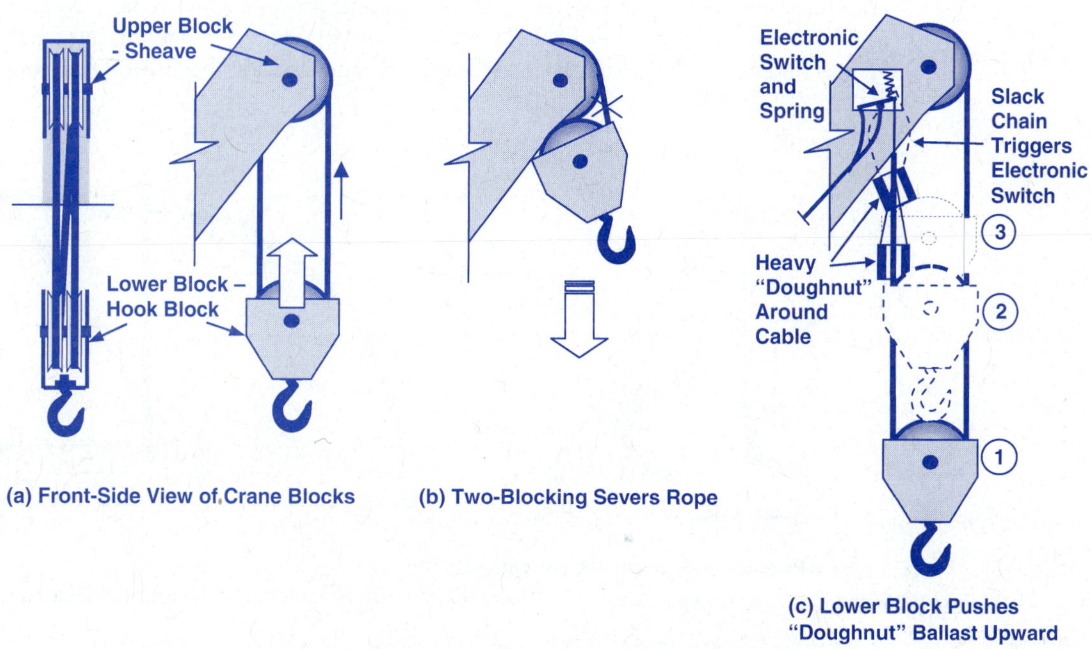

Figure 7.40 The Mechanics of Two-Blocking

Figure 7.40b illustrates the situation that leads to deaths due to the rupture of the hoist line after the operator does not realize that the two blocks are locking up. This situation may also occur in a situation where the telescopic boom is being extended without the simultaneous lengthening of the hoist line. The outcome of two-blocking is a ruptured wire rope whipping through the air, possibly slicing through the operator cabin. Simultaneously, the heavy hook block drops or is catapulted from the crane, hitting bystanders.

A system to warn and prevent such accidents is referred to as an anti-two-block device. Figure 7.40c depicts the key component of one technology that is being used. As the hook block gets close to the upper block (state 2), it hits a "doughnut" weight surrounding one of the lines. That ring ballast is hanging on chains connected to an electronic switch inside a control box mounted on the crane boom. As soon as the weight of the ballast is sitting on the lower block instead of pulling on the chains (state 3), a spring inside the control box pulls the switch away from the contact. The lack of contact stops the current inside the communication cable and will trigger the warning signals that have been installed, or even stop the hoist winch automatically. Newer models will use wireless communication instead of cables, thus reducing the danger of a faulty line.

7.9.2 POWERLINE DETECTION SYSTEM

Between 1984 and 1994, OSHA investigated 502 deaths in 479 incidents involving cranes in the construction industry. Electrocution was the largest category, with 198 deaths (39%) reported. The most frequent accident scenario involves a crane working close to a high voltage line that comes in direct contact with one of the wires and turns into a conductor. As Figure 7.41 illustrates, it is easy to forget about the existence of the dangerous wires when everybody's attention is directed at activities lower down. As soon as a metallic crane element touches the active wire, the crane boom and load line become conductors.

The U.S. Army Corps of Engineers prohibits storing materials under powerlines unless absolutely necessary. They even require that electric power or distribution lines be placed underground in areas where there is extensive use of equipment having the capability of encroachment on the clear distances specified, and ask for the use of non-conductive

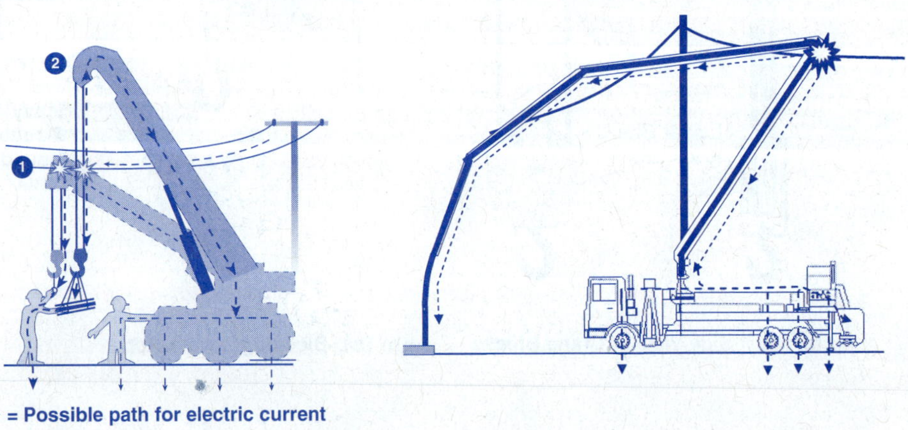

(a) Booming Up or Down in Vicinity of Power line

(b) Articulated Crane Boom Reaching Back

Figure 7.41 Examples of Craning Situations that Led to Fatal Accidents

taglines. ANSI standards define a 10 ft (3 m) cylindrical zone around powerlines that shall be kept clear of any parts of a crane. Since visual perception is easily "tricked" by differing light conditions, relying on such protection methods seems ineffective.

A more reliable, albeit not perfect, measure is an approach that uses the capabilities of sensors that detect the change in the magnetic/electric field caused by an electric current. The sensors' most critical feature is the fact that it senses current without touching the powerline. One such system is SigAlarm, which provides both audible and visual warning signals to alert the operator and attendant ground personnel when the boom comes within proximity of an energized high-voltage powerline. By attaching a longitudinal sensor to the entire boom, any part of the boom that enters the warning zone will set off the alarm. The attentive reader will realize that case 1 in Figure 7.41a represents a case that, SigAlarm cannot always cover, since the touching load line might be too far from the sensor on the boom.

7.9.3 INSTRUMENTS THAT MEASURE INCLINATIONS

OSHA has two rules that specify the need to measure angles:

- Part Title: **Safety and Health Regulations for Construction**
- Subpart Title: 1926 Subpart N:

Cranes, Derricks, Hoists, Elevators, and Conveyors

. . . 1926.550(g)(3)(i): **Operational criteria (for Crane and Derricks)**: The crane shall be uniformly level within 1% of level grade and located on firm footing. Cranes equipped with outriggers shall have them all fully deployed following manufacturer's specifications, insofar as applicable, when hoisting employees.

. . . 1926.550(g)(3)(ii): ***Instruments and Components***

. . . 1926.550(g)(3)(ii)(A): Cranes and derricks with variable angle booms shall be equipped with a boom angle indicator, readily visible to the operator. (OSHA)

The traditional approaches to fulfill these requirements used gravity as guidance. Figure 7.42 illustrates the two main concepts.

Figure 7.42a presents a simple angle-measuring device attached to the crane boom. The arrow rotates freely around a pin and points to the center of the earth at all times. As soon as the operator lifts the boom up, the circular scale rotates about the hinge-pin of the boom while the arrow still points in the original direction, vertically down. From his

(a) Boom Angle Measurement Using Free Rotating Arrow

(b) Carpenter Level Indicate Levelness

Figure 7.42 Traditional Technologies to Measure Angles on Cranes

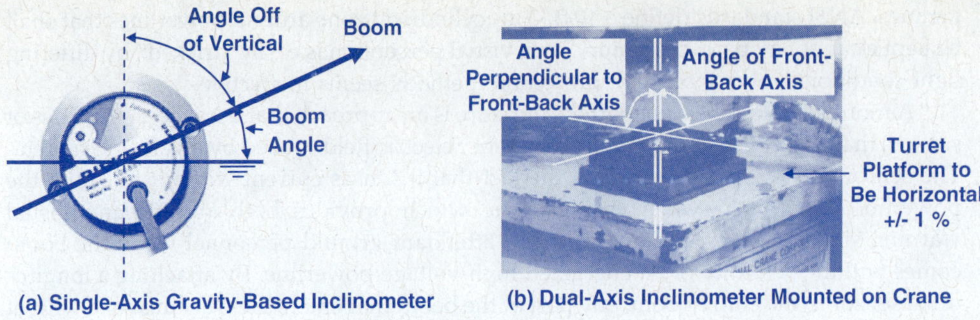

Figure 7.43 Electronic Inclinometers Provide Real-Time Data

vantage point, the operator is able to see both, the angle measuring device and the load chart which tells the maximum loads according to angles and boom extensions. Figure 7.42b shows a carpenter level to measure the levelness of the chassis. The vial features a slightly curved glass tube that is incompletely filled with ethanol. These levels are also known as a tilt meter, tilt indicator, slope meter, slope gauge, gradient meter, gradiometer, level gauge, level meter, or pitch & roll indicator.

The desire to calculate load-moments using a microprocessor in real time as well as having a device that controls the levelness at all times, as the soil underneath outriggers may move, led to the introduction of ruggedized inclinometers. An inclinometer is an instrument for measuring angles of slope and inclination of an object with respect to its gravity. It is also known as a tilt sensor.

These types of inclinometers are able to detect the change in the gravity pull when the center-line is off the vertical. They provide either an analog micro-voltage (mV) or micro-amperage (mA) output that is calibrated to degrees. Figure 7.43a depicts such a small device mounted on the side of a crane boom. As it moves away from a vertical orientation, the sensor will provide electronic data about the changes in real time. Some inclinometers, like the one shown in Figure 7.43b, are combined with single-axis inclinometers and thus have two outputs representing the tilting of a plane, such as a crane carrier platform. The dual axis in the example is mounted on the turret platform of a telescopic truck-crane measuring the levelness in two main axes, which can be used to calculate the levelness of the truck base before, during, and after an operation. If this is connected to a display or a warning device, the operator will be able to receive instantaneous information in his cab. These kinds of tilt sensors can be also found on manlifts, aerial work platforms, and boom lifts.

7.9.4 INTELLIGENT ELECTRONICS PREVENT CRANE TIPPING

It was discussed earlier that more than one situation can result in a catastrophic failure of the crane. Causes for overloading are mostly due to someone's improperly using the crane due to an error, lack of training, or negligence. The physical phenomenon that causes a crane to tip is an insufficient counter-moment against the total moment created by the load, the weight of the rigging, and the boom itself. The load charts attached to every crane should give the allowable load as a function of boom length and boom angle. Unfortunately there are several factors that make the load chart ineffective:

1. Crane carrier is not leveled.
2. Load line is not vertical.

3. Weight of load is not known exactly.
4. Load is stuck (in mud).
5. Weight of rigging is not considered as part of a load.
6. Counterweight is mounted wrong.

The key to preventing an overload condition is to either provide the operator with accurate and real-time information about the condition of the crane or to create a feedback system into the crane control that locks/overrides the control input from the operator if the condition reaches a dangerous point.

The electronic crane monitoring system to protect against overload has five main components: (a) electronic sensors, (b) crane configuration parameters, (c) monitoring software, (d) microprocessor, and (e) output(s). Similar to electronic inclinometers there exist devices that measure load, called load cells, or hydraulic pressure, distance, and so on, while some simply signal if a switch is on or off. One application of the latter is at the outriggers. If positioned properly, a switch is only activated if the outrigger is fully extended and locked. The output of hydraulic pressure sensors, one on each side of a cylinder with known dimensions, can be used to calculate the load it is experiencing. By reading the sensory data into a digital converter linked to a microprocessor that has access to information about the crane's characteristics, the actual load can be automatically calculated and compared with the allowable load for that situation. If the load is getting close to or enters the range of maximum capacity, not only can the system be programmed to display the results on the monitor in the cab, but different alarms can be started or the operator commands overridden to allow only actions that reduce the overload condition.

United States Patent 6496766, http://www.patents.com/us-6496766.html. This crane monitoring system and method is characterized by a plurality of sensors mounted to a crane, which communicate data to an on-board control unit. The control unit is characterized by its "black box" functionality. A tip switch is mounted in parallel relation to the vertical portion of a hoist cable below the boom tip. The control unit processes and stores input data from the sensors, which will indicate unsafe crane conditions, defined as alarm events such as dragging and extrication events. Each stored event has a time and date stamp. Upon detection of an alarm event, the control unit logs data from the sensors into a non-volatile memory, along with the data from a period of time prior to the alarm event. Data is stored continuously until the alarm event clears. Additional data is stored for a period of time after the conclusion of the alarm event. This results in a discrete event log residing on the non-volatile memory, which can subsequently be accessed by authorized personnel for analysis and identification of the crane operator who caused the alarm event.

7.9.5 ELECTRONIC NETWORKS PROTECT AGAINST TOWER CRANE COLLISIONS

By nature, cranes are only able to serve a circle area defined by the maximum radius that the boom can reach. On large construction sites, where one crane is not able to serve the entire footprint, multiple cranes are necessary. In order to avoid dead-zones, zones that can't be served, the circular areas they cover have to overlap, which of course creates the danger that the crane booms and load lines may interfere with each other. Furthermore, existing buildings or other pedestrian areas may possibly be declared as restricted zones where no load can be "flown" overhead. Figure 7.44 describes the problem graphically and shows the electronics controls that are necessary to create a safeguard against mid-air collisions.

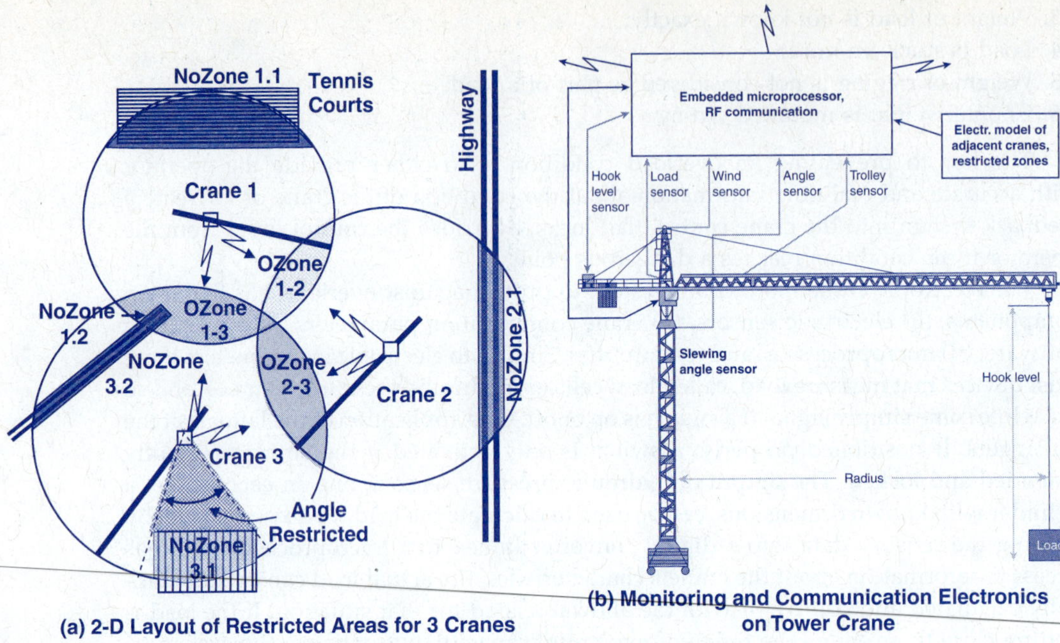

Figure 7.44 Concept of Spatial Interference Prevention for Tower Cranes

Figure 7.44a distinguishes four different zones. The white areas are free of any limitations to crane movements while the NoZones represent sectors where the crane is not allowed to swing a load, such as over tennis courts or a highway. OZones are regions where two or more crane jibs overlap, while the restricted angle is related to a zone describing its edges in degrees. Also indicated is the fact that modern systems rely on radio communications between the cranes to establish a network for letting the others know position and direction of present motion.

Figure 7.44b portrays the electronics used by each crane, which again depends on distance sensors but now also includes a load sensor, a wind sensor, and a slewing angle sensor for the jib. Although the latter also measures an angle, this time the rotation is horizontal, which makes the inclinometer inapplicable. This sensor counts the revolutions of a small wheel as the jib slews above the stable tower, thus providing the angular position of the jib relative to a starting position. While inside a restricted angle section, the sensor that measures the trolley position becomes important, since it has to stay outside the NoZone. As soon as the operator swings the jib close to either side of the restricted zone, the controllers have to verify that the trolley position will not violate the NoZone parameters if the jib keeps turning in the same direction.

The area where two or more cranes overlap requires a more sophisticated control algorithm, in that the decision making involves a second active "player," not just a static model. The situation in Figure 7.44a portrays crane 1 in the OZone 1-2 and crane 2 in OZone 2-3. The simplest rule that could be applied says that if one crane occupies the OZone, then no other crane is allowed to enter. A more sophisticated rule could stipulate: if two cranes are within the same OZone, crane speeds are automatically drastically reduced. This would allow both cranes to continue working but drastically reduce the possibility of a collision, because both operators are forewarned and the effect of inattentiveness mitigated. One can easily see that the key to implementing safety procedures for tower cranes are (a) accurate sensory data from position sensors, (b) a model of the site layout with the restricted zones, and (c) a controller observing that the programmed rules are being followed. Unlike the previous overload protection sensor, this system has to directly interfere with the operation of the crane, since a simple warning could be overlooked by the operator. Independent of which rule is chosen, the goal is always to slow down the slewing of the approaching crane(s) until it comes to a standstill before colliding with either the hook line or the boom.

7.10 Traits of the Master Crane Operator

It should not be surprising that an operator who has the smallest number of accidents also has the highest productivity. As a consequence, the causes that lead to accidents can be used to identify qualities of a master crane operator. Table 7.12 lists the main causes of accidents and characteristics of an excellent operator:

Table 7.12 Crane Fatal Events by Proximal Cause and Contributing Physical Factors (according to 125 cases analyzed in Beavers et al., 2006)

Proximal Cause	% of cases	Contributing Factors	Events
Struck by the load	32	Rigging failure	24
		Unbalanced load	3
		Load dropped	10
		Accelerated movement	1
		Equipment damage	5
Electrocution	27	Failure to maintain required clearance	34
		Boom contact	15
		Cable contact	12
		Headache ball/sling contact	5
		Jib contact	1
		Load contact	1
Crushed during assembly/disassembly	21	Improper assembly	3
		Improper disassembly—pin removal	10
		Improper boom support	6
Failure of boom/cable	12	Boom buckling	2
		Boom collapse	5
		Overload	6
		Equipment damage	5
		Incorrect assembly	3
		Cable snap	3
		Two blocking	1
Crane tip over	11	Overload	5
		Loss of center-of-gravity control	3
		Outrigger failure	2
		High winds	2
		Improper maintenance	1
Struck by cab/counterweight	3	Intentional turntable turning	3
		Bridge crane in motion	1
Falls	2	Missing hand rails	1
		Improper operation	1
		Improper maintenance	1

Table 7.13 Seven Routines of a Safe and Productive Crane Operator

		Factors that Trigger Accidents	Operator Activities Leading to Safe Productive Operation
Preparation	1.	**Lack of planning and supervision**	Plans critical lifts regularly and has knowledgeable supervisors
Preparation	2.	**Incorrect setup of cranes**	Knows and implements the rules to set up different crane types
Preparation	3.	**Failure to correctly calculate/estimate load**	Is capable of estimating the weight of objects and considering adverse effects of environmental conditions (rain, mud, etc.)
Preparation	4.	**Use of wrong rigging gears**	Understands how to use the different rigging methods and slings that are appropriate for lifting a load
Preparation	5.	**Faulty machine and devices**	Is capable of identifying when a crane is damaged, wire rope defective, slings worn out, or hooks ruined
Preparation	6.	**Lack of proper maintenance**	Is competent in performing a crane inspection and in pinpointing items/mechanisms that require maintenance, insists on regular maintenance
Operation	7.	**Lack of sufficient skill training**	Extensive hands-on training with the particular crane, certified by an accredited training agency such as the National Commission for the Certification of Crane Operators (NCCCO).

On the other hand, a cause-and-effect analysis of highly competent operators led to seven categories of safe activities. OSHA has an agreement with the National Commission for the Certification of Crane Operators (NCCCO) that its national certification program meets agency requirements for crane operator proficiency. The NCCCO certification program is accredited by the National Commission for Certifying Agencies and is formally recognized by the National Skills Standard Board. This program, or its equivalent, establishes an OSHA-approved standard for crane operator training and certification for both general industry and construction. An NCCCO-certified operator will carry a card that states the training he or she has received. Other qualification requirements have been set up by the ASME B30.5 section 5-3.1.2 of the American Society of Mechanical Engineers Safety Standards.

The legal framework for the certification and training of crane operators is provided by the Occupational Safety and Health Act from 1970 when **OSHA** became the main federal agency charged with the enforcement of safety and health legislation.

OSHA Regulation with Part Number 1926 addresses Safety and Health Regulations for Construction, while its Subpart N covers specifically Cranes, Derricks, Hoists, Elevators, and Conveyors and the section with the standard number: 1926.550 on Cranes and Derricks.

7.10.1 WRITTEN OPERATOR EXAMINATIONS

The Army Corps of Engineers has its own regulations that are valid for its own people as well as for contractors working for them. The main articles related to construction are published with the Publication Number: EM 385-1-1, containing 32 sections. Appendix G presents the procedures for the examination of crane operators and stipulate:

OSHA, through ANSI/ASME B30.5, requires crane operators to be qualified. Paragraph 5-3.1.2 of ANSI/ASME B30.5 requires that operators pass a written or oral examination and a practical operating examination unless able to furnish satisfactory evidence of qualifications and experience. Contractor crane and derrick operators are required to be designated as qualified operators by a source that qualifies crane and derrick operators. This includes independent testing and qualifying company or agencies; unions; governmental agencies; or qualified consultants.

The following types of cranes and derricks are treated as different machines:

(1) Mobile crane, lattice boom
(2) Mobile crane, telescopic boom crane
(3) Articulating boom (boom truck) crane
(4) Floating crane or floating derrick
(5) Gantry crane
(6) Portal or pillar crane
(7) Overhead crane
(8) Tower crane
(9) Derrick
(10) Monorail or underhung crane

The topics of the written or oral test that a crane operator working for USACE has to pass mirror closely the seven factors that trigger accidents. In particular, the operator will be tested about:

a. Responsibilities of operator, rigger, signalpersons, and lift supervisor
b. Knowledge of USACE crane safety requirements and the crane's operator manual
c. Ability to determine the crane configuration, compute the size and shape of loads, and determine the crane's capacity using the load chart
d. Use and limitations of crane operator aids
e. Inspection, testing, and maintenance requirements
f. Determination of ground conditions and outrigger and matting requirements
g. Crane setup, assembly, dismantling, and demobilization procedures
h. Requirements for clearance from power sources
i. Signaling and communication procedures
j. Factors that reduce rated capacity
k. Emergency control skills

7.10.2 PRACTICAL OPERATOR EXAMINATIONS

The practical test examines the operator's ability to perform the following:

a. Inspecting the crane
b. Establishing a stable foundation and leveling the crane
c. Raising, lowering, extending, retracting, and swinging the boom
d. Raising and lowering the load line
e. Attaching the load, holding the load, and moving the load
f. Reading load, boom angle, and other indicator devices

The hands-on test for operating a crane includes several tasks that test the motor skills of the operator handling a particular type of crane. Here is an example that describes how the test of negotiating a zigzag corridor with a suspended load is executed:

Optimum Time

Small Telescopic Crane —	four (4) minutes each direction
Lattice Boom Crane —	three (3) minutes each direction
Large Telescopic Crane —	three (3) minutes each direction

This task is divided into two tasks A and B. A requires you to go through the corridor in a forward direction. 4B requires you to go through the corridor in a reverse direction. (This is the only task in which the time allowed differs depending on the crane type you are operating.)

At the Examiner's direction to start (at which point timing will begin), lift the Test Weight into the air and swing, boom up or down, hoist up or down, as you judge necessary to guide the load through the zigzag corridor without touching the ground with the Test Weight, or raising the Test Weight so high that the chain leaves the ground, or touching or knocking over any part of the PVC barrier.

Points will be deducted for the following:

— Knocking ball off pole
— Moving pole base off line
— Knocking pole over
— Chain leaving ground
— Load touching ground

7.11 Signaling with Hand Signs and Voice—Danger of Miscommunications

In 2000, CalOSHA published the result of an analysis of Crane Accidents between in 1997–1999 (http://www.dir.ca.gov/dosh/CraneAccidentReport.html). It was found that the second-highest cause of crane accident, behind tipping and loads falling from the hook or sling, was miscommunication between the crane operator and a signal person. According to OSHA 1926.1419, a signal person is needed whenever the operator loses a direct view of the load, lacks the ability to judge the accurate movement of the load due to distance or light conditions, or when the crane operates close to powerlines or electrical equipment. In fact, if both the pickup and the point of placement are out of direct view of the operator, two signal persons need to be in place.

Today's large booms may even make the traditional hand-signal insufficient since the operator may not be able to distinguish between signals that are similar or not clearly executed. In such situations, phone and radio communications are used. Unfortunately, OSHA has not yet developed a standard for verbal signaling as it has for signaling by hand.

OSHA 1926.1420 further establishes rules for the use of radio, telephone, or other electronic transmission of signals, requiring the testing of any electronic device before the operations and the installation of a dedicated channel, especially when multiple cranes are involved. Not surprisingly, the operator needs to be equipped with a hands-free system.

OSHA 1926.1421 covers the use of voice signals. Not having developed a standard, OSHA states only that the operator, signal person, and lift director need to get together and agree on voice signals. Interestingly, a special requirement is made in that everybody is supposed to "effectively communicate in the language used." Still, the danger is that the people involved select verbal terms that can be easily misunderstood such as "Stop" and "Up." The American Society of Mechanical Engineers (ASME) recognized the need for standardization in the terms used when audio devices are used (ASME B30.5 2004).

Finally OSHA 1926.1430 makes the employer responsible for providing the training for the signal, competent, and qualified persons.

Definitions in OSHA 1926.1419

Competent person means one who is capable of identifying existing and predictable hazards in the surroundings or working conditions which are unsanitary, hazardous, or dangerous to employees, and who has authorization to take prompt corrective measures to eliminate them.

Qualified person means a person who, by possession of a recognized degree, certificate, or professional standing, or who by extensive knowledge, training and experience, successfully demonstrated the ability to solve/resolve problems relating to the subject matter, the work, or the project.

Dedicated channel means a line of communication assigned by the employer who controls the communication system to only one signal person and crane/derrick or to a coordinated group of cranes/derricks/signal person(s).

7.11.1 STANDARD HAND SIGNALS

The universal hand signals shown in Table 7.14 apply for conventional cranes, but not for some of the new cranes such as articulated cranes. In order to close those "holes," individual trades have established their own signals, causing major confusion for the operator who has never worked with a certain trade. Let us review some to the key hand signal from OSHA 126.1419.

Table 7.14 Universal Hand Signals

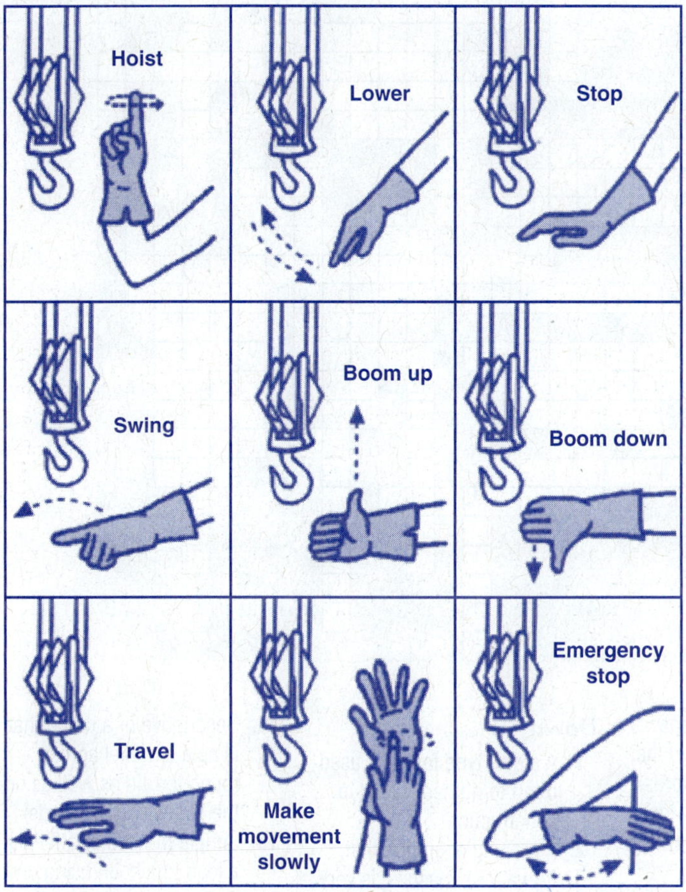

7.12 Chapter Seven Review

THE EQUIPUZZLE

The circled letters spell the name of a Roman engineer who published several books on the state of the art in civil engineering design and construction.

Across

2. A tracked crane, usually with a lattice boom, found especially on highway or road jobs (two words)
3. The configuration with which a rope is fed around drums and sheaves
4. An activity needed to secure a load to the crane hook
5. The smaller boom section of a tower crane holding counterweights and hoisting winches (two words)
6. Lasting deformation of wires and strands
9. Aerial conveying system for transporting single loads along a suspended track cable
10. A specialist who is competent and experienced in a subject area (two words)
11. Boom section adding reach to a crane
12. It is attached to the load and used to guide it into place (two words)
13. Attachment to a mobile crane that, when deployed properly, stabilizes the sub-structure
16. Formed metal piece protecting the eye of a wire rope
19. Rubber-tired highway truck providing the base for mounting a crane
26. A person able to identify and authorized to remedy dangerous situations on a construction site (two words)
29. Center section of a wire rope that provides flexibility and base support (two words)
30. The segment of the rope between the winch and the fastest sheave (two words)
31. A crane on wheels having the support structure of a sawhorse (two words)
32. The element of the rigging connecting the load to the hook
33. Crane mounted on a mobile carrier. Commonly with telescopic boom (two words)
34. A vertical drum with horizontal levers providing a means to lift a heavy load
36. Section of a crane boom that contains the upper sheaves

Down

1. A clevis type fastener used often to rig heavy load in construction
2. A craning operation that, for various reasons, is very risky (two words)
7. Element of a rope consisting of wound wires
8. Mechanism consisting of several parts to pull in a rope
12. Descriptive term for truck crane boom that reminds us of an invention that allowed Galileo to write the book Sidereus Nuncius
14. Descriptive of a crane that is able to travel across rough terrain as well as on the highway (two words)
15. Simple method of attaching a load to a hook using one or several slings (two words)
17. When you let the crane freely rotate in the wind you . . . . the crane
18. Created by a wire rope around a drum
20. That segment of the rigging that runs from the hook to the winch drum
21. Invented by Archimedes, this system is used by basically every crane today (three words)

22. A steel piece added to hook to provide a base tension in the line (two words)

23. A crane with a long history. It was critical in the successful construction of the Empire State Building in New York

24. This apparatus is located at the boom head and prevents a lethal accident caused by the upper and lower block (three words)

25. Value indicating the inclination of a boom (two words)

27. A load limit defined by the manufacturer of a crane, hoist, etc. (two words)

28. Loading condition for which no crane is ever certified (two words)

35. A rotating device guiding a wire rope

JOURNALING QUESTIONS

1. Crane technology has a long history and has influenced many historic developments and achievements. Besides tower construction by the Romans, what do you consider to be the three main contributions of this technology over time?

2. Other technological advancements had major impacts on the use and usefulness of cranes. One example is the concept of mechanical advantage discovered by Archimedes. Discuss the other examples that contributed heavily to today's cranes.

3. Wire ropes and bridge cables have much in common. What are some of the differences?

4. Excessive wind can have catastrophic consequences. Discuss the mechanics of why winds cause cranes to collapse. What really happens before and during the accidents?

5. During the practical test for all-terrain crane operators, the chain hanging from the hook is not allowed to leave the ground during the zigzag corridor test. Explain why this might be a test of the operator's motor skill.

6. Is the load capacity of a tower crane different for a 2-part and a 4-part line? Present a logical argument to support your answer.

BASIC CONCEPTS AND DEFINITIONS

Provide brief definitions or descriptions for the following questions.

1. OSHA standards 1926.550 in the Code of Federal Regulations 29 (CFR 29) is important for anybody who operates a crane in construction. Why is that? Which issue, in your opinion, is most important?

2. What are advantages and disadvantages of the six different types of slings?

3. Wire ropes are made of wires and strands laid in a helical manner. What are the differences of left lay and right lay ropes? Which one is more common in construction?

4. Non-running sheaves don't move much and are used to equalize the tension in opposite parts of a rope. Can you list some applications that you have seen?

5. Describe the parts and their functions of a trolley used by a tower crane. What is needed to keep the hook horizontal when the trolley is moving?

6. Describe and sketch the danger zones of a crane working close to powerlines? What do you think of the effectiveness of the 10-ft rule?

7. How does an anti-two-blocking device inhibit an accident from happening? How do people get killed if a crane does not have one installed?

8. What are the skills, responsibilities, and authorities of a competent person for cranes?

9. What components of a crane should a qualified crane operator inspect before starting the workday? What specifically should he look for?

LESSONS FROM THE ACCIDENT FILE

1. Accident #1: Working on barges that are not secured by piles will reduce the crane's lifting capacity. Sketch the situation when tipping started. Explain the reason for tipping although the load was still within the crane capacity. (A small experiment might be helpful here. Make a small model with two pieces of wood and a clothes hanger. Hang something from the hook the almost makes the crane tip. After that, repeat the experiment in the kitchen sink

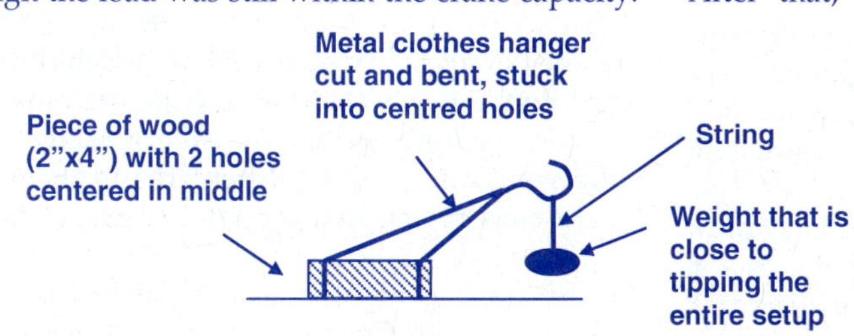

Step 1

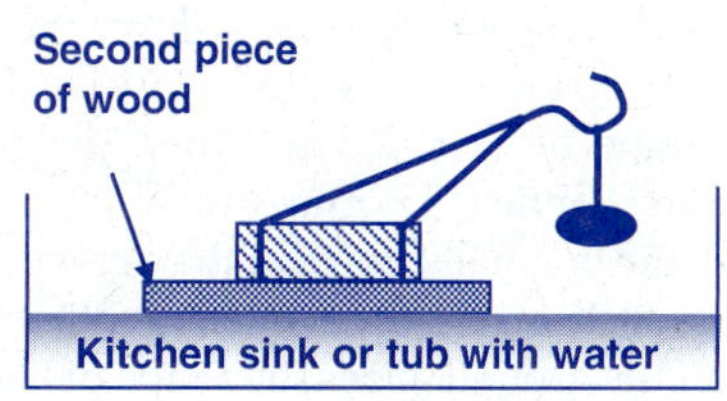

Step 2

half-filled with water and with a larger wood piece representing a barge. What are some rules for operating cranes on barges?)

2. Accident #2: Sketch the situation that led to the laborer being electrocuted. What could the laborer have done to protect himself from receiving the deadly electric shock? What OSHA rules were violated? How would having a critical lift plan have avoided the accident?

3. Accident #3: Sketch how the pipes were rigged to the hook. Draw what would be a "competent" method to rig the pipes. What simple recommended gear would have saved the foreman's life? Draw how it should have been used.

4. Accident #4: What happened that severed the hoist line? Sketch the apparatus that would have prevented this accident?

5. What is the possible reason why the scissor lift fell over, even though the foreman had tested it first? Why did the scissor lift not use the outriggers? Why would outriggers possibly have prevented the accident?

CALCULATION PROBLEMS

1. The Romans and people all through the Middle Ages used the treadwheel to power the hoists of cranes used at harbors and in construction. Consider the following arrangement of an A-frame crane designed to build a bridge abutment.

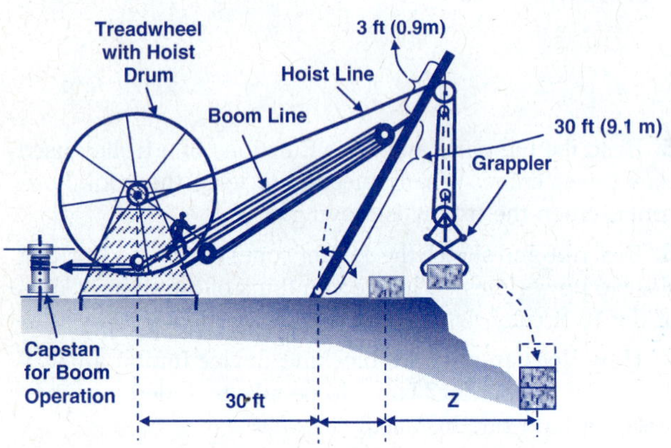

A piece of granite weighs about 4,000 lbs (1.8 t). The person operating the treadmill steps onto the wheel at a distance of $^1/_2$ radius from the wheel center and weighs 166 lbs (75.3 kg). Assume the diameter of the hoist line drum to be 2 feet. The minimum distance for X = 9 ft (2.7 m). Friction force in each pulley = 15%.

(a) What is the minimum radius of the treadwheel to lift a granite block if no lower block is used (i.e., just a single load line)?

(b) What are the minimum treadwheel radii for 2-, 3-, and 4-part lines?

(c) Assuming that the treadwheel is also used to place the blocks, how long will it take to lower one granite piece 20 ft (6.1 m) using a 4-part block and tackle, a treadwheel with a radius of 24 ft (7.3 m) when the person walks at about 25 min/mile (15.6 min/km)?

(d) Draw the boom line force at the capstan as a function of Z if the pulley at the capstan redirecting the line is 4 ft (1.2 m) off the ground. Does the shape of the function make sense?

(e) What happens with the force at the capstan if the treadwheel (with the last boom line pulley) is moved closer to the A-frame?

2. As the concrete placement operation proceeds, the crane has to turn all the way to 180°, at which time the operator could slew in the opposite direction. Let's assume that this is impossible until the electricians are finished installing the electrical conduits. This means that the crane will have to keep turning in the same direction until it reaches 270°, at which point it can change direction and slew only 90° or less.

(a) Draw a top view of the situation as the concrete operation continues around the crane to 270° and continue to 360°.

(b) Draw a bar chart of process tasks for a slewing angle of 270° and 120 ft (36.6 m) between filling and emptying radii, assuming that an experienced operator is on the controls.

(c) What is the estimated hourly production at 270° if all standard deviation values stay the same and the minimum duration will be 90% of the mean task durations?

(d) Draw the hourly production as a function of the slewing angle as the crane slews from the 130° in the example to 270° and from 90° to 0°. Why does the graph make sense to you? What is the maximum hourly production that you *lose* because the electricians did not finish their job yesterday?

3. The book site contains pdf files for the 50-ton All Terrain crane from Link-Belt RTC-8050.

(a) It is sold with either a standard ¾-in (2 cm) RB type or a ¾-in ZB type rope. What wire construction does each use, and what are their lifting capacities? Discuss the difference.

(b) The rated lift capacities differ when the crane operates over front between the tracks. What are possible reasons for this? Assuming that the sole reasons are that the outriggers, calculate the differences in their positioning (ft/m) considering that the rated loads are 85% the tipping load.

(c) The crane is allowed to lift without having the outriggers installed. Compare the rated loads and calculate

(1) the effect of eliminating support from the outriggers, (2) reducing the rated load to 75% of tipping load.

(d) Assume that you need to put a piece of equipment onto the roof of a 90-ft (27. m) high building. The closest that you can get to the building is 70 ft and the rigging with the load will be at least 20 ft (21.3 m) high.

What are the tradeoffs that you have to discuss with the client that exist between the weight of the load and the distance from the edge of the roof that you can place the equipment? What is the maximum weight of a load that you can lift onto the roof, and what is the farthest you can reach in (possibly with less weight)?

4. Section 3.2.3 presented force diagrams for five prominent crane types. Establish equations for each that need to be fulfilled in order for the crane boom to be stable (i.e., in equilibrium).

OPEN-ENDED PROBLEMS

1. SigAlarm is an electronic system mounted on the boom of a crane in order to prevent the crane from accidentally touching an overhead powerline. What are the basic components of this system, and what physical principle(s) are employed to make it work? Why might the alarm not work in cases where the load line, instead of the crane boom, touches the powerline? Design a system that would detect when any part of the crane breaks through a safety zone around the powerline. (Hint: You might consider using a stand-alone laser-based approach!)

2. The Space Station utilizes several articulated crane booms to move cargo in the weightlessness of space. The requirements of the job demand that the arm be extremely flexible while allowing precise movements of the load during construction. Most important, it has to be fail-safe or at least redundant in case something breaks.

(a) Does the weightlessness create any special limitations and/or opportunities to designing a crane boom?

(b) What motors and gears are used to move the arm?

(c) What mechanisms are used to grab, hold, and guide the load?

(d) How does the operator control the boom (hand-controls and feedbacks)?

(e) What are limitations in what space station crane booms can do? What are their causes?

3. United States Patent 6496766 has a "black box" functionality while also detecting dangerous dragging and extrication (shock-loading) operations. Show in a graph why dragging a load and extrication are unsafe (pick the right dimensions for the axis). How does Patent 6496766 prevent dragging while allowing the load line to swing freely when no load is attached? How does it detect that the crane is being shock-loaded?

BIBLIOGRAPHY

Al-Hussein, M., S. Alkass, and O. Moselhim. "Optimization Algorithm for Selection and On Site Location of Mobile Cranes" *J. Constr. Engrg. and Mgmt* 131, no. 5 (2005).

Ali, M. S., N. R. Babu, and K. Varghese. "Collision Free Path Planning of Cooperative Crane Manipulators Using Genetic Algorithm." *J. Comp. Civil Engrg.* 19, no. 2 (2005).

ASCE Manual Nr. 93. *Crane Safety on Construction Sites.* Task Committee on Crane Safety on Construction Sites. ASCE, Reston VA, 1998.

Beavers, J. E., J. R. Moore, R. Rinehart, and W. R. Schriver. "Crane-Related Fatalities in the Construction Industry. *J. Constr. Engrg. and Mgmt.* 132, no. 9 (2006).

Bernold, L.E. "A Cable Based Lunar Transportation System." *J. Aero. Engrg* 7, no. 1 (1994).

Bernold, L.E, "Spatial Integration in Construction." *J. Constr. Engrg. and Mgmt.* 128, no. 5 (2002).

Bernold, L.E., S.J. Lorenc, and E. Luces. "Intelligent Technology for Truck Crane Accident Prevention." *J. Constr. Engrg. and Mgmt.* 123, no. 3 (1997).

Bernold, L.E., S.J. Lorenc, and E. Luces. "On-line Assistance for Crane Operators." *J. Comp. in Civil Eng.* 11, no. 4 (1997).

Frolov, V.A., G.P. Arumov, V.P. Bogomolov, O.P. Klishev, R.A. Myllyla, and I.M. Sidorov. "A Self-Guided Platform for an Automation of Construction and Assembly Work Involving the Use of Helicopter." *Proc. Robotics 2000.* February 27–March 2, Albuquerque, NM.

Hasan, S., M. Al-Hussein, U. H. Hermann, and H. Safouhi. "Integrated Module for Mobile Crane Dynamic Instability Analysis and Supporting System Design." *Proc. Constr. Res. Congress.* April 5–7, 2009. Seattle, Washington.

Ju, F., and C.J. Sang. "Dynamic Analysis of Tower Cranes." *J. Engrg. Mechanics* 131, no. 1 (2005).

Kang, S.C., and D. Miranda. "Computational Methods for Coordinating Multipole Construction Cranes", *J. Comp. Civil Engrg.* 22, no. 4 (2008).

Lee, J., and L.E. Bernold. "Ubiquitous Agent-Based Communication in Construction." *J. Comp. Civ. Engrg.* 22, no. 1 (2008).

Liu, J., L. Ma, and Y. Yan. "Coordinate FEM Formulation for Boom of Crawler Cranes Dynamic Analysis." *Proc. Inter. Conf. Transp. Engrg.*, July 25-27 2009. Southwest Jiantong University, Chengdu, China.

Mahalingam, A., H.S. Nair, and K. Varghese. "A Computer-Aided Heavy Lift Planning Model." *Proc. Int. Conf. Comp. Civil Building Engrg.*, August 14-16, 2000. Stanford University, CA.

McCarthy, P., and F. Vazifdar. "Securing Cranes for Storm Wind: Uncertainties and Recommendations," *Proc. Ports 2004*, May 23–26, 2004, Houston, TX.

Peraza, D.B., and J.A. Travis. "Crane Safety—An Industry in Flux." *Proc. Fifth Cong. Forensic Engrg.*, November 11–14, 2009. Washington, DC.

Richter, T., S.J. Lorenc, and L.E. Bernold. "Cable Based Robotic Work Platform for Construction." *15th Int. Sympos. Auto. Rob. Constr.* Munich, Germany. March 31–April 1, 1998.

Russo, C.J., and E.A. Puchek. "Anatomy of a Tower Crane Collapse—An Opportunity to Learn." *Proc. Fifth Cong. Forensic Engrg.* November 11–14, 2009. Washington, DC.

Shapira, A., and B. Lyachin. "Identification and Analysis of Factors Affecting Safety on Construction Sites with Tower Cranes." *J. Constr. Engrg. and Mgmt.* 135, no. 1 (2009).

Shapira, A., Y. Rosenfeld, and I. Mizrahi. "Vision System for Tower Cranes." *J. Constr. Engrg. and Mgmt.* 134, no. 5 (2008).

Shapira, A., and M. Simcha. "AHP-Based Weighting of Factors Affecting Safety on Construction Sites with Tower Cranes." *J. Constr. Engrg. and Mgmt.* 135, no. 4 (2009).

Tam, C. M., W. T. Leung, and D. K. Liu. "Nonlinear Models for Predicting Hoisting Time of Tower Cranes." *J. Comp. Civil Engrg.* 16, no. 1 (2002).

Tam, C. M., T.K. Tong, and W.K. Chan. "Genetic Algorithm for Optimizing Supply Locations around Tower Crane." *J. Constr. Engrg. and Mgmt.* 127, no. 4 (2001).

Tamate, S., N. Suemasa, and T. Katada. "Analyses of Instability in Mobile Cranes due to Ground Penetration by Outriggers." *J. Constr. Engrg. and Mgmt.* 131, no. 6 (2005).

Tantisevi, K., and B. Akinci. "Simulation-Based Identification of Possible Locations for Mobile Cranes on Construction Sites." *J. Comp. Civil Engrg.* 22, no. 1 (2008).

Zhang, C., and A. Hammad. Towards Supporting Construction Equipment Operation Using Collaborative Agent-Based Systems." *Proc. Inter. Workshop Comp. Civil Engrg.*, July 24–27, 2007. Pittsburgh, PA.

OSHA Regulations (Occupational Safety and Health Administration), U.S. Department of Labor http://www.osha.gov/pls/oshaweb/owasrch.search_form?p_doc_type=STANDARDS&p_toc_level=0/.

29 CFR (Code of Federal Regulations) Standards,

1910, General Industry

- 1910.27, "Fixed Ladders"
- 1910.178, "Powered Industrial Trucks"
- 1910.179, "Overhead and Gantry Cranes"
- 1910.180, "Crawler Locomotive and Truck Cranes"
- 1910.184, "Slings"

1926, Construction

- 1926.106, "Working Over or Near Water"
- 1926.251, "Rigging Equipment for Material Handling"
- 1926.550, "Cranes and Derricks"

ANSI Specifications (American National Standards Institute), http://www.ansi.org/

- 10.28, Safety Requirements for Work Platforms Suspended from Cranes or Derricks for Construction and Demolition Operations.

ASME Specifications (American Society of Mechanical Engineers), http://www.asme.org/

- B22, Articulating Boom Cranes
- B56.10, Manually Propelled High Lift Industrial Trucks
- NOG-1, Rules for Construction of Overhead and Gantry Cranes (Top Running Bridge, Multiple Girder)
- NUM-1, Rules for Construction of Cranes, Monorails, and Hoist (with bridge or trolley or hoist of the underhung type)
- B30.2, Overhead and Gantry Cranes Top Running Bridge, Single or Multiple Girder, Top Running Trolley Hoist
- B30.5, Mobile and Locomotive Cranes
- B30.9, Slings
- B30.10, Hooks
- B30.11, Monorails and Underhung Cranes
- B30.16, Overhead Hoists (Underhung)
- B30.17, Overhead and Gantry Cranes (Top Running Bridge, Single Girder, Underhung Hoist)
- B30.20, Below-the-Hook Lifting Devices
- B30.21, Manually Lever Operated Hoists
- B30.23, Personnel Lifting Systems
- B56.1, Safety Standard for Low Lift and High Lift Trucks
  - 7.5, "Nameplates and Markings"
  - 7.35, "Platforms"
- B56.6, Safety Standard for Rough Terrain Forklift Trucks

chapter eight

Creating and Securing a Deep Dig

Many structures that a contractor has to build are located in flat areas that are close to oceans, lakes, or rivers filled up with soft and unstable soil. Furthermore, the cost of land in city centers requires, for economic reasons, building downward as well as upward. Both of these situations require not only extensive site-investigations and careful engineering designs, but also special methods and technologies to dig deep to construct the foundation. As this chapter will demonstrate, contractors have developed a variety of fascinating techniques, tools, and materials that allow them to build soil retaining structures and foundations in areas that were until recently out of reach. While the two issues are covered in two separate chapters, they have become intertwined to the point where retaining walls fulfill double functions as both soil support and permanent foundation for the final structure.

Table 8.1 Topics Covered in Chapter 8

Construction			Planning & Control		
Equipment	**Methods**	**Materials**	**Managerial-Engineering Tools**	**Rules & Standards**	**Modern Tools**
✓ Trucks ✓ Centrifugal pump ✓ Impact hammers ✓ Axial turbine pump ✓ Wellpoint ✓ Sheet pile threader ✓ Pile driver ✓ Vibratory drivers ✓ Jacking driver ✓ Auger drill ✓ Trench cutter ✓ Hydrocyclone ✓ Slurry desander ✓ Rakers ✓ Mobile anchor drill ✓ CFA ✓ Progressive cavity pump ✓ Pipe tremie	✓ Slurry walls ✓ Grouting ✓ Dewatering ✓ Dynamic pipe head ✓ Pump curve ✓ Retaining walls ✓ Wall shoring ✓ Earth anchoring ✓ Tieback Anchoring ✓ Jet grouting ✓ Grout batching	✓ Bentonite Slurry ✓ Cement grout ✓ Colloidal mixture ✓ Sheet piles ✓ Chemical grouts	✓ Pump performance analysis ✓ Equivalent pipe length ✓ Hazen-Williams friction loss ✓ Navier-Stokes equation ✓ Anchor load ✓ Marsh Funnel Viscosity ✓ Standard penetration test ✓ Pascal ✓ Piezometer	✓ ASTM A36 Steel ✓ ASTM A82, A615 and A884 ✓ ASTM A416 A421, and A882 ✓ FHWA-IF-99-025 ✓ ASTM D1586	✓ PDA ✓ Strain gauge ✓ Load or Pressure cell

8.1 The History of Digging Deep and Retaining the Earth

3800 BC	**Retaining Walls**	Large dry stones are used to build walls held together by interlocking the stones. The walls are backfilled to create terraces for agriculture or defenses.
500 BC	**Deep Drilling**	Chinese develop deep-drilling using oxen, wheels, and ropes to power rotary and percussion drills. The goal is to mine rock salt from the surface.
230 BC	**Archimedes' Screw**	Greek mathematician Archimedes invents the screw pump after visiting Egypt, where he sees the compartmented rotating vertical wheel used to lift water.
170 BC	**Piston Pump**	Ctesibius of Alexandria is thought to have invented the piston pump with lever and valves to pump water. The invention is still used today.
1126	**Deep Drills in France**	French Carthusian monks in 1126 operate a mechanism with a thin rod and a hard cutting tip to drill a deep hole. The rod is struck with a hammer, breaking up the bottom of the hole. Artesian wells are named after Artois, France, where this took place.
1773	**Calculation of Earth Pressure**	The French civil engineer Charles Coulomb observes that retaining wall failures occurred along a distinct plane. He suggests that the earth pressure could be calculated using the soil cohesion and friction forces along the plane with friction angle.
1833	**Screw Pile**	The blind Irish engineer Alexander Mitchell invents the screw pile. A helical blade at the end of a pile allows it to be screwed into muddy rivers or sandy banks to help create foundations for lighthouses or bridges.
1851	**Centrifugal Pump**	During England's Great Exhibition, John Appold enters a competition for water pumps. His pump has curved impellers and has three times higher efficiency than anybody else's.
1875	**Steam Hammer Pile Driver**	The Vulcan steam hammer begins when the company manufactures hammers under the patent of Thomas T. Loomis in 1875. This hammer uses many of the main features of the Nasmyth hammer from 1839.
1902	**Sheet Piles**	TryggveLarssen from Bremen, Germany, invents the use of long deformed steel profiles to be rammed into the ground supporting each other. The original walls are still functioning.
1912	**Screw Anchor**	Albert Bishop Chance invents the first practical earth anchor. Instead of a deadman, the anchor depends on a short helical screw that is turned into the ground with minimal disturbance of the soil. Thus, the holding capacity of the anchor is close to maximum.
1930	**Bentonite**	Sodium bentonite, a clay, swells in water, absorbing several times its dry mass. This property is found useful in creating subsurface seals, and first patents are established.
1949	**Vibratory Pile Driver**	The Russian D.D. Barkan is the first to test a new method to install sheet piles as retaining walls. Rather than hammering them, a motor turns several eccentric weights inside a frame on top of the pile. The spinning weights create sinusoidal down forces.
1950	**Slurry Walls**	The first vertical narrow trenching supported by a bentonite-water-mix slurry during excavation is tested in 1940s in Italy. The first full scale slurry wall is built in Italy in 1950 as a cut-off wall.
1955	**Benoto Piles**	The French Benoto piles launch the success story of bored piles used as building foundations and retaining walls. A double-walled casing is pressed into the ground during alternate turning (corkscrew motion). The soil inside is removed, and the space filled with concrete while the casing is slowly removed.

8.2 From the Accident File

Case 1: During the mid-afternoon, a 100 ft (30 m) high retaining wall of a cut-and-cover subway construction failed. The diaphragm walls, approximately 60 ft (20 m) apart, were made of reinforced concrete and supported by a steel bracing system with king posts, walers, and struts. Four workers were killed with three more injured, and two cranes were swallowed up. The large cave-in also impacted the highway that ran parallel to the retaining wall, closing most of its lanes for traffic for several months.

Case 2: Three sections of a 158-ft (48-m) long concrete retaining wall was being built to widen a road. Previously, a warehouse had been built into the hillside above the road and construction debris mixed with soil dumped at the location of the retaining wall. When the accident occurred, the first two sections had been already built, and the space for the third had been excavated. While three workers were laying rebar for the footing, the 18 ft (6 m) free-standing soil wall suddenly collapsed, spilling 75 tons (68 t) of mixed material onto the ground. One worker was able to run away, one was covered up to his chest but lived, and one was buried and died.

Case 3: A piling crew was installing a sheet-pile wall alongside a highway. A dock builder was assigned to help guide a second sheet into position to be joined with the first one that was held by a horizontal cantilevering H-beam frame 9 ft (2.75 m) above ground. The first sheet was not driven into the ground but sank about 18 in (0.46 m) into the dirt under its own weight. The goal was to vibrate two sheets together after they were set properly and bolted to the H-beam. For this purpose, the dock builder hitched a ride to the top of the first sheet pile by standing on the ball of the crane line. Using stirrups (a piece of flat stock steel formed into foot rests), he sat on top of the first sheet, which was an accepted practice in the industry. Apparently, the interlock joint between the two sheets did not align properly. Thus the worker started to rock and push the second sheet when the bottom of the first sheet he was sitting on kicked out of the ground, pivoting on the point where it was attached to the horizontal H-beam. As the sheet began to slowly tip over, the victim jumped from the stirrups, falling and striking his head on the supporting H-beam frame before landing on the ground.

Header Problem 8.1 / Planning a Deep Excavation and Special Foundation

You have been asked to study and assess the best method for excavating an approximately 60 ft (18 m) deep pit in groundwater. The pit is needed to build an underground garage. Removing 1 m (3.3 ft) of saturated sandy clay for a building with a footprint of 600 m^2 (6,480 ft^2) reduces an average weight of 600 m^3 × 1,800 kg/m^3 = 1,080 t (1,210 tons). Naturally, the buoyancy produced by the groundwater will reduce the actual weight. Nevertheless, replacing the soil material with air space and concrete walls able to support the building above might be a good solution to avoid consolidation. Figure 8.1 and Table 8.2 provide the basic information about the subsurface conditions.

While the excavation of the sandy clayey material to the required depth should not provide any major difficulties, preventing such a deep pit in a "lake" of groundwater from collapse looks a little more difficult.

Table 8.2 Representative densities for various materials

	Dry Densities		Saturated Densities	
Materials/Soil	**lb/yd^3**	**kg/m^3**	**lb/yd^3**	**kg/m^3**
Styrofoam	50–200	30–120		
Plastics	1430–2350	850–1400		
Silts	1940–2760	1150–1640	2360–3030	1400–1800
Sandstone	3370–4380	2000–2600	3710–4380	2200–2600
Limestone	3200–4380	1900–2600	3540–4550	2100–2700
Sand	1850–2700	1100–1600	2530–3370	1500–2000
Clay	1900–2800	1130–1660	2190–3540	1300–2100
Clay & Organics	1180–1520	700–900	1690–2190	1000–1300

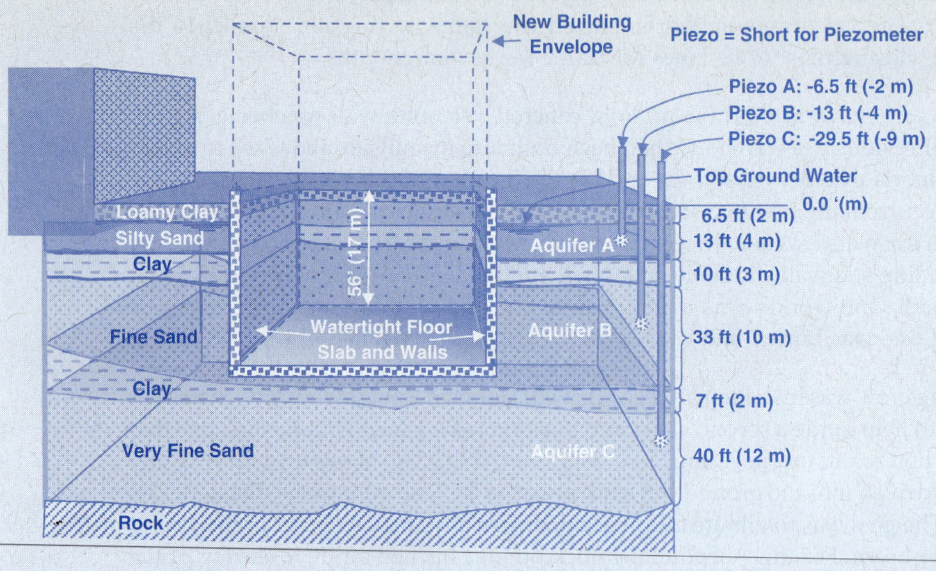

Figure 8.1 Result of the Geotechnical Investigation and the Design of an Open Dig

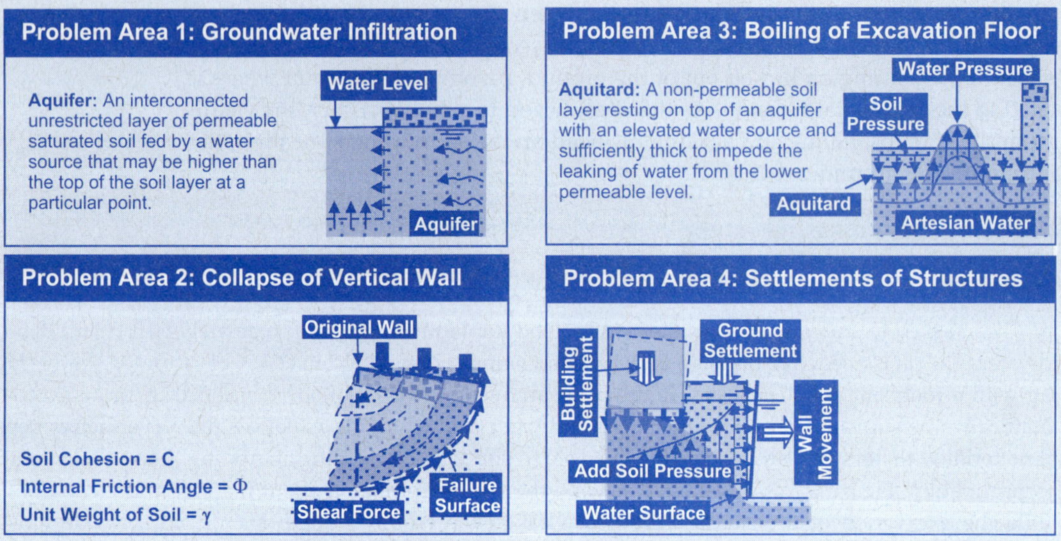

Figure 8.2 Risky Situations During Deep Excavation

Furthermore, identifying the most economical and efficient deep foundation might also take some time. In fact, the geotechnical engineer has already created a short list of four subproblems that are most important to study, presented in Figure 8.2.

8.3 Building on Soft Ground

What follows are presentations of a wide range of technologies some of which might be useful in solving Header Problem 8.1. Please keep it in mind when you study the material, as you will be asked to propose and argue at least one solution to the many subproblems.

8.4 Dewatering the Ground to Excavate in the Dry

Raising water from the underground became a necessity when people started to switch from a nomadic to an agrarian life far from mountainous regions. People and the domesticated animals needed the water to survive, while the farmers depended on it to irrigate their crops. It is thus not surprising that the history of technology related to elevating water is global, as people all around the world studied on how they could improve the primitive approach, a bucket lowered via a rope into a shaft. As we learned earlier, the origin of today's crane was the *shaduf*, a beam with counterweight at one and a rope or rod with water bucket at the other end. The Egyptians synergized the waterwheel with a camel-powered bucket wheel, thus integrating power generation and water-lifting into one wheel with diameters from 10 to 30 m (33–99 ft), known today as *norias* (Spanish for waterwheel with buckets).

Figure 8.3 illustrates not only the ingenuity of ancient engineers, but also the effect of science on construction technology. Modern pumps received a first boost from Evangelista Torricelli, an Italian physicist and mathematician (1608–1647) who invented the barometer by demonstrating the suction effect of "underpressure" in a glass tube filled with mercury, which is 14 times as heavy as water.

The second major contribution came from Daniel Bernoulli, a Dutch-Swiss mathematician (1700–1782) who experimented with liquids and established a basic equation for incompressible fluids:

$$\frac{v^2}{2} + gz + \frac{p}{\rho} = \text{constant}$$

where

v = fluid speed
g = acceleration due to gravity
z = elevation of the point above a reference plane
p = pressure at the point
ρ = density of the fluid at all points in the fluid

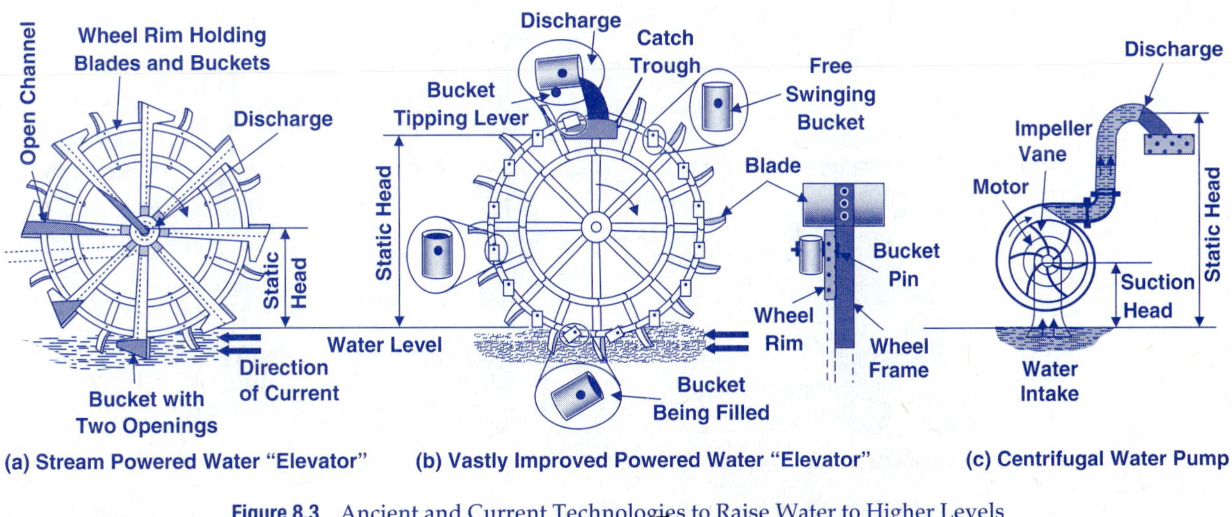

Figure 8.3 Ancient and Current Technologies to Raise Water to Higher Levels

Finally, the concept of the centrifugal impeller originated with the French physicist Denis Papin (1647–1712) and perfected by John George Appold (1800–1865). Appold presented his amazing pump during the London's World Exhibition in 1851 after he had experimented with many different vane shapes and observed that the amount of water increased when they were slightly bent. Without realizing it, he had tapped into the Venturi effect formulated by Bernoulli, in that curved vanes increased the negative (= low) pressure on the backside, pulling an increasing amount of water into the pump to be "pushed" forward by the front side. Appold's pump showed an efficiency of 68%, more than three times better than any of the other pumps participating in a competition during the Exhibition.

Today's construction in areas with high groundwater is unthinkable without a variety of pumps able to keep the excavation area temporarily dry and safe, a process known as dewatering. Let's take a close look into three basic pump technologies used today, all made possible by the work by Torricelli, Papin, Bernoulli, and Appold.

8.4.1 PRINCIPLES OF DEWATERING PUMPS

The two main families of pumps are the so-called rotodynamic and jet-eductors (eductor = Latin for "one that leads out"). While the latter employs the Venturi principle to pull water into a low pressure void, the first depends on spinning propellers or impellers to accelerate the water. The pump propellers operate like those used by ships to create a thrust for propulsion while the impellers accelerate the water away from the rotating shaft, creating centrifugal velocity that can be channeled into a pipe that leads away from the impellers. Figure 8.4 compares the mechanical assemblies as well as the physical principles of three most frequently used pumps in construction.

Without a doubt the volute centrifugal pump, or just centrifugal, is the favorite since it is able to create the highest pressure and with it the highest lift or better, static head. A volute casing is shaped so that the high-speed "twirling" water is quickly "straightened" out, thus reducing the loss of energy. At the tip of the impellers the water reaches the highest speed and with it the lowest pressure (according to Bernoulli) pulling more water from the inlet. In the growing cross-section inside the casing, the speed is decreasing and the pressure increasing, pushing the water out of the discharge.

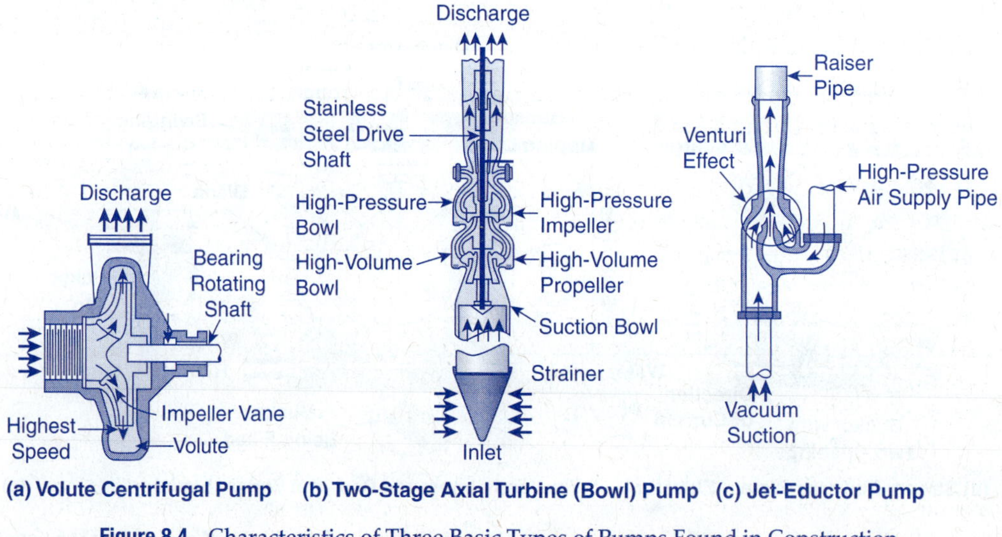

Figure 8.4 Characteristics of Three Basic Types of Pumps Found in Construction

Figure 8.4b presents the inner world of a pump that pulls the water, unlike from the centrifugal pump, parallel along the drive shaft. Its elongated shape makes it a perfect candidate to be deployed in narrow shafts, even submerged if a waterproof electric motor or long drive shaft is used. The pump is typically multistage containing a propeller, to pull in a large volume, and one or more impellers (depending on head requirements) housed in a bowl assembly.

8.4.2 PRESSURE LOSSES DUE TO FRICTION INSIDE PIPE

Water being pushed through long pipes, even those that run horizontally, faces resistance similar to the rolling resistance of wheels discussed in Chapter 5. Here, however, the causes are related to the friction between the fluid in the pipe walls and restrictions that impede its free flow from the intake and the outflow at the other end such as bends or valves. It is similar to losing speed and energy when a biker faces a headwind while trying to climb a hill without touching the pedals. Without sufficient speed at the bottom, the bike will slow down bit by bit not only because of the energy needed to ascend but also because of the headwind. In dewatering pumps, the energy that drives the water is called pressure or head pressure, not hp.

There are two different pressures:

(a) **Static or better hydrostatic pressure** that is caused by the water column above the point of interest. The following function can be used:

$$P = \frac{H * \rho_{Water} * g}{2.31}$$

✓ 1 ft column of water exerts 0.433 psi
✓ 1 m column of water exerts 9.8 kPa or 0.098 bar
✓ 1 psi is created by 2.31 feet of water (head)
✓ 1 psi = 6.9 kPa

where

P	= Static pressure in lb_{force}/in^2 (psi) or N/m^2 (Pa)
H	= Head or height of water column (ft or m)
ρ_{water}	= Fluid density (62.3 lb/ft^3 or 9.8 kN/m^3)
g	= Gravitational acceleration (32.174 ft/s^2 or 9.807 m/s^2)

and

(b) **the dynamic pressure or dynamic head** is the kinetic energy needed to keep a liquid in motion inside a pipe or hose. As we will learn, there is no simple function, as liquids experience many different kinds of friction causing "headwind." Figure 8.5 illustrates the differences between the static or head pressure while the pump is turned off and the dynamic pressure during pumping.

As a visual means to demonstrate the effectiveness of the two types of pressure, consider three risers installed along the pipeline. A riser consists of a long hollow pipe that is connected to the pipe "rising" vertically above it. In Chapter 3 we talked about Pascal's experiment with the three vases that were connected at the bottom, causing the water in all three to rise to the same level. In Figure 8.5, the vases are the three risers, and the connecting tube is replaced by the pipe filled with water to the top. The difference of the water levels with the pump off and the pump on, as in Figure 8.5b, is apparent. Pushing the water uphill requires enough pressure not only to overcome the height difference, but also to overcome the friction along the pipe. The amount of friction is a function of its length, as the riser in the middle shows a lower amount of friction loss to reach the top of the pipe, which shows 0 pressure at the outflow into the canal.

Before installing a dewatering system, we should have a good idea about how large the head losses will be, since this will be one key dimension for installing and operating a pump with sufficient performance. Let's first review the different causes of friction or head losses of a liquid moving inside a long pipe, such as the one shown in Figure 8.5.

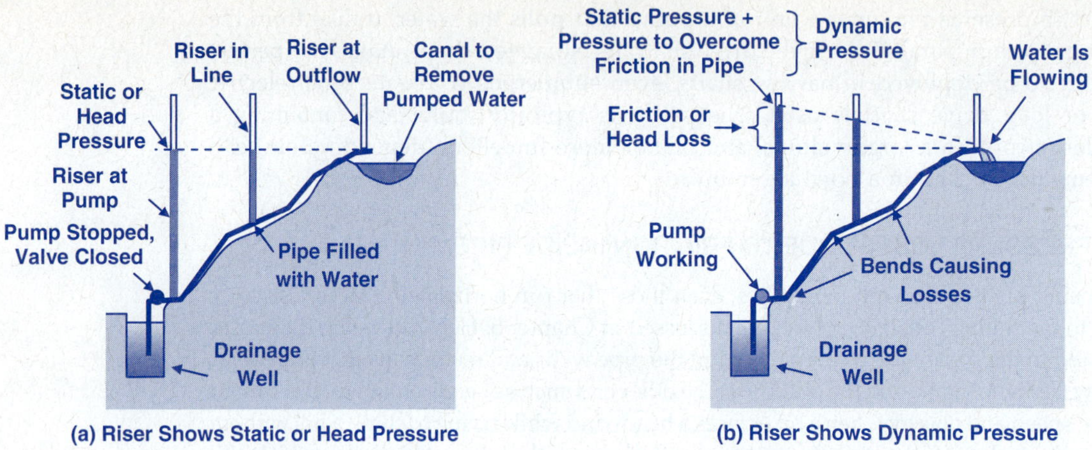

Figure 8.5 The Difference Between Static and Dynamic Pressure

1. **The velocity of a fluid.** As the velocity increases, the flow increases from a laminar to turbulent status, raising the drag, which results in larger pressure losses.
2. **The size (inside diameter) of a pipe.** According to Bernoulli's principle, changing the diameter of a pipe with a constant flow will impact the pressure and the speed of the fluid. Decreasing pipe sizes increases the velocity and, at the same time, decreases pressure in the pipe. As shown above, higher velocity leads to higher losses.
3. **The friction at the pipe walls.** Surface roughness of the pipe causes friction loss.
4. **The length of the pipe.** The friction losses are cumulative, as the water travels through the length of pipe. The greater the distance, the greater the friction losses will be.
5. **Special head losses caused by**
   - Valves and fittings
   - Changes in direction of flow path
   - Obstructions in the flow path
   - Sudden or gradual changes in the cross section and shape of the flow path

There are two common approaches to calculating the amount of normal friction losses:

(a) The empirical Hazen-Williams equation:

$$\mathbf{f = 0.2083\,(100/c)^{1.852}\,q^{1.852}/d_h^{4.8655}}$$

f = friction head loss in feet of water per 100 feet of pipe (ft_{H20}/100 ft pipe)
c = Hazen-Williams coefficients
q = volume flow (gal/min)
d_h = inside hydraulic diameter (inches)

The Hazen-Williams "c" coefficients have been developed for different materials. The higher the friction, the lower the c value, leading to a higher friction head f.

(b) The more complex D'Arcy-Weisbach equation. While it considers more factors, it is also more generally applicable to fluids, even for laminar and transient phases.

For the case of calculating friction losses for dewatering, the Hazen-Williams formula has been found sufficient.

The graph in Figure 8.6 plots the effect of pipe throughput in gal/min (liters/sec) on the friction loss for three pipe types and give diameters using the Hazen-Williams equation.

The combined effect of flow and pipe size is apparent. The friction losses for 3-in (7.5-cm) pipes that have to carry 300–400 gal/min (19–25 l/sec) increase dramatically,

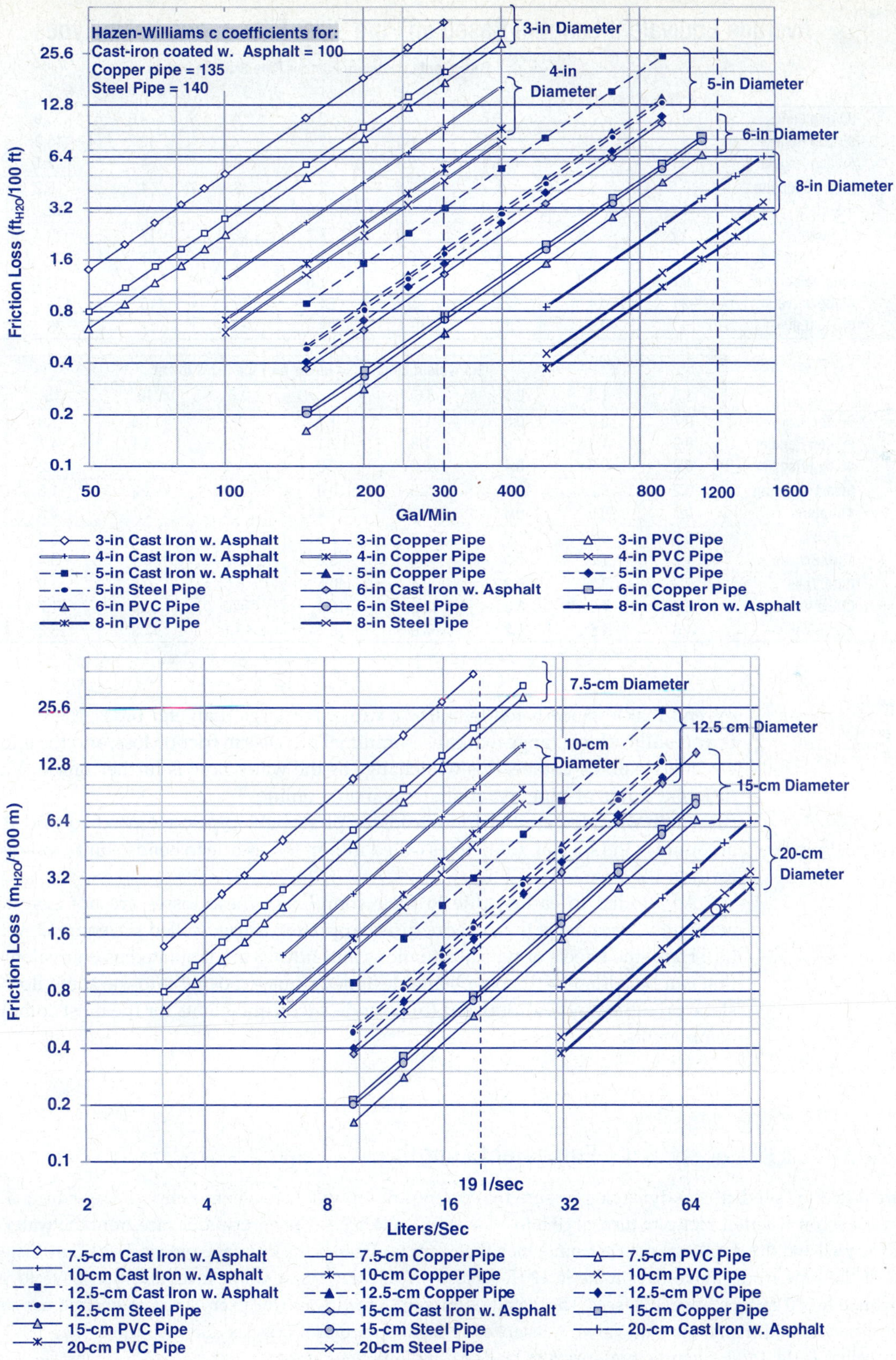

Figure 8.6 Friction Losses as a Function of Pipe Material, Diameter, and Flow Capacity

Table 8.3 Average Equivalent Lengths Based on Pipe Size, Fitting, and Valve Types

		Equivalent Lengths in Feet for Pipe Sizes in inches								
		1/2	3/4	1	1 1/2	2	3	4	6	8
Elbows	90 deg Long	2.2	2.3	2.7	3.4	3.6	4	4.6	5.9	6.7
	90 deg Regular	3	4.4	5.2	6.1	7	7.7	10.2	15.3	25.2
	45 deg Regular	0.75	0.9	1.3	2.1	2.8	3.6	4.7	7.1	12.0
	Insert Coupling	0.5	0.75	1	1.5	2	3	4	5.2	5.9
	Adapters	1	1.5	2	3.5	4.5	6.5	9	12	15.2
Tee	In Flow	1.7	1.4	3.2	5.6	7.7	12	17	19	24.2
	Thru Branch	4	5	6.6	9.9	12	17	21	31	38
Valves	Gate Valve	0.6	0.7	0.8	1.2	1.5	1.9	2.1	3.2	4.3
	Globe Valve	22	24	29	42	54	79	110	170	250
	Angle Valve	15	15	17	23	29	43	57	75	92

		Equivalent Lengths in Meter for Pipe Sizes in cm								
		1.3	1.9	2.5	3.8	5	7.5	10	15	20
Elbows	90 deg Long	0.7	0.7	0.8	1.0	1.1	1.2	1.4	1.8	2.0
	90 deg Regular	0.9	1.3	1.6	1.9	2.1	2.3	3.1	4.7	7.7
	45 deg Regular	0.2	0.3	0.4	0.6	0.9	1.1	1.4	2.2	3.6
	Insert Coupling	0.2	0.2	0.3	0.5	0.6	0.9	1.2	1.6	1.8
	Adapters	0.3	0.5	0.6	1.1	1.4	2.0	2.7	3.6	4.6
Tee	In Flow	0.5	0.4	1.0	1.7	2.3	3.6	5.2	5.8	7.4
	Thru Branch	1.2	1.5	2.0	3.0	3.6	5.2	6.4	9.4	11.6
Valves	Gate Valve	0.2	0.2	0.2	0.4	0.5	0.6	0.6	1.0	1.3
	Globe Valve	6.7	7.3	8.8	12.8	16.4	24.0	33.4	51.7	76.0
	Angle Valve	4.6	4.6	5.2	7.0	8.8	13.1	17.3	25.8	28.0

a trend that is exacerbated by a low c value. For large-diameter pipes, the higher water flow is still within a range that causes minimal increase in friction loss, and the effect of the c is still small. Of course, this will change as the water flow is further raised. Values for different kinds of pipe material are available online.

After we calculate the friction losses of straight pipes, we need to find a way to account for the special, sometimes called minor, losses due to bends, joints, valves, and so forth in the pipe.

As alluded to earlier, due to their complexity these losses are not easy to assess accurately. Because of this, a more direct approach is used called the equivalent length L ($_{eq}$). Here, the effects of typical situations and fittings are tabulated as equivalent lengths that can be added to the pipe being designed, thus exposing it to the same flow rate, or static pressure. Tables 8.3(a) and (b) provide such equivalents for the most common pipe components.

Worked-Out Problem 8.1 / Equivalent Length

You are asked to "predict" the dynamic pressure that a sump-pump will face when pumping water from a deep well inside a construction pit up to discharge into a large concrete pipe. The vertical distance from the water level inside the well and the discharge nozzle from which the water pours into the drainage pipe is 75 ft (23 m). The total length of the pipe from intake to outtake is 120 ft (36.6 m). The pump is at water level and will have to pump approximately 300 gallons/min (19 l/sec). The installed smooth steel pipe is 6 in (15 cm) in diameter and has several built-in flow "impediments": 1 gate valve, 6 insert couplings, 6 regular 45 degree elbows, and 2 long 90-degree elbows. What is the total dynamic pressure loss or head this pipeline?

Assumptions:

The Hazen-Williams equation will provide a sufficiently accurate result.

Calculations:

(1) Static pressure:
The total height of the water column = 75 ft (22.8 m).

(2) Pressure loss due to friction in a straight pipe (see Figure 8.6):
For a 6-in steel pipe with 300 GPM = $0.71\ ft_{Water}/100\ ft$ of pipe
For a 15-cm steel pipe with 19 l/sec = $0.71\ m_{Water}/100\ m$ pipe

(3) Equivalent lengths are summarized in Table 8.4.

(4) Total equivalent length of the pipe = Straight length + Added L_{eq} = 120 ft + 88.8 ft = 208.8 ft
Total equivalent length of pipe = 36.6 m + 27 m = 63.6 m

(5) Dynamic pressure loss = static pressure + friction loss = Total Dynamic Head (TDH)

$$TDH = 75\ ft_{Water} + 208.8\ ft\ of\ pipe * 0.71\ ft_{Water}/100\ ft\ of\ pipe = 76.5\ ft_{Water}$$
$$TDH = 76.5\ ft_{Water} = 76.5\ ft_{Water}/2.31\ ft_{Water}/psi = 33.11\ psi$$
$$TDH = 22.8\ m_{Water} + 63.6\ m * 0.71/100 = 23.25\ m_{Water} = 23.25\ m_{Water} * 9.86\ kPa/m_{Water} = 229\ kPa$$

Table 8.4 Equivalent Lengths for 6-in and 15-cm Steel Pipes Based on Table 8.3

Type	L_{eq}/Unit	Number	Total L_{eq}
	ft		ft
Gate Valve	3.2	1	3.2
Insert Couplings	5.2	6	31.2
Regular 45° Elbows	7.1	6	42.6
Long 90° Elbow	5.9	2	11.8
	Added Total L_{eq}		88.8

Type	L_{eq}/Unit	Number	Total L_{eq}
	m		m
Gate Valve	1.0	1	1.0
Insert Couplings	1.6	6	9.6
Regular 45° Elbows	2.2	6	13.2
Long 90° Elbow	1.8	2	3.6
	Added Total L_{eq}		27.4

8.4.3 DESIGNING A DEWATERING SYSTEM

The two main design parameters are the amount of water that has to be moved per minute, called the flow rate or discharge, and the static height, defined as the vertical distance between the lower water level and the height of the discharge. Most other parameters are interdependent, which demands that the engineer make several choices, each leading to a different branch along the decision tree. Some of the choices include (a) size and material of pipes, (b) type of pump, (c) location of pump (e.g., submerged), (d) RPM of pump, and (e) impeller/propeller size of pump. The foremost problem that has to be solved is to assess the internal losses that have to be overcome and compensated by pump power. The source for internal losses are (1) thermal and mechanical losses by the motor, (b) mechanical losses of the drive system (e.g., shaft bearings), (c) hydraulic and mechanical losses inside the pump (e.g., impellers), and (d) friction losses along the piping system (e.g., suction losses). Each pump has its special characteristics, which can be

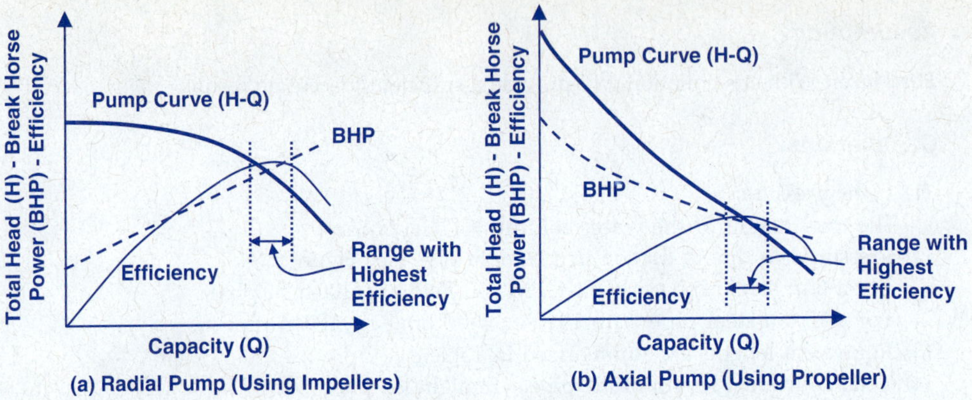

Figure 8.7 Generic Pump Performance Curves with a Fixed Speed

represented as curves in a chart with the GPM on the x- and the total head on the y-axis, each speed and impeller/propeller size having its own curve established by the pump manufacturer through experimental tests. Figure 8.7 compares the performance relationships of a radial impeller and an axial propeller pump.

It is apparent that each pump performs with the highest efficiency within capacity range. Since the total head H cannot be changed, the only variable that is available to "steer" the pump toward the highest efficiency for a given speed is to control the intake of the pump, using valves. But most importantly, one has to find the pump that best fits the overall requirements. It is best to experience the decision-making process, which might not necessarily lead to the absolute best solution, on a worked-out example.

Worked-Out Problem 8.2 / Design of a Deep Well Pumping System

Beth Veach, an excavation contractor, has to lower a groundwater table and would like to get a first idea about configuring a deep well pump as displayed in Figure 8.8. Based on the hydraulic situation, she expects an average inflow capacity of 1,200 GPM (75.7 l/sec) with a drawdown of approximately 27 ft (8 m) while the static head will be about 80 ft (24.2 m) and the horizontal distance to the discharge into a nearby canal is 330 ft (100 m). She has found that in this kind of situation a vertical turbine pump is most reliable and economical. Two types of pipes are available: 6-in (15.2 cm) and 8-in (20-cm) smooth steel. We are charged finding an efficient configuration including the power requirement for the pump.

S8.1 Goulds Vertical Pump
S8.2 Goulds Pump Selection

Layout Sketch:

Pump: The booksite includes a brochure on vertical turbine pump model VIT from ITT Goulds Pumps, matching the capacity requirement of 1,200 GPM (75.7 l/sec) at 1760 RPM.

Background and Assumptions:

Total Head = Static head + Losses due to friction inside pipe system + Velocity head
Static Head = Suction head (from water surface to Impeller) + Discharge head (from impeller to discharge)

- Friction losses depend on type and size of pipe and are available for 100 ft and 100 m
- Figure 8.6 indicates that both available pipe sizes could deliver the necessary 1,200 GPM (75.7 l/sec)
  However, the 6-in (15 cm) is a little small; thus, the resulting pressure or head loss caused by friction is approximately 4 times higher than the 2.3 ft_{Water}/100 ft (2.3 mWater/100 m) of the 8-in (20 cm) pipe.

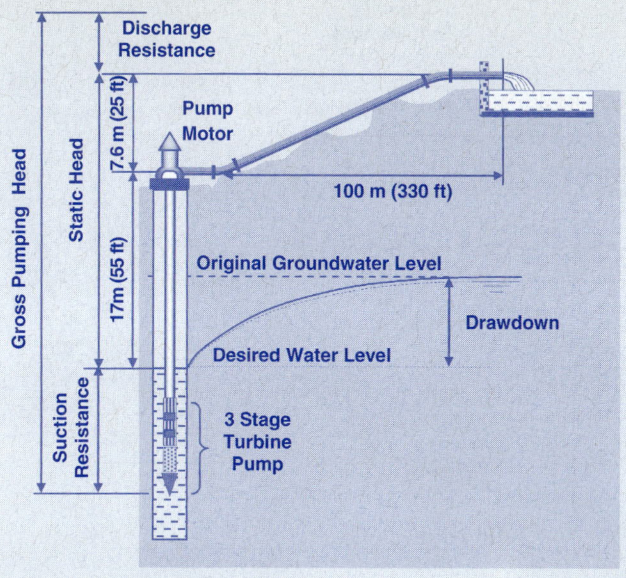

Figure 8.8 Configuration of Pumping Station

- L_{eq} for a regular 45° elbows is available in Table 8.3:
  For an 8-in pipe the equivalent length $L_{eq} = 12\,\text{ft}$
  For a 20-cm pipe the $L_{eq} = 3.6\,\text{m}$

Calculation of Total Head (H) = Total Dynamic Head (TDH):

$$\text{Static Head} = 55\,\text{ft} + 25\,\text{ft} = 80\,\text{ft}_{\text{Water}} = 34.6\,\text{psi}$$
$$\text{Static Head} = 16.7\,\text{m} + 7.6\,\text{m} = 24.3\,\text{m}_{\text{Water}} = 240.6\,\text{kPa}$$
$$\text{Length of pipe} = \sqrt{(330^2 + 25^2)} + \text{intake} = 331\,\text{ft} + 20\,\text{ft} = 351\,\text{ft}_{\text{Pipe}}$$
$$\text{Length of pipe} = \sqrt{100^2 + 7.6^2} + \text{intake} = 100.3\,\text{m} + 6.1\,\text{m} = 106.4\,\text{m}_{\text{Pipe}}$$
$$L_{eq}\text{ for 2 elbows} = 2 * 12\,\text{ft}\,(2 * 3.6\,\text{m}) = 24\,\text{ft}_{\text{Pipe}} = 7.2\,\text{m}_{\text{Pipe}}$$
$$\text{Total equivalent length} = 375\,\text{ft} = 113.6\,\text{m}_{\text{Pipe}}$$

$$\text{Head loss for a 375 ft steel pipe} = 375 * 2.3\,\text{ft}_{\text{Water}}/100\,\text{ft}_{\text{Pipe}} = 8.6\,\text{ft}_{\text{Water}}$$
$$\text{TDH} = 8.6\,\text{ft} + 80\,\text{ft} = 88.6\,\text{ft}_{\text{Water}}/2.31\,\text{ft}_{\text{Water}}/\text{psi} = 38.36\,\text{psi}$$
$$\text{TDH} = (2.3\,\text{m}_{\text{Water}}/100\,\text{m}_{\text{Pipe}} * (106.4 + 7.2\,\text{m})) + 24.6\,\text{m}_{\text{Water}}) * 9.86\,\text{kPa}/\text{m}_{\text{Water}} = 264.\,\text{kPa}$$

Work or hydraulic horsepower (Hhp) is the horsepower needed to pump the liquid.

$$\text{Hhp} = Q\,(\text{gallons/minute}) * \text{TDH}\,(\text{psi})/1{,}714)\,\text{hp}$$
$$\text{Hhp} = (1{,}200\,\text{GPM} * 38.36\,\text{psi})/1714\,\text{hp} = 26.5.\,\text{hp}$$
$$\text{Hhp} = (9.81 * 26.77\,\text{m} * 75\,\text{l/sec}/1000) * 1.341\,\text{hp/kW} = 26.5\,\text{hp}$$

Assuming an efficiency factor of 80% for the pump = PHP = 26.5 hp/0.8 = 34 hp
Assuming efficiency factor of 90% for the drive system and 100% for the electric motor
EngineEHP = 34/0.9 = 38 hp

Discussion of Results:

By plotting the two key parameters, H and Q, into the composite performance graph, we can show that the resulting intersection is well inside the area that describes high efficiency at an RPM of 1760. The pump manufacturer will be able to equip the pump with impellers that provide a pump curve that crosses the desired performance, such as the dashed curve shows.

In addition, if more water seeps into the well, the pump is able to handle it efficiently.

The required EHP of 38 hp provides a guide to select the proper electric motor.

In summary, the selected pump is ideal for the pumping requirements given.

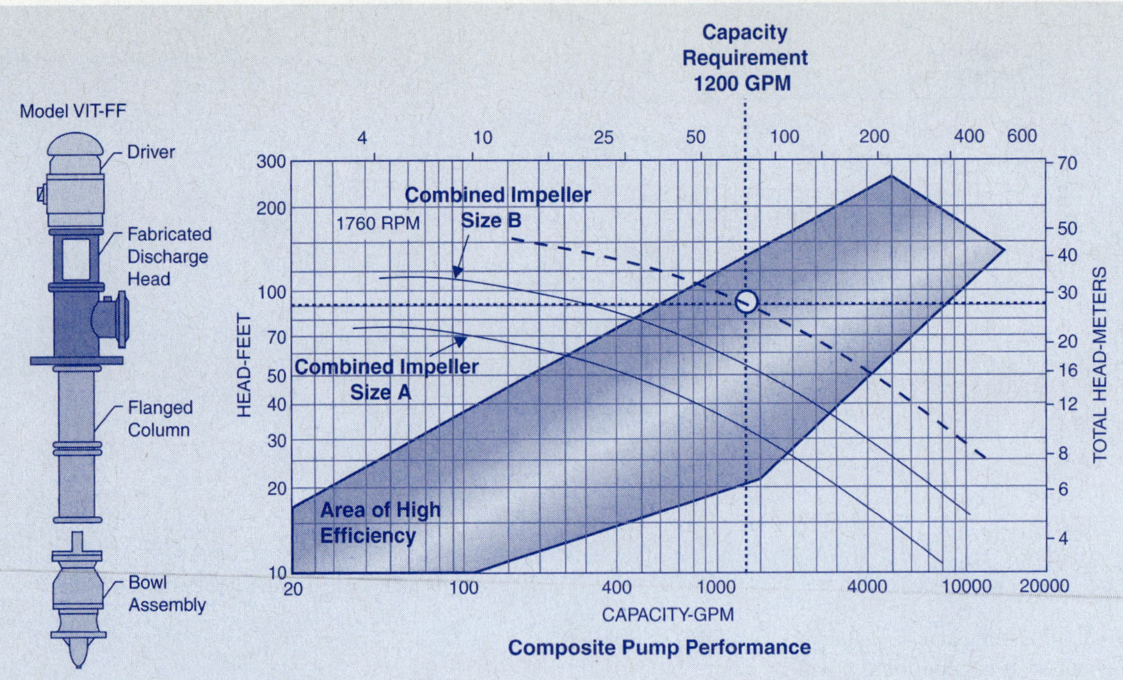

Figure 8.9 Vertical Turbine Pump Model VIT-FF by ITT Goulds

8.4.4 INSTALLING A CONVENTIONAL WELLPOINT SYSTEM

In situations where silty soil reduces the permeability and where large drawdowns impact large areas, a conventional wellpoint system offers an economical alternative. The backbones of this dewatering method are a centrifugal pump and one or more vacuum pumps that are connected to a large header pipe, which encircles the area to be dewatered. As Figure 8.10 shows, 1 to 2-in (2.5–5.0 cm) diameter riser pipes are connected to a header pipe 3–10 ft (1–3 m) apart, after they have been inserted into the ground, using the jetting method.

Wellpoint tips are made of perforated plastic pipes to form a screen that holds off sand and silt. They are slightly bigger than the riser tube, 2 to 5 ft (1.5 m) in length, and may include a water-jetting tip that will be closed automatically under vacuum with the help of a floating ball. They may or may not be surrounded with a filter, depending upon the type of soil drained. Due to the physical constraints of vacuum, the ability to lift water vertically is limited to 20 ft (6.1 m). Thus, if the water table has to be lowered by more than 15 ft (4.6 m), wellpoints need to be staged as depicted in Figure 8.11.

Each wellpoint level is able to lower the water table by approximately 4.5 m (15 ft) as the piping hardware relative to the surface and so on accounts for about 1.5 m (5 ft) per stage. As a result, a two-stage system is able to lower the water by about 9 m (30 ft).

While it is necessary to create a dry bottom to build a solid foundation for a new building, it is equally important to ensure that the benches, slopes, or walls of the excavation are secure and safe during the construction phase. The next section will elaborate on the key issues related to constructing vertical retaining walls that are able to support the forces created by the surrounding soil.

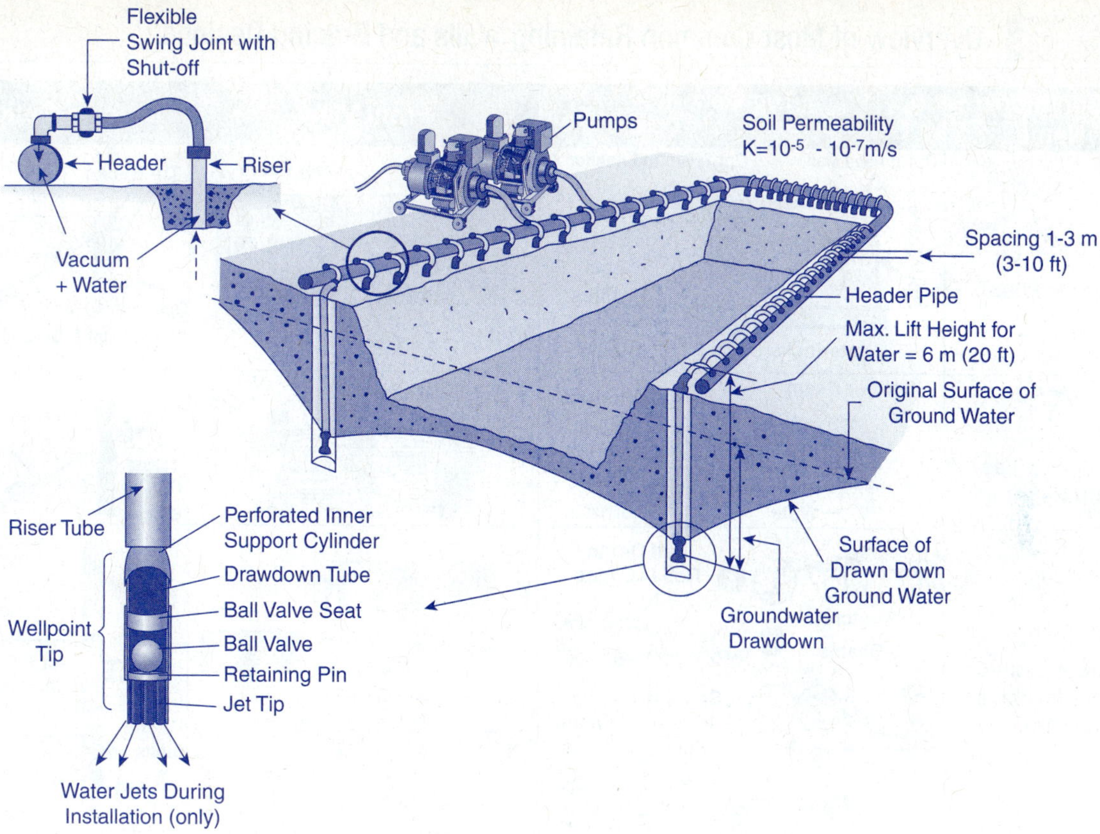

Figure 8.10 Components of a Conventional Wellpoint System

8.5 Preventing Collapse of a Vertical Earth Wall

Creating vertical enclosures that support the surrounding earth and adjacent structures is one approach to creating a space that is accessible to build a foundation. One can think of creating a vertical tunnel in soft soil that requires a structure strong enough to resist the weight of the soil above the crown. In addition, it would be nice to create an enclosure that is watertight in situations where the excavation reaches into the groundwater. Table 8.5 provides an overview of the most commonly used methods that have been developed over the years in response to both geotechnical conditions and economic efficiency.

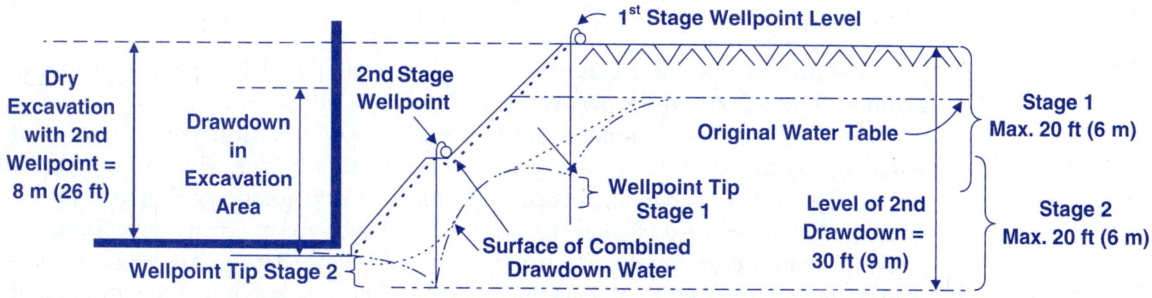

Figure 8.11 Two-Stage Wellpoint to Create a Dry Excavation to 8 m (26 ft)

Table 8.5 Overview of Most Common Retaining Walls and Bracing Designs

Common Name	Top View with Several Bracing Options	Front View	Bracing Options
(A) Sheet Piles	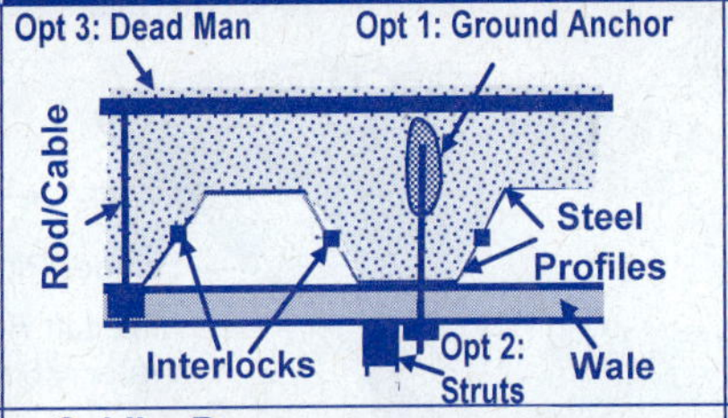	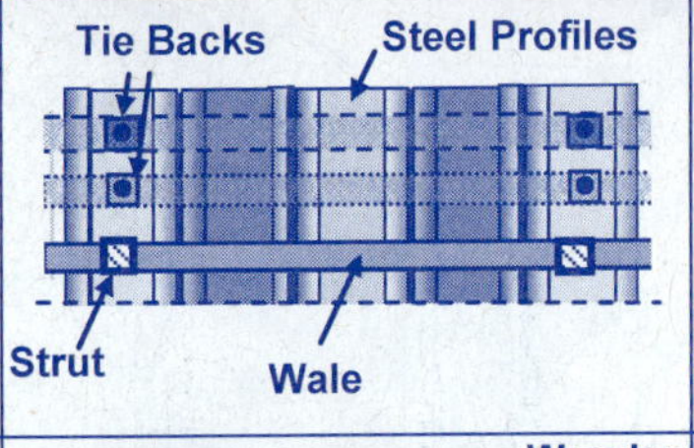	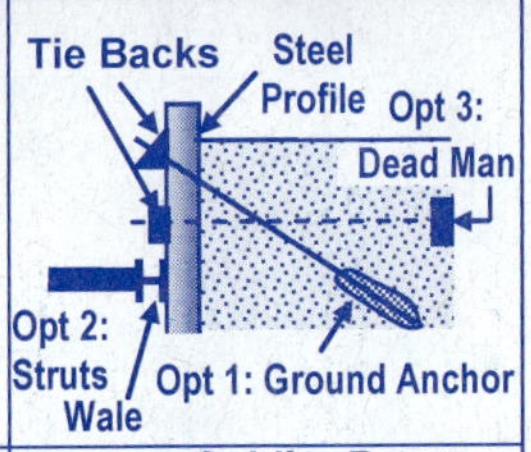
(B) Soldier Piles with Lagging	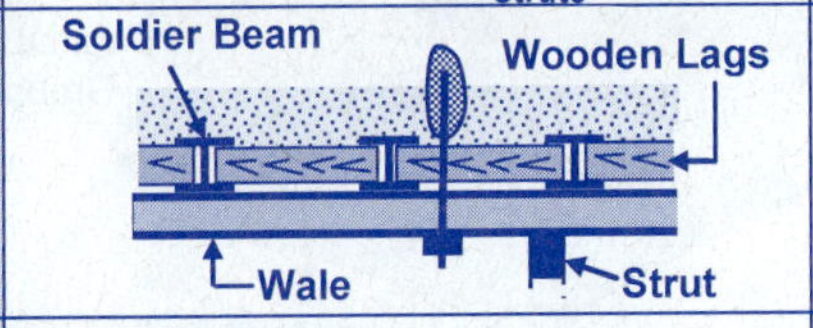	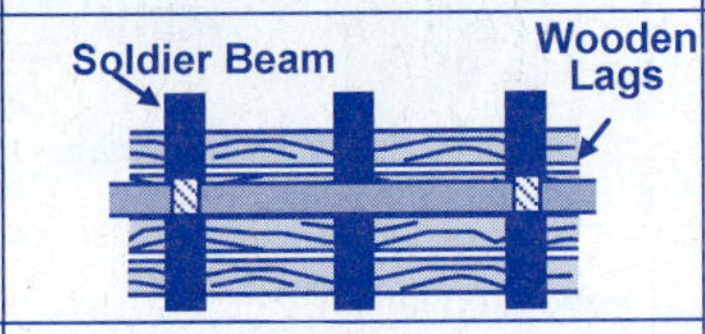	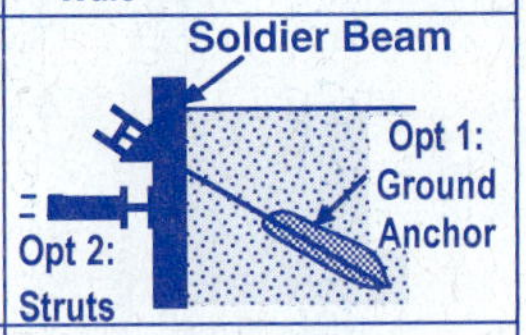
(C) Auger Drilled Cast-in-Place Piles (Soldier Piles)	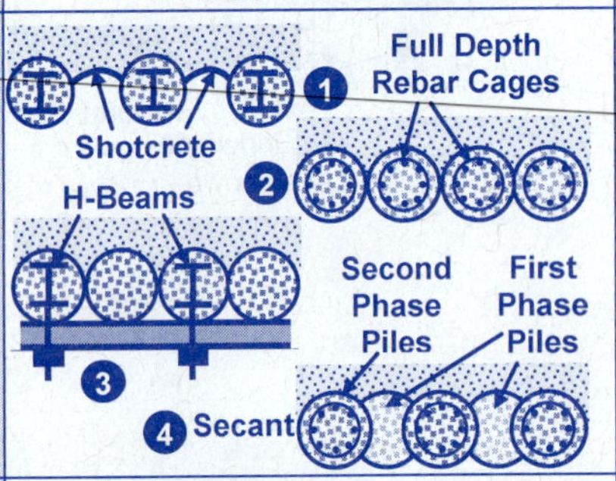	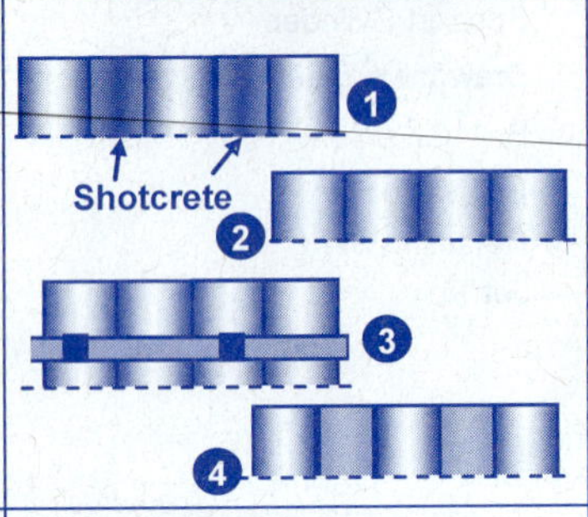	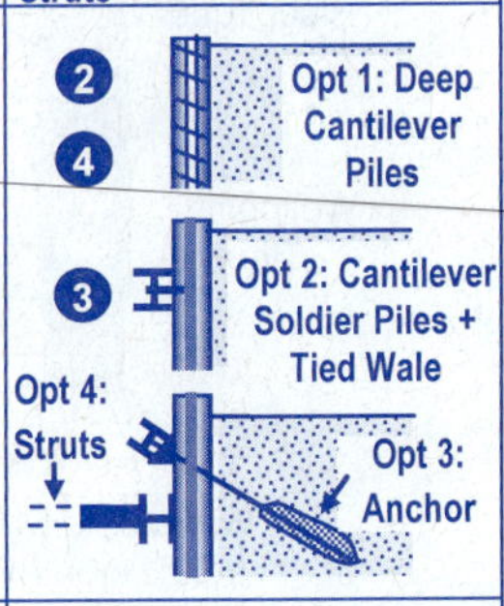
(D) Soil Mixed Wall	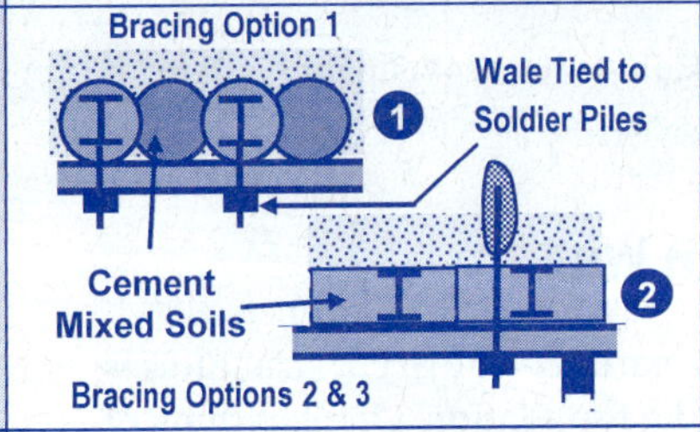	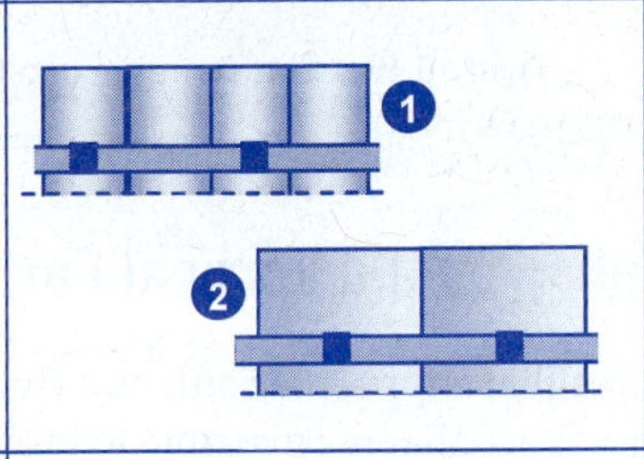	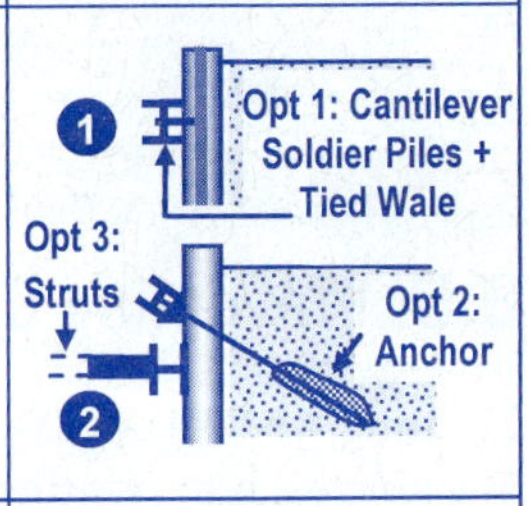
(E) Diaphragm/ Slurry Walls	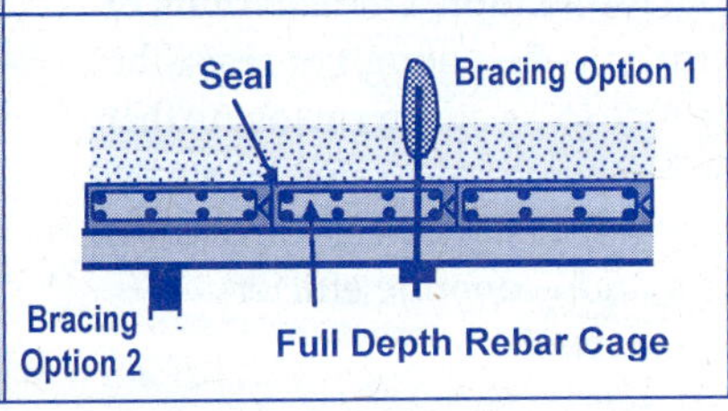	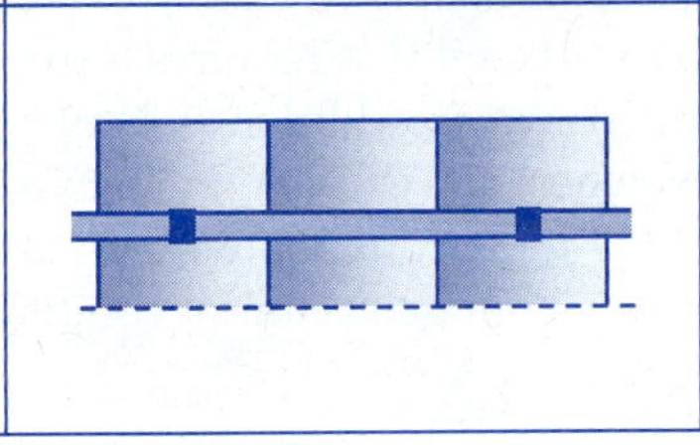	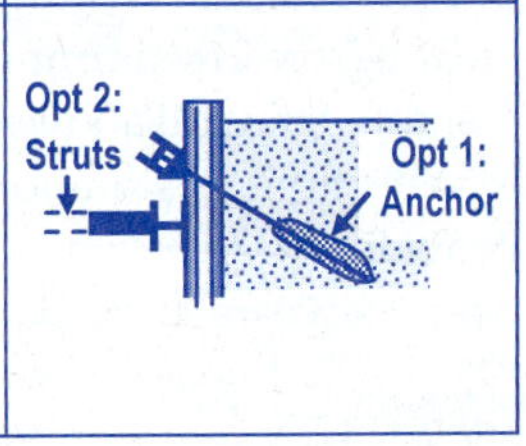

Let's first have a closer look at how to build retaining walls using soldier-piles with lagging (B) and study the more specialized methods afterward.

The name *soldier-pile* reflects the importance of the vertical pile in that it carries the active earth load when completed. The installation begins with driving steel piles (H-beam or pipe) or drilling deep shafts for cast-in-place or precast concrete piles. Commonly, the space between the piles is 5–10 feet (1.5 m 3.0 m) depending on the soil conditions. In soft soil the piles are closer together since the space in between will have to stay unsupported while the lagging is being built. As shown in Table 8.5, one of the materials used for lagging are wooden boards or lags that are 3.0–6-in (7.6 cm–15.2 cm)

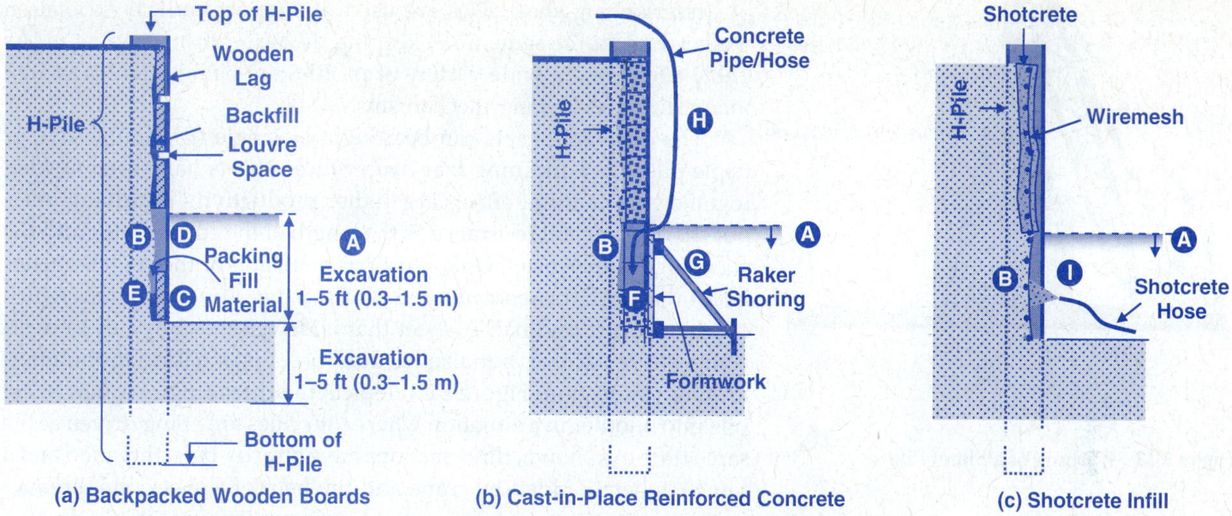

Figure 8.12 Construction Methods for Soldier Pile Laggings

thick depending on the pile spacing and the soil conditions. Also shown is the purpose of the piles. For example, the flanges of the H-piles serve as the end-supports for the lags. Commonly, the piles are shored by horizontal wales held in place by struts or earth anchors. Figure 8.12 depicts the process of installing three different lagging systems.

With the soldier piles in place, a cycle begins with excavating a layer of soil Ⓐ. Before the wooden lags are installed between two piles, the earth is cut back Ⓑ as smoothly as possible, with just enough room for the lags to fit Ⓒ. Wedges between the board and the H-pile flange can be used to push the board against the soil. Almost certainly, small overcuts will emerge and need to be filled in Ⓓ to avoid future collapse of the soil behind the lagging, which could lead to settlement on the surface behind the retaining wall. The fill material needs to packed tightly so that the lagging is able to provide the needed counter-pressure Ⓔ. The next lag is then installed with a small space, called a louvre, separating it from the last to allow backfilling and sufficient drainage of the soil behind the lagging.

Figure 8.12b presents a lagging method that is using cast-in-place concrete. As with the wooden lags, the cycle begins with excavation followed by a careful cutting back of the open section Ⓑ to make space for the reinforcing cages Ⓕ and a sufficient concrete width. Subsequently, the vertical formwork between two soldier piles can be put in place and the raker shoring setup Ⓖ. Concrete is supplied by a hose-pipe combination and filled into the space between the formwork and soil.

Finally, Figure 8.12c shows how shotcrete fulfills the lagging function. Here the soil is cut back in slight arches between the piles. After the affixing of the wiremesh, shotcrete is shot at the surface Ⓘ until the mesh is totally covered and a desired thickness is reached.

Soldier piles and lagging walls are considered the most cost-effective temporary retaining walls. On the other hand, some of these features make them unfit for certain situations, such as (a) when soil conditions allow a large inflow of ground water, and (b) when the soil is not stiff enough for certain applications, leading to possible soil movements.

8.5.1 INSTALLING SHEET PILES

Sheet piles are thin interlocking sheets of steel that can be driven into soft ground that is void of large obstacles such as boulders, solid debris, and buried utilities. In 1902,

TryggveLarssen from Bremen, Germany, invented the use of steel profiles to be rammed into the ground supporting each other.

Interlocking sheet piles are used to support vertical excavation walls while, at the same time, keeping the water from seeping in. As Table 8.6 shows, a wide variety of profiles are being offered, as well many different locking mechanisms.

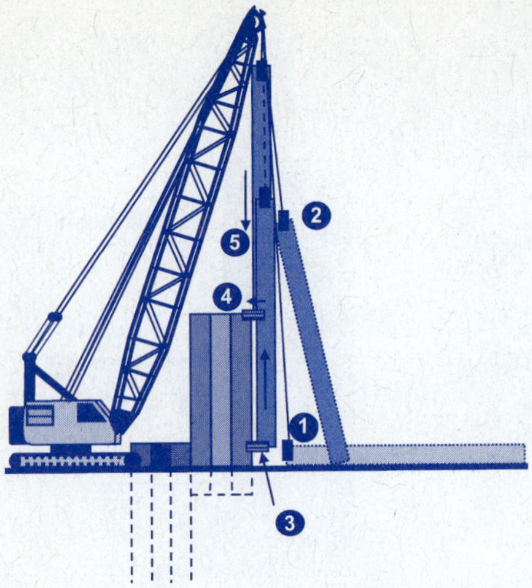

Figure 8.13 Pitching of a Sheet Pile

The different sheets can be bought as single (E), double (D), or triple piles (Dr), meaning that two or three sheets have been welded together. The natural effects are higher productivity in driving piles, but also greater soil resistance as the length of the cutting edge and the sheet area is increased. Only a proper matching of the soil characteristics that will be encountered during the drive will tell which combination is optimal. A sheet that is driven not only has to overcome soil resistance, but also friction forces originating in the interlocking connection. Figure 8.13 depicts the threading of a new sheet pile into another in a situation where four piles are being driven at the same time. As shown, three sections have already been threaded and a fourth is being added by crane and the help of a sheet pile threader mechanical device that eliminate the need for someone to personally guide it into the lock of the previous piece.

❶ Crane hook is connected to next sheet pile on the ground. ❷ Crane lifts new pile into vertical position. ❸ The sheet pile threader is attached to the lower end and locked to the fixed pile. Cranes lifts the new pile with the steel pile threader to the top of the fixed pile.	❹ When the top is reached, a spring inside the threader pushes the tongue of the new pile into the groove of the fixed steel pile. ❺ New pile is lowered, interlocking the two piles on the way down. On the bottom, pile threader is removed.

The booksite includes a piece that exists in more detail several safe threading procedures. As the insert shows, there exist a wide variety of tongue-and-groove lock designs for the many sheet pile configurations and purposes. In order to minimize water seepage through the locks, special seals are available as well.

Sheet piles are available as hot rolled or cold rolled profiles. Generally, the cold rolled sheets are thinner and have a higher strength-to-weight ratio. On the other hand, the cold rolling does not result in the same tight interlocks as hot rolling. Thus, more water will seep through at the beginning, but possibly slow down when the tight space fills up with silt.

One of the many benefits of sheet piles is their quick installation as well as the possibility of extracting and reusing them at the end of the construction project. For the insertion of the sheets several technologies, referred to as pile drivers, have been developed over time, although all depend on mass and gravity to overcome three sources of resistance.

Figure 8.14 presents three different technologies to driving piles into the ground: (a) Impact hammers, (b) vibrators, and (c) hydraulic jacks.

Impact hammers are powered by compressed air, steam, hydraulic, or diesel rams that range in size from light air hammers up to nine-ton (8.2 t) rams. The air hammers are usually used for the lighter work done by low-maintenance tools. Hydraulic and hydraulic impact hammers have the ability to operate at a much higher capacity and are more suitable for large-scale piling operations. A sheet piling operation is usually performed with one of the many types of vibrators. The basic down force is provided by

Table 8.6 Samples from a Large Selection of Available Larssen/Hoesch Sheet Profiles

LARSSEN sections.
HOESCH sections.

Section (E = Single pile, D = Double pile, Dr = Triple pile)		Elastic section modulus [5] W_y	Plastic section modulus [5] W_y	Weight	Cross sectional area	Circum-ference [6]	Coating area [7]	Static moment S_y	Second moment of inertia I_y	Radius of gyration i_y	Classification nach to ENV 1993-5, Steel grades S 240 GP	S 355 GP	S 430 GP
		cm³	cm³	kg/m	cm²	cm	m²/m	cm³	cm⁴	cm	For further steel grades see page 57		
LARSSEN 600	per m wall	510	580	94.0	119.7	225	2.25	290	3825	5.65	3	4	4
	per E	130		56.4	71.8	156	1.44	–	850	3.45	–	–	–
	per D	614		112.8	143.6	291	2.79	–	4590	5.65	–	–	–
	per Dr	712		169.2	215.4	426	4.14	–	6400	5.45	–	–	–
LARSSEN 601	per m wall	745	836	78.0	98.3	245	2.45	418	11520	10.83	2	3	3
	per E	251		46.8	59	172	1.60	–	2650	6.70	–	–	–
	per D	890		93.6	118	319	3.07	–	13830	10.83	–	–	–
	per Dr	1050		140.4	177	466	4.54	–	19320	10.44	–	–	–
LARSSEN 43	per m wall	1660	2184	166	212	280	2.80	1100	34900	12.80	2	2	2
	per E	483		83	106	167	1.55	–	6230	7.67	–	–	–
	per D	1660		166	212	308	2.96	–	34900	12.80	–	–	–
	per Dr	1990		249	318	449	4.37	–	48670	12.40	–	–	–

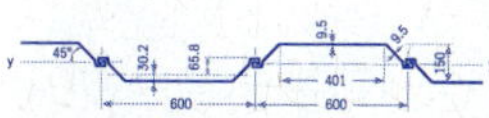
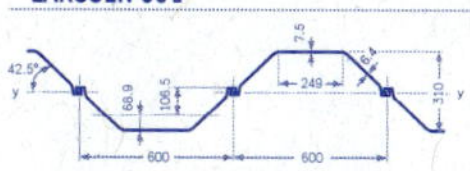
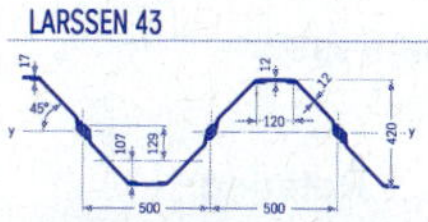

New Z Profiles

	W_y [1] cm/m Wall	cm³/ Single pile	kg/m² Wall	kg/ Single pile	I_y cm⁴/m Wall	t mm	s mm	h mm	b mm
HOESCH sections (LARSSEN interlock)									
HOESCH 1807	1800	1260	109.3	76.5	37800	9.2	9.0	420	700
HOESCH 2507	2525	1768	141.4	99.0	55550	11.5	11.5	440	700
HOESCH 2607	2600	1820	146.2	102.3	57200	12.0	12.0	440	700
HOESCH 2707	2705	1894	153.0	107.1	59400	12.7	12.6	440	700
HOESCH 2807	2765	1936	156.7	109.7	60830	13.1	13.1	440	700

Selection from the complete range	Pile section [9]	Section modulus W_y	W_z	Weight	Dimensions B	H	t	s	Circum-ference Devel-oped [10]	Area Steel cross section	Included [11]	Second moment of inertia I_y	I_z	Radius of gyration i_y
		cm³	cm³	kg/m	mm	mm	mm	mm	cm	cm²	cm²	cm⁴	cm⁴	cm
LP	LP 23	2310	2330	155	536	470	11.5	10.0	175	197	1960	54300	62400	16.6
	LP 24	2840	2400	175	536	470	15.6	10.0	175	222	1960	66700	64300	17.3
	LP 704	2540	3170	161	738	484	10.2	9.5	205	205	2650	61520	116800	17.3
	LP 25	3440	2720	206	536	470	20.0	11.5	175	262	1960	80700	72800	17.6
	LP 605 K	2830	3020	173	637	470	12.2	10.0	192	221	2160	66400	96200	17.3
	LP 606 n	3410	3080	188	636	485	14.4	9.2	196	240	2430	82800	97900	18.6

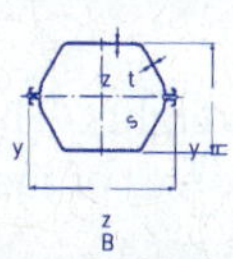

Interlock designs

LARSSEN section
Interlock design to DIN EN 10248-2 and E 67 of EAU 2004

LARSSEN 43, 430
Interlock design to DIN EN 10248-2 and E 67 of EAU 2004

HOESCH section (Finger-and-socket interlock)
Interlock design to DIN EN 10248-2 and E 67 of EAU 2004

PEINE locking bar/ PEINE sheet piling
Interlock design to DIN EN 10248-2 and E 67 of EAU 2004

UNION straight-web section
Interlock design to DIN EN 10248-2 and E 67 of EAU 2004

Lightweight section KL
Interlock design to DIN EN 10249-2

Source: Copyright ThyssenKrupp Steelcom

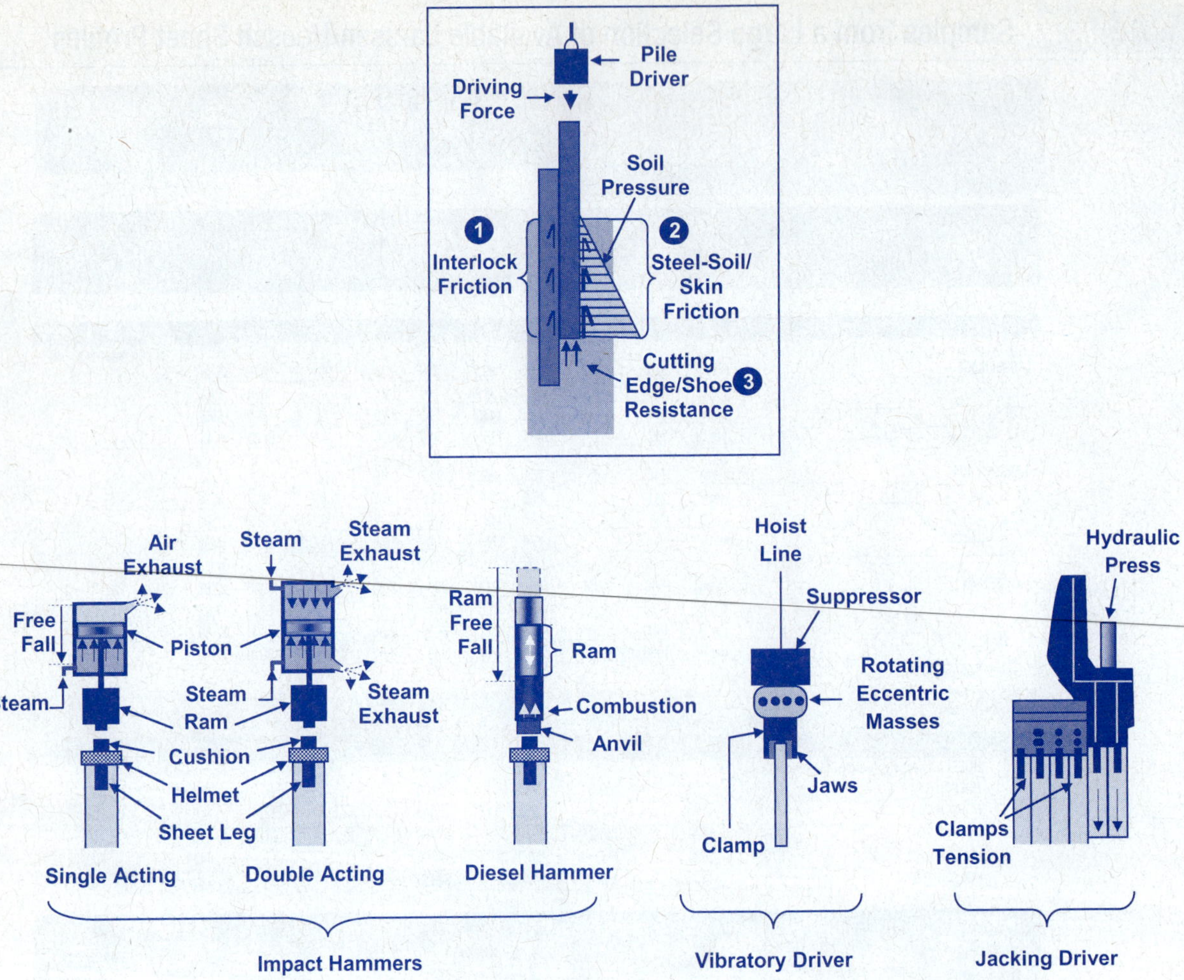

Figure 8.14 Pile-Driving Technologies and the "Work" They Do

the weight of the equipment, and possibly by additional weights. They fulfill a double function, since they are also needed to counter the upward forces of the vibrating masses, as presented graphically in Figure 8.15.

As indicated above, resistance to pile driving is a sum of friction and cutting forces, of which the friction along the surface is most significant (friction function $F_f = \mu_o * N, \mu_o =$ friction coefficient steel-soil). The friction function indicates that F_f varies with N, the normal force, which increases with penetration. Thus, it is not surprising that, as shown in Figure 8.15a, the forces needed to drive a hypothetical steel sheet increase with driving depth from 200 to 1200 kN (22.5–135 tons). As soil layers also change, linearity clearly simplifies the combined effect. Nevertheless, the main observation remains, the needed driving forces increase over time, making it desirable to have a vibrator that can change the force output in real time. Figure 8.15b shows the basic concept of timing the rotation of two eccentric weights.

The kinetic centripetal/centrifugal force F_c caused by a rotating mass can be calculated with

$$F_c = m * a_c = m * v^2/R = m * R * \omega$$

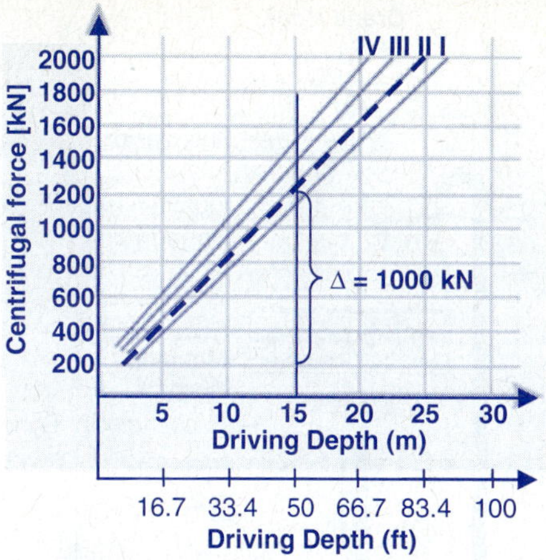

Curve	Sand&Gravel	Silty Soils
I	Loose	Soft
II	Medium	Plastic
III	Dense	Hard
IV	Very Dense	Very Hard

(a) Changing Centrifugal Force Requirements

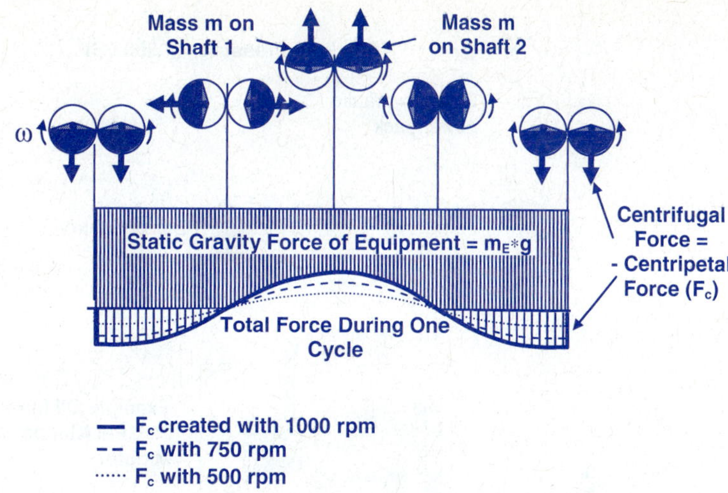

(b) Variable Pattern of Driving Forces

Figure 8.15 Changing Driving Force Needs Met by Vibratory Mechanism

where

R = distance between shaft to the center of mass of the eccentric mass
ω = rate of rotation (rpm)
m = rotating mass

The maximum eccentric moment of a vibrator is calculated by summing up all the moments created by the individual masses: $M_{ecc} = \Sigma(m * r)$ kgm.

The function shows that the amount of force can be adjusted by changing either the *mass*, the *radius*, the *rate of rotation* or a *combination*. It is not surprising that the most direct approach is to modify the rate of rotation (rpm), as hydraulic power offers motors and valves that can be opened and closed linearly. Synchronizing the two identical eccentric masses, the centrifugal force vectors equal each other out when the center of mass reaches 45° and 315°, respectively. As soon as the vectors align vertically, however, the forces sum up downward or upward. Their magnitude, also depicted in Figure 8.15, changes with the rpm. A hypothetical 1000 rpm is reduced by one-quarter, and subsequently by one-half, each time resulting in smaller amplitudes. Figure 8.16 presents an overview of an example vibrator system equipped with two sets of synchronized eccentric masses.

Vibratory hammers operate by spinning eccentric weights at high speed in a gear case. The crane/excavator is isolated from these vibrations by the suppressor housing, which is mounted on springs. The vibrations are transmitted to the pile via a hydraulic clamp. During startup and slow-down, high-frequency hammers, which reach 3500 rpms, have to pass through a phase where they are in dangerous resonance with the surrounding soil. This effect can be avoided by counterbalancing two or four masses, which is possible with the four eccentrics shown in Figure 8.16. Resonance-free vibrators have to synchronize the rotation so that all the masses counterbalance each other, not

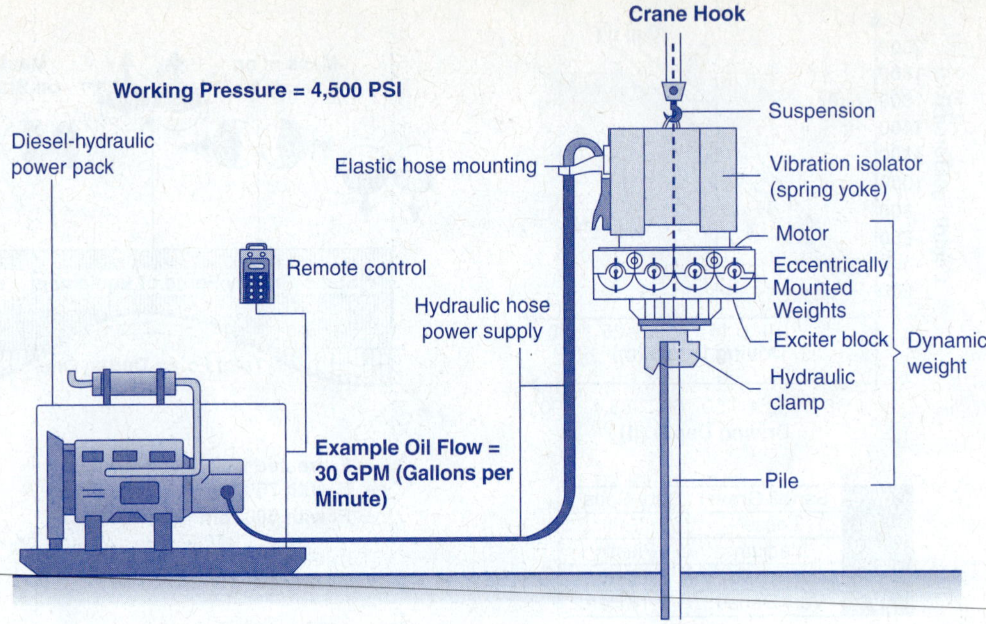

Figure 8.16 Main Components of a Vibratory Pile Driving System

only when horizontal but also when vertical. Thus, during the startup they are able to turn the eccentric moment off until the hammer has reached its designated operating speed.

Instead of needing a crane, some vibrators are built so they can be handled by excavators in place of a bucket. This allows rapid and accurate pile positioning and insertion/extraction. Their high frequencies ensure a high power-to-weight ratio, while their effective piling and extraction force is increased by the excavator's weight and the down force created by the hydraulic arm.

Thanks to the strong clamp, the same vibrators can also be used to extricate the piles when not needed anymore. It is apparent that the available force will be reduced as the gravity force now works against the operation. On the other hand, the cutting force is eliminated.

The hydraulic jacking method simply pushes the pile into the ground. This is a perfect solution for situations where vibrations and noise must be avoided. Figure 8.17 presents schematically the core mechanism of a piler that sits atop three or four already piled sheets by clamping itself to the top. In fact, this reaction-based approach depends on the already driven pile to push against. The world's first such equipment was introduced by the Japanese company Giken Seisakusho Co. in 1975. Since the skin friction of several piles exceeds the resistance force of a new section, enough counterforce (tension) can be expected to press it, using a hydraulic jack or ram.

The operating cycle begins with sliding a new sheet pile into the clamping mechanism. After an initial amount of down pressing enough to support the weight of the entire system, the four clamps holding on to the reaction piles, the power pack is lifted and inches forward by one pile. After the four clamps gripped the new set of pile-heads (see Figure 8.17), the jacking operation can continue until the desired depth, at which time the cycle starts over.

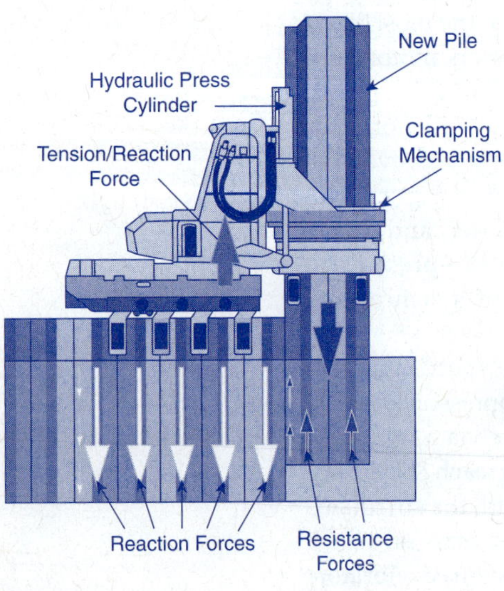

Figure 8.17 Reaction-Based Pile Jacking

8.5.2 DRIVING SOLDIER PILES

We learned earlier that the soldier piles and lagging make the most cost-effective retaining wall system. While we discussed three different lagging systems in Figure 8.12, we did not discuss the method of driving the soldier piles into the ground. First, the conditions have to be appropriate for this method, in particular the very low likelihood that a pile might "meet" a boulder when at half-depth.

Driving H-beams is similar to driving sheet piles. As we have focused so far on the vibratory and jacking approach to driving piles, we will now look at impact hammers. A major difference, of course, is the driving apparatus. As every pile is independent and unattached to a previous element, establishing and keeping a proper inclination require a support system. This is accomplished with the lead, a vertically suspended truss beam, depicted in Figure 8.18.

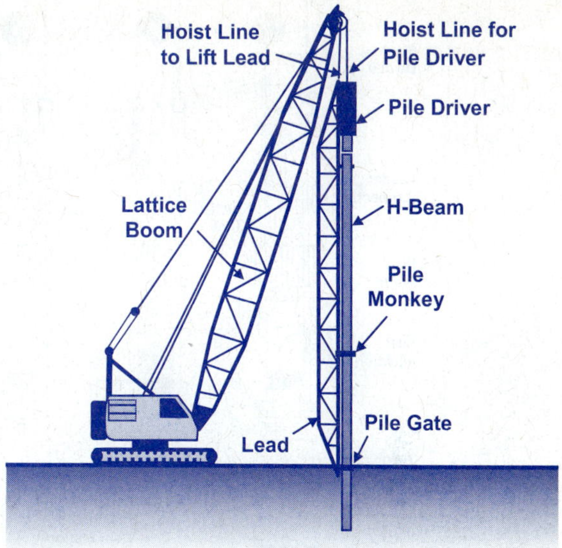

Figure 8.18 Pile-Driving Equipment with Swinging Lead

The pile-driving system shown in Figure 8.18 uses a crawler crane as its hoisting apparatus. As shown, the lead serves as a physical conduit for both the impact hammer and the pile during the driving process. In particular, the pile driver uses the lead as a linear guide as it hits the top of the long H-beam that is being held in place by the pile gate at the bottom.

Two kinds of pile drivers are common: (1) externally powered and (2) internal combustion hammers. While the first group depends on a power pack to generate compressed air, steam, or hydraulic pressure to operate single- or double-acting pistons that impact the top of the pile, the internal combustion hammers are self-reliant. In fact, they depend heavily on gravity to accelerate a mass of steel, the ram, turning static into kinetic energy. Some systems are also double-acting, configured with two combustion chambers at each end of the driver to accelerate the ram vertically upward and downward. Figure 8.19 presents the basic elements of a diesel hammer with an open end at the top.

Happenings Along a Hammer Stroke Cycle

1. After the anvil of the pile driver is set atop the top plate of the H-beam, the starting mechanism lifts the ram to its initial position and fuel is injected into the combustion chamber.
2. The release of the ram grab causes the piston to accelerate downward until the top plate decelerates it, thus causing a growing action-reaction force followed by a motion reversal. At this time, the air-fuel mix reaches its maximum compression, at which time it combusts. The resulting pressure pushes the ram back up.
3. During the upward motion the exhaust vent opens and the pressurized combustion gas escapes and new air is sucked in, while the hammer still moves upward.
4. After reaching the maximum height, the ram accelerates downward, closing the air vent and causing a new injection of fuel before it reaches the top plate for the start of a new cycle.

Closed-end diesel hammers operate similarly except that a compression or vacuum chamber is included on top of the piston to "push" the ram downward. This speeds up the blow rate of the hammer. It is not possible for the striking end of the ram of an impact hammer to directly adapt itself to all shapes of piles; therefore, it is necessary to insert driving accessories of various types between the bottom of the hammer and the pile to

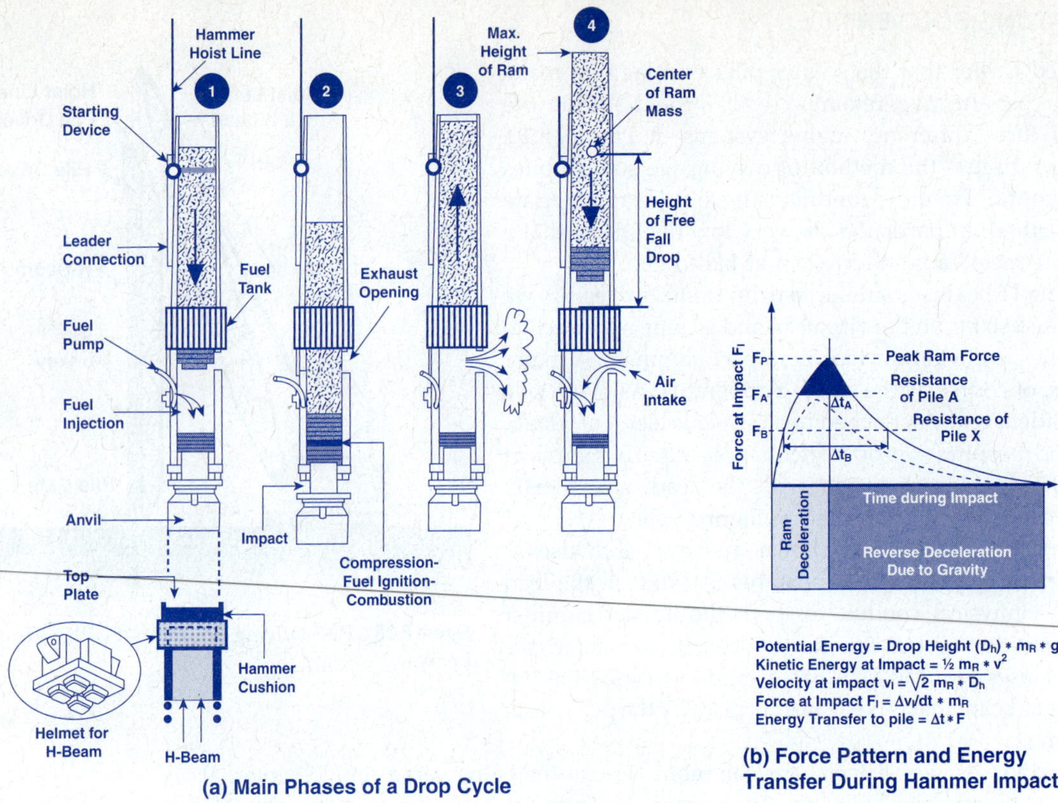

Figure 8.19 Mechanics and Physics of the Open-End Diesel Hammer

both mate the two geometrically and transmit the force of impact from hammer to pile. Impact hammers have a ***cushion*** under the strike plate that protects the battered parts from damage; it also modulates the force-time curve of the striking impulse and can be used to match the impedance of the hammer to the pile, thus increasing the efficiency of the blow. The actual material of the cushion and its configuration will vary, depending upon the hammer configuration and the cushion material. Of course, any hammer cushion should be installed and used in accordance with the recommendations of the hammer manufacturer. Internal combustion hammers need an *anvil* to trap the combustible mixture and thus allow it to build pressure. A *helmet* distributes the driving force more evenly onto the top of the pile.

In difficult situations, the H-beam is equipped with a small pipe to provide pressurized fluid to the shoe of the pipe that temporarily loosens the bond between pile and soil. This process is called jetting, as water or another fluid is injected via small jets, thus reducing the resistance of the pile at the bottom and on its skin surfaces. Of course, jetting will change the soil conditions around the pipe and may lead to later complications. Thus, jetting of pipes should be executed with caution and supervised by engineers experienced with this method.

8.5.3 AUGER-DRILLED CAST-IN-PLACE PILES (CIPP)

An alternative to soldier piles with lagging are augered cast-in-place piles (ACIPP). Although ACIPP walls are not totally impermeable to ground water, they can be easily excavated without a problematic change to the surrounding soil. If a large amount of water penetrates the wall, the specific location can be grouted later or the water captured

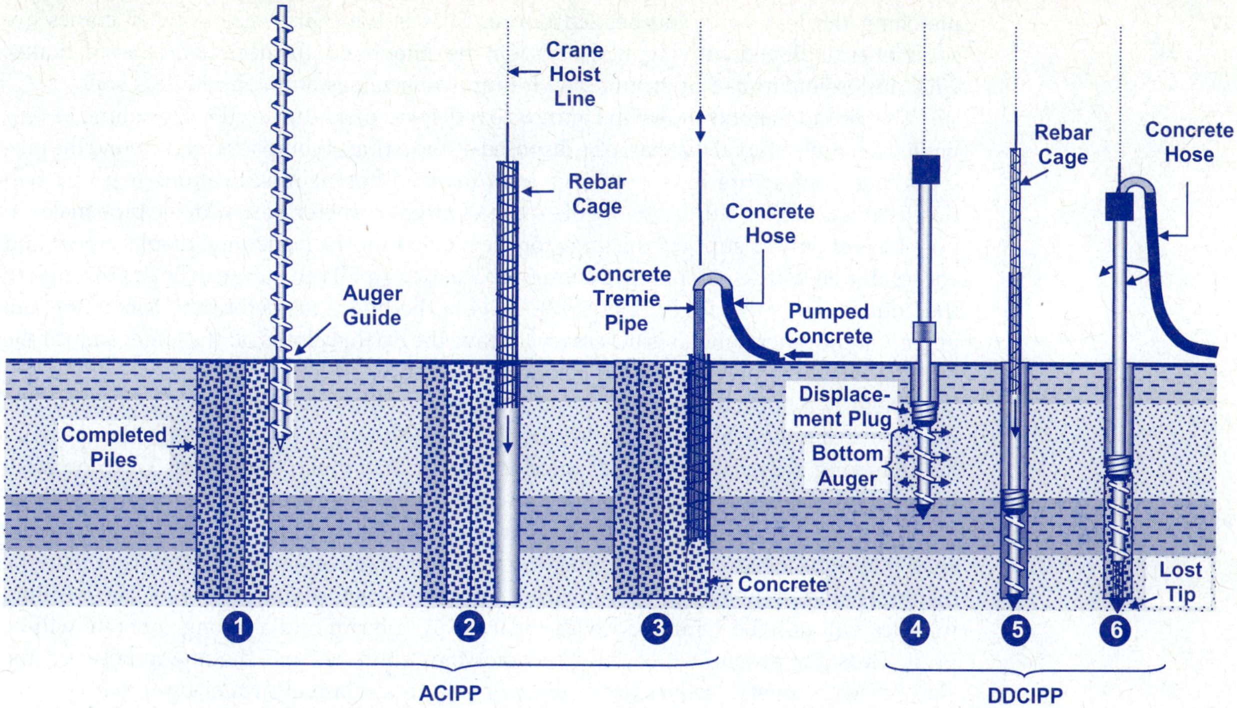

Figure 8.20 Main Tasks for Cast-in-Place Piles Using Auger and Displacement Drilling

and pumped to the surface. Figure 8.20 presents the main steps of constructing a new pile. The flight auger as shown in the figure is most commonly selected as the method of excavation. Augers consist of flights of helical or spiral fluting attached to a torque bar or stem that is either solid or hollow.

Because of the large torque forces needed to drill and to overcome friction with the shaft wall and lift the soil from the bottom to the top, the diameter and length of flight augers are kept at around 0.3 m (1 ft) and 18 m (60 ft), respectively. The torque bar is rotated by a hydraulic motor stabilized by the crane's lead (see Figure 8.18), driving the auger on its own weight. The tip or bottom of the auger has a fitting that allows the mounting of different cutting tools that best match the geotechnical material (e.g., drill bit for hard material). Following the principles of Archimedes' screw, the spiraling flutes lift the loosened material. To save time, the contractor will select an auger length that does not require the adding or removing of extension elements. This, of course, will require

Process Sequence of CIP Pile Construction

1. Alignment of continuous flight auger (CFA) drill powered by hydraulic motor attached to leader; auger drilling and spoil removal.
2. Lowering of prefabricated rebar cage into drilled shaft.
3. Placement of tremie pipe and pumping of concrete to fill the shaft from bottom up without segregating the concrete. Tremie pipe is continuously raised as the casing is installed.
4. Displacement drill (DD) method with hollow drill stem. Displacement bulge forces the drilled soil to compact and force itself into the surrounding area.
5. Small-diameter rebar cage is lowered through the center.
6. Concrete is pumped through the stem, forcing the auger tip to be lost. Auger is slowly pulled back and rotated to compact the concrete at the bottom of the pile.

matching the lead with the needed depth. That is why large-size crawler cranes are sometimes deployed with gigantic leads. On the other hand, the need to lift the soil makes CFA impossible to use in grounds with high water tables and cohesionless soil.

The second method shown in Figure 8.20 is displacement drilling (DD), resulting in very limited amounts of spoil that has to be disposed of and a ring of compacted soil around the pile.

While contractors have come up with many different modifications (e.g., partial-displacement drills), the basics are the same. A large-diameter hollow drive pipe makes it possible not only to support the large torque needed for the horizontal displacement and consolidation, but also to provide enough space for a small rebar cage to be fed through to the bottom. Thus the auger does not to have to be lifted out before the concreting can begin. On the other hand, it is necessary to leave the tip that closes off the auger stem at the bottom, which is forced out by the concrete pressure.

There is a special approach to organizing the piles with the goal to decrease the possibility that the wall will permit water leakage. Shown in Figure 8.20 ❹, the specialty lies in having the secant wall piles interlock rather than touch each other. To accomplish this, piles are drilled with an open space between them that is smaller than the diameter of a pile. Before the concrete in two adjacent piles is fully hardened the drill augers a new shaft in the space between them. Since the diameter of the auger is slightly larger than the distance between the two plies, a small section from each is removed as well. These (filling in) piles will now be reinforced with rebar or an H-beam and a strong concrete will be used. Thus, the second set of piles becomes the "soldiers" and the one in between the "lag." A well constructed secant wall can result in a relatively water-tight wall.

8.5.4 WALLS MADE BY TREATING THE SOIL IN-SITU

The concept of reinforcing mixed soil with soldier H-beam piles was first tested and used in 1970s Japan. The concept belongs to the family of deep mixing methods where the soil is not excavated but instead treated to improve one or several of its undesirable characteristics. In the case of the soil-mixed wall (SMW), the soil is blended with cementitious materials, called binders, usually in slurry form, that are produced in a plant on the top and pumped down the hollow stem shaft to the mixing tools. At that point, jets are installed that inject the slurry into the surrounding soil to be immediately mixed in by rotating tools such as auger flights, wings, blades, or paddles. The result is a wall made of in-situ soil material, which provides not only an excellent water barrier, but also a vertical wall with significant stiffness, perfect for retaining walls, causing no vibration during installation, and only very little spoil material that has to be disposed of.

Figure 8.21 presents three basic technologies that have been developed out of a wide variety of special applications used around the world. The mixing apparatus consists of a crane-mounted drive table or a drilling rig that turns a hollow shaft with the mixing tools attached at the lower end. The top of the shaft is attached to a "slurry swivel" that connects the slurry hose to the rotating mixing shaft, which is not necessary for the horizontal cutter-mixer system, as the shaft does not rotate. They all follow the same basic procedure to reach the desired depth, starting with preparing the slurry that has been found to be most effective in combination with the existing soil. This is accomplished with an on-site plant equipped to measure the appropriate amount of cement for a batch of slurry blended inside a tank from which it can be pumped down the hollow shaft, as depicted in Figure 8.21d.

After the slurry is prepared and the mixing tool is properly located, the mixing tools are engaged and slowly pushed in (e.g., 10–20 rpm) and advanced at 0.5–1.5 m/min (1.64–4.9 ft/min). As the mixing tool proceeds downward, the cement slurry is pumped through the hollow shaft(s) and jetted out through the specially designed openings. Here, the slurry immediately becomes part of the mixing process, lubricating the soil-tool interaction and helping to breaking up the larger soil aggregates into smaller and smaller pieces to be later

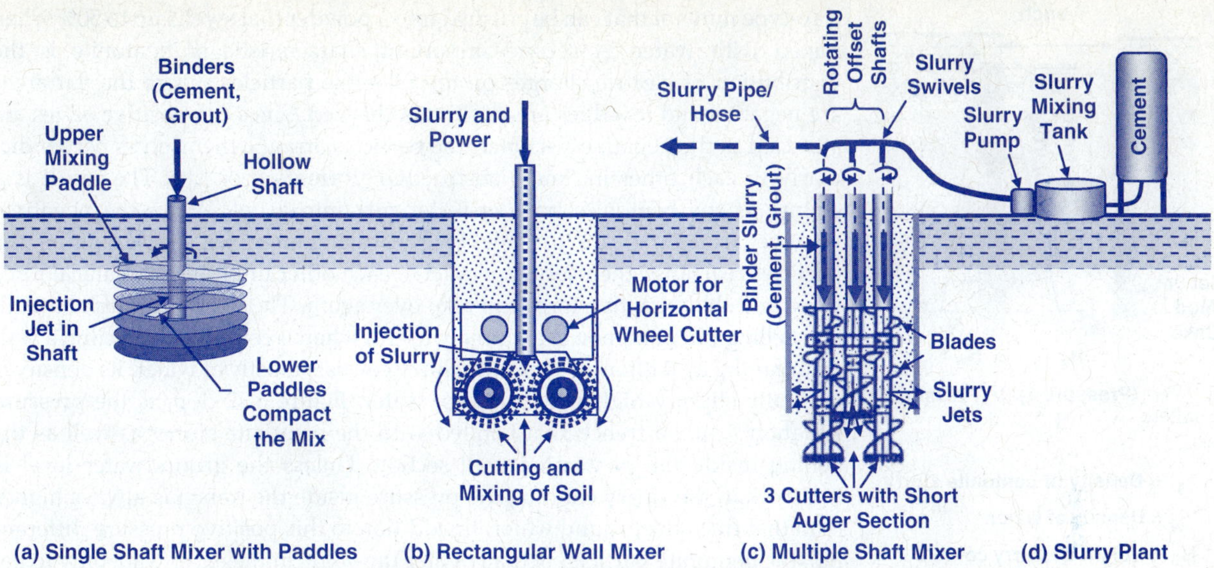

Figure 8.21 Three Technologies to Construct In-Situ Soil-Mixed Walls

bound together by the curing cement. After final depth is reached, the mixing tool is slowly retrieved to the top while a much reduced amount of slurry is pumped down.

Eventually, the soldier piles have to be lowered into the soil mix before the cement has started to cure. As the soil mix will behave much like a liquid at that time, the H-beam is suspended from a rigging above the wall segment, or the beam will sink into the mix and might even come to rest at an inclined angle.

The exhaustive literature about SMW includes some rules of thumb that recommend installing the soldier piles at 1 to 1.5 m (3 to 5 ft) apart, and the soil-cement compressive strength should reach 0.7 to 1.0 MPa (100 to 145 psi).

8.5.5 SOLID CONCRETE DIAPHRAGM OR SLURRY WALLS

In a human body, the diaphragm is the thin membrane made up of muscles and tissues that separate the chest from the digestive system. In much the same way, a diaphragm wall is constructed as a means to partition off an area in the ground. This technique was introduced in Europe in the 1950s, mainly to stabilize retaining walls for open excavations in narrow downtown areas that had been destroyed during the war. Experience with this technology led to the observation that the walls were effective not only in retaining the soil, but also in groundwater cut-off, and very important as temporary or permanent load-bearing supports. Thus, the wall could be become part of the building of itself where traditional retaining walls were stand-offs leaving space for the building of the basement wall formwork. However, diaphragm walls could be built right next to the adjacent building foundation without any space wasted. The virtually vibration-free operation minimizes the potential for ground movement and its resulting detrimental settlements in surrounding buildings. The multipurpose use made it a very economical solution in urban construction, which in turn spurred the contractors to experiment with innovative new equipment, some of it tailored to solve especially difficult situations such as lack of available height. In addition, while the early depth in reach of the excavation equipment was 20 m (66 ft), today's machines reach up to 120 m (400 ft).

What made the invention possible was the ingenious use of a peculiar mix of bentonite and water that had a very important characteristic. Bentonite is a natural

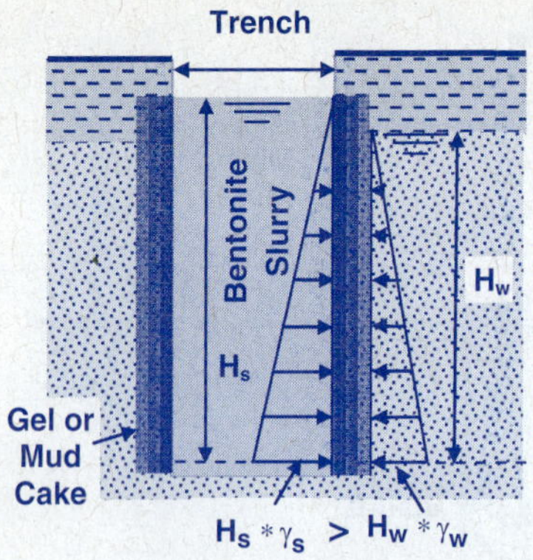

Figure 8.22 Characteristics of Bentonite Slurry

clay-type mineral that can be crushed into a powder that swells up to 50% when mixed with water. A second important characteristic of bentonite is the distribution of electric charges on the platelike particles where the flat areas are negative and its edges are positively charged. Since the positive edges are attracted to the negative flat areas, the particles arrange themselves perpendicularly to each other in "most obstructive" formation of |-|-|. The result is a gelling of the bentonite powder-water mix into a gel. When the negative-positive attractions are forcibly broken, using a mixer, for example, the particles rearrange themselves parallel to each other until the mechanical force stops and the attractive force can take over again. The density combined with this gelling phenomenon is extremely useful when it comes to protecting a wall from caving in. Although bentonite slurry consists mostly of water, its density is naturally slightly higher than that of water. Figure 8.22 depicts the pressure situation inside a trench that is filled with the bentonite slurry as well as the gelling inside the penetrated wall section. Unless the groundwater level is higher than the slurry column, the pressure inside the trench is always higher than that from the ground water. In addition to this positive pressure differential, the bentonite particles actually enter the voids in the earth wall, only to get stuck in the smaller crevices. However, this has an important effect in that it stabilizes the soil particles inside the border region of the soil and, with it, of the entire face inside the trench.

As mentioned above, the use of bentonite slurry was critical in the development of this technology. What is missing, of course, are the equipment and tools able to build the large underground walls. Figure 8.23 introduces a selection of the most common types of machines and excavation tools used.

The carrier technology graduated from large crawler cranes to specialized and compact rigs that were easily deployable as well as able to quickly track from one location to another. Hydraulic systems enable the rapid adjustments of the telescopic mast to fit under low obstacles such as bridges. Both clamshell-type grabs and wheel cutters are used, the latter allowing the control of its verticality. The grab adopted the clamshell mechanism, two hinged jaws actuated by rods, and integrated it into a heavy grab frame to provide the jaws gravity force during the closing of the jaws. Sufficient down force (= weight of grab – buoyancy) is especially needed to dig in hard material. The grabs can be closed either by a separate closing line from the crane or by a hydraulic piston designed to operate the jaw rods up or down (see Figure 8.23b).

The trench cutter, taking a totally different tack in digging and spoil handling, allowed the record depths that the industry has become used to. One key component that is built into the cutter frame is a guidance system that senses the verticality of the frame along two dimensions (2-D) using built-in electronic inclinometers. The active control of the direction of the cutter is enabled by four sets of 2-D steering plates, two on the top and two at the bottom, that can be hydraulically extended out, thus increasing the distance between the frame at the point and the wall. Even the cutter heads can be pivoted by the operator, who is able to monitor and control the large system from his cabin. The result of this guidance system is that very small tolerances can be achieved.

The second key difference from the traditional trench excavators is the fact that the bentonite not only stabilizes the walls but has become the transport media for sand and stones that are loosened by the rotating cutters. Figure 8.23b shows a slurry pump mounted right above the cutters. The intake of the pump is designed to most efficiently "collect" the cuttings and to push them through the slurry hose to the desanding plant, where the solids will be separated from the slurry, which in turn is returned back to the trench. Thus, the desanding operation takes on a key role in the overall operation. Figure 8.24 provides a graphic overview of the hydrocyclone, the technology that cleans the slurry.

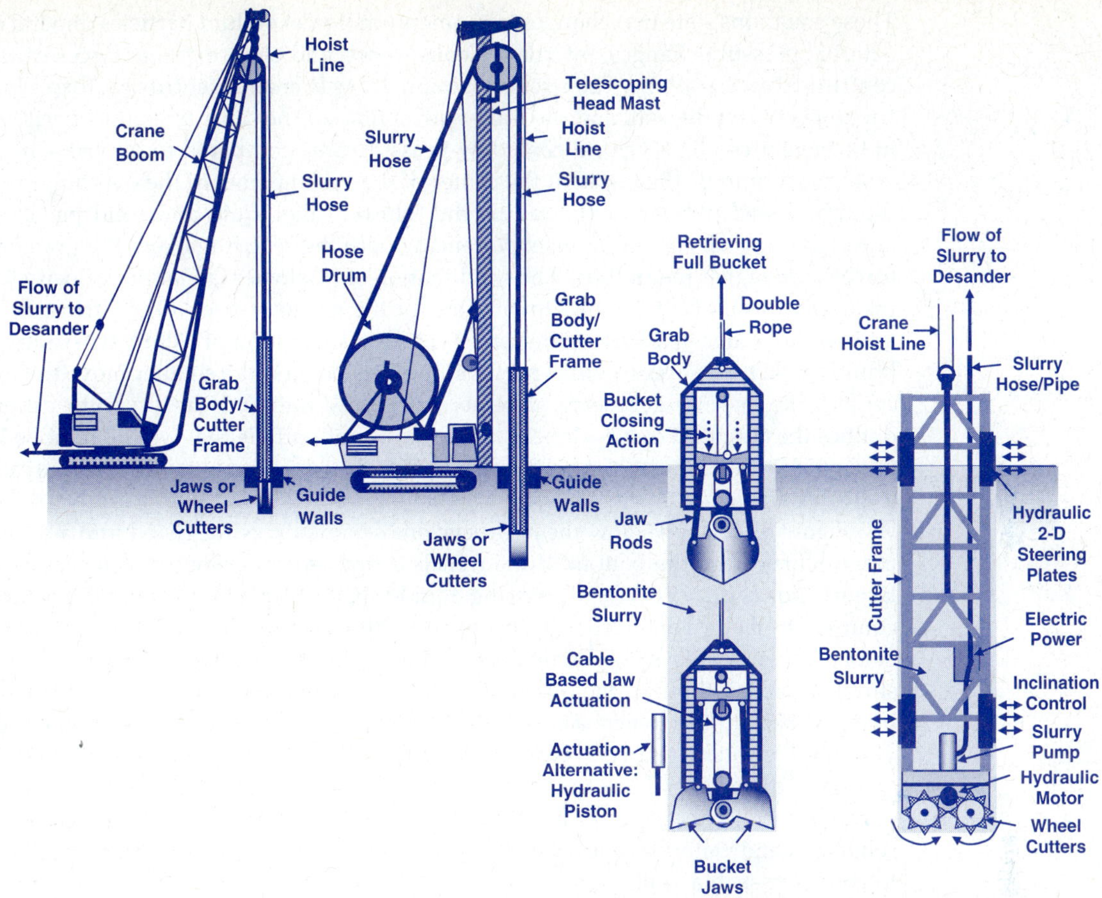

(a) Carrier Models for Handling Grabs and Trench Cutters (b) Grab and Steerable Trench Cutter

Figure 8.23 Hardware to Excavate Narrow Slurry Wall Trenches

The flow conditions in a hydrocyclone are characterized by a three-dimensional, turbulent two-phase flow, a complex process that is still not adequately understood. The graphic representation in Figure 8.24a shows how the pressurized slurry-mix is pushed tangentially into the upper end of the conic cyclone, forcing the liquid into a rotation. The Navier-Stokes equations describe the balance of forces acting at any given region of the fluid.

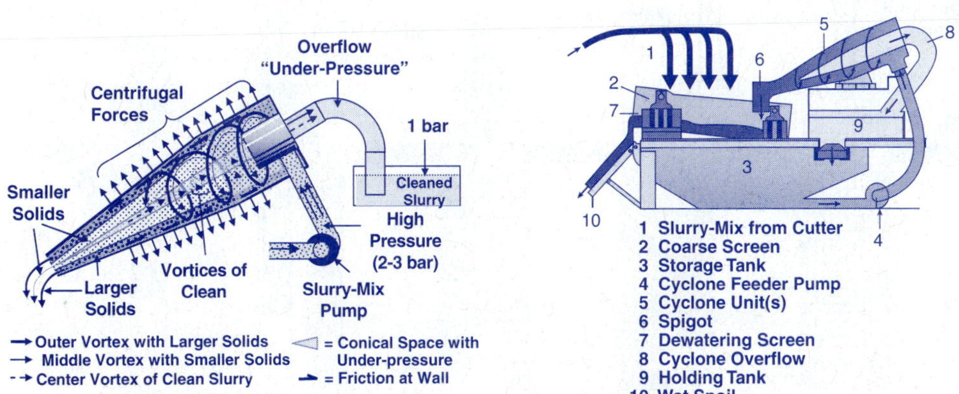

(a) Model of Hydrocyclone "Mechanics" (b) Basic Components of a Slurry Desander

Figure 8.24 Principles of Cyclonic Separation and Desanding Slurry

These equations state that changes in momentum (force) of fluid particles depend on the external pressure changes, internal viscous forces, and other external forces caused by centrifugal accelerations. The circular motion not only creates centrifugal forces but also friction between the wall and the fluid with solids. At the same time, the overflow pipe in the center creates a central conic low-pressure area compared to the pressure at the wall, "vacuuming" the liquid at the center of the cyclone toward the overflow pipe. On the other hand, the larger the radius the "slower" the liquid and solid particles, the latter, due to their higher mass, experiencing both the centrifugal and the gravitational forces with higher magnitude. The combination of (1) slower tangential velocity toward the cyclone wall, (2) the slow-down effect of the friction forces, (3) centrifugal forces, and (4) the under-pressure at the center causes a separation of solid particles from the liquid. In fact, the larger solids end up at the outside, and the smaller move toward the center. The cyclone geometry, together with flow rate and the pressure difference, defines the cut point of a cyclone. This is the size of particle that will be removed from the liquid with a 50% efficiency. Particles larger than the cut point will be removed with a greater efficiency.

Figure 8.24b shows how the hydrocyclone separation is integrated into a desanding plant. Slurry from the bottom of the trench is first fed to a sieve/screen removing the larger solids (e.g., > 5 mm). The passing liquid-mix is collected in a large storage tank that connects to the cyclone feeder pump pushing the slurry with 2–3 bars into the hydrocyclone above. While the solid-mix leaves the cyclone at the lower spigot, the cleaned slurry is sucked through the overflow pipe into the "cleaned slurry" tank, ready to be pumped back to the trench. The solid-mix, however, is drained for the last time on the vibrating dewatering screen before being disposed down the chute to the spoil pile.

After discussing the main resources needed to excavate and secure the deep but narrow trench, we are ready to follow the process of actually building a cast-in-place reinforced and watertight slurry wall. Figure 8.25 provides schematic graphs for six tasks to construct such a wall.

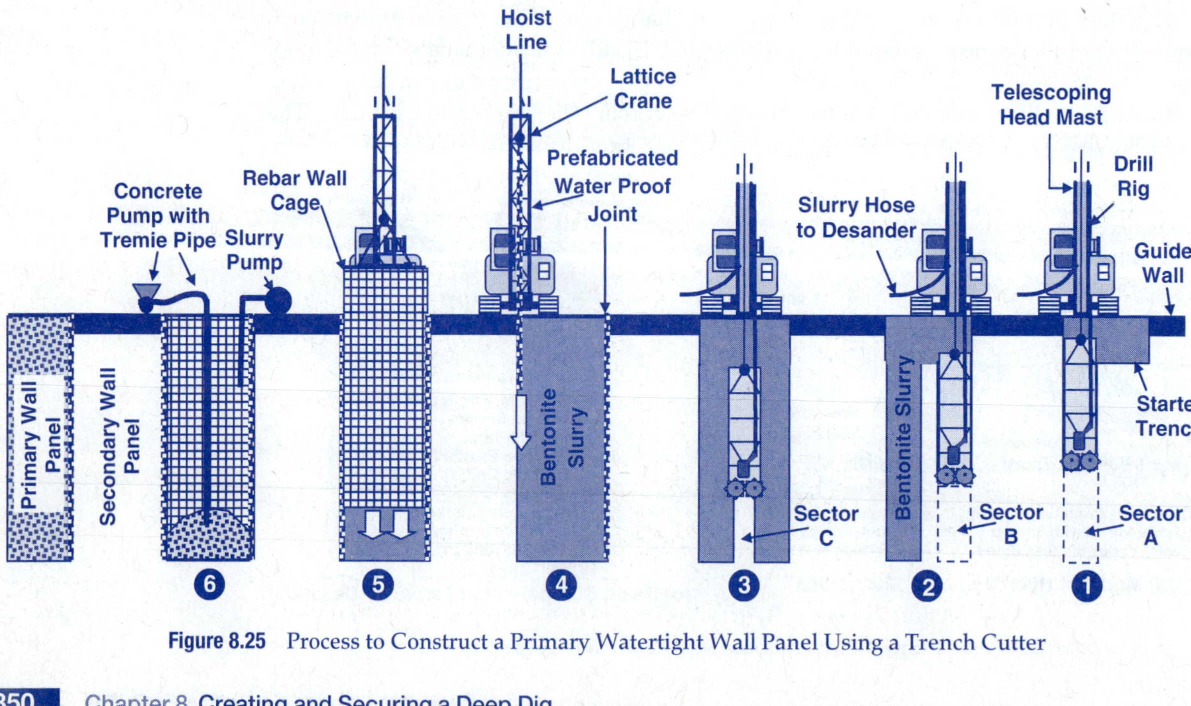

Figure 8.25 Process to Construct a Primary Watertight Wall Panel Using a Trench Cutter

Before starting step ❶ of a segment, the geotechnical investigation as well as experience from digging the adjacent primary wall section must have revealed what can be expected during the excavation. In addition, a sufficient quantity of bentonite slurry of acceptable quality (e.g., 5% bentonite in water with a minimum density of $64\,lb/ft^3$ ($1.03\,gm/cm^3$)) has to be ready inside a mixing tank. Finally, the guide walls on both side of the trench and the starting "channel" are needed to properly direct the cutter.

Phase ❶ The width of one wall panel or section has been divided into three sectors A–C, with C being the center. In this phase, sector A is dug by the trench cutter suspended from a drill rig. The slurry pump sucks the cuttings and pushes the mix into the desander. The pumped-out slurry mix is replenished with cleaned slurry from the desander.

Phase ❷ Phase 2 continues the excavation of the panel by moving the cutter over to sector B, leaving a pile in the center.

Phase ❸ In this phase, the center sector will be removed by the cutter, thus leaving a cleaned-out panel filled with bentonite slurry.

Phase ❹ While the drill rig moves on to excavate the next panel, a lattice crane starts to install prefabricated concrete elements at the two boundaries of the trench. Figure 8.26 presents three common joint configurations including the prefab-element that joins two adjacent cast-in-place panels. This configuration has a cast-in dumbbell rubber-type waterstop on both sides along the entire length of the element.

Phase ❺ A prefabricated rebar cage is lowered and set inside the slurry-supported wall section. Figure 8.25 shows that the lifting and turning of large cages requires special lifting tools and may involve more than one crane.

Phase ❻ The last phase of constructing a primary wall panel is concrete placement with a pump and tremie pipe so that the concrete is pushing up from the bottom. In order to prevent the slurry from mixing with the concrete, its sand/silt content (at the bottom) should be kept below 4%. Simultaneously, the excess slurry is removed via an overflow or a pump.

The joints between two adjacent panels represent a weak point for creating a watertight retaining wall. Consequently, different sealing mechanisms have evolved, starting with the stop-end casing to prefabricated concrete elements that have cast in rubber-based waterstops.

The *stop-end casing* is only a hollow steel pipe with a diameter matching the wall thickness. It needs to be anchored at the top and the bottom so that concrete is not able to

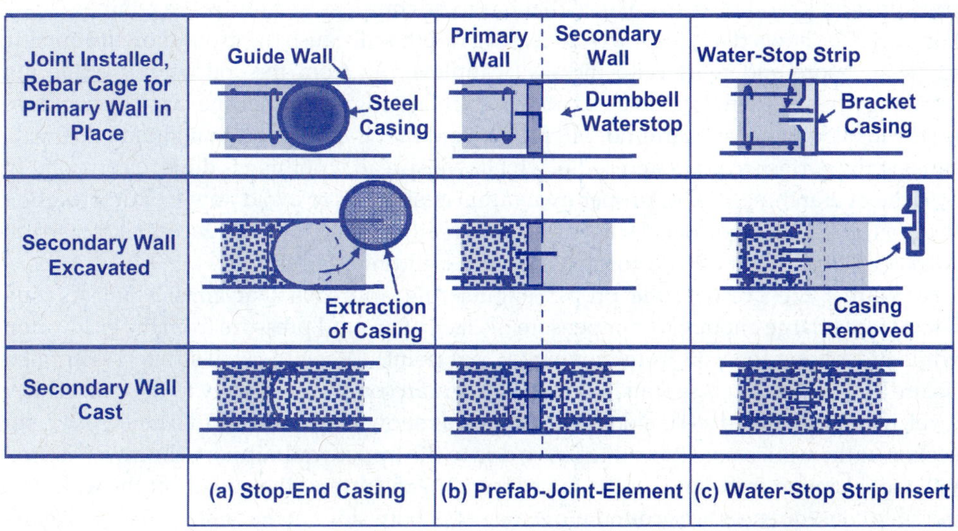

Figure 8.26 Three Joint Designs for "Watertight" Cast-in-Place Diaphragm Walls

mix with the soil. After the concrete has cured and possibly after the secondary panel has been excavated, the casing is extracted, leaving the concave surface to be jointed with the fresh concrete. From the three options, this option is the least effective in preventing water from penetrating the joint although it lengthens its path when compared to a straight cross-section. Figure 8.26b illustrates the use of a prefabricated joint-element with rubber dumbbells extending into both wall sections. The element is positioned inside the section after the excavation is complete, either before or after the lowering of the rebar cage. Due to the weight of the long, slender element, special care has to be taken during handling in order not to put an excessive stress on the element. The concave shape of the element is able to protect the waterstop from damage during the excavation of the secondary wall. Finally, Figure 8.26c combines the positive traits of (a) and (b). Instead of casting the waterstops into a large prefabricated element, this method uses a shaped metallic end-casing to position a strip insert holding the water-stop from top to bottom. The casing will be removed only after the secondary wall has been excavated, protecting the rubber stops, but before the concrete is poured encasing the dumbbells.

8.6 Bracing Systems to Support Retaining Walls

Although diaphragm walls are extremely strong and in direct contact with the soil, leaving no opportunity for the soil to move, deep walls bound into the ground but unsupported on top will experience such large stresses from the soil pressures that deformations will most likely occur. One has to consider as well that the joints, although watertight, are hinge points in the wall, since the rebar does not continue across the joints. As a result, it is necessary to brace the earth retaining wall in order to minimize its horizontal movements and accompanying settlement of the surrounding soil and any structure that is adjacent to the wall.

The last section, about the construction of retaining walls, will discuss the installation of bracings to minimize the possible horizontal movements during excavation.

As we reviewed earlier, the main function of a solid retaining wall is to prevent the earth behind the wall from moving or settling due to loss of counterpressure from the excavated soil. When excavating below the groundwater level without the availability of dewatering option, a watertight wall may also have to withstand the water pressure. Figure 8.27a illustrates how the earth pressure in its natural condition increases with depth in an approximately linear manner. Removing the soil on site, with the help of a loader or excavator, will therefore reduce the earth pressure on one side, creating a pressure and lateral force imbalance that has to be compensated for by the retaining wall. Figure 8.27b shows that the slowly increasing depth will cause the ground on the outside to "lean" more and more onto the wall cantilevering from the soil below. While not breaking, the wall will start to bend and even rotate slightly under the large lateral force on the top. Of course, the ground will fill the void that is created and will follow the wall, causing the surface to crack and settle, which will spread if nothing is done to correct it. In fact, the overarching goal of proper excavation design has to avoid any wall deformation that could result in soil displacement as this really is a sign of a failure that is hard to correct and can have drastic consequence if a building is nearby.

Figure 8.27c presents the proper engineering approach that aims at introducing lateral forces large enough to compensate the lack of ground pressure from the excavation while preventing the wall from moving (at that point). When the wall support is applied before a critical depth is reached, the excavation can continue, but only to the next critical level, in order to avoid a second type of failure, as shown in Figure 8.27d. While the upper section of the wall is not allowed to move due to the upper wall support, the lower section will face a similar imbalance of the forces from the inside and the outside of the wall. The possible consequences are ground movements on both sides of the wall, as shown. Again,

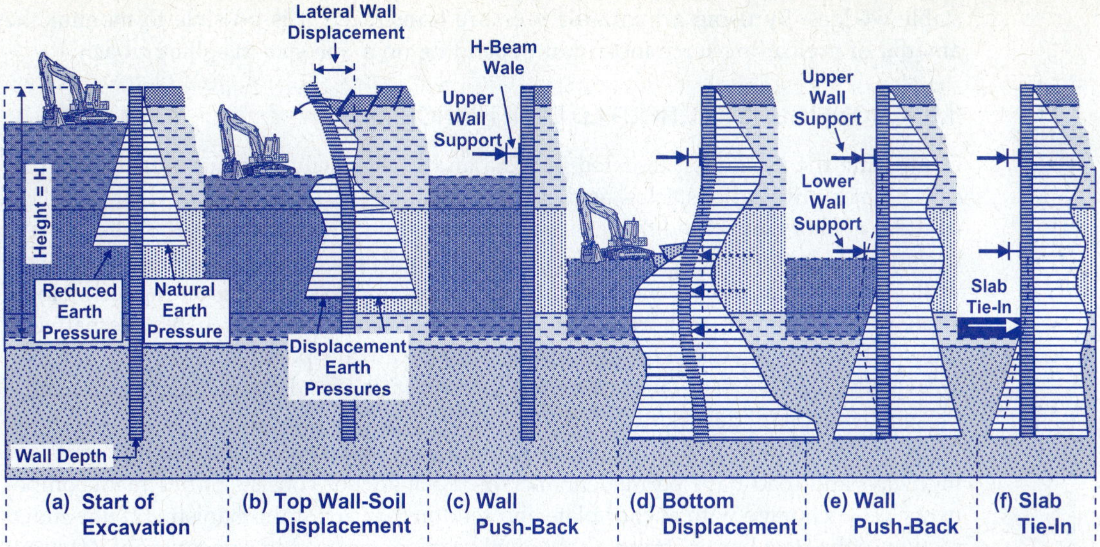

Figure 8.27 Possible Wall Movements and Earth Pressure Changes During Excavation and Wall Stabilization

the proper engineering approach is to install wall supports before the critical excavation depth is reached. If the wall depth has been properly designed, the foundation slab should be at a safe depth and, after being tied into the wall itself, provide a solid frame-type connection.

We hope that the discussion about leaving out the wall support has been convincing enough that the reader has began to wonder about how to apply those large stabilizing forces. This is the topic of the next section. Figure 8.28, in effect, shows five common methods in three different situations.

The first wall support system portrayed is probably the oldest, albeit not necessarily the simplest to build and maintain. It consists of compression struts, made of timber or steel beams/pipes, using the earth (and water) pressures on one wall to support the same on the opposing walls. In order to distribute the point forces from the struts, they are tied to horizontal force-distributing H-beams, called wales, that in turn support the soldier-piles or retaining wall sections. With the goal of countering the small displacements of the wall that may have occurred already, the struts may be preloaded. One way to accomplish this is to use two hydraulic jacks that fit between jacking plates, welded onto the strut, and the wale. The space that is opening up between wale and strut can be filled with solid and

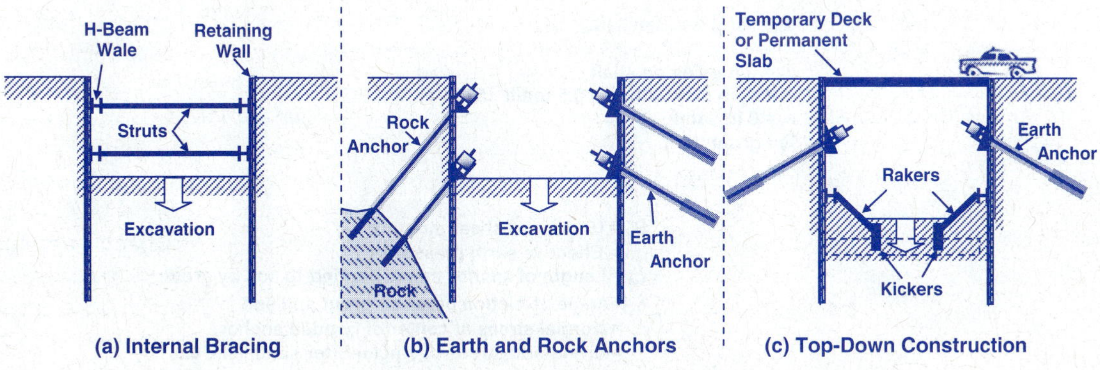

Figure 8.28 Approaches to Supporting Retaining Walls During Excavation

stable wedges. By using a hydraulic pressure transducer, it is possible to monitor the amount of preloading force instead of depending on a one-size-fits-all approach.

8.6.1 CONFIGURING GROUTED EARTH ANCHORS

Comparing the methods presented in 8.28a and b one easily recognizes that the struts cause major physical obstacles, as they are spaced approximately 4.5 m (15 ft) apart. In contrast, anchors set behind the retaining walls provide wide-open spaces. Tiebacks have become popular for exactly this reason, although they may have to penetrate a watertight wall, and they cost more than the struts. Advances in pressurized grouting technology, allowing the setting of anchors outside of rock, enabled its use as a reliable alternative to internal bracing. Nevertheless, anchors may not be always possible, especially in urban areas with many underground obstacles and sensitive foundations. Figure 8.28b and c present five types of tiebacks constructed both for temporary or permanent use. Retaining walls for excavation are generally of a temporary nature, while retaining walls for highways and road have permanent anchors. Each anchor consists of five main components: (1) anchorage with anchor plate and tendon/bar arrest mechanism, (2) unbonded section of the steel bar or strand, (3) bonded segment encased in cement grout, (4) grout tube to inject cement grout, and (5) tendon/bar corrosion protection. The length and size of the bonded part are key to how much force the anchor is able to hold without being extracted. To calculate the ultimate anchor load, the Army Corps of Engineers has established the simple formula presented in Figure 8.29.

As the formulas show, there is a large amount of uncertainty that is caused by the wide variety of soil conditions one can expect. A key input variable is the effective earth pressure caused by the soil resting above. The term *effective* means that the uplift created by groundwater has to be recognized in the calculation. This means for the soil that is submerged, the density of water has to be subtracted (10 kN/m^3 (62.4 lb/ft^3)) from the dry soil density. Thus, if the anchor is under 20 ft (6 m) of fine sand with 18 kN/m^3 and the water table is at 10 ft (3 m), the effective pressure is calculated as follows: $\sigma_{v'} = 3\,m * 18\,kN/m^3 + 3\,m * (18 - 10)\,kN/m^3 = 78\,kN/m^2$ (11.3 psi).

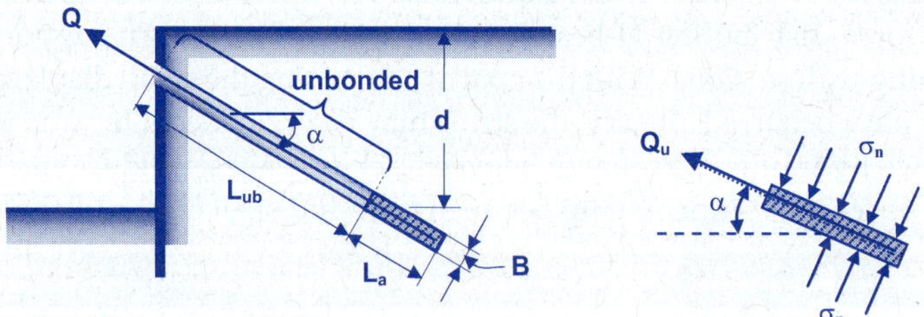

$Q_u = B(c_a + \sigma_n \tan \delta)\, L_a$

c_a = adhesion on shaft
$c_a = s_u$ in clay with $s_u < 0.5$ ton/ft^2 (50 kPa)
$c_a = 0$ in sand
= grout

S_u = Undrained shear strength
$\sigma_{v'}$ = Effective earth pressure
L_a = Length of anchor that is bonded to soil by grout
δ = Angle of friction between grout and soil
σ_n = Normal stress at center of bonded anchor
$= \sigma_{v'} *$ in-situ stress coefficient (for sand 0.4 – 0.5)

Figure 8.29 Calculating Maximum Load of a Grouted Anchor

It is critical that the unbonded section not move and stay corrosion-free. There are several different means to select from. One supplier encases the strands in plastic pipes with grease between the steel and plastic. The spacing between tiebacks cannot fall below approximately 3 m (10 ft), as the pressure bulb around each anchor needs sufficient space to dissipate.

Figure 8.28c illustrates two more concepts that are sometimes used in combination with each other. As the label indicates, in the top-down method of construction, the top built first offers a strategy to keep a highway operational while the construction is proceeding underneath. After the diaphragm wall is completed, either a temporary deck or a permanent concrete slab is built on top of the wall. If the span is too wide, columns or piers can be drilled as additional supports.

Now excavation can start underneath while the side walls are being supported by the deck struts or a solid slab. To keep the space free, tiebacks are installed as a second tier of support, while the third tier uses rakers that brace the wales against kickers. When the basement slab is being built, the kicker can be replaced with inserts into the concrete.

After the removal of the struts and rakers and the detension of the tiebacks, the basement and elevated slabs will have to be able to provide the permanent lateral supports to resist the earth pressure that has built up. This transfer of the stresses from the temporary to the permanent supports has to be carefully planned.

Because of the increasing importance of tiebacks, we should spend a little more time in figuring out how they are actually installed, commonly by specialty contractors. The excavator has to prepare a flat area in front of a wall section, and one or several drilling rigs have to be set up. Figure 8.30 portrays a small but mobile and fully hydraulic drill. Not shown are a small plant to mix the cement grout and a place to store the tendons or bars to be inserted.

The tracks allow the drill to maneuver easily within the excavation pit, and the many hydraulic articulators and motors provide the strength to position the drilling feed boom at the desired height and angle. In order to reach further than the length of the feed, it is possible to extend the boom length or to couple drill rods together. Drill rods fulfill two important functions. First, they connect the drill motor with the drill bit in order to transmit thrust and hammer blows while rotating the drilling tool. Secondly, the tube-like construction of the rods is essential in pumping gas or fluids to the drill bit in order to clean out the borehole and to cool the grinding bit.

Most often drillers utilize rotary-percussion drills that combine chipping and grinding actions to penetrated layers of sand-silt, sandstone, and even rock. Figure 8.30 depicts the rotary-percussion hammer, also called a drifter, located at the end of the rod on a frame that slides along the feed when the pulldown motor activates either a steel rope or chain connected to the feed slide. Here is a set of specifications for a small drifter:

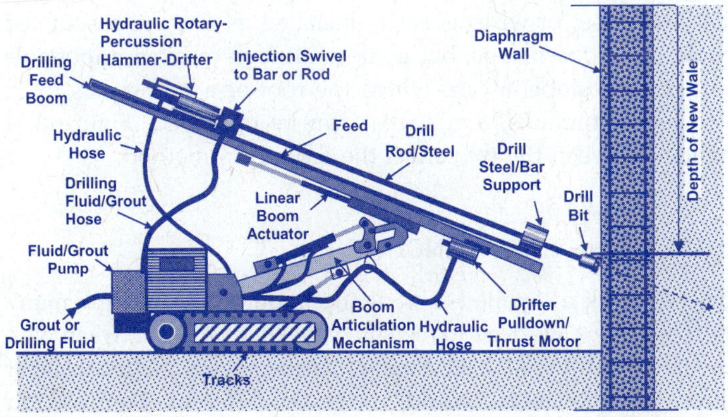

Figure 8.30 Small Mobile Drill Perfect for Installing Earth Anchors

Hammer Specifications:	Rotation Specifications:
• Hole sizes 1 3/8″ to 1 3/4″ • 18 GPM @ 1850 PSI Hydraulic Flow • 80 FT/LB Blow Energy • 3500 Blows per Minute	• 200 FT/LB of Torque • Up to 250 RPM • 35 Revolutions per Gallon of Hydraulic Flow • Maximum 7 GPM of Hydraulic Flow

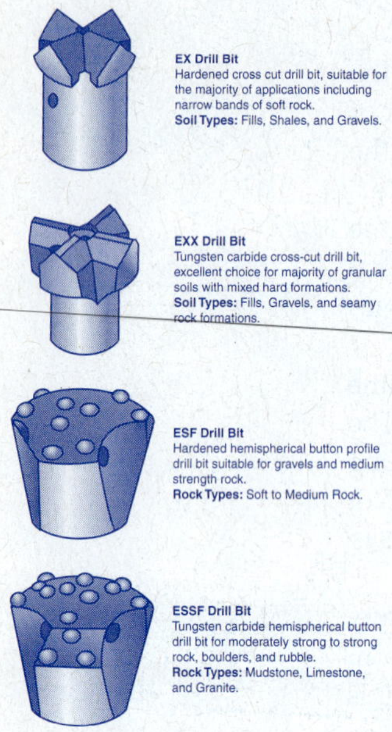

Figure 8.31 Sample Drill Bits

Right in front of the drifter is a swivel mechanism that allows the injection of drilling fluid, compressed air, or grout while the drill steel is rotating. The first two are used in removing the chipping from the drill bit while, at the same time, cooling the bit (especially in rock). If compressed air is used, only when no groundwater is encountered, a lot of dust can collect around the borehole, which can be remedied by a dust collection system. At the front end of the feed, the drill steel/bar support keeps the drill bit pointed into the correct direction while letting the drill steel rotate freely and slide forward and backward. Figure 8.30 also shows that the excavation needs to reach a level below the position of the tied-back wales so as to allow the drill rig the necessary flexibility to position and direct the drill boom. One important decision is the selection of the drill bit to match the ground that has to be penetrated as well as the purpose of the drilling (e.g., sample coring for geotechnical investigation, simple borehole for blasting or anchor installation). What has to be balanced is the cost of the type of drill bit with the speed of penetration. Drill bits differ not only in their shapes but also in the material used for their fabrication. The base material consists of hardened metal to which tungsten carbide alloy or diamonds are added. One common "family" used for ground anchor drilling in soils and soft rock is the drag bit (fishtail, bladed, cross-cut). The tips of fishtail and bladed bits are usually made of a tungsten carbide alloy for wear resistance.

Replaceable blade bits have insert blades, which are individually replaceable. Figure 8.31 portrays two types of cross-cut bit, the first made of only hard steel while the second has cutting edges made of the extremely hard tungsten carbide, guaranteeing that the bit stays sharp for a longer period even in rocky material. Larger-size boreholes can be drilled with other bits. The example bits in Figure 8.31 illustrate that each contains several holes in the center or on the sides through which the pressurized drilling fluids or air exit, after passing through the hollow rods, to return to the entry of the borehole carrying along the chippings abraded/loosened by the drill bit. If hollow bars are used as the tension-carrying element, even the grout will be injected through these holes.

Before we can study the drilling process in more detail, we need to take a closer at the different tieback hardware, at least comparing the two main options. The key difference is the material to be tensioned between the bonded end and the anchorage. One approach applied the technology developed for cable suspension bridges, tendons consisting of twisted wire strands. This flexible set of wires is easily installed as one pre-assembled piece as long as the borehole has a downslope, but is more difficult or even impossible when the borehole has an upward slope, a case where the roof of a tunnel has to be secured with anchors. For such situations, a rigid bar can be deployed. Figure 8.32 highlights how the differences between the two effect the anchoring method.

8.6.2 INSERTING AND GROUTING EARTH ANCHORS

In Chapter 7, about cranes, we took a detailed look at the importance, and the many different makeups, of wire ropes. Here we need to recognize the distinction between wire ropes and wire strands, the later used to make tendons for earth anchors. Both consist of twisted high-strength wires, but the inner "world" is vastly different. Figure 8.32a

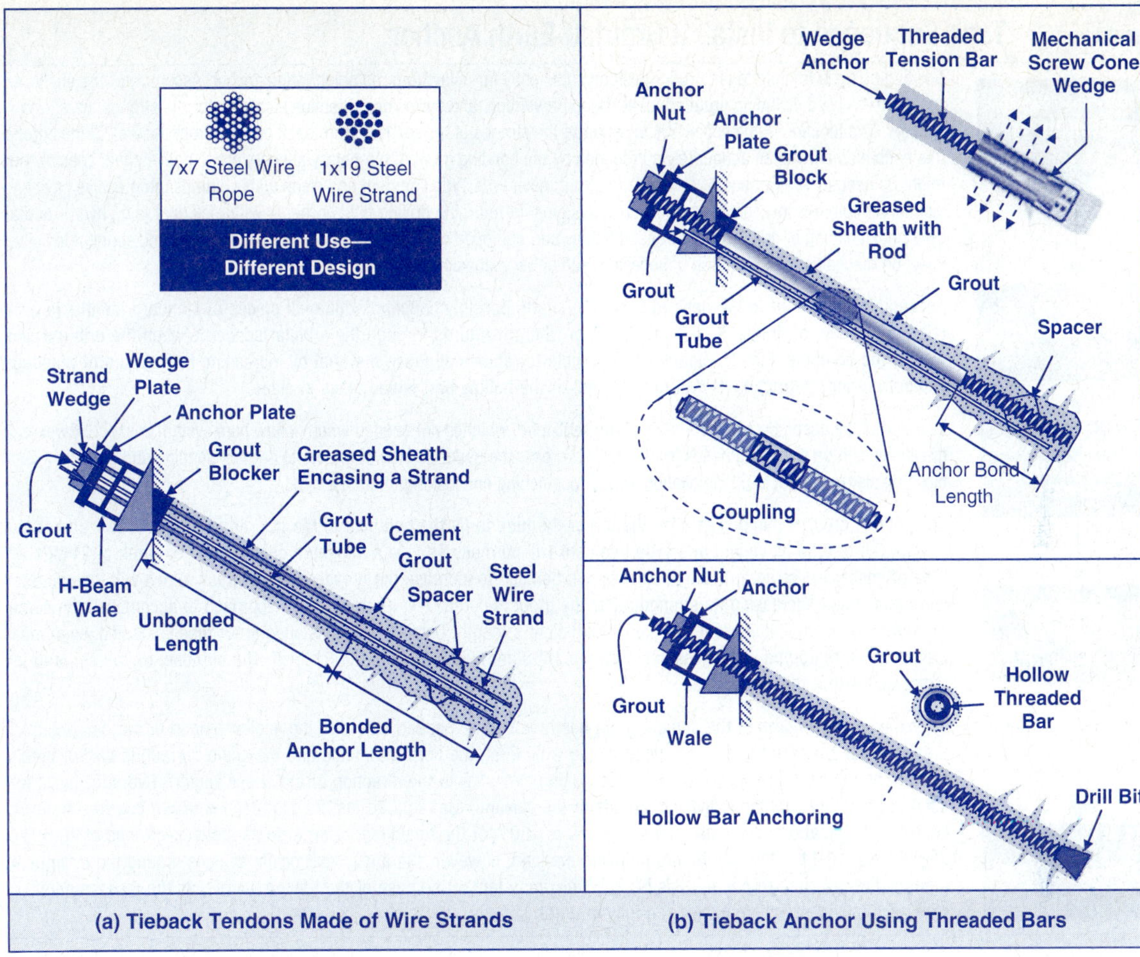

Figure 8.32 Grouted and Mechanical Anchoring Technologies to Tie Back Retaining Walls

presents cross-sectional views of the two, illustrating the use of cores for the ropes. The origin of this, of course, is the fact the ropes experience constant bending around pulleys, creating internal friction, which would grind down the strands over time without the use of a "soft" core material (for more detail please see Chapter 7). Since anchor cable only needs to bend during installation, the friction-reducing cores are not needed.

When should we use grouted anchors, and when should we use mechanical anchors, as shown in Figure 8.32b? The answer lies behind the dashed arrows representing the bracing forces caused by the expanding wedges when the rod is turned. The anchor depends on reaching significant bracing forces to activate the friction between the wedges and the borehole walls that are needed to apply the necessary tension to the anchor head. What will happen when the material surrounding the borehole is soft like butter? Of course, the wedge will just push the soft material away, never creating any major bracing, and with it no friction force. So, in order for the mechanical wedge anchor to work, the ground has to be hard enough for it to develop sufficient friction forces. As was shown in Figure 8.28b, solid rock would be best, but even sandstone or hard clay might hold the needed tension force. However, most earth is too soft and needs improvement with a grout before applying tension to the cable or the bar.

Let's take a closer look at the the installation cycle for earth anchors, broken down into five steps.

Table 8.7 Task Sequence to Install a Grouted Earth Anchor

Step	Description
1. Setup	The drilling rig is tracked to the proper position sufficiently far away from the neighboring anchor. Assembly of the anchoring system to be used including grouting tubes, bond prevention, and corrosion protection elements for the strands, bars, and spacers. The function of the spacers is to separate the strands of the tendons from each other or from the wall of the borehole so that each element will be adequately surrounded by the bonding grout. If temporary anchors are to be installed, special couplers might be needed. Preparation of drilling fluid, grout mixer with Type I cement and clean water, pumps (progressive cavity, piston), air compressor, hoses, and collection systems for returning drilling fluid or the air with rock dust (e.g., hydro- and air-cyclones). Loading of drill steel or threaded hollow bar, equipped with the proper drill bit, into the feed and connect to swivel injection mechanism. Preparation of necessary rod or bar extensions with couplers.
2. Drilling	Advancing the drill bit to the desired borehole length, possibly coupling additional pieces of standard lengths (e.g., 3 m (10 ft)). Pumping of drilling fluid to the bit from where it returns through the annular space between the drill rod and the wall of the borehole. Finally, cleaning of borehole to remove all loose material by increasing the amount of pumped fluid without cutting. Retraction of drill rods but not of the hollow bar, which stays in place.
3. Installation of Anchoring Hardware	Insertion of the assembly of tendons-spacers-grout-tube, which is not needed when hollow bars system is used. However, a grease-filled tube sheathing needs to be "screwed" onto the upper section of the bar to prevent bonding and corrosion. Grout blocking sleeves might need installation. Proper positioning and locking of anchor head.
4. Grouting-Regrouting	The cement grout is pumped through the bar or the tube to fill the borehole to the desired length. A water/cement (w/c) ratio of 0.4 to 0.55 by weight and Type I cement will normally provide a minimum compressive strength of 21 MPa at the time of anchor stressing. If needed, sleeves can be used to increase the pressure of the grout, which will thus penetrate the surrounding soil, increasing its strength after the grout is cured. The more common approach to accomplish the same is regrouting after the initial grout has gained sufficient strength. Using the pre-installed grout tubes, specific, even localized, areas can be regrouted with pressures between 1400 and 3600 kN/m^2 (200–522 psi). The purpose, of course, is to increase more, and with it the anchor load.
5. Pre-Stressing	After the curing process of the cement grout has reached the needed strength, the anchor tendon or bar can be stressed to the level that is needed to support the retaining wall. First, the horizontal wale and the angle-correcting anchor bracket need to be in place (the anchor plate has to be perpendicular to the direction of the stress vector). Hydraulic jacks are then used to grasp and pull the individual strands or the bar until the required test load is reached, according to standards or contract specifications, normally between 120% to 150% of the final load, referred to as the lock-off load of 40,000–250,000 lbs (18.1 t – 113.4 t). Before reducing the load, however, the anchor nut or the wedges around the strands are tightened or inserted. Of course, this is especially important in the case of the tendon anchor, as the wedges need some backward movements before they can fully grab the strands.

Solving Header Problem 8.1 / Planning a Deep Excavation and Special Foundation

The depth and breadth of the list of methods, equipments, tools, and materials that were covered in this chapter bear evidence to the complexity of building on or inside a difficult subsurface. Moreover, the uncountable number of different conditions that an excavation contractor faces not only provide technical challenges, but also increase the risks associated with dealing with the ground adjacent to large building. Not surprisingly, owners are not inclined to spend large sums for unnecessary costly measures. Thus, experience with modern equipment, creative thinking, and an appreciation of how the different soils behave is the basis to continued success in this exciting area of construction. While it is impossible to address all the detailed issues "buried" in the header problem, let's at least cover the main underlying concepts and take note of all the interrelationships that link one decision to another or many other decisions.

Basic Assumptions and Rules

(1) Open cuts with sloped or benched sides are usually the most economical solutions to build a flat foundation, even at a greater depth than just 10 ft (3 m). However, the creation of a stable slope will require space beyond the footprint of the building, at least during construction. In addition, the backfill material is

best kept on site to avoid transportation cost, requiring even more space away from the excavation. On the other hand, this material provides an opportunity to build a ramp for vehicle access to the bottom.

(2) Watertight slurry walls are able to serve as a foundation for the building and may also become the outer wall of the underground segment of the building. This makes them a very cost-effective alternative to sheet-piles and traditional walls.

Priority Problem: Protection of Adjacent Building

There are two issues that have to be solved: (a) Retaining wall that will not move under the earth pressure, and (b) dewatering between retaining wall and adjacent building.

Retaining Wall. There is a wide variety of methods to choose from. The presence of such a high groundwater suggests selecting a "near" watertight option, like sheet piles or slurry wall with horizontal bracing or ground anchors.

Dewatering. Any lowering of the groundwater level below the building has to be avoided, as loss of buoyancy will result in settlement (see Figure 8.33). The three main options to accomplish are to (1) create a watertight barrier, (2) lower the water level close to the wall and replenish the water underneath the building, and (3) build a watertight retaining wall.

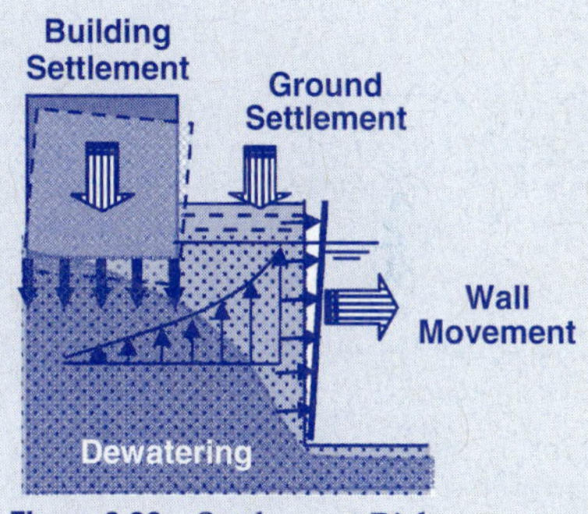

Figure 8.33 Settlement Risks

Synergistic Solution. The watertight diaphragm wall would offer a solution that addresses both problems. The contractor could offer this as an economical solution if the structural engineer would be willing to use the wall as part of the underground building. This would eliminate having to excavate outside the footprint of the building, which is necessary when building a traditional formwork for the wall. The depth of the wall could reach the rock surface and thus eliminate the need to install support underneath. In an added benefit, this would, in fact, cut off the water infiltration from the third aquifer. In order to extend the water cut-off effect, retractable sheet piles could be used to extend the slurry wall on both sides. This would probably eliminate any drawdown of water around the adjacent building. Figure 8.34 presents the envisioned solution to protect one side of the open pit during excavation.

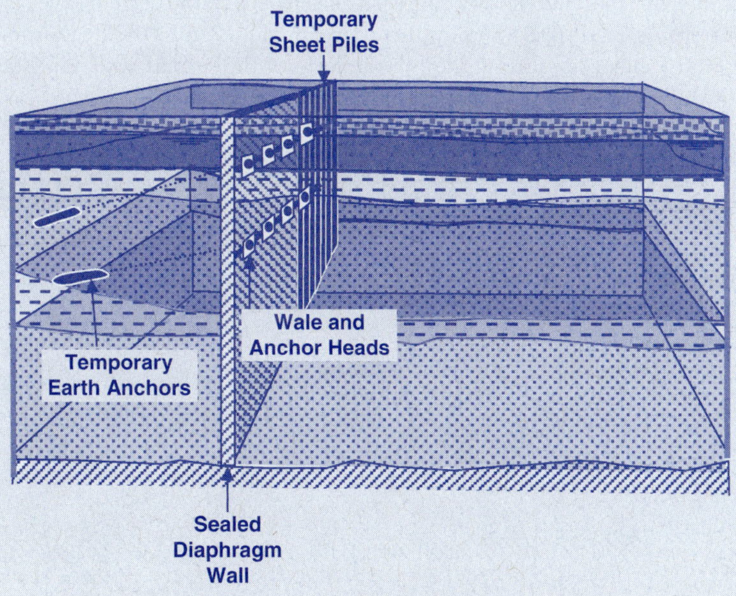

Figure 8.34 Sealed Slurry Wall with Earth Anchors

Table 8.8 Permeability of Sands

Type of Sand	Grain Size Range (mm)	Coefficient of Permeability k $\times 10^{-4}$ cm/sec	$\times 10^{-4}$ ft/min
Sandy silt	0.006–0.02	5–20	10–40
Silty sand	0.02–0.0625	20–50	40–100
Very fine sand	0.0625–0.125	50–200	100–400
Fine sand	0.125–0.25	200–500	400–1,000
Fine to medium sand	0.25–0.4	500–1,000	1,000–2,000
Medium sand	0.4–0.6	1,000–1,500	2,000–3,000
Medium to coarse sand	0.6–1.0	1,500–2,000	3,000–4,000
Coarse sand and gravel	1.0–6.0	2,000–5,000	4,000–10,000

Dewatering Strategy

One important decision concerning dewatering the pit has already been made, in that the diaphragm wall will be made watertight reaching down to the rock layer. In addition, the wall will be extended with sheet piles to serve as a water cut-off wing during excavation so that the drawdown toward the adjacent building is minimized. While it might be very reasonable to extend the concrete wall all the way around the footprint of the building, we want to explore how the rest of the pit walls would be sloped or benched with an access road leading to the bottom of the excavation. The three basic options that are available to dry out the excavation area temporarily are (1) deep wells with large drawdown, (2) wellpoints around the periphery of the excavation, and (3) a combination of the two. Relying only on deep wells seems undesirable, as the water table in a large semicircular area around the site would be lowered. Also, the pressurized fine sand might not allow a quick enough gravity-based drawdown of the water table. The rate at which water can be pumped from dewatering system is directly proportional to the coefficient of permeability of the formation being dewatered (see Table 8.8); thus, this parameter should be determined reasonably accurately prior to the design of any drainage system. Further soil investigations should address this potential problem if a deep well solution is selected. On the other hand, the use of a staged wellpoint at least for aquifers A and B seems to offer a cost-effective solution. The 10 m (33 ft) of fine sand is a perfect to take advantage of the vacuum effect of a narrowly spaced wellpoint system.

As we learned, the maximum lift capacity of a conventional wellpoint is 6 m (20 ft). Since the entire excavation has a depth of 17 m (55 ft) with the bottom only about 1 m (3.3 ft) above the second layer of clay, three wellpoint stages should be sufficient. As Figure 8.35 shows, the approximately 3 m (10 ft) thick upper clay layer between aquifer A and B offers an excellent opportunity to put the head of the second wellpoint stage at its bottom, thus allowing the tip of the well to be lowered by about 2 m (7 ft).

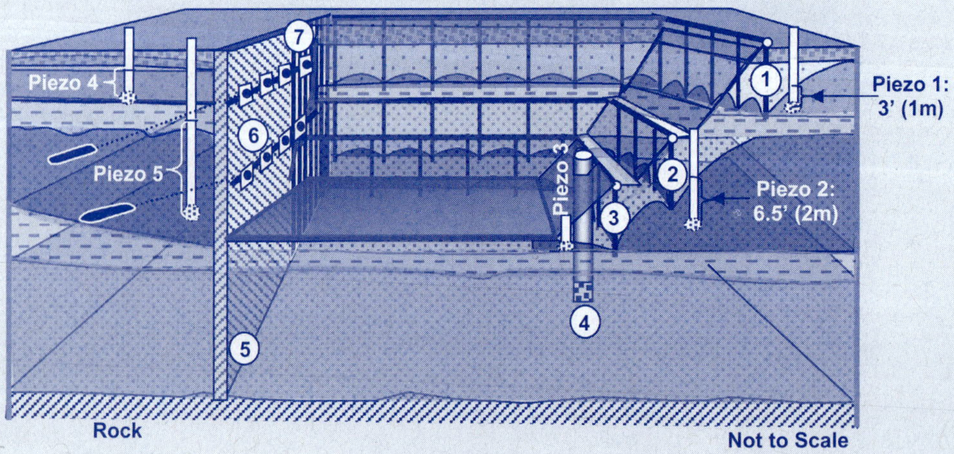

Figure 8.35 Various Dewatering Schemes

The two wellpoint stages are built into the terraced sides of excavation pit. The narrow paths simplify the supply of the wellpoint pipes and rods as well as making inspection and possible repair more efficient. Each stage is accompanied by one or more piezometers to monitor the drawdown level. Since it is not necessary to reduce the water table in aquifer C, it should be sufficient to operate with a few encased deep wells equipped with a piezometer and a water pump. Their function is to prevent boiling of the bottom by keeping the water pressure under the second clay layer at a level that has to be determined with the geotechnical engineer after the jet-grouted pile option has been approved. This should be made easier by the cut-off walls that reach to the rock layer, thus sealing the influx of ground water from one side. During the anchor installation, the specialty contractor must be aware of the water pressure behind the retaining wall and be required to fit the anchor head with proper seals. Finally, additional piezometers have to be driven to the first and second aquifer behind the wall. Their sole purpose is to monitor and document any changes in the water pressure that could cause any settlement of the surrounding. Of course, should anything unusual occur, countermeasures could be immediately undertaken.

8.7 Chapter Eight Review

THE EQUIPUZZLE

The encircled letters make a word that stipulates a piece of equipment that is commonly used during the construction of a slurry wall.

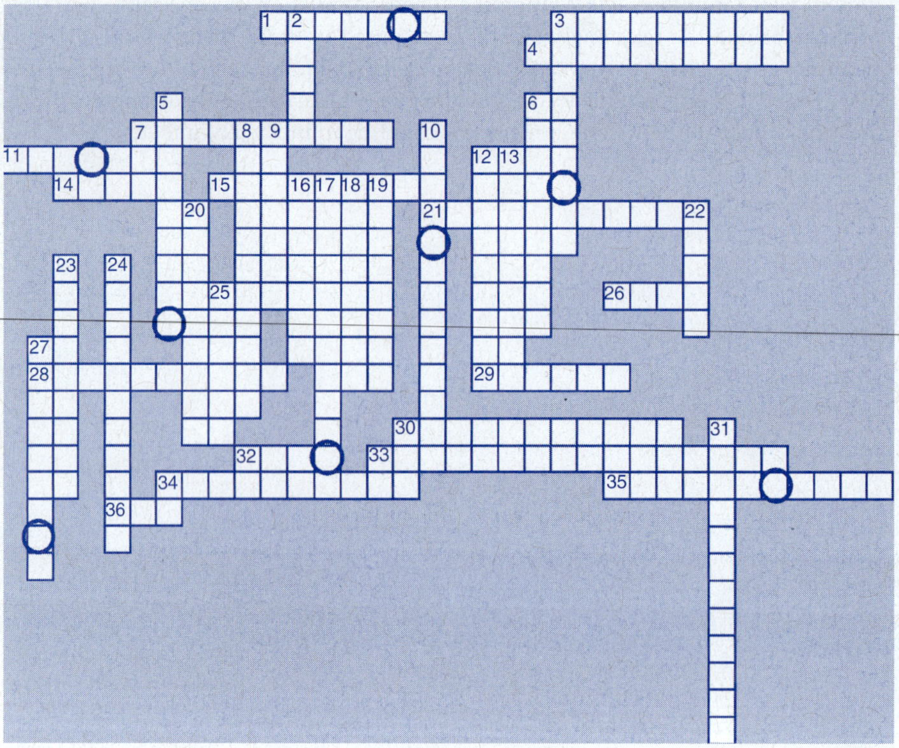

Across

1. Family name of a German inventor who rammed deformed steel profiles into the ground supporting each other

3. Looks like a solution but consists of two separate phases

4. 19th-century pioneer of the centrifugal pump (first and family name together)

7. A mechanism to prohibit sand and silt from entering the suction pipe of a dewatering

11. An object that is buried, used to anchor a tension rod or strands needed to stabilize a wall

14. A liquid mix of materials that hardens after it is pressure-injected into the ground

15. A dewatering system for soils that consist of fine sand and silt

21. An anchor technology that reverses the direction of the axial forces created by Alexander Mitchell's screw piles (two words)

26. One structural element of a waling

28. Lowering of the groundwater table to obtain a "dry" space for excavation

29. A hopper with steel pipe used to place concrete underwater

30. A force associated with a rotating body

32. Short for Net Positive Suction Head Available.

33. A plot that relates several characteristics of a liquid pump to the total head (two words)

34. Method used to create tension force in a horizontal strut during installation in order to push against the ground

35. The difference in elevation between the liquid level of the fluid source and the centerline of the pump (two words)

36. Short for a perforated grouting pipe

Down

2. A helical rotary tool for drilling a cylindrical hole in soil

3. A dam-like structure built around a construction site to keep the water out

5. An absorbent aluminum silicate clay formed from volcanic ash. It swells in water to several amounts of its dry mass

6. Descriptive name of a reinforced concrete wall that is built in sections within a deep slurry-filled trench

8. An economical method used to create a vertical earth-retaining structure that also blocks ground water from entering the site. The individual elements are installed with the help of a threader

9. A liquid composed of finely divided particles that do not settle out of suspension

10. Ground water beneath a layer of clay and connected to water source that is at higher level (two words)

12. The state of a streaming fluid producing agitated flow patterns and eddies

13. An inverted air hammer with yoke to pull piles out of the ground

16. Name of the structural pieces used to close the space between two soldier piles

17. Energy created by a water pump (two words)

18. Name of the inventor of the steam pile driver (first and family name)
19. The rotating element of a centrifugal pump that consists of a disk with curved vanes
20. A curve plotting the total head vs. the flow for a specific pump model (two words)
22. Name of the pipe that has to be jetted into the ground when building a wellpoint
23. Minerals that are mixed with water, fly ash, cement, etc., to produce a grout
24. A key characteristic of grout. It tells how easily it will pass through pipes under pressure
25. A steel pipe pushed into the ground to prevent the collapse of a borehole
27. Additive to a fluid to achieve a desirable characteristic
31. A vertical cylindric hole in the earth that is filled with reinforced concrete (two words)

JOURNALING QUESTIONS

1. Both the secant pile wall and diaphragm walls provide great opportunities to construct retaining walls that are "fairly" watertight. What does it mean to have a retaining wall that is fairly watertight? How do these two systems make it hard for water to seep through?

2. Figure 8.28 presents three different technologies to support the retaining walls of a deep excavation. What are the advantages and disadvantages of using earth anchors?

3. Figure 8.3 shows different methods that have been used over time to lift water to higher elevations. Discuss which one of those methods you consider most elegant and/or most energy-efficient.

HOMEWORK PROBLEMS

Question 1. The header problem requires the anchoring of the diaphragm and the sheet pile wall. In order to decide on the spacing between the straight shaft tendon anchors, we should know how much one anchor will hold. As both rows will be anchored in silty or fine sand ($18\,kN/m^3$), no adhesion to the bonded length will occur. Assume the angle of the anchor to be approximately 30° and friction angle F of the cohesionless soil 33°. We assume that the clay has a saturated density of $2{,}000\,kg/m^3$ or $3400\,lb/yd^3$ (see Table 8.9).

From experience we know that the ultimate transfer load for smaller earth anchors and this kind of cohesionless soil is between 700 and 1300 kg/m due to variability in density. Postgrouting of an anchor is expected to increase the load capacity by 30%.

Using the function developed by the Army Corps of Engineers, calculate the borehole diameter that would provide the ultimate transfer load that is expected.

Question 2. The US Army Corps of Engineers developed a simple function to calculate the radius of influence of a natural water table lowered by either a well or a line of wellpoints:

$$R = C * (H - h_W) * \sqrt{k}$$

where

R = Radius or distance from wellpoint line (in feet)
C = 3 for artesian and gravity wells, 1.5 to 2.0 for single line wellpoints
H = Piezometer height before drawdown, natural (in feet)
h_W = Piezometer height after drawdown (in feet)
k = Permeability in 10^{-4} cm/sec

Your task is to estimate the radius of influence for deep wells and wellpoints for this chapter's header problem. Calculate and compare the radius for these two options for aquifer A and aquifer B.

Table 8.9 Soil Densities

	Dry Densities		Saturated Densities	
Material/Soil	lb/yd³	kg/m³	lb/yd³	kg/m³
Silts	1940–2760	1150–1640	2360–3030	1400–1800
Sandstone	3370–4380	2000–2600	3710–4380	2200–2600
Limestone	3200–4380	1900–2600	3540–4550	2100–2700
Sand	1850–2700	1100–1600	2530–3370	1500–2000
Clay	1900–2800	1130–1660	2190–3540	1300–2100
Clay & Organics	1180–1520	700–900	1690–2190	1000–1300

Question 3. Consider that a contractor is installing Cast-in-Place Concrete Piles (CIPs) to build a retaining wall using the method described in Figure 8.36.

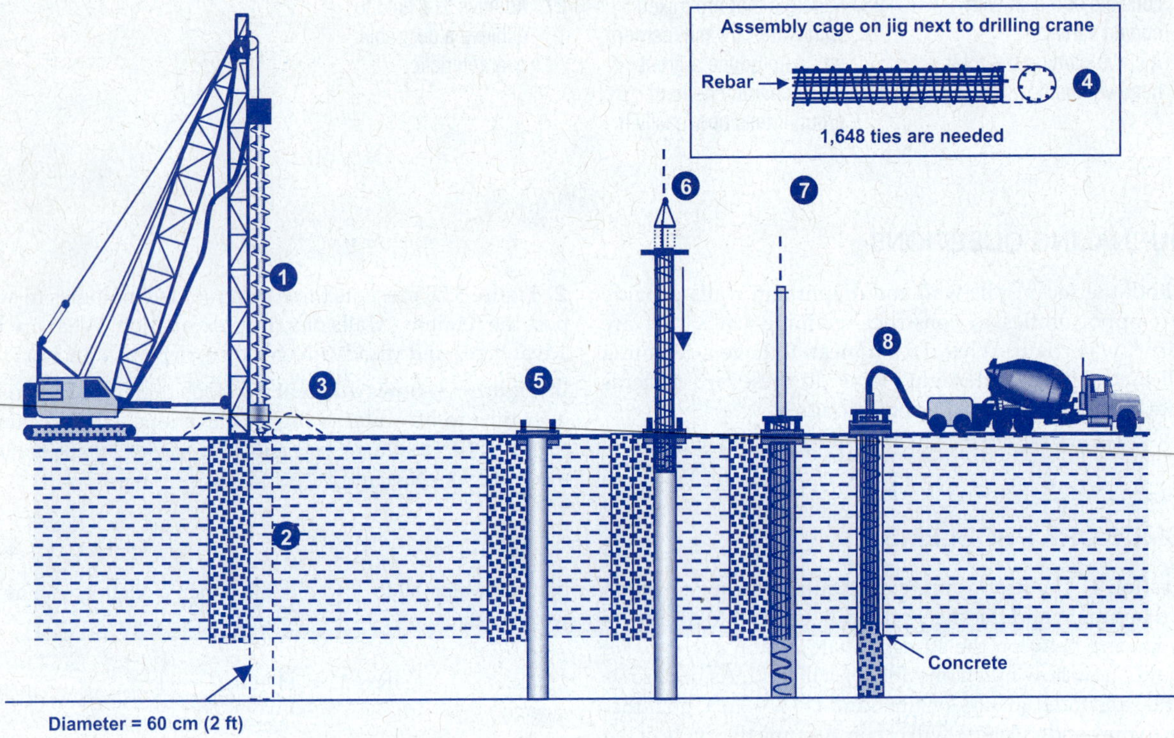

Figure 8.36 Cast-in-Place Pile Construction

No.	Description of Tasks	Dimension	Duration	Needs Crane
1	Installation of drill rig and positioning of auger at correct location	1 unit	10 min	▲
2	Pile drilling with auger, diameter = 60 cm	21 m	1.66 min/m	▲
3	Removing spoil from auger by a front-end-loader with 0.5 m^3 bucket	0.5 m^3	3.5 min/cycle	
4	Assembly of rebar cage with 9 × 21 m rebar, each has to be tied - Preparation of long rebar on jig - Making 1,648 rebar ties per rebar cage	 1 unit 1,648 units	 15 min 0.1 min/tie	
5	Installation of rebar cage positioning bracket	1 unit	5 min	▲
6	Lowering of rebar cage into drilled hole with drill rig	1 unit	10 min	▲
7	Lowering of concrete tremie with drill rig	1 unit	4 min	▲
8	Pumping concrete - Preparation of pump and hose - Pumping concrete from trucks - Cleaning pump and tremie	 1 unit 6 m^3 1 unit	 10 min 15 min 10 min	▲

(a) Attach the proper labels to all the arrows.
(b) Model the crane/drill rig cycle as an I/O process.
(c) Add the other tasks and other cycles that are needed:

- The spoil from the auger needs to be removed before ❺ can be done!
- The laborers working on the cage need to also do ❻ + ❼ + ❽

(d) Draw the balance point to find the number of laborers you need for assembling the rebar cage right next to the drilling operation. The crew needs to keep up with the augering operation (remember that one laborer needs 0.1 min/tie). How many laborers will you hire for the rebar tying?
(e) What would you recommend to do if your boss allows the hiring of only 50% of the number you need to reach balance point?
(f) What would result from the drilling operation if you gave the front-end loader a larger bucket (> 0.5 m^3)?

LESSONS FROM THE ACCIDENT FILE

1. Accident case 1 had tragic results, killing four workers and shutting down a major highway. What method/technology/precaution would you have recommended to the contractor that would have showed that the retaining walls were at the brink of collapsing?

2. The death and injuries of the laborers in case 2 could have easily been avoided. What were key indicators that the soil embankment was very unstable? What protective measures would you have implemented in order to prevent the sudden collapse or at least to avoid workers getting killed?

3. The situation leading to the fatality described in case 3 was extremely dangerous from the beginning. Sketch the situation when the sheet pile the worker was sitting on gave away. What technology would you recommend should have been used instead?

OPEN-ENDED PROBLEMS

1. In the 1980s–1990s the Up-Down method of constructing buildings with large underground space for garages, similar to the Header Problem of this chapter, became common. The trick is to build the subspace and the superstructure at the same time. Sketch the Up-Down method for the Header Problem 8.1 that allows for the simultaneous construction of the sub- and superstructure

2. The jacking of sheetpiles is an interesting method. What are its benefits considering the environmental impact when compared to the two alternative installation methods? Sketch and describe how you would apply a sheetpile jacking system to build a containment wall in a protected (swampy) area that does not allow heavy equipment to drive around. Design the start-up system. How would you supply the sheets when most of the wall is outside the reach of a crane?

BIBLIOGRAPHY

Boehm, D.W. "The Utilization of Jet Grouting and Soil Mixing Methods to Repair and Support Bulkhead Structures." *Proc. Conf. Ports 2004. Houston*, TX, May 23–26, 2004.

Brown, D.A., S.D. Dapp, W.R. Thompson, and C.A. Lazarte. *Design and Construction of Continuous Flight Auger*, Geotechnical Engineering Circular (GEC) No. 8, Federal Highway Admin. Washington, April 2007.

Bruce, I.A., *An Introduction to the Deep Soil Mixing Methods as Used in Geotechnical Applications*, Rep. FHWA-RD-99-138. USDOT. Washington, DC. March 2000.

Brunner, W.G., R. Fiorotto, E. Stötzer, and M. Schöpf. "The Innovative CSM-Cutter Soil Mixing for Constructing Retaining and Cut-Off Walls." *Proc. GeoCongress*. Atlanta, GA. February 26–March 1, 2006.

Brunner, W.G. "Development of Slurry Wall Technique and Slurry Wall Construction Equipment" *Proc. Geo-Support*. Orlando, Florida. January 29–31, 2004.

Chow, H.L., and C.Y. Ou. "Boiling Failure and Resumption of Deep Excavation." *J. Perform. Constr. Fac.* 13, no. 3 (1999).

Davis, A. G. "Nondestructive Evaluation of Existing Deep Foundations." *J. Perform. Constr. Fac.* 9, no. 1 (1995).

Dunker, K.F., and D. Liu. "Foundations for Integral Abutments." *Prac. Per. Structural Design Const.* 12, no. 1 (2007).

Elwood, N.J., and J.W. Gaythwaite. "The DeLong Pier Repair Project: The Unique Challenges of Designing Repairs to a 1950 Vintage Marine Structure in the Arctic." *Proc. Conf. Ports 2007*. San Diego. March 25–28, 2007.

Hsieh, H.S., C.C. Wang, and C.Y. Ou. "Use of Jet Grouting to Limit Diaphragm Wall Displacement of a Deep Excavation." *J. Geotech. Geoenvironmental Engrg.* 129, no. 2 (2003).

Larsson, S. "State of Practice Report—Execution, Monitoring and Quality Control." *Proc. Conf Deep-Mixing*. Stockholm, Sweden. May 23–25, 2005.

Lee, H.S., J.Y. Lee, and J.S. Lee "Non-Shored Formwork System for Top-Down Construction." *J. Constr. Engrg. Mgmt.* 125, no. 6 (1999).

Liou, D.D. "Thermal Effects in Large-Sized Diaphragm Wall." *J. Perform. Constr. Fac.* 13, no. 1 (1999).

Liu, G. B., C.W. Ng, and Z. W. Wang. "Observed Performance of a Deep Multistrutted Excavation in Shanghai Soft Clays." *J. Geotech. Geoenvironmental Engrg.* 131, no. 8 (2005).

Mansur, C.I., and S.G. Durrett. "Dewatering Cofferdam for Construction of Olmsted Locks." *J. Geotech. Geoenvironmental Engrg.* 128, no. 6 (2002).

Mayne, P.W., B.R. Christopher, and J. DeJong. *Subsurface Investigations—Geotechnical Site Characterization*, Rep. FHWA-NHI-01-031. USDOT. Washington, DC. May 2002.

Poh, T., A.T. Goh, and I.H. Wong. "Ground Movement Associated with Wall Construction: Case Histories. *J. Geotech. Geoenvironmental Engrg.*, 127, no. 12 (2001).

Robinson, H.D., J.E. Gómez, C.M. Englert, and A.W. Cadden. "New York City Harbor- Remediation of Seawalls on the Historic Liberty and Ellis Islands." *Proc. Ports 2007*. San Diego March 25–28, 2007.

Ryan, C.R., and C.A. Spaulding. "Strength and Permeability of a Deep Soil Bentonite Slurry Wall." *Proc. Geo-Congress 2008*. New Orleans LA. March 9–12, 2008.

Schmieg, H., and P. Vielsack. "Transmission of Shear Forces in Sheet Pile Interlocks." *J. Geotech. Geoenvironmental Engrg.* 128, no. 4 (2002).

Stuart, D.M. "Project-Specific Steel Sheet Piling Applications." *Pract. Period. Struct. Design Constr.* 9, no. 4 (2004).

Yang, M.P., S.J. Chen, and S.T. Chen. "Innovative Central Opening Strut System for Foundation Excavation." *J. Constr. Engrg. Mgmt.* 132, no. 1 (2006).

chapter nine

The Many Ways to Construct Deep Foundations

But if a solid foundation is not found, and the site is loose earth right down, or marshy, then it is to be excavated and cleared and remade with piles of alder or of olive or charred oak, and the piles are to be driven close together by machinery, and the intervals between are to be filled with charcoal. Then the foundations are to be filled with very solid structures. Let double-walled formwork be set up in the designated spot, held together by close-set planks and tie beams, and between the anchoring supports have clay-packed down baskets made of swamp reeds. When it has been well tamped down in this manner, and is as compact as possible, then have the area bounded by the cofferdam emptied and dried out by means of water-screw installations and water wheels with compartmented rims and bodies. The foundations are to be dug there, within the cofferdam.

From *De Architectura*, written by the Roman Engineer Vitruvius, first century B.C.

We learned in Chapter 8 that the objective of deep excavation is to remove soft material to a depth that has soil or rock sufficiently strong to support the structure that will eventually rest on it. The Romans already found, however, that many times the depth at which appropriate material can be found renders excavation technically infeasible or economically not viable. In the first century BC engineer Marcus Vitruvius wrote in his book De Architecture that there exists an alternative, namely piles standing on solid rock. In Roman times, such piles were made of "alder or of olive or charred oak" wood and driven into the ground by some kind of a machine. Subsequently, the heads of the piles were covered with a solid structure, thus serving as the base of a building such as a bridge column or a tower. While this technology has evolved dramatically over time, the basic premise has not changed. Deep foundations offer an economical solution to build in areas where the subsurface consists of soft soils to a great depth. This chapter will review the many innovative approaches that contractors and geotechnical engineers have developed to respond to the many different ground conditions that they face in different parts of the world.

Table 9.1 Topics Covered in Chapter 9

Construction			Planning & Control		
Equipment	**Methods**	**Materials**	**Managerial-Engineering Tools**	**Rules & Standards**	**Modern Tools**
✓ Trucks	✓ Franki Pile	✓ Bentonite Slurry	✓ Navier-Stokes Equation	✓ ASTM A36 Steel	✓ PDA
✓ Impact Hammers	✓ Pressurized Caisson	✓ Precast Piles	✓ Anchor Load	✓ ASTM A82, A615 and A884	✓ CAPWAP
✓ Pile Driver	✓ Cast-in-Place Piles	✓ Cement Grout	✓ Safe Pile Load	✓ ASTM A416 A421, and A882	✓ Accelerometer
✓ Vibratory Drivers	✓ Deep Mixing Method	✓ Load-Bearing Piles	✓ Engineers News Formula	✓ FHWA-IF-99-025	✓ Strain Gauge
✓ Swinging Lead	✓ Underreaming	✓ Fiber Reinforced Polymers	✓ Case Method Capacity	✓ ASTM D1586	✓ Pile Integrity Tester
✓ Auger Drill	✓ Jet Grouting	✓ Colloidal Mixture	✓ Marsh Funnel Viscosity		
✓ Trench Cutter	✓ Grout Batching	✓ Sheet Piles	✓ Pile Toe Resistance		
✓ Hydrocylcone	✓ Open Caisson	✓ Chemical Grouts	✓ Standard Penetration Test		
✓ Rakers			✓ Pascal		
✓ Mobile Anchor Drill			✓ Piezometer		
✓ Kelly Bar					
✓ CFA					
✓ Core Barrels					
✓ Rock Bucket					
✓ Rock Auger					
✓ Pipe Tremie					

9.1 From Wooden Stilts to Jetted Piles

Humans have tried to build structures on grounds that were too weak to support their weights. One of the most famous examples of what can happen in such cases is the 55.9 m (183 ft 3 in) high and 14,500 t bell tower of Pisa that is now leaning 3.9 m (12 ft 10 in) at the top. Built off and on over 177 years starting in 1173, it apparently allowed one of his most famous citizens, Galileo Galilei, to conduct his famous drop experiments in 1590 while professor at the university. Although usually not with such dramatic results, building on weak ground still creates havoc when not undertaken with professional expertise and skill. The following short review marks some of the key innovations that allow us today to create massive supports unseen by the unsuspecting public.

4000 BC	**Houses Built on Stilts**	A Neolithic tribe called the Swiss Lake Dwellers built its houses on wooden piles in lakes to be safe from attack. Those piles are preserved still rammed into the bottom of some lakes in Switzerland.
500 BC	**Piles for Pons Sublicius**	Egyptians, Greeks, and Romans built buildings, bridges, roads and viaducts on wooden piling. The Romans built the first bridge across the Tiber River, the **Pons Sublicius,** on timber piles (around 500 B.C.). (Sublicius = "resting on pilings.") Buildings in the cities of Venice and Ravenna were built on piles from 100 B.C. to 400 A.D., as well as the first bridge across the Thames River in London in 60 A.D.
500 BC	**Deep Drilling**	The Chinese developed deep-drilling using oxen, wheels, and ropes to power rotary and percussion drills. The goal was to dissolve rock salt in water and pump the resulting brine to the surface.
230 BC	**Archimedes' Screw**	The Greek mathematician Archimedes invented the screw pump after visiting Egypt, where he saw the compartmented rotating vertical wheel used to lift water.
1126	**Deep Drills in France**	French Carthusian monks in 1126 operated a mechanism with a thin rod and a hard cutting tip to drill a deep hole. The rod is struck with a hammer breaking up the bottom of the hole. Artesian wells are named after Artois, France, where this took place.
1450	**Piling Rigs**	Sophisticated drop-hammer piling rigs were invented. Francesco di Giorgio was an inventor.
1825	**Erie Canal Locks**	The Erie Canal locks in New York were constructed with one- and two-ton blocks on the floor (against uplift) supported on a system of 6-foot (1.8 meter) timber piles. Each lock was supported by 700 piles.
1839	**Steam Hammer**	James Nasmyth, born in Edinburgh, invented the steam hammer. This heavy machine allowed large steel pieces to be forged with great accuracy. A four- or five-ton hammer was lifted by one steam piston and then dropped, accelerated by a second steam piston.
1851	**Pressurized Caisson**	The first pressurized steel caisson was built for constructing the foundations for a bridge over the Medway at Rochester, in Kent, England.
1875	**Steam Hammer Pile Driver**	The Vulcan steam hammer began when the company began to manufacture hammers under the patent of Thomas T. Loomis in 1875. This hammer used many of the main features of the Nasmyth hammer from 1839.
1887 (1925)	**Chemical Soil Stabilization**	Jeziorsky received a patent for a method to solidify soils with liquid glass (sodium silicate), also known as water glass. Two holes had to be drilled, one for sodium silicate and the other for a coagulating agent. In 1925 von Joosten developed the concept for use in the field.
1908	**Franki Pile**	The Belgian Edgard Frankignoul invented an alternative to piling and drilled shafts that offered load capacities that surpassed traditional methods. He pressed the concrete at the bottom of a vertical shaft outward, lowered a rebar cage, and filled it with concrete.

1939	Precast Piles	First prestressed precast concrete piles used in Sweden.
1950 (1970)	In-Situ Soil Mixing	In the early 1950s a method was invented in the United States of mixing soil using an auger driven down into the soil via a high torque turntable and pneumatically fed with a chemical. When the chemical is thoroughly mixed with the in-situ soil, the auger is pulled back. On November 30, 1970, a first patent was initiated in Japan for a method to create subsurface piles/columns referred to as CCP (chemical churning pile).
1972	PDA	The Pile Capacity Computer was introduced (later renamed the Pile Driving Analyzer (PDA)) in 1972. Initially, these PDA units used analog computation with digital readout.
1976	WEAP	In 1976 George Goble and Frank Rausche produced a wave equation analysis program (WEAP), the first public-domain software to predict pile capacity modeling the dynamics of diesel hammers.

9.2 From the Accident File

Case 1: A vibratory hammer rigged to a 70-ton lattice boom crawler crane was being used to install steel sheet pilings. The hammer was powered by a power pack, which consisted of a diesel engine that provided the energy required by the hydraulic motors and hammer clamps. The victim's usual location during the steel sheet pile-driving task was next to the power pack and outside of the crane's swing radius operating the vibratory hammer. There had been an extended delay in the steel sheet piling installation task due to an obstruction in the ground. During the general contractor's attempt to remove the obstruction, the pile driving crew's downtime led into their scheduled work break. The crane operator spent the downtime and the coffee break inside his cabin, and the victim went to his car but had walked over to the crane, which did not have a swing radius barricade around the rear, apparently to get something out of the toolbox. Outside the crane operator's line of sight, he entered the crane's swing radius when the operator suddenly swung the superstructure without advance notice, crushing the hammer operator between the left rear track and the superstructure.

Case 2: A construction worker was in a 15-ft-(4.6 m)deep trench setting the bottom of a 50-foot (15.2 m) long pile weighing 1,500 lbs (0.7 t) suspended within the leads of a pile driver. The crane used two hoist lines, one to lift the pile into position and a second to move the hammer in place. Only after a pile is set into its proper position should the hammer be lowered onto the pile held on top by a sleeve. A spotter outside the trench signaled the operator to lower the pile into its final place by calling for "line 1," which indicates the hoist line. Instead of lowering "line 1," the operator lowered "line 2," which lowered the hammer. While the pile was still suspended and manually pushed into place by the laborer in the trench, the 5 ton (4.5 t) hammer hit the top, causing it to swing into the laborer and pushing him into the wall of the steel trench box, where he was crushed.

Case 3: A specialty foundation company was contracted to drill and insert 63 pilings to underpin the basement and foundation of a building. To perform the work the company employed a drilling machine, which was powered by a diesel hydraulic system. The auger used to drill through the concrete floor and into the earth was approximately eight inches in diameter and spun at 200 revolutions per minute. A laborer was employed to assist the machine operator during drilling processes, and was using a shovel to clear the earth brought up by the auger. He had to work immediately next to the revolving auger in order to perform the job and had been warned of the danger of wearing loose clothing around the auger. The laborer had taken precautions that morning to tape his rain slicker close to his body. Nevertheless, as the 34th hole was being drilled, an appendage on the auger caught the arm area of the rain slicker, pulled and spun him around the auger multiple times before the machine operator could disengage the machine. The machine operator was unable to immediately disengage the auger, because he had walked away from it while it was running, and had to make his way back to the control panel in order to hit the emergency stop button.

9.3 Problems with Building on Soft Ground

9.3.1 DISASTROUS ENGINEERING THAT BECAME A WORLD HERITAGE

Any man-made structure needs a solid foundation (Latin *solidus* = safe, sound, reliable). Not surprisingly, early structures were not built close to rivers and lakes but on hills and mountains strong enough to carry the load (weight) of the building. However, commerce and industry depended on rivers for transport and power to drive the many machines. Unfortunately, plains that had been created by rivers and oceans depositing layers of fine soil, sand, gravel mixed in with organic material do not provide the same sturdy ground as mountains. The most renowned building to display what can happen to buildings built on the layered deposits of rivers and oceans is the Leaning Tower of Pisa, whose foundation was constructed in 1173 on the delta deposits of the River Po. Ignoring, or not aware of, Vitruvius' book *De Architectura*, the foundation engineers did not use piles, but a flat solid plate. Tower construction lasted 177 years, as it was repeatedly interrupted for long periods of time (up to 50 years) because of wars and/or a lack of money. Today we know that the long spans in between construction were the main reasons why there is still a standing tower today, attested by Figure 9.1. The person who helped the engineering community in understanding the principles behind this phenomenon was Karl Terzaghi, who was born in 1883 to an Austrian military family. Through careful experiments with scientific tools that he invented singlehandedly, he established what is known today as soil mechanics. He found that the settlement due to an added load consolidated the different soils in different amounts.

Figure 9.1 The Tower of Pisa Today

Every soil consists of small grains or, in case of clay, flakes that touch each other, leaving some empty voids in between. In case of the tower of Pisa, located in a river delta, the voids were filled with ground water, creating a saturated soil. While the water was relatively quickly squeezed out of the sand layers during the construction of the tower, the flakes of clay offered only tiny "pores" for the water to leave, which raised water pressure in the pores. As liquids can't be compressed easily, the water will take on the stresses added by the load, thus reducing the stress that would compact the flakes (referred to as the effective vertical stress). (Only after water is squeezed out of the clay will the effective stress that causes the settlement become active.) The consolidation, however, will be slowed down immediately as the pore pressure will become active again. As a result, the settlement of clay was a much slower process than the settlement of sand underneath the tower. Figure 9.2 shows that the layered clay underneath the tower had a total height 11 m + 5 m + 13 m = 29 m (36 ft + 18 ft + 43 ft = 95 ft).

Figure 9.3 presents graphical views of the behavior of clay for the first two construction phases starting in 1173. The solid line shows the theoretical effect of adding load L1 all at once, while the dashed line represents the results of a slow construction process of that time. It is apparent that the primary consolidation process slows down significantly when compared to the settlement curve related to instantaneous loading. The long interruption drawn logarithmically, leads to a secondary settlement of the clay as a result of creeping clay and the decay of organic material. This long phase, of course, reduced the void ratio and created a new balance between the soil and the tower, while the creep widened the area of consolidated soil, reducing the maximum effective vertical stress before the second construction phase began, which repeated the slow process. One can now easily understand that the long breaks between the construction reduced the speed consolidation and gave the clay the time to achieve a new equilibrium. If this had happened, the tower would most probably not be standing today. Later we will discuss what is being done now to actually reduce the leaning with a method that was not available even 30 years ago.

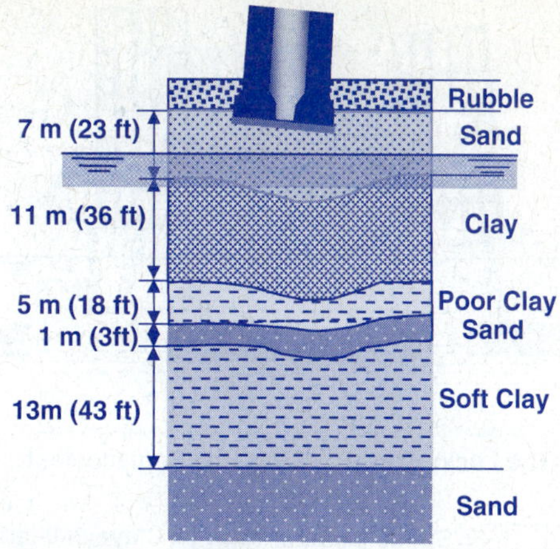

Figure 9.2 Stratified Clays Under the Tower of Pisa

Karl Terzaghi, 1883-1963

1904 Diploma as mechanical engineer
1906 Work for Austrian construction company
1910 Ph.D. from Tech. Univ. Graz
1913–1916 Army engineer during WWI
1916–1925 Professor in Istanbul, Turkey
1925 Publication of *Soil Mechanics Based on Soil Physics*
1925–1929 Visiting Prof. at MIT
1929–1938 Prof. Tech. Univ. in Vienna
1939–1956 Prof. of Soil Mechanics at Harvard

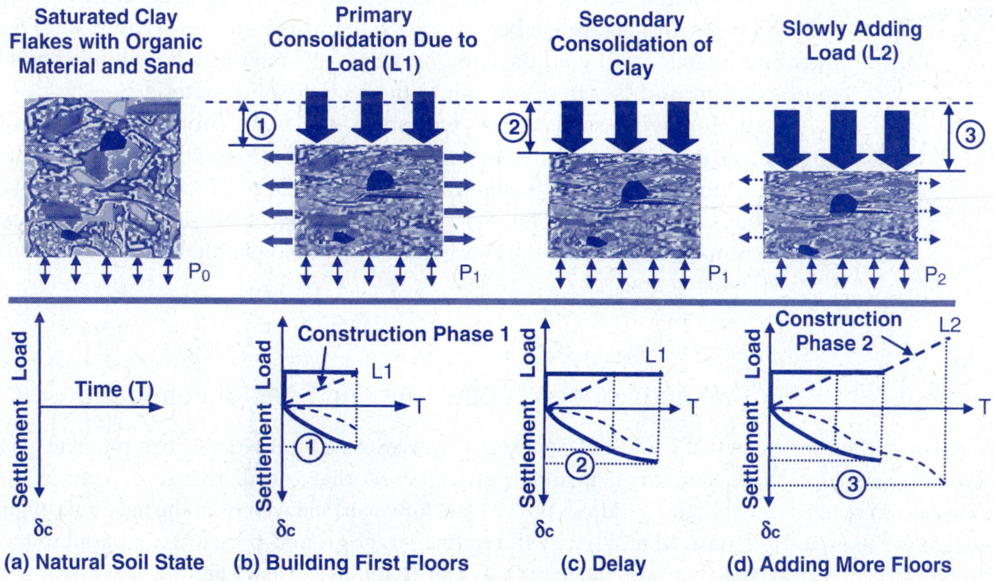

Figure 9.3 Consolidation Mechanisms of Clay Under the Tower of Pisa

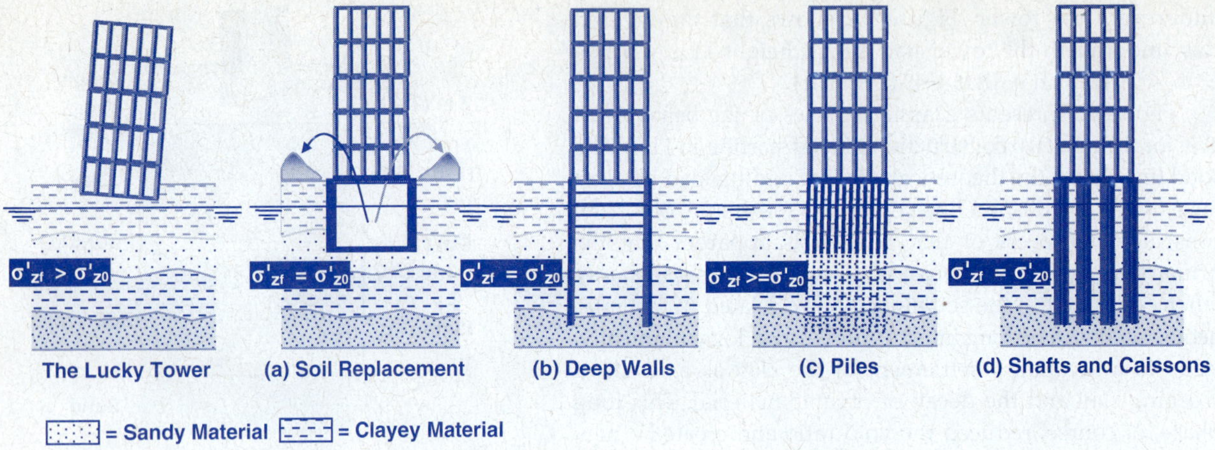

Figure 9.4 Main Approaches to Minimize or Avoid Consolidation

9.3.2 METHODS TO AVOID EXCESSIVE SETTLEMENTS OF NEW BUILDINGS

Terzaghi showed that settlement has four main causes:

1. Compression index of the soil
2. Void ratio of the soil
3. Thickness or height of the stressed soil strata
4. Ratio of effective stress caused by the new load

As we will learn, modern deep foundation methods are geared toward improving one or more of the root causes depending on the design parameters and the surrounding conditions. Let's see how some of the most used approaches relate to Terzaghi's function.

Figure 9.4 compares schematically four engineering designs that all have one goal in common: to add little to no stress to the initial or natural stress that has been established over 100s or even 1000s of years. This is achieved in different ways: (a) excavating a volume of soil that matches the weight of the new building, (b) excavating walls down to solid grounds and creating a box on which the upper structure can rest (c) driving piles, like the Romans, until the tips stand on stronger material, (d) drilling shafts or caissons until solid ground is found and filling them with solid material.

In the following sections, we will study the main problems that each of the four approaches faces during construction. As we will learn, each approach requires a unique set of technologies and understanding of engineering principles mainly related to geotechnology and hydraulics. But first, you will be faced with a complex problem requiring you to comprehend the rich and advanced technologies that have been developed over the last 3,000 years.

Header Problem 9.1 / Planning a Deep Excavation and Special Foundation

In Chapter 8, we studied methods that could be used to safely excavate a 60-ft (18 m) deep pit. One geotechnical benefit of removing such a large amount of material, of course, is that it will reduce or even eliminate the consolidation caused by the weight of the building. In our case, however, the weight of the new building is greater than the weight of the excavated material and thus will create a larger ground pressure compared to the original status. Thus, a series of piles that support the building was found necessary. Figure 9.5 provides the basic information about the subsurface structure.

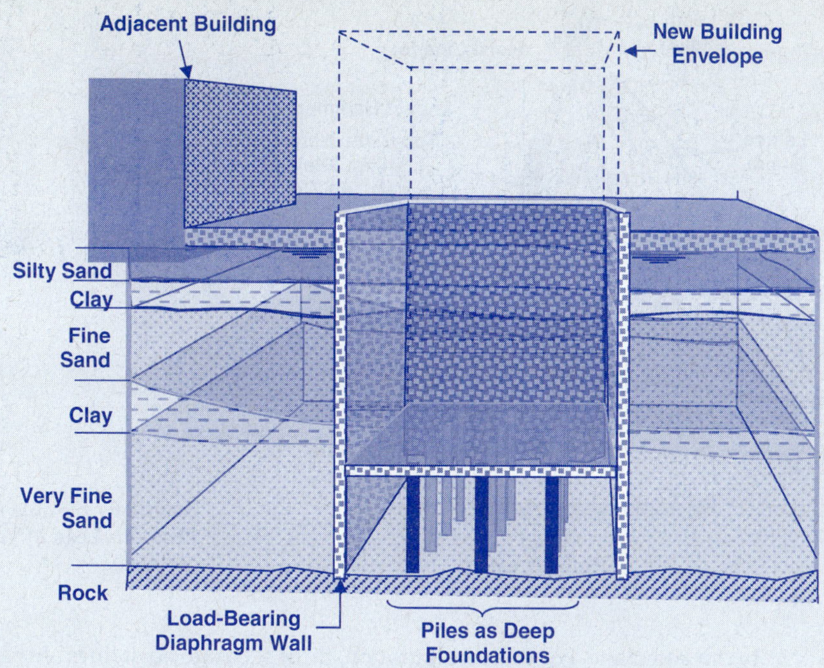

Figure 9.5 Sketch of the Engineer's Deep Foundation Concept

9.4 Pile-Driving Technology

If the architect of the Tower of Pisa had been able to study Vitruvius' *De Architectura* before the construction began, there would be no leaning tower and no UNESCO World Heritage site. In fact, Vitruvius' work had been lost but was resurrected by Leonardo da Vinci and Galileo in the 1500s. Since then, driven-pile technology has advanced, not only in the materials used to make piles, but also in the use of sophisticated equipment, and lately in quality-control devices such as real-time monitoring of the bearing capacity.

9.4.1 DRIVING EQUIPMENT

Section 8.5.2 introduced the use of the impact hammer to drive soldier piles needed for the construction of retaining walls. Four different impact hammers were shown. However, piles are also useful to shore up horizontal forces, such as from a breaking train on a bridge. For this the lead needs to be positioned at an angle also referred to as batter. Figure 9.6 introduces a crawler crane manipulating a sliding lead, thus adding the capability to reach surfaces not in the same plane as the crane.

Of special interest in Figure 9.6 is the sliding guide on the boom point that allows linear movement. This movement kicks in when the lead hoist rope, attached to the base of the lead, is winched in or out. The angle or batter can be reached by activating the hydraulic cylinder in the telescopic brace that pushes or pulls the lead off its vertical position. As shown, the telescopic spotter is able to create positive or negative batters, sometimes called fore or aft batters. In fact, a so-called moonbeam can be mounted on the end of the brace, which allows side batters to the left and the right of the plane made by the

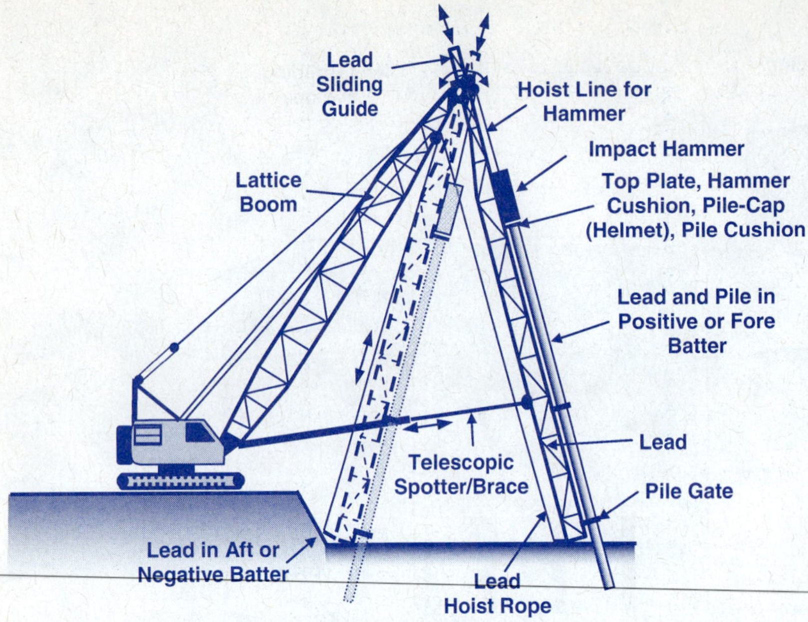

Figure 9.6 Flexible Pile-Driving System Positions Lead in Various Batters (Angles)

boom and crane. For this configuration, three separate hoist lines are needed for lifting the (1) hammer, (2) pile, and (3) sliding lead.

9.4.2 TAXONOMY OF DRIVEN PILES

S9.1 USACE Pile Driving Engineering Instructions
S9.2 USACE Pile Foundations

Figure 9.7 presents a paradigm for organizing the many different load-bearing pile types used in construction. The main grouping is according to the material a pile is made of (steel, timber, concrete, or a composite). In the following some of the types will be briefly introduced.

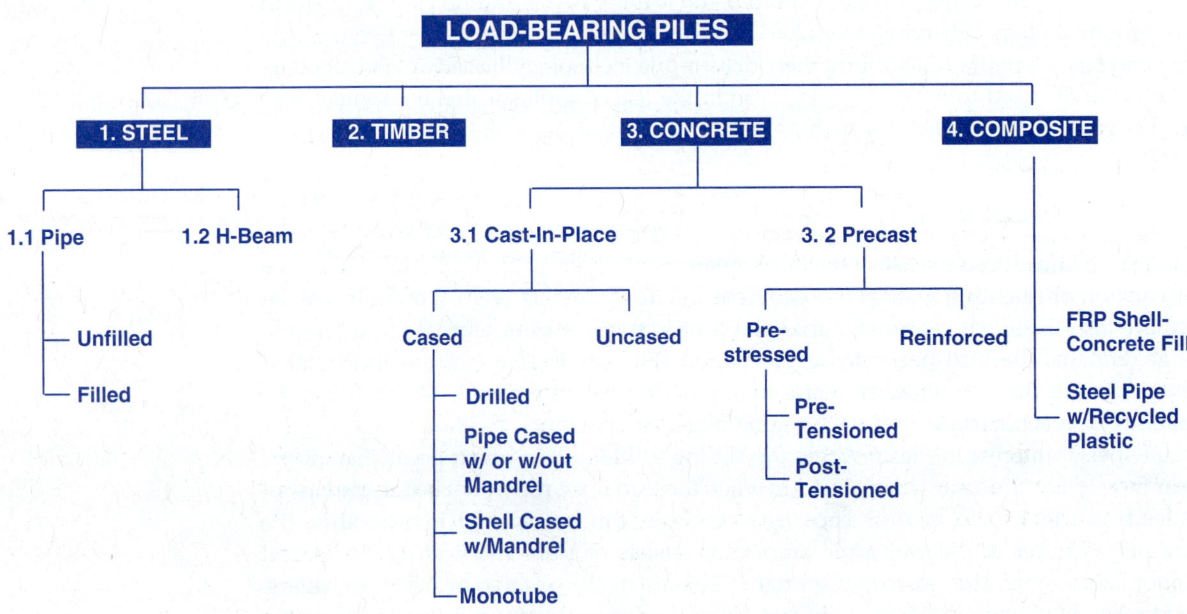

Figure 9.7 Overview of Common Foundation Piles

1. Steel Piles: Steel piles are produced in many forms, but they predominately consist of H-section or pipes sections that can be welded to reach 12–231 ft (36–70 m).

1.1 Steel Pipe Pile: The main specification for pipes is ASTM A252-98 (2007), Standard Specification for Welded and Seamless Steel Pipe Piles. Both the open pipe and the H-section are low-displacement piles as their cross-section does not require the soil to move aside, possibly causing a lifting on the top. Thus, they can be driven close together. Both can be designed as end or tip-bearing piles working similarly to free-standing columns that transfer the load from the top to the bottom or the tip resting on a solid stratum. A newly driven pipe is either left as is, or its center or core is excavated and filled with concrete, possibly combined with a steel beam. Of course, the acceptable design load varies widely. For example, the Army Corps of Engineers specifies the design load for a pipe pile with a concrete core as between 500 and 1500 tons (455 and 1,365 t), while without a core the allowable load is only 80–120 tons (73–109 t).

Alternatively, one end of hollow pipe can be welded shut with a flat or conical tip, making it a closed-end pile. A hardened steel tip is extremely helpful in soils that include boulders that might damage the rim of a hollow pipe or, equally undesirable, deflect the tip from the intended alignment. To improve piercing capabilities further, some tips are shaped for particular conditions. For example, welded-on pointed teeth support driving through obstructions and keep the pile on line when the tip reaches a sloped stratum. Like every pile type, pipe piles have advantages and disadvantages. For example, closed pipes will also result in soil displacement, which, of course, is minimal with open pipes. Also, steel pipes are costly compared to other material. On the other hand, they can reach extremely deep strata when welded together and are able to carry high loads when the core is full of concrete. Corrosion resistance can be obtained with cathodic protection or the application of a coating made of coal-tar epoxy, metalized zinc, or phenolic mastics.

1.2 Steel H-Pile: This is probably the most common bearing pile in areas without large boulders. With sufficient bearing capacity at the tip, Grade 50 steel allowing stresses up to 50,000 psi (3.5 t/cm^2) can be selected. H-piles will normally have to meet ASTM A36 with a strength of 36,000 psi (2.5 t/cm^2). The design loads lie between 40 and 200 tons (36–182 t). Like the hollow pipes, they displace a small volume of soil and thus can be driven at relatively close spacings. Sometimes, however, the soil can jam up between the flanges and the web. As a result, the cross-section of the pile will mirror that of a closed-end solid pile. Referred to as plugging, the effect is increased driving resistance and possible soil heaving around the top of the pile.

As with steel pipe, H-piles can be easily spliced together to extend their reach. In the same vein, piles that extend beyond the final grade can be easily cut with a torch. Naturally, the presence of boulders and a sloped bedrock ask for special treatment of the tip to prevent damage and deflection by a hard sloped surface. In fact, tip reinforcement is sometimes specified when it is known that boulders or thin layers of rock will be encountered. Corrosion protection is similar to that of steel pipe piles.

2. Timber Piles: More than 2,000 years before the Romans used timber piles to build solid foundations in areas with soft underground, lake dwellers in other parts of Europe built their houses on poles over water. Many of those driven piles have survived for 4,300 years with little damage. Thus, timber piles can be considered the oldest deep foundation structures. They are made of round, straight, undamaged tree trunks trimmed of all branches and bark. Most common trees used to make piles are southern pine and Douglas fir with allowable stresses of 1,200 psi. ASTM standard D25, Specification for Round Timber Piles, gives the minimum timber dimension. Because the timber is slightly tapered, with the tip between 5 and 9 inches (12–23 cm) and the butt 12 to 20 inches (30–50 cm), these piles are extremely hard to splice. As a consequence, timber piles

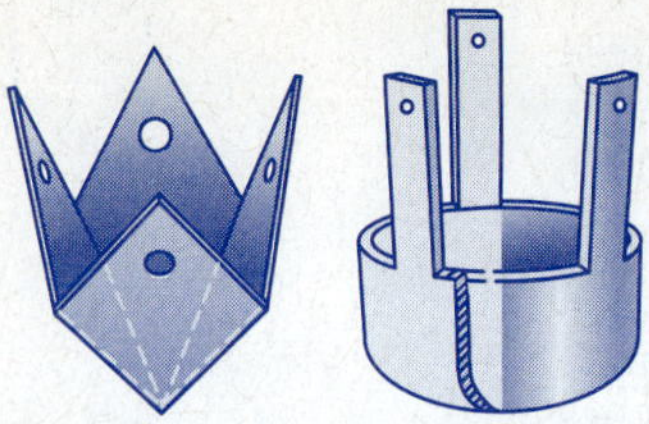

Figure 9.8 Timber Pile Point and Boot

are normally restricted to a depth of 66 ft (20 m), although Douglas fir can reach 120 ft (36 m).

Weak points are the soft tip and butt that will be impacted by a heavy hammer. The potential problem is splitting and even breaking of the pile body. Like a pipe end, the tip of a timber pile can be reinforced with a pointed shoe or a boot, shown in Figure 9.8. The use of heavier and more productive hammers subjects the piles to higher compression forces when the pile reaches obstructions. The metal point fits on the tip and can be nailed and ben to fit the diameter of the pile. Fitting the round boot may sometimes require trimming. On the other hand, the point requires a careful orientation so that the tip of the "pyramid" is perfectly centered and aligned with the centerline of the pile. Otherwise, the pile will be easily deflected away from its vertical axis. To protect the top of the pile from splitting in heavy driving in a dense subsurface, it is recommended that a metal band be applied at about 1.5 ft (45 cm) from the top.

Another method to reduce the driving forces needed is the use of a small water jetting nozzle at the tip of the timber pile. We will learn later that jetting will cause the liquefaction of the surrounding soil and, as a consequence, reduce its resistance to the advancing pile.

S9.3 USACE Driven Piles

As we learn from historical artifacts, timber survives a long time if permanently covered with water but decays easily when water and air can intermittently get access to the wood. Naturally, dry wood can easily burn. While pressure treatment with creosote will not preserve the wood forever from decay and wood borers, it will significantly extend its life.

3.1 Cast-in-Place Concrete Piles: Figure 9.7 lists two options for cast-in-place piles, cased and uncased. The later was added just for completeness since it requires drilling, a topic covered in the following section. That said, one recognizes the importance of a cylindrical shell that is driven to a desired depth before it is filled with concrete. In that procedure, driving a shell that is closed at the tip end is very much a displacement pile with the same potential for heaving. Driving of the shell may be done with the help of a shaped mandrel that is inserted during driving and removed when done. Shells driven with a mandrel can be made of much thinner steel (e.g. $\frac{1}{8}$ to $\frac{1}{4}$ in) compared to 1.0–2.5 in (25–63 mm). The mandrel-driven shells are sometimes corrugated, which increases the frictional characteristic of the completed pile.

Of course, in soil that contains boulders, driving the shell will face problems such as deflection that are akin to driving pipes or H-piles. On the other hand, the empty shell can be inspected via a light and camera prior to filling it with concrete. The visual inspection will focus on damage or distortion of the shell as well as the cleanliness and dryness of the inside, especially the bottom.

3.2 Precast Concrete Piles: Again, Figure 9.7 shows two types of piles, prestressed or plain reinforced. One major advantage of precast concrete made in a plant is the consistency and quality of the concrete in combination with an all-around perfect concrete cover not achievable when cast-in-place. The final product is a high-strength pile that is corrosion resistant. With the help of special cements and coatings, these piles will also resist chemically or organically polluted water.

It is common that precast piles, prestressed or reinforced, are made with a hollow core. The outside can be round, square, or octagonal. While the concrete has to meet ACI Specification 318 and the rebar ASTM A82, A615, and A884 for reinforcing steel, pre- and post-tensioning cables must conform to ASTM A416 A421, and A882.

In prestressed piles (prestressing is extensively discussed in Chapter 11), the heavy longitudinal bars are replaced with high-tension rods or cables. Before or after the pouring of the concrete, the tensioning steel is stressed, causing the concrete to be in constant compression across the entire cross-section. This in turn will lead to lower moment deformations during transport to the construction and thus avoid cracking of the concrete. As a result, the use of high-stress tension cables or rods makes it possible for pre-stressed piles to be made with thinner walls, leading to lighter and longer piles. For example, piles with standard reinforcement reach a length of 50 ft (16 m), while prestressed piles can be made with double that length. On the other hand, a pretensioned pipe is very hard to shorten on site if it has been cast too long. Piles that use the post-tensioning method are able to circumvent this major disadvantage. Instead of creating long, slender piles that have to be cast as one piece, post-tensioning allows piles to be cast in segments to be assembled and stressed only when it is known how large they need to be.

Another benefit of using precast concrete is the large—400-ton (364 t)—loads that can be put on an end-bearing pile. On the other hand, the tension loading makes tips, especially those of prestressed piles, more vulnerable to driving damage from boulders during driving. As protection, a steel H-section or a "stinger" can be attached or cast into the tip. To improve the footing of the end-bearing pipe in the rock surface of the solid underground, a special rock shoe made of solid steel can be cast into the end preventing any future lateral movement.

4. Composite Piles: The objective of composite piles is to allow the use of different materials along the pipe in order to address conditions that differ from the top to the bottom. For example, the lower end of a pile could consist of an H-pile, offering low soil displacement, while being protected against corrosion. This H-pile is then cast into a precast concrete pipe for the top part, providing the benefit of low corrosion. Obviously, the load capacity is tied to the lowest element in the link.

More recently, the resin used in fiber-reinforced polymers (FRP) or recycled plastic has attracted interest since it can be used to protect piles from decay and corrosion. For example, the FHWA-HRT-04-043 report from 2006 (http://www.tfhrc.gov/structur/pubs/04043/index.htm#toc) presents three examples suitable for load bearing. The first is a steel **Pipe Core Pile** where a normal steel pipe has been covered with a thick shell of recycled plastic. While the steel core still provides the structural strength, the plastic shell functions as a coating.

The second example mentioned in the FHWA report is the **Structurally Reinforced Plastic Matrix Pile** where recycled plastic takes over the place of concrete that is reinforced with either FRP rods or a steel rebar cage. This composite pile type uses approximately 240 recycled milk jugs per linear foot (0.305 m) of a 12-inch (0.305-m) nominal diameter pile. The **Concrete-Filled FRP Tube Pile** simply replaces the steel of a pipe shell with RFP and fills it with concrete, with or without reinforcement. The RFP can be first filled with concrete and driven after it is cured.

9.4.3 PILE PERFORMANCE EVALUATION METHODS

How can a contractor know that a pile that is driven 50 ft (15 m) through various layers of silt, sand, clay, gravel, and so on has reached the capacity required by the design? How can he be assured that the pile is still in good condition? For a long time, the only true measure one could have was the amount of settlement after a blow, established by the number of blows it took to advance a meter or a foot, and the type and size of the hammer. One famous and long-standing way was proposed by Arthur Mellon Wellington on December 29, 1888 in an article in *Engineering News*. It became known as the Engineering News formula and had the following form:

$$\textbf{Safe load L} = \mathbf{F}\frac{\mathbf{w} * \mathbf{h}}{\mathbf{s} + \mathbf{c}}$$

where

F = "constant determined from experience"
w = weight of ram
h = height of ram drop (assumes single-acting hammer)
s = penetration of pile per blow
c = "constant added to s"

In addition to the extremely empirical nature of the formula, its application was limited to timber, used exclusively at that time, and a sandy and silty underground. Later, the following refinements to the formula were made (for more information please download from the booksite a digital copy of the original book of 1893, *Piles and Pile Driving*, published by the Engineering New Publishing Co., New York):

$$\textbf{For drop hammers } \mathbf{L(lb)} = \mathbf{F}\frac{\mathbf{2\,w * h}}{\mathbf{s+1}}$$

$$\textbf{For steam hammers } \mathbf{L(lb)} = \mathbf{F}\frac{\mathbf{2\,w * h}}{\mathbf{s+0.1}}$$

w = weight of drop hammer or striking parts of the steam hammer (lb)
h = drop height of ram or striking parts (lb)
s = settlement of pile as a result of the last blow (in)

The scientific approach to soil behavior led by Terzaghi in the early twentieth century, the introduction of steel and concrete pipes, and the introduction of standard testing of the load capacity after the completion of the drive—all these developments led to increasing dissatisfaction with the Engineering News formula. Today, the FHWA asserts that except where extensive data have been collected to fit the empirical formula in an area with uniform conditions, this and other dynamic formulas, as they are called, should not be used. New methods include many of the factors that could not be measured at the time when the dynamic formulas were propagated, such as the elastic deformation of the pile under impact and the dynamic resistance of different soil conditions.

Let us jump from the past to today dominated by three science-based tools: (1) Pile Driving Analyzer with (2) Case Method Capacity and (3) Case Pile Wave Analysis Program (or CAPWAP).

Pile Driving Analyzer (PDA): This system is used to collect and analyze data during the driving of a pile that has been equipped with two types of sensors, an accelerometer and a strain transducer. The data is processed in real time to obtain velocity and force waves as they travel up and down the length of the pile after the strike by a hammer. Today, the sensor output can be transmitted wirelessly to the geotechnical office for immediate analysis with the CAPWAP software presented below. In the field, the PDA uses the Case Method to calculate the static capacity of the pile and also evaluates pile integrity and establishes driving stresses and hammer energy during pile installation.

The dynamic testing starts with the attachment of the four sensors three pile diameters below the head of the pile, on opposite sides. The reusable gauges are bolted onto the pile to be removed after the end of the test. The electronic cables are bundled, left hanging from the pile, and connected to the PDA, which collects the data and processes analogue information into digital data. The result after each blow is stored and presented on a screen for immediate review. Figure 9.9 offers three different graphs that indicate different pile-driving situations. Before we can interpret the curves, it is necessary to understand the behavior of a wave, such as a stress wave created by a hammer blow, and what an accelerometer will measure over time. The speed of a wave speed = C, the cross-sectional area = A, and the elastic modulus = E. When the hammer hits the head of the pile, the surrounding soil is compressed, creating a reaction force at the point of contact.

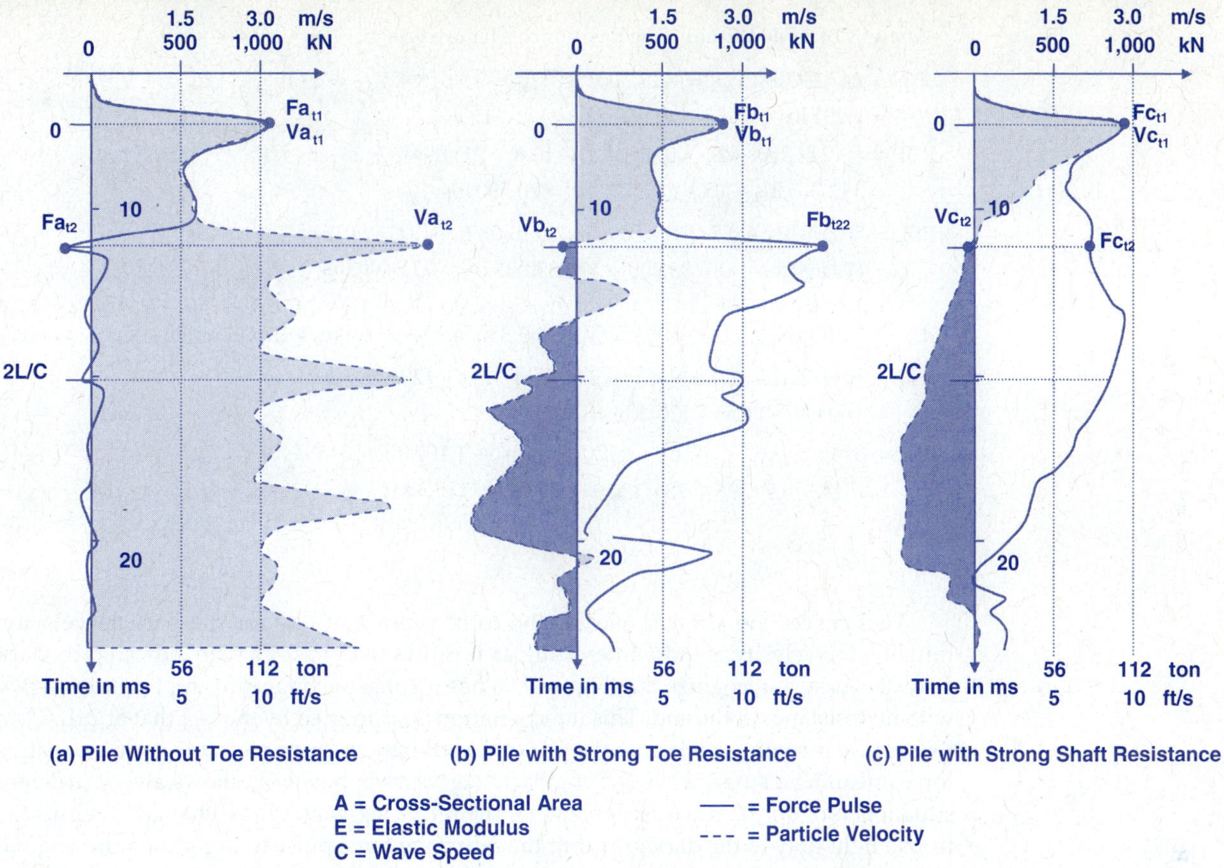

Figure 9.9 Force and Particle Velocity Measurements for Various Piling Situations

The initial acceleration caused by the hammer causes the material on its top to be accelerated at a velocity V, called a particle velocity. In theory, the force pulse and the particle velocity in a pile that experiences no outside resistance will follow the same wave pattern over time. The force at any time should equal the particle velocity times a constant, $E*A/C$.

There are two important features of how a wave behaves when it reaches the end of a solid rod, such as a pile. At the two extremes, the end of a rod could be either totally free to move, such as the prong of a tuning fork, or held rigidly fixed. When a wave reaches a free end, it returns in the same phase as it arrived. When the end is held, however, the force will cause a phase change, and the wave returns back in the opposite phase. Figure 9.9a and b show the results of those two extreme situations.

The vertical axis shows that time in milliseconds after the hammer force has reached its maximum ($= 0$). The first major spike after the initial blow indicates the return of the wave and the force pulse that had traveled along the length L of the pile, down and back up. The time is when the first wave arrives, which lets us calculate the speed of the wave, since the travel time of the wave $= 2 * L/C$. Naturally, the time $4*L/C$ is the time the first and largest wave traveled twice to the end of the pile and back. The size of the force, as measured by the strain gauge, measures a complementary set of data. If the acceleration hits the free end, there will be no action and no reaction force created. Thus, the acceleration "dies" or attenuates. On the other hand, if the pile feels resistance, the acceleration creates a dynamic force, which can be larger than the force created by the hammer. Let's review what this theory means by looking at the data.

Analysis of a Pile Without Toe Resistance—Figure 9.9a

$F_{at1} = 247{,}250\ \text{lbs}, F_{at2} = -22{,}500\ \text{lbs}, V_{at1} = 10.8\ \text{ft/s}, V_{at2} = 21\ \text{ft/s}, EA/C = 8{,}231\ \text{lbs s/ft}, J = 0.7$

$$RTL = \tfrac{1}{2}(247{,}250 - 22{,}500)\ \text{lbs} + (\tfrac{1}{2}(10.8 - 21)\ \text{ft/s}) * 8{,}231\ \text{lbs s/ft} = 112{,}375\ \text{lbs} - (5.1 * 8{,}231)\ \text{lbs} = 70{,}590\ \text{lbs}$$

$$RSL = 70{,}590\ \text{lbs} - 0.7((10.8\ \text{ft/s} * 8{,}231\ \text{lbs s/ft}) + 247{,}250\ \text{lbs} - 70{,}400\ \text{lbs}) = (70{,}590\ \text{lbs} - 0.7(88{,}900 + 176{,}850)\ \text{lbs} = \underline{\mathbf{-115{,}400\ lbs}}$$

$F_{at1} = 1{,}100\ \text{kN}, F_{at2} = -100\ \text{kN}, V_{at1} = 3.3\ \text{m/s}, V_{at2} = 6.4\ \text{m/s}, EA/C = 120\ \text{kNs/m}, J = 0.7$

$$RTL = \tfrac{1}{2}(1,100 - 100)\ \text{kN} + (\tfrac{1}{2}(3.3 - 6.4)\ \text{m/s} * 120\ \text{kNs/m} = 500\ \text{kN} - (1.55 * 120)\ \text{kN} = 314\ \text{kN}$$

$$RSL = 314\ \text{kN} - 0.7((3.3\ \text{m/s} * 120\ \text{kNs/m}) + 1{,}100\ \text{kN} - 314\ \text{kN}) = (314 - 0.7(396 + 786)\ \text{kN} = \underline{\mathbf{-513\ kN\ (115.3\ kip)}}$$

Two curves are shown, one for the force pulse and one for the particle velocity. Initially, the velocity is more interesting as it shows the phase-change indicated by dark gray. It is easy to recognize that Figure 9.9a has no phase change, thus representing a pile with no resistance at the end. This interpretation is supported by the fact that at 2∗L/C no force pulse is coming back from the pile end as the strain gauge measures even a negative force, meaning a stress wave coming back. Figure 9.9b, however, shows a very different situation. Not only can we recognize that the returning waves have changed their phase from a light gray to the dark gray, but the returning force pulse is larger than the original at time 0. The particle velocity and the force pulse are inverse, as one would expect. When the force reaches the free end, it simply dissipates, as there is no resistance, while the velocity doubles, as shown in Figure 9.9a. On the other hand, if a pile hits hard rock, the hammer force reaching the tip will be met, resulting in a force that is even larger than the initial blow due to the dynamic response of the pile itself. As shown in Figure 9.9b, the force F_{bt2} returning to the strain gauge at 2∗L/C is large then F_{bt1}, the initial blow, while the velocity V_{bt2} has turned to negative, indicating that the pile tip did "bound" backward. These two simple cases are rare in the real world. Usually the resistance that a pile experiences comes from both skin friction and toe. Let us look at a such a case.

Figure 9.9c shows that the recognizable wave forms of the force pulse and particle velocity have disappeared. In other words, both the velocity and the force graphs show no oscillations. This means that at any time waves at various amplitudes arrive back at the sensor "smoothing" each other out. The results are time-based measurements that are averages of many waves. The only possible interpretation of this pattern is a situation where the pile experiences resistance all along its length, not just at its tip. Friction resistance at the circumference of the pile also resists the acceleration and the force pulses, thus reflecting waves that "overlap" each other. Thus, the graph in Figure 9.9c represents a case of a pile with strong friction resistance along its skin.

Case Method Capacity: While the graphical representation of the waves helps us understand the conditions surrounding the pile, a contractor needs more specific information about the pile capacity, in particular the precise static load that a specific pile would be able to carry if the driving equipment were turned off right now. The establishment of such an approach was the longstanding topic of research at Case Western Reserve University in Cleveland, Ohio, which resulted in the Case Method Capacity method. It takes advantage of the PDA measurements and a model

Analysis of a Pile with Toe Resistance—Figure 9.9b

$F_{at1} = 179.8$ kip, $F_{at2} = 337$ kip, $V_{at1} = 8.85$ ft/s, $V_{at2} = -1$ ft/s, EA/C = 8,231 lbs s/ft, J = 0.05

$F_{at1} = 800$ kN, $F_{at2} = 1{,}500$ kN, $V_{at1} = 2.7$ m/s, $V_{at2} = -.3$ m/s, EA/C = 120 kNs/m, J = 0.05

$$\text{RTL} = \tfrac{1}{2}(800 + 1{,}500)\,\text{kN} + (\tfrac{1}{2}(2.7 + 0.3)\,\text{m/s}) * 120\,\text{kNs/m} = 1{,}150\,\text{kN} + (1.5 * 120)\,\text{kN}$$
$$= 1{,}330\,\text{kN}\ (299\,\text{kip})$$

$$\text{RSL} = 1{,}330\,\text{kN} - 0.05\,((2.7\,\text{m/s} * 120\,\text{kNs/m}) + 800\,\text{kN} - 1{,}330\,\text{kN})$$
$$= (1{,}330 - 0.05\,(324 - 530))\,\text{kN} = \underline{\mathbf{1{,}340\,kN\ (301\,kip)}}$$

Analysis of a Pile with Toe and Shaft Resistance—Figure 9.9c

$F_{at1} = 202.3$ kip, $F_{at2} = 157.4$ kip, $V_{at1} = 9.5$ ft/s, $V_{at2} = -.66$ ft/s, EA/C = 8,231 lbs s/ft, J = 0.4

$F_{at1} = 900$ kN, $F_{at2} = 700$ kN, $V_{at1} = 2.9$ m/s, $V_{at2} = -.2$ m/s, EA/C = 120 kNs/m, J = 0.4

$$\text{RTL} = \tfrac{1}{2}(900 + 700)\,\text{kN} + (\tfrac{1}{2}(2.9 + 0.2)\,\text{m/s}) * 120\,\text{kNs/m}$$
$$= 800\,\text{kN} + (1.55 * 120)\,\text{kN} = 986\,\text{kN}\ (222\,\text{kip})$$

$$\text{RSL} = 986\,\text{kN} - 0.4\,((2.9\,\text{m/s} * 120\,\text{kNs/m}) + 900\,\text{kN} - 986\,\text{kN})$$
$$= (986 - 0.4\,(348 - 86)\,\text{kN} = \underline{\mathbf{881\,kN\ (198\,kip)}}$$

of the pile as linearly elastic and a constant cross-section to calculate the **T**ota**L** Resistance **R**TL and the **S**tatic **R**esistance of a **P**ile, RSP:

$$\mathbf{RTL} = \tfrac{1}{2}\,(\mathbf{F_{t1}}+\mathbf{F_{t2}}) + (\tfrac{1}{2}(\mathbf{V_{t1}} - \mathbf{V_{t2}})) * \mathbf{EA/C}$$
$$\mathbf{RSP} = \mathbf{RTL} - \mathbf{J} * (\mathbf{V_{t1}}(\mathbf{EA/C}) + \mathbf{F_{t1}} - \mathbf{RTL})$$

where

J is a dimensionless damping factor reflecting the soil type near the pile toe (0.1 for clean sand and 0.7 for clay)

The RSP function works best with piles experiencing a large shaft resistance. For piles with a large toe resistance, a different function should be used (see booksite FHWA Design and Chapter 18, "Construction of Driven Pile Foundations").

From Figure 9.9 we learn that t1 refers to the time where the blow of the hammer reaches its peak while t2 is the time when the first wave returns from the tip of the pile at 2∗L/C. Let us apply this function to the different cases in Figure 9.9.

It is interesting to see that the results of the Case Method Capacity calculations tell us that the pile sitting on a tough layer has, with 1,340 kN (301 kip), clearly the highest static capacity. It also shows very plainly the small negative dynamic contribution to the RTL of 10 kN (2.25 kip). This stands in stark contrast with the free-end pile, which shows a low RTL of 314 kN (70.6 kip) and a negative static capacity, which is obviously not possible. This makes it apparent why the Case Method should only be used on piles that have significant toe resistance, such as those in Figures 9.9b and c. In fact, the PDA output for the last pile, shown in Figure 9.9c, computes into an RTL of 986 kN (222 kip) and a RSL 881 kN (198 kip), indicating a "healthy" dynamic contribution of 986 kN – 881 kN = 105 kN (23.6 kip) to the total static load resistance RTL.

Case Pile Wave Analysis Program (CAPWAP): This program takes PDA data collected on site to conduct a more thorough analysis with the goal of refining the Case Method results. The program is also based on the wave equation and the elastic pile and soil models. The final objective of using this program is to match the collected data with that of a model for soil and pile. In this iterative method, the factors representing possible soil conditions are changed until they match the PDA data as closely as possible,

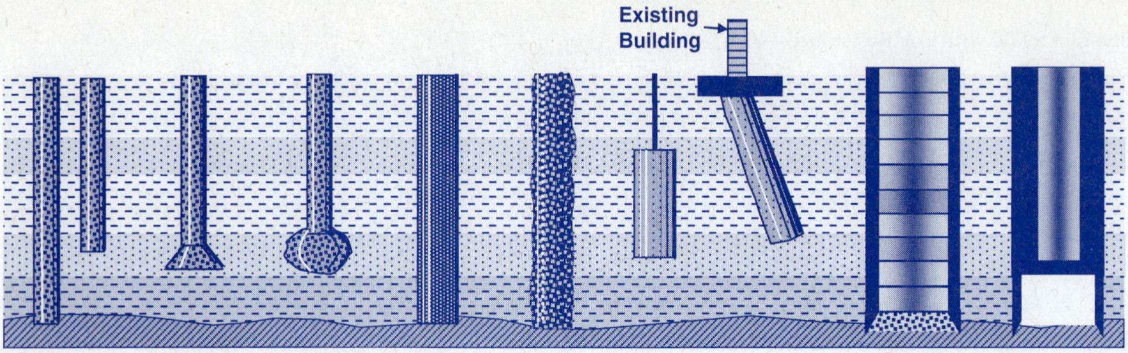

Figure 9.10 Overview of Common Deep Foundation Technologies

representing the best estimate for the static pile capacity, soil resistance on the shaft, and its damping characteristics.

9.5 Non-Driven Load-Carrying Piles, Columns, and Caissons

This section will introduce a wide range of technologies that avoid the noise and vibration created by driving piles and sheets through boring, drilling, displacing, mixing, jetting, vibrofloating, and sinking of boxes and rings. As with piles, each method has its advantages and disadvantages, making them suitable in situations where others are not. Figure 9.10 provides a graphical product layout of those technologies, which will be discussed below. Again, the presentation first reviews each process before studying the mechanics of the main equipment and some unique tools.

9.5.1 DRILLED AND BORED CAST-IN-PLACE CONCRETE PILES

As with driven piles, the objective is translate the stresses created by the weight of a structure built on the surface down into more solid soil layers or, if necessary, all the way to rock. Despite this simple objective, the almost indefinite number of possible subsurface conditions has led to a diverse set of equipment, tools, and attachments.

S9.4 MoDOT Drilled Shafts
S9.5 USACE Driven Piles

Drilling and Concreting Methods

We looked at the method for construction with drilled cast-in-place piles for building vertical walls to allow vertical excavation. While those were needed to carry horizontal forces coming from the surrounding soil, the piles we are discussing now are built to transfer heavy vertical loads to a lower stratum or, if possible, down to solid rock. As a result, they are larger in diameter and require much more care in terms of pile integrity and the quality of the contact between the bottom/shoe of the pile in the soil/rock. While it is sometimes possible to deploy a continuous auger, most often the size and soil condition require much a shorter and more solid auger, including a set of additional tools not needed for wall piles. Figure 9.11 presents the more common method without the use of a continuous long auger.

While the sequence of these nine steps has gained wide acceptance, innovative contractors have developed many modifications to improve their productivity where the job conditions allow it.

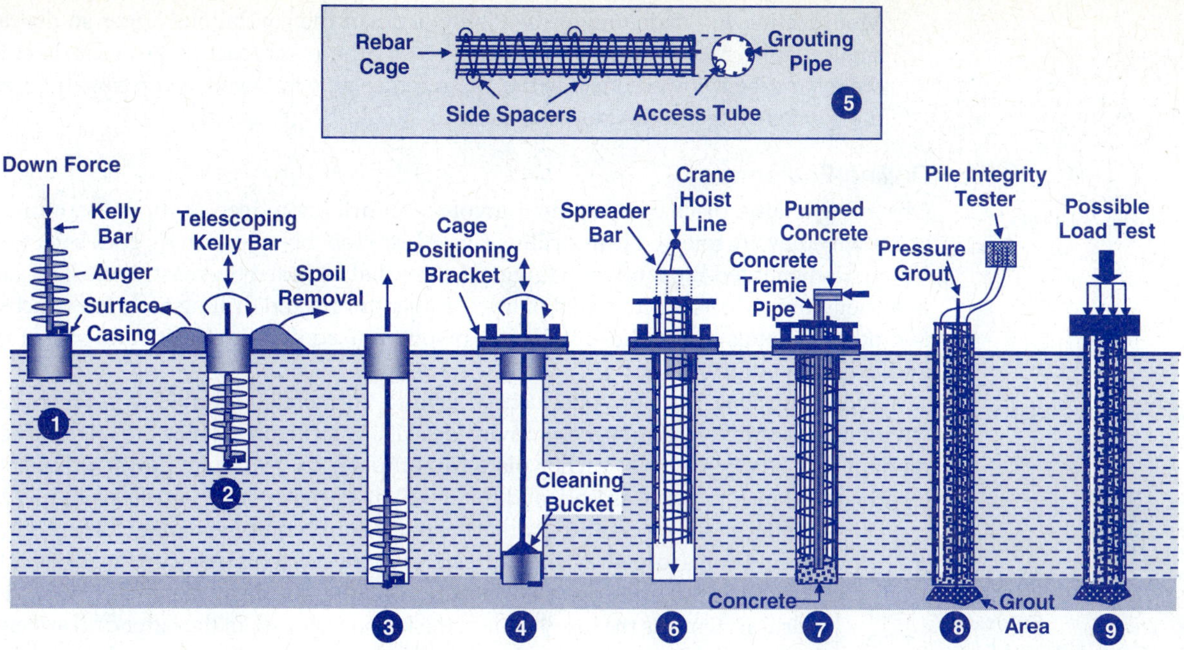

Figure 9.11 Basic Steps for the Construction of a Large Load-Carrying Pile

1. Installation of surface steel casing to prevent collapse of topsoil. Alignment of auger drill and Kelly bar powered by rotary drive and Kelly bushing.
2. Auger drilling and soil removal by lifting and cleaning the auger piece. Lowering of a casing if needed.
3. Removal of auger after desired depth is reached.
4. Operation of cleaning bucket to remove loose soil at the bottom. Setting up of steel bracket needed to position/suspend rebar cage inside drilled shaft until concrete has set.
5. Pre-assembly of rebar cage and possibly the insertion of access tube (to be later used to test pile integrity) and grouting pipe.
6. Lowering of prefabricated rebar cage using a spreader bar until cage collar sits on bracket.
7. Placement of tremie pipe and pumping of concrete to fill the shaft from bottom up without segregating the concrete. Tremie pipe is continuously raised as is the casing if installed.
8. Integrity testing of pile to identify possible large holidays caused by soil collapse during concrete placement. Possibly grouting of area below the tip of the pile.
9. Load testing of pile if planned.

Modification A: The casing is extended down into the shaft in order to support the walls or to cut off groundwater from layers serving as aquifer.
Modification B: During the drilling, bentonite is pumped into the shaft in order to keep groundwater out and to support the walls of the shaft.
Modification C: The short auger with Kelly bar extension is replaced with a continuous auger piece. Now the screw-mechanism brings the soil directly to the surface.
Modification D: The stem of the continuous auger is hollowed to serve also as concrete tremie pipe. After the required depth is reached, the auger is slightly raised, and concrete is pumped to the bottom. Auger is lifted as the shaft fills with concrete, interrupted by down-movements to compact the concrete already cast. Rebar cage is lowered or vibrated into the concrete of the already filled shaft. Grouting of the area around the bottom of the pile is needed since the bottom was not cleaned prior to concreting.

Modification E: Widening of the hollow stem of the continuous auger so that a small-diameter rebar cage will fit through. After the auger reaches the desired depth, the rebar cage is lowered inside the auger center pipe followed by a tremie to pump the concrete. Again, this pile bottom area should be grouted.

Drilling Equipment

S9.6 USACE Drill Riggs

VG9.1 Drilling Equipment

Observing a large piece of equipment involved in drilling a deep shaft, one can understand why they are referred to as drilling rigs. Most probably, the term *rig* goes back to the Vikings, who raided England with "rigged" ships that included the mast, spars, and sails. In ancient times, drill rigs consisted of large timber towers and drilling tools to extract salt and other minerals before the oil industry revolutionized the complexity and size of rigs and drilling platforms.

Today's drilling rigs for construction have to be mobile and able to reach rugged environments or low-ceiling spaces even inside existing structures. Thus, the "rigging" of a modern drill consists of a (1) carrier platform, (2) plant for hydraulic power production, (c) articulated mast to serve as lead, and finally (d) drive motors and winches. Figures 9.12 and 9.13 shows graphically that contractors and equipment manufacturers have found ways to let the same carrier deploy various tools based on the needs of the job. In general, one rig can drill using a continuous flight auger (CFA) or a Kelly drill system. As the insert shows, the key mechanism in the latter drill system is a square or any non-round rod or bar that is rotated by a round drive wheel with an opening in the center through which the bar can move freely up and down. During drilling, a drill piece is attached to the bottom of the bar that is subsequently turned by the Kelly drive powered by one or two hydraulic motors. The drive mechanism includes also protective bushings that can be easily adjusted to fit different bar sizes.

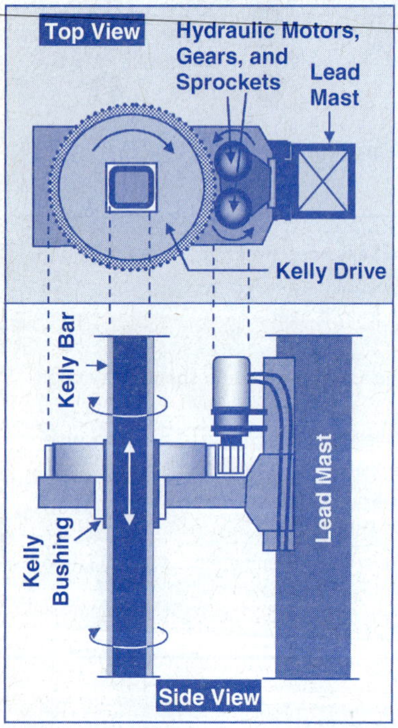

Figure 9.12 The Kelly Drill Drive System

Naturally, the efficiency of drilling deep shafts is linked to several factors such as the appropriateness of the auger head to extract the soil and rock, the available torque from the drill rig, and last the number of times an auger is brought up to spin off the soil from the screw. Should the length of the pile extend the height of the lead mast, extra nonproductive time has to be spent on decoupling Kelly bars or continuous lift auger elements. Thus, it should not come as a surprise to see drilling rigs that are 100 ft (33 m) high.

Crawler-mounted rigs offer more maneuverability and require less overhead clearance than the other rigs, making them the rig of choice for restricted work areas.

There is a variety of tools utilized by contractors when drilling shafts. The wide assortment includes drilling augers, for rock and soil, core barrels to casings and cleanout tools. Regardless of how powerful the rig is, if the wrong tool, or a poor-quality tool, is used, the result can be costly or even fatal.

Earth and Rock Augers (See also Figure 9.15)

Earth augers, like the one shown in Figure 9.14, are typically used in hard sands and cohesive materials. As the drill rig on the surface turns the Kelly bar or the auger extension, the head teeth scratch the soil and lift it onto the flight of the screw. The head is coupled to the first auger piece, pushing the loosened soil further up.

Rock augers are designed to overcome significant resistance to cutting due to buried boulders or layers of harder material. Having to take more abuse and exert more force, they are constructed of heavier material than the earth augers. The flat teeth of an earth auger are replaced by ferocious-looking conical hard-steel "dinosaur daggers." As these teeth wear more quickly than the auger, they are set into sockets to make them replaceable.

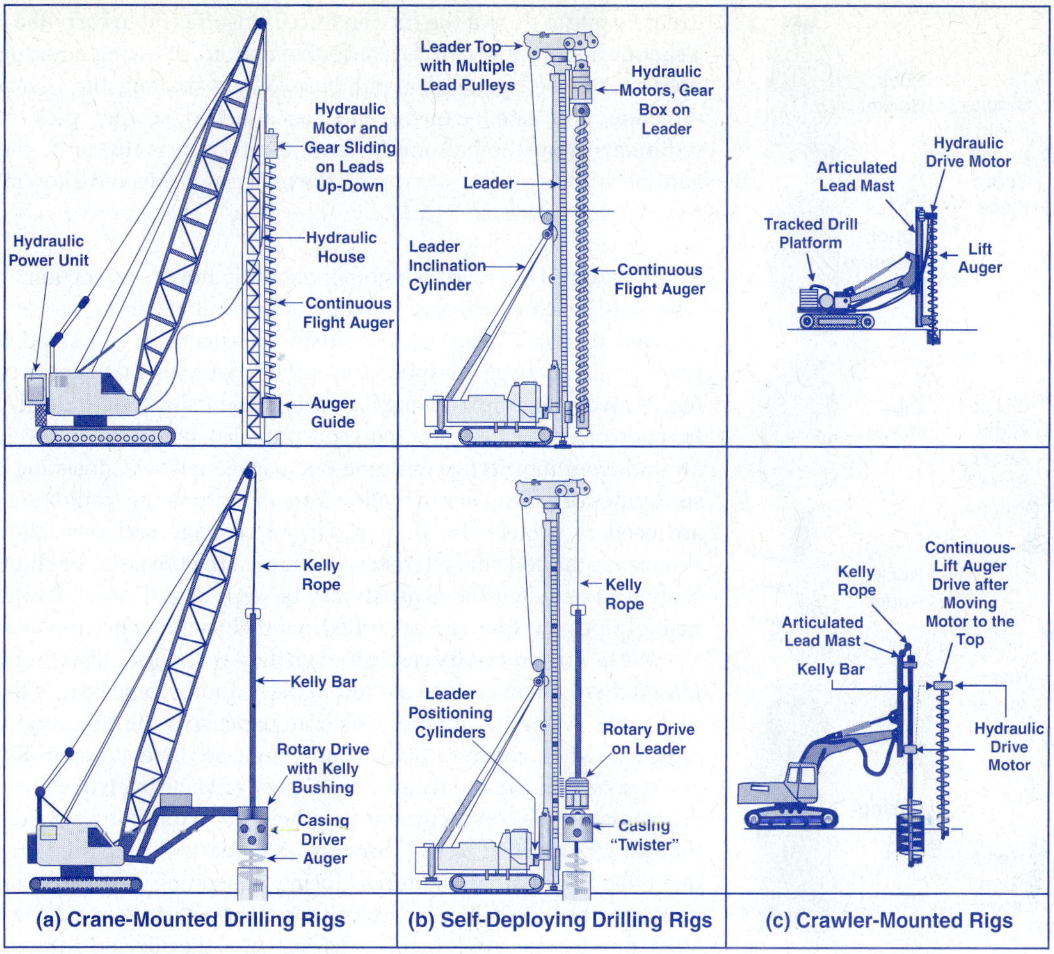

Figure 9.13 Major Drill Rig Configurations Used in Deep Foundation Construction

Core Barrels. Breaking or grinding through hard rock that may be encountered on the way to deeper depths can slow down progress. A more effective alternative to slowly fracturing the entire rock is to cut out large pieces and retrieve them. Borrowing from the concept of core drilling, where drills cut only along the perimeter of the circle, thus leaving the center intact for investigation, shaft drilling contractors switch to core barrels. After pulling up the auger piece, a hollow barrel is mounted and lowered into the shaft. When the bottom edge of the barrel hits the rock, its hardened teeth will cut along the perimeter while leaving the core intact. When a joint or discontinuity is encountered, the core breaks off and can now be removed as one piece.

Buckets. Buckets come in three types, earth, rock, and cleanout buckets. As the names imply, each has a designed use for either advancing the shaft or for cleaning the bottom. Like the augers, the buckets are attached to a Kelly bar and cut into the bottom of the shaft. The cleanout bucket fulfills a critical need in that it removes all the loose material that collects at the bottom of the shaft before being filled with concrete. Excessive amounts of unconsolidated soil left at the bottom will cause the finished

(a) Single Flight Auger Pieces

(b) Hard Soil Auger Head

Figure 9.14 Earth Auger

concrete pile to drop if the friction force is insufficient to carry the load. The cleanout bucket normally has a double bottom, allowing a cutting gate to open and close openings at the base. When rotating the bucket in one direction, the gate is open, and the attached scraper picks up loose sediments from the bottom. When the direction is reversed, the scraper gate closes the openings, trapping the material inside and allowing its safe removal to the top.

The Casing. This word is another example how construction adopted a Norman-French word, *casse*, having its root in the Latin word *capsa* meaning case or box. Today, casing outside construction can stand for many types of covers (e.g., computer casing) or materials that encase or enclose (e.g., window casing). Casings used in drilled shafts do indeed fulfill the function of a protective box, like the Latin *capsa*, in that they prevent the soil around from interfering with the open space inside. Consisting of round steel pipes of a diameter that allows an auger to fit the trough, short pieces are used to protect the shaft rim from damage and possible flooding. Whenever the soil characteristics encourage the break-off of chunks in the shaft wall or when the shaft should be kept dry of groundwater, longer casing pipes are lowered or drilled parallel to the main operation.

We talk of temporary casings when they are pulled out as the concrete is placed. Permanent casings are left in place and become part of the cast-in-place pile. A condition where such an expensive solution might be warranted is when concrete placement cannot be done successfully without such protection. Since only an integrity test after curing will show if the pile is acceptable, the contractor has no other option for fixing a weakened pile than to replace it or to dig down to the area to be repaired. Figure 9.16 demonstrates that soil failures during concreting can cause significant volumes of soil to fall onto the rising concrete surface and to be encapsulated, thus leaving voids in the cross-section of the pile. In addition, concrete will be lost as it fills in the space left by the collapsed wall area.

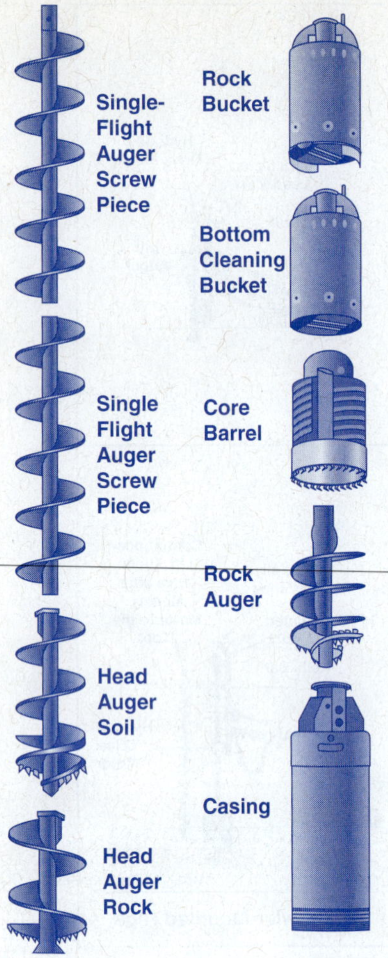

Figure 9.15 Augers, Buckets, and Barrels

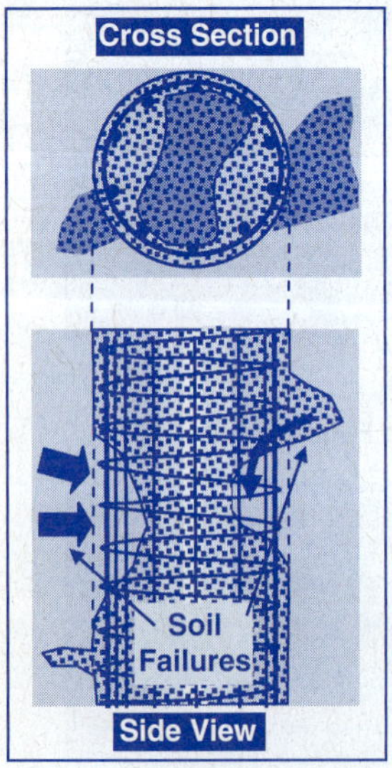

Figure 9.16 Pile Weakened by Soil Intrusions

Percussion Tools

While the rotary tools presented so far are most commonly used for drilling, in certain instances a brute-force technology is needed. One such example is a sloping hard-rock surface. When a barrel tends to bind, the tungsten tips of a rock auger slide sideways and push the auger off the vertical. To avoid the costly consequences of losing a tool at the bottom of the hole, the contractor may opt to use a rock breaker that can be dropped inside the shaft, acting as a slow-moving pneumatic hammer as it is raised up via a hoist line from the crane. After exchanging the rock breaker for a clamshell or grab bucket, the broken-up rock pieces can be retrieved before the drilling operation continues. The time needed to switch from percussion to retrieval tools and back again slows down the drilling operation.

Slurry to Support the Shaft Walls

Similar to the diaphragm-wall construction discussed in Chapter 8, contractors were looking for a method to protect the wall shafts of drilled piles without having to install a steel casing. To no one's surprise, the solution was bentonite slurry. Even more important than in the construction of retaining walls, which were eventually exposed during excavation, contractors had to ensure that no cavities could develop during both the excavation and concreting face. Thus, it was critical to better understand the mechanics of the interaction between the earth wall and the bentonite. Figure 9.17 highlights the effect of the bentonite slurry on the shaft "edge" area. As shown, the slurry changes from a liquid to a gel as it is pushed deeper into the soil due to the higher pressure inside the shaft filled with slurry ($H_\sigma * \gamma_s > H_w \cdot \gamma_w$). The reason for this soil-clogging property is that the electrically charged bentonite particles "square up" when left undisturbed. Shown in Figure 9.17 is the gelling effect causing the jelly zone, referred to as a filter or mud cake. It is capable of sealing the shaft against influxes and outfluxes of water that weaken and eventually erode vertical wall areas lacking the cohesion and shear strength necessary to counteract the combined pressure from groundwater and soil.

An alternative to the original bentonite are polymers consisting of chain-like hydrocarbon molecules. Like the bentonite plates, the chains are electrically charged and act similarly; in particular, they can be pressured into sand or silty soil where the long chains get lodged inside the pores of the soil. Eventually, the long polymer strands begin to hold soil particles together and to clog the flow of the slurry, resulting in the same sealing effect that bentonite exhibits.

Experience has shown that prevalent soil characteristics have to be considered when mixing the slurry. Most important is the viscosity of the slurry measured in poise or centipoise, named after the French physician Jean Louis Marie Poiseuille (April 22, 1799–1869), who studied the viscosity of blood inside the artery. A quicker but less accurate measure is the Marsh funnel viscosity or MV (see insert). Adopted from the oil industry, it was found that clay, silt, or sand need a slurry with a MV of 32 seconds (32 MV with a 946 ml funnel volume), while gravel needs up to 50 MV to create a sufficient filter cake. Slurry with a larger MV is very hard to desand and can create problems during concrete placement as it may stick to the rebar and thus create large "caverns" inside the completed pile.

While drilling at the bottom of the shaft filled with bentonite, soil particles will enter the slurry where they stay in suspension, only to settle when excavation ends. Of course, the heavier sand and fine gravel will accumulate at the bottom, which can create a problem when concreting starts. A rule of thumb is that a slurry with a sand content exceeding 4% should be desilted and/or desanded before concreting begins. Of course, an alternative to desanding is removal and replacement with fresh slurry with a sieve for larger objects and cyclones to remove the sand particles.

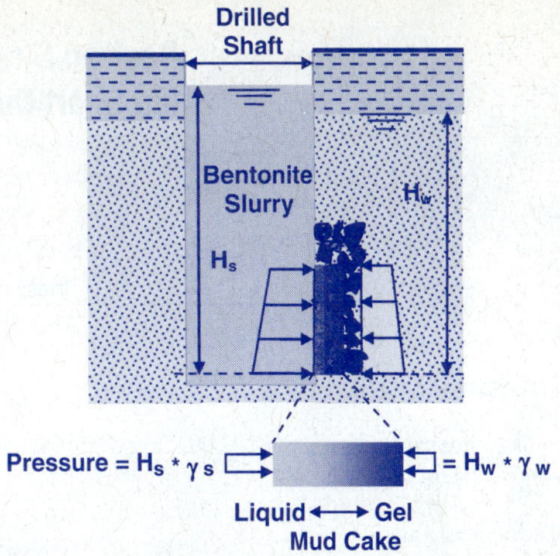

where

γ_s = Density of bentonite slurry

γ_w = Density of water

H_s = Height of slurry column

H_w = Height of water column

and:

$\gamma_s > \gamma_w$ and $H_s > H_w$

Figure 9.17 Wall Support Mechanism of Bentonite Slurry

Marsh Funnel Viscosity (MV)

Viscosity is defined as the shear stress in the slurry liquid divided by the shearing rate. The viscosity of slurry (mud) can be measured with the Marsh funnel and translated into poise or centipoise. The modern funnel consists of a cone 6 inches (152 mm) across and 12 inches in height (305 mm) to the apex of which is fixed a tube 2 inches (50.8 mm) long and 3/16 inch (4.76 mm) internal diameter.

While blocking the exit with one finger, the liquid to be measured is poured through a mesh into the cone holding about 1.5 liter. To take a measurement, the finger is released as a clock is started. The time in seconds is recorded as a measure of the viscosity (MV).

Worked-Out Problem 9.1 / Bentonite Slurry Mixing

Assume that you need to prepare a bentonite slurry inside a tank 24 hours before it is needed for drilling a 24-m (78.7 ft) deep shaft with a diameter of 2 m (6.6 ft) and a water table at 1.5 m (5 ft). Expect an overbreak/added depth of 15%. For the existing silty-sandy soil conditions, it is recommended to use average values for viscosity and density. How many 50 lb (0.22 kN) bags of bentonite and how many gallons of water are needed to create a workable slurry for the next day?

Table 9.2 Desired Slurry Properties Slurry Is Desanded by Pumping It a Short Distance to Special Equipment

Property	Range During Drilling	Range Before Concreting
Density	$9.95 - 10.3\,\text{kN/m}^3$ $63.0 - 65.3\,\text{lb/ft}^3$	$9.95 - 10.4\,\text{kN/m}^3$ $63.0 - 66.0\,\text{lb/ft}^3$
Viscosity	28 – 50 MV for 0.946 ml	28 – 50 MV for 0.946 ml
These values apply to mineral bentonite only.		

Assumptions

The desired density is $9.99\,\text{kN/m}^3$ or $64.15\,\text{lb/ft}^3$. The density of water is $998\,\text{kg/m}^3$ at 20 °C or $9.8\,\text{kN/m}^3$ ($62.1\,\text{lb/ft}^3$ or 8.3 lb/gal).

Calculation

$$\text{The theoretical volume of the shaft} = 24\,\text{m} * 3.14 * 1.0\,\text{m}^2 = 75.4\,\text{m}^3$$
$$= 78.7\,\text{ft} * 3.14 * (3.28\,\text{ft})^2 = 2,660\,\text{ft}^3$$

Volume that will fill overbreak and possible holes $= 1.15 * 75.4 = 87\,\text{m}^3$ ($2,059\,\text{ft}^3$)

Water needed (liters) $= 87\,\text{m}^3 * 1,000\,\text{l/m}^3 = 87,000\,\text{l}$

Water needed (gallons) $= 87,000\,\text{l} * 0.2642\,\text{gal/l} = 22,985\,\text{gal}$

In order to increase the density of water from 9.8 to 9.99 kN/m3, we need to add $0.19\,\text{kN/m}^3$ ($1.21\,\text{lb/ft}^3$) or 0.162 lb/gal of bentonite.

Amount of 0.222 kN bags $= (87\,\text{m}^3 * 0.19\,\text{kN/m}^3)/0.222\,\text{kN/bag} = 74.45$ **bags**

Amount of 50 lb bags $= (22,985\,\text{gal} * 0.162\,\text{lb/gal})/50\,\text{lb/bag} = 74.45$ **bags**

Discussion of Results

In order for the bentonite to activate, it is important to leave it 24 hours in the mixing tank. It is not necessary to fill the shaft all the way to the top, since the water table is at 1.5 m. However, the top of the slurry should not sink below 1 m. The viscosity should be measured with the Marsh funnel from time to time to make sure that it stays between 30 and 40 MV. Check density and viscosity of samples from the bottom before concreting in order to avoid costly repairs later due to sand and debris that settled and became encased.

Profiling the Shaft

S9.6 FHWA Drilled shaft installation plan
S9.7 FHWA Drilled shaft log
S9.8 FHWA Drilled shaft soil excavation log
S9.9 FHWA Drilled shaft concrete log

Drilling into the multi-layered subsurface does not create the intended smooth cylindrical opening. Drilling most often means breaking up and yanking large pieces out of the immediate surroundings of the shaft wall. This can create big surprises by the time that concrete is being pumped into the cavity. For example, a shaft with a theoretical volume of $20\,\text{yd}^3$ ($15.3\,\text{m}^3$) may need $26\,\text{yd}^3$ ($20\,\text{m}^3$) of concrete to fill. Knowing the actual shape of the created concrete pile is also important when interpreting the data of pile integrity and load testing. To solve this problem, two methods have been developed. The first uses a mechanical or electronic caliper device that is lowered into the shaft (caliper = instrument having two adjustable arms or jaws to measure the diameter or thickness of round objects). Most recently, a 360-degree sonic "radar" is used to create a 3-D "as built" image of the drilled shaft, showing the surface relative to the vertical centerline in real time. This data, of course, allows the instant calculation of the shaft volume as a basis for the contractor to order ready-made concrete from the batch plant.

The second method uses data from a concrete flow meter or the hopper volume and the height of the raising concrete column inside the shaft. The surface of the concrete can be easily measured with a weighted tape lowered into the shaft; this works even with slurry. When the weight at the end of the lowered tape meets the concrete the tape slacks off, telling the observer, who is able to read the tape instantaneously, that it reached the bottom. By plotting the theoretical with the actual volume of pumped concrete, the observer gets an immediate overview of the situation and is able to predict the amount of concrete needed to fill the drilled shaft. Figure 9.18 presents the graph of a hypothetical situation.

The imaginary shaft is 24 m (80 ft) deep, needing 20 m^3 (26 yd^3) of concrete and resulting in the theoretical linear fill line. One always needs to expect a modest overbreak of 3–6%, but, as the development of the actual fill line shows, three events result in the contractor needing almost 40% more concrete than planned.

Potential Problems: FHWA-IF-99-025

- Shaft off location or out of plumb
- Base of shaft not in proper founding stratum
- Crack in the shaft when hit by equipment early in curing process
- Bulge or neck in the shaft (soft ground zones that were not cased)
- Cave-in of the shaft walls
- Excessive mud cake buildup
- Temporary casing that cannot be removed
- Horizontal separation or severe neck
- Horizontal sand lens in concrete
- Quarter-moon-shaped soil intrusion on the side of the shaft
- Soft shaft bottom
- Voids outside of rebar cage
- Honeycombing, washout of fines or water channels in the concrete
- Folded-in/encased debris

Event 1: When the concrete reaches a depth of 20 m (66 ft), it stops rising until 6 m^3 (8 yd^3) are added. At 11 m^3 (14.3 yd^3), the top of the concrete is still at 19 m (63 ft) instead of the expected 12 m (40 ft). As the sketch indicates, one has to suspect that concrete entered into a rather large side cavern that was created or had existed there before drilling began.

Event 2: At 13 m (36 ft) the concrete surface suddenly drops 2 m (7 ft) even though more concrete is pumped in. Unlike event 1, there was an instant drop, not just a smooth transition. This indicates an abrupt creation of an opening in the shaft leading to an empty cavern that was so far untouched. The cause of such an event must have been the pressure of the concrete combined with a lack of water inside the cavern.

Event 3: At a depth of 10 m (33 ft) the concrete starts to rise quicker than predicted, meaning that the cross-sectional area suddenly got smaller. At a depth of 6 m (20 ft) the fill slope returns to what it should be. This can only mean that debris from cave-ins had accumulated at this depth and was subsequently encased by the concrete, thus reducing its needed volume.

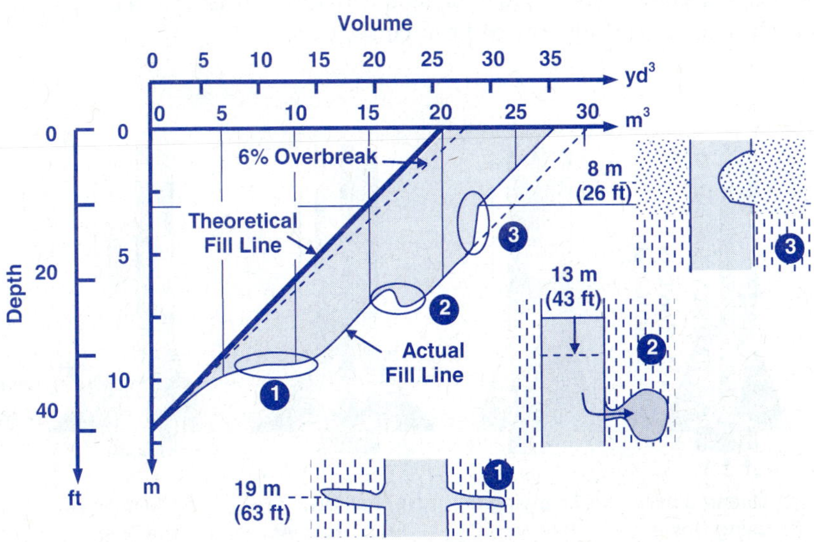

Figure 9.18 Concrete Filling Measurements for Hypothetical Shaft

While the vast majority of drilled shafts are built without any difficulties, the Federal Highway Authority (FHWA) has put together a list of problem areas that a contractor has to watch out for.

9.5.2 UNDERREAMED PILES

The bearing capacity of most piles depends strongly on the amount of load that the base or point of the pile can take after the initial settlement. The two main factors, of course, are the area and the soil strength below the base of the pile. Widening or belling the base is one method to take advantage of this equation since the area increases with the square of the radius (area $= \pi * r^2$).

> **To ream:**
>
> - To enlarge, taper, shape, or smooth out (a hole) with a reamer
> - To enlarge a bore or a shaft by removing material

Not surprisingly, foundation contractors have found a way to take advantage of the open shaft to insert an expandable "reaming tool" similar to an angioplasty balloon that opens up a clogged artery. The tool, referred to as a belling bucket, underreamer, or simply the reamer, is presented in Figure 9.19. After the drilling auger or bucket reaches a desired depth, it is replaced with the reamer and lowered into the hole at the end of the Kelly, as shown in Figure 9.19b. A two-link mechanism forces the wings of the reamer outward when the drilling rig exerts a downward force, since the round tip component does not turn with the Kelly and thus acts as the stable arm of a c-clamp (Figure 9.19c). Rotating the Kelly will result in the cutting teeth of the reamer digging into the soil of the shaft wall, which collects at the bottom of the shaft. Naturally, reversing the down-pressure will cause the reamer wings to close again, trapping the loosened soil inside, which is ready to be removed by raising the reamer to the top through the shaft (see Figure 9.19d). Repeating this process will increase the underream angle until a mechanical stop that dictates the maximum (see Figure 5.19d). Finally, the bottom of the shaft can be cleaned with a separate bucket, thus making it ready for installing the rebar cage.

One necessary condition for the bellying operation to work as designed is a dry shaft. As we learned earlier, a possible method to keep groundwater from seeping in is the use of a casing that can be properly sealed, as depicted in Figure 9.20 ❸.

In the hypothetical situation depicted in Figure 9.20, a gravel-sand layer below a layer of clay carries the water of an aquifer that has to be held back by a casing that is being drilled through it and into the clay layer underneath. In order to prevent the groundwater from entering the shaft, a seal has to be created between the casing and the clay. If this can be accomplished successfully, the bell can be kept dry.

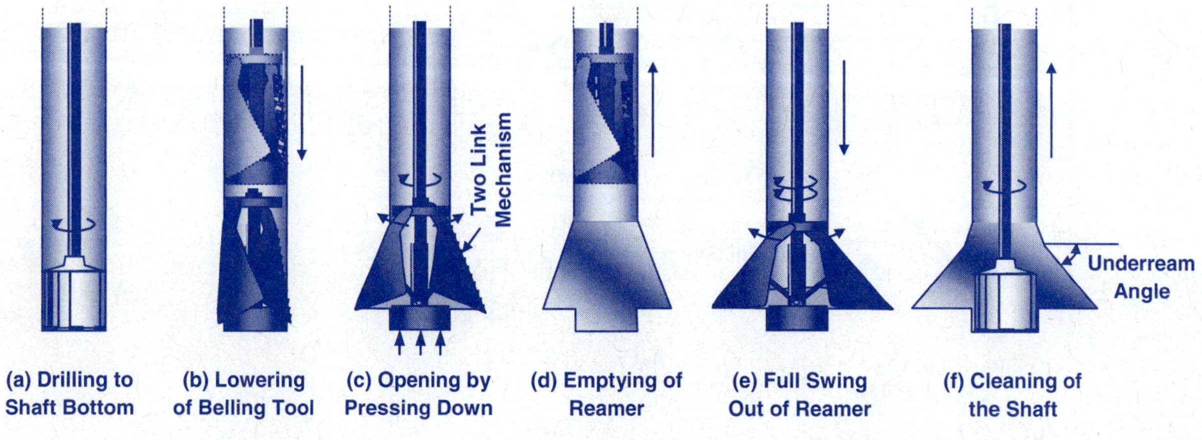

Figure 9.19 Process Steps to Ream Out the Bottom of the Drilled Shaft

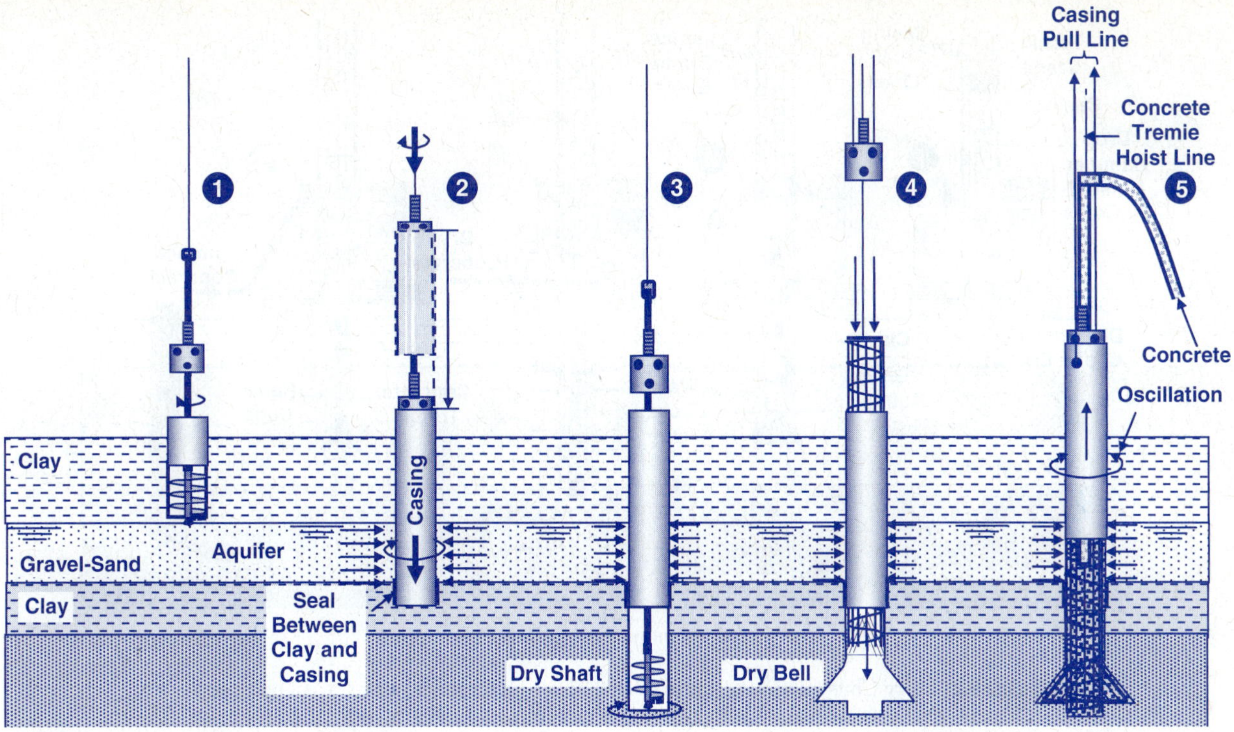

Figure 9.20 Main Steps in Constructing an Underreamed Pile

9.5.3 THE FRANKI PILE

In the 1910s the Belgian contractor Edgard Frankignoul came up with a method that speeded up the casting of a pile in place to a variable depth. Further, he found a way to expand the point of a pile, thus drastically increasing the bearing capacity of such piles. It may have been the first time that a steel pipe was used as temporary casing, but rather than "piling" the case into the ground, causing a lot of energy loss and vibration, a drop-hammer is operated inside the case to hit a bottom plug that can be pushed out when the desired depth is reached. Figure 9.21 demonstrates the main steps for installing a Franki pile.

The unique features of this cast-in-place pile construction are clearest in steps 2 to 4. The use of the end-plug keeps the casing dry inside while the drop energy is most effectively applied to displace the soil ahead of the cap. At the preferred depth, some concrete is pumped to cover the plug, and the casing is locked in place before the hammer is dropped again, pushing the plug out of the case at the bottom. The added concrete will ensure that no water is able to enter. By repeating the cycle of pumping and hammering the concrete is forced out of the case into the surrounding soil, which is densified in the process. It is clear that the purpose of this step is to achieve a benefit that is similar to the "bellying" out of drilled shafts, creating a large increase of point load capacity. Early experiments by loading the completed

Steps to an Underreamed Pile

1. Drilling dry shaft to groundwater level
2. Advancing casing into the clay layer, creation of seal around casing
3. Drilling dry shaft to desired depth
4. Reaming bell, cleaning, and installing rebar for bell and shaft
5. Concreting bell and shaft with a coordinated retrieval of casing

Steps to Build a Franki Pile

1. Positioning lattice crane with casing plugged at the bottom and a drop hammer inside
2. Displacement of soil and water as hammer drops onto the cap
3. Hammering out of plug after adding some concrete to the bottom of the arrested casing
4. Cycles of concrete pours and hammering of concrete
5. Lowering of rebar cage and tremie to fill concrete while pulling casing

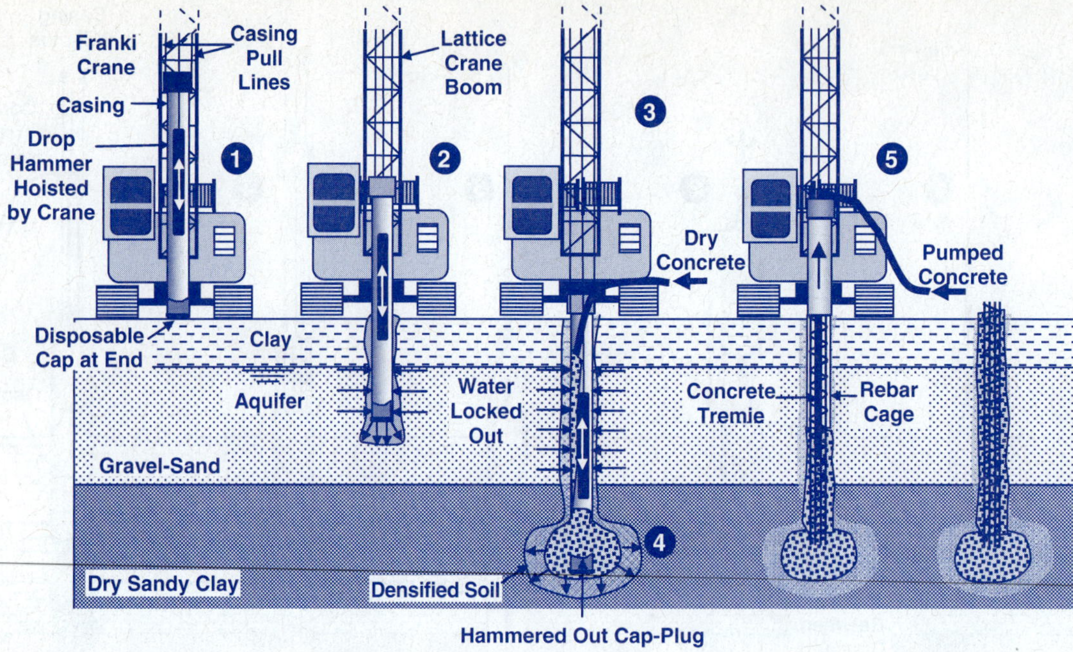

Figure 9.21 Dry Construction of a Franki Pile Through Groundwater Carrying Layers

Franki piles did indeed verify the drastic effect of such "bulbs." Finally, a rebar cage is lowered through the still dry casing followed by a tremie (if desired) and concrete. The pulling of the case, of course, needs to be coordinated with the height of the concrete column to ensure that the groundwater is kept out at all times. In other words, the bottom of the casing follows the top of the concrete at a distance that secures a sufficient sealing effect.

9.5.4 IN-SITU-SOIL OR DEEP MIXING METHOD (DMM)

"If you can't beat them, join them" is a well known saying that summarizes the basic idea behind in-situ soil mixing. Instead of spending so much effort dealing with difficult soil conditions and the groundwater, this method just uses the soil and inserts material that makes it stronger, lowers permeability, or reduces compressibility. Typical applications include soil or sludge stabilization, waste treatment, retaining walls, soil improvement, dewatering, foundation improvement, and so on. Column diameters typically range from 0.6 to 1.5 m (2–4.9 ft) and may extend to 40 m (131 ft) in depth.

Improving the in-situ material has a long tradition in agriculture and was perfected by Roman road engineers. The core idea is to add ingredients that synergistically react with the soil to achieve the desired outcome. In order to create a ground that is able to carry more weight without settling, contractors have found a way to force a tool into the subsurface able to inject various binders, and then blend and knead the resulting mix before compacting the treated in-situ material. Figure 9.22 illustrates how the different components work together flexibly so as to adjust to the many different soil conditions around the world.

S9.11 FHWA-Deep Soil Mixing

It is clear that the main carrier of the mixing tool is the crane or drill rig that is being used to drill shafts. Figure 9.22b shows how the entire system is integrated. The lead mast and the motor drives have been modified to handle more than one Kelly (up to eight have been used). Three different injection substances, grout, air, and water, mean that the contractor has to switch from one to the other or combine them. Figures 9.22a and c

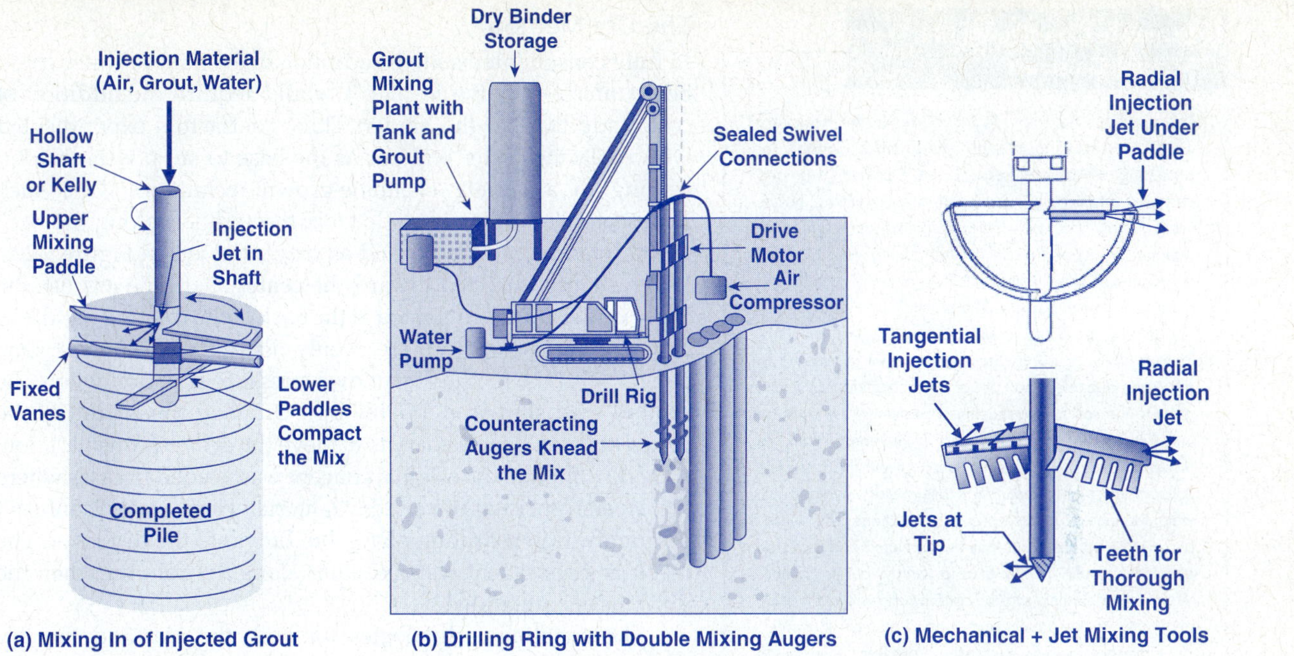

Figure 9.22 Overview of Main Soil Mixing Hardware

provide a better insight into the different shapes of mixing tools as well as the various strategies for injecting new material into the loosened soil. While one system may employ one or several straight paddles, including fixed vanes between them to improve the mixing action, other use half-rounds, or one or several short paddles with teeth. The injection jets are placed at various locations. The straight mixer has a hole in the shaft right above the upper paddles while the half-round includes a horizontal pipe nozzle underneath the upper-cross paddle and injects radially. The toothed tool has three different options, including one at the tip, a radial jet, and a series of tangential jets built into the paddle. We can identify at least two main classes of mixing operations, the dry and the wet method. Let's look at the different benefits and uses.

The Wet Method

Wet mixing is the more commonly used method for creating deep columns. It got its name from the fact that it uses water as a base liquid to inject different kinds of binders such as lime, cement, or fly-ash. While drilling downward to the bottom, the paddles break up the soil, loosening it into "lumps"; water and air are injected, with or without a binder component or chemical additive. On the way up, the main binder is injected while the paddles compact the mix. A main concern is the even and thorough distribution of the binder along the cross section, which depends on good disaggregation of the original soil. The danger is that large chunks may remain together and thus stay untreated as they are encased in the column. If left in significant amounts, they will cause a settlement or a leak. Of special concern is the bottom, as some mixing tools have their injection jets above the lowest paddle. One approach to improve disaggregation is the use of fixed vanes close to a rotating paddle, which prevents the soil material from rotating along with the mixing tool. A second method is the use of two or more close-mounted augers or offset paddles that rotate against each other, resulting in a tight kneading action as depicted in Figure 9.22b. Still another modification that helps break up large agglomerates is the addition of teeth at the edge of the paddles, as shown in Figure 9.22c.

Example Conditions in a Deep Mixing Contract

The grout mixing plant shall include all the necessary equipment including a high shear mixer capable of producing a *colloidal suspension* of cement/lime and chemical additives in water and pumps, valves, hoses, supply lines, and all other equipment as required to *adequately supply grout* to the mixing tool. *Positive displacement grout pumps* shall be used to transfer the grout to the mixing/auger. The plant shall be equipped to accept *dry or liquid additives* in measured amounts. Storage tanks or ponds shall be provided as needed to allow an adequate supply of *batches or continuously mixed grout* to the soil mixing machine. Grout shall be *agitated* until fully mixed and recirculated in the storage tanks to *maintain a homogeneous mix* and prevent flash set. The *proportioning equipment* may use meters, weights, or weight-volumes to ensure proper proportions. The density and viscosity of the grout shall be monitored and recorded as per the *quality control plan* to verify grout proportion. Weighing equipment shall be *calibrated* to within 2% at the beginning of the project and verified monthly thereafter. The material formed by mixing the grout with the in-situ soil shall have an unconfined *compressive strength* of 25 to 150 psi minimum at 28 days and a *permeability* of $1 \times 10-6$ cm/sec.

The Dry Method

In fully consolidated soil, the addition of even more water could be counterproductive in that it would require the addition of costly additives to the binder. Thus, contractors experimented with replacing water with air as the base to supply the binder, leading to a variety of home-grown techniques. One such approach injects the air-binder mix through jets along the back of the rotating paddle, labeled as tangential jets in Figure 9.22c. While slightly directed toward the center shaft, the jets use the slight under-pressure created at the back of the rotating paddle as a synergistic means to more evenly distribute the binder. One danger with the method is an over-pressure situation where the ground may shift (e.g., bubble up). Draining the air out of the mixed soil will be necessary to achieve the desired consolidation. A modification to the original concept was needed in cases where soil layers consisted of very dry material. Here, water is infused during the down-drilling phase, but only into the dry layer. The result is a consistent water content along the column when the dry binders are added.

The enabling technologies for both the wet and the dry methods are still in development as contractors are asked to build foundations in ever-more-difficult soil conditions.

Grout Mixing and Delivery

The inset provides an example of stipulations that can be found in a construction contract about how a contractor is to run the grout mixing plant. It not only lists the required equipment components of a plant but also their target functions, intended use, and even validated quality/calibration. This level of detail indicates the importance of the grout plant as it is key to producing, supplying, and injecting the exact amount of the binding ingredients according to a recipe that may change rapidly. In the same vein, it is important that we take a closer look into what it takes to fulfill all the conditions listed in the hypothetical mixing contract.

Figure 9.23 presents the main components of a common batch-mixing system. When visiting a deep mixing operation, one immediately recognizes the large storage hoppers for the dry binders and additives, which have to be readily available. The storage in elevated towers allows a simple gravity feed into the mixers. In automated systems, the quantity of dry powder is controlled by screw conveyors that can also serve as weigh stations. If properly calibrated, each turn of the screw provides a fixed amount of material. An alternative is to combine the screw with a sensor able to weigh in real time the total amount of material in the screw—data that can be combined with the speed of the calibrated screw to calculate not only the volume, but also the weight of the material fed to the mixing tank.

Colloidal Mixture

Colloids and suspensions are heterogeneous mixtures that differ in the size of the particles dispersed in the continuous medium. The bits of a colloidal substance are typically between 10 to 1000 angstroms, so small that they can't be filtered out, unlike suspended particles. Whipped cream is an example of a gas dispersed in a liquid, while paints or grouts consist of solids dispersed in a liquid.

As Figure 9.23 indicates, each of the stored binders, liquid additives, and water can be delivered individually according to a preset recipe. In the case of water, a pressure pump can be turned on if necessary. The function of the mixer, of course, is to blend all the ingredients prescribed for a certain batch into a colloidal suspension. As the box about colloids shows, the goal is to disperse and distribute the tiny binder particles uniformly within the water without leaving chunks, which could clog up jets or fail to bind with the material as intended. The mixing operation is all

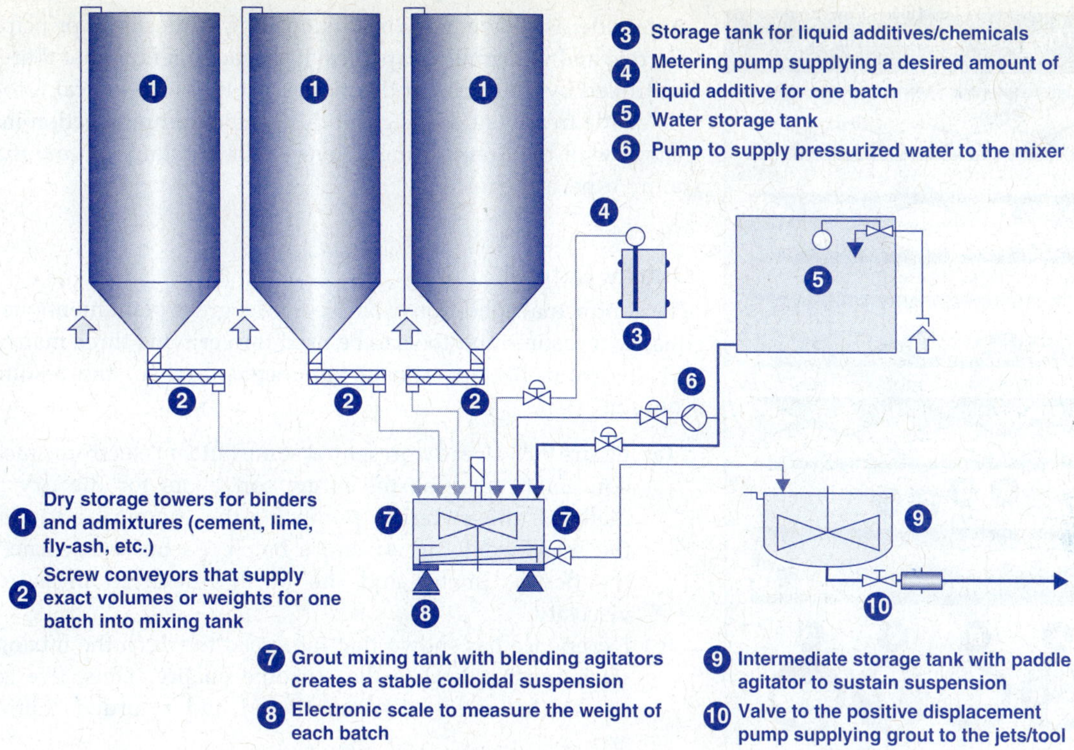

Figure 9.23 Schematic of a Grout Batching Plant

the more critical in that the dry state of such particles is not finely sized as desired, but instead comes in the form of clusters and lumps. Consequently, the task of colloidal mixing technology is to separate these particles from one another and uniformly distribute them in the dispersion medium during the mixing process. To avoid reactions between the tank and the mixture, various materials are used such as aluminum, stainless steel, plastic, or plastic-coated as appropriate.

One critical issue of batched mixes as compared to continuous mixers is timing, in particular, the time between the mixing and the gelling of the mixture. If a grout is gelling while being pumped for injection, it could easily lock up the pipe due to its changing viscosity. Proper timing and the avoidance of delays after the mixing has started are obviously key to success.

S9.11 USACE Chemical Grouting

One condition incorporated into the example requirement for the contractor is the use of a positive displacement grout pump. What is so special about such a pump? In fact, there is more than one reason. For one thing, displacement pumps treat the fluid much more gently than the bladed impellers of a rotodynamic centrifugal pump used in dewatering of construction pits. This feature is important when working with colloidal mixtures, which might become disaggregated by the blade actions. Displacement pumps trap a fixed amount of the fluid in a cavity and force the volume into an area with higher pressure. One example is the reciprocating piston pump that pushes concrete to high elevations. If the pulsating of a piston pump is undesirable, as in a grout injection system (since exact volumes need to be applied), one is able to choose between a rotary vane/lobe or the helical twisted rotor pump called the progressive cavity pump, as shown in Figure 9.24, or eccentric screw pump. It has only one basic moving part—a metal rotor gyrating within an elastomeric lined stator. As the rotor turns, small cavities between the rotor and stators open and progress from the low-pressure intake to the high-pressure

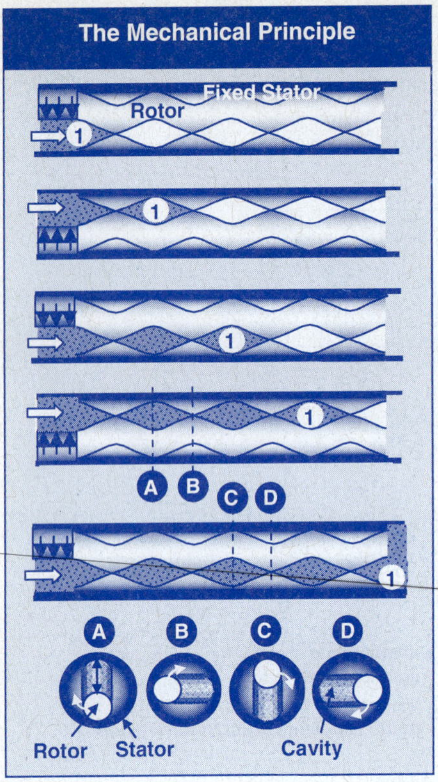

Figure 9.24 The Progressive Cavity Pump

outlet side, as shown in the insert depicting its mechanic principles. The discrete and uniformly shaped cavities lead to a flow rate that is easily controlled by the speed of the motor that turns the helical rotor. Thus, the gentle treatment of the grout during the pumping action in combination with the non-pulsing delivery of a constant volume make this pump a perfect choice.

Quality Control

The contractual condition example for the grout plant mentions several quality measures and tools to be used for verifying three main aspects: (a) the grout, (b) operational parameters, and (c) the resulting soil column.

(a) Figure 9.23 already presented some built-in electronic measuring equipment such as the screw conveyors for the dry powder material, the metering pump for the chemical additives, and the scale weighing an entire batch. Earlier we learned about the density meter and the Mash funnel to measure grout viscosity.

(b) Experience has shown that the speed to which the mixing tool is set is most critical to the outcome quality. Thus, five common values that need to be monitored and recorded continuously are

1. Penetration and retrieval speed of mixing tool
2. Rotation speed of the rotating unit(s) of mixing tool
3. Air pressure (in case of dry mixing)
4. Grout injection pressure
5. Feed rate of grout

At the same time, it is imperative that the contractor document the depth of the mixing cycle, start and end time of each cycle, and possibly the pore water pressure around the column being mixed.

(c) Not unexpectedly, compressive strength and permeability are two quality measures of the finished product, as they are closely related to the expected settlement under load and the amount of voids present. However, two other important parameters should be included in the quality control plan, namely shear strengths and density along the entire column, each depending on a different set of measuring tools. For shear strength, the Vane test has been found most useful. Here, two standardized blades are attached to a hub that is pulled through the completed column while the required force is recorded. For the second, a common tool is the standard penetration test (SPT). Here a thick steel tube is driven into the ground at the bottom of a borehole using a standard hammer weight and drop height. The number of blows it takes to advance a fixed distance, called the blow count, provides a measure that can be correlated to density. This SPT procedure can be repeated for different depths by extending the borehole. The test procedure is described in the British Standard BS 1377-9:1990 and ASTM D1586.

9.5.5 VIBRATORY STONE COLUMNS

Cohesionless sandy grounds allow the use of a more economical method that exploits the densification effect of applied forces. As we learned from the Roman road builders aggregate with the right mix can be densely compacted by applying force, both static and

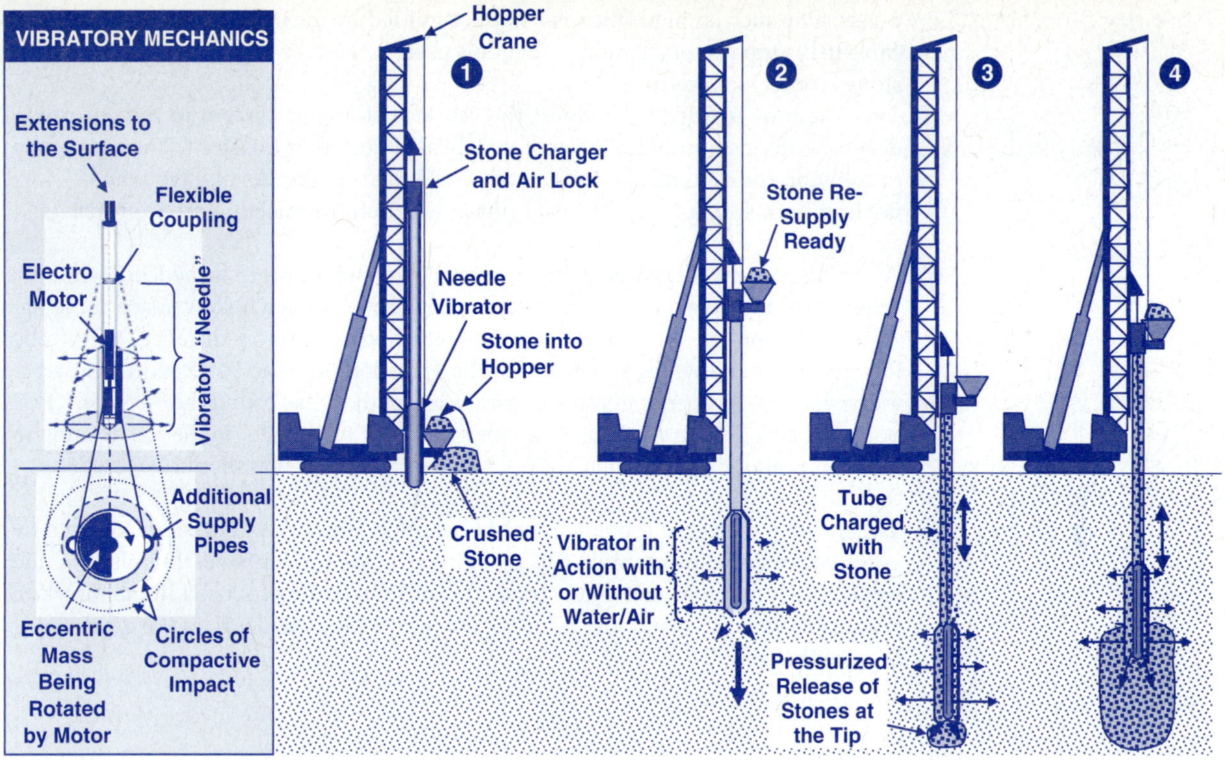

Figure 9.25 Deep Vibratory Soil Compaction and Insertion of Crushed Rock

dynamic. Natural in-situ grounds without clay are perfect candidates to apply vibro-compaction, where, similar to concrete, a needle vibrator is forced into the ground, thus densifying the area around the needle. The void that is created in the process, as the soil particles force out the air in between, is replaced with new material that is added from the top. Unfortunately, this simple method is only effective when the percent of fine material is below 10–15%. At higher percentages, the next-best solution is to replace a column of the silty-sandy material with stones using basically the same needle vibrator. Figure 9.25 presents the main steps of this cost-effective ground improvement technique.

Like the common vibrators used in heavy construction, the heart of the needle vibrator is a motor that rotates an eccentric mass attached to a central shaft. Changing the motor speed allows the modification of the vibrator's frequency, which is especially important when interacting with non-cohesive soil. A hollow tube connects to the needle vibrator via a flexible coupling through which service hoses and cables can be run. For example, the power cable for the motor, compressed air, and even water may be needed to enhance the process. The method presented includes one more capability. Different from the frequently used vibratory approach, where fill-material is fed to the top of the created cavity, stone (GP or GW) is fed to the bottom tip of the vibrator as soon as it reaches the desired depth. The following provides some specific comments to each of the four phases depicted in Figure 9.25.

1. To allow an effective process without having to add extensions to the "needle," tracked carriers are equipped with a high mast with the necessary crane cables and supply lines. A separate mechanism is necessary to supply stone to the top of the extension tube that can be closed on the top and pressurized during the retraction

phase. One such configuration is a hopper guided by the mast and lifted via a hoist cable to the top rim of the tube. A loader is used to refill the hopper with the crushed stone from a storage pile.

2. After the tube is filled with stone, the vibrator starts its descent to the required depth, with or without the help of water/air injected at the front. The frequency of the vibrator dictates the time needed to lower a tube. As always when dealing with soil, a "what-if" trial phase will help in selecting an appropriate setting.
3. When the desired depth is reached, water or air injection is turned off while the pressurized insertion of the gravel stone begins by creating a hollow space at the tip. After restarting the vibrator it is further retracted for a short distance, only to be lowered again. Now, the frequency of the vibration needs to be changed so that it is able to exert its energy into the densification of the area around the "needle." This point is reached if the vibration frequency finds its "resonance" in the soil, leading to its amplification. Again, trials are needed, but the frequency should be lower than 20 Hz.
4. As indicated, the cyclic retraction and re-insertion forces the stone sideways into the granular surrounding material. The resulting densified column will thus also include some of that material as it enters the air voids if no water was used in the insertion phase. On the other hand, the resulting column is much wider than the diameter of the needle itself.

Worked-Out Problem 9.2 / Gravel Stone

The company that you are working for has been awarded a contract to construct 800 stone columns with an average length of 10 m (33 ft) in an area it had worked before. The assigned construction manager, Paul Posco, expects that the average diameter of a column will need to be 1.1 m (3.6 ft). The site investigation showed the following data:

Table 9.3 Soil Layers

Layer	Depth (m, ft)		Soil Description
	(m)	(ft)	
1	0–1	0.0–3.3	Loose silty sand, poorly graded (SM)
2	1–6	3.3–19.8	Average dense, well graded clean sand (SW)
3	6–10	19.8–33.0	Dense, poorly graded clean sand (SP)

It is required to reach a dry density equivalent to 95% of Standard Proctor for the stone column.

The question is how much stone should be ordered all at once in order to get a good price. Paul wants you to do a quantity-take-off for the cheaper GP. He also asks to add 5% in order to provide contingencies for surprises. Is it necessary to add water to the inserted stone?

Base Data and Assumptions

The volume of one column = 9.5 m^3 (336 ft^3)

The density of the stone will decrease from the center to the outer edge. Density at the center will be slightly above the standard range.

The new stone will be delivered with loose dry density and with a water content of 4%.Water table is unknown and will be neglected. Loss of stone volume during handling = 2% and 5% should be added to cover contingencies.

It can be expected that the achieved density will be 95% Standard Proctor.

10% of the column volume will be filled with the in-situ sand (SM-SW-SP).

The answer should be given in tons.

The following table provides basic soil properties of the stone for filling and the underground in the construction area. We can assume that the vibrator will, on average, reach the energy required for Standard Proctor at the optimal water content applicable to the treated soil.

Table 9.4 Standard Soil Properties

Soil Type		Dry Density (Min–Max)		Optimal Moisture	Shrinkage	Swell	Permeability	
		lbs/yd^3	kN/m^3	%	%	%	ft/min	m/min
GW	Well graded clean gravel, gravel-sand mix	3,360–3,630	19.6–21.2	11–8	5–9	14–18	$5 * 10^{-2}$	$1.5 * 10^{-2}$
GP	Poorly graded clean gravel, gravel-sand mix	3,090–3,360	19.0–19.6	14–11	8–12	11–13	10^{-1}	$0.3 * 10^{-1}$
GM	Silty gravel, poorly graded gravel-sand-silt	3,225–3,630	19.8–21.2	12–8	8–10	12–15	$> 10^{-7}$	$>.3 * 10^{-7}$
SW	Well graded clean sand, gravelly sand	2,960-3,500	17.3-20.4	14–9	10–14	12	$> 10^{-3}$	$> 0.3 * 10^{-3}$
SP	Poorly graded clean sand, sand-gravel	2,690–3,630	15.7–21.2	21–12	12–16	10–12	$> 10^{-3}$	$> 0.3 * 10^{-3}$
SM	Silty sands, poorly graded sand-silt	2,960–3,360	17.3–19.6	16–11	10–14	12	$5 * 10^{-5}$	$1.5 * 10^{-5}$

Calculations

GP has a maximum dry density at 12% moisture content of 3,360 lbs/yd^3 (19.6 kN/m^3)

Water to be added to delivered GP (4%) = 12% − 4% = 8%

Weight of Water to be added = Weight of dry solids * Water content = 3,090 lbs/yd^3 * 0.08 = 247 lbs/yd^3 (19.0 kN/m^3 * 0.08 = 1.5 kN/m^3)

The GP delivered to the construction site is assumed to have a loose density of 2,990 lb/yd^3 (17.2 kN/m^3)

Swell% = 12 and a Shrinkage% = 8

Bank density of delivered GP = 2,990 lb/yd^3 * (1 + .12)

= 3,349 lb/yd^3 (17.2 kN/m^3 * 1.12 = 19.26 kN/m^3)

Compacted density of GP = 3,349 lb/yd^3 * (1 + 0.08) = 3,616 lb/yd^3 (20.8 kN/m^3)

Empty volume of one column = 9.5 m^3, 336 ft^3, 12.44 yd^3

Volume to be filled with GP = 90%, 8.55 m^3, 11.2 yd^3

CCY (Bank Cubic Yard) GP needed to fill one column = 11.2 yd^3, CCM = 8.55 m^3

LCY = CCY * 1.08 * 1.12 = 13.5 yd^3, LCY = CCM * 1.08 * 1.12 = 10.3 m^3

Weight of loose GP per column = 2,990 lb/yd^3 * 13.5 yd^3 = 40,365 lb = 20 tons

(= 17.2 kN/m^3 * 10.3 m^3 = 177.2 kN = 18 t)

Number of columns to be filled = 800

Loss due to handling = 2% : Contingency = 5%

GP to be ordered = 800 * 1.02 * 1.05 * 20 tons = 17,130 tons = 15,540 t

In order to reach the 95% Standard Proctor, we will have to add water to the stone in the amount of 247 lbs/yd^3 = 247 lbs/yd^3/8.33 lbs/gal = **29.6 gallons/yd^3** = (1.5 kN/m^3/ 0.00981 kN/l = **153 l/m^3**.

Discussion: The calculations are based on several assumptions, most importantly on the accuracy of the average diameter and depth of one column. Adding water could be best accomplished right before the stone is being loaded into the hopper.

9.5.6 JET GROUTED PILES

This is a perfect place to address the soil-mechanical behavior and conditions that are being manipulated with "forceful" injection of grouts.

Injecting Grouts into Soil and Rock

The deep mixing method injects grout into the soil that has been mechanically loosened and blended to improve its strength and reduce its permeability. Jet grouting, however, tries to accomplish similar results without the use of mechanical paddles and augers. Therefore, the main problem of this method is to inject grout into voids, fissures, and cavities in soil (or rock) formations in order to improve the treated soil as desired. The base material of ***cementitious grout*** is water, cement, and sand, but many admixtures are used, such as lime, fly-ash, or superplasticizer (SP), to achieve specifically desired modifications in its handling. During the hydration of cement, water-containing compounds are formed that create among other by-products, calcium silicate hydrates and crystals that grow in a longitudinal shape. Eventually, the grout begins to harden, and the crystals link the cement particles solidly, including the soil particles in between. Superplasticizers, accelerators, antifreeze, air-entraining agents, and many others are generally used to achieve specific performance characteristics (e.g., pumpability). Latex (acrylic resin) has been used for improving the properties of cement-bentonite grouts used in soil mixing. It is important to monitor their impact on the desired outcomes: strength, durability, impermeability, and resistance to chemical erosion of the grouted soil or rock mass.

> **L=Low, M=Moderate, H=High**
>
> **Rheology**
>
> **Rheology** *(from the Greek word* rheos, *meaning stream)* relates to the study of liquids and other matter that flow such as water, blood, solutions, grouts, and ketchup. One key measure is viscosity, which is a fluid's tendency to resist deformation. Higher viscosity means "thick" fluids, while low viscosity refers to "thin" fluids like water. Ketchup belongs to the group of fluids, referred to as non-Newtonian, that can change viscosity.

Although the average particle size of cement clinker is 10 – 20 μm (394 – 788 micro-inches), the particles do not pass through fissures of soil or rock that needed to be closed: for example, around and under water dams. This problem was solved with **chemical grouts** solutions with particle sizes below the minimum of colloids of 10 nm or 0.01 μm. Chemical grouting utilizes chemically reactive solution components, called activators, inhibitors, and catalysts, that after time form gels, semisolids, and solids. In order that the gelation and curing process not start before the grout is injected, the timing of the mixing is critical. What makes this even more complex is the fact that many different types of chemical grouts have different reactivities and thus need to be carefully proportioned and monitored. Furthermore, chemical grouts are sensitive to changes in temperature, dilution by groundwater, chemistry of groundwater, including pH, and contact with undissolved solids that may be in pumps or pipes. Table 9.5 shows how some of the main chemical grouts qualitatively compare with the cementitious grouts.

Generally, chemical grouts are more expensive but will develop a greater tensile strength, a better bond, and a higher compressive strength. Most importantly, the solution-based chemical grouts penetrate smaller openings than the colloidal cementitious grouts.

Table 9.5 Comparison of Cementitious and Chemical Grouts

	Cement	Silicates	Acrylates	Lignins	Urethanes	Resins
Penetration	L	H	H	H	M	L
Durability	H	M	M	M	H	H
Ease of Use	M	H	H	H	M	M
Relative Cost	L	L	H	H	H	H

L = Low, M = Moderate, H = High

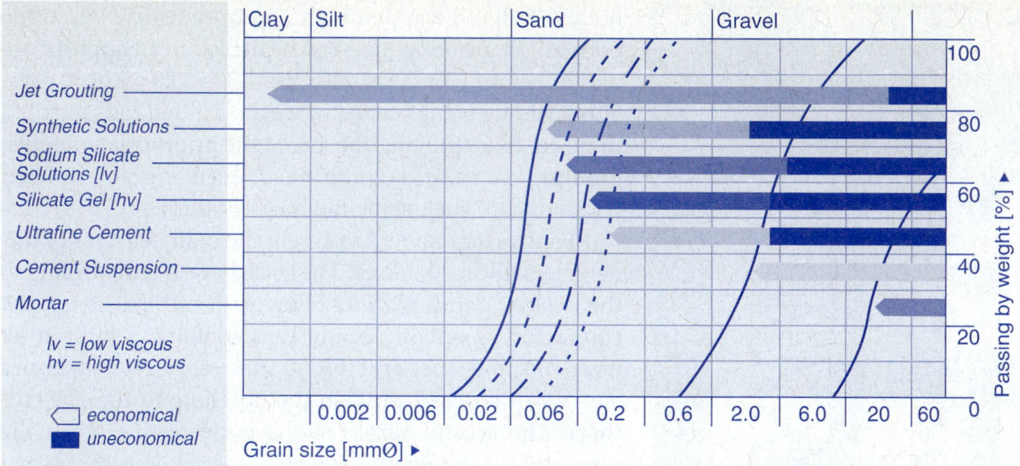

Figure 9.26 Application Limits of Grouting Techniques (based on Keller Grout Injection)

Superstars Between Chemical Grouts (see also Figure 9.26)

Sodium silicate grouts are the most popular grouts because of their safety and environmental compatibility. A silicate solution forms a colloid that polymerizes further to form a gel that binds soil or sediment particles together and fills voids. Sodium silicate solutions are alkaline, and when neutralized will aggregate to form a gel, if 1 or 2% of sodium silicate is available.

Acrylate grout gels through the polymerization of acrylates (usually magnesium acrylate). Methylene-bis-acrylamide is used as a crosslinking agent. Potassium ferricyanide is used as an inhibitor if long setting times are required.

Urethane grouts form a rubbery polymer. The one-component grout will gel or foam depending on the amount of water available. The two-component grouts employ a direct reaction between an isocyanate liquid and a polyol and produce either a hard or flexible foam depending on the formulation.

Lignin grouts forms an insoluble gel a short time after being combined with an oxidizer. Lignin is a by-product of the sulfite process of making paper and can be mixed into solutions with various viscosities.

Resin grouts are solutions of resin-forming chemicals that combine to form a hard resin upon adding a catalyst or hardener. Injection is by the one-solution process. The principal resins used as grouts are epoxy and polyester resins. The terms *epoxy* and *polyester resins* apply to numerous resin compounds having some similarity but different properties.

After we achieve a general understanding of the important characteristics of grouts and how they behave after they have been activated, we are prepared to gain a good understanding about the technologies that inject the different grouts to improve the soils in the subsurface.

Technologies for Jet Mixing or Jet Grouting Piles

The reader must wonder by now how jet grouting is able to treat soils without the use of mechanical mixers when even low-viscosity chemical grouts fail. The simple secret is high-pressure gases and fluids that are "blasted" at the soils, like gold miners using pressurized water. In fact, the technique replaces the paddles of a mixer with a pressurized agent able to unlock the soil particles for a period long enough to inject and mix in the grouts that will bind them closer together afterward. By using jet nozzle openings in a

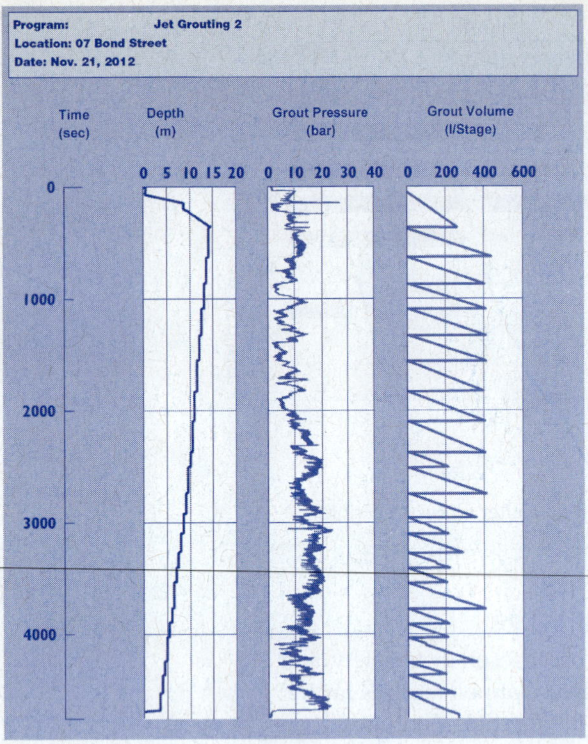

Figure 9.27 Output of Sensory System on Jet Grouting Drill (based on Keller Grout Injection)

hollow drill rod (capped with appropriate drill bits at the bottom), jet grouting tools can be drilled to an appropriate depth before, for example, the 430 – 720 psi or 30 bar – 50 bar jets are being started while the drilling rod is rotated in place. By deploying the jets at the appropriate depths, grouting is easily accomplished even around isolated zones of soil, such as below buried utilities.

Figures 9.27 and 9.28 present the main features of this very versatile technology. The cycle begins with drilling to the desired depth with a hollow drill rod. At its tip, the contractor mounts a section that contains a built-in jet nozzle(s) and the drill bit. If necessary, low-pressure slurry can be used to flush the borehole to remove cuttings. The second phase of the cycle begins when the powerful 600 hp engine is activated to begin the jetting with high-velocity slurry injection.

The goal of this phase is to loosen the soil at the bottom of the column sufficiently by moving the rod up and down several times before starting the grouting phase. As shown in the figure, the excess slurry soil mix escapes through the space between the drill rod and the borehole wall, and is collected on the top to be recycled.

Maximum penetration and mixing are sought by choosing the appropriate mixture of cementitious or chemical grouts (which are of course more expensive). As discussed above, the viscosity of the grout, measured in centipoise (cP), determines the ability of a grout to flow into and through the pore spaces in a soil. The distance (= radius of jetted column) and rate of penetration will depend on three factors: (1) permeability (= hydraulic conductivity) of the soil, (2) viscosity of grout, and (3) pressure of injection pump. There exists a direct relationship between viscosity and pressure, in that higher viscosity grouts require higher pressure to obtain an equal penetration.

By plotting in real time the volume of grout injected, the pressure, and the retrieval speed, as shown in Figure 9.27, the operator is able to monitor the performance of the operation and to readily detect anomalies.

If specific strata need special treatment, the operator may be able to use packer or "tubes à manchettes" (French for collar tubes). Only if the soil above and below is strong enough can rubber sleeves or manchettes be installed and pressurized. Control valves can be opened and closed to direct grout through individual jets that are positioned at the end or between two inflated packers. In this manner, only the targeted layer will be grouted, possibly with a different grout or a different pressure or both.

In summary, a contractor setting up its equipment for jet grouting will have to deploy several pieces of equipment and organize the supply and storage of various bulk materials and chemicals. Some of the main pieces include

- ✓ A drill rig specially equipped for jet grouting with a mast high enough not to require adding rod section during drilling.
- ✓ A grout plant able to continuously produce grout mix for the pump.
- ✓ A high-pressure grout pump, 350–600 HP.
- ✓ Air compressor and water pump in the case of the use of double or triple jet-grouting.
- ✓ Accessories such as a high-pressure grout hose, special swivel, special single or double or triple rods, monitor, nozzles, etc.

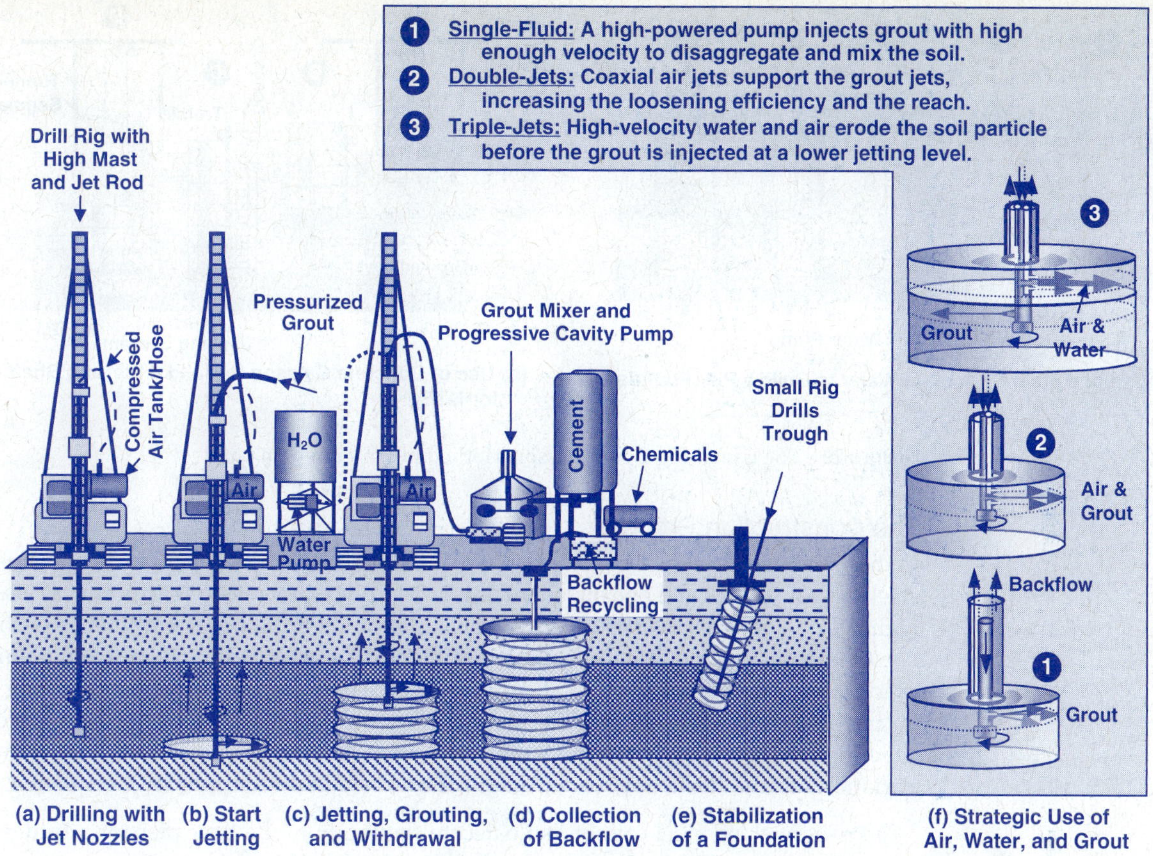

Figure 9.28 Equipment and methods for soil improvements using jet grouting technology

Most, importantly the contractor needs to carefully plan his slurries, grouts, pumps, and mixers, as well as the timing of each step. Here is a quick summary of the parameters to be considered:

- ✓ Soil particle distribution, permeability of soil, void ratio
- ✓ Grout mix composition, viscosity
- ✓ Grout pump pressure
- ✓ Air and water pressure, in case of double and/or triple jetting
- ✓ Diameter and number of the nozzle(s)
- ✓ Penetration depth and speed
- ✓ Withdrawal speed of the jet rod
- ✓ Volume of grout mix per m (ft) of withdrawal
- ✓ Rotation speed of the rods/nozzle(s)

9.5.7 OPEN CAISSONS

Caissons (from French meaning big boxes or chests) are used when large concentrated "footprints" have to be created at significant depths. These footprints may be needed to establish a sufficient foundation for a large bridge or heavy structure or an opening to install a horizontal tunneling system. The unique characteristic of a caisson is that the "box" normally sinks by its own weight into the ground while the soil inside the box is safely removed. As the name of this technology indicates, the caisson is open at both top and bottom.

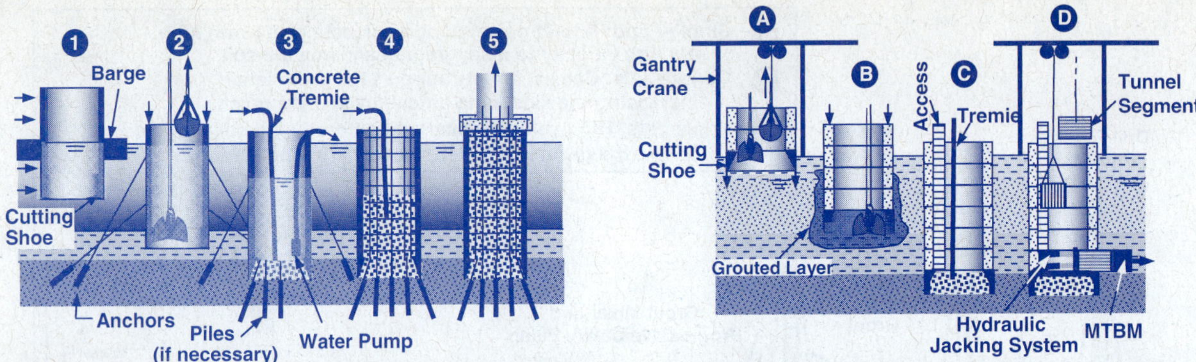

(a) Use of a Steel Caisson in Water to Build a Pier Foundation

(b) Use of Concrete Caisson to Build a Starting Shaft for MTBM

Figure 9.29 The Use of the Open Caisson Method in Water and on Land

The Construction Process

In the following we take a closer look at two different uses of the caisson, one in water using steel to create a shell and one on land using cast-in-place concrete to "slip-form" a caisson. Figure 9.29 highlights the basic steps of each application fulfilling two different purposes.

One option is to fabricate the lower section of a steel caisson in a more convenient place someplace on-shore and to float it into place using large barges. The floating can be facilitated with the temporary installation of a horizontal deck. Also, instead of a circular caisson, designs with rectangular cross-sections are used. Despite the many variations to the basic approach, what follows is a common five-step sequence:

1. Pre-assembled steel caisson is floated to the location of a new pier for a bridge or other project. The leading edge or "cutting shoe" of the caisson is sloped outwards and slightly larger than the steel wall to reduce its friction during the sinking. The floating can be facilitated by the use of false bottoms or temporary domes that provide substantial uplift without creating an unstable situation that might cause the caisson to overturn.
2. After that, the caisson is lowered to the exact place where it will be anchored, and is properly aligned. Anchors have to sustain maximum flow forces (if built in river). Begin excavating with clamshell bucket, adding further caisson rings if necessary. Caisson should sink due to its own weight. If necessary, injection of slurry at the cutting edge could be used to reduce the build-up of excessive friction. A large boulder might require that a diver remove the obstacle.
3. The risk of high up-thrust forces due to water pressure, an insufficient bearing capacity of the finished caisson, or the risk of high lateral forces may require the installation of piles below the bottom of the caisson. After the pouring of tremie concrete for the plug, the inside area can be dewatered. At this point, the piles may serve as anchors to hold down the concrete plug, avoiding boiling of the bottom.
4. Installation of rebar cage, if necessary, and filling of caisson with concrete.
5. Capping of caisson and casting of pier to the proper height.

Microtunneling boring machines (MTBM) find more and more use in installing underground water and sewer lines without the need for open trenches. For deep installations in groundwater, sheet-piles and bracings are being replaced with concrete caissons. As Figure 9.29b demonstrates, the basic principles are the same, but the opening is used to install the horizontal boring machine and the hydraulic cylinders for the jacking system that provides the necessary thrust for the drill to engage the soil. The starting opening in the caisson can be opened and secured ahead of time; otherwise the machine will drill through the concrete.

A. For vertical and horizontal transportation an adequate gantry crane is erected first. The steel cutting shoe can now be installed into a prepared trench before a first concrete section of the caisson is formed and cast in place, providing the weight to sink under its own weight after the clamshell excavation begins.

B. In case of groundwater, three options are available. The first is to lower the water table using a well and pump. If this option creates too much risk for the surrounding buildings, the water-carrying layer may be selectively grouted to lower the permeability around the caisson as it works itself through the layer. The third is to excavate underwater until a denser layer is reached, at which time the water can be safely pumped out. If the sinking of the caisson is impeded, several countermeasures are possible. Friction can be reduced via slurry injections. Excessive resistance to cutting can be remedied by digging or drilling directly underneath the cutting edge. On the surface, concrete sections are continuously formed and cast, matching the progress of excavation.

C. At the appropriate depth, the tremie concrete can be poured to dry out the caisson if necessary. A safe access for personnel has to be installed next. It might consist of an elevator or stairs that are properly secured. For safety reasons, at least the installation of stairs is necessary. Next, the foundations for the jacking system and the launching pad for the tunnel boring can be poured with concrete. Power lines can be installed as well as a transfer mechanism to remove the muck from the boring machine through the caisson. Several different muck-removal methods can be used. Finally, the gantry crane is now able to lower the mechanical components for the jacks, and the boring machine is ready to be installed.

D. The jacking system includes hydraulic rams that extend the length of a concrete segment. After the boring machine has advanced this length, the gantry crane lowers a new segment to be installed after the rams are retracted. During the next advance, the muck is removed with a mucking container or other method.

9.5.8 PRESSURIZED CAISSONS

Although the concept of using an overturned "cup" to dive into the ocean or a river has long been practiced, in particular by Alexander the Great, it was only in the 1850s in England that this concept was used to repair a foundation underwater. The French called it the *caisson perdu* (lost box). Throughout Europe, engineers started to deploy and further develop the method. Its first uses in the United States were on the Eads Bridge in St. Louis (1867–1874), at 6,442 feet (1,964 m), then the longest arch bridge of the world, as well as with the Brooklyn Bridge (1870–1883), at 5,989 feet (1,825 m) the longest suspension bridge at the time. Even the foundation of the two legs of the Eiffel Tower (1887–1889) closest to the River Seine used the pressurized caisson method. It was crucial to building the bridges in that it allowed their pier foundations to be built inside the river, and it provided dry access for digging down to the rock through the river bottom. A simple experiment with a see-through glass

Short Review of Key Pressure Laws

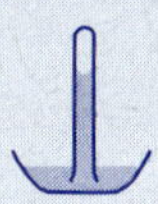

The Italian physicist Evangelista Torricelli (1608–1647) was the first scientist who observed that air had weight. His famous experiment was to fill a glass tube with mercury and hold it upside down in a bowl that was also filled with mercury. What kept the mercury up above the level in the bowl? As we now know, the secret is the weight of the air pressing on the liquid in the bowl combined with the vacuum at the top of the glass tube. The barometer was born when he observed that the mercury level moved up and down according to the air pressure. **760 mm** of mercury (abbreviated mmHg) came to be the standard atmospheric pressure at sea level (= **14.7 psi** = **1 atm** = **760 torr**).

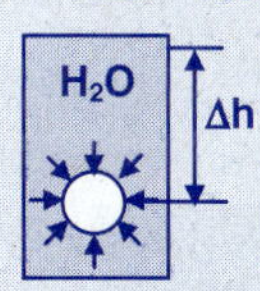

Blaise Pascal (1623–1662) formulated the hydrostatic pressure for fluids as: $\Delta P = \rho * g\,(\Delta h)$. He showed that the pressure of a fluid at rest is isotropic. As a consequence of this phenomenon, any change in pressure applied at any point of the fluid is transmitted undiminished throughout a continuous fluid.

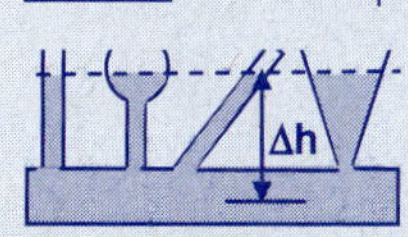

Much earlier the ubiquitous inventor Archimedes had already defined a principle that says that the buoyancy of an object immersed in a fluid will have an upward force acting on it that is equal to the weight of displaced fluid.

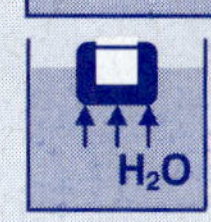

Figure 9.30 Air Trapping

pushed upside-down into water will allow you to visualize both the advantage and the problems that this method creates. As Figure 9.30 demonstrates, the oxygen (20% of the natural air and most important to survive) is kept inside the cup. On the other hand, the volume of air becomes smaller the deeper the cup is pressed. From thermodynamics we know that there exist relationships among pressure, volume, and temperature of a gas defined as $p * V = R * T$ where $p =$ pressure of the gas, $V =$ volume, $R =$ gas constant, and $T =$ temperature of the gas. If we assume that the temperature for the cup in the experiment is constant like R, it becomes evident that p and V are inversely related: when the pressure increases, the volume has to decrease and vice versa. We then conclude that the pressure of the air inside the cup increases with its depth. This phenomenon occurred to the child prodigy Blaise Pascal, who devised the famous function $\Delta P = \rho * g(\Delta h)$, known as the hydrostatic pressure function (see box on previous page). Pascal's function says that the increase in pressure inside a fluid is related to the density of the fluid, to gravity, and to depth.

He added another important fact that we need to consider, namely that pressure inside a fluid is isotropic, acting in all directions. Now we are able to connect the dots. The all-directional pressure for water increases with the depth of the cup, squeezing the air inside the cup more and more. Higher air pressure causes its volume to shrink, and thus the water will enter the cup from the open bottom until the hydrostatic water pressure is counter-balanced by the air pressure (ΔP of air inside cup $= \rho * g *$ Depth of water level inside cup). Now we are ready to reveal the secret about how this technology works.

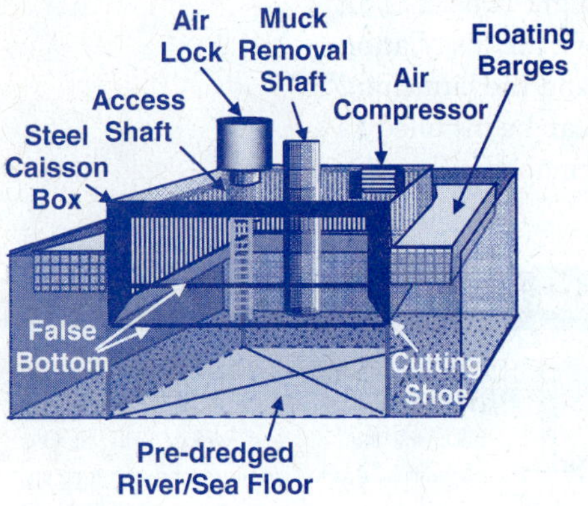

Figure 9.31 Floating and Positioning of Caisson

The sequence commonly starts some distance away from the location of the construction site, with the fabrication of the open-bottom caisson box. While the cutting shoe is steel, the box itself can be made of concrete or even timber, as in the case of the Brooklyn Bridge. To improve its buoyancy during the "floating" operation, the caisson sometimes has false bottoms, which are removed after the submerging of the caisson. Figure 9.31 shows a cross-cut of the steel caisson as it is when barges position it right above the prepared location of the future pier. The prefabrication of the box includes four important elements: (1) air lock, (2) personnel access shaft, (3) muck removal shaft, and (4) compressed air pump. You must be asking what all that extra stuff is for when the box alone looks sufficient. Their need becomes clear as we continue along the installation process by submerging the box so it rests on the bottom of the river/sea by overcoming the buoyancy force, depicted in Figure 9.32. The box has been weighted down with four lifts of cast-in-place concrete. Each lift is cast above the water surface but causes the box and with it the previous lift to sink. Held in place by strong moorings, the cutting edges of the caisson eventually come to rest where the reaction forces from the soil and the buoyancy are larger than the downward forces that can be created with concrete. Concrete is supplied from the river's edge or via barges. Consequently, the shafts are still tightly shut, and the air pressure pipes have to be extended so as to keep the tops above the high-water level. The attentive observer will surely notice that the space inside the box, with removed false bottoms, is void of water, a fact that contradicts our air-trapping experiment. There should be water at least at the bottom of the box. However,

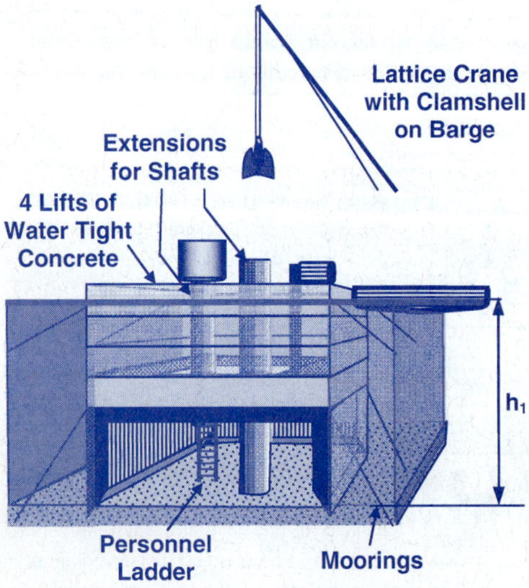

Figure 9.32 Submerging and Preparation to Excavate

Worked-Out Problem 9.3 / Pressure to Ensure Dry Excavation

What pressure is needed to keep an excavation 33 ft (10 m) under water dry? The problem at hand is to create enough pressure to counterbalance the pressure with which the water forces itself in through the open bottom. The question can be rephrased as follows: What is the absolute pressure inside the water column at the depth we desire to keep it?

Qualitative Answer

The absolute pressure at the depth h_1 is comprised of the atmospheric pressure on the surface of the water plus the hydrostatic pressure created by the water.

Calculation

p_1 = Air pressure at the surface $+ h_1 * g * \rho$

Approx. air pressure at ocean level = 14.5 psi = 1 atm = 1 bar = 750 torr = $100\,kN/m^2$

$p_1 = 100\,kN/m^2 + (h_1 * 9.81\,m/s^2 * 1{,}000\,kg/m^3)$, 1 Newton = $1\,kg * m/s^2$ At $h_1 = 10\,m$ (33 ft) we need following counter-pressure inside the caisson:

$p_1 = 100\,kN/m^2 + h_1(m) * (9{,}810\,(kg * m/s^2)/m^3 = 100\,kN/m2 + 10\,m * (9.81\,kN/m^3) = 198.1\,kN/m^2$

$\mathbf{p_1 = 198\,kN/m^2 = 198\,kPa = 28.7\,psi}$ **or approximately 2 atm**

Discussion of Results

The 10 m of water doubles the air pressure on its surface created by the atmosphere. Although water densities change slightly with temperature and salinity, it is a good rule of thumb that the water pressure increases by 1 bar or 1 atm every 10 meters, a fact that will become important when we start digging.

the contractor remembered the formula that links gas pressure and air volume for a constant temperature: $p * V$ = constant. While the hydrostatic water pressure causes the water inside the box to rise, compressing the air volume, the contractor figured out that he only had to turn on the air compressor supplying air through a pipe. The only question was how much pressure he had to create to dry out the bottom.

In order to make sure that the water will stay below the bottom, the contractor might slightly increase the pressure above 2 atm. The next issue is the removal of the muck from the pressurized caisson box. While it is possible to use an air lock approach, similar to the one through which the laborers enter the caisson, the procedure of opening and closing the hatches of the lock as well as the loss of air make this a cumbersome method. Figure 9.33 presents a clever solution. It is clear that opening the muck removal shaft would simply allow the air inside the box to escape to the outside. However, by combining Torricelli's experiment with the heavy mercury with Pascal's observation about the uninterrupted pressure distribution within a continuous fluid, we are able to create a simple solution. All we have to do is to dig a pit into the soil that reaches below h_1 or h_2, the top of the water plane underneath the caisson, to create a pond of standing water. Inserting the bottom of the muck shaft into the created basin and opening the lock on the top, we observe that the water inside the shaft will rise exactly to the level of the river or sea. Pascal's law does work and

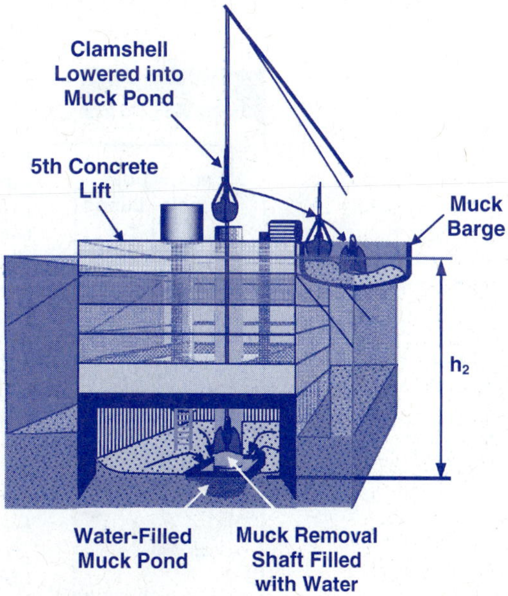

Figure 9.33 Muck Removal Without Air Locks

creates a superb opportunity to improve efficiency! While the dry soil inside the caisson is being filled into the muck pond, a crane with a clamshell bucket operates in parallel to remove the muck through the vertical shaft. Figure 9.33 shows a barge being loaded with the muck from the bottom, but there are also many other options for depositing the excavated material. For instance, it could be dropped into the water outside the caisson or deposited on a conveyor belt reaching the embankment or the shore. One will also notice that a fifth concrete lift was added, accompanied by more shaft elements as the caisson slowly sinks deeper into the ground. The plumbness of the entire "pack" can be secured by shifting the excavation around the cutting shoes to a location that resists an even settlement. Most importantly, boulders or wooden logs that might be encountered by the edge of the caisson can be broken up using appropriate tools. As Washington Roebling so impressively demonstrated at the Brooklyn Bridge, even the blasting of boulders is possible.

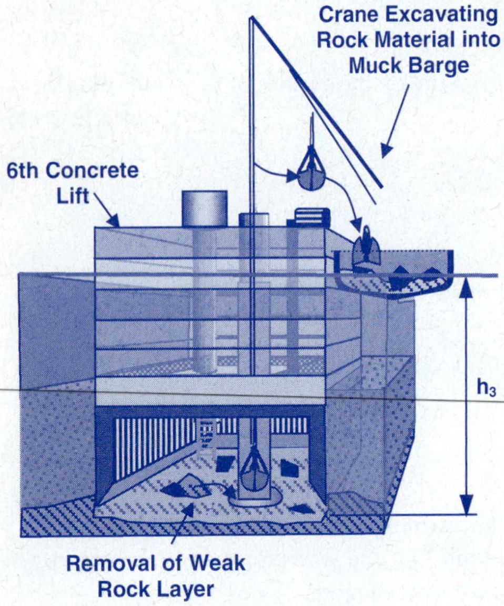

Figure 9.34 Removing Weak Rock

As the caisson reaches the rock below the soft soil deposited by the river or the ocean current, the excavation tools have to be changed. What are needed now are rock-breaking mechanisms such as percussion hammers or rippers able to break up the soft rock layers on top. In addition, a pond is needed for the removal of the broken-up rock by the clamshell. However, as indicated in Figure 9.34, the depth of the caisson has now increased to h_3. While this might not be much of an issue if remotely controlled equipment were used, humans working under higher pressure may face physiological effects. As a matter of fact, the physicians at the Eads and the Brooklyn Bridge reported that laborers emerging from a shift inside the caisson sometimes had to be hospitalized; several even died. It was observed that this occurred after a depth of 50 ft (15 m) had been reached. An effective treatment eluded the physicians for some time. Later, the malady facing the workers was termed "caisson disease" or "the bends." The latter name emerged

The Bends

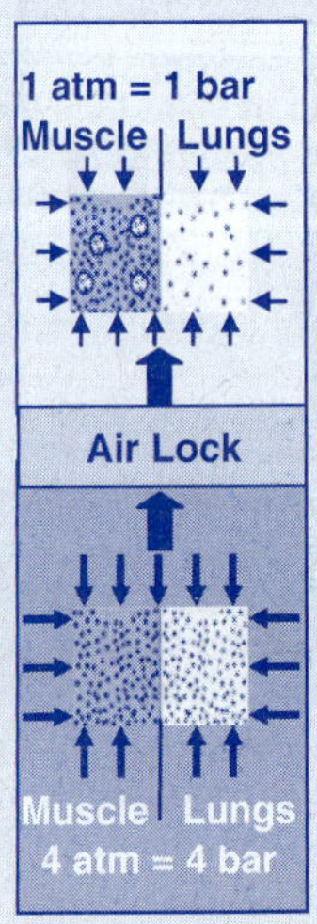

As the deeper caisson at the Brooklyn Bridge reached a depth of 50 ft (with a projected depth of approx. 90 ft), workers started to complain about severe pains in muscles and joints. As the depth increased, so too did the severity of the effects on the laborers, ranging from deafness, vomiting, partial locking up of muscles, to fainting. Dr. Leonard Hill, employed by the Brooklyn Bridge project, suggested slow decompression as a remedy but was unsuccessful in convincing the management. At a depth of about 76 ft, two workers died, and at 80 ft a third, causing Washington Roebling to stop excavation. As we now know, the secret can be traced to a law defined by the English chemist William Henry (1775–1836), which says: "*. . . the amount of a given gas dissolved in a given type and volume of liquid is directly proportional to the partial pressure of that gas in equilibrium with that liquid.*" For example, the breathing human body absorbs 1125 ml of N_2 at 1 atm. At 4 atm, the imbalance between the pressure in the lung and the body will result in a slow increase of N_2 to 5000 ml. This does not create major problems for working humans until they return back to the surface through the air lock. While the lungs are able to instantly adjust to the decrease in atmospheric pressure, the body tissues take hours to release the extra gas safely through the lungs. If sufficient time is not allowed to slowly bring the body from 4 to 1 atm, the 3875 ml nitrogen will form bubbles inside body tissues, causing the excruciating painful caisson disease, or the bends.

because the afflicted laborers bent over with pain, like ladies performing the fashionable dance the Grecian Bend (1880).

Employing all the decompression technologies that have been developed since the Brooklyn Bridge allows humans to reach 120 ft (36 m) requiring 50 psi (345 kPa) of air pressure to keep the water out. Hopefully before this limiting depth is reached, solid rock is reached, terminating the caisson excavation process. Reversing the removal operation, concrete is used to plug the bottom of the caisson box so that it can be kept dry without the use of pressure. Figure 9.35 presents one possible configuration for filling the space with concrete. A barge with a concrete pump is moored at the pier together with a batch plant that is fed with aggregates and cement from a second barge. Again, many different setups can be designed that would fit local conditions and availability of equipment.

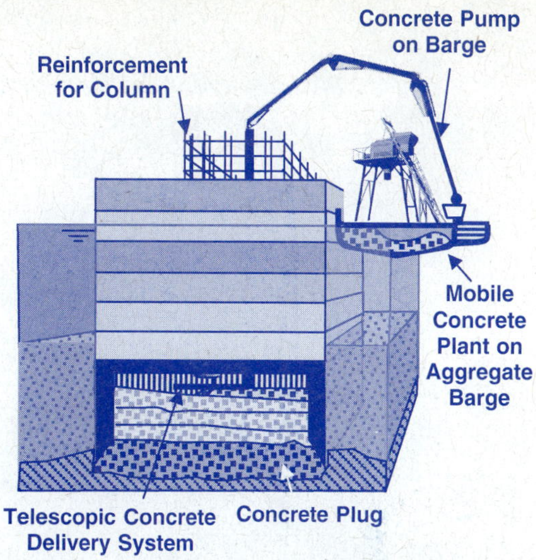

Figure 9.35 Completion of Pier Foundation

Solving Header Problem 9.1 / Planning a Deep Foundation

When solving the Header Problem in Chapter 9 it was decided to use a concrete diaphragm wall to retain the water and earth on one side but to bench the rest of the excavation. However, considering the weight of the building in combination with the poor soil underneath, the foundation engineer decided to design load-bearing diaphragm walls all around. While the walls probably will cost more to construct, the benefits in reduced dewatering costs and the "free" load bearing structure make it an overall economical solution. Still, as Figure 9.5 shows, three rows of piles are needed.

Strategies to Build the Deep Foundations

While we understand how to build slurry walls (see Chapter 8.5.5), the installation of piles below the bottom of the excavation looks significantly more difficult. There are three basic options:

1. Drill vertical shafts from the top to the bottom, install rebar, and fill the shaft with concrete up to the bottom of the pit. Excavate.
2. Excavate to bottom, lower driving or drilling rig, and install the piles.
3. Jet-grout the piles from the top before excavation starts.

Option 1 will require tight control of depth and concrete level in order to avoid a lot of rework to shape the piles when the flat foundation plate is constructed. Also, drilling through the upper level to reach the bottom will be costly.

Option 2 needs a heavy crane to lower and lift the piling equipment. In addition, work will be needed to provide extra mats for the rig to travel across the soft bottom of the pit. The pile crew will stall any other work until they are done.

Option 3 does not need drilling shafts and can be done from the top, thus avoiding the negative aspects of the other two options.

Jet-Grouting Deep Piles: This phase should be completed as quickly as possible as long as the potential of boiling from water through the thin layer of clay hangs over the operation.

We learned that jet-grouting offers an alternative if it provides sufficient support. Figure 9.36 illustrates how this could work.

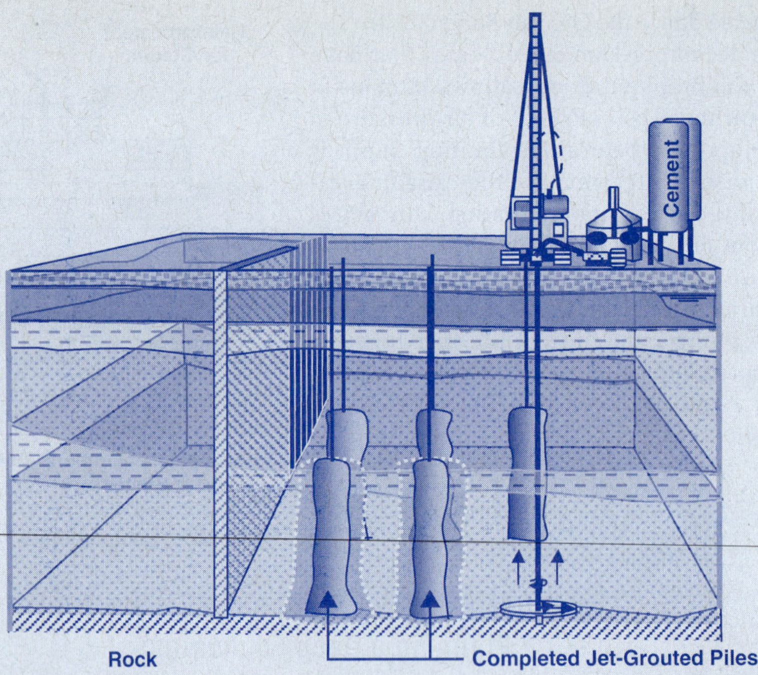

Figure 9.36 Jet-Grouted Piles as Alternative Recommendation for Deep Foundations

The jet pile alternative offers several advantages that could result in significant time and cost savings while reducing risks. First, the work would not require any pile-driving operations that cause noise and ground vibration. Most importantly, the piles can be built before the start of the excavation from the surface without any hindrance and even before mobilization of the excavation contractor. Another problem that the jet piles would help alleviate is the risk of a boiling floor. Because of the friction forces between the layer of clay and the grouted area, additional forces will be created to counter the water pressure from the third aquifer. On the other hand, puncturing the clay layer might create weak spots that could actually allow water to seep through much more easily. Nevertheless, the many benefits that jet-grouted piles offer should encourage the contractor to submit an alternative to the engineers' design.

9.6 Chapter Nine Review

THE EQUIPUZZLE

The circled letters make a word of an important tool for placing concrete in a deep shaft.

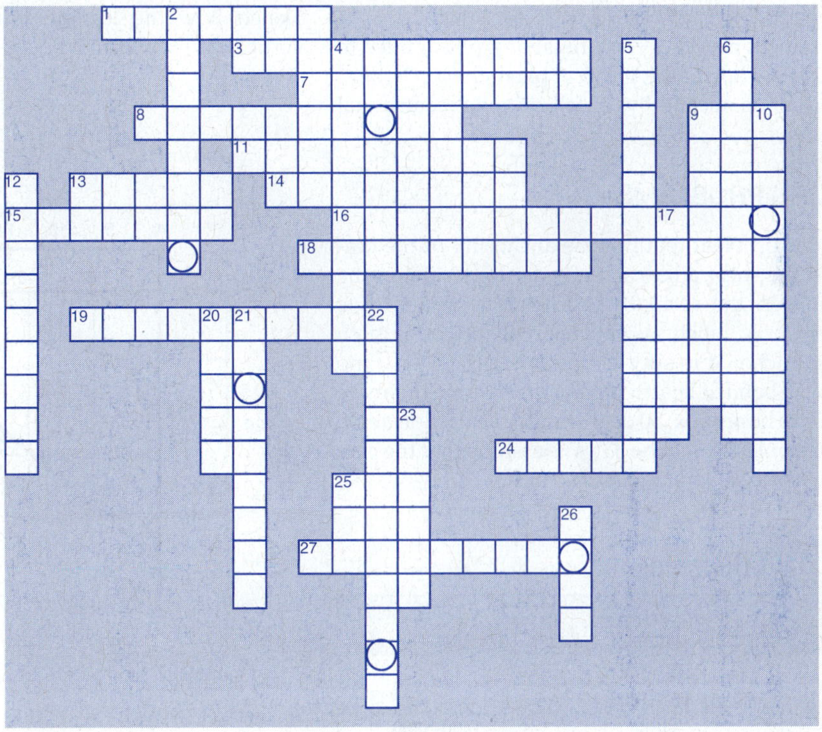

Across

1. A "p . . . box" first used in 1851 to build the foundations for the bridge over the Medway at Rochester, Kent, England

3. The difference in elevation between the liquid level of the fluid source and the centerline of the pump (two words)

7. Looks like a solution but consists of two separate phases

8. Lowering of the groundwater table to obtain a "dry" space for excavation

11. An inverted air hammer with yoke to pull piles out of the ground

13. A hollow bar attached to the top of the drill column

14. Name of the inventor of the the steam pile driver (first and family name)

15. Family name of a German inventor who rammed deformed steel profiles into the ground supporting each other

16. A cylindrical tool used for removing "cuttings" from the shaft bottom (two words)

18. A natural substance that swells in water to several amounts of its dry mass

19. 19th-century pioneer of the centrifugal pump (first and family name together)

24. A disease that was first observed in the 1870s during the construction of the Eads Bridge in St. Louis, where 15 workers died of it

27. Additive to a fluid to achieve a desirable characteristic

Down

2. The rotating element of a centrifugal pump that consists of a disk with curved vanes

4. Last name of a scientist who first thought about employing mercury to create vacuum inside an inverted tube

5. The name of the first Roman bridge built on timber piles (two words)

6. A reinforced or unreinforced concrete foundation element formed by drilling a hole in the earth and filling it with concrete (two words)

9. An auger-type drilling tool, equipped with hard metal teeth to enable it to drill in soft or weathered rock (two words)

10. A force associated with a rotating body

12. An undesirable condition where soil falls into an uncased shaft during or after the drilling

17. A liquid composed of finely divided particles that do not settle out of suspension

20. A helical rotary tool for drilling a cylindrical hole in soil

21. A curve plotting the total head vs. the flow for a specific pump mode (two words)

22. Electronic device to measure in real time the density of a fluid passing through (two words)

23. A hopper and drop pipe used to place concrete underwater

25. Abbreviation for a shaft inspection device that allows the inspector to see the bottom of a drilled shaft

26. An "under-ream" to increase the bearing area of a drilled shaft

JOURNALING QUESTIONS

1. The tower of Pisa became a global tourist attraction although its engineering was deeply flawed. The key problem was that the responsible foundation engineer did not know about Roman engineering principles. Explain what the foundation engineer should have known and what soil characteristics, combined with events of the time, saved the tower. What other tall buildings in the world are leaning, and why are they still standing?

2. The static suction head of centrifugal pumps is limited to about 6 m (20 ft). Explain the physical phenomenon behind this fact. What is meant by a self-priming centrifugal pump?

3. Artesian aquifers create special problems when dewatering the ground to create a dry excavation pit. What is the cause of artesian aquifers, and what is the effect on gravity as well as wellpoint dewatering systems?

4. Jet-grouting the soil underneath existing buildings provides an opportunity to reverse settlements that have already occurred. Explain the principles behind this effect. Sketch how the leaning tower of Pisa could be made vertical. Why will this never happen?

HOMEWORK PROBLEMS

1. The header problem requires the anchoring of the diaphragm and the sheet pile wall. In order to decide on the spacing between the straight-shaft tendon anchors, we should know how much one anchor will hold. As both rows will be anchored in silty or fine sand (18 kN/m^3), no adhesion to the bonded length will occur. Assume the angle of the anchor to be approx. 30 degrees and the friction angle F of the cohesionless soil to be 33°. We assume that the clay has a density of 20 kN/m^3.

Table 9.12 Densities

Material/Soil	Dry Densities (kg/m^3)	Saturated Densities (kg/m^3)
Styrofoam	30–120	
Plastics	850–1400	
Silts	1360	1400–1800
Sandstone	2000–2600	2200–2600
Limestone	1900–2600	2100–2700
Sand		1500–2000
Clay		1400–2200
Clay & Organics	700—900	1000–1300

From experience we know that the ultimate transfer load for smaller earth anchors and this kind of cohesionless soil is between 70 and 130 kN/m due to variability in density. Post-grouting of an anchor is expected to increase the loading capacity by 30%.

Using the function developed by the Army Corps Engineers, calculate the borehole diameter that would provide the ultimate transfer load that is needed.

2. You are pouring concrete with a tremie to fill up a drilled shaft. During the pumping operation you use a flow meter to record the concrete entering the shaft. In addition, you measure the level of the concrete as it rises inside the shaft. After the cast is finished you get the graph shown in Figure 9.37, plotting volume vs. depth of concrete surface.

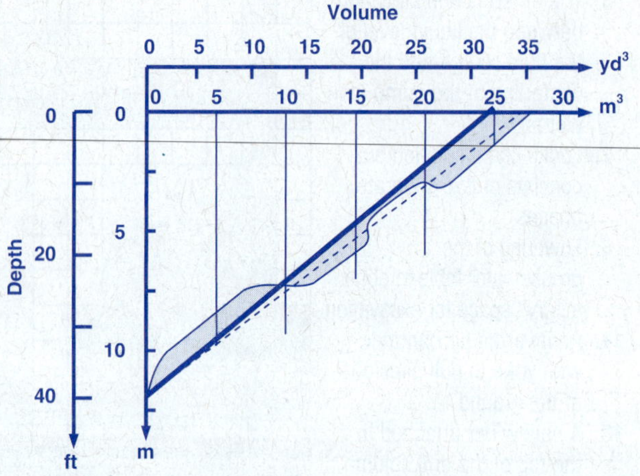

Figure 9.37 Concrete Use During Concrete Pouring

1. Are you happy with what you see?
2. Is the amount of concrete that you needed within an expected range?
3. Sketch the possible causes for the deviations at 24 m (79 ft), 15 m (49 ft), 10 m (33 m), and 7 m (23 ft). Explain briefly.
4. Your boss is interested in calculating the eccentric moment, centrifugal force, and the amplitude at 1000 and 2000 rpm of a new vibratory hammer she is intending to buy. Here are its basic characteristics:
   Mass of the hammer incl. clamp = 3.3 t or 3.64 tons
   Number of eccentric hammers = 4
   Distance from center of rotation to center of gravity of one eccentric weight = 20 cm (0.65 m)
   Mass of one eccentric weight = 300 kg (661.4 lbs)
   The amplitude is calculated so: A = rpm∗ (eccentric moment/total dynamic weight)

For the total weight consider a sheet pile weighing 1.19 t, or 1.31 tons.

(a) What is the maximum eccentric moment that can be achieved?

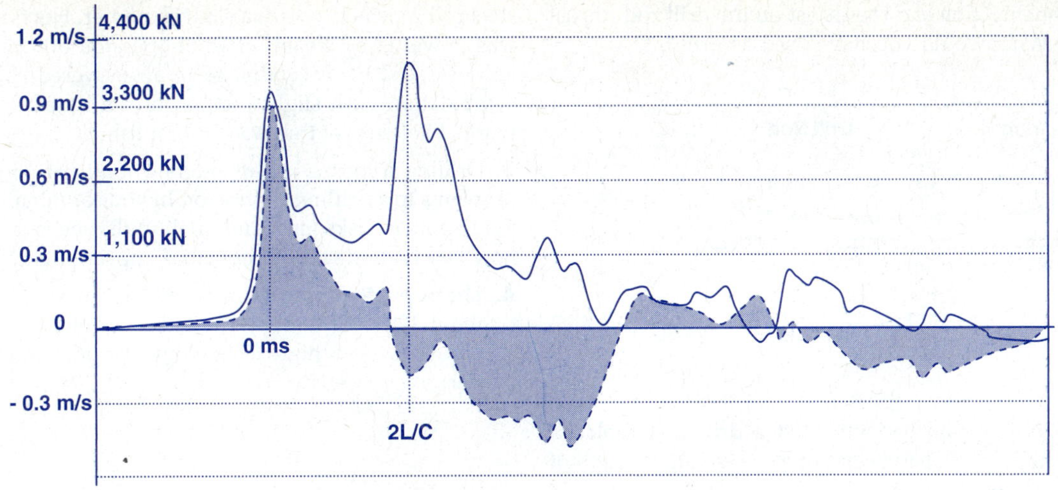

Figure 9.38 PDA Output for H-Pile

(b) Sketch the four masses in a situation where exactly ½ the maximum M_{ecc} is achieved. Will it change if you reduce the rpm?
(c) What are the centrifugal forces at 1000 and 2000 rpm?
(d) What are the amplitudes at 1000 and 2000 rpm?

3. Figure 9.38 presents the force and velocity graphs from an HP 14 × 89 (360 × 132) H-pile with a cross-sectional area of 26.1-in^2 (168 mm^2) made of ASTM A-36 steel.

The elastic modulus E of the pile = 210,000 MPa (30,500 ksi)

Wave speed in the pile C = 5,122 m/s (3.18 mile/s)

The AASHTO (American Association of State Highway and Transportation Officials) recommends that the maximum driving stress for an A-36 steel pile be 223 MPa (32.3 ksi).

1. Describe qualitatively the resistance profile of the H-pile.
2. Calculate the total resistance RTL.
3. What is the static resistance at the tip of the pile?
4. Is the present driving force safe? Should anything be changed?

LESSONS FROM THE ACCIDENT FILE

1. Case 1 could have easily been prevented. Sketch the situation in which the laborer got crushed. One method that would have avoided the accident would have required the laborer to tell the operator (face-to-face) that he was going to step behind the crane to access the tool box. What are three other methods of communication that could have prevented the accident?

2. Case 2 is yet another example how miscommunication or, maybe better, inappropriate action in response to proper communication, may result in death. Come up with at least two measures that would have avoided this accident.

3. A terrible and avoidable death is described in case 3. While it is easy to focus on the apparently still loose rain slicker, although it was taped tight, the operator seemed to have enough time to stop the operation to prevent the accident. But he had stepped away from the controls. Discuss two technical preventions that would have prevented this accident (even with the operator far from the controls).

4. The accident in case 4 could have been avoided. Describe what method you would use that would have accomplished the same as the "pile monkey" who was killed.

OPEN-ENDED PROBLEMS

1. The Egyptians used a so-called bowdrill to bore into hard rock. One purpose was to drill into granite along a straight line to split a long piece in order to fashion an obelisk. The splitting was actually done by wooden wedges that were wetted with water. Build a simple replica and show how the back of the drill rod could have been built to control the direction and the thrust on the drill rod (thrust bearing). What would you have used as drill bit?

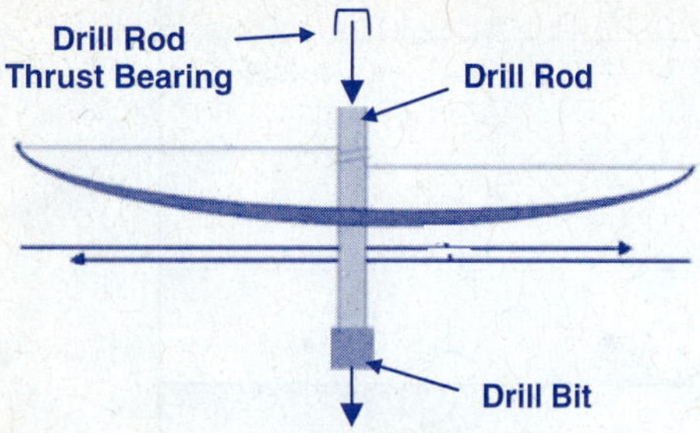

2. By 1870, Ingersoll had invented a drill that replaced hand drilling, which had been done by striking a length of steel rod repeatedly with a sledgehammer. His first design was powered by steam but later operated on compressed air. How did his invention of an air-powered hammer in combination with slight rotation of the bit after each hit work? Sketch how this was accomplished.

3. Drilling a vertical shaft offers a long list of potential problems to a drilling contractor. In your opinion, what are the five main risks (cost and quality) that he is facing, and how can he mitigate the potential impact of each?

4. The depth that pressurized caissons can reach is limited to about 120 ft (36.4 m). Why is that? Sketch and discuss a solution where high technology would eliminate this barrier.

BIBLIOGRAPHY

Abu-Farsakh, M.Y., and H.H. Titi. "Assessment of Direct Cone Penetration Test Methods for Predicting the Ultimate Capacity of Friction Driven Piles." *J. Geotech. Geoenvironmental Engrg.* 130, no. 9 (2004).

Armour, T., P. Groneck, J. Keelev, and S. Sharma. *Micropile Design and Construction Guidelines, Implementation Manual.* FHWA Pub. No. SA–97–070, USDOT. Washington, DC: June 2000.

Boehm, D.W. "The Utilization of Jet Grouting and Soil Mixing Methods to Repair and Support Bulkhead Structures." *Proc. Conf. Ports 2004.* Houston, TX. May 23–26, 2004.

Brown, D.A., S.D. Dapp, W.R. Thompson, and C.A. Lazarte, *Design and Construction of Continuous Flight Auger Piles,* Rep. FHWA, USDOT. Washington, DC. April 2007.

Bruce, I.A., *An Introduction to the Deep Soil Mixing Methods as Used in Geotechnical Applications.* Rep. FHWA-RD-99-138, USDOT. Washington, DC. March 2000.

Brunner, W.G. "Development of Slurry Wall Technique and Slurry Wall Construction Equipment." *Proc. Geo-Support.* Orlando, Florida. January 29–31, 2004.

Brunner, W.G., R. Fiorotto, E. Stötzer, and M. Schöpf. "The Innovative CSM-Cutter Soil Mixing for Constructing Retaining and Cut-Off Walls." *Proc. GeoCongress.* Atlanta, GA. February 26–March 1, 2006.

Canale T.S., G. Drohen, and J.L. Kaufma. "Design & Construction of the Foundations for the Watauga Raw Water Intake Facility in Karstic Limestone Near the City of Johnson City, TN. *Proc. Conf. Karst 2005,* San Antonio, TX. September 24–28, 2005.

Chakrabarti, S.K., P. Chakrabarti, and M.S. Krishna. "Design, Construction, and Installation of a Floating Caisson Used as a Bridge Pier." *J. Waterway, Port, Coastal, Ocean Engrg.* 132, no. 3 (2007).

Chow, H.L., and C.Y. Ou. "Boiling Failure and Resumption of Deep Excavation." *J. Perform. Constr. Fac.* 13, no. 3 (1999).

Davis, A. G. "Nondestructive Evaluation of Existing Deep Foundations." *J. Perform. Constr. Fac.* 9, no. 1 (1995).

Dunker, K.F., and D. Liu. "Foundations for Integral Abutments." *Prac. Per. Structural Design Const.* 12, no. 1 (2007).

Elwood, N.J., and J.W. Gaythwaite. "The DeLong Pier Repair Project: The Unique Challenges of Designing Repairs to a 1950 Vintage Marine Structure in the Arctic," *Proc. Conf. Ports 2007,* March 25-28, San Diego, 2007.

Fernandez, A.L., M.A. Pando, and P.G. King. "Load Test Program to Validate Model for Post Grouted Drilled Shafts." *Proc. Geo-Denver.* Denver, CO. February 18–21, 2007.

Harbin, B., M. Wallin, D. Bennett, and K. Staheli. "Starting Off on the Wrong Foot—Problems with Microtunneling Shafts." *Proc. NO-DIG 2005.* Orlando, FL. April 24–27, 2005.

Hartman, J.J., R.J. Castelli, and S. Malhotra. "Design and Installation of Concrete Cylinder Piles." *Proc. Geo-Denver Congress.* Denver, CO. February 18–21, 2007.

Hussein, M. and G.G. Goble. "A Brief History of the Application of Stress-Wave Theory to Piles." In *Spec. Cur. Pract. Future Trends in Deep Found. GSP 125,* edited by M.H. Hussein and J.A. DiMaggio. Reston, VA: ASCE, 2004.

Hsieh, H.S., C.C. Wang, and C.Y. Ou. "Use of Jet Grouting to Limit Diaphragm Wall Displacement of a Deep Excavation." *J. Geotech. Geoenvironmental Engrg.* 129, no. 2 (2003).

Hwang, J.H., N. Liang, and C.H. Chen. "Ground Response During Pile Driving." *J. Geotech. Geoenvironmental Engrg.* 127, no. 11 (2001).

Kaplan, H., A. Elburg, and I.D. Tommelein. "Analysis of the Variability in Precasting and Installation of Pile Foundations." *Proc. Constr. Res. Cong.* 2005. San Diego, CA. April 5–7, 2005.

Larsson, S. "State of Practice Report—Execution, Monitoring and Quality Control." *Proc. Conf Deep-Mixing.* Stockholm, Sweden. May 23–25, 2005.

Lee, H.S., J.Y. Lee, and J.S. Lee. "Non-Shored Formwork System for Top-Down Construction" *J. Constr. Engrg. Mgmt.* 125, no. 6 (1999).

Liou, D.D. "Thermal Effects in Large-Sized Diaphragm Wall." *J. Perform. Constr. Fac.* 13, no. 1 (1999).

Liu, G. B., C.W. Ng, and Z. W. Wang. "Observed Performance of a Deep Multistrutted Excavation in Shanghai Soft Clays." *J. Geotech. Geoenvironmental Engrg.* 131, no. 8 (2005).

Lopez-Anido, R., A.P. Michael, S. Thomas, C. Sandford, and B. Goodell. "Repair of Wood Piles Using Prefabricated Fiber-Reinforced Polymer Composite Shells." *J. Perform. Const. Fac.* 19, no. 1 (2005).

Mansur, C.I., and S.G. Durrett. "Dewatering Cofferdam for Construction of Olmsted Locks." *J. Geotech. Geoenvironmental Engrg.* 128, no. 6 (2002).

Massarsch, K.R., and B.H. Fellenius. "Deep Vibratory Compaction of Granular Soils." In *Ground Improvement-Case Histories*, eds. B. Indranatna and C. Jian. Elsevier Publ., 2005.

Mayne, P.W., B.R. Christopher, and J. DeJong. *Subsurface Investigations—Geotechnical Site Characterization*. Rep. FHWA-NHI-01-031, USDOT. Washington, DC. May 2002.

Massoudi, N. "Rock Socketed Micropiles." *Proc. Geo-Denver*. Denver, CO. February 18–21, 2007.

Mullins, G., D. Winters, and S. Dapp. "Predicting End Bearing Capacity of Post-Grouted Drilled Shaft in Cohesionless Soils." *J. Geotech. Geoenvironmental Engrg.* 132, no. 4 (2006).

Nam, M.S., and C. Vipulanandan. "Roughness and Unit Side Resistances of Drilled Shafts Socketed in Clay Shale and Limestone." *J. Geotech. Geoenvironmental Engrg.* 134, no. 9 (2008).

O'Neil, M.W., and L.C. Reese. *Drilled Shafts: Construction Procedures and Design Methods*. FHWA, DTFH6 1-96-2-0005 1, USDOT. Washington DC, 1999.

Poh, T., A.T. Goh, and I.H. Wong. "Ground Movement Associated with Wall Construction: Case Histories." *J. Geotech. Geoenvironmental Engrg.* 127, no. 12 (2001).

Robinson, H.D., J.E. Gómez, C.M. Englert, and A.W. Cadden. "New York City Harbor—Remediation of Seawalls on the Historic Liberty and Ellis Islands." *Proc. Ports 2007*. San Diego, CA. March 25–28, 2007.

Ryan, C.R., and C.A. Spaulding. "Strength and Permeability of a Deep Soil Bentonite Slurry Wall." *Proc. Geo-Congress 2008*. New Orleans LA. March 9–12, 2008.

Schmieg, H., and P. Vielsack. "Transmission of Shear Forces in Sheet Pile Interlocks." *J. Geotech. Geoenvironmental Engrg.* 128, no. 4 (2002).

Stuart, D.M. "Project-Specific Steel Sheet Piling Applications." *Pract. Period. Struct. Design Constr.* 9, no. 4 (2004).

Vipulanandan, C., O. Guvener, and M. McClelland. "Monitoring the Installation and Curing of a Large Diameter ACIP Pile in Very Dense Sand." *Proc. Geo-Denver Congress*. Denver, CO. February 18–21, 2007.

Yang, M.P., S.J. Chen, and S.T. Chen. "Innovative Central Opening Strut System for Foundation Excavation." *J. Constr. Engrg. Mgmt.* 132, no. 1 (2006).

Zayyat, M.M., A.W. Jackson, T. Tanaka, T. Hino, and Y. Matsuba. "Foundation Soil Improvement Using Vibro Compaction Combined with Geo Grids." *Geotech. Spec. Pub. No. 136*. Eds.V.R.

chapter ten

Forms and Temporary Structures for Material and Laborers

This chapter introduces a variety of pieces of equipment and mechanisms that don't make many headlines in construction but are vital to productivity, quality, and safety. While these modern marvels are essential in constructing, they are all gone by the time the topping out is celebrated. In fact, they are mostly eco-friendly, sustainable production systems that are used again and again such as the shores and falsework that carry the heavy elevated decks under construction. No structure can be constructed without scaffolds that provide safe access at great heights for laborers and material. Finally, no concrete wall can be poured without strong formwork that contains the wet and heavy concrete until it is strong enough to support itself. All in all, this chapter will bring to bear a large number of devices and systems that are overlooked by the casual observer but are essential to constructing almost anything.

Table 10.1 Topics Covered in This Chapter

Construction			Planning & Control		
Equipment and Mechanisms	**Methods**	**Essential Components**	**Managerial-, Engineering Tools**	**Rules, Laws, Standards**	**Control Elements**
✓ Formwork ✓ Shoring ✓ Work Platform ✓ Tube & Coupler Scaffold ✓ Tubular Frame Scaffold ✓ Mobile Scaffold ✓ Bamboo Scaffold ✓ Aerial Work Platforms ✓ Climbing Scaffold ✓ Scissor Lift ✓ Falsework ✓ Modular Formwork ✓ Mechanized Formwork ✓ Pre-Cast Formwork ✓ Formwork Tables ✓ Crane Fork ✓ Gang Form ✓ A-Frame Wall Form ✓ Slip Form ✓ Cylinder/Drum Finisher ✓ Paving Train ✓ Slipform Paver ✓ Stringline Sensor ✓ Second-Lift Paver ✓ Texturing Machine	✓ Shoring ✓ Trench Shoring ✓ Bracing ✓ Tower Shoring ✓ Pump Jacking Scaffold ✓ Backshoring ✓ Pouring ✓ Crane-Handled Forming ✓ Self-Climbing Forming ✓ Vertical Slip-Forming ✓ Horizontal Slip Forming	✓ Shore ✓ Prop ✓ Post ✓ Ledger Frame ✓ Shore ✓ Screw Jack ✓ Caster ✓ Stringer ✓ Tower Shore ✓ Outrigger ✓ Swiveling Tube Coupler ✓ Base Plate ✓ Concrete Masonry Unit ✓ End Guard Rail ✓ Plywood with Protective Overlay ✓ Joists ✓ Raker Bracing ✓ Wales ✓ Form Ties	✓ Pre-Engineered ✓ Pre-Fabricated ✓ Pre-Tested ✓ Allowable Load ✓ Triangle Bracing ✓ Factor of Safety ✓ Tipping Point ✓ Stability ✓ Rated Working Load ✓ De-Energizing Power Lines ✓ Formwork Deflection ✓ Stiffness of Forms ✓ Maximum Load Capacity ✓ Hydrostatic Effect of Concrete ✓ Compressive Strength ✓ Perimeter Beam ✓ Slab Beam ✓ Rate of Pouring ✓ Automatic Bar Inserter ✓ Crack Control	✓ OSHA 29CFR 1926, Subpart Q ✓ OSHA 29CFR 1926.501(b) ✓ OSHA 29CFR 1926.451 ✓ OSHA 29CFR 1926.550 ✓ ANSI A10.9 ✓ ACI 347 ✓ ACI 347-04 ✓ ASTM A-153 ✓ Federal-Aid Highway Act of 1956	✓ Self-Retracting Lifeline ✓ Full Body Harness ✓ Automatic Track and Screed Control

10.1 The Family Tree of Temporary Supports

4000 BC	Tunnel Centering	The history of **falsework,** as far as we know today, is tied to the use of the arch, also called a vault (barrel vault–tunnel vault). The earliest known example of an arch is a tunnel vault found under the Sumerian ziggurat at Nippur in Babylonia, ascribed to about **4000 BC**. It was built with burnt bricks amalgamated with clay mortar. Because of the circular nature of the arch and the use of bricks, one has to assume that the Sumerians used some kind of falsework system, also referred to as **centering**.  Falsework/Centering
2000 BC	Crib Shore	One theory about how the Egyptians moved large blocks of rock onto the top of a pyramid involves a rocking cradle and a type of **crib shore**. Heavy loads can be "rocked" upward by inserting wooden logs or shims underneath the free side before rocking the cradle back. At the end of an incremental lift, the cradle would be skidded onto a new horizontal ledge to begin another lift. 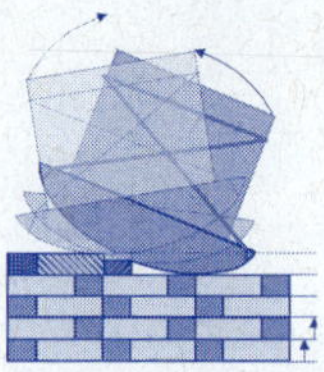
50 AD	Scaffolding	Finished in 50 AD, the Pont du Gard Aqueduct (49 m/161 ft high and 275 m/902 ft long) was built by the Romans in Southern France with precut stones that fit together without the need for mortar. The heavy blocks had to be lifted by treadmill cranes (see Chapter 3), which were supported by falsework. A separate falsework/centering was needed to support the arched "formwork" until the keystone closed the arch. One can still see the protruding **scaffolding for the masons'** supports and ridges on the piers.
Medieval Europe	Jumping, Centering and Scaffolds	From the twelfth to the eighteenth century, Europe carried out a large (re) building boom for cathedrals, churches, monasteries, castles, palaces, and city walls to protect against armies. Both Gothic and Baroque architecture used the arch/vault structure to create high, wide open spaces. Following the concept of creating falsework that rested on the already existing part of the building, the centering was built high above the ground. In contrast, scaffolding for the masons were built on cantilevered timbers inserted into openings in the wall at predefined levels. For one type of jumping scaffold used at the time, see Chapter 3.
Nineteenth Century	Deep Excavation and Trench Shoring	The rapid growth of cities such as London and New York put higher demands on basic underground infrastructure to provide more water for the growing population and to remove sewage to ensure a healthy environment. However, the available space underneath roads required the excavation of narrow trenches with vertical walls that tended to collapse. Trench shoring slowly became a crucial safety element of any trench operation. Thus, trenching operations at that time used both vertical and horizontal falsework to prevent the collapse of the earth structure. Vertical planks provided the sheeting in between the wooden beams and stringers organized into layers. The sewer pipe included an arch that was built, as the Romans did, on an arch formwork.
Early Twentieth Century	Heavy Falsework for Concrete Bridges	The revival of the Roman Puzzolana, now called cement, and the introduction of the reinforcing steel (rebar) brought new challenges for heavy falsework that carried the moulds, the rebar, and the heavy concrete before the cement started to harden. One cost-effective use of concrete was employed in the bridges for heavy trains that were built in Switzerland at the beginning of the twentieth century. Richard Coray became an expert in building temporary scaffolding for railroad bridges built in stone and later with reinforced concrete.

1918	Universal Tubular Scaffolds	 In 1909 Daniel Palmer-Jones received a patent for Scaffixer, a safe, simple way to couple crossing scaffold elements. In 1918, the universal coupler was invented, and in 1920 the use of tubular steel scaffolding took off. This flexible concept depends on steel or aluminum tubes of various lengths, coupling devices, ties, base plates, and boards to stand on. The flexible concept has since extended to create scaffold systems with modular pieces and to construct mobile towers.
1950s–1960s	Mobile Falsework and Formwork Mechanism for Bridges	The economics of cheap reinforced and post-tensioned concrete led to the development of several innovative falsework systems to carry either prefabricated concrete elements or the formwork needed for cast-in-place concrete. While spans were increased across rivers, valleys, ocean inlets, and inhabited areas, the surrounding conditions did not allow traditional falsework that reached the ground. Two examples of new falsework systems were the launching gantry and the balanced cantilever method. Both systems required building of piers, which were used either as a support for the gantry or the launching of deck-supported form-travelers. Several modifications to the basic concepts were developed while applying the falsework structures to transporting and erecting pre-cast segmental elements.
1960s	Climbing and Slip-Forms for Concrete	Both the high-rise buildings in growing city centers after World War II and the high piers to support the large bridges built in steel or concrete led to innovations in formwork for concrete walls. Two examples are self-climbing formwork and vertical slip-forming. The lower part of the multitier climbing formwork is anchored to the hardened concrete, thus propping up the formwork used to cast the next higher level. The slip-form approach is a continuing 24-hour process in which the form is jacked upward in incremental segments. Climbing rods are inserted into the hardening concrete and used by the jacking system to push the formwork system upward without leaving any visual horizontal marks. Both methods require precise surveying to stay plumb.
1960s	Heavy Mobile Formwork for Tunnels	While the need to put roads above ground led to ever-longer bridges, hills and mountains had to be tunneled. The large diameters and the necessary flexibility to adapt to changing soil conditions challenged the industry to develop large segmental formwork systems that could be easily moved (on rails) and could be used again and again. Heavy falsework in steel, rather than timber, not only had to keep the formwork in place during concreting but also had to allow sufficient space for the construction traffic to pass underneath.
1990	Mobile Slip-Formers for Highways	The mechanized construction of concrete surfaces for highways, traditionally built in sections, was replaced by slip-formers traveling at slow speed on a sturdy sub-base. Laser beams or string lines provided the vertical guidance for the self-adjusting form suspended by a hydraulic cylinder from the moving platform. A conveyor continuously moved concrete supplied by concrete trucks into a hopper on top of the machine.
2000	Powered and Ergonomic Climbing Scaffolds	Anyone looking at a pump jack scaffold supported by poles and platforms that can be raised by jacks can easily understand that a mechanized version using strong frames could provide many benefits. Today, motorized mast climbers can be found on many sites. Installed in 30% of the time it takes for conventional 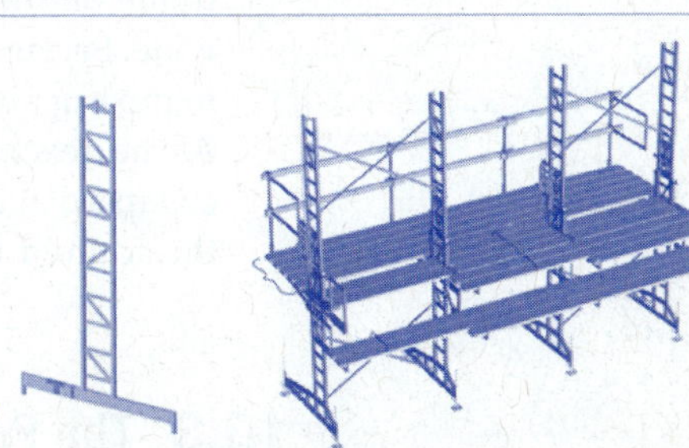 scaffolds, they can be lowered and raised with the push of a button, carrying both people and material (e.g., masonry units). Because of its ease of use, laborers pay more attention to safe and ergonomic operation—always working at the optimal height.

10.2 From the Accident File

The following accidents are examples of what can go wrong. The underlying problems provide focal points for many of the discussions and the reader will again be challenged to design solutions at the end to the chapter.

1. While repositioning a launching gantry on a bridge project, the front leg of the overbridge slipped off its pier, causing the entire gantry to collapse to the ground, killing four ironworkers. Three ironworkers were on a catwalk around pier 13 preparing to bolt down the back leg of the underbridge when the accident occurred. The front leg of the overbridge, only held by friction, slid forward and off pier 13. At the time of the collapse, ironworker victims #1 and #2 were in position on top of the underbridge, and victims #3 and #4 were in position on top of pier 14, directly under the underbridge, waiting for the front leg of the underbridge to come into position so that they could bolt it to the pier. The operator, who suffered serious injuries, was on a catwalk above and behind the front leg, and rode the overbridge down as it fell to the ground.

S10.1 Launching Gantry Collapse

2. A crew consisting of six erectors and two foremen was completing the last tier of a scaffold. After completing the handrails, the foreman instructed the victim to climb down one level with him to get more boards for the floor. The foreman was climbing down the scaffold when he heard a rush of air and turned to see the victim falling from the scaffold.
3. Two laborers were installing metal frames on the exterior wall of a residential building. One of them, working from a ladder on the scaffold, was trying to pull a nail out of the wall. The scaffold moved away from the building, causing the victim to fall 10 m (33 ft), hitting his head on the second story before crashing onto the concrete floor.
4. A construction worker was killed when he was struck by concrete formwork that fell from an overhead crane. Plywood and steel forms were rigged to the crane hook with a two-leg wire rope bridle attached to two nylon slings, which were choked around the load. The load was level and was raised without problems; however, in the middle of the swing, one sling end slipped or twisted out of the hook, and the entire load of forms fell to a work area underneath. The victim was struck by the falling load, which caused fatal head and neck injuries. The man who rigged the load, as well as the crane operator, were experienced workers and did not notice anything unusual prior to or during the lift. The area from which the forms were lifted was out of sight of the crane operator, and the rigging man used a radio for communication.
5. An ironworker installing a roof for a 20-m-(66 ft) diameter, 15 m-(50 ft) high tank used scaffold towers made of tubular steel frames to work from. A scaffold tower was built of six frames, each of which measured five feet in height, plus a three-foot guard rail frame on the top. Each bay was five feet, ten inches long and five feet, two inches wide. Because the third roof section misaligned with the second, the victim attempted to push against it using both hands. He stood on one plank at the top of the scaffold. As he reached with his arms the unstable scaffold tower tipped backward and collapsed against the inside wall, causing the victim to fall onto the steel base. When the scaffold tipped, two of the four legs came out of the base plates.

10.3 The Principles of Shoring

The reader has undoubtedly felt that construction terms are often totally confusing. For example, a bridge shoring seems to fulfill vastly different function from a trench shoring, while a centering looks really like a falsework or even a scaffold. Like many other English construction terms, the word *shoring* has its roots in shipbuilding and sailing (other terms include a *shore*, a *stay*, a *mainstay*, a *bulwark*, a *post*, a *brace*, an *upbear*, an *underpin*, a *support*), while the German equivalent, *das Gerüst*, was used in the Middle Ages to describe the armor of a knight on horseback, including hooks to support his long lance, as well as the high timber structure built to break the defenses of a walled-in town. The German word for *trench shoring* is *der Verbau*. Its origin is in military engineering, where it was used to indicate a defensive wall system (*ein Bau*) to prevent an enemy access to a town. It also describes the "defenses" against falling rock in underground mining, in this case the wooden structure (*Bau*) preventing the ceiling from collapsing and killing the miners.

Considering its historical background, the following best describes the basic meaning of the term *shoring* in construction:

> Shoring is a temporary horizontal or vertical support system that prevents the failure or collapse of a permanent or temporary structure under construction or undergoing a change causing a modified force distribution within. The verb shore denotes to the process of creating a support structure.

Accidents caused by failed shoring have led to several standards and codes that a contractor needs to follow. They include the United States Federal Occupational Safety and Health (OSHA) standards, state standards, (such as California (Cal/OSHA), American National Standards Institute (ANSI) standards, and the American Concrete Institute (ACI) standards. The following are some applicable examples:

S10.2 Safe Practices in Frame Shoring

- OSHA 29CFR1926, Subpart Q
- ANSI A10.9, *American National Standard for Construction and Demolition Operations—Concrete and Masonry Work-Safety Requirements*
- ACI 347, Guide to Formwork for Concrete

Of course, compliance with standards alone does not guarantee an accident-free use of shoring.

10.3.1 COMMON CAUSES OF CATASTROPHIC FAILURES

With one exception, shoring systems utilize props, posts, struts, frames, and trusses in combination with braces, ties, and jacks to transfer load from one end of the shore to the other. For example, the deck shoring carries the load of formwork and fresh concrete to a lower level that is strong enough to sustain it. Of course, one tries to keep the shoring as lightweight as possible in order to minimize its use. As we will discuss later, each system has an allowable load capacity as a function of some basic parameters. Frequent accident reports, mostly nonfatal, attest that such structures, albeit temporary, require careful planning, communication, construction, and inspection. A review of 250 fatal accidents with shoring by Flint (1981) revealed that almost all could have been avoided with proper care (Table 10.2).

Table 10.2 Main Sources of Fatal Accidents Involving Shoring

Cause	% of Cases
Inadequate scaffolding or ladders	21
Inadequate technical supervision	20
Plant, equipment, fittings and details defective, overloaded, or unsuitable for purpose	16.4
Inadequate signals, instructions, or briefing	14
Inadequate inspection and maintenance	12
Inadequate planning	5
Inadequate coordination between contractors	4
Total	92.5

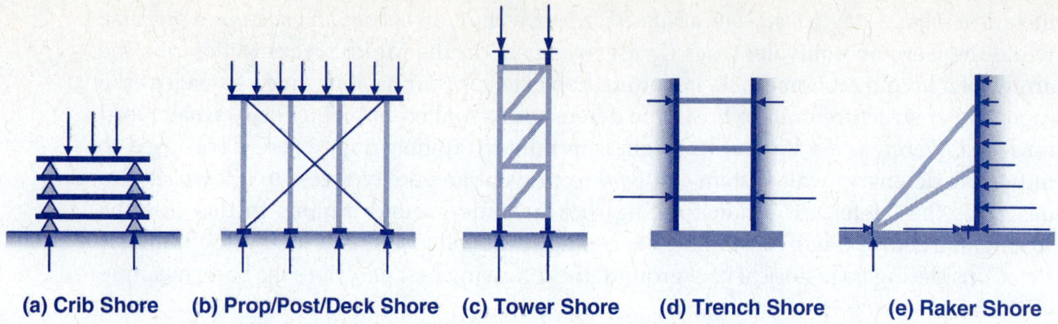

Figure 10.1 Forces Acting on Heavy Shoring Systems

It is apparent that the lack of technical expertise required for proper installation, including the use of appropriate material, was the main contributor to 69.5% of shoring accidents. Therefore, every engineer involved in any way with shoring must have a solid understanding of the underlying force system and an appreciation of the basic mistakes that can cause a large structure to tumble and crash.

10.3.2 FORCE MODELS FOR MAIN SHORING SYSTEMS

Shoring encompasses supports that act vertically as well as horizontally. Accordingly, we can create the following five force models for the shoring systems that are most often used in today's construction (Figure 10.1).

10.3.3 THE PROP/POST/DECK SHORE

As the title of this section indicates, this most common of all shores has many different names and labels. They all indicate that an elevated platform, such as a concrete formwork, is being supported temporarily. Figure 10.2b indicates that such shores are not used alone, but rather several are configured in such a way as to evenly carry a distributed load resting on a beam from the head to the bottom plates. While the props or posts stand separately, their connections to the beam on the top and a proper cross-brace make them work as a system. As we will learn later, their flexibility is an advantage when it comes to shoring uneven platforms but the one-by-one approach of assembly is very labor intensive. Fabricators, not surprisingly, offer a remedy for this drawback.

Let's quickly review some of the most common behaviors and problems that can lead to a catastrophic collapse of prop shores.

While scaffolds embraced the use of the universal couplers and fixed aluminum tubes, heavy loads of fresh concrete required a much sturdier solution for the shoring of formwork. A flexible and sturdy solution was found in the form of telescoping posts depicted in Figure 10.2a. Two mechanisms allow quick length adjustments along the main axis. By inserting and locking the pin into one of the pre-drilled holes in the smaller pipe segment, quick adjustments of the telescoping length can be achieved. A threaded collar with a handle permits fine height adjustments over a limited range. Due to the telescoping action, where the smaller-size tube telescopes into the outer one, prop shores are manufactured to operate between a lower and an upper height limit. Figure 10.3 presents length-load graphs for five shores that flexibly cover a span from total retraction to full extension.

Post/prop/deck shores are available in many different models even by the same fabricator. Of course, the load that each model is allowed to support depends on its cross-sectional area, the material used, and, most visibly, its length. As the tendency of a long

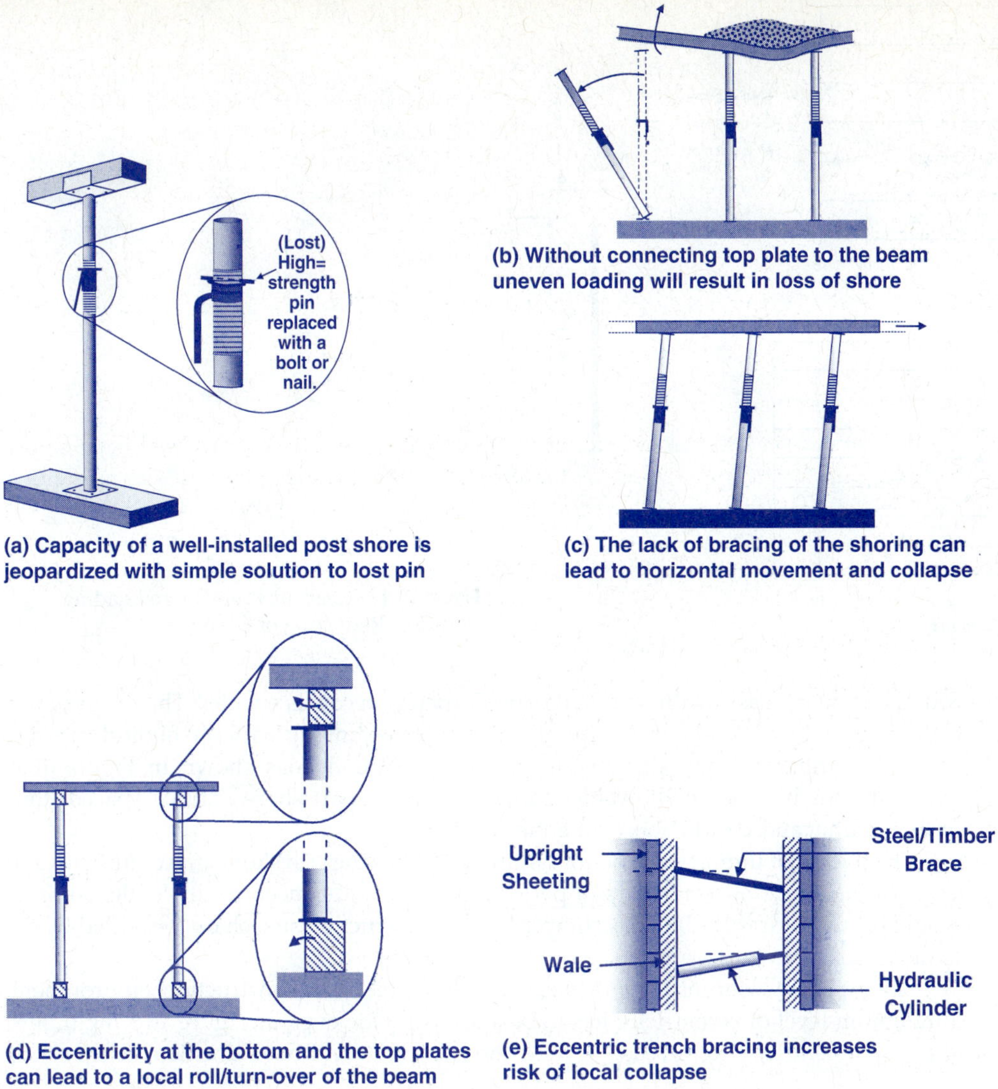

Figure 10.2 Common Problems with Heavy Post and Trench Shores

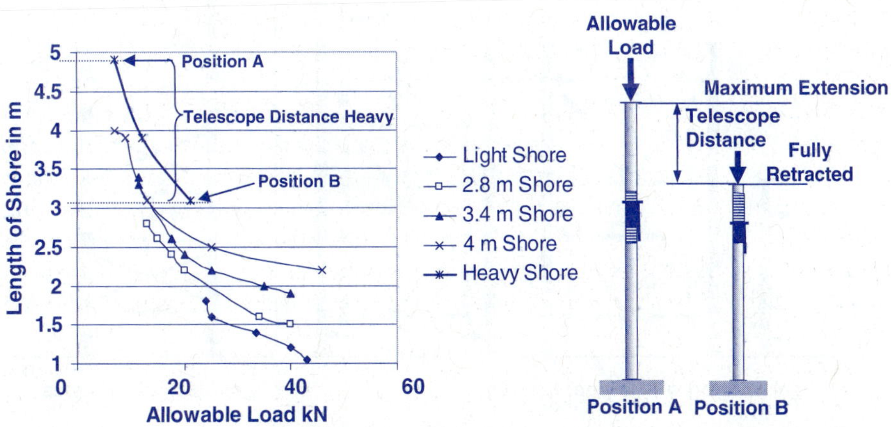

Figure 10.3 Load Capacities as a Function of Size and Extension

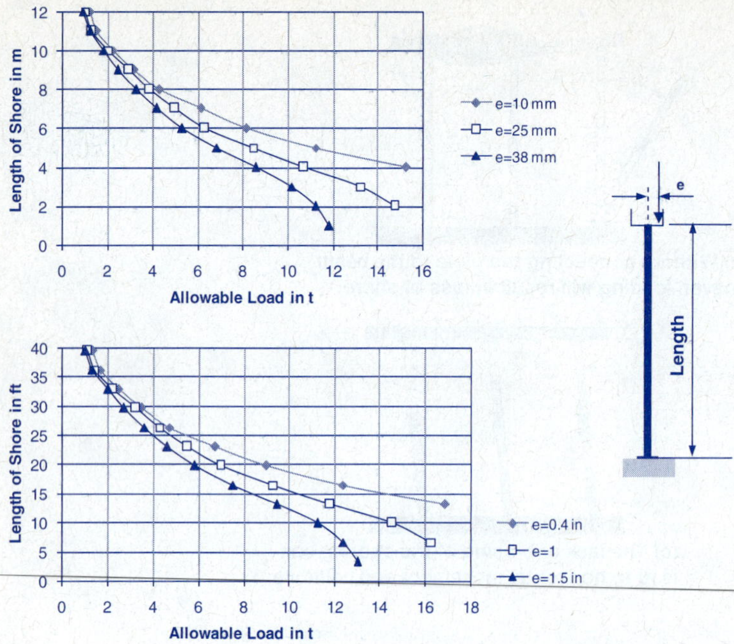

Figure 10.4 Eccentric Prop/Shore Loading Requires Reduction of Load

rod to buckle increases with length, the load capacity becomes smaller. The result of this effect is a nonlinear load curve for each shore type, beginning with the highest capacity when fully retracted and the lowest when fully extended as shown in Figure 10.4. Naturally, the higher the allowable load, the further apart shores can be spaced, thus reducing time and cost to shore an entire deck.

The two most important types of failure for all but the crib shore are eccentricity and overloading. Since the setting up of perfectly vertical and concentrically loaded shoring would be almost impossible to accomplish on a construction site, charts provided by the manufacturer provide for some amount of imperfection.

Since perfect concentricity is close to impossible in a construction environment, a minimum level of eccentricity has to be assumed. However, as Figure 10.5 highlights, allowing eccentricity to increase beyond the minimum will have drastic consequences.

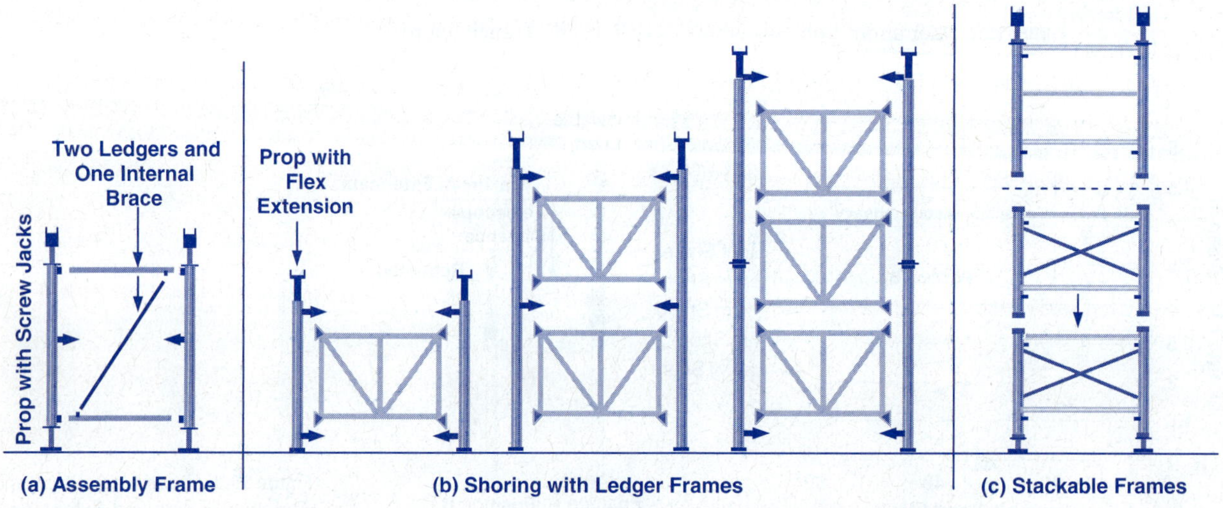

Figure 10.5 Three Common Techniques for Building Framed Shoring

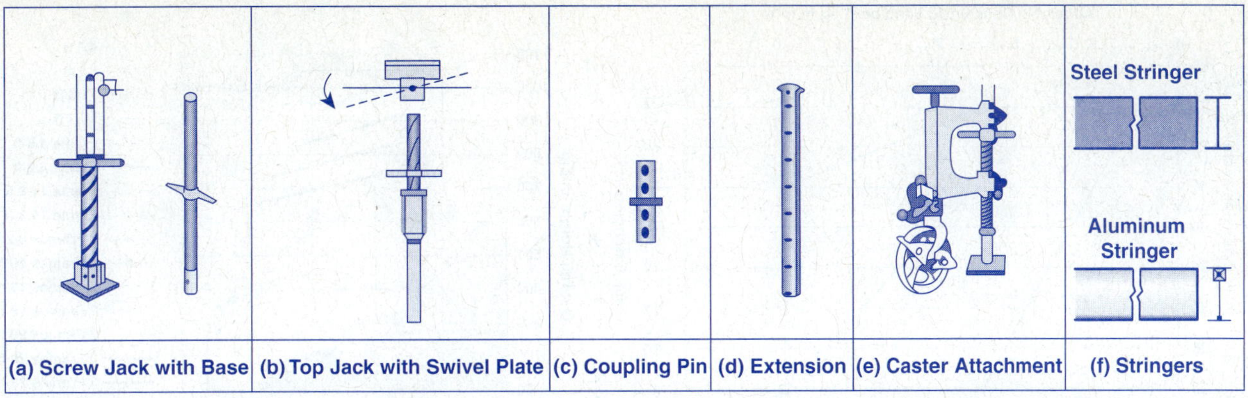

Figure 10.6 Common Accessories for Framed Shoring

For example, the shore at 4 m (13 ft) looses 4.5 t (5 tons) or 30% of its allowable load if the eccentricity increases from the minimum 10 mm (.4 in) to 25 mm (1.5 in).

To increase the stability of post shores fabricators have developed frames that can be used to connect and brace individual props into 2-D or space assemblies. There are a multitude of combinations to accomplish this, including the use of individual steel pieces as ledgers and cross-braces or, increasingly, the use of frames that either connect two props or include them in the frame.

There are three main reasons for a contractor to use shoring frames: (1) building efficiency, (2) loading capacity, and (3) calculation of allowable loads. As in housing construction where modularization and pre-fabrication of roof-trusses has drastically speeded up roof construction, shoring technology has benefited from similar principles. As Figure 10.6 indicates, pre-engineered modules come in many different subassemblies, but all are designed to be quickly fit together using built-in connection mechanisms. Variations in sizing the elements, such as ledgers and cross-bracing, provide the needed flexibility in both width and height. Finally, the screw jacks on top and bottom or other flex extension mechanisms make it possible to fine-tune shoring height, within limits.

The second benefit from frames is added strength through welded joints. While ledger elements provide great flexibility, the allowable load is impacted as the joints have to be considered as pinned connections. Welded ledger-frames, shown in Figure 10.5b, add stiffness to the shoring assembly and thus increase the allowable load despite being limited by the screw jacks, as discussed below.

As it is the responsibility of the contractor to build the necessary temporary structures, the company and its insurance have to assume the associated risks, which are potentially large. Pre-engineered and pre-fabricated frame-modules not only reduce the amount of individual pieces but also the opportunity to make mistakes. Furthermore, elements can be pre-tested in the laboratory, making it possible to establish standardized loading charts that can be easily understood by a contractor. Predictability reduces risks. Having available pre-tested shoring elements that can be put together, following pre-determined procedures, provides a great opportunity for the contractor to lower risks while, at the same time, increasing safety and efficiency.

While single-prop shores have built-in telescoping capabilities to be fitted to the required length (= height), frames require a two-step approach. First, a quantity and length of individual frame elements have to be calculated so as to match the desired height as closely as possible. Secondly, jacks or flexible extensions at the two ends, bottom and top, are used to close the final gap.

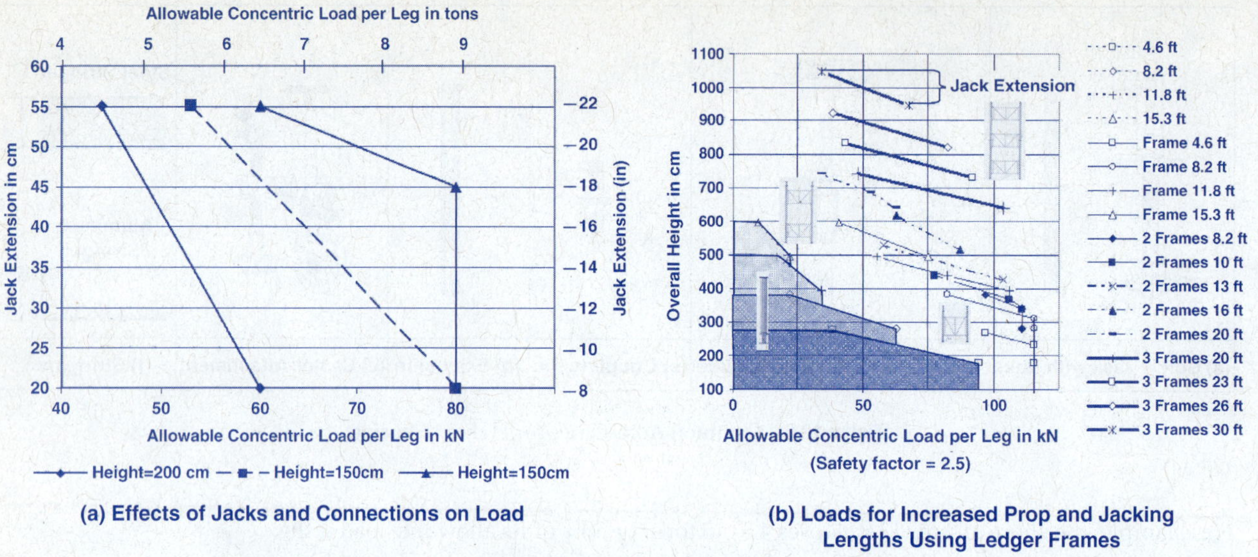

Figure 10.7 Charts of Allowable Load as a Function of Frame Size and Screw Extension

By turning a handle nut fitted to a solid screw, as shown in Figures 10.6a and b, the upper bracket can be lifted and lowered. Both bottom-end and top-end plates can swivel in cases where the stringers are not horizontal. Stringers are made of lightweight aluminum, timber, steel, or even composite material (see Figure 10.6f). Other add-ons include a connector piece for the vertical frame standards, the telescopic extensions and the casters. The latter becomes especially useful in cases where several frames are connected to reusable tables or towers.

As mentioned above, the load limits of the jacks have to be considered when calculating the number of frames needed to support an expected load. The fabricator will provide load tables for the most common cases. Figure 10.7 shows two such graphs.

These charts are best explained through an example.

Worked-Out Problem 10.1 / Shoring Using Frames

You are asked to provide a preliminary assessment of the spacing of shoring frames presented in Figure 10.5b for supporting the formwork for a reinforced-concrete deck 20 cm (7.9 in) thick and a floor-to-ceiling height of 5.2 m (17 ft) off the ground as depicted in Figure 10.8.

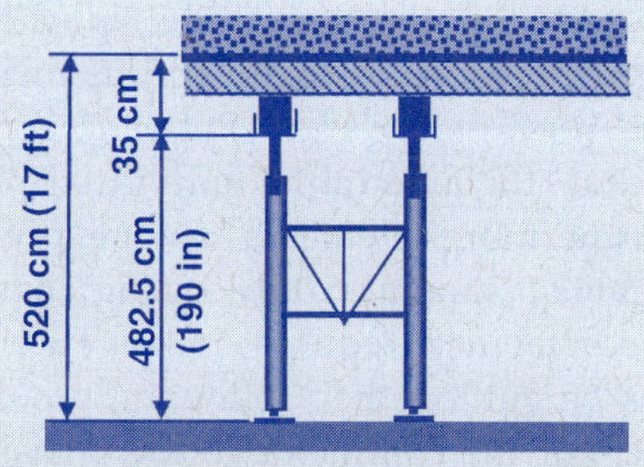

Figure 10.8 Shoring Frame

Assumptions

Weight of the reinforced concrete = 25 kN/m^3 (158 lb/ft^3)
Dead load of formwork = 0.48 kN/m^2 (10 lb/ft^2)
Live load = 2.4 kN/m^2 (20 lb/ft^2)
Total load = Concrete load + Dead load of formwork + Live load
Height of stringer = 15 cm (6 in)
Height of joists = 20 cm (7.7 in)
Formwork plywood sheathing = 2.5 cm (1 in)

Calculations

$$\text{Concrete load} = 0.2\,\text{m} * 25\,\text{kN/m}^3 = 5\,\text{kN/m}^2$$
$$= (7.9\,\text{in} * 158\,\text{lb/ft}^3 * 1\,\text{ft}/12\,\text{in}) = 104\,\text{lb/ft}^2$$
$$\text{Total load} = 5\,\text{kN/m}^2 + 0.48\,\text{kN/m}^2 + 2.4\,\text{kN/m}^2 = 7.88\,\text{kN/m}^2$$
$$= 104\,\text{lb/ft}^2 + 10\,\text{lb/f}^2 + 20\,\text{lb/ft}^2 = 134\,\text{lb/ft}^2$$
$$\text{Required length of shore} = 520\,\text{cm} - 20\,\text{cm} - 15\,\text{cm} - 2.5\,\text{cm} = 482.5\,\text{cm}\ (190\,\text{in})$$

Graph Figure 10.7b: At total height of 482 cm (14.7 ft) the following shore frames can be chosen:

Option A: 1 × 11.8 ft frame (360 cm) with an extension of 2.8 ft (122 cm) jack extension

$$\text{Allowable load} = 62\,\text{kN per leg}$$

Option B: 2 × 6.5 ft frames = 13 ft (396 cm) with an extension of 1.7 ft (86 cm) jack extension

$$\text{Allowable load} = 79\,\text{kN}\ (13{,}938\,\text{lb})\ \text{per leg}$$

For this preliminary calculation we assume a loading situation as presented in Figure 10.9. with shore props organized in a square pattern with a spacing of x.

Thus,

$$x^2 * 7.88\,\text{kN/m}^2 \Leftarrow 79\,\text{kN/shore}$$

$$x = \sqrt{(79/7.88)\,\text{m}} = \sqrt{10} = 3.16\,\text{m}\ (10.36\,\text{ft})$$

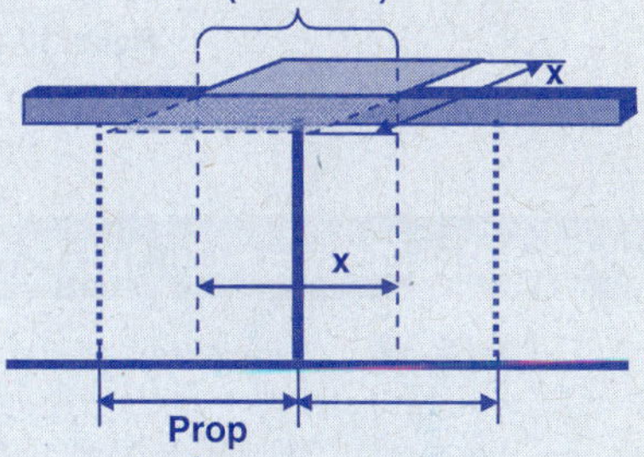

Figure 10.9 Load Impact Area

Discussion

The assumption of an organization of the props in a square can provide only preliminary results since the available frame sizes will dictate a pattern that might be used and most often will not include those spacing distances. Nevertheless, this calculation gives us an important first indicator of an appropriately sized spacing pattern.

10.3.4 THE TOWER SHORE

As depicted in Figure 10.10, a tower shore is a three-dimensional version of the framed shore system. Employing a pre-engineered and modular approach, frames and bracings are tested for allowable loading capacities. By following the instructions from the fabricator, a contractor is able to build multi-tiered towers to reach great heights.

Again, the complexity of calculating allowable loads of space frames/trusses has led to the adoption of pre-engineered systems. As with individual shore frames, working loads depend on the type and lengths of frames, the bracing, and especially the lengths of the screw jacks on the base and on top. High-strength aluminum alloy has also become the preferred material for fabricating the frames, primarily because of its lighter unit weight. Tower shores systems are usually divided into categories based on allowable leg loads to include 6.7 ton, 9 ton, 13.5 ton, or 60 kN (80 kN or 120 kN). Ultra-high-load shoring towers can provide allowable loads up to 39.3 ton (350 kN) per leg, all with a factor of safety of 2.5.

The incentives for pre-assembling tower-sections on the ground and using a crane to stack the tower-elements are safety and productivity, which are related in this case. In general and especially when erecting steel structures such as towers, laborers can be

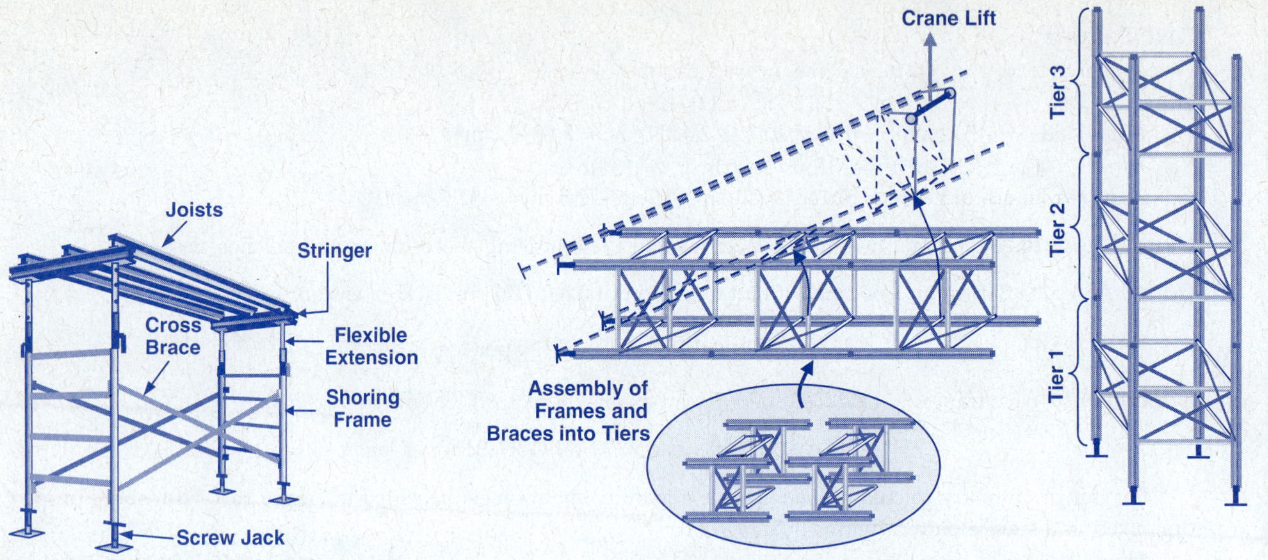

(a) Shoring Tower with Frames and Cross Bracing **(b) Pre-assembling Tower Horizontally and Lifting into Place**

Figure 10.10 Typical Tower Shores Used for Various Applications

> **OSHA Regulation 1926.501(b)**
>
> 
>
> "Unprotected sides and edges." Each employee on a walking/working surface (horizontal and vertical surface) with an unprotected side or edge which is 6 feet (1.8 m) or more above a lower level shall be protected from falling by the use of guardrail systems, safety net systems, or personal fall arrest systems.
>
> The self-retracting lifeline.

protected from the consequences of a fall only by wearing personal fall protection gear. The most commonly found technologies are the fall-arrest system and the self-retracting lifeline (SRL). Both require that a full body harness be worn, and properly placed lifelines have to be installed. Thus, working on a large tower shore may require not only the installation or use of work platforms, but also the wearing of fall protection equipment. While keeping the workers safe, such equipment can slow down the overall operation, which can be counteracted, however, by careful planning and preparation of the erection sequence.

It may not always be possible or even desirable to employ a crane. Individual circumstances have to be considered, including crane availability during the day, space needs, cost of crane, and cost of labor. Figure 10.11 presents the results of on-site time-measurements and curve fitting on erecting tower shores with the traditional pulley-hoist method without pre-assembly published by A. Shapira (2004). By entering the results of repeated field observations, two curves were plotted that related the worker-hours per tier for assembly and disassembly with the height of the tower.

$$\text{Assembly: Worker-hrs/Tiers}_{1-x} = 0.0035x^3 + 0.024x^2 + 0.36x + 0.30 \quad (1)$$

$$\text{Disassembly: Worker-hrs/Tiers}_{1-x} = 0.0028x^3 + 0.016x^2 + 0.15x + 0.01 \quad (2)$$

where

x = number of tier from the ground.

Because of lack of data for extremely high towers, the two functions (1) and (2) are limited to a maximum height of 30 m (91.5 ft). These two functions demonstrate the impact of tower height on the efficiency of assembly operation.

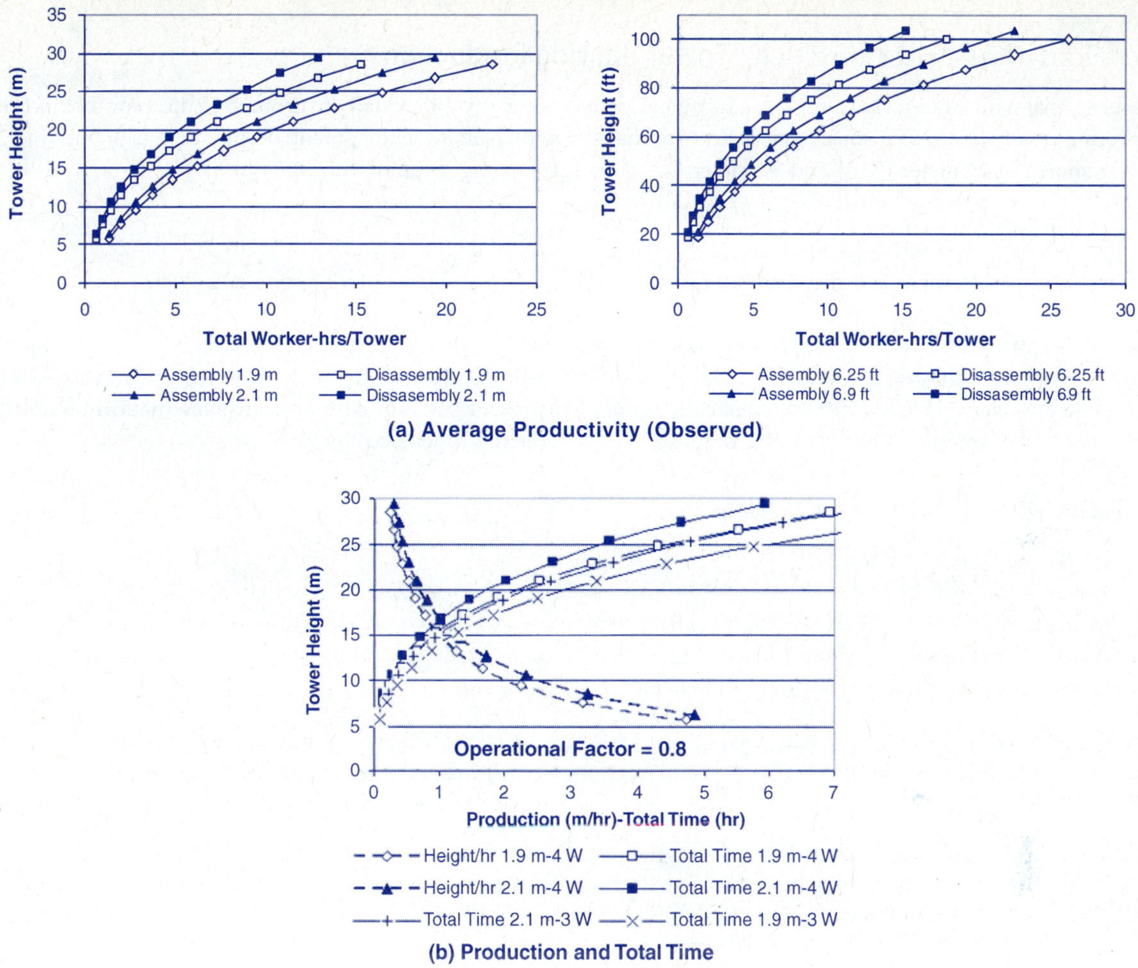

Figure 10.11 Crew Performance as a Function of Tower Height (Based on Aviad Shapira, 2004)

For this analysis, frame sizes of 1.9 m (6.2 ft) weighing 19.5 kg (42.9 lbs) and 2.1 m (6.9 ft) weighing 21 kg (46.2 lbs) are used. Figure 10.11a clearly shows the drastic effect of tower height on the time to assemble and disassemble towers. According to the function, the time it takes a 3-person crew to build a 5-tier tower with 1.9 m (6.2 ft) frames is approximately 20 minutes, while the construction of a tower with 10 tiers (19 m) takes 190 minutes, or more than 9 times longer. Furthermore, the needed times for disassembly, indicated by the lines with the square data point labels, follow the shape of the assembly durations but are approximately 35–45% lower. This fact indicates that the hoisting and lowering of the frames and braces as well as the additional safety measure to prevent a fall impact both operations in equal amounts.

The plots in Figure 10.11b are based on an operational factor of 0.8 assuming a 48 min/hr and crew sizes of 3 and 4 laborers to calculate the total time to build one tower at differing heights. The dashed curves compare the production in m/hr with 4 in a crew while the solid lines compares the total erection time for towers of variable heights. As can be expected, the larger frame sizes allow the crews to make more progress per hour, thus leading to lower total times of construction.

Worked-Out Problem 10.2 / Tower Shoring Production

Assume that you are working for an equipment rental company interested in promoting its new boom-type elevating work platform for construction. You are charged with evaluating the potential time savings in building 20 tower shores of 15 meter (50 ft) and 30 meter (100 ft) heights, using a common frame type.

Assumptions

We can use the assembly and disassembly functions (1) and (2) established by Shapira as a basis.

We can assume that the work above five meters consists of two separate elements: (1) fixed time for connecting the frames and braces for one tier, and (2) variable time for hoisting the tower elements. The hoisting of frames and bracing using the mechanical lift will drastically reduce the present amount of time because the platform is able to carry the hardware of an entire tier at a time, so that the hoisting time is cut by 70%.

Calculations Using 1.9 M (6.25 ft) Frames:

Worker-hrs/tower (w-h/twr)

Assembly 5.7 m or 18.7 ft ($x = 3$ tiers) $= 0.0035x^3 + 0.024x^2 + 0.36x + 0.30 = 1.4$ w-h/twr

Disassembly 5.7 m or 18.7 ft ($x = 3$ tiers) $= 0.0028x^3 + 0.016x^2 + 0.15x + 0.01 = 0.67$ w-h/twr

Assembly tier/time = 1.4/3 = 0.47 w-h/tier

Disassembly tier/time = 0.67/3 = 0.22 w-h/tier

Assembly 15 m (x = 8) = 6.2 w-h/twr Fixed time = 8 × 0.47 w-h/tier = 3.76 w-h/twr

Disassembly 15 m (x = 8) = 3.7 w-h/twr Fixed time = 8 × 0.22 w-h/tier = 1.76 w-h/twr

Assembly 30 m (x = 16) = 26.24 w-h/twr Fixed time = 16 × 0.47 w-h/tier = 7.52 w-h/twr

Disassembly 30 m (x = 16) = 18.0 w-h/twr Fixed time = 16 × 0.22 w-h/tier = 3.52 w-h/twr

	Tower-Shores Trad.			Tower-Shores with Lift				
	Fixed w-h	Variable w-h	Total w-h	Fixed w-h	Variable w-h	Total w-h	Differ. w-h/twr	Differ. for 20 w-h/twr
Assembly 15 m	3.76	2.44	6.20	3.76	0.73	4.49		
Disassembly 15 m	1.76	1.94	3.70	1.76	0.58	2.34		
TOTAL 15 m			9.90			6.83	3.07	61.4
Assembly 30 m	7.52	18.72	26.24	7.52	5.61	13.13		
Disassembly 30H m	3.52	14.48	18.00	3.52	4.34	7.86		
TOTAL 30 m			44.24			20.99	23.25	465.0

Discussion

As expected, the time savings increase rapidly with height. For a 3-member crew the difference at 15 m is merely 1 hour/tower, but almost 8 hours at 30 m. For 20 towers the time reductions overall are 20 (2 days) and 155 hours (16 days) respectively. It is apparent that the time savings will also result in cost savings, which will counter-balance the rental cost for the equipment.

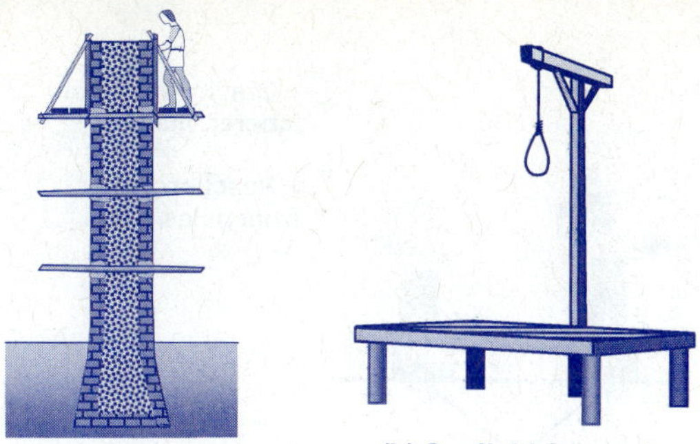

(a) Roman Scaffold to Raise High Walls

(b) Scaffold Supports a Gallows

Figure 10.12 Scaffolds to Build and to Destroy for Everybody to See

10.4 The Ubiquitous Scaffold

It is unlikely that another term related to an ubiquitous piece of construction equipment carries with it such an etymologically discomforting past. A review of dictionaries shows that they agree that the basic meaning of a scaffold is "a raised wooden platform," but they list two vastly different applications: (1) support for workmen and materials performing tasks above the ground, and (2) framework used for hanging or beheading condemned prisoners. The German word *Schafott,* meaning blood scaffold, and the French *échafaud* both seem based on the Latin *scaffaldus* but specifically relate to its use for executing people in front of large crowds.

For the following discussion, however, we will confine ourselves to its use in construction, where it has mushroomed into a wide variety of applications and configurations far beyond the cantilever beam-and-planks used by Babylonian and Roman masons, as depicted in Figure 10.12a.

10.4.1 UBIQUITOUS SCAFFOLDING SYSTEMS

Scaffolds, as non-consumables, must have several high-priority characteristics that are significantly different from those of consumable building materials. Temporary in nature, they should be (a) made of low-cost material, (b) easy to fabricate, (c) fast to assemble and disassemble, (c) light in weight, (d) reliably strong for the duration, and (e) spatially adaptable. From a mechanics point of view, to create an elevated platform to temporarily support both a laborer and materials, two approaches are available as shown in Figure 10.13: (1) suspension, and (2) elevation from the ground. In addition, such systems need to be spatially stable and sustain external forces.

S10.3 USACE Scaffolds
S10.4 OSHA Scaffolding

The suspended work platform requires an existing building from which to attach a rope or cable overhead; the majority of scaffolds in use today are elevated platforms. Worldwide, the following types can be found in great numbers:

(1) Bamboo scaffolds
(2) Tube and coupler scaffold
(3) Tubular frame scaffolds
(4) Mobile scaffolds
(5) Mast-climbing work platforms
(6) Climbing scaffold
(7) Aerial work platforms

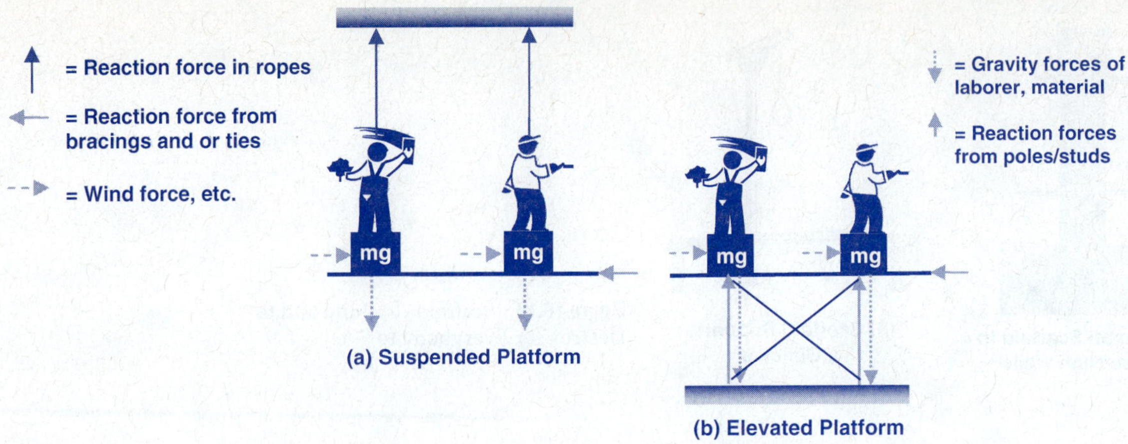

Figure 10.13 Stable Elevated Work Platforms

10.4.2 THE BAMBOO SCAFFOLD

Bamboo is the fastest-growing plant on Earth. This fibrous grass has found a wide variety of uses that include serving as an eco-building material. After mature bamboo is properly treated it forms a very hard wood that is both lightweight and exceptionally durable. As this prehistoric plant covered large areas of South East Asia, it became an effective building material for erecting structures, or as shown in Figure 10.14, as sustainable guardrails in parks. It is believed that the Chinese have been using bamboo for over 5,000 years, and it is still heavily used to build scaffolds all across Asia. Its continued success with contractors is particularly based on the speed with which the 20 ft (6 m) long bamboo poles can be assembled. Some experts claim that bamboo scaffolds are erected 6 times faster and dismantled 12 times faster than metal scaffolds with the same number of workers.

Buildings Department
The Government of the Hong Kong Special Administrative Region

Guidelines on the Design and Construction of Bamboo Scaffolds

- The commonly used bamboo types are Kao Jue and Mao Jue. They should be 3 to 5 years old and air-dried in vertical positions under indoor conditions for at least 3 months.
- The scaffolds should be effectively braced and tied back to ensure overall stability.
- The bracings should extend from the base to the top of the scaffolds.
- Knottings for posts/standards and ledgers, ledgers and transoms, bracings/rakers, etc., should be tied by nylon strips.
- The minimum ultimate strength of the nylon strips should not be less than 0.5 kN per strip, while the nominal width is 5.5 to 6.0 mm.
- All knots should be tightened with at least 5 rounds of nylon strips. The ends of the nylon strips should be crossed and twisted to form a single twisted end which passes through the knot twice to give one round turn for proper anchorage.
- The distance between two knottings on the overlapping portion of the bamboo members should not be greater than 300 mm, and the tail of one bamboo member should be connected to the head of the other.
- The competent person should check the strength and stability of the scaffold and ascertain there are no defects and deterioration. Defects found during the inspection should be rectified immediately.
- Removed bamboo members must not be thrown, tipped or shot down from a height.

Main Structural Support: Bamboo scaffolds use the double pole structure. This means that a double row of vertical bamboo (*mao jue*) provide the vertical supports. Cross-beams, ledgers, and braces are tied to the uprights to provide lateral stability. Wooden planks or metal catwalks rest on the cross-beams.

Figure 10.14 Bolted Bamboo Guardrail

Bamboo Knotting: Fastening or joining bamboo to extend their length, by overlapping them by approximately 6.5 ft (2 m), or to create lateral supports is done through knotting them together with the help of sufficiently strong string. In Hong Kong the building department's guideline stipulates the use of 0.4 inch (610 mm) wide nylon strips with a minimum ultimate strength of 112 lb (0.5 kN) to be wrapped tightly four times.

Fixed Lateral Bracing: As it is being erected, the scaffold needs to be rigidly

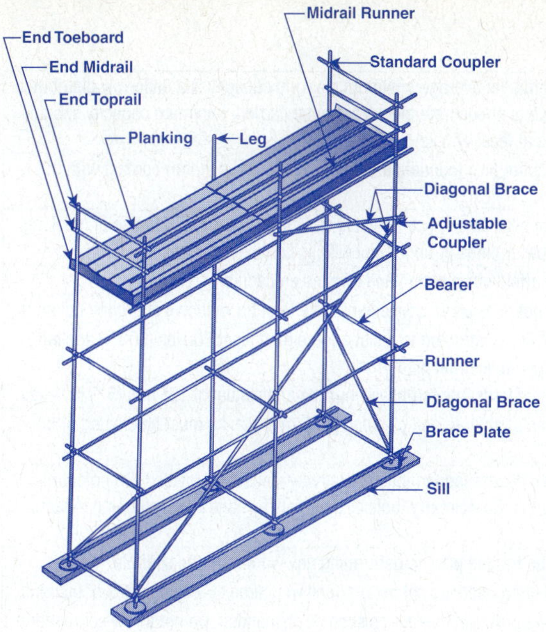

Figure 10.15 Tubular Steel Scaffold

braced against the existing building structure. In cases where the outside of the building consists of solid walls made of brick, block, or concrete, anchors or putlogs have to be installed. Putlogs, installed by the masons while building the wall, are metal brackets that can later be removed. For the brace, a steel bar or a combination of a wire tied to the anchor or putlog and a bamboo strut can be employed.

Metal-Bamboo Scaffold: The goal of this type of scaffold is to reduce one major drawback of the bamboo scaffold: safety for the laborers after erection. A particular concern is the sudden breakage of wooden boards due to overloading or to deterioration by weather as well as to the lack of staircases requiring the laborers to climb the scaffold. Thus, prefabricated and multi-use metal elements are being integrated into the bamboo structure. Of particular use are metal catwalks with toe boards and staircases.

10.4.3 TUBE AND COUPLER SCAFFOLD

As we learned from the historical review, 1920 was the birth year of the tubular steel scaffold, slowly replacing the cumbersome wooden systems. Similar to the bamboo, the T&C scaffold represents a braced double-pole or double-leg structure, as shown in Figure 10.15. While the hollow tubes, today made of aluminum, reduced the weight, the universal coupler, presented in Figure 10.16, had a drastic impact on the time for assembly and disassembly. With only a few operations, the coupling devices connect legs/uprights, braces, bearers, and runners/ledgers. Although tube and coupler (T&C) scaffolding is the most labor-intensive, scaffold systems offer important benefits, such as ease of use with a minimal of pieces, unlimited adaptability, access to even the most unique jobsite.

Due to its wide use and the many accidents caused by improper *installation or use*, many governmental agencies, contractor associations, and suppliers publish a wide variety of safety guidelines (OSHA Regulations 29 CFR-1926.451 General Requirements for Scaffolds). Table 10.4 points out some key issues.

Table 10.4 Scaffold Hazards

Overloading scaffold platforms is one of the most frequent violations of good scaffold practice. An average minimum platform capacity is a uniformly distributed load of 50 lbs/sq ft for construction-related work. For masonry construction where large pallets of concrete blocks are to be carried, minimum capacity should be at least a uniformly distributed load of 150 lbs/sq ft. This means that scaffolds should be at least double-planked.
No Base Plate: All scaffold legs require the use of a base plate **and** a mudsill or other adequate firm foundation. Base plates must be in firm contact with the sills/foundation and scaffold legs, and should be centered on the sills.
Uneven Ground: Compensate for uneven ground by using screw jacks and base plates with sills. Do not use unstable objects or materials. The use of blocking or packing such as bricks, short pieces of lumber, or scrap material, either under scaffold base plates or under mudsills, is prohibited.
Plumbness: Plumb and level scaffold until connections can be made with ease. Be sure scaffold stays plumb and level as erection progresses.
Tie Backs: Wall tie brackets are an alternative tying method, and they are used where it is not possible or convenient to tie through windows or around columns.
Guardrails must be used on all open sides and ends of scaffold platforms. Guardrails must be constructed to resist a force of at least 200 lbs (900 N) applied anywhere on the guardrail. Failure to provide guardrails is one of the main reasons for falls from scaffold platforms.
Toeboards must be installed whenever people are required to work or pass under a scaffold platform. The toe-board must be a minimum of 125 mm (5″) in height.
Slipping: Scaffold platforms can be either sawn or laminated lumber planks or prefabricated aluminum and plywood platforms. Planks must be cleated or be otherwise secured to prevent slipping.
Access must be provided to all work platforms. If access is not available from the structure, access ladder units or stairways must be provided. Do not climb bracing, guardrails, or vertical posts. Both hands must be free to hold guardrails or ladder rails. Do not carry tools or materials by hand when climbing ladders. Wear a tool belt and pouch, and move material up or down by rope.
A **Competent Person** must check the entire scaffold assembly before use. All users must be trained prior to performing any work from the scaffold.
Do not erect scaffold on **wagons, trucks,** or other **wheeled vehicles**. Stand only within the platform area; do not try to extend work area by leaning out over guardrail.
Dismantling: Do not remove ties until scaffold above has been removed. Lower components in a safe manner as soon as dismantled. Do not throw components off scaffold. Stockpile dismantled equipment in an orderly manner.
Wind: Scaffolds that are going to be enclosed (hoarded) will be subjected to significantly larger wind loads. Wind can lift lighter platform materials from the scaffold if they are not secured. Where severe wind conditions are anticipated or where high scaffolds are involved, platform materials such as aluminum/plywood panels should be secured to the scaffold.

S10.5 Inspection of System Scaffold
S10.6 Visual Inspection of Scaffolds

10.4.4 TUBULAR FRAME SCAFFOLD

Similar to shoring frames, scaffolding is meant to replace flexibility with higher reliability in creating solid joints. This is accomplished by replacing couplers and clamps with welded joints and tight coupling pins. The rigidity of the pre-engineered frames reduces the chances of making mistakes during assembly, while at the same time opening the door for creating standard elements that easily fit into the frames built in standard widths and heights. Nevertheless, fabricators address the special needs arising from common situations by offering a variety of frames and special accessories. Figure 10.17 provides a limited set of frames that can be mixed because of their standard sizes.

The assembly of frame scaffold begins, as with the T&C scaffold, with a solid foundation for the base plates on top of solid boards/mudsills. In addition, uneven grounds has to be compensated with screw jacks so that the first frames are plumb and level. The tubular frames fit the screw jacks and have to be connected to avoid possible uplift. Horizontal and diagonal braces come with various end brackets to allow quick connection to the uprights of the frames. Figure 10.18 shows a system that is based on studs welded to the frames that allow easy insertion of the end bracket when the distance to the corresponding stud on the neighboring frame meets the requirement. A simple pin or quick lock can be engaged to prevent the brace from slipping off the stud. The upper cross-beams of the frame provide the supports for either wooden planks or prefabricated platforms made of an aluminum

Couplers (90 degrees or swiveling)

- Joins scaffold tubes.
- Can be used to resist tension or compression loads.
- The two halves of a coupler can connected with a swivel pin.
- Replaceable T bolts, nuts, and washers,
- Made of forged steel, hot dip galvanized, (ASTM A-153).
- Collar nuts, T bolts, and rivets are mechanically galvanized.

Figure 10.16 Universal Coupler

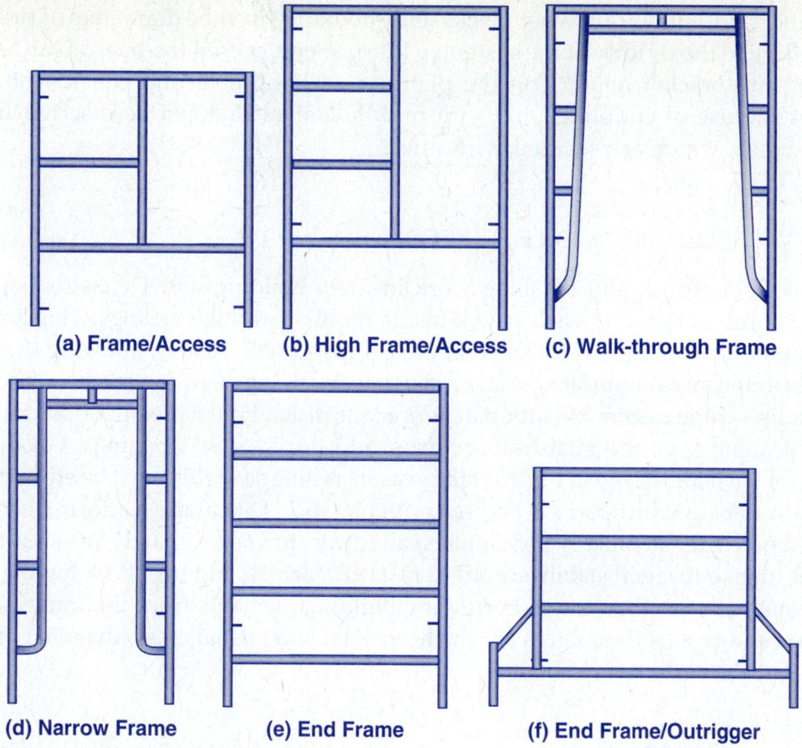

Figure 10.17 Common Scaffold Frame Configurations

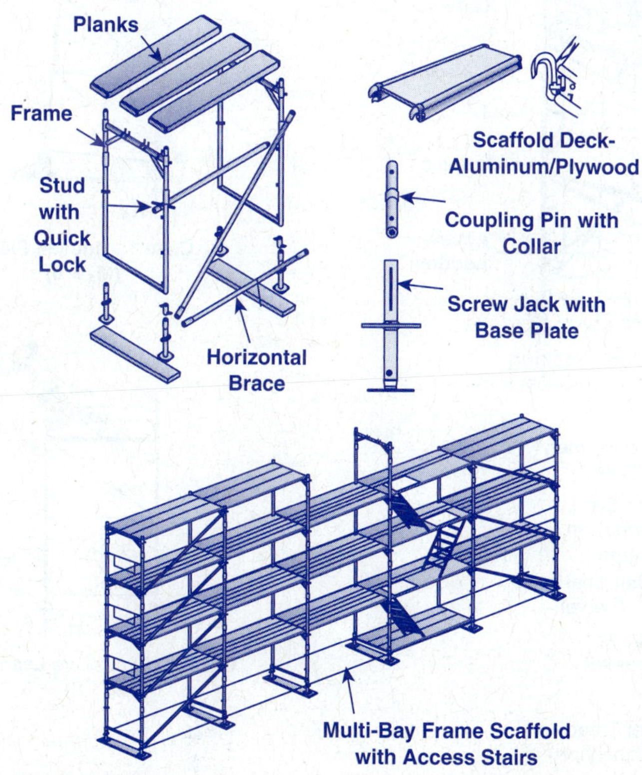

Figure 10.18 Elements of a Basic Frame Scaffold

frame and aluminum or plywood decks. End-hooks fit the tube diameters of the frame, thus adding to the rigidity of the system, while pre-engineered toe-boards can be easily inserted into bracket holders on the platform. Adding a second echelon of frames requires the use of coupling pins with middle collars that can be inserted into the hollow tubes, which are secured with pins.

10.4.5 WHEELED OR MOBILE SCAFFOLDS

Mechanical, electrical, and finishing work inside a building under construction many times requires workers to reach places inside rooms with high ceilings. This led to the invention of mobile scaffold towers that are equipped with casters, allowing them to be pushed around an even surface such as a concrete floor. The main danger of such scaffolds is instability, as the casters can lift up during eccentric loading of the upper platform (e.g., worker leaning over the guardrail) or drop into unprotected openings. Good housekeeping of the floor is a must. For a similar reason, rolling scaffolds must be erected so that the height-to-base width ratio is no greater than 3 to 1. This limits platform height with standard outrigger stabilizers and single-span towers to approximately 9.6 m (30 ft). This assumes that outrigger stabilizers are 3 m (10 ft) across. Figure 10.19 highlights the components of a one-bay tower borrowing building elements from the frame scaffold. The main and crucial differences are the horizontal and spatial cross/diagonal braces to ensure that the casters stay aligned.

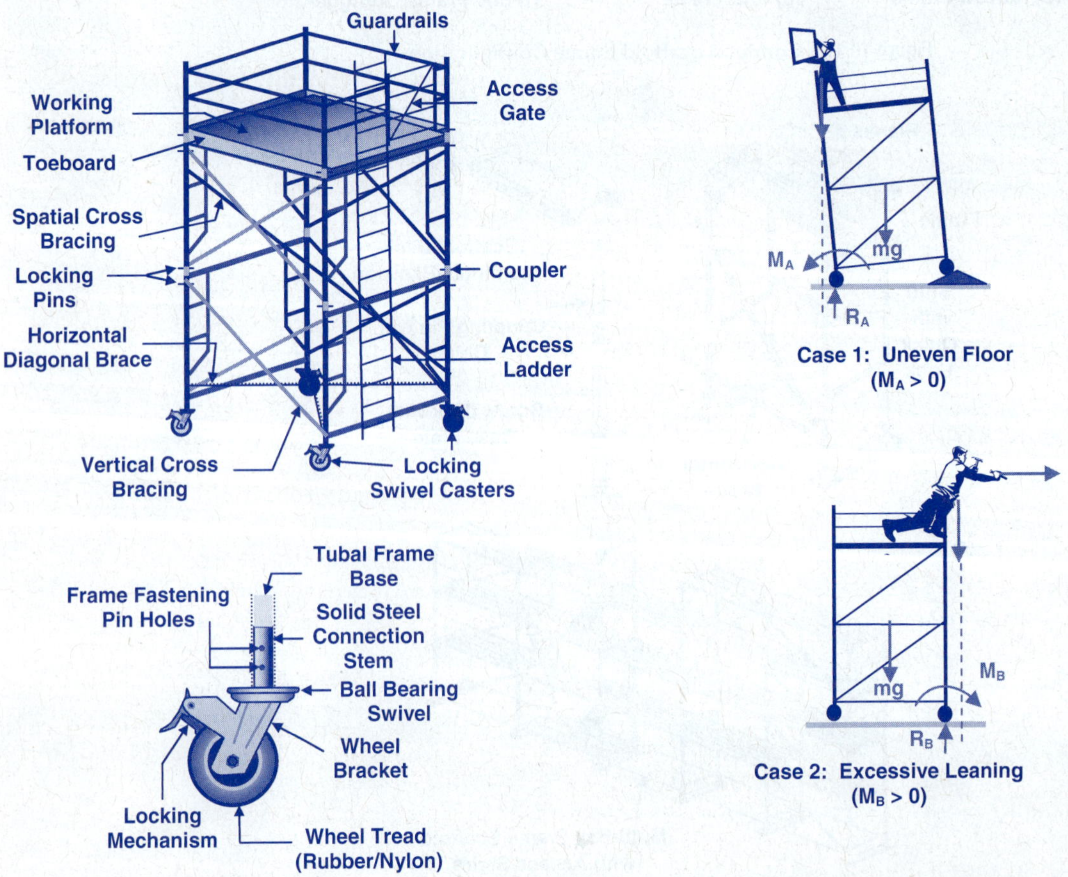

Figure 10.19 Structural Elements and Force Diagrams of Common Turn-Over Accidents with Mobile Scaffolds

Similar to shoring, wheeled scaffolds are organized into three classes according to their capacities:

Light duty: Up to 2.2 kN (496 lb) per full-width platform suitable for painting, electrical work, and other light tasks.
Medium duty: Up to 4.41 kN (992 lb) per full-width platform per bay.
Heavy duty: Up to 6.6 kN (1,488 lb) per full-width platform per bay. This is what is generally needed for bricklaying, concreting, demolition work, and most other work tasks involving heavy loads or heavy impact forces.

These safe load limits include the weight of people, taken to be a nominal 176 lb (0.78 kN) per person, plus the weight of any materials, tools, and debris on the platform.

Therefore, a properly constructed scaffold with a light duty working platform can safely support 1 worker and 320 lb (1.4 kN) of tools and materials, or 2 workers and 143 lb (0.64 kN) of tools and materials.

> **DOs and DON'Ts for Operating a Mobile Platform**
>
> Review *OSHA*'s General Requirements for Scaffolds *1926.451*.
> Ensure that the surface on which the scaffold rolls is level and without holes or obstructions.
> Brace all rolling scaffolds horizontally and diagonally.
> Install the appropriate guardrails.
> Prevent joints from separating by inserting and "locking" all the necessary bolts and pins.
> Cleat or secure all planks or plywood sheets.
> Fasten access ladders to the scaffold.
> Ensure that each wheel or caster is equipped with brakes.
> Lock the caster brakes before climbing onto scaffold.
> Tie down or remove all material, equipment, and personnel from platform before moving it.
> Do not stay on the scaffold when it is being moved.
> Do not allow the working platform height to exceed three times the base width, unless it is guyed and equipped with outriggers or otherwise stabilized.
> Do not bridge between separate rolling towers.
> Do not pull or push rolling towers from the top, push or pull only at the base level.
> Do not stand on platform overhangs. Stand only within the platform area; do not try to extend work area by leaning out over guardrail.
> Do not extend working heights by standing on planked guard rails, boxes, ladders, or other materials.

10.4.6 CLIMBING SCAFFOLDS

Masons depend on strong scaffolds, especially masons laying heavy concrete blocks referred to as **concrete masonry units** (CMU). A typical load-bearing concrete block weighs 38–43 lb (0.17–0.19 kN), which puts a lot of stress not only on the scaffold but also on the mason's back. The increased use of the CMUs in the 1950s and 1960s led to another critical innovation that had a synergistic impact on both safety and productivity, namely the elevating-adjustable or climbing scaffold. This type of scaffold allowed the platform to be continuously adjusted upward along preinstalled poles or towers. The immediate beneficiaries of this technology were the brick or block masons who were able to work continuously at a comfortable waist-high level, as the platform could keep pace with the growing wall. This, in turn, had a first dramatic impact on laying productivity because the amount of bending and lifting needed to laying the block was minimized, reducing fatigue. A second significant productivity booster was not having to interrupt work in order to add another scaffold level and to relocate tools and boards. The combined effect of the adjusting capability was estimated to increase productivity by 20%.

Pump Jack Scaffolds

One of the early systems that supported pole climbing was called the pump jack scaffold. The key elements of this system consist of vertical poles that are braced to a solid structure on the top and a pump jack hoisting mechanism embracing the poles. While the poles are made of straight-grained solid lumber, the foot operated pump jack, support braces and the platform brackets are made of steel. Figure 10.20 shows the original as well as the present aluminum-based concept. As shown, the scaffold consists of two elevations, the platform for standing and the work bench to keep tools and/or material. The common maximum working load is 500 lb (2.2 kN), but the scaffold is capable of supporting at least four times the maximum intended load without failure. Like all the other adjustable scaffold systems, the pump jack hoisting allows the work platform to be easily raised or lowered, keeping the top of the masonry wall between knee and shoulder height. Furthermore, the material needed can be stored

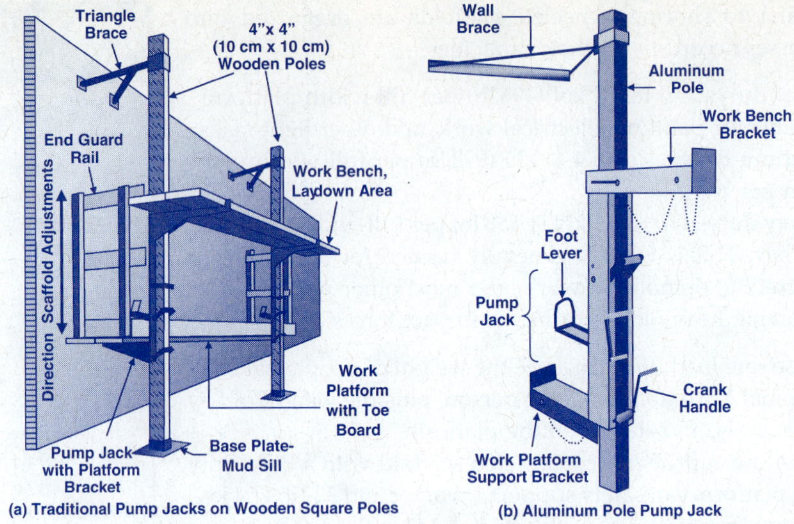

Figure 10.20 Components of a Pump Jack Scaffold Hoisting System

above the platform level on the work bench, eliminating the need for bending during material handling.

Workers often perform more than 1,000 forward-bending tasks per shift. Working at waist level reduces bricklayer fatigue and reduces the risk of back and shoulder problems associated with bending at low courses and reaching for courses above the chest (Figure 10.21). Studies have reported a decrease in lower back pain and stress among those working with materials kept at or near waist height. As mentioned earlier, the work platform and bench, including the stored material, climb together. This further reduces the risk of developing work-related lower back pain by decreasing heavy lifting and manual material handling, and reduces the risk of injury from falling objects. Despite all the many benefits, the pump jack poles require a triangular brace on the top, which, in turn depends on the availability of a supporting structure. In many large single-story projects, such a structure may not be available. This gave rise to a freestanding scaffolding system: the tower or mast climbers.

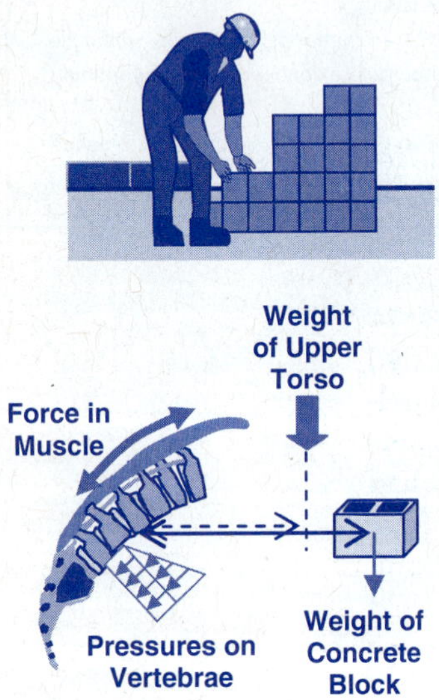

Figure 10.21 Load on Back

Powered Tower and Mast Climbers

Many tower or mast climbers are mounted on a trailer that is easily deployable. Outriggers are setup on solid mud sills before the tower base can be leveled. Both towers and mast climbers come either with manual or powered hoisting mechanisms, commonly a cable and winch system. For jobs higher than 30 feet, mast climbers usually need to be braced to a building; otherwise, they can be used as freestanding structures. Mast climbers are useful for tall and short structures. Their modular design allows contractors to tailor the platform to most building shapes. Figure 10.22 presents three basic configurations, where, similar to the pump jack scaffold, the platform can be equipped with a lower level of scaffold planking that projects toward the building face as well as a higher main materials platform. Alternatively, the work platform of a mast climber can be modified to include a raised area for mortarboards and bricks.

As one would expect, the work load capacities vary depending on the strength of the structure, especially the towers and masts. Some of the bigger platforms are able to lift 6,500 lb (28.9 kN). So called rack-and-pinion lifts, depicted in Figure 10.21b and c, replaced the hand-cranks with an electric motor. The freestanding mast climbers can be

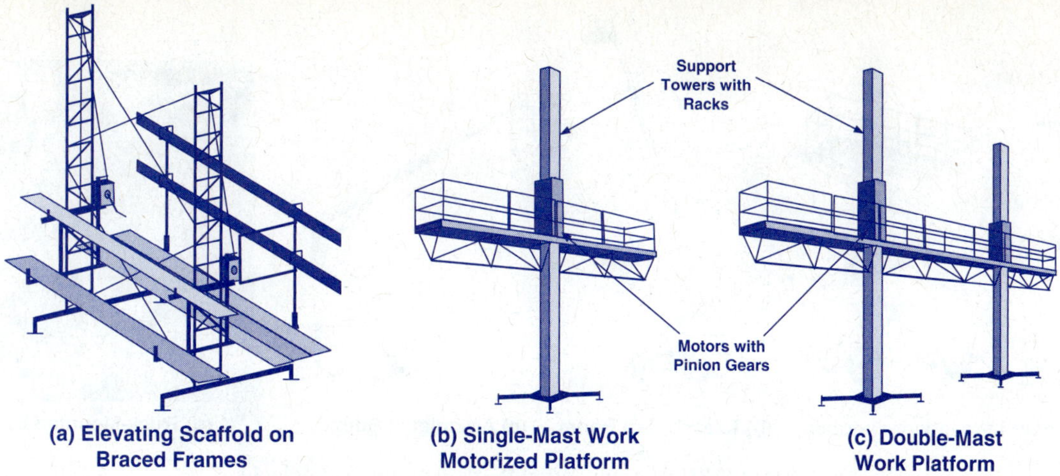

Figure 10.22 Powered Climbing Work Platform Technologies

extended to contour the face of the building with platform lengths for single masts reaching 63 ft (19.2 m).

10.4.7 AERIAL WORK PLATFORMS

Aerial work platforms combine the technologies and benefits of the mobile/rolling with the climbing scaffold. As Figure 10.23 indicates, aerial work platforms are self-propelled wheeled devices equipped with (a) a power-plant, (b) an extendable lifting mechanism, and (c) a work platform.

As the different undercarriages or bases indicate, such systems are designed for a range of conditions, which opens the door for mistakes. For example, some scissor lifts are specifically meant to be used on horizontal concrete slabs. However, many accidents report indicate that such devices have been deployed outside on rough terrain. Another invitation to abuse is to load or extend the platform beyond capacity or its tipping moment. Still another highly dangerous practice is to extend the reach of the platform by equipping it with ladders or planks, which again can cause the system to tip.

Scissor-Type Work Platforms

Scissors are an interesting mechanical device that can be found in many lecture books on mechanics. The platform is raised by extending one (or two) hydraulic cylinders connected to two linkages, as depicted in Figure 10.23a. Naturally, this type of equipment operates only along the vertical axis and thus is very susceptible to tipping if the ground is unstable. Indoor units are commonly equipped with DC battery power, eliminating the exhausts from gas or diesel engines, and solid rubber tires. Outdoor or rough-terrain units are built on a much larger undercarriage with a wider wheel base and pneumatic tires. They usually are powered by an internal combustion engine. Finally, scissor lifts are manufactured in different sizes and range in capacity from 500 lb (2.2 kN) to several thousand lb.

Boom-Supported Platform

This highly flexible type of equipment has become as ubiquitous on construction sites as hard hats. Self-propelled and mounted on a rough-terrain undercarriage with lots of counterweight—or even a truck with outriggers—the boom extensions allow an operator to reach almost any height. Unlike scissor lifts, however, they are generally not designed

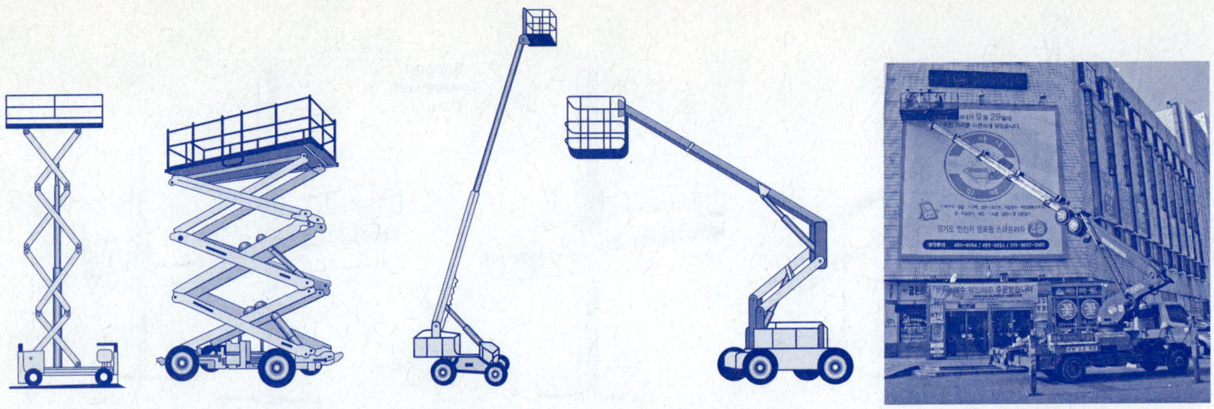

(a) Scissor Lifts Indoor-Outdoor (b) Telescoping Boom (c) Articulated Boom (d) Truck-Mounted Boom

Figure 10.23 Various Aerial Work Platform Technologies

to lift heavy loads in addition to two workers with a combined weight of 500 lb (2.2 kN) as the platform reaches beyond the wheel base. Figures 10.23b and c depict the two main mechanisms found on boom-supported aerial platforms, the telescopic and the articulated boom, which sometimes are combined in one unit and mounted on a truck as in Figure 10.23d.

Stability and Tipping

The safe operation of a boom-supported aerial platform requires a basic understanding of the effect of both static and dynamic forces on the equipment. Nevertheless, the most basic requirement is clean, even, horizontal, and solid ground for the wheels to stand on. Figure 10.24a shows forces that provide stability and forces that are able to overturn the system. Like any crane (see Chapter 7), a central issue is the location of the tipping point. In the cross-sectional view, that point is between the center of the front wheel and the surface it is standing on. It is not so much the total of the forces that is the deciding factor but rather the moment (**Moment = Force × Distance to perpendicular force vector**) that those forces create around the tipping point. As long as the moment created by the forces that stabilize the system are larger than those that tend to tip it, nothing will happen. Thus we can establish the basic function establishing the critical tipping moment (M_{TP}):

$$M_{TP} = M_{(Stability)} - M_{(Boom\text{-}Load)} = 0$$
$$M_{(Stability)} = (\text{Counterweight} * D_C) + (\text{Weight of Undercarriage} * D_U)$$
$$M_{(Boom\text{-}Load)} = (\text{Weight of Boom} * D_B) + (\text{Weight of Load} * D_L)$$

It is apparent that one has to avoid letting the M_{TP} reach 0. As a matter of fact, manufacturers provide a capacity chart or safe-reach envelope that include a safety factor that considers the overall risk. If humans are riding on the work platform, the Factor of Safety (F_S) is generally between 1.5 and 2.0. This translates into the following condition:

$$\text{Factor of Safety (FS)} = \frac{M_{(Stability)}}{M_{(Boom-Load)}}$$

Unfortunately, there are additional forces that can impact the equipment. We distinguish between the static and dynamic forces. The forces in Figure 10.24a are considered static as they are caused by the static acceleration g, called gravity.

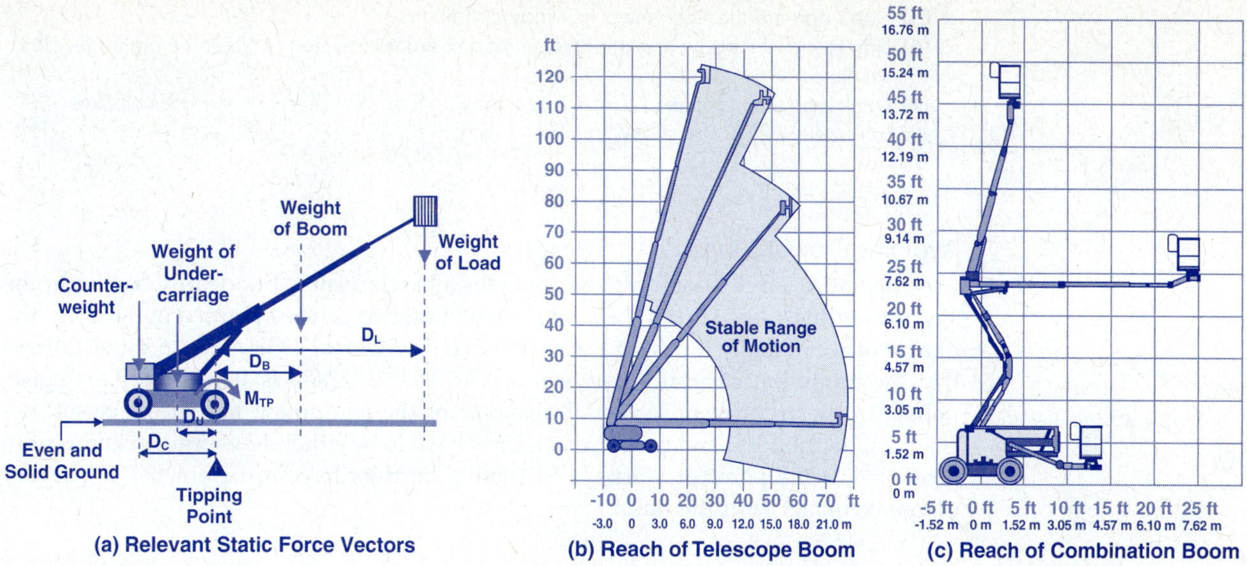

Figure 10.24 Stable Range of Operation

Any time that the boom experiences additional accelerations, additional forces are created (Force = Acceleration $*$ Mass), causing significant additions. Causes of dangerous dynamic forces include

- Jerking movement of the unit or parts of the unit when elevated
- Sudden stopping when elevated
- Abrupt release after a component got stuck on an obstacle (possible whipping)
- Dropping one wheel into a hole in the floor surface
- Operating the equipment in windy conditions

Most of the advice listed in operating manuals is related to avoiding tipping. Here is a partial list:

Absolute DOs for Boom-Supported Work Platforms:

(1) Check tires for proper pressure.
(2) Check hydraulic hoses, lift cylinder(s), and connections for leaks or loose connections.
(3) Inspect all structural components for damage, broken parts, cracks in welds, including outrigger arms and pads.
(4) Secure loads and tools on the platform so that machine movement won't dislodge them.
(5) Make sure that extension cords are long enough for the full platform height and won't get stuck, pinched, or severed.
(6) Wear a full body harness and tie off to a designated tie-off point while moving.
(7) Ensure that guardrail gate is securely closed before moving the platform.
(8) Deploy stabilizers or outriggers according to the manufacturer's instructions.
(9) Position the boom in the direction of travel where possible.

Absolute DON'Ts for Boom-Supported Work Platforms:

(1) Don't remove guardrails while the platform is raised.
(2) Don't access the platform by walking on the boom.
(3) Don't try to push or move the machine by telescoping the boom.
(4) Don't use a boom-supported platform as a crane.

(5) Don't operate the equipment in windy conditions.
(6) Don't place the boom or platform against any structure to steady either the platform or the structure.
(7) Don't load the platform above its rated working load (RWL). Wherever possible, keep the load below 2/3 of the RWL.

Avoiding Power Lines

Last, operating large booms always includes a barely visible but deadly barrier in the sky, power lines. As many of the aerial boom platforms can be operated in the open, the danger of accidentally touching hot power lines is very real. OSHA §1926.550 stipulates that the minimum clearance between for 50 kV and below is 10 ft (3 m). For higher voltages, this distance increases. If any part of the equipment has to get closer, the electrical distribution and transmission lines have to be either de-energized and visibly grounded at the point of work, or insulating barriers have to be erected to prevent physical contact with the lines.

OSHA Tips for Working Around Overhead Power Lines (Refer also to Figure 10.25)

✓ Don't operate equipment around overhead power lines unless you are authorized and trained to do so.
✓ If an object (scaffolds, crane, etc.) must be moved in the area of overhead power lines, appoint a competent worker whose sole responsibility is to observe the clearance between the power lines and the object. Warn others if the minimum distance is not maintained.
✓ Never touch an overhead line if it has been brought down by machinery or has fallen. Never assume lines are dead.
✓ When a machine is in contact with an overhead line, DO NOT allow anyone to come near or touch the machine. Stay away from the machine and summon outside assistance.
✓ Also, never touch a person who is in contact with a live power line.
✓ Be trained in cardiopulmonary resuscitation (CPR).
✓ If you should be in a vehicle that is in contact with an overhead power line, DON'T LEAVE THE VEHICLE. As long as you stay inside and avoid touching metal on the vehicle, you may avoid an electrical hazard. If you need to get out to summon help or because of fire, jump out without touching any wires or the machine, keep your feet together, and hop to safety.
✓ When mechanical equipment is being operated near overhead power lines, employees standing on the ground may not contact the equipment unless it is located so that the required clearance cannot be violated even at the maximum reach of the equipment.
✓ To maximize his or her own safety, an employee should always use tools that work properly. Tools must be inspected before use, and those found questionable removed from service and properly tagged. Tools and other equipment should be regularly maintained. Inadequate maintenance can cause equipment to deteriorate, resulting in an unsafe condition.
✓ Tools that are used by employees to handle energized conductors must be designed and constructed to withstand the voltages and stresses to which they are exposed.

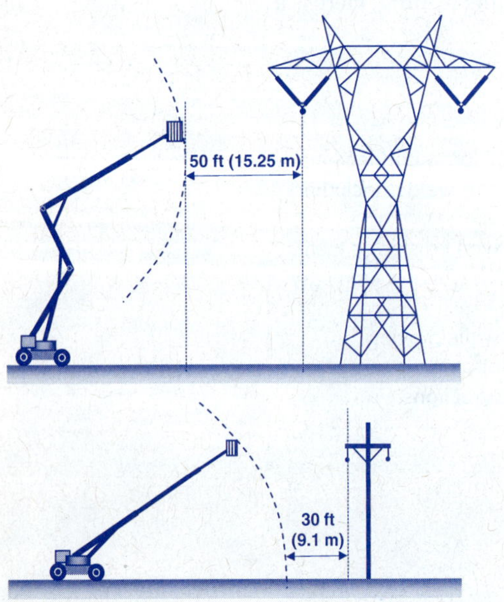

Figure 10.25 Keeping Clear of High Voltage Power Line

- ✓ Use the personal protective equipment appropriate for the job that is performed. This equipment may consist of rubber insulating gloves, hoods, sleeves, matting, blankets, and so on. These items must be inspected prior to each use and tested annually.
- ✓ When working near overhead power lines, the use of non-conductive wooden or fiberglass ladders is recommended. Aluminum ladders and metal scaffolds or frames are efficient conductors of electricity.
- ✓ Avoid storing materials under or near overhead power lines.

10.5 The Large World of Concrete Formwork

The success story of concrete as one of the preferred construction materials in the twentieth century is linked to the advances in cost-effective and high-quality formwork. In other words, the success of concrete would have been impossible without the tremendous progress that formwork has made. How come? As the term implies, formwork has two functions: (1) **form**ing the concrete and (2) providing the false**work**. Because of the large masses and high density of concrete used to build walls, decks, tunnel walls, bridges, and so on, keeping the formwork from deforming under its weight represents the key objective for both design and fabrication. The consequences of failure to do so are costly, as the hardened concrete requires time-consuming work if it has to be removed and redone. Furthermore, the forms are also responsible for leaving an aesthetically pleasing surface in areas that will not be covered. As a result of the large differences in conditions where concrete is used, a virtually infinite number of formwork types and accessories have been developed over time. For this reason, this section is only able to provide a limited insight into the main categories and uses of formwork.

10.5.1 MAIN FORMWORK SYSTEMS AND THEIR APPLICATIONS

The literature offers various approaches to grouping formwork, to include the main material used to build them (e.g., timber), the location of fabrication, the sizes of elements, etc. In this book, we distinguish among five main categories of formwork technologies defined in Table 10.5. As indicated in Table 10.6, each of the five major formwork technologies is used for more than one area of application.

S10.7 Deck Formwork

Table 10.5 Overview of Main Formwork Technologies

Formwork Technology	Main Features
1. Traditional Cut-to-Fit Formwork	Predominant until the 1970s and very flexible. Solid wood, plywood, or particle board is cut and joined to establish the mold. Still used today where other systems are not flexible enough to fit the required design.
2. Engineered Modular Formwork	Pre-engineered and prefabricated modules of small- and large-size panels with reusable sheathing and the necessary support structure. Supplied in various sizes, panels are easily joined into tables or walls using aligning connectors, clamps, props, walers, and braces.
3. Mechanized Formwork	Pre-engineered and prefabricated formwork actuated by mechanical systems and able to move on their own from one pour to another.
4. Stay-In-Place Formwork	Prefabricated panels made of different throwaway materials that are able to support the wet concrete. Examples are hollow tubes made of plastic, corrugated sheet-metal, or slim precast elements. After concrete is cured the formwork is not removed (stay-in-place).
5. Pre-Cast Steel Formwork	Fixed but flexible casting systems that are reusably made of steel. Forms mounted on mechanized tables and serviced by gantry cranes.

Table 10.6 Overview of Most Common Formwork Systems Today

		Traditional	Pre-Engineered-Modular	Mechanized Vertical	Mechanized Horizontal	Stay in Place	Pre-Cast Steel Forms
1.	Walls						
2.	Columns						
3.	Continuous Shafts						
4.	Building-Deck/Slab						
5.	Bridge-Deck						
6.	Bridge Piers						
7.	Tunnel Walls						
8.	Dams						
9.	Road-Tarmac						
10.	Road Curbs						
11.	Pipes						

10.5.2 UNDERSTANDING THE MASSIVE FORCES ACTING ON FORMWORK

As discussed earlier, shoring supporting the formwork has to be able to sustain both dead and live loads. For example, the average unit weight of the reinforced concrete is 158 lb/ft^3 (25 kN/m^3), the dead load for slab formwork is 10 lb/ft^2 (0.48 kN/m^2), and the live load due to workers and the dropping of concrete onto the formwork is generally given as 20 lb/ft^2 (2.4 kN/m^2). For walls, however, these simple distributed load assumptions are not sufficient.

Before reviewing the acting forces, we should first understand how the concrete and the formwork interact (Figure 10.26). While the wooden plywood panels can be replaced with steel or even thin precast concrete panels, the force models for horizontal slabs are

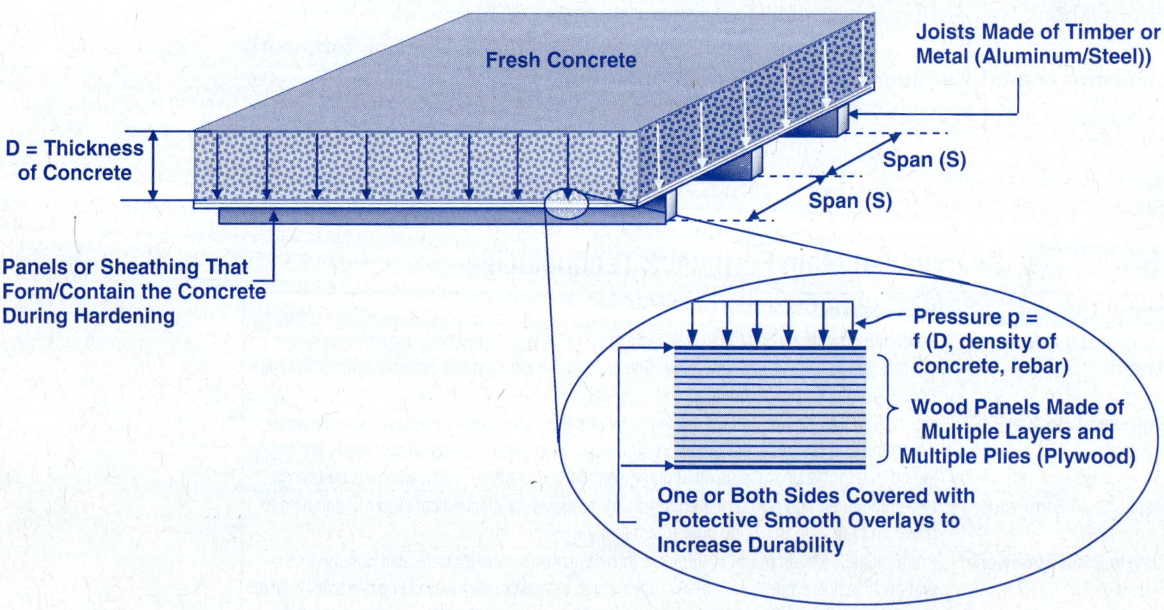

Figure 10.26 Concrete-Formwork Force Interaction for Common Horizontal Wood Panels

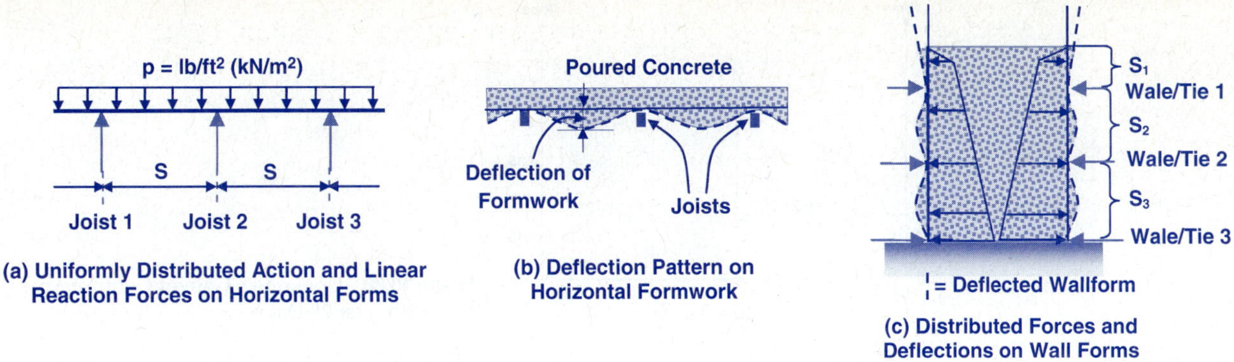

Figure 10.27 Concrete Pressure Causing Deflections of Formwork

the same while vertical wall forms are additionally exposed to hydrostatic pressures of the fresh semi-liquid concrete.

The drawings in Figure 10.27 show that several interdependent issues have to be considered when designing or selecting formwork for a particular concrete operation. In an iterative design process one can either begin with a preferred joist size, a form, or a span width S. One needs to consider following variables:

(1) Safety factor
(2) Weight of concrete, rebar, and forms on joists = (p ∗ S) lb/ft (kN/m)
(3) Live-load to be expected during concrete placement (e.g., local dropping of concrete batch)
(4) Acceptable deflections between joists
(5) Type of form panel (wood, steel, aluminum)
(6) Stiffness of forms in the direction of the spans (e.g., number and orientation of wood plies)
(7) Max. load capacity on joists due to shoring limitations (stringers, posts, frame shores)
(8) Allowable load capacity of (selected) joist
(9) (Desired) span S between the joists

For walls and columns, as shown in Figure 10.27c, the hydrostatic effect of wet/liquid concrete has to be considered in the design of the ties and wales. But there are other significant factors that put pressure onto the forms. They include such things as (a) vibration of concrete, (b) consistency of concrete, (c) method of filling forms, and (d) aggregate mix and additives.

> The American Concrete Institute (ACI) provides extensive and up-to-date information about formwork design standards that are continually being improved (see ACI 301, ACI 317 and ACI 347). Students who are interested in learning more about sizing formwork are referred to this excellent resource (http://www.concrete.org).

10.5.3 ELEVATED CONCRETE SLAB FORMING

One of the most risky states of an elevated concrete slab is during the placement and finishing of fresh concrete because it needs to be supported until the cement has reached sufficient "binding strength" to support itself. Until then, the heavy concrete depends on the temporary support provided by the formwork, by the shoring below, and ultimately by a solid foundation for the shores and props such as an already completed floor in a building or the ground.

A most unfortunate characteristic of concrete is the slow development of its compressive strength, a main factor for a concrete slab to carry itself let alone additional loads originating at the formwork at a higher level. (See Figure 10.28.) While the engineer creates designs using a compressive strength of mature concrete, the same concrete after 4 days is

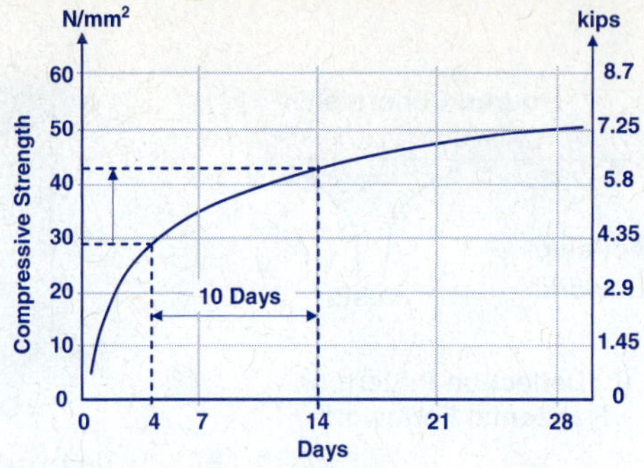

Figure 10.28 Concrete's Gain of Compressive Strength over 28 Days

maybe 40% of mature strength. Even this estimate is a maybe, since the ambient temperature, protection, and treatment of the concrete play a significant role in its development. A floor that has been poured 7 or 10 days prior to pouring concrete onto a formwork above might not be able to carry any additional load. It is therefore necessary to transfer some of the load to a floor further below, a process that is called reshoring. Consisting of either single posts or towers, the process transfers loads to lower and therefore older slabs of concrete, which in turn might be reshored as well. Preshoring, also known as backshoring, is used when the formwork is removed from the freshly placed concrete before the concrete has achieved sufficient strength to support itself totally. Vertical supports are installed through the formwork while the formwork is carefully removed so it can be reused on the next-higher floor, utilizing new shores. Similar to the shoring technology discussed in the previous section, the formwork for elevated slabs has made significant progress due to the increasing cost of labor and ever-higher buildings with identical floor layouts in each floor. The obvious trend was to replace labor-intensive tasks such as cutting, nailing, fitting, and stripping of joists and sheathing with technologies that were more costly but time-savers for crews doing the work.

Singapore: *On 27 June 1999 at about 10:45 P.M. the formwork for a ramp on the third level of a parking garage collapsed as the workers were pouring concrete. In previous cases involving collapses of formworks, the Ministry of Manpower had found that one of the most common causes of such accidents was inadequate bracing.*

$101 Million Settlement in Deadly Parking Garage Collapse That Killed Five Workers

In April 2004, the U.S. Occupational Safety and Health Administration (OSHA) issued citations to and fined four contractors after its investigation revealed that the construction site—and particularly the installation of the crucial floor-to-wall elements—was an accident waiting to happen. One of the citations charged . . . failure to erect, support, and maintain steel formwork capable of supporting all possible vertical and lateral loads.

Failure of Cold-Formed Steel Beams During Concrete Placement

During concrete placement on the second story of a building under construction, the supporting cold-formed steel beams collapsed. Four workers were injured. The collapse occurred while concrete was being placed onto steel decking on the second floor of the structure.

The complexity involved in designing, building, and maintaining shores and reshores that are strong enough to stand up to the many shifting loading conditions during the pour of fresh concrete and a solid understanding of safe methods and technologies are most important. The inserted short accident reports should provide the necessary impetus to reread this section extra carefully. Figure 10.29 introduces several such inventions that have found rapid adoption by contractors.

Formwork Tables

These systems consist of pre-engineered and pre-assembled modules of shores, posts, frames, cross-bracings, stringers, joists, and sheathing that can be transported as a unit. On multiple-story buildings that maintain the same floor height, not having to dismantle and reconstruct the formwork piecemeal will lead to substantial productivity increases. The assembled sections are lowered using jack-screws, equipped with casters for horizontal movements, lifted by elevator or "flown" with the help of a crane fork from one level to the next. Once in position, the gaps between the tables or table and wall are bridged with "fillers."

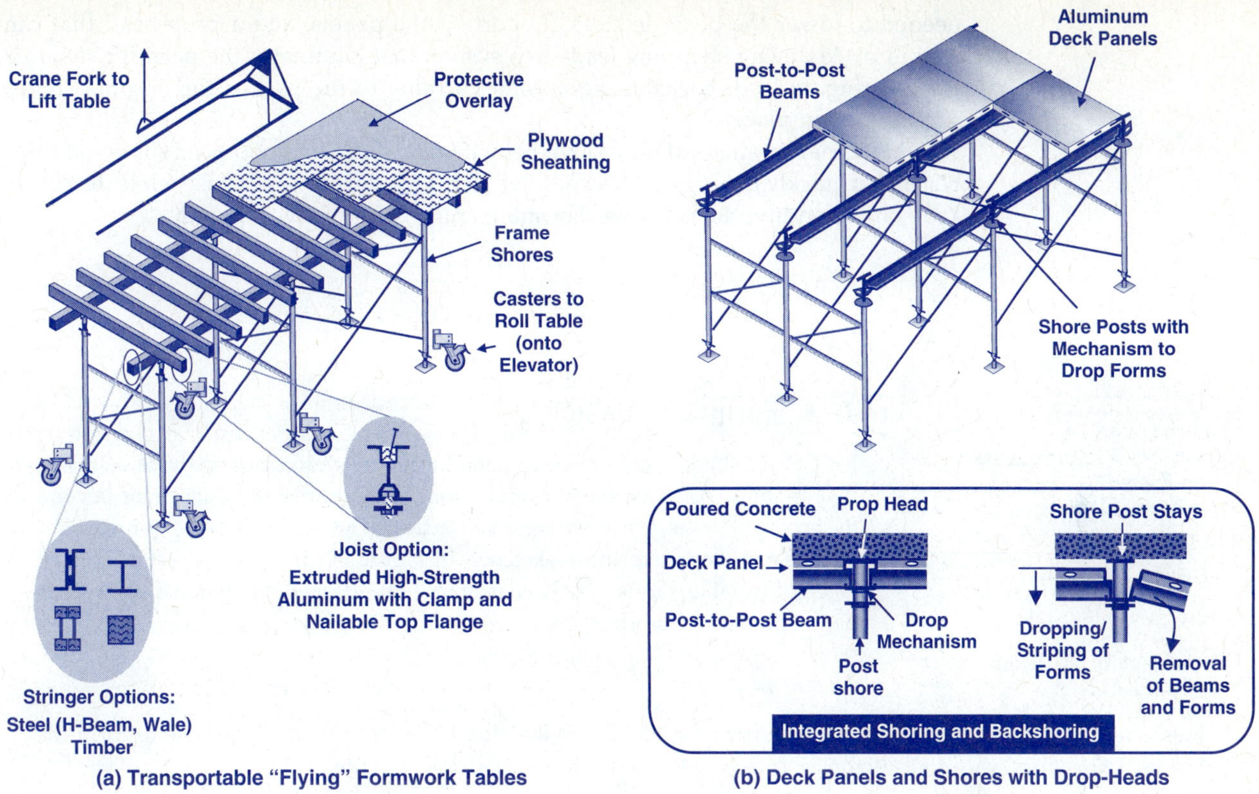

Figure 10.29 Two Highly Productive Elevated Slab Formwork Systems

Figure 10.29a indicates that both stringers and joists are made of various materials, including timber, wooden beams or trusses, steel, and aluminum, and temporarily connected to each other using bolts and clamps as they have to be lifted as a whole. Again, the most common sheathing is plywood, but steel and fiberglass are also being used.

Not surprisingly, the tables moved by the ubiquitous cranes are nicknamed "flying" tables as they are flown from deck to deck. In order to pick up ever-larger table sizes, the use of a fork is essential. Newer systems combine use of the fork and casters to pick up extremely long tables from within a building.

Panel Forms with Drop- and Prop-Head Shores

This innovative concept synergizes the advantages of lightweight and low-cost aluminum with the demands for concrete formwork to provide structural support for an extended period of time while being of productive use. One important step was the introduction of aluminum panels of a size that can be carried by one laborer, at least when clean. Figure 10.29b highlights that the panels are of sufficient strength to allow the elimination of the joists, which contributed significantly to the ease of stripping after the concrete has cured. A second step was the replacement of the long stringers with shorter simple beam elements reaching from post to post. This added flexibility allows very small areas to be stripped at once by dropping four shore heads at a time, and the deck in that area to be immediately backshored. The third important advance is the introduction of the prop-head that can be inserted into the top of prop-shores. From Figure 10.29b, we can see that the combination of a drop-head, needed to lower the beam-ledgers supporting the panels, and a prop-head that can stay in place during stripping leads to a system that eliminates the need for separate backshoring. Indeed, backshoring becomes a thing of the past as the original shores simply stay in place.

In summary, panelized forms together with drop-head and pop-head shores offer the benefits of quickly freeing up the expensive forms for cleaning and reuse while the shores keep on supporting the concrete slab still gaining compressive strength.

Worked-Out Problem 10.3 / Formwork Design

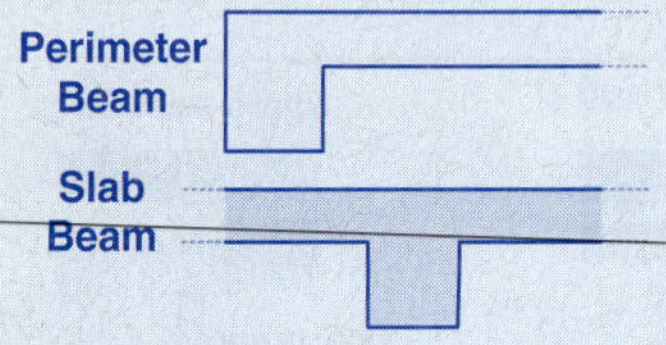

Figure 10.30 Beam Forms

Irregularly shaped slabs challenge the formwork designer to provide solutions that are both quick to assemble and strong enough to avoid deflections beyond the accepted limits. We are being asked to design a solution for the perimeter and slab beams similar to those sketched in Figure 10.30. The contractor decided to employ table forms for the flat areas. The beams run parallel as well as perpendicular to the stringers.

Assumptions

Frame tower shores, aluminum H-beam stringers, and short wooden H-beam joists are available to form the beams. Other hardware that might be necessary to form the beams can be rented.

Principles

1. The forms for the beams have to perform in unison when the concrete is being poured.
2. The shores for the beam will have to carry not only the concrete for the beam but also a portion of the slab.
3. The vertical form elements will be exposed to the hydrostatic effect of liquid concrete and require careful bracing of the bottom edges.

Design Step 1: Shoring Options (see Figure 10.31)

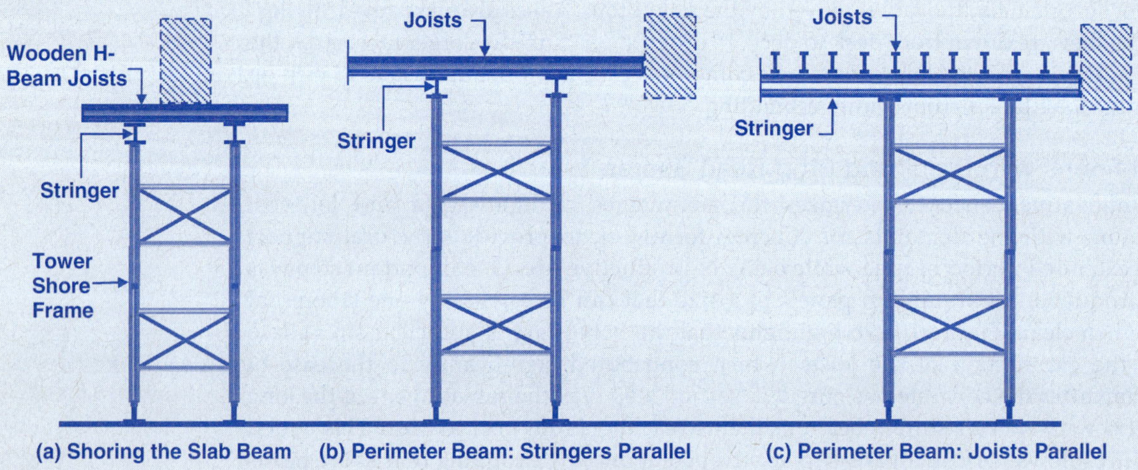

Figure 10.31 Beam Shoring Alternatives

Design Step 2: Forming the Beams Parallel to Stringers

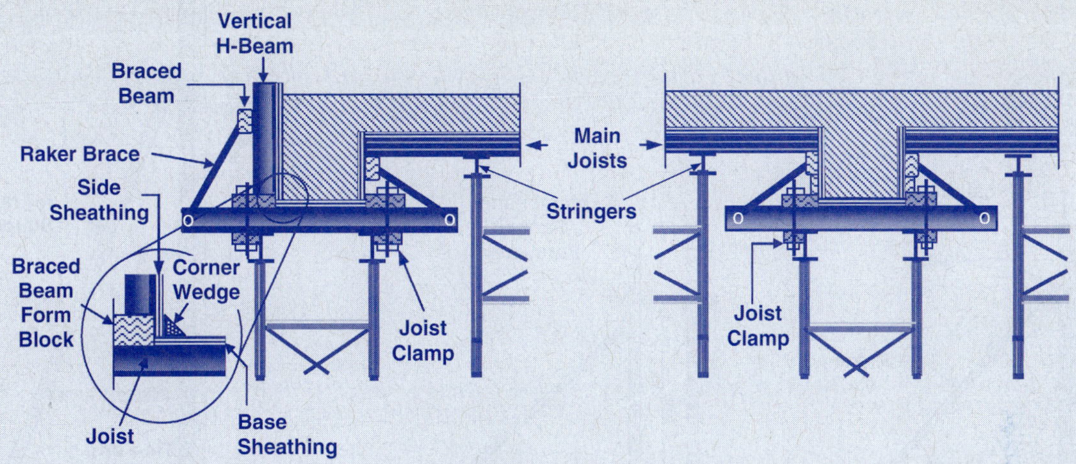

Figure 10.32 Form Designs for Beams Parallel to Stringers

Design Step 3: Forming the Beams Parallel to Joists

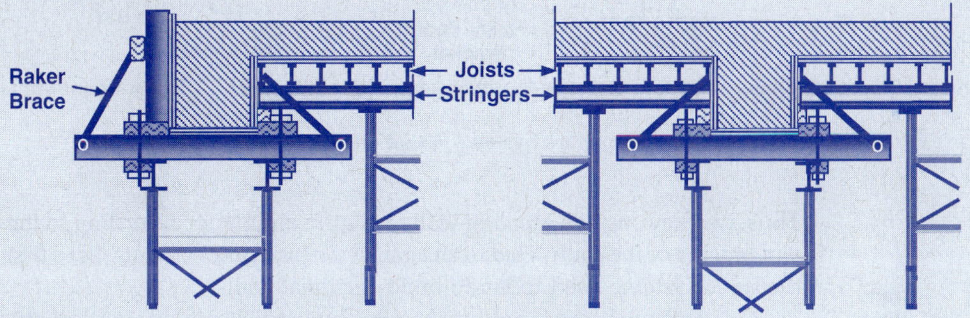

Figure 10.33 Form Designs for Beams Parallel to Joists

Discussion

Figures 10.32 and 10.33 indicate that the joists underneath the beams support both the side and the base sheathing of the beam forms. Two types of bracings are used to keep the vertical forms from bulging out. The bottom is kept in place by blocking beams on both sides. Raker braces provide support for horizontal beams and vertical H-beams. In order to avoid uneven settlement of the sheathing, it is critical that the stringers and joists be properly connected to the beam shoring systems.

10.5.4 SECTIONAL WALL AND COLUMN FORMS

The vertical orientation of concrete could lead to the misconception that walls are much simpler to form than elevated slabs. However, stories about collapsed wall forms are abundant, and the money spent on wall formwork can easily outstrip the cost for slab formwork. The following section will provide some basic principles for forming wall and columns.

The two sources of forces that impact walls that have to be considered more than anything else are the lateral pressures created by the fluid concrete and wind that can reach 0.21 psi (1.44 kPa) and is able to create extensive damage to unsecured wall forms.

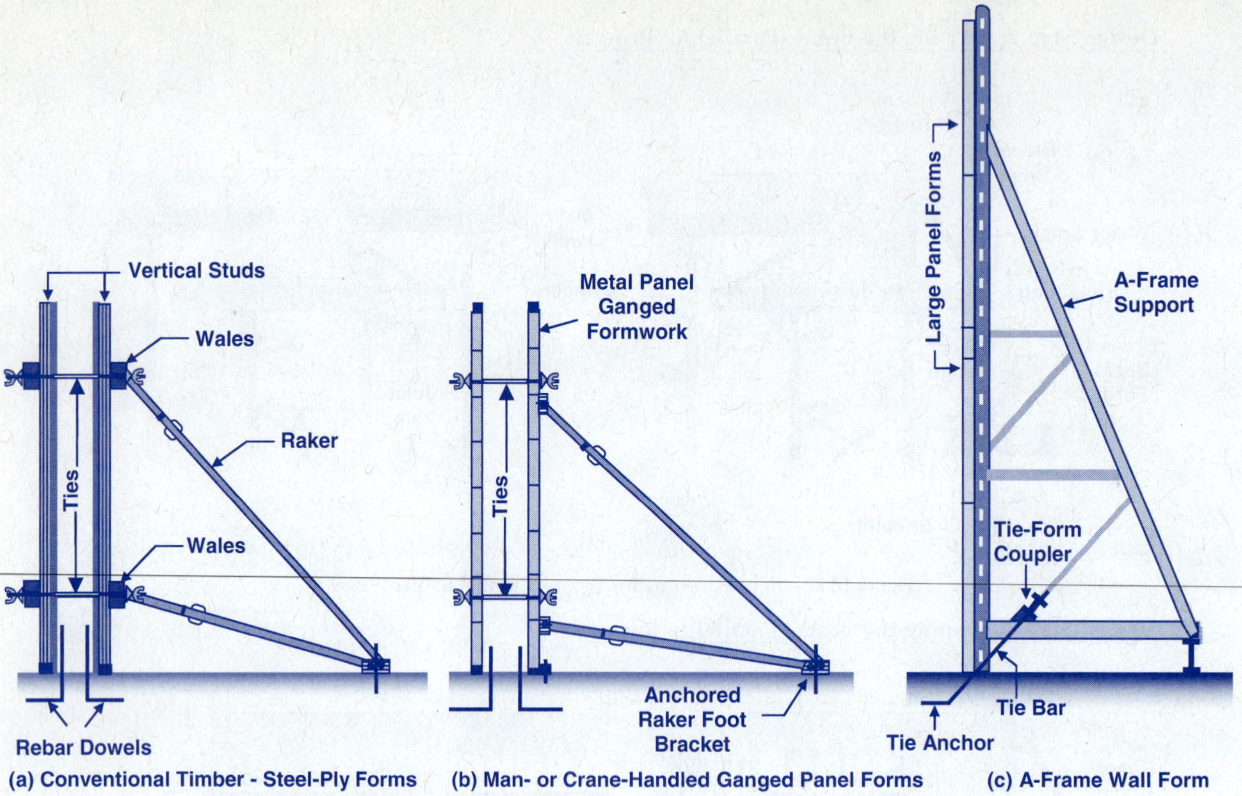

Figure 10.34 Common Wall Formwork Systems Found on Construction Sites

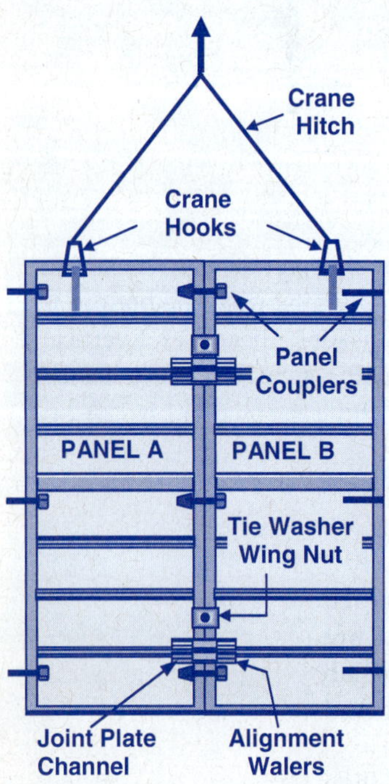

Figure 10.35 Crane-Handled Gang Panels

Thus, two key steps in erecting wall forms are alignment according to the design and bracing of the completed form against excessive movements. Let's first look at the basic systems used today to build sectional walls.

Conventional forms are build with plywood sheathing, vertical studs, and horizontal wales on both sides, which are kept from separating by metal ties, shown in Figure 10.36, that pass between the vertical forms. Figure 10.34a displays the need for rakers to align the forms vertically and to provide lateral support in case of wind. Assembled on the ground, the forms end up quite heavy, requiring the help of cranes for setting them up. Having a crane makes it possible to make the wales and beams out of steel, and, in addition, several forms can be ganged together on the ground and lifted in one piece, as illustrated in Figure 10.35. The length of a gang form for walls is often determined by the distance between control joints or by the capacity of the jobsite crane. Another advantage surfaces during stripping, cleaning, and resetting of the forms—all done in large ganged elements. Not having to strip and rejoin individual form components leads to significant time saving relative to conventional forms.

Similar to deck forms, the panelized forms are made out of lightweight aluminum which weighs between 7 and 15 lbs/ft^2 (0.33 kN/m^2 – 0.7 kN/m^2) or steel. Man-handled panels are smaller and naturally lighter but are able to withstand 9.96 psi (68.6 kPa) of lateral concrete pressure. Figure 10.35 illustrates how larger panels are ganged together using couplers, clamps, and aligners to be lifted by cranes. These sturdy aluminum forms can withstand concrete pressure up to 12 psi (82 kPa), while steel gang forms take as much as 18 psi (124 kPa).

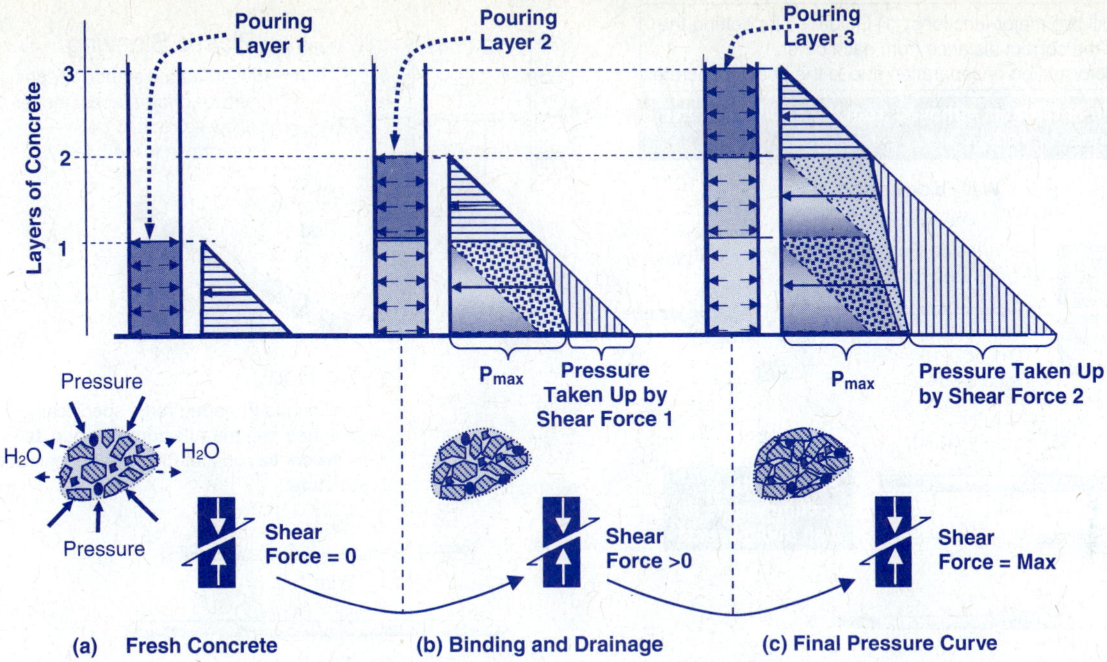

Figure 10.36 Simplified Model of Lateral Force Buildup Inside Wall Forms

As mentioned above, the consistency of fresh concrete causes it to exhibit hydrostatic behavior, meaning a linear increase of lateral pressure with increasing depth. The actual amount of pressure at the bottom of a high wall is of course of prime importance, since the forms and the braces used have to be able to sustain the developing forces. Since concrete does not behave like a true liquid, while other factors also play a role in the development of pressure, a series of experimental studies have been performed over time. They have found that many other factors impact the result: (a) rate of placing the concrete, (b) temperature of the concrete, (c) proportion of the concrete mix, (d) consistency of the concrete, (e) consolidation method of the concrete, (f) impact during placing, (g) size and shape of the formwork, (h) amount and distribution of the reinforcing steel, (i) unit weight of the concrete, (j) height of the concrete, (k) ambient temperature, (l) smoothness and permeability of the formwork, (m) pore water pressure, and (n) type of cement. One of the most important indicators is the pore-pressure of the setting concrete. For example, dry and highly permeable forms allow the water in the mix to drain slightly, thus increasing the coefficient of friction between the aggregates and hence the shear-strength of the concrete itself. The sequence of filling in 3 ft (.9 m) lifts of a 9 ft (2.7 m) concrete wall is modeled in Figure 10.36.

The effect described here depends on the reduction of the pore pressure in fresh concrete through lateral drainage mainly and possibly through some chemical binding during the filling process. Resulting from such a drainage process is the direct contact of more and more concrete aggregates, which in turn builds up the shear-strength of the concrete itself. Interestingly, while adding the concrete mass of the second layer increases the pore pressure in layer 1, its additional weight also leads to additional drainage, thus furthering its shear-strength. Finally, the maximum pressure P_{max} reached at the bottom end of the wall will be less than what would be possible under hydrostatic conditions. The apparent benefit of the drainage effect is that the formwork can be built with less structural support, and so the forms will be less heavy and less costly. Naturally, this will also affect the needed form ties shown in

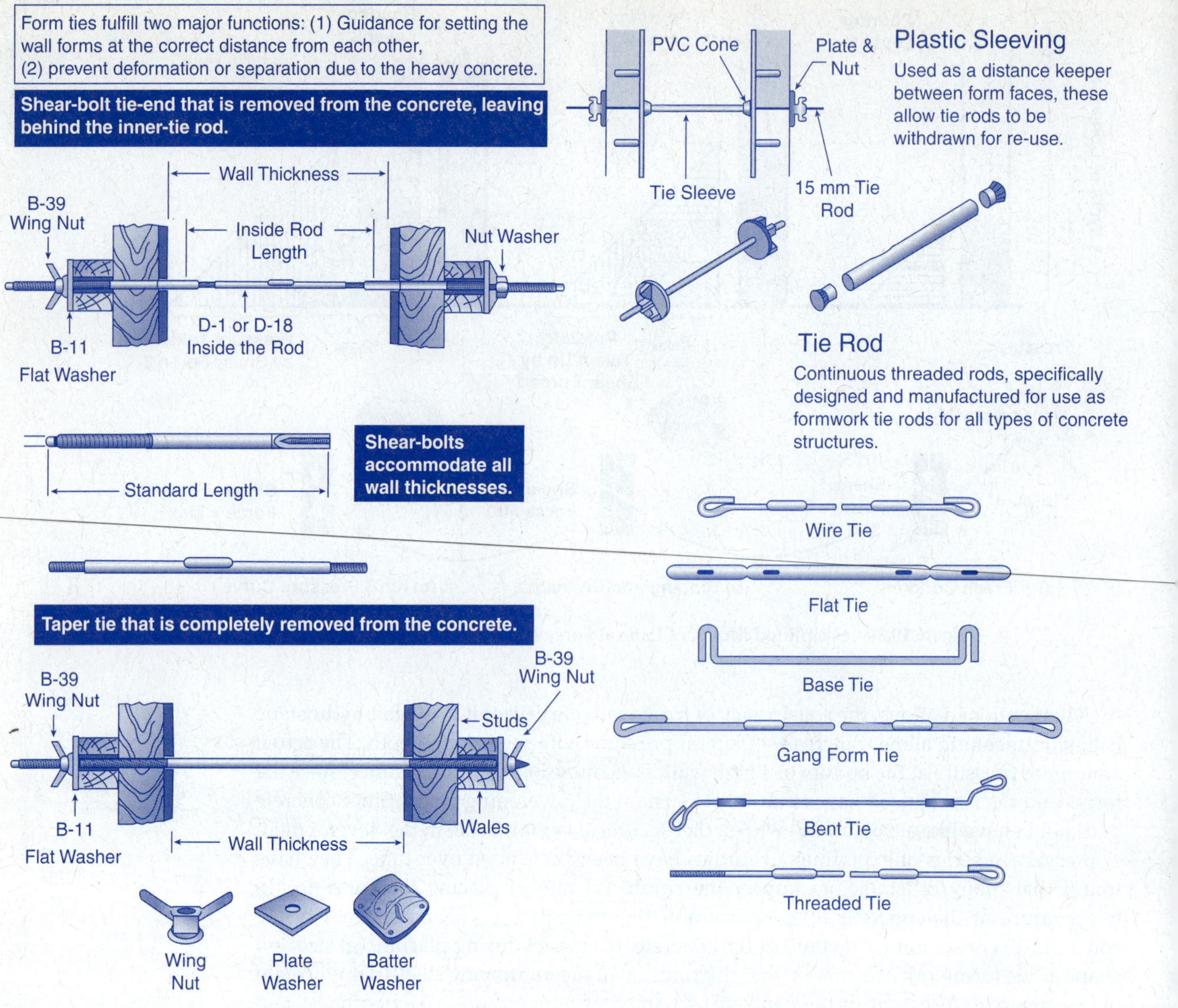

Figure 10.37 Design and Functions of Major Form Ties

Figure 10.37. It is apparent that there exists a large variety of ties or tie rods with washers and nuts that have to be supported by wales to distribute the large forces from the semi-liquid concrete supported by the vertical walls. Each design has a cost associated with its use and needs to be matched with the requirements of the job.

The attentive reader will observe that the discussion of wall forms was done from a qualitative point of view providing very little quantitative guidance. In fact, the inherent complexity of formwork is still the topic of research. Still, design standards and codes are available that cover most situations. One such important standard is **"ACI 347,"** a design guidance for **"Lateral Pressure on Wall Forms"** from the American Concrete Institute (ACI).

ACI 347-04 (see insert on the next page) gives two equations for wall form pressure applicable to two different wall heights and speeds of filling the formwork. However, both are applicable to concrete with a slump of 7 in (175 mm) or less and vibration to a depth of 4 ft (1.2 m) or less and the unit weight for concrete ($C_w = 1$) is 150 lb/ft^3 or 7.2 kN/m^3. Note that the pressure can never get less than 4.17 psi (28.7 kPa).

ACI 347-04: Guide to Formwork for Concrete

This guide offers ways to meet the specification requirements for concrete formwork!

Publisher: ACI (American Concrete Institute)

Objectives of safety, quality, and economy are given priority in these guidelines for formwork. A section on contract documents explains the kind and amount of specification guidance the engineer/architect should provide for the contractor. The remainder of the report advises the formwork engineer/contractor on the best ways to meet the specification requirements safely and economically. Separate chapters deal with design, construction, and materials for formwork. Considerations specific to architectural concrete are also outlined in a separate chapter. Other sections are devoted to formwork for bridges, shells, mass concrete, and underground work. The concluding chapter on formwork for special methods of construction includes slipforming, preplaced-aggregate concrete, tremie concrete, precast, and prestressed concrete. 32 pp.

The following equation applies to walls with a rate of placement less than 7 ft/hr (2.1 m/hr) and a placement height of 14 ft (4.2 m) or less:

$$\textbf{P-Lo-Wall:} \quad p_{max} = C_w C_c\,[150 + 9000\,R/T]\ \text{psf}$$

where

p_{max} = Maximum pressure in psf
C_w = Concrete unit weight coefficient
C_c = Chemistry coefficient
R = Rate of concrete pour (ft or m per hour)
T = Temperature of the concrete at the time of placing (degree Fahrenheit or Celsius)

Case 1

$T = 60\,°F\,(15.5\,°C), R = 6\,ft/hr\,(1.8\,m/hr), C_c$ for types I, II, III cement without retarder $= 1.0, h = 10\,ft\,(3\,m)$

$p_{max} = [150 + 9{,}000 * (6/60)] = 150 + 900 = 1{,}050\,psf\,(7.3\,psi\ or\ 50.3\,kPa)$

Adding retarder to the same cement extends the fluid state of the concrete and, because of the higher pore pressure, should result in a higher p_{max}. According to ACI 347-4, the value for C_c for types I, II, III cement with retarder = 1.2.

Case 2

$T = 60\,°F\,(15.5\,°C), R = 6\,ft/hr\,(1.8\,m/hr), C_c = 1.2$

$p_{max} = 1.2\,[150 + 9{,}000 * (6/60)] = 1.2 * 1050 = 1{,}260\,psf\,(8.75\,psi\ or\ 60.3\,kPa))$

The second equation, P-High-Wall, applies to all walls with a placement rate of 7 to 15 ft/h (2.1 to 4.5 m/h), and to walls placed at less than 7 ft/h (2.2 m/h), but having a placement height greater than 14 ft (4.2 m).

$$\textbf{P-High-Wall:} \quad p_{max} = C_w C_c\,[150 + 43{,}400/T + 2{,}800\,R/T]\ \text{psf}$$

Case 3

$T = 60\,°F\,(15.5\,°C), R = 6\,ft/hr\,(1.8\,m/hr), C_c = 1.0, h = 16$

$p_{max} = [150 + 43{,}400/60 + 2800\,(6/60)] = 150 + 723 + 280 = 1{,}153\,psf\,(8\,psi\ or\ 55.2\,kPa))$

> **The MOST Important Thing to Learn About Pouring Walls:**
>
> The two most critical issues that surface when reading articles concerning the safe pouring of concrete into wall forms (***safe = no destruction of the forms***) is the rate of pour (ft/hr) and the even placement along the wall and around corners. An uneven distribution leads to pressure concentrations. In addition, vibration will become difficult, leading to air pockets, aggregate segregation, or honeycombing (PUBLICATION #C850516, 1985).

10.5.5 MECHANIZED FORMWORK: VERTICAL

Different from sectional wall formwork, mechanized formwork is designed to move with the wall upward by attaching itself to the fast-setting concrete and using it as a "step-ladder." Today, we differentiate between two basic systems: (a) the jump/climbing and (b) the slip(ping) forms. The economic benefits of this concept contributed to the rise of concrete as a main construction material, and vice versa. Its capabilities is one of the main factors that allows the construction of super-tall buildings like the Burj Dubai. Its emergence as a key enabling technology, however, depended on many other elements coming together in a synergy—of course, all closely linked to reinforced concrete.

Possibly the first use of concrete to build high walls was in 1899 when Frank H. Peavy, called the "elevator king," hired the civil engineer Charles Haglin to design a cylindrical storage silo as an experiment on the outskirts of Minneapolis. Peavy was looking for a better way to protect grain from fire, as the silos had traditionally been made of metal-clad wooden planks. To everybody's surprise, the 12-inch (0.3 m) walls, tapering off to 8 in on the top, rose to 125 ft (38 m) without collapsing or breaking apart when full with grain. They still stand today having a respectable insider diameter of 20 ft (6 m). Since it was just an experiment, the structure was never really used as a grain silo, but the proven concept of the concrete grain elevator spawned a building boom that reached across the entire globe. The French architect LeCorbusier praised it as "the magnificent First Fruits of the new age."

For pouring of the construction, steel-braced round forms made of wood were used. The concrete was placed in layers. After the concrete of one layer had dried the forms were pulled off, lifted, realigned, and braced before concreting the next layer. This procedure pretty much describes today's use of jumping formwork. The 15-story Ingalls Building in Cincinnati, Ohio, is considered the first concrete "skyscraper". It was built in 1903 by A. O. Elzner using Ransome's patented reinforcing bars consisting of square, twisted steel bars inserted in concrete.

> **History of the "Climbing" Formwork**
>
> 1824—English mason Joseph Asplin patents his Portland cement
>
> 1849—French gardener Joseph Monier patents flower pots made of concrete with inserted wire nets
>
> 1884—Ernest Ransome patented twisted bars for greater bond
>
> 1899—Peavy-Haglen concrete silo
>
> 1903—15-story Ingalls Building with reinforced concrete structure
>
> 1908—American Concrete Institute published "Requirements for Reinforced Concrete or Concrete-Steel Constructed Buildings"

Slip forms and jump forms are the terms given to self-climbing formwork systems specifically intended to construct concrete walls and columns in high-rise buildings and other concrete structures, such as chimneys. In slip forms, the climbing is usually carried out continuously during the concrete pour. With jump forms, the climbing is done in steps, following the concrete pour. The term "climb form" is also used to describe a slip form or jump form. The power for the climbing operation can be provided in a variety of ways, but usually by means of hydraulic rams or electric motors connected to climbing feet or screw shafts.

Modern Jump or Climbing Forms

For buildings higher than five stories, the use of jump forms, shown in Figure 10.39, is more economical than conventional forming. One can find the two terms, *jump* and *climbing*, used interchangeably. However, the two terms are also combined to describe two sub-groups: (a) self-climbing and (b) crane-handled jump forms. Both systems comprise three main elements: (a) the steel structure attached to the already cured concrete providing the supports for (b) the forms for the next higher stage, (c) and several working platforms for the setting of the forms, rebar installation, pouring operation, and formwork stripping and cleaning. Systems are modular and can be joined to form long lengths, rectangular or circular shafts, or even varying construction geometries.

The following will describe the common features for each system, since both types have been modified by the various fabricators and vendors.

Crane-Handled Jump Forms The following discussion explains how such a system is used.

Figure 10.38 presents a cross-sectional view of the cast cycle of a jump formwork arrangement for a tall building or a bridge pier or even a round silo. The structural frame is made either of steel or aluminum. The main assembly as shown, which is put together on the ground, consists of three work platforms each with guard rails, shoe boards, ladders, and wind-shields. The opposite form consists of a conventional wall system that is attached to the wall at the foot of the form via bolts that are held by cast-in anchors. Cast-in anchors on the opposite side of the wall, offset to avoid overlaps, support the main work platform in addition to the retractable wall form, here pivoting around a bottom hinge. While the top deck facilitates concrete placement and rebar placement serves the second deck for stripping the wall form, affixing anchor bolts and aligning the form before the next pour can begin. The alignment has to consider the designed profile

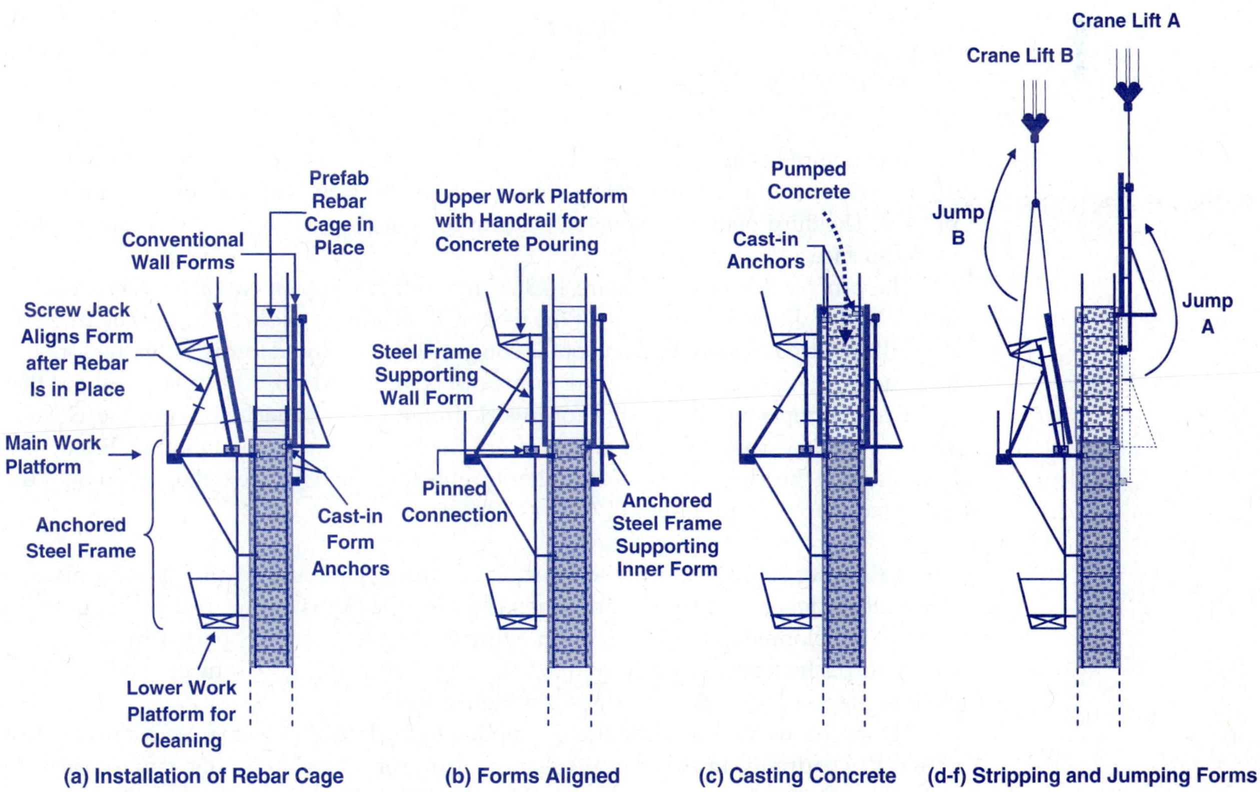

Figure 10.38 The Sequence of Jumping the Inside and Outside Forms for a Concrete Wall

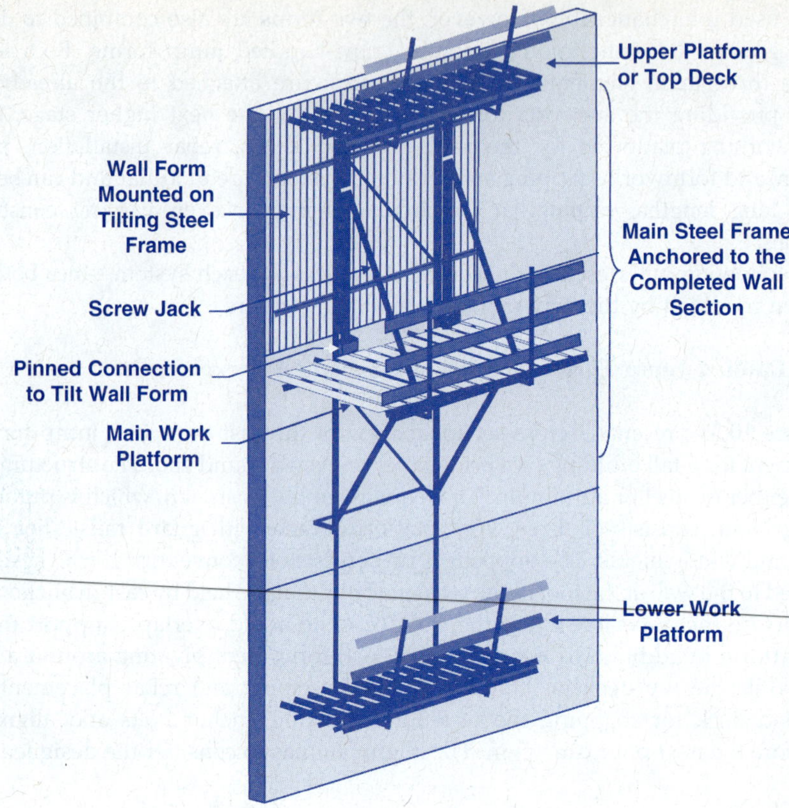

Figure 10.39 Jump Form in 3-D

of the structure, its height, radius, curvature, and thickness. As indicated in the figure, screw jacks serve as flexible struts for each wall segment and support the alignment process. The third platform allows safe access for smooth finishing and possible painting of the surface.

The process depicted in Figure 10.38 starts with the placement of the rebar and the installation of the cast-in anchors for the next wall section (a) followed by the alignment of the wall and (b) the layered casting of the concrete into the closed forms. After the concrete has reached the proper strength, the forms can be retracted (d). At this stage, shoes that assist in the repositioning of the formwork frame can be attached using the cleaned anchors. One by one, the form elements are hooked to the crane, unbolted, lifted, and lowered onto the shoes of the next-higher level before the bolts are tightened to hold the frame structures, as indicated in (e) and (f).

Self-Climbing Jump Forms. The objective of adding climbing gear to jump forms is to eliminate its dependence on a crane when resetting the forms for the next-higher pour. In a recent development, the forms of an entire casting level are supported by one frame structure pushed up vertically by one or several climbing apparatuses. Both of these climbing technologies are introduced in Figure 10.40.

The power needed for climbing is supplied by hydraulic power packs connected to a network of hydraulic lines and hoses supplying one or several hydraulic rams or jacks. In order to minimize the hydraulic supply lines, the hydraulic pump rides with the form and is powered by electricity. While comparing Figures 10.40a and b, one recognizes that the

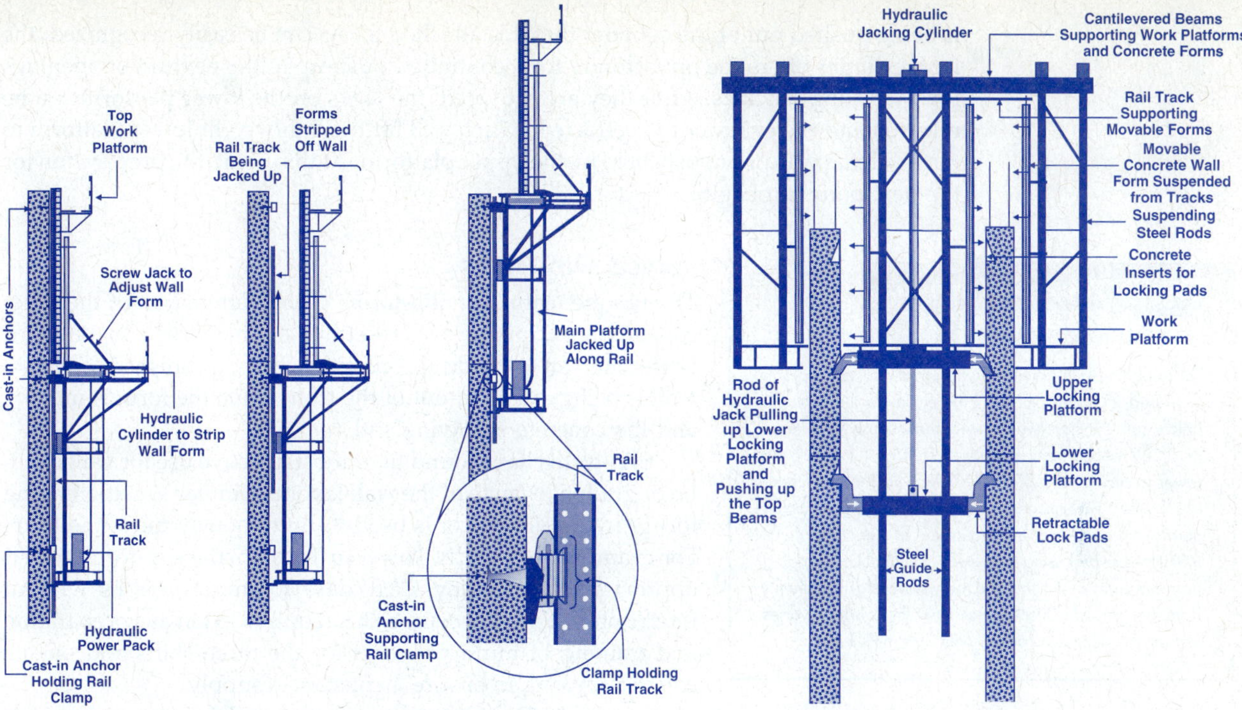

(a) Three-Level Self-Climbing Wall Form **(b) Climbing Multi-Form Platform Frame**

Figure 10.40 Self-Climbing Jump Form Systems

actual climbing mechanism differs in the two systems shown. The self-climbing jumpform employs the help of a rail track that can be "jumped" or moved up independently from the frame structure. By leapfrogging the rails ahead and fastening them to the anchor bolts on the top and the bottom, they serve as an interim structural support during the push-up of the main frame. Depicted in the insert, mechanical clamps that can be opened and closed are used to connect the rails to the anchors. In the same vein, clamps connect the platform brackets to the rails. During the push-up of the rails, the platform is connected to additional anchors allowing the release of the rail clamps. After the rails are reattached at the new anchor locations, the platform clamps can now be closed and used as fixed points for the jacking rams. Commonly, the jacking is semiautomated to ensure that the opening and closing of clamps are coordinated with the operation of the rams, since it will take several cycles of pushing and reclamping. It is feasible to operate 20 or more small hydraulic jacks and activators from one hydraulic pump.

The multiform platform frame shown in Figure 10.40b utilizes a concrete shaft (e.g., for an elevator) as the structure to climb and stabilize the upper-structure. Here, one or two long hydraulic jacks provide a stroke length sufficient to push up the entire structure in only one cycle. Instead of individual clamps, it depends on two horizontal platforms inside the shaft, upper and a lower, each with bracket mechanisms that lock themselves into cast-in openings in the concrete walls. Multiple vertical struts and rods serve as guides as well as the support for the cross-beams the cantilever out on the top.

Opposite to the rod pushing the boom of a backhoe upward, the rod points down toward the lower locking platform. During a climbing cycle, the locks of the upper platform are released and the hydraulic cylinder activated. Guided by the steel rods, the upper platform underpinning the entire superstructure is lifted by the hydraulic rod that is being pushed out of the cylinder that it is attached to. As can be easily recognized, the long cylinder offers the opportunity for a continued push-up to the next set of openings for its locking brackets. After they are activated, the locks on the lower platform can be released and the cylinder rod retracted, which will lift the lightweight lower platform to the level that was previously held by the upper platform. At this point, the preparation for the next pour can begin.

The CN Tower, in Toronto, Canada 553.3 m (1,815 ft)

To build the main tower, a hydraulically raised slipform was used, climbing 6 m (19.7 ft) per day as the concrete below set. Concrete was poured continuously by a team of 1,532 people until February 22, 1974. In total, the tower contains 40,500 m^3 (52,972 yd^3) of concrete, all of which was mixed on site in order to ensure batch consistency. The vertical accuracy of the tower was maintained by using massive plumb-bobs hanging from the form and controlled with surveying instruments from the ground. Over the height of the tower, it varies from true vertical accuracy by only 29 mm (1.1 in).

Vertical Slip-Forms

The unique feature of slip-forms is the elimination of the time-consuming stripping and resetting of the forms. In effect, it represents an extruding process where a finished concrete wall is being extruded out of the form while the form is moving and the concrete is staying still.

Slip-forms have found their most effective use for walls with no or gradual change of the wall layouts. Similar to self-climbing forms, hydraulic power is used to continuously move the form. For example, for the CN Tower in Toronto, the slip-form moved approximately 6 m/day (20 ft/day) for three months without interruption, 24/7. Of course, this required extensive preparation and training of multiple crews for the three shifts and on-site concrete mixing to ensure a constant resupply.

Figure 10.41 presents a generic model of a slip-forming assembly that has been launched on a vertical wall. While the climbing jump forms attach themselves to the outside of a concrete wall, the slip form straddles the wall while supporting itself via steel rods, called climb rods, that are continually extended upward. In order to increase the stiffness of the rod section that is not surrounded by hardened concrete, it is encased in a pipe sleeve section that moves with the form.

The slipping structure consists of the yoke frame that is fastened to the hydraulic clamp and jacking mechanism that uses the climb rods as support. The yoke in turn holds the two main work platforms on both sides and the suspended lower-level work platform from where the fresh concrete can be treated (e.g., cooled during a hot summer). Between the vertical brackets are the two form elements between which the horizontal rebar can be installed before pouring the concrete and small layers. Form braces provide the necessary support and allow the wall thickness to be adjusted when necessary.

Also shown in Figure 10.41 are trailing screens that, like perimeter screens, provide fall protection for the laborers and prevent materials from falling.

When designing and operating these types of form, the following issues should be addressed:

- Provision for safe access to all areas and work platforms.
- Control of risk associated with working at height.
- Provision of sufficient working space and head room.
- Specification of appropriate manual handling methods for shutters and other components. As an example, the use of cranes to lift shutters and provision of safe access for persons attaching slings to shutters.
- Provision for retrieval of an injured or incapacitated person from any work area located within the structure.

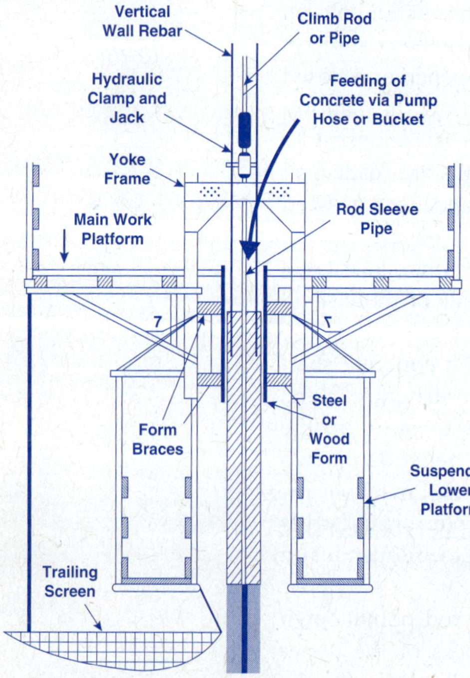

Figure 10.41 Cross-Section of a Slip Form

Workplace health and safety regulations have to be followed and cover common issue, such as the following

- All penetrations should be covered by plywood that is secured in position or should be provided with leading or perimeter edge protection.
- When placing steel or pouring concrete into a form, controls should be provided to prevent a person falling into or off the form. This becomes more of an issue when the width of the form is greater than 225 mm (8.8 in) prior to placement of reinforcing steel. A person working in proximity to projecting reinforcing steel or other objects must be protected from the risk of being impaled by these objects. Suitable controls may include edge protection on the internal side of the working platform or sheets of steel reinforcement mesh temporarily placed on top of the form, so that the steel can be fed through the gaps in the mesh.
- All platforms that may be affected by uplift should be positively secured so that uplift cannot occur. Uplift can occur from wind loading or when a platform is out-of-level and unsecured. In the latter instance, the platform can seesaw when a load is applied to the end of the platform and then drop a person off the platform if one of the platform supports moves. This may be the case where work platforms are supported by a trailing screen on one side and a parapet of the building on the other.
- Where individual cells of the form are climbed at different times, edge protection should be provided on each of the raised cells.

10.5.6 HORIZONTAL SLIP-FORMING

As of 2010, the total length of the U.S. Interstate Highway System was 47,182 miles (75,361 km), or twice around the Earth. What began on June 29, 1956, when President Eisenhower signed the Federal-Aid Highway Act of 1956 turned into a gigantic public works project. According to the American Road & Transportation Builders Association, the record paving day in 1950 was 4,700 ft (1,432 m) of pavement 11 ft (3.35 m) wide. By 1974, the American Concrete Pavement Association's Mile-A-Day Club included 70 contractors. The record was 4.7 miles (7.5 km) by 24 ft (7.3 m) wide, placed on I-80 in Idaho by a contractor that worked 23 hours nonstop. These accomplishments would have been impossible without mechanized pavement technology that was able to extrude rather than "place" white- and blacktop onto a prepared road base.

A Historic Event for Paving Technology

One can argue that tighter specifications for concrete surfaces that were implemented in the early 1960s led to the success of the precursors for today's concrete slip-form pavers. Experience from previously built highways taught the design engineers that a smoother final surface of a concrete road extended its life significantly. Harold Godbersen, a contractor in Iowa, was tinkering with a mechanical device that was able to produce a more uniform bridge deck. It seemed to have worked, since he used the finishing system for himself and other contractors were renting it from him. The economic breakthrough came when Iowa's state specifications incorporated the use of the finisher and regional contractors began buying the machines. His new company, Gomaco, was joined by others, like Bid-Well, to offer what is referred to as cylinder finishers that operated a cylinder on an "x-y table" provided by a traveling truss frame, similar to the more modern screed and finisher shown in Figure 10.42. Later versions added the capability to create skewed surfaces, thus adding a third dimension.

As Figure 10.42b illustrates, today's cylinder pavers serve three distinct tasks: (a) screeding with an auger and a vibrating drum, (b) finishing with the cylinder finisher and (c) surface treatment. These three operations are all executed simultaneously from the traveling carriage that moves back and forth between the two ends. The truss frame upon

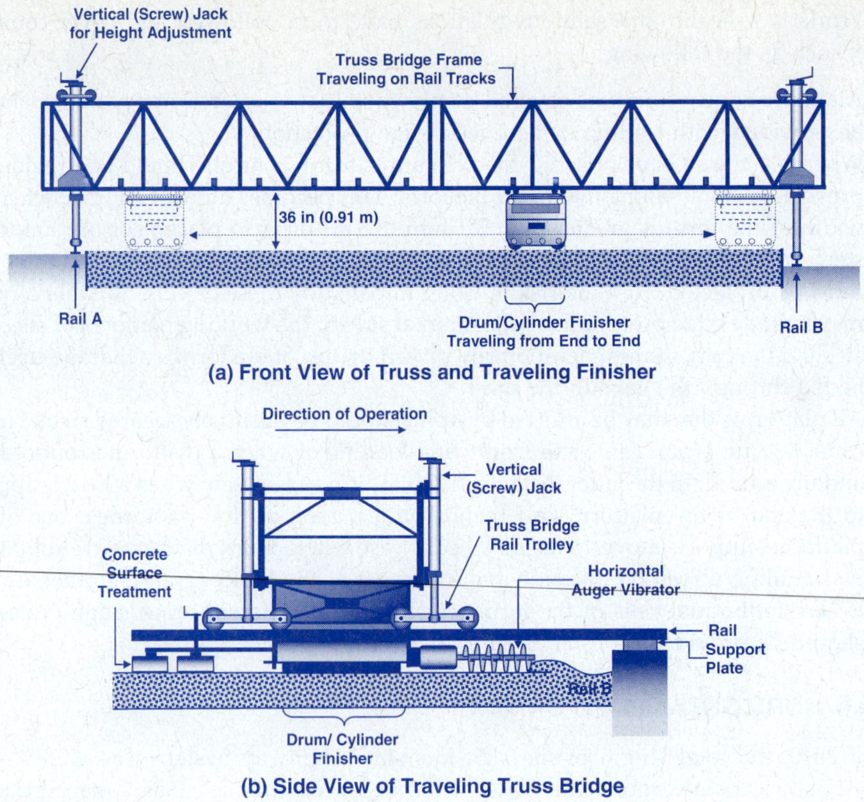

Figure 10.42 Components of a Modern Screed and Finisher Used for Fixed Form Paving

which the carriage travels can be set to provide flat, parabolic, crowned, super-elevated, and tapered surfaces.

Reviewing the main elements of a cylinder finisher, one realizes there is but a small step from this technology to the slip-form pavers of today. While the development of slip-form technology did begin in the early 1950s, the success of the cylinder finishers in the industry helped them to be adopted in the 1960s. Undoubtedly the main driver for the successful launch was their high productivity.

The Paving Train

Large jobs require the placement and finishing of massive amounts of concrete. Truck after truck has to be available to supply concrete in order to keep the slip-former moving. However, trucks may be delayed or the paver may have to stop for many reasons, which can cause a logistics nightmare. To remedy this problem, the placement and spreading of the concrete has been separated from the finisher. In addition, the floating, texturing, and application of a curing membrane have also been mechanized. As a consequence, three sets of systems are sequenced thus, constituting the paving train:

1. Placing and spreading machine
2. Slipform paver
3. Curing and texturing machine

Placing and Spreading Machine As the name of the first component of the train indicates, its job is to receive the concrete from rear-dump trucks not allowed onto the base of the concrete slab. Not having concrete trucks with turning drums deliver means lower delivery cost but requires short travel distances or additives that delay the setting of concrete.

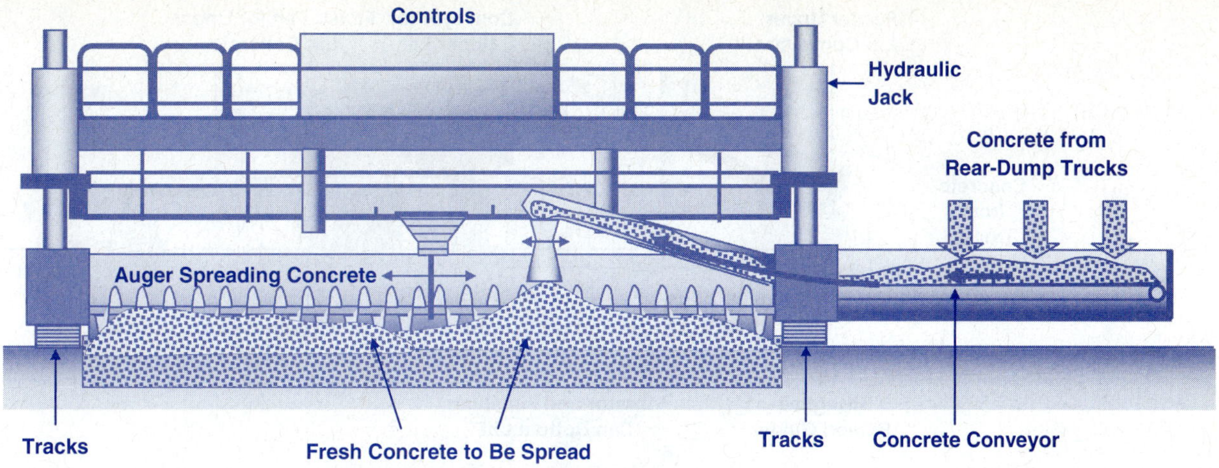

Figure 10.43 Front View of a Placing and Spreading System

Figure 10.43 presents a tracked placer that is receiving concrete on its left side (the right side from the reader's point of view) and moving it via a wide conveyor to the center for interim storage. However, the conveyor system can be switched, and some systems have conveyors that can be fed from both sides simultaneously. Occasionally, the trucks are allowed to travel onto the base, eliminating the need for conveyors.

Reversible and split augers are able to rapidly spread the concrete to the required depth matching the following slipform paver.

The Slipform Paver

Depicted in Figure 10.44, slipform pavers normally use one or two augers to spread the concrete before striking off the concrete at the correct elevation using a simple strikeoff plate, or screed.

Depending upon mix design and slab depth, vibrators are usually set to 7,000–9,000 vibrations per minute (VPM) range. Vibrators are positioned next to one another so that their influence zones overlap by about 50–75 mm (2–3 inches) at normal paver speed (ACPA, 1995). Initial finishing is accomplished by squeezing the concrete through the

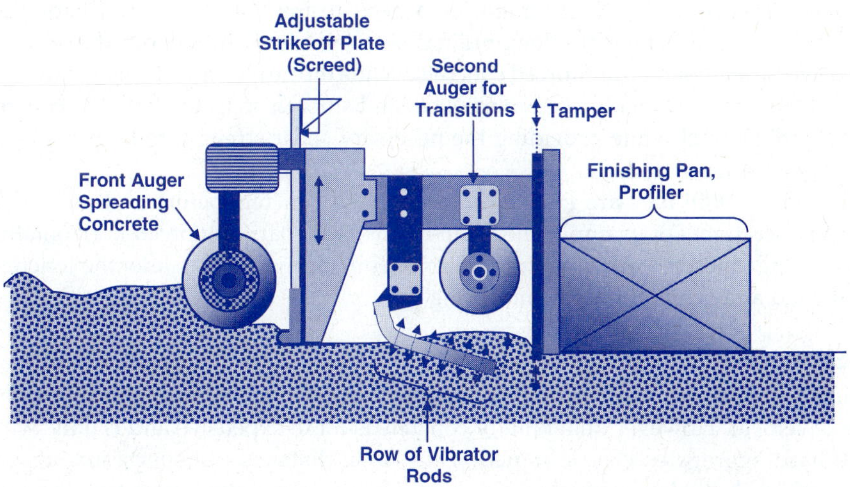

Figure 10.44 Side View of Basic Components for Screeding, Vibrating, and Initial Finishing

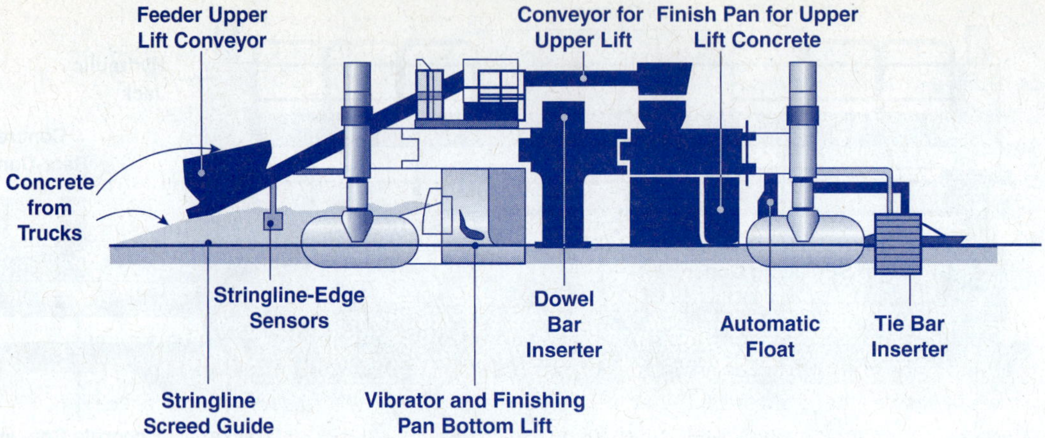

Figure 10.45 Advanced Two-Lift Paver with Automatic Bar Inserters

slipform made up of the base of the road, the side forms, and the finishing pan. This extrusion provides the final slab dimensions and also serves to imbed larger aggregate particles below the surface, which results in a smooth finish. Some pavers are also equipped with a hydraulic tamper bar just behind the vibrators. In certain situations, the tamper bar may assist in the consolidation of the concrete by pushing large aggregate particles below the slab surface. Well-designed concrete doesn't require this operation, which also might cause the mortar to gravitate to the top, which could later scale. On the other hand, it might be effective with low-slump concrete.

Recent models offer an automated floating apparatus attachable to the paver. The "floating" pan traverses the surface of the fresh concrete at a slow speed perpendicular to the direction of the traveling paver. This operation further smooths the surface of the fresh concrete and the still opened pores are sealed. If necessary, the floats can also apply water to the surface during hot and dry conditions. Dehydration of the concrete may begin very soon, causing surface cracks.

Figure 10.45 presents a cross-sectional view of an advanced paving machine including several unique technologies:

Automatic Bar Inserters There might be a need to insert two types of bars into the concrete: (a) tie bars along the longitudinal side joint and (b) smooth dowel bars for the transverse joint of non-reinforced concrete. While the tie bars are of small diameter, the sizable dowel bars are made of stainless steel and smooth so as to allow the concrete to expand and contract while providing the necessary shear strength to keep the concrete slabs aligned during traffic.

As shown, both bars are inserted into the fresh concrete before and after the final finishing operation. For example, the small-diameter tie bars are forced horizontally into the side of the concrete slab via a hydraulic or pneumatic mechanism after the concrete has been floated and while the final smoothing occurs.

The Second-Lift Paver

Research has shown that providing an extra smooth and hard surface will extend the life of the pavement. However, that kind of concrete is more expensive and is only available where hard aggregates can be mined. Thus, long-distance transportation costs have to be added. A "value-engineering" approach was the development of the two-lift pavement where the bottom layer can be poured using the locally available cheaper concrete while the thin second lift consists of the more durable expensive material. Figure 10.45 shows an approach where the two paving operations are combined into one piece of equipment, thus minimizing equipment cost. The space between the two screeds is dictated by the timing needed to prevent the mixing of the two layers while still creating the desired bonding. The upper lift concrete is delivered to the front of the paver where it is poured into a feeder that loads a conveyor supplying the upper lift screed toward the back of the paver. Of course, this strategy only works if the area in front is not already filled by a concrete spreader. In such a case, the concrete feeding has to be done from the side.

Automatic Track and Screed Level Control A key performance criterion of any paver is providing the designed cross-slopes and an accurate line. As the slopes, especially in curves, change, the control of the screed and the tracks of the pavers ask for a very complex set of maneuvers that have to be executed simultaneously. Thus, it is not surprising to learn that paving contractors were early adopters of electronic technologies. An early system is the stringline screed guidance mechanism. Here, stringlines were installed on both sides of the paver using surveying tools. The stringline followed an imaginary line exactly parallel to the upper edge of the future concrete slab. By attaching two touch sensors, one for the horizontal and one for the vertical plane, the hydraulic controls were able to sense if the paver was on track or if the screed was at the proper height. Discrepancies could be easily corrected by activating the appropriate cylinder responsible for steering the tracks or moving the screed up or down. Figure 10.46 depicts two guidance mechanisms utilizing sonar and touch sensors following the surface of an already existing slab or a stringline. Both technologies are now very common.

Texturing and Curing Machine

Curing and providing the necessary texture to create sufficient friction while keeping the noise level at a minimum is typically done once finishing of an area is complete and the original wet sheen has nearly disappeared. The wet sheen is created by the bleeding

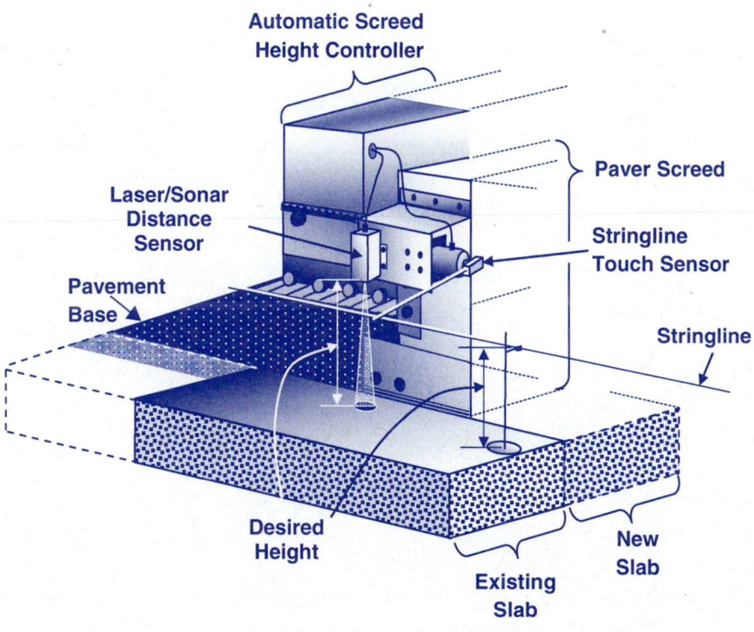

Figure 10.46 Sonar and Touch Sensors Provide Tracking Data

of the concrete, which forces the water under pressure at the bottom through capillaries to the surface of the concrete. This natural phenomenon might be inhibited by a concrete that is dehydrating too quickly. It was found that evaporation rates of as little as $(0.3\,N/m^2)/hr$ (9.64 ounces/$ft^2{*}hr$) may cause critical losses from some concrete if not protected. As a consequence, the operator of the texturing and curing machine has to monitor not only humidity and temperature, but also the concrete surface itself. In particular, applying the curing membrane too early might lead to the spalling of the top concrete caused by water that was not able to bleed out.

Before covering the concrete with the curing compound, the concrete will be textured according to the requirements set in the contract. The machine can be equipped to apply a macro or a microtexture. Tined bars are dragged across the surface of the slipped concrete to create a large or macro texture, while brooms, burlap, or artificial turf sections create much smaller micro ridges in the concrete.

Finally, the machine is able to travel back and spray the compound consisting of a liquid. It consists of a liquid membrane-forming chemical compound. Once the solvent evaporates, the curing compound forms a near-impermeable cover. Pigments are often added to curing compounds in order to reduce (white pigment) or increase (dark pigment) heat absorption. Additionally, pigments allow workers to see where the curing compound has been applied, which helps to ensure complete coverage. This now completes the main paving operation.

Crack Control

Although there are a number of ways to create joints in PCC pavements, the most common method is by sawing the hardened PCC to an established depth, typically one-quarter to one-third of the slab thickness; this creates a weakened plane, thus encouraging cracking at that location in a controlled manner (see Figure 10.45). This sawing must be done neither too soon (which will cause raveling of the concrete) nor too late (which may result in random cracking due to the buildup of residual stresses). The latter is of particular concern, because the longer that sawing is delayed, the greater the chance that random cracking may develop (ACPA 2002).

10.6 Chapter Ten Review

THE EQUIPUZZLE

The letters in the circles can be put together to a word used in a European country when they need a structure to support masons and material.

Across

4. A device for locking together the tubes of a tube and coupler scaffold

5. Name used for the steel or aluminum pipe used to build scaffolds

8. Short for Portland Cement Concrete

9. Temporary structures used in construction to support spanning or arched structures in order to hold the component in place until its construction is sufficiently advanced

11. Term used for water that rises to the surface of fresh concrete

13. A deformed steel bar or connector embedded across a longitudinal joint in a PCC slab to prevent separation of abutting slabs (two words)

15. Stops a plank from slipping of a scaffold

22. Also known as backshoring

24. The deterioration of the upper 3 to 12 mm (0.125 to 0.5 in) of the concrete surface, resulting in the loss of surface mortar

25. Either single posts or towers that transfer construction loads to as many lower floors as required to ensure the partially completed building is neither overloaded nor overstressed

26. Process of supporting a structure in order to prevent collapse

27. A person who can identify existing and predictable hazards in the working environment or working conditions that are dangerous to personnel and who has authorization to take prompt corrective measures to eliminate them. (two words)

28. Last name of civil engineer who designed and oversaw the first vertical slipforming operation

Down

1. Short for Jointed Reinforced Concrete Pavement that has no reinforcement

2. Main danger when operating a mobile boom work platform

3. A plain round steel bar which extends into two adjoining PCC slabs so as to transfer shear loads

6. Plant that is widely used to build scaffolds

7. A powered or unpowered, portable, caster, scaffold, or wheel-mounted supported scaffold (two words)

10. Common formwork solution for situations with high clearance (two words)

12. Also called single prop (two words)

14. A support sloping between the face of a structure to be supported and the ground (two words)

16. Last name of Frenchman who patented the first steel reinforced concrete system

17. Rope connecting a personal fall arrest system to a solid anchor

18. A supported scaffold consisting of a platform supported by brackets attached to formwork

19. Descriptor for a joint that runs perpendicular to the pavement

20. Formwork that climbs continuously up a highrise building during construction (one word)

21. A manual or power-operated mechanical device to raise or lower a suspended wallform

23. The primary horizontal force on scaffolds and vertical formwork

JOURNALING QUESTIONS

1. The Pantheon in Rome was built in 125 AD. At 43.3 m (142 ft) in both height and diameter, this concrete structure was probably the first time that formwork, falsework, and scaffolds were heavily used. Discuss how today's technologies differ from what the Romans probably employed to build the Pantheon.

2. Self-climbing formwork is very similar to slipform technology. Explore where each concept has been used successfully in construction. What are the benefits that the users of the two mention?

3. Visit some construction sites in your town, and describe the used of different scaffold technologies (fixed, climbing, mobile, and boom-based). Would you recommend different systems from what is being used? What improvements to safety should be made?

HOMEWORK PROBLEMS

1. The two-lift paver in Figure 10.45 uses a cumbersome concrete feeding mechanism for the second lift. What is the main reason why the manufacturer built it in the present configuration? Sketch an alternative mechanism, and discuss its benefits.

2. Controlling the development of cracks in concrete pavements depends on creating artificial weaknesses. How does this process work? When exactly should it be scheduled? What happens if one is too early or too late with the process? Create a plot (with two axes) that you could use to explain the best timing to a paving supervisor.

3. Calculate the maximum lateral pressure inside a wall form considering the following conditions:

Concrete unit weight = 150 lb/ft^3 (23.5 kN/m^3)
Chemistry coefficient = 1.1
Rate of concrete placement = 5 ft/hr (1.5 m)
Temperature of the concrete at the time of placing = 60 °F (15.5 °C)
Height of concrete wall = 12 ft (3.6 m)

4. Say that you are trying to decide which shoring system to pick for a new job that requires shores that are 2.4 m (7.9 ft) to support a slab formwork. Presently you are reviewing the plot in Figure 10.3. Which shore would you recommend that should be rented? Argue the benefits in case your solution is a little bit more expensive than the next best solution.

5. Cost comparison of a tube and coupler T&C scaffold with a climbing scaffold:

Job description:
Panel installation on an outside wall area 13,000 sq. ft. = 130′ length × 100′ height

Option 1. T&C (Duration of installation: 4 months)

Transportation costs: 3 trucks × $1,000
Erection & Dismantle costs: $1.11/ft^2 (11.9 $/m^2)
Monthly rental costs: $.28/ft^2 (3 $/m^2)
Customer's labor costs: $300/day/person × 12 persons × 4 months × 20 days

Option 2.2. CLIMBER units (Duration of installation: 3 months)

Transportation costs: 2 trucks × $1,000
Erection & Dismantle costs per unit $600
Crane rental for re-positioning $5400
Supervising fee of assembling/dismantling $2000
Rental costs: $4,000/month/unit
Customer's labor costs: $300/day/person × 6 persons/2 units × 3 months × 20 days
Compare the two options and describe which of the two you would recommend to be used.

LESSONS FROM THE ACCIDENT FILE

1. Name and describe two different technologies that would have prevented the laborer from falling to his death in Case Number 2.

2. Which basic requirement for installing a high two-pole scaffold has not been followed in Case Number 3?

3. In Case Number 4, what apparently went wrong when the concrete forms were rigged to the crane? If you have difficulty in finding the answer, please review the basic approaches to sling-rigging a set of elongated material without the use of a crate.

4. What indicates that the tower in Case Number 5 needs to be considered a rolling tower that was locked? What basic rule was broken when the tower was built? What requirement did the laborer break when he tried to re-align the roof section? What installations would you recommend for the described situation inside?

OPEN-ENDED PROBLEMS

Reviewing the historical development of formwork, one can observe some general trends.

1. How did the development in concrete influence formwork technology?

2. How did the development in other areas influence the productivity of formwork installation?

3. How was the issue of backshoring addressed by different systems?

4. What future developments do you expect next?

BIBLIOGRAPHY

Flint, A. R. "Risks and Their Control in Civil Engineering." *Proc. of the Royal Society of London*, Series A, Mathematical and Physical Sciences 376, No. 1764 (1981) 167–179.

Shapira, A. "Work Inputs and Related Economic Aspects of Multitier Shoring Towers." *J. Construction Engineering and Management* 130, No. 1 (2004): 134.

Shapira, A., and Y. Raz. "Comparative Analysis of Shoring Towers for High–Clearance Construction." *J. Constr. Engineering and Management* 131, No. 3 (March 1, 2005).

chapter eleven

How Do They Build These Magnificent Bridges?

Figure 11.1 On the way to work

A search for the meaning of the word *bridge* will bring up a long list of subject areas outside engineering and construction. For example, *bridge* is a card game with a variety of specialties, such as *rubber bridge*. The captain of a large ship uses a *flybridge* high up to steer with controls on both side of the ship. A dentist anchors a *bridge* between two healthy teeth, while the optician makes sure that the new glasses fit perfectly onto the *bridge* of the nose. A *bridge* piece is able to connect two sections of a music composition or the two lenses of eyeglasses. One can measure the resistances, inductances, or capacitances of a circuit with the help of a *bridge*, while an individual might apply for a *bridge* loan or the violinist check the seat of the *bridge* that raises the strings. International treaties are designed to create bridges between nations while politicians sometimes want to build "bridges to nowhere." Another large field where the word has found a home is in psychology or sociology. One of the most famous quotes that belongs to this genre is by a man who is familiar to every engineer, Sir Isaac Newton, who supposedly said: "We build too many walls and not enough bridges."

This final chapter, however, concerns itself with the original use of the word to describe those large structures that pass over rivers and deep valleys to make travel for people and animal safer and faster. Looking at one of those gigantic structures, as depicted in Figure 1.1, able to withstand hurricanes and earthquakes, one can only wonder: "How do they build these magnificent bridges?"

Table 11.1 Topics Covered in Chapter 11

Construction			Planning Control		
Equipment	**Methods**	**Materials**	**Managerial-, Engineering Tools**	**Rules & Standards**	**Bridge Designs**
✓ Launching Gantry	✓ Flame Cutting	✓ Ordinary Portland Cement	✓ Cantilever	✓ ASTM C 143	✓ Beam/Girder
✓ Multi-Strand Cable Tensioning/Jacking	✓ T-Joint Welding	✓ Cable Stay	✓ Arch	✓ AASHTO T-23	✓ Rigid Frame
✓ Slipformer	✓ Oxy Fuel Gas Welding	✓ Concrete	✓ Box Girder	✓ AASHTO **T-119**	✓ Truss Through
✓ Gantry	✓ Arc Welding	✓ Structural Steel	✓ Elastomeric Bearings	✓ AASHTO T-126	✓ Top Deck Arch
✓ Movable Formwork	✓ Rebar Tying	✓ High-Performance Steel	✓ Expansion Joints	✓ ASTM C 31/31M-95	✓ Suspension
✓ Internal Formwork	✓ Concrete Casting	✓ Reinforced Concrete	✓ Dead Load	✓ ASTM A709	✓ Cable-Stayed
✓ Underslung Launching Gantry	✓ Prestressing	✓ Steel Bearings	✓ Balanced Cantilever	✓ ACI 318	✓ Continuous Girder
✓ Self-Propelled Mobile Transporter	✓ Cast-in-Place	✓ Elastomeric Bearings	✓ Bending Moment	✓ ACI 222R-01	✓ Open Box Girder
✓ Form Traveler	✓ Precast	✓ Superplasticizers	✓ Tension/Compression	✓ OSHA 29 CFR 1926	✓ Plate Girder
✓ Cantilevered Beam Hoist/Crane	✓ Post-Tensioning	✓ Ultra-High-Perform. Conc.	✓ Moment of Inertia	✓ Grade HPS 70W	✓ Bailey Bridge
✓ Hydraulic Lift Jacks	✓ Cable Spinning	✓ Acetylene	✓ Concrete Slump Test	✓ AASHTO SBB-1	✓ Pier Table
✓ Floating Cranes	✓ Falsework Shoring	✓ Rebar grades 40, 60,or 75	✓ Concrete Sampling	✓ ASTM A416	✓ Pylon
✓ Luffing Crane	✓ Unshored Construction	✓ Epoxy Coating	✓ Yield Stress	✓ ASTM A325	✓ Stay Cables
✓ Mobile Crane	✓ Floating Barge	✓ Cable Wire/Strand	✓ Cable Stressing Force	✓ CFR 1926.550(f)(1)(iv)	✓ Deck Anchor
✓ Launching Nose	✓ Deck Panel Construction	✓ Drift Pin	✓ Suspender/Hanger Cable		✓ Arch Strut
✓ Launching Station	✓ Segmental Precast Girder	✓ Production Bolt	✓ Composite Deck		✓ Spandrel Pier
✓ CC Total Station	✓ Balanced Cantilever Precast Elements	✓ Deck Panels	✓ Back Stay Anchoring		✓ Hollow Arch Rib
✓ Track Cable	✓ Span-by-Span Precast	✓ Bent/Pier Caps			
✓ Guy Cable	✓ Full-Span Segmental	✓ Shear Stud			
✓ Temporary Lift Cables	✓ Incremental Launching				
	✓ Reverse Circulating Drilling				
	✓ Orthotropic Deck				

11.1 History of Bridge Construction

8000–1000 BC	Prehistoric Bridges	Most civilizations have built structures to provide safe passage to people, vehicles, and animals; bridges have been made of various materials, such as bamboo, wood, vines, ropes, stone, and iron. Because of the characteristics of the different materials, four main bridge forms evolved. 
900 BC	Chinese Pontoons	Chinese armies use temporary pontoon bridges.
480 BC	Greek Pontoon	Herodotus reports a bridge across the Hellespont made of 600 pontoons and white-flax cables. The Persian king Xerxes planned to move his army across, but the bridge was destroyed in a storm.
200 BC	Chain-Link	The first chain-link suspension bridge, the Panhogiao or Panho Bridge, was built by Chinese General Panceng of the Han Dynasty.
200 BC	Cofferdams	The Romans used cofferdams made of two rows of driven timber piles with the space in between filled with stone and "Pozzolana mortar" that created a water barrier. Bad soil was removed by buckets down to rock for a solid pier foundation.
18 BC	Pont du Gard	Construction of the multistory Roman aqueduct near Nimes in France using semicircular arches built with stones.
600 AD	Zhao Zhou Bridge	Chinese engineers built the partial circular stone arch with a span of 37 m (122 ft). The arch openings not only lighten the load on the arch ribs, but they also make the bridge look slender and delicate.
1345	Ponte Vecchio	The "Old Bridge" across the Arno replaces the many that had been destroyed before by floods. Use of flat arches to achieve larger spans. The only bridge in Florence not dynamited by the Germans during World War II and still standing today.
1779	Iron Bridge	Abraham Darby succeeded in building the first iron bridge in Coalbrookdale, England, across the Severn. Unlike other attempts before, this 30 m (100 ft) bridge was strong enough and still survives.
1806–1869	Johann August Roebling	Born in Thuringia, Germany, emigrates in 1831 to the United States. Invented machines to manufacture wire ropes and ran a rope factory with 8000 employees. Another important development of his is the mechanized in-situ cable spinning method, in which wires are carried by a wheel back and forth between the anchorages. The method ensured that all wires have the same amount of tension. Designs Brooklyn bridge.
1832-1923	Gustave Eiffel	His company designed steel bridges across the world. He was the first engineer to design steel structures with detailed drawings showing every element, down to single rivets. His most famous works were the Eiffel Tower (1889) and the Statue of Liberty in New York (1886).
1823	Pont St. Antoine Wire Cable Suspension	In Geneva, Switzerland, engineers Dufour and Séguin built the first permanent wire cable suspension bridge. Six cables of 90 wires each were used.
1834	Grand Pont Cable Suspension	In Fribourg, Switzerland, French engineer J. Chaley completes a 273-m (896 ft) cable suspension bridge, the longest in the world for 15 years. Four cables, each with 1056 prestressed wires, as Dufour had done before him. The cables were laid out on the bottom of the valley and lifted up onto the top of the towers.
1848	Warren Truss	James Warren and Willoughby Monzani obtained a British patent for a configuration of repetitive equilateral triangles that could support a roadbed on either its top or bottom chord. 

1850	**Britannia Bridge**	Robert Stephenson finished a railway bridge in the shape of two rectangular tubes (each 4.4 m (14.43 ft) wide and 9 m (29.5 ft) high), each with a railway track. The bridge was designed after extensive experimentation on models in the scale 1:6 and functioned well for 120 years.
1850	**Vertical Lift Bridge**	Robert Hood builds a lift bridge across the Grand Surrey Canal for the Branch of the London, Brighton and South Coast Railway.
1859	**Steel Tubes**	Opening of the Royal Albert Bridge designed by Isambard Brunel with steel tubes as the main support for the bridge.
1864	**Clifton Suspension Bridge**	Near the English town of Clifton, Isambard Brunel creates a beautiful chain bridge with a span of 214 m (706 ft).
1866	**Cable Suspension**	Johann Roebling finishes the longest cable suspension bridge across the Ohio, at 322 m (1,062 ft)
1867	**Reinforced Concrete**	French inventor Monier receives a patent for reinforcing concrete with steel wires and bars inserted into fresh concrete.
1874	**Iron-Steel Arch Bridge**	James Eads finishes the longest arch bridge with a maximum span of 159 m (525 ft). He used tubular members partly of iron and partly of steel to form the latticed arch.
28. Dec. 1879	**Tay Bridge Collapse**	During a storm the mail train, including several spans of the Tay bridge, disappeared with approximately 80 people into the water below. The immediate inquiry into the cause of the collapse brought to light a long list of problems that contributed to the failure, including bad design, lack of material testing, poor construction, and lack of structural maintenance.
1883	**Brooklyn Bridge**	With a span of 486 m (1,600 ft), the bridge was completed by Johann Roebling's son and his wife (a lawyer) after he died at the start of construction. It was 50% longer than any previous suspension bridge. The two masonry towers were 107 m (353 ft) in height supporting 4 cables made of 5,358 wires stiffened by deep trussed deck girders and a large number of diagonal cable stays.
1885	**Truss Bridge at Henderson**	Albert Fink completes a 158 m (522-ft) span truss bridge using two Warren-type truss structures stacked on top of each other.
1890	**Firth of Forth Bridge in Scotland**	Cantilever truss bridge, built without falsework, had two main spans of 521 m (1,720 ft). This sort of truss construction had been pioneered by the German Heinrich Gerber. By increasing the depth of the truss girders at the piers, it could cantilever into the center, where it could meet the opposite side or a prefabricated center element could be lifted in.
1897	**Muengstener Railbridge**	First large bridge built using cantilevered construction methods. The steel truss arch had a span of 170 m (561 ft).
1919	**Steel Truss Cantilevered Bridge**	The Quebec Bridge (Pont de Québec) with a world-record center span of 549 meters (1800 ft) remains the longest cantilevered bridge span and is considered a major engineering feat. It was completed after large construction accidents
1928	**Prestressed Concrete**	French engineer Eugene Freyssinet patents the technique of prestressing concrete elements (to be used for bridge construction).
1930	**Salginatobel Bridge**	Robert Maillart completes the acclaimed three-hinged concrete arch bridge in Switzerland.
1931	**George Washington Bridge**	Othmar Amman, an emigré Swiss engineer, set a new record for suspension bridges with a span of 1,067 m (3,521 ft). It used 4 cables with over 20,000 wires each. It carried the greatest live load of any bridge, consisting of two traffic decks and 14 lanes.
1932	**Sydney Harbor Bridge**	Arch bridge by Freeman with a span of 503 m (1,660 ft).
1937	**Golden Gate Bridge**	Austrian engineer Joseph Strauss completes the most famous suspension bridge with a span of 1281 m (4,230 ft).

1940	Tacoma Narrows Bridge Collapse	Due to insufficient stiffening, a heavy wind through the Narrows caused the slender 853 m (2,815 ft) suspension bridge to oscillate heavily and eventually break apart. The failure was recorded on film.
1957	Theodor Heuss Bridge	First cable-stayed bridge is completed. Crossing the Rhine at Düsseldorf, it has a center span of 260 m (860 m).
1962	Seine River	The first precast segmental concrete bridge, built in 1962, crossed the Seine River in France.
1964	Verrazano Narrows Bridge	Othmar Amman completes the largest suspension bridge in New York harbor with a span of 1,298 m (4,280 ft).
1968	Incremental Launching	First concrete bridge built using the incremental launching method across the River Inn at Kufstein in Austria. Concept was pioneered by Leonhardt and Andra.
1998	Akashi Kaikyo Bridge	Suspension bridge with a central span of 1,990 m (6,560 ft) between Kobe City and Awaji Island.
2003	Lupu Bridge	Built in Shanghai, the 550 m (1,815 ft) long arch bridge replaces the New River Gorge bridge as the world record holder.
2004	Millau Viaduct	At the town of Millau, at 270 m (891 ft), the highest and most spectacular multi-span cable-stayed highway bridge closes a major gap in the French highway system.
2008	Sutong Bridge	At 1,088 m (3,570 ft), the longest cable-stayed bridge spans the Yangtze River between Nantong and Changshu.

11.2 From the Accident File

1. **Unanchored Rear Leg of Launching Gantry Slips:** Four male ironworkers were killed. The incident occurred during the moving of a launching gantry (LG) made of long horizontal steel trusses that span the distance between two elevated bridge piers.

   Figure 11.2 displays how both the underbridge and the overbridge leapfrog from pier to pier resting on movable legs where the underbridge serves as the support during the installation and post-tensioning of the precast elements for an entire span. When the accident occurred the underbridge was being moved to the next pier at a 1% down grade and a slight curve to the left. Despite the manufacturer's instructions, the rear legs were not anchored to a completed bridge section. Supported by the driving legs, the rear legs of the underbridge had just been lowered and readied to be connected to the new middle pier when the unanchored rear legs of the overbridge slipped off the last pier and caused the entire launching gantry to crash 65 feet to the ground.
2. **Crane Tips Over on Barge:** A crane operator working for a contractor building a bridge across a river was involved in driving piles for a cofferdam. His large crawler crane had a 90-ft (27.5-m) boom and a 33,500 lb (149 kN) counterweight and was

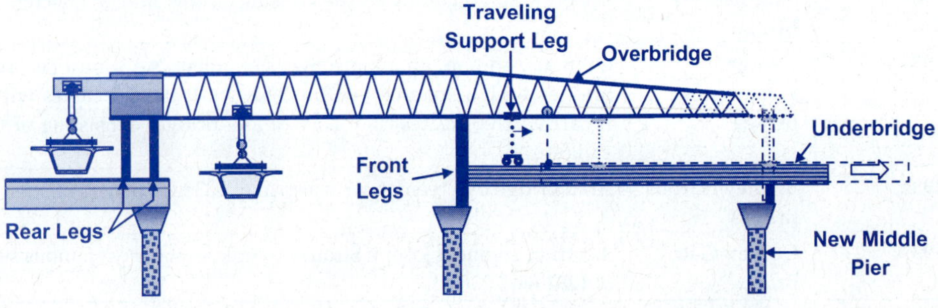

Figure 11.2 Basic Structure and Operation of the Launching Gantry

stationed on a barge, located next to the cofferdam. At the time of the accident, the boom was supporting a set of 70-ft (21 m) leads and a diesel hammer with a combined weight of 15,500 lb (69 kN). The victim was in the process of rotating the crane boom past the starboard bow of the barge when the crane began tipping, raising the left track off the barge. The crane continued tipping and turned on its side, impacting a pile installation machine located on the starboard side of the barge. The operator was crushed in the cab of the crane. A witness said it appeared the boom was at 45 degrees or a little more during the time the crane was turning with the pile driver. Assuming a conservative angle of 52 degrees, the maximum load (with side frames extended, which was the case), according to the load chart in the crane at the time of the incident, is 10,550 lbs (47 kN). The crane was not positively secured to the barge at the time of the incident.

3. **Collapse of Bridge Falsework:** Twelve workers died and 17 were injured when portions of a highway ramp under construction collapsed. The formwork for the cast-in-place, post-tensioned, concrete superstructure rested on shoring towers, longitudinal stringers supported by transverse crossbeams. The collapse most likely began with the cracking of concrete pads underneath the falsework towers. It was felt that the collapse would not have occurred if the tower footings had rested on properly reinforced concrete slabs.

11.3 Characteristics of Modern Bridges

Anybody who travels by car, train, or ship will observe the wide variety of shapes and forms exhibited by bridges across the world. Their key characteristics, length and height, are the result of improved engineering and building materials. In the early twentieth century, increased confidence in suspension bridges pushed the length of main spans over an amazing 3,930 ft (1,200 m) mark before the collapse of the Tacoma Narrows Bridge. After learning from this failure, new design rules were established, and the push for even longer bridges continued. Today we are able to build bridges with a center span of 6,557 ft (2,000 m).

But length is only one criterion with which to characterize bridges. For construction, the more important considerations are the structural materials that are selected, horizontal alignment, project duration, environmental and geotechnical conditions and restrictions. For example, building a bridge in a downtown area with no space for cranes and trucks is markedly different from building one over water or across a deep valley or canyon.

11.3.1 PRINCIPLES OF STRUCTURAL DESIGN

While the Romans were sophisticated bridge builders, they did not have the materials that we have today. The Romans, however, taught us to look at the bridge as a system made of several parts that work together. In Chapter 9 we learned how effective they were in first establishing a solid foundation for piers and abutments. In fact, what we see of a bridge is matched by what we don't see underground. Please review Chapters 8 and 9, since solid foundations are a prerequisite for constructing the superstructure.

The Romans used stone, brick, and timber to build their bridges. At the same time, timber was critical in other ways, as it was the main material with which to build the falsework to support the bridge under construction. Stone, brick, and timber have since been replaced with steel, concrete, and tension cables. Although not as strong as cables, steel is able to sustain high tension forces while concrete, consisting mainly of bonded aggregates, has excellent capabilities to sustain high compression forces.

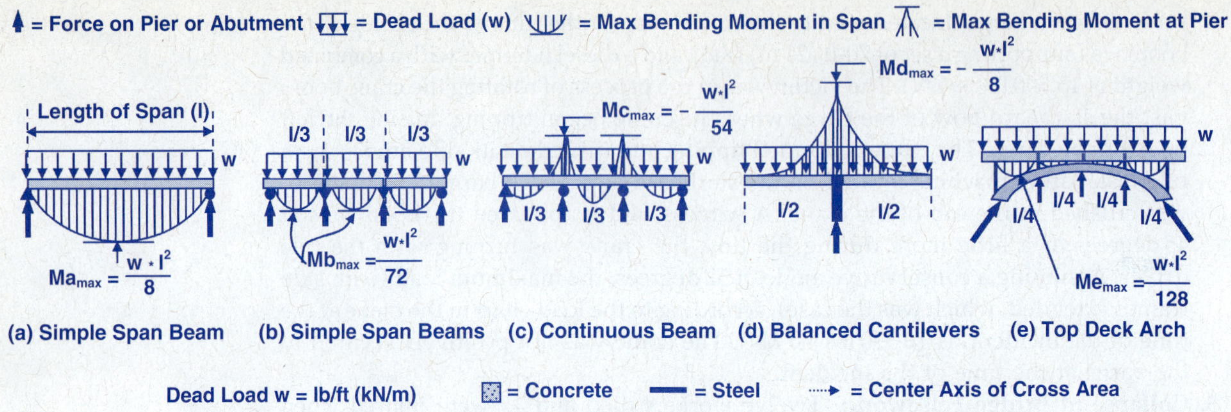

Figure 11.3 Load and Bending Moment Diagrams for Some Common Bridge Spans

Engineering mechanics-statics allows us to study the invisible forces and moments that a bridge will have to be able to sustain. Figure 11.3 provides a qualitative comparison of some common bridge structures that demonstrate why we find so many different bridge shapes and forms.

For this demonstration only dead loads are introduced. The vertical arrows indicate that this load originates with the weight of the bridge deck itself, including asphalt and other materials that rest on the abutments, columns, or piers even without any traffic. One readily recognizes that the two maximum bending moments have the same function ($\frac{w \cdot l^2}{8}$). In Figure 11.3a Ma_{max} is positive and in the center of the span, while in Figure 11.3d Md_{max} is negative and at a center pier. Figures 11.3b, c, and e all have smaller spans (e.g., one-third) and smaller bending moments, as they are quadratically linked to the length of the span.

You may ask: Why are we talking about bending moments? There are two main reasons, the problems of bending and failure during construction. The reader is encouraged to run a small experiment by creating a small bridge with a ruler as shown in Figure 11.3a. Loading it with a cell phone, one will observe that the ruler will in fact bend similar to the moment diagram. Adding more weight will result in a larger bending as the moment increases. It is readily understandable that the heavy beams necessary to support trucks or trains driving across will bend from their own weight. Failure in bending will occur when the material tears open on the underside or crushes on the top of the beam. When a material cracks open, we talk about a failure due to tensile stress and when it is squashed, it's a failure due to compressive stress. Bridge design and construction have to consider the compressive and tensile strength of the material as the bridge evolves.

Figure 11.4 presents some examples to highlight the key relationships that need to be understood in order to avoid a catastrophic collapse.

The largest diameter of a tree is close to the roots, because it is where the tree will experience the highest bending moment during a storm. Thus, we might infer that the size and the strength of the cross-sectional material have something to do with the bending moment it can sustain. In fact, structural engineers use a simple formula to calculate stresses (σ) parallel to the beam's center line:

$$\sigma = \frac{\text{Moment}_x * y}{I_x}$$

where

I_x, called the moment of inertia, indicating the cross-section's resistance to changes in its geometry due to an outside bending moment.

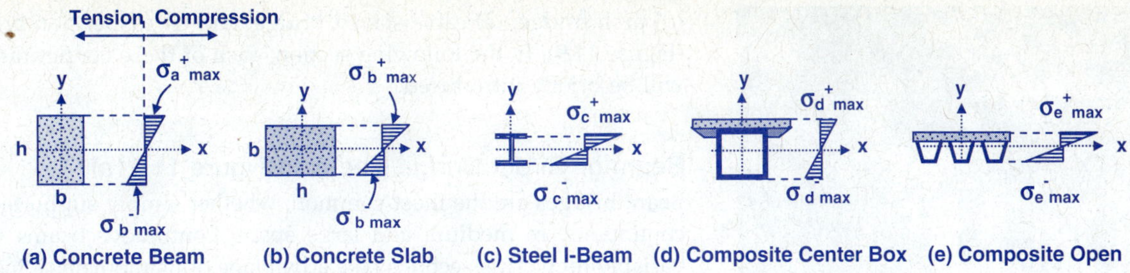

Figure 11.4 Comparing Stresses as a Function of Cross-Section Designs

y signifies how far the point of interest is from the neutral axis of the cross section. The formula for rectangular cross section, as in Figures 11.4a and b is: $I_x = \frac{Area*Height^2}{12}$ If we assign a value of 737, 373 lb-ft to the moment, b = 3.28 ft, and h = 6.56 ft, y = h/2

$$\sigma_{max} = 737{,}373\,\text{lb-ft} * 12 * (6.56/2)\text{ft}/(3.28\,\text{ft} * 6.56^3\,\text{ft}^3)$$
$$= 29{,}023{,}001\,\text{lb-ft}^2/926\,\text{ft}^4 = 31{,}344\,\text{lb/ft}^2 \text{ or } 218\,\text{psi (compression)}$$

Moment = 1,000 kN-m, b = 1 m and h = 2 m. We can calculate for Figure 11.4a:

$$\sigma_{max} = 1{,}000\,\text{kN-m} * 12 * (2/2)\,\text{m}/(1\,\text{m} * 2^3\,\text{m}^3) = 12{,}000\,\text{kN-m}^2/8\,\text{m}^4$$
$$= 1{,}500\,\text{kN/m}^2 \text{ (compression)}$$

For the configuration in Figure 11.4b we get:

$$\sigma_{max} = 737{,}373\,\text{lb-ft} * 12 * (3.28/2)\,\text{ft}/(6.56\,\text{ft} * 3.28^3\,\text{ft}^3) = 14{,}511{,}500\,\text{lb-ft}^2/231\,\text{ft}^4$$
$$= 62{,}688\,\text{lb/ft}^2 \text{ or } 436\,\text{psi (compression)}$$

$$\sigma_{max} = 1{,}000\,\text{kN-m} * 12 * (1/2)\,\text{m}/(2\,\text{m} * 1^3\,\text{m}^3) = 6{,}000\,\text{kN-m}^2/2\,\text{m}^4$$
$$= 3{,}000\,\text{kN/m}^2 \text{ (compression)}$$

Because the two rectangles are symmetrical along their central axis, the magnitudes of the tension stresses are the same, but negative, because y is negative.

Assuming that the two rectangular cross-sections in Figures 11.4a and b are made of OPC (Ordinary Portland Cement) concrete, we can expect a compressive strength of 2,844 psi (1.96 kN/cm^2) and a tensile strength of 284 psi (0.196 kN/cm^2). Based on the assumed values, a design engineer will have to conclude that even the thinner beam in Figure 11.4b will not reach the compression strengths that an OPC will achieve, but on the other hand, its tensile strength is too small to carry the load without cracking. In Section 11.4.3 we will discuss what is done to overcome this weakness of concrete.

Without going into more detail on how the design engineer considers the many different structural effects and material constraints, this review will close with pointing to the three final cross-sections where I-beams of steel and concrete maximize the moment of inertia by having large areas of resistance as far from the neutral axis as possible. The two steel boxes in Figures 11.4d and e are combined with top decks made of concrete creating a composite structure of steel and concrete. This takes advantage of the compressive strengths of concrete, high tensile strength of steel and the high moment of inertia created by the high steel boxes.

11.3.2 MOST COMMON BRIDGE DESIGNS

An observant traveler will recognize the basic features of six common designs such as the (a) beam bridge (Figure 11.5), (b) frame bridge (Figure 11.6), (c) truss bridge (Figure 11.7),

Figure 11.5 The Beam or Girder Bridge

Figure 11.6 The Rigid Frame Bridge

Figure 11.7 The Truss-Through Bridge

Courtesy FHWA

Figure 11.8 The Top Deck Bridge

(d) arch bridge, (e) cable stayed bridge, and (f) suspension bridge (Figure 11.9). In the following section, each of these configurations will be briefly introduced.

Beam or Girder Bridge (see also Figure 11.11a)

Beam bridges are the most common, whether simply supported or continuous. In medium and large spans, continuous beams with variably high cross-sections take advantage of its moment of inertia and the strengths of the main materials, steel and concrete. The beam, or girder, is made of steel I-beams welded together from single plate (plate girders), steel, or concrete boxes. In continuous concrete box girders, post-tensioned steel cables encased in concrete are used in addition to rebar to sustain the large tensile stresses in the open span.

Rigid Frame Bridge (see also Figure 11.11b)

The term *frame* indicates a key characteristic of this bridge type. It depends upon the ability of structures like picture frames to provide a stiff and sturdy structure to a painting or fabric. Of particular importance is rigidity of the corners, which remain rigid even if the vertical elements are cut in half. By extending the horizontal element on both sides and sloping the "legs" outward but making the connection between the two equally stiff, a common form of a frame bridge develops.

Truss-Through Bridge (see also Figure 11.11c)

Truss bridges used to be omnipresent because of their light weight, ability to create cross-sections with high moment of inertia, and available skilled manpower to assemble and connect the pieces of timber or steel. While steel has outlasted maintenance-intensive and flammable timber, the rising price of steel and of qualified manpower has caused a slow retreat from truss bridges. Another factor might be the large cost for the mills to roll great big beams and plates that allow the fabrication of huge girders. However, the truss-through bridge is still alive, especially for heavy trains, as the large moment of inertia results in lower deflections in the free span.

Top Deck Arch Bridge (see also Figure 11.11d)

Of course, arched structures have been built for a long time even before being used for bridges. The "secret" of the arch is that it only creates compressive stresses and experiences very small bending moments. Thus, arches can be built with material that is strong in compression, like stone or brick in early times or concrete today. The arch, however, provides only the "frame," or ribs, on top of which to build something or from which to hang a bridge deck. In the example shown in Figure 11.8, two arch ribs support a deck that is built on the top, opening up spandrels on both sides. When the deck is lowered below the crown we talk about a through-deck arch. While the Romans filled them in with stone, today the spandrels are left mostly open, making the structure look lighter and weigh much less.

Figure 11.9 The Suspension Bridge

Suspension Bridges (see also Figure 11.11e)

The emergence of the modern suspension bridge is directly related to the development of high-strength steel in the nineteenth century. Of course, much earlier cables for footbridges were made of sisal or other fibers and were in heavy use by the Inkas in the Andes. The first slender bridge decks were suspended from chains or pinned plates made of steel (Clifton 1864). But Roebling, a German immigrant running a rope company in Trenton, NJ, came up with a new concept to create steel cables made of high-tension wires that could be "spun" from one end of the bridge to the other and back again over the towers in between, anchored on both sides, and finally bundled and protected against water and snow. The "inverted arch" created by the bundled wire cables serves as the platform for hangers, shorter vertical cables of variable lengths, that supported the bridge deck. The wires to spin the main suspension cables have an allowable tensile strength of 600 MPa or 87,000 psi or 60,000 kN/m^2. Some of the advantages include that no falsework during construction is necessary, and the pleasing aesthetic. One problem was the small moment of inertia horizontally. This weakness could be countered by building very stiff deck structures. Today, this is the only bridge design that will support spans longer than 1,100 m (3,600 ft), requiring large structures in order to anchor the large tension forces in the suspension cables.

Figure 11.10 Pylon of Cable-Stayed Bridge

Cable-Stayed Bridges (see also Figure 11.11f)

This type of cable-based bridge, depicted in Figures 11.1 and 11.10, does not depend on anchorages at both ends. A stay defines a nonvertical direct support, such as individual cables that are tied to the top of a vertical pylon and to a segment of a continuous beam, thus creating a triangle where the stay is in tension and the other two edges are in compression. By creating two equivalent "triangles" on the sides of a pylon, a balanced system emerges, only needing a strong support in the middle to carry the vertical load. The horizontal compression forces that are introduced to the structure by the inclined cables anchored to the deck differ from the suspension bridge design. Wind forces on both the cables and the deck are mitigated by minimizing drag, with the help of aerodynamic-efficient designs. Some designs use only one stay cable anchored in the center of the bridge, mostly for smaller bridges. Most common is the arrangement in pairs anchored on both sides of the bridge. These are shaped like a harp, a fan, or a mix of the two. The pairs of cables can be supported by two separate straight pylons or one single one looking like an inverted Y (see Figure 11.11f).

11.3.3 FURTHER CRITICAL ISSUES

With the exceptions of the rigid frame and cable bridges, the superstructure rests on special supports called bridge bearings. These bridge bearings not only concentrate the weight of the bridge deck, but also allows it to move and rotate slightly. This capability is most critical in longer bridges since the temperature changes between summer and winter result in expansions and contractions that would otherwise create large forces inside the deck. Appropriate bearings are accompanied by expansion joints in the deck itself,

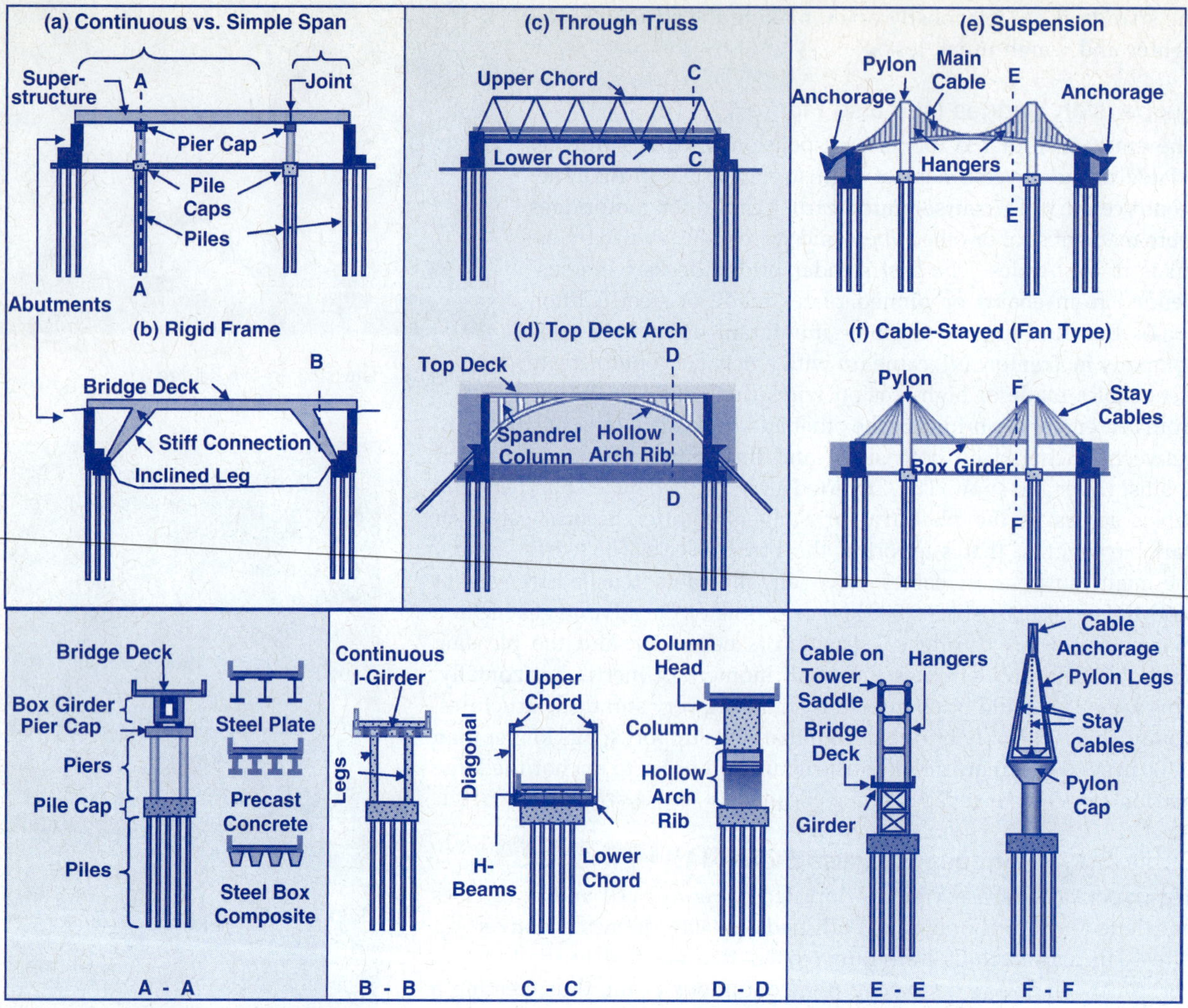

Figure 11.11 Overview of the Most Common Bridge Designs

providing space for the bridge to expand while limiting physical shocks when traffic drives across the joints. As will be discussed later, expansion joints come in various forms and shapes and are installed by specialized companies.

Large weights that travel across the bridge, such as two heavy trucks one behind the other, will cause slight vertical deflections in the center-span, accompanied by a slight rotation at the pier. Therefore, continuous beams bridges need bearings that allow slight rotations.

Many different types of bridge bearings are used by bridge designers. Many different types of bridge bearings, such as metal rollers and laminated rubber blocks, are used by bridge designers. Each type has to be specially designed for the bridge and requires special care and skill to be installed precisely.

The inclined cable stays are exposed to vibrations in the bridge deck that are caused by traffic, wind, and earthquakes. Thus, so-called bridge dampers are installed at each anchor point to absorb abrupt load changes.

Each bridge also has at least two abutments that fulfill three basic functions: (a) provide vertical support for end spans, (b) accept horizontal forces (e.g., arches requiring a special foundation), and (c) stabilize the embankment.

Finally, each bridge construction has to consider a variety of laws and regulations, both federal and state, that will require special measures taken by the contractor. Here are some examples:

- National Environmental Policy Act
- Department of Transportation Act
- National Historic Preservation Act
- Clean Air Act
- Wild and Scenic Rivers Act
- Noise Control Act
- Fish and Wildlife Coordination Act
- The Endangered Species Act
- Water Bank Act

Header Problem 11.1 / The Ocean Bay Connector

The state DOT has just finished its design of the Ocean Bay Connector (see below). The city of Orkin expects to be heavily impacted by the largest project for 30 years in this area. Although it is looking forward to having the main traffic bypass the city, it is concerned about the impact on its citizens and on businesses that have a sizable number of tourists during the summer months enjoying the pristine Orkin Bay area. The city has hired you to review the design, not from a structural, but from a construction, point of view. What has to be expected in terms of construction traffic, road closings, shipping restrictions, and so on? What can the city do or require from the bidding contractors to speed up construction while mitigating its negative effects on its citizens? Figure 11.12 indicates that the connector has been broken up into eight separate sections, each built by a separate contractor. Thus you are asked to treat each section separately as well to elaborate on possible methods of construction, possible transportation routes, main equipment, noise and dust to be expected, and so on. You are also asked to point out construction solutions that would best meet the key goals of the city, supported by your expert opinion.

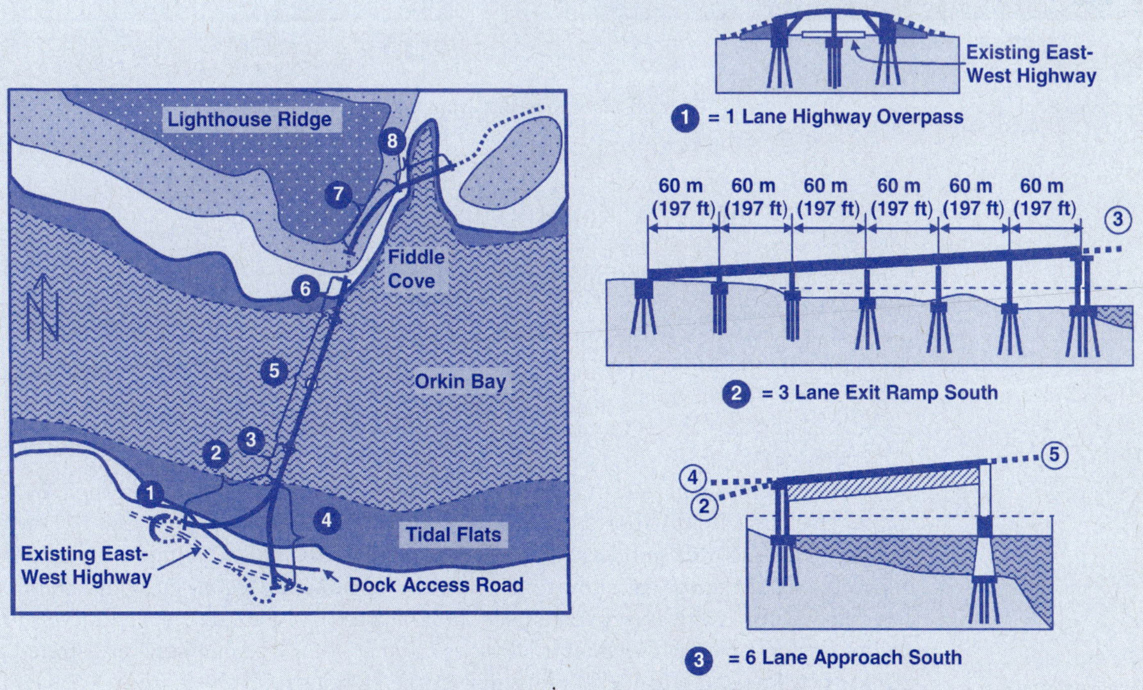

Figure 11.12 Bridges on the Ocean Bay Connector

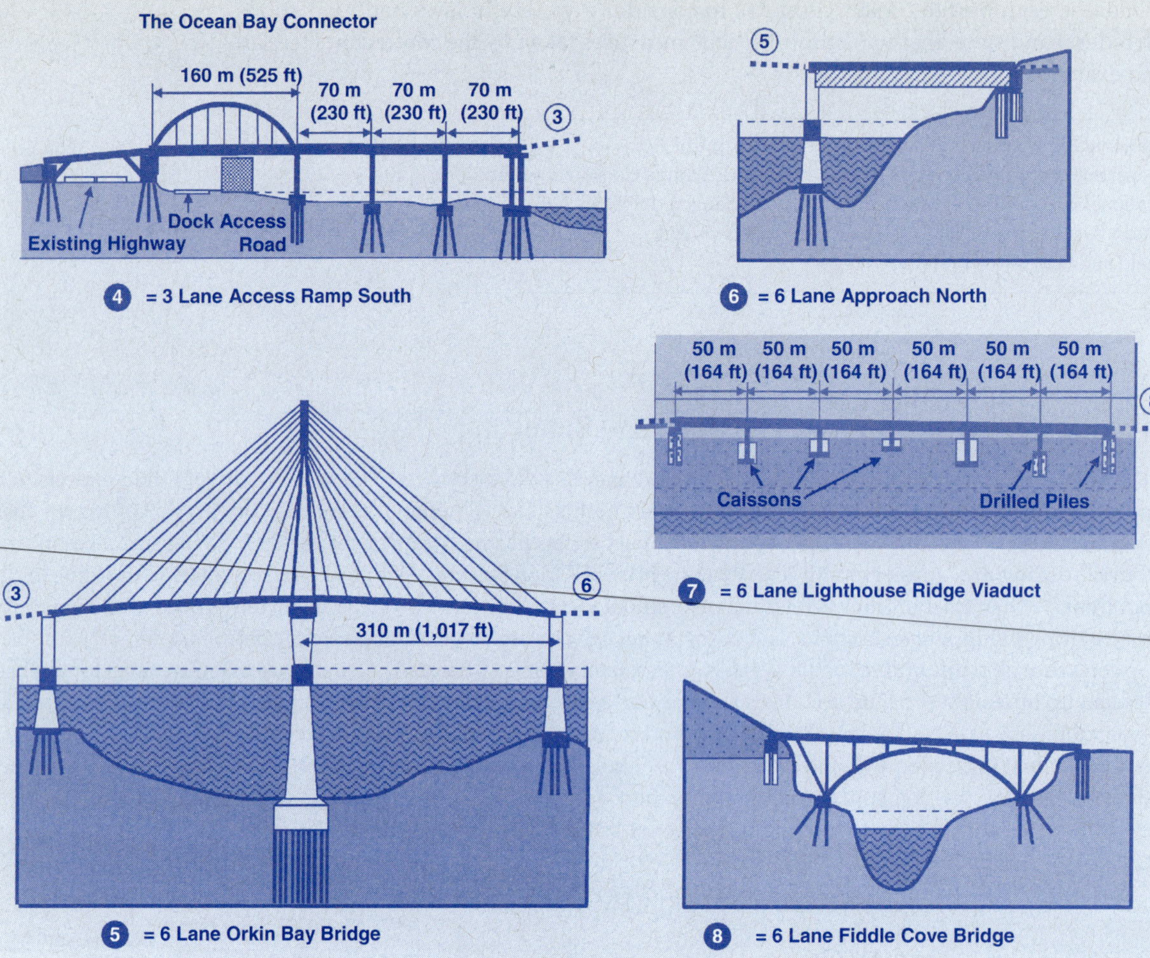

Figure 11.12 Bridges on the Ocean Bay Connector (Continued)

11.4 Properties of Key Building Materials

Three materials share the bridge market: timber, concrete, and steel, with concrete and steel making up the vast majority. While cement as a binder has a long history going back to the Assyrians, modern concrete only goes back to 1756, when the British engineer, John Smeaton, made the first concrete, and 1824, when Joseph Aspdin patented Portland Cement.

11.4.1 THE AMAZING CONCRETE

Average concrete consists of 41% coarse aggregates (gravel-crushed stone), 21% fine aggregates (sand), 110% Portland cement, 16% water, and 5% air (by weight). However, modern concrete mixes add or replace portions with fly-ash and admixtures to improve concrete performance (e.g., flowability.) Recently, cements, techniques, and composites have also made drastic improvements in the compression strengths that engineers can use in the design of bridge elements. It is anticipated that average concrete strengths of 207 to 276 MPa (30,000 to 40,000 psi) will become standard.

S11.1 FHWA High Performance Concrete
S11.2 TRB Concrete Sustainable

Several weeks must pass for concrete to develop its advertised strength. It could be extremely costly if the concrete does not achieve its strength, as it may have to be removed

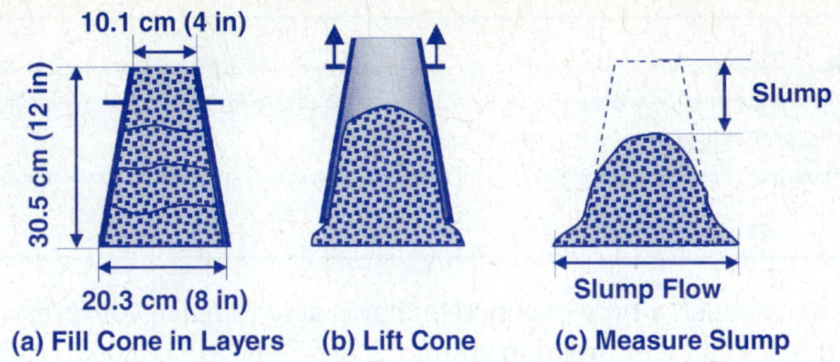

Figure 11.13 Steps to Conduct Slump Test ASTM C 143, C 192, AASHTO T-23, T-119, T-126

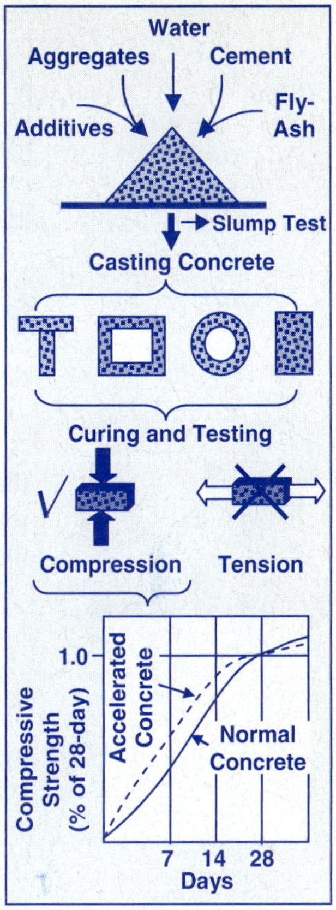

or strengthened. In order to avoid such failures, a series of quality testing procedures have been developed over time, primarily for the cement and aggregates (clean, hard, and durable particles). A crucial process is the apportioning of the different components. The water-cement ratio is the chief factor that controls the cement's final strength. Also affecting strength and durability is the grading of the aggregates, an indicator of how well the distribution from the largest to the smallest aggregate matches a desired recipe. A first measure indicating the appropriateness of the water-cement ratio is the concrete slump test, depicted in Figure 11.13, conducted with fresh concrete directly at the point of delivery. While this test is not able to indicate compressive strength, stable values show that the deliveries contain a consistent concrete mix. The test further determines the workability of concrete, which provides a scale of how easy it is to handle, compact, and cure. For example, a low slump indicates too much stiffness, which makes it difficult or impossible to fill the concrete into the forms and around reinforcing steel. On the other hand, too fluid a mixture is also detrimental. The ASTM C 143 standard describes in detail the exact steps to be followed when conducting a slump test.

For example, for bridge decks the maximum and the minimum slump might be 4 in (11 cm) and 3 in (7.6 cm), respectively. A mix with unacceptable plasticity and cohesiveness crumbles, segregates, and falls apart.

Again, concrete strength grows over time, following a curve that can be predicted after seven days. For this reason, fresh concrete is sampled in multiples, cured and tested at predefined time intervals. When the compression test is properly conducted, a maximum force is obtained at the point at which the concrete breaks. With this force, the compressive strength in megapascal (MPa) or pounds per square inch (psi) is calculated. In order to provide uniformity, ASTM has established appropriate standards such as ASTM C 31/31M-95, *Standard Practice for Making and Curing Concrete Test Specimens in the Field*, and ASTM C39, *Compressive Strength of Cylindrical Concrete Specimens*.

Admixtures for Concrete

Water-Reducers decrease the required water content for a concrete mixture by about 5–10%. Consequently, concrete containing a water-reducing admixture needs less water to reach a required slump than untreated concrete. The treated concrete can have a lower water-cement ratio, leading to a higher-strength concrete.

Retarders counteract the accelerating effect of hot weather on concrete hardening, which make placing and finishing difficult. Most retarders also function as water reducers and may entrain some air in the concrete.

Accelerators increase the rate of early strength development, which is especially useful in cold weather.

Superplasticizers make high-slump flowing concrete that can be placed with little or no vibration or compaction. The effect of superplasticizers lasts only 30 to 60 minutes, followed by a rapid loss in workability. Therefore, they are added at the jobsite.

Corrosion-Inhibitors slow the corrosion of reinforcing steel in concrete, which causes concentrations of chloride. Corrosion inhibitors can be used as a defensive strategy for concrete structures, such as marine facilities, highway bridges, and parking garages, that will be exposed to high levels of salt.

Shrinkage-Reducers minimize the development of "hair cracks," thus creating a surface that is not penetrable by water.

In an effort to establish a type of concrete that is able to reach compressive strengths that far exceed the normal compression strengths of 55–65 MPa (8,000–9,400 psi), High Performance Concrete (HPC) has been developed. The National Concrete Bridge Council defines HPC as "concrete that attains mechanical, durability or constructability properties exceeding those of normal concrete," and the American Concrete Institute (ACI) specifies, "Concrete that meets special combinations of performance and uniformity requirements that cannot always be achieved routinely using conventional constituents and normal mixing, placing, and curing practices" (ACI 116R). Presently, concrete that meets one of the following three criteria would fall under the definition of HPC:

1. Very early strength of 21 MPa (3,000 psi) within 4 hours after placement
2. High early strength of 34 MPa (5,000 psi) within 24 hours
3. Very high strength 69 MPa (10,000 psi) within 28 days

Concrete with higher compression strength means that bridges can be (a) lighter, (b) with more clearance, (c) with longer spans, (d) more durable, and (e) cheaper and with less maintenance cost.

Another high-strength concrete has been termed Ultra High Performance Concrete (UHPC) with a mix as shown in Table 11.2. The unique characteristics are the steel or glass fibers that are added to the concrete to increase its tensile strength, which impacts compressive strength as well. In fact, tests show that the latter is able to consistently reach 124 MPa (18,000 psi), with potential to go even higher.

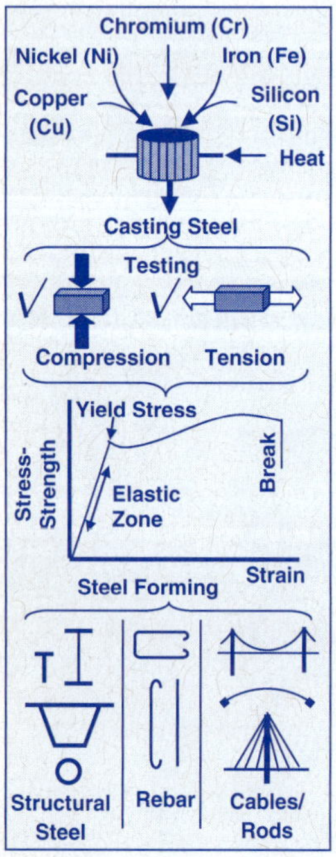

Table 11.2 Typical Mix Recipe for an Ultra High Performance Concrete (UHPC)

Material	Amount (kg/m^3 (lb/yd^3))	Percent by Weight
Portland Cement	712 (1,200)	28.5
Fine Sand	1,020 (1,720)	40.8
Silica Fume	231 (390)	9.3
Ground Quartz	211 (355)	8.4
Superplasticizer	30.7 (51.8)	1.2
Accelerator	30.0 (50.5)	1.2
Steel Fibers	156 (263)	6.2
Water	109 (184)	4.4

11.4.2 STRUCTURAL STEEL

Steel, the second most important material, is much more expensive than concrete but makes up for its higher cost per ton with its targeted use. For example, while concrete has little tensile strength, steel is able to withstand large tension forces. One example is the steel ropes used by hosting cranes or the metal rods and pins used to build backhoe's excavators. The three main applications of steel in bridge construction are structural steel, reinforcing bars and steel cables for pre-stressing, post-tensioning, suspending, and staying.

Due to its different uses, steel can be tailored to specific needs. Again, standards have fulfilled a critical role in ensuring that the many different types of steel that have evolved over time meet minimum specifications that the design engineer and contractor can

(a) Automated Flame Cutting of Steel (b) Spot Welding of Stiffening Plates (c) Semi-Automated Final T-Joint Welding

Figure 11.14 Shaping Flat Steel into Desired Components and Structures

depend on. A common steel used for building the superstructure of a bridge is called Grade 50, which has to meet the standard established by the American Society for Testing and Materials (ASTM) or M270 created by the American Association of State Highway and Transportation Officials (AASHTO). The number after the grade refers to the yield stress of the steel. Thus, Grade 50 is a type of steel with a yield stress of 50,000 psi (345 MPa).

VG11.1 In the Steel Fab Shop

An additional characteristic of special importance to bridge construction is workability to create individual parts, tiny assemblies, or large structural components. For building bridges in steel, the most important metalworking operations are cutting, drilling, grinding, and welding. Figure 11.14 presents pictures that exemplify the workability of steel in manufacturing a steel bridge.

The capability to drill holes very precisely and the manufacturing of bolts made of steel with high tensile strengths are responsible for the widespread use of bolt connections. Specifications for most bolted structural joints call for the use of high-strength steel bolts tightened to a specific tension to create large friction forces between the splice plates, web, and flange. In contrast, a sophisticated set of welding methods has been developed that create strong joints when properly executed.

The earliest known form of welding, called forge welding, dates back to 2000 BC. Forge welding is a primitive process of joining metals by heating and hammering until the metals are fused (mixed) together. Although forge welding still exists, it is mainly limited to the blacksmith trade. The technologies for welding of steel elements, mostly done in manufacturing plants, continue to develop rapidly. Creating even the common weld joints, shown in Figure 11.15, requires extensive training, and welding of load-bearing

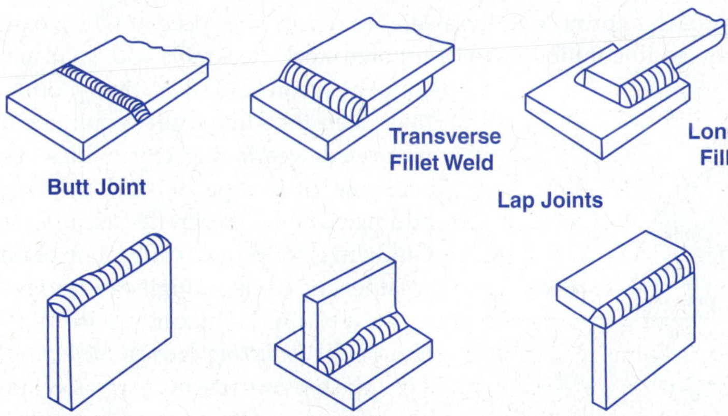

Figure 11.15 Common Weld Joints

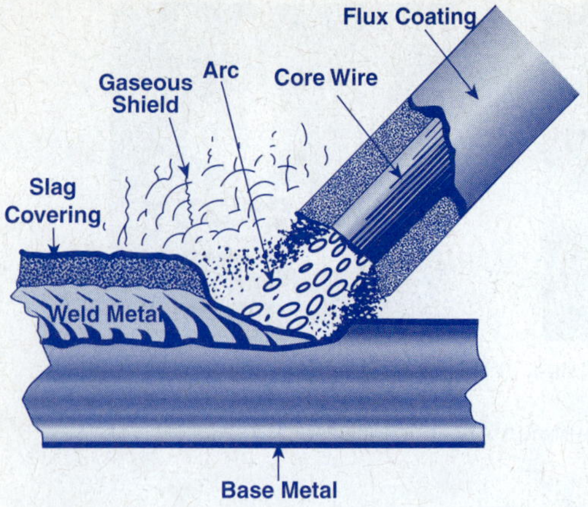

Figure 11.16 Principle of SMAW

parts of a structure should be performed only by qualified personnel. The American Welding Society (AWS) defines welding as "a materials joining process which produces coalescence of materials by heating them to suitable temperatures with or without the application of pressure or by the application of pressure alone and with or without the use of filler material." The two principal welding processes used on structural steel are arc welding (AW) and oxy fuel gas welding (OFW). For structural steel, the so-called **shielded metal arc welding** (SMAW) process is used extensively. AWS defines the arcing mechanism presented in Figure 11.16 as "an arc welding process which produces coalescence of metals by heating them with an arc between a covered metal electrode and the work-piece. Shielding is obtained from decomposition of the electrode covering. Pressure is not used and filler metal is obtained from the electrode." The covering of the electrode or core wire, when reaching high temperature, turns into a gas of hydrogen and/or carbon dioxide that shields the air from the arc, thus minimizing oxidation of the molten metal.

AWS defines the OFW as "a group of welding processes which produces coalescence by heating materials with an oxy fuel gas flame or flames with or without the application of pressure and with or without the use of filler metal." Heat is obtained by a chemical reaction between two gases, acetylene (or hydrogen) with oxygen, as it is discharged from a torch designed for this purpose. Figure 11.14a indicates that the OFW concept of heating two pieces of steel to be joined is being used for cutting steel (oxy fuel cutting) where the torch is guided by a computer-controlled arm able to automatically follow a predefined path such as a circle.

S11.3FHWA Steel Fabrication Europe Japan
S11.4 FHWA Prefabricated Steel Bridge

While steel has several characteristics that are very useful in building safe bridges efficiently, it comes with a major disadvantage, corrosion. Steel exposed to water and oxygen begins to oxidize, a chemical process that eats away at the iron, creating tiny holes that grow to large openings. This anodic reaction produces, in addition to rust (= ferrous ions), electrons that are looking for an "attractive" new partner, which they find inside the water, as shown in Figure 11.17. The cathodic reaction closes the destructive circle by consuming the electrons created during the oxidation process producing hydroxides (OH^-). Eliminating the flow of electrons from the destructive anode is the goal of cathodic protection of rebar used for building bridges. This is accomplished by inducing stronger currents of electrons to flow to the cathodes provided by sacrificial anodes connected to the steel.

Another approach to protecting the steel is to reduce the speed at which oxidation of iron takes place by adding inhibitors to the molten steel. Steel mills add small amounts of copper, silicon, and so on, thus producing steel that meets the same grade requirement but has improved weathering capabilities. This characteristic of a type of steel is indicated by adding a W to the grade, as in Grade 50W. Guidelines for proper application of unpainted weathering steels in highway bridges appear in the FHWA Technical Advisory T 5140.22, *Uncoated Weathering Steel in Structures*.

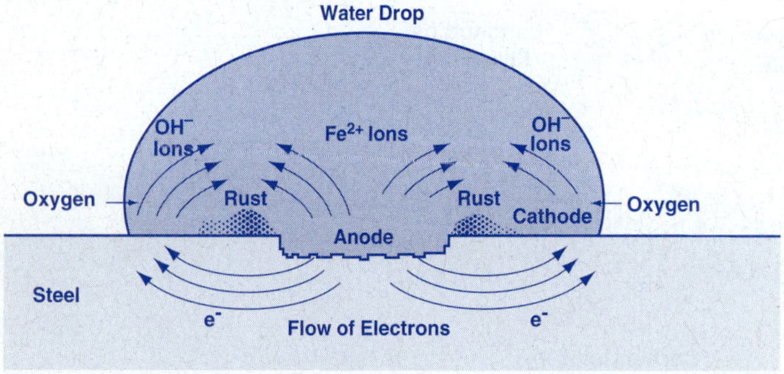

Figure 11.17 Schematic of Corrosion Process

Like the concrete industry, steel producers have raised quality and performance by creating a High Performance Steel (HPS), containing less carbon and sulfur, thus improving weldability and strength. The Grade HPS 70W (485 MPa) is now used for many steel bridges, resulting in lighter and/or smaller sections that can be welded more efficiently, thus reducing construction costs.

S11.5 FHWA Corrosion Protection of Rebar

11.4.3 REINFORCING CONCRETE

As we saw above, concrete has only a minimal capacity to sustain tension forces. In fact, the slight shrinking of concrete during the curing process, when water evaporates, causes small tensile forces inside. Unfortunately, curing is also the time when the concrete is building up its strength, including the minimal tensile strength. To prevent fine cracks from developing, called hair cracks, it is necessary to either eliminate the evaporation of water early in the process and/or counteract the developing stresses. The French inventor-gardener Joseph Monier observed in the mid-1800s that a wire mesh encased into the fresh concrete would serve this function. Lacking a background in mechanics his patent from 1867 focused mainly on the use of wire mesh to create small concrete objects such as flower containers, light posts, or guardrails. Nevertheless, Monier's demonstrations during the 1889 World's Fair in Paris (also featuring the Eiffel Tower) triggered great interest in his concept, especially from bridge builders across Europe. Bridges, of course, have dramatically larger dimensions requiring a large amount of heavy and expensive steel to be added. The two inserts demonstrate the principles that were responsible for the rapid embrace of reinforced concrete by bridge designers all over Europe.

Reinforcing a Simple Concrete Beam

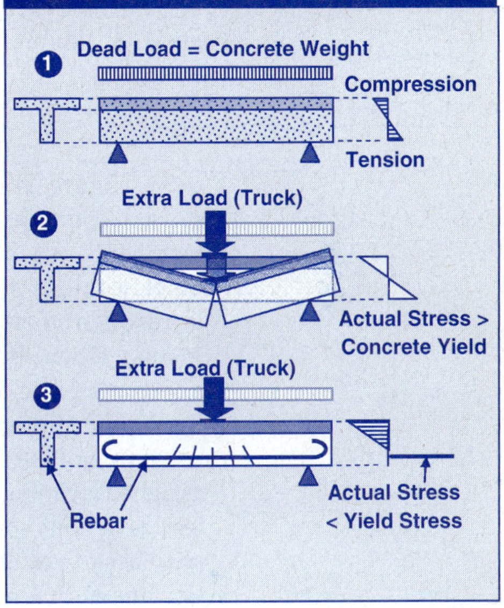

While hair cracks are a natural phenomenon of curing concrete, larger cracks are a sign of overstress, a situation that may lead to an instant failure of a concrete element, such as a beam ❶, ❹. What we call rebar today had its origin in the idea to use steel bars to reinforce concrete at its weak "side," the area where it experiences tensile stresses that lie above its yield strength ❷, ❺. By inserting and anchoring steel bars in those "endangered" places, engineers found a way to take advantage of cheap and compressively strong concrete to build bridges with larger and larger spans. The two inserts show how simply supported and cantilevered beams are reinforced at the bottom and at the top, respectively ❸, ❻.

VG11.2 Reinforcing Bars

Of course, the amount and the location of rebar are related to the total amount of tension that has to be expected when the bridge is completed. The establishment of those forces and the distribution of rebar within the concrete are the subject of the design process. The designer can choose from different grades and specifications that vary in yield strength, chemical composition, and surface deformations needed to improve the bonding between the bar and the surrounding concrete. A detailer will indicate the number of bars, bar lengths, types, bends, and positions. As for structural steel, the grade designation of rebar, typically manufactured in grades 40, 60, and 75 (275, 412 and 515 MPa respectively), gives the minimum yield strength in 1,000 psi. As Figure 11.18 illustrates, there is a wide range of problems that a contractor has to consider when working with concrete reinforcement.

Reinforcing a Cantilevered Beam

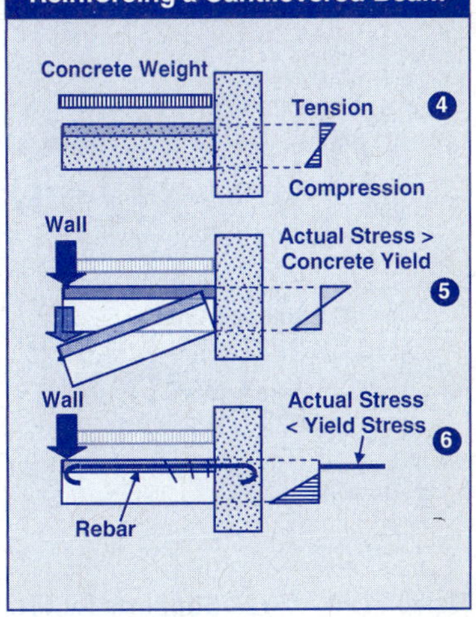

Commonly, rebar is placed right into the formwork after it has been set up. One class of large exceptions, of course, are deep foundations like drilled shafts for which cages of rebar are fabricated and lowered as one

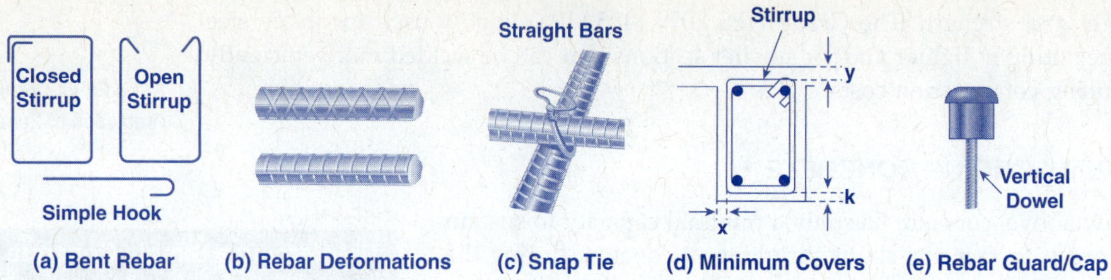

Figure 11.18 Features Associated with the Proper Use of Reinforcing Bars

piece. Figure 11.18d indicates that a beam consists of both straight bars and cross-sectional stirrups. It's the steel crew's problem to figure out how to assemble them. The fact that the shapes, sizes, and weights can vary for every part of a bridge can turn this quickly into a strategic decision that has a major effect on the productivity of the ironworker. For example, a stirrup can be ordered as close or open (see Figure 11.18a). Open stirrups allow the longitudinal bars to be easily placed inside; however, a separate piece of bent rebar closes the opening on the top. The closing bar needs to be tied to the stirrup in at least at four places, requiring extra work.

Tying of the rebar with tie wires is necessary so that the reinforcing steel stays in the assigned location during concrete placement. Incorrect positioning of reinforcing steel can lead to serious concrete structural failures. A variety of tie types (see Figure 11.18c) from snap ties to saddle ties are used in the concrete reinforcing industry. Proprietary snap-on ties are also available, such as tools that help in this extremely repetitive task. For preassembled cages and mats, the ties have to make the assembly rigid enough that it will sustain the strain caused by lifting it with the crane or by hand. Although bars should be placed as close to the specified position as possible, there will always be a little bit of variation. The tolerances on rebar positions are defined by ACI 117, *Tolerances for Concrete Construction and Materials.*

A second reason for tying rebar is to ensure that a sufficient distance to the surface of the concrete is kept. Referred to as concrete cover (see Figure 11.18d), it fulfills two purposes: (1) protection against corrosion by preventing water from reaching the steel and (2) allowing the rebar to bond with the concrete. Bonding, helped by the surface deformations (see Figure 11.18b), allows the concrete to benefit from the high yield strength of the rebar, avoiding the development of large cracks.

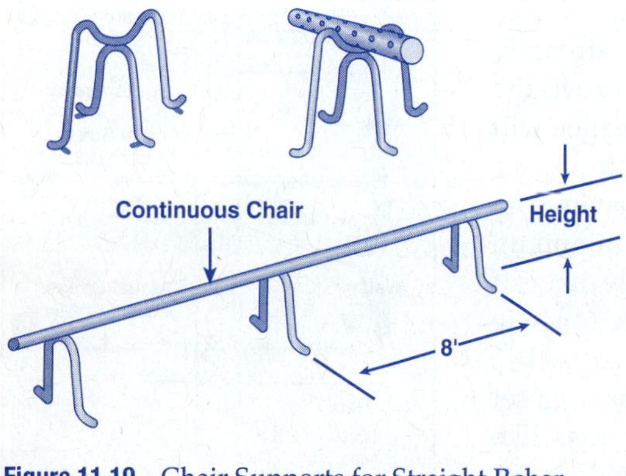

Figure 11.19 Chair Supports for Straight Rebar

The requirements for minimum cover are usually listed in the project specifications or shown on the drawings. Where not specified, the minimum cover for cast-in-place concrete is specified by ACI 318, *Building Code Requirements for Structural Concrete* Section 7.7 "Concrete Protection for Reinforcement." Generally, for concrete cast against earth, the cover should be 3 inches (7.5 cm), for wall exposed to the weather 2 inches (5 cm), beams and columns 1.5 inches (3.7 cm), and for indoor slabs and walls $^{3}/_{4}$ inch (2 cm).

A special problem is to keep the rebar off the formwork, as laborers walking on top of the bars, stored material, and even equipment may cause it to deform or ties and supports to break apart. For this reason, prefabricated chairs for different slab thicknesses and loads are available, made of steel, precast concrete, or plastic. Figure 11.19 illustrates the

Table 11.3 Partial List of Rebar Sizes, Dimensions, and Grades

Size US	Size Metric	ϕ (inch)	ϕ (mm)	Weight (lb/ft)	Weight (N/m)	Availability (Grade)
#3	#10	0.375	9.525	0.376	5.25	60, Epoxy
#4	#13	0.500	12.7	0.668	9.7	40/ 60, Epoxy
#5	#16	0.625	15.875	1.043	15.1	40/ 60, Epoxy
#6	#19	0.750	19.05	1.502	21.9	60, Epoxy
#7	#22	0.875	22.225	2.044	29.8	60, Epoxy
#8	#25	1.000	25.4	2.670	38.9	60, Epoxy
#9	#29	1.125	28.65	3.400	49.6	60, Epoxy
#10	#32	1.270	32.26	4.303	62.7	60, Epoxy

variety of wire chairs with feet that are shielded with a plastic cover, thus avoiding the developing corrosion marks from the exposed steel after the formwork is removed.

The standard concrete cover is not always sufficient to protect the rebar. In particular, bridges that are treated with de-icing solutions have an increased risk of highly corrosive water finding cracks and becoming destructive very quickly. Thus, protective measures used include corrosion-inhibiting admixtures, coating of reinforcement (for example, with an epoxy resin), and use of sealers and membranes on the concrete surface (see ACI 222, *Protection of Metals in Concrete Against Corrosion*, ACI 222R-01). Table 11.3 indicates that rebar can be ordered with the epoxy coating already applied.

The heavy weight of long, large-size rebar has had a negative impact on the health of ironworkers, especially those involved in carrying and placing. For example, one 20 ft (6 m) #10 (#32) weighs 86 lb (0.38 kN). Combined with continuous bending to tie the bars, a human spine is easily overloaded. A sore elastic disk between two vertebrae may be so overstressed that a small section breaks off, as sketched in Figure 11.20. The freed disk pieces will rub and inflame the main nerve system controlling the legs. A deadly risk is rebar that is installed in a vertical orientation, where the bars become spears. Figure 11.18e illustrates one way to guard against the hazard of impalement, as required by the OSHA standard, 29 CFR 1926, Subpart Q, *Concrete and masonry construction.*

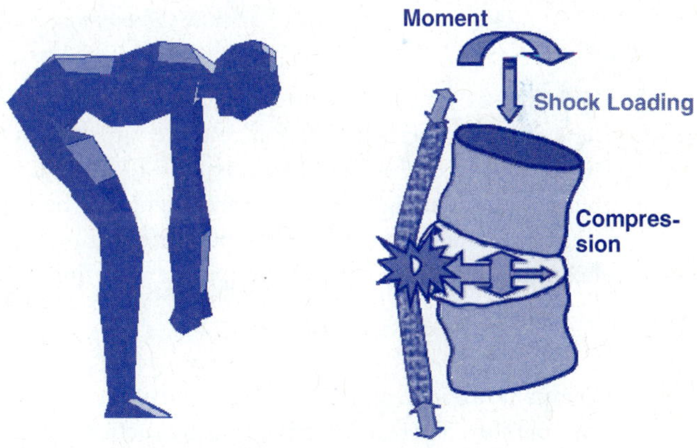

Figure 11.20 Mechanics of Back Injury

11.4.4 BEARINGS AND EXPANSION JOINTS

A bridge superstructure reacts to external forces such as traffic moving across, columns that settle, or changing temperatures that influence the elongation of each material. For example, 100 ft (30.5 m) of concrete or steel elongates approximately 0.01 in for every degree Fahrenheit (0.45 mm/°C). Figure 11.21 presents a hypothetical box girder bridge with four bearings and expansion joints at each abutment. Each of the four bearings has to transmit static and dynamic forces from the superstructure to the substructure while accommodating horizontal movements as well as rotations due to expansion-contraction and loading of the structure they support.

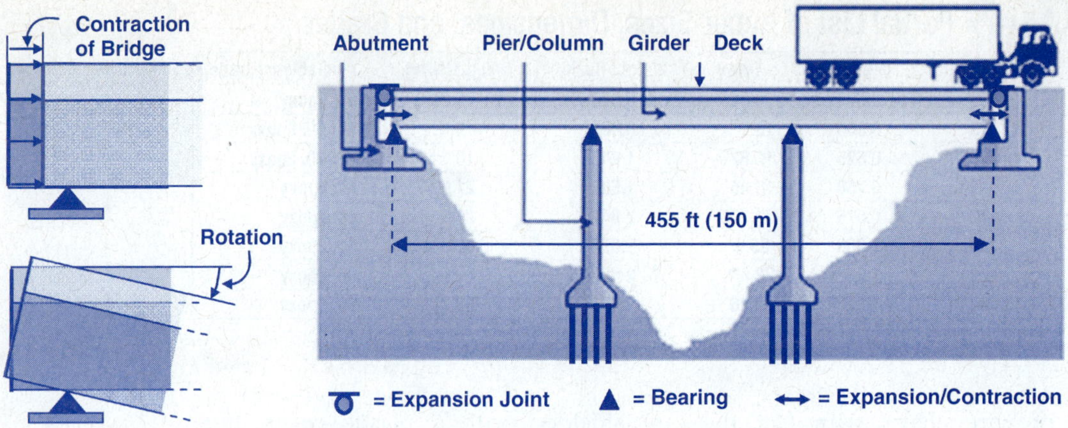

Figure 11.21 A Box Girder Bridge that Is "Alive"

Worked-Out Problem 11.1 / Estimating Expansion-Contraction

You are working for a bridge contractor who wants to check the designer's estimate of how much reinforced concrete will expand and retract at the abutments. You are told that the expected average summer temperature during construction will be 60 °F (15.5 °C). The bridge deck is supported by one continuous cast-in-place concrete girder of 455 ft (150 m) resting on two piers and two abutments.

Assumptions

1. The bridge is simply supported at four points similar to the one shown in Figure 11.3c.
2. The expansion coefficient of concrete is $8 * 10^{-6}$ in/in °F ($14.4 * 10^{-6}$ cm/cm °C)
3. The expansion coefficient of steel is $7.3 * 10^{-6}$ in/in °F ($13.1 * 10^{-6}$ cm/cm °C)
4. The maximum elongation at one abutment will be that caused by one half of the bridge
5. Minimum long-term concrete temperature during winter = 37 °F (2.8 °C)
6. Maximum long-term concrete temperature during summer = 90 °F (32 °C)

Calculation

Temperature increase relative to construction = 90 °F − 60 °F = 30 °F (16.5 °C)

Temperature decrease relative to construction = 60 °F − 37 °F = 23 °F (12.75 °C)

Maximum elongation of concrete = 30 °F * ($8 * 10^{-6}$ in/in °F) * 1,200 in/100 ft = 0.024 * 12 in/100 ft = 0.288 in/100 ft = 23.7 mm/100 m

Maximum elongation steel = 0.288 * 7.3/8 = 0.263 in/100 ft = 9.1 mm/100 m

Maximum contraction concrete = −23 °F * ($8 * 10^{-6}$ in/in °F) * 1,200 in/100 ft = −0.018 * 12 in/100 ft = −0.22 in/100 ft = 18.34 mm/100 m

Maximum contraction of steel = −0.22 * 7.3/8 = −0.2 in/100 ft (15.3 mm/100 m)

Maximum elongation of total bridge (concrete) = (455 ft * 0.288 in)/100 ft = 1.31 in (33.3 mm)

Maximum contraction of total bridge (concrete) = −455 ft * 0.22 in/100 ft = 1 in (25.4 mm)

Total elongation of steel = 4.55 * 0.263 in = 1.2 in (30.5 mm) = minor difference compared to concrete

Total contraction of steel = −4.55 * 0.2 in = −0.94 in (24 mm) = minor difference compared to concrete

Needed opening at one abutment due to elongation and contraction of concrete = (1.31 in + 1 in)/2 = 1.15 in (29 mm)

As the rebar is encased in the concrete the two will expand as one.

Discussion

The joint at one abutment might be blocked, which would result in an expansion/retraction concentrating on one abutment only. In order to add a margin of safety one might choose a joint opening of 2.31 in, which would also consider the possibility of larger temperature variations as the ones assumed.

Bearings and joints are discontinuities in the bridge structure that provide openings for foreign objects such as water, dust, or even pieces of steel or rock. When high-strength steel is used, corrosion has a free hand. It is therefore not unexpected that bearings that do not depend on steel and that are exchangeable have been on the increase. The document AASHTO SBB-1, *Steel Bridge Bearing Design and Detailing Guidelines,* put together by the ASSHTO/NSBA Steel Collaboration, lists three main groups of bearings: (1) elastomeric, (2) high load multi-rotational, and (3) steel.

Elastomeric Bearings: This bearing consists of a solid square or cylindrical piece of rubber sandwiched between sub- and superstructure. It is a maintenance-free bearing involving no moving parts and having a design life in excess of 50 years. Substantial deflections can be avoided through lamination, intermediate metal plates that are bonded into the block. Laminated elastomeric bearings allow horizontal movement by shear deflection and allow rotation by angular deformation, as demonstrated in Figure 11.22.

S11.6 AASHTO Steel Bridge Bearing Design
S11.7 NSBA Bearing Design

In general, the elastomeric bearings are used for simple-span, prestressed concrete beams, simple-span steel girders, and some continuous beams. As shown, movements and rotations are accommodated by compressing or shearing layers of rubber, so these bearings are always deforming elastically. At the same time, the rubber damps vibrations and minimizes the effect of impact loading. If larger longitudinal movements will occur, the bearings can be fitted with a vulcanized PTFE (Polytetrafluoroethylene = Teflon) sheet and possibly a slider plate that accommodate large translations, as shown in Figure 11.22e. Also indicated is the fact that these types of bearings are normally vulcanized to a top or sole plate in order to avoid any movement. In order to allow easy replacement, the bearing can be left unattached but held in place by a guard or a special bridge seat.

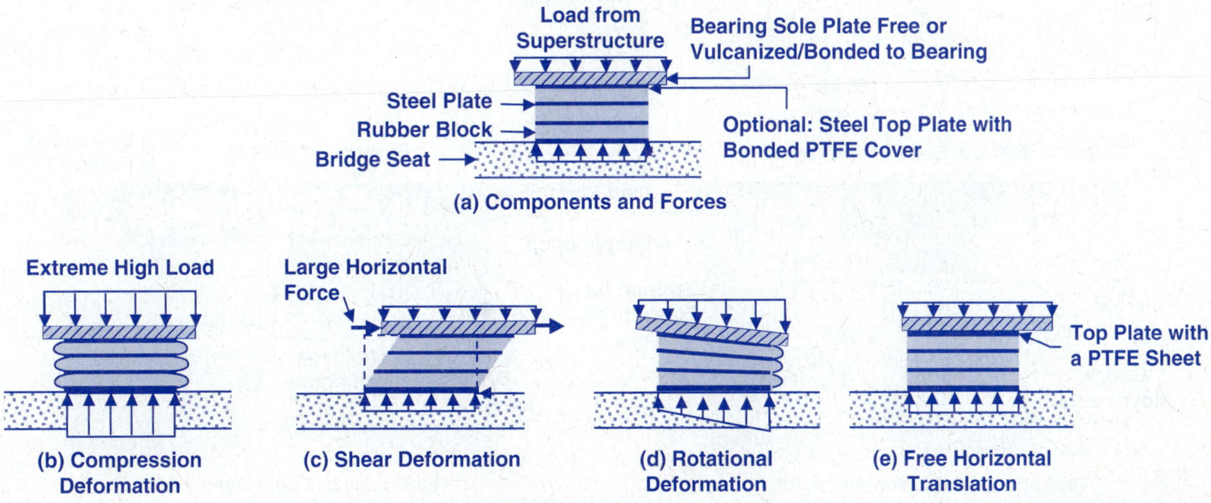

Figure 11.22 Basic Design and Functions of the Laminated Elastomeric Bearing

High Load Multi-Rotational Bearings (HLMR): In cases where the vertical loads are large, as with large steel bridges with a minimal number of girders, elastomeric bearings are not sufficiently strong. HLMR are normally manufactured specifically to fit the characteristics of a new bridge. There are three basic HLMR bearing types currently used: elastomeric pot bearings, polyurethane discs, and spherical bearings. Standard bearings offer a range of vertical load capacities of up to 3,372 tons (3,059 t) and allow for horizontal movements of more than 2 in (50 mm). Unlike the laminated "block" bearings, the rubber or polyurethane is supported by a rim (e.g., pot), while the higher-compression forces are distributed using a thick masonry steel plate on the bridge seat and a top plate. Horizontal translation is achieved by a sliding bearing on one external face.

Steel Bearings: These types of bearings use high-strength steel elements, such as pins and rollers, to allow rotation and translation in one direction only. While their construction is straightforward, corrosion and debris are more difficult to manage. It is therefore not surprising that many bridge owners have experienced long-term problems with steel expansion bearings. Scheduling regular inspection-maintenance and the needed repairs of damaged steel parts also raises the cost. It is, therefore, not surprising that the AASHTO/NSBA committee on Steel Bridge Bearing Design writes: "Where practical, steel bearings should only be considered for fixed bearing types."

Expansion Joints: While bearings support the entire bridge superstructure, the expansion joints support the wheels of the moving traffic when they pass from the "living" bridge to the "solid" abutment. As we learned from the solved problem, it is possible that the gaps between those two objects easily reach 1 to 2 inches (25 – 50 mm). Special mechanisms are used to pass over those gaps, providing continuity to the road surface. The sources of horizontal and rotational movements to be accommodated are identical to the normal bearings. Therefore, expansion joints and bearings for a particular bridge span should be compatible. As with bearings, there are many different joint designs available, which vary in the opening they span and the allowable load. Figure 11.23 presents two designs appropriate for larger joint gaps.

The two examples show the measures to prevent corrosion of the joint hardware and to drain water and debris that is able to get through the unavoidable openings. Vulcanized rubber is used not only to protect against corrosion, but also, as in the bearings, to offer elasticity for elongation and rotation of the bridge deck. For safety reasons, the surface of the finger-type joint is grooved to improve tire grip and drainage.

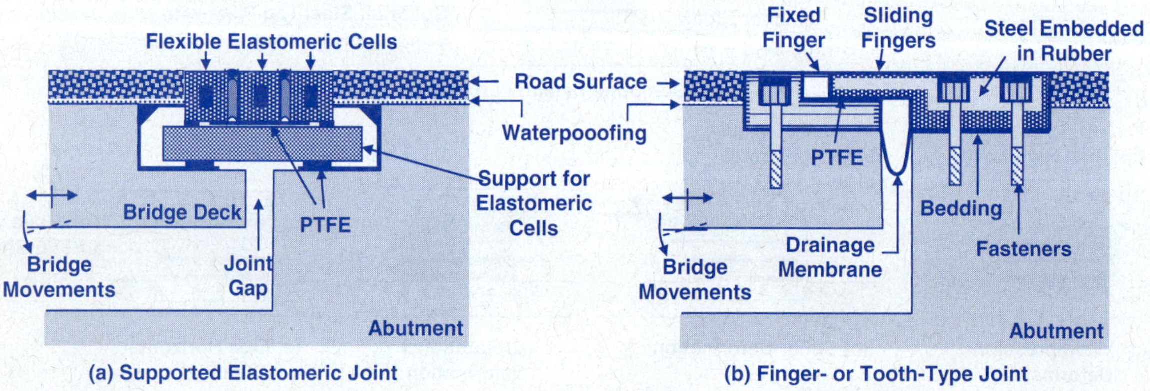

Figure 11.23 Two Common Expansion Joint Designs

11.4.5 PRESTRESSED CONCRETE

Prestressing is a method of reinforcing (strengthening) concrete or other materials with high-strength steel strands or bars, also referred to as tendons. In Chapter 7 we studied the history and use of ropes to lift heavy objects. Cables used in bridge construction differ from steel ropes as they don't have to constantly bend when wrapping around pulleys and winches. On the other hand, their purpose is the same, namely to carry large tension forces. Let's first investigate how cables can improve one of the weaknesses of concrete, its tensile strength.

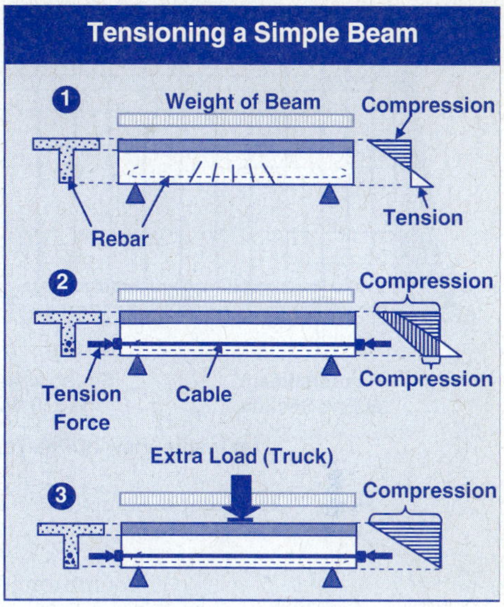

The insert, Tensioning a Simple Beam on the right, shows a concrete beam that is strengthened with rebar or a cable, at the bottom. In order for the rebar to activate its high yield strength, it needs to develop a strain that will require the stiff concrete to crack first, thus creating tiny openings for water to seep in and corrode the rebar (1). As step (2) illustrates, a cable that stretches the entire length of the beam applies, if tensioned with a tightening nut outside the beam, a compressive force that is larger than the tension in the center of the beam (see insert). Adding extra load, such as a truck (3), will create additional tension forces on the bottom, possibly not large enough to counterbalance the compressive effect of the cable.

Figure 11.24 introduces another benefit of using flexible cables hooked together or following a tight curve. The question is how much vertical force we can create if the cable is angled instead of kept parallel to the beam. For this primitive experiment, one cable is suspended at angles between 10 and 45 degrees amid two sidewalls of a steel box. In the middle of the cable rests a weight of 100 kN (11.2 tons). At different angles the tension force is noted and plotted in Figure 11.24b. As expected, the larger the angle, α, the closer the tension force gets, 50 kN or 5.6 tons. At 30 degrees, however, the pulling force reaches 115 kN or 12.9 tons and climbs rapidly to 285 kN (32 tons). The nut that keeps the cable taut pushes against the outside of the box, creating a compression force in reaction to the tension needed to support the weight. As the dashed line shows, its horizontal component is of course slightly lower than the tension force in the cable. From this primitive experiment, we learn that when tension cables are inserted into a concrete beam following a smooth curve, two helpful forces result, a vertical (lift) force and a compression force.

Pre-Tensioning Precast Concrete

Fabricating concrete structures in a factory, rather than on site, provides the opportunity to employ specialized installations that could not be used in the field. On the other hand, prefabricated concrete elements have to be transported. Moving them not only results in transportation costs but also exposes the elements to forces they do not experience after they are put into the bridge. For those reasons, prestressing technology has been embraced by precast companies because it allows them to fabricate long girders that are smaller and lighter. In fact, factory-type production fostered a special type of prestressing called pre-tensioning. Here the wire strands are tensioned before the concrete is placed and released only when the concrete is hard enough to provide sufficient friction and compressive strength to sustain the forces from the tendons.

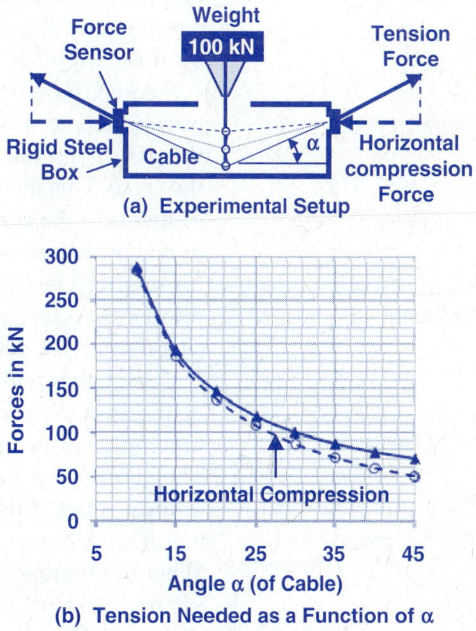

Figure 11.24 Effect of Angled Cables

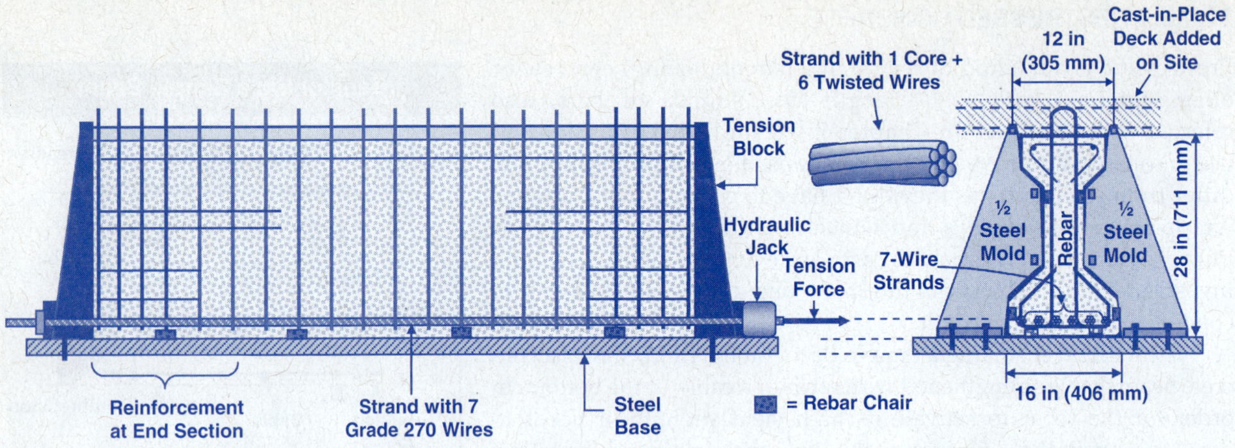

Figure 11.25 Sample Setup to Pre-Tension a Beam with AASHTO Type I Cross-Section (Not to Scale)

A common prestressing tendon used is a seven-wire strand 0.6 in diameter (15.2 mm) made of Grade 270 steel (ASTM A416). Figure 11.25 presents a typical installation for casting bridge beams to a desired length (e.g., 110 ft or 33.5 m).

While Figure 11.25a is not to scale, the setup for casting a prestressed concrete beam shows a common arrangement. Because of repetitive use of the concrete forms, precasters generally prefer molds made of steel. In order to speed up the stripping of the mold, large elements are bolted together and can be lifted with the help of a crane. The two end pieces are important because they have to carry the developing tension forces when the hydraulic jacks pull on the strands, stretching them to 70–80% of their ultimate strength. Before the concrete is poured, the necessary rebar has to be placed and tied with the appropriate supports and chairs to ensure sufficient concrete cover. Once the concrete reaches the required strength, the stretching mechanisms at the end are released, causing the tendons to contract minimally and gain in circumference. Because of the bond between concrete and the strands, the remaining tensile stresses (pre-stressing up to 80% of the ultimate stress) are translated into a compressive stress in the concrete.

S11.8 CPCI Precast Structures

After the curing process has sufficiently advanced, the beam can be lifted and transported to a storage area and to the construction site thereafter. The removal of the piece from the floor will create a suction force that can be 30 to 50% of the weight of the piece. This fact becomes important when designing concrete inserts needed to rig the elements to the crane hook. Naturally, the forces caused by lifting and transporting have to be considered in their design. Figure 11.26 provides a set of different arrangements for rigging a beam, accompanied by the localized stresses that each will cause. As a rule, fewer connection points will cause higher stresses and larger deformations. In fact, the single-point lift will require extreme measures to avoid cracks or a breakup. In addition, inclined ropes will create horizontal forces, which normally will require special hooks to be inserted in the concrete to prevent damage. In the case of the AASHTO bridge beam, looping rebar extends out of the top flange to connect to the cast-in-place concrete deck. Their configuration provides excellent hooking points for option (d) with four or more vertical ropes connected to a spreader beam.

Also, different loading schemes are available during road transport via a trailer-truck. When comparing the three examples shown in Figures 11.26e to g, one sees that the end-supported setup in (f) resembles the eventual situation of a simple beam sitting on two pier caps. The other two scenarios, with cantilevered sections, will require extra rebar in order to survive the transport without cracks in the concrete.

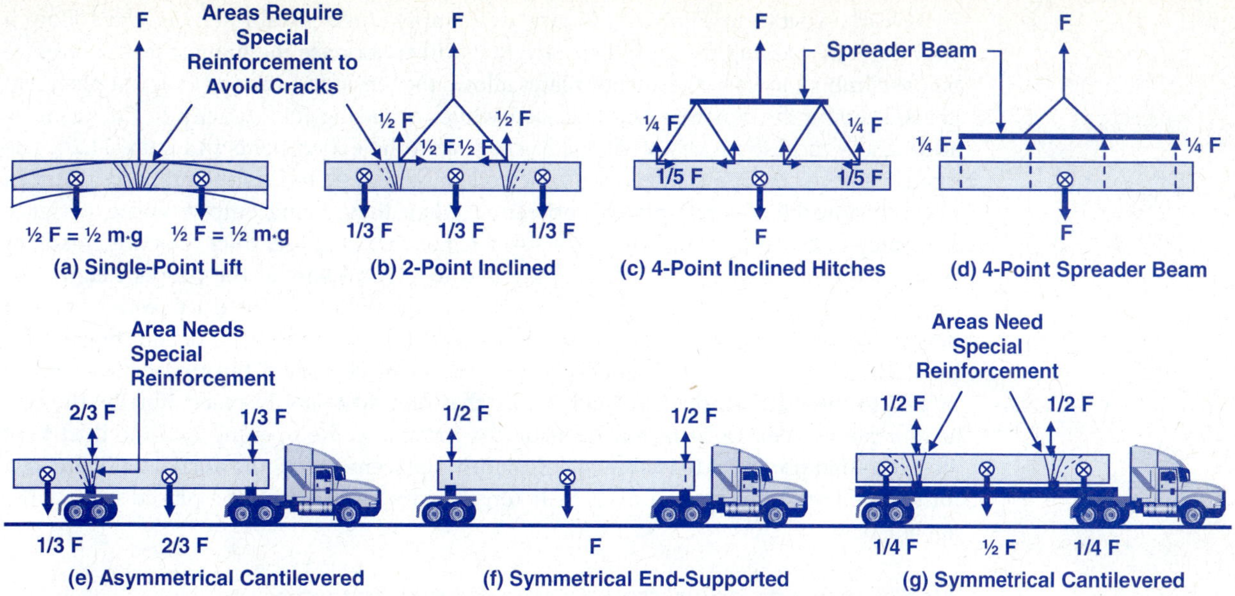

Figure 11.26 Options for Rigging and Transporting a Prestressed Concrete Beam

Post-Tensioning Precast Concrete

The name of the second prestressing method indicates that the tensioning of the wire strands is done after the casting and curing of the concrete. This method arose from contractors' desire to stepping up the tension parallel to the concrete hardening. A core step that enabled this was to separate the installation of the steel strands from the casting operation and instead "block out" a space by fitting an empty duct between the rebar. They serve as conduits for the strands to be inserted and tensioned whenever needed. This flexibility offers several opportunities such as (1) tensioning in phases, (2) installing the ducts in many different shapes instead of a straight line, (3) bonding or unbonding the post-tensioning tendons, and (4) using special grout to protect the steel strands. Let's use the pre-tensioned AASHTO beams discussed above to investigate what would change if post-tensioning is used.

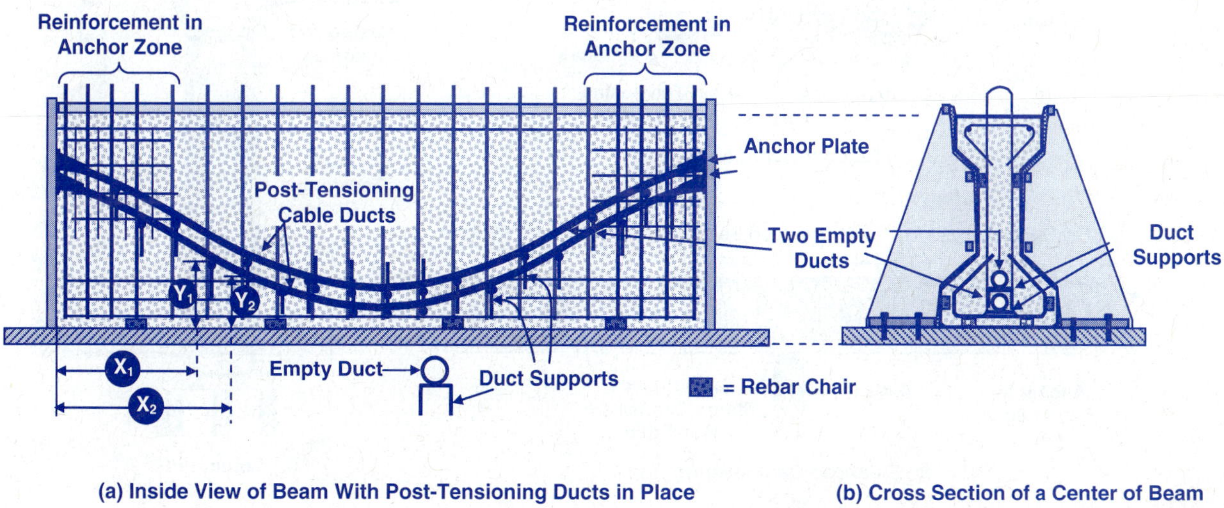

Figure 11.27 Schematic of Casting Prestressed AASHTO Beam Using Post-Tensioning

Clearly visible in Figure 11.27 are two empty ducts flexible enough to follow a curvature. By passing through the space that will experience the highest tensile stresses before climbing toward the anchor plates allows the tensioned strands to accomplish two goals, keeping the concrete cracks closed while increasing the capacity of the beam by adding a vertical force vector, as we learned from the experiments in Figure 11.24. The ducts, made of corrugated steel or plastic, follow a predesigned curve and need to stay in place during the concrete-placing operation when they, being empty, will experience buoyancy forces and may float to the surface. For this purpose, duct supports, made of rebar bent into an L, U, or Z-shapes, are tied to vertical bars at the proper height. For example, at a distance X_1 or X_2 from the left end, the bottom of the duct needs to be at a height Y_1 or Y_2, respectively. Supports should be at intervals of no more than 2 to 3.3 feet (0.6 to 1.0 m). Subsequently, the ducts can be pulled through or lowered to be tied to the duct supports in such a way that they will stay in place. Finally, the two anchor plates need to be fastened inside the form, and the opening into the duct kept closed so that no concrete is able to enter during placement. As the anchors will create a lot of localized compression and shear forces, extra rebar has to be placed behind the anchor plate.

11.4.6 IN-PLACE POST-TENSIONING OF BRIDGE SEGMENTS

S11.9 AASHTO Prefabrication 2004
S11.10 FHWA Prefabricated Bridges
S11.11 FHWA Post-Tensioned Bridges
S11.12 FHWA Testing Post-tensioned Cables

Bridge construction makes great use of post-tensioning technology on the construction site, as no tensioning frames are necessary. As in the precasting application, empty ducts are secured into the prepared formwork with the rebar in place. After the concrete has gained sufficient strength, the tendons are fed into the ducts before hydraulic jacks are used to tension the cables all at once or in phases. Figure 11.28 presents three examples where post-tensioning finds important uses.

All three applications take advantage of the capabilities offered by post-tensioning cables. Instead of supporting the column head of a highway with a wide wall, the cables reinforce the cantilevering beam, enabling it to be supported by a slender column. In the second example, it serves the construction of bridges without the need for shoring

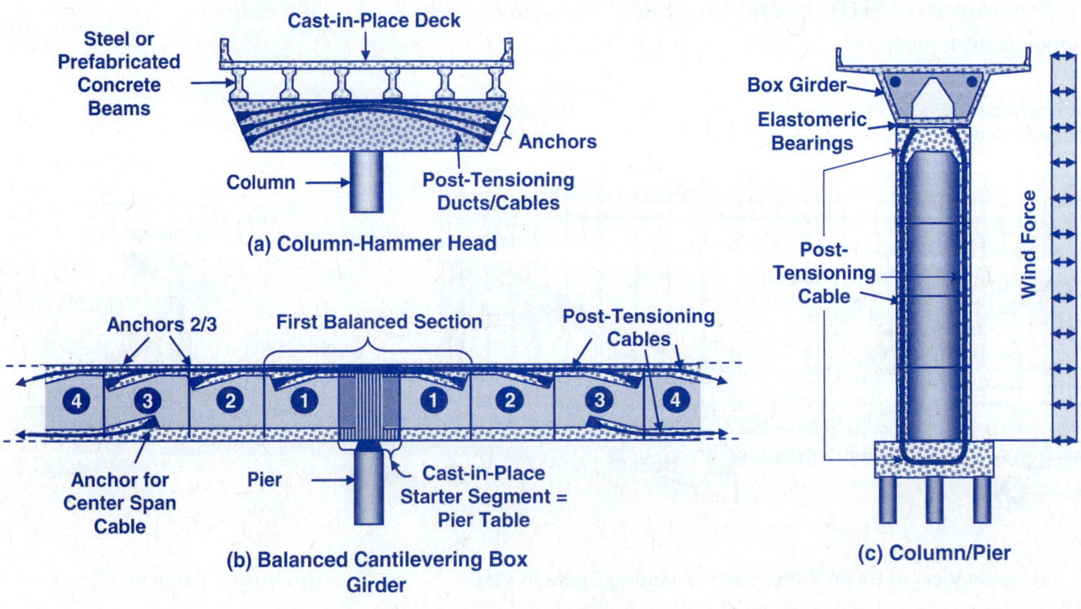

Figure 11.28 Key Applications of Post-Tensioning in Bridge Construction

between the columns. The shoring is replaced with a clever use of tension cables or bars that are placed symmetrically from one side of the column to the other. After the starter segment on top of the column is complete, sections 1 will be built at the same time while supported by a temporary structure sitting on top of the starter segment. After completion, strands are inserted into the duct, connecting the balanced sections, anchored and tensioned. Subsequently, the temporary support structures are positioned on segment 1 to build 2, and so on. In the following section, the various methods for this fascinating concept will be discussed in full.

A totally different use of post-tensioning is presented in Figure 11.28c. Here a pier is built as a hollow vertical shaft. In a strong wind, the shaft will experience forces similar to horizontal cantilevering beams. With the goal of keeping the shaft as slender and light as possible, post-tensing cables are used to provide sufficient tensile strength in the concrete from top to bottom. Since the wind can come from all directions, the cable is placed on all sides, tying into the pile cap.

A vital process of the system is the tensioning of the cables at one or both anchoring points. In the following section, we take a closer look at the equipment and methods used for this purpose.

Post-tensioning hardware consists of two main components: (1) anchorage and (2) duct with wire strands. Figure 11.29 illustrates two basic anchor designs, stressing and dead-end anchors, each with various designs offered by different manufacturers. If both ends are accessible, the seven helically wound wire strands will be cut and inserted only after the concrete has cured. Again, the strand should be of Grade 270, 270 ksi (1,860 MPa), conforming to the requirements of ASTM A 416, "Standard Specification for Steel Strand, Uncoated Seven Wire Strand for Prestressed Concrete." Also shown are two drains and a vent at the low and middle points of the duct. Encased in concrete, they are drained and cleaned out before being filled with cement-based grout bonding the strands with the structures while protecting them from corrosion. According to FHWA's "Post-Tensioning Tendon Installation and Grouting Manual," adding fly-ash to the grout mix improves corrosion resistance in aggressive environments, and dry silica fume (micro-silica) improves resistance to chloride penetration. The water-cementitious material ratio should be limited to a maximum of 0.45 to avoid excessive water retention and bleed, and to optimize the hydration process. Any temptation to add water to improve fluidity on site must be resisted at all times. Fluidity may be enhanced by adding a high range water-reducer, HRWR (Type F or G).

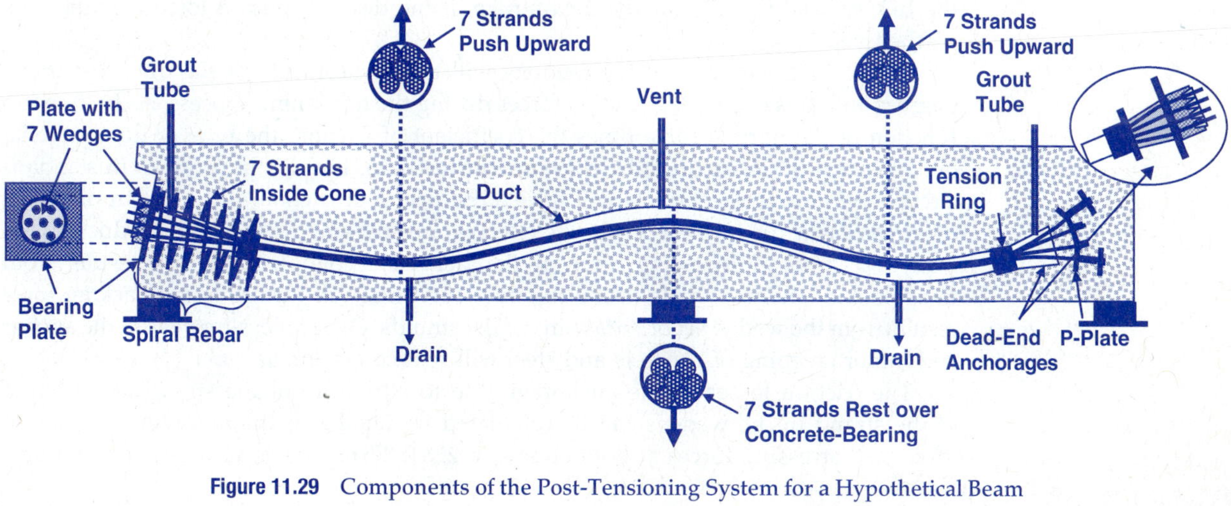

Figure 11.29 Components of the Post-Tensioning System for a Hypothetical Beam

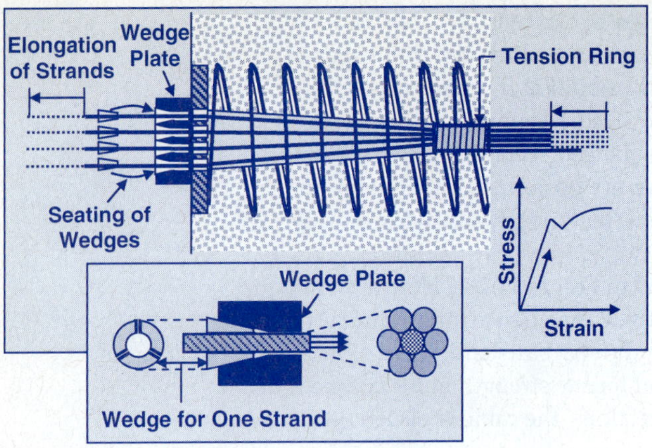

Figure 11.30 Principle of Strand Wedging

Of particular importance is the proper construction and installation of anchors to include the space between the tension ring and bearing plate where the individual strands are separated wide enough to feed through the cylindrically openings in the wedge plate. As shown in Figure 11.30, the area is strengthened with a spiral reinforcement to resist the large stresses introduced by the steel bearing plate that would crush the concrete locally. The stressing of a multi-strand cable requires wedges with serrated teeth that grip each strand without letting it slip during pulling. A wedge assembly, made of case-hardened low carbon or alloy steel, typically consists of two- or three-part wedges with a retainer clip in a groove to keep them together. In order for the wedge to grip the tendon firmly, it has to move slightly with the strand, a process that is called wedge seating. FHWA requires that the wedge plates meet a test requirement, "Acceptance Standards for Post-Tensioning Systems."

Pulling one strand will strain each wire creating tensile stresses inside the steel (see strain-stress curve). As we learned, steel responds elastically until it reaches its yield stress. This fact is used to control the tensioning process, as in the above case, which includes seven strands of standard diameters of either 0.5 in (13 mm) or 0.6 in (15 mm). Knowing that the yield strength of the wires used in post-tensioning is 270 ksi, it is simple to calculate the maximum pulling force a hydraulic jack should apply. Figure 11.31a depicts a common setup used on site to tension a multi-strand bridge cable. The bearing plate, which distributes the final tension force to the concrete, has a sufficient number of holes for each strand to pass to and through the wedging plate. In fact, the tendons have to be long enough to reach through the hollow section of the hydraulic stressing jack and the jacking wedge plate. Important elements are the wedge springs that will be activated to push the wedges after a jacking stroke has been completed, securing the tendons from contracting again. The first jacking sequence begins with the seating of the jacking wedges after each has been pulled to a common starting tension. Next, hydraulic pressure is slowly increased, which activates the ram that moves one stroke length ΔL while pushing the jacking plate with the wedged-in strands. At this time, the wedge springs can be activated to take hold of the tendons at the bearing plate. Then the hydraulic ram can be reversed by the length of a stroke allowing the resetting of the jacking wedge plate for the next stroke, if the desired tension force has not been reached yet.

Figure 11.29 shows that the tendons will touch the inside of the duct to various degrees and thus experience friction forces during the tensioning process. As friction is a function of the normal force times the coefficient of friction, the type of duct material becomes important, as does the curvature of the duct. In fact, the tendons in standard circular galvanized steel ducts will have a coefficient of friction $\mu = 0.20$, but only 0.14 when in plastic ducts. As a result, the effective prestressing force along the duct differs from the initial prestressing force at the jack. A second friction loss results from unavoidable tendon wobble inside the duct curvatures. A further loss of jacking force results from the wedge set or "draw-in" of the strands. Over time, slippages in the anchor and minor creeping of concrete and steel will reduce tension as well.

The friction losses in the anchorage due to curvature of the strand and friction of the strand in the wedges can be calculated in detail. For the following estimation of needed stressing forces at both ends of a 228 ft (75 m) beam, we assume a uniform loss.

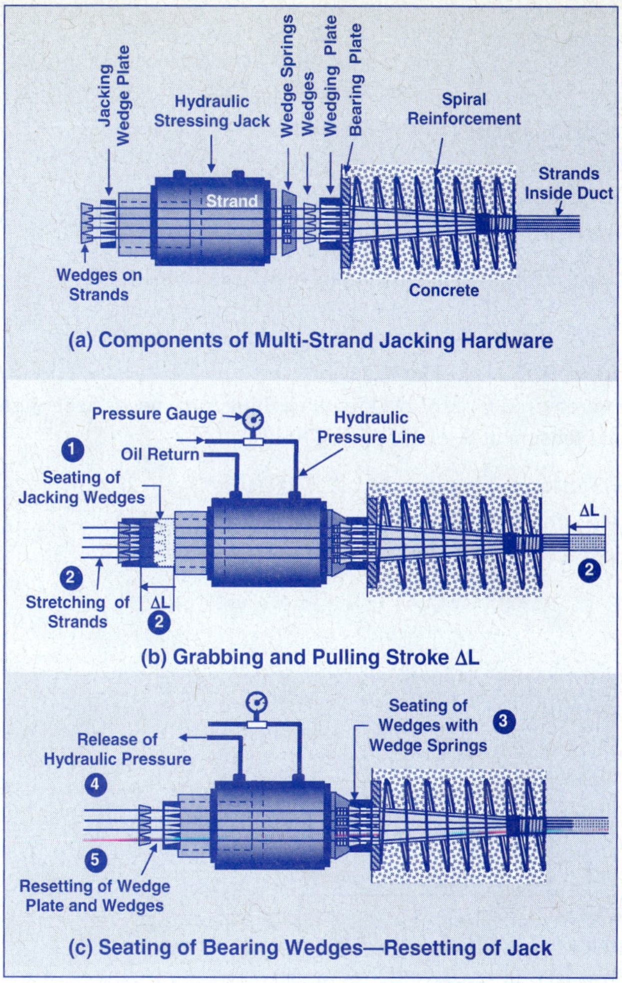

Figure 11.31 Process of Tensioning Multi-Strand Cable

Worked-Out Problem 11.2 / Calculating Stressing Forces

The minimum tension force in a 300 ft (91 m) symmetrical bridge with two anchor points A and B is 175 kip (778 kN). What are the necessary jacking forces at the two anchor locations A and B? What is the final distribution of tension forces inside the duct before grouting?

Assumptions

Only one jacking system is available.
The hydraulic gauge pressure is accurate and can be used to calculate the jacking force.
Friction between tendons and duct is uniformly distributed, 0.24 kip/ft (3.5 kN/m).
Both anchors are accessible for the installation of the jacking system.
The wedge seating, after jacking results in a drawback of the tendons, is about 3/8 in (10 mm).
1 in (25 mm) of cable elongation/length reduction will change the force by 100 kip (445 kN).
The drawback during wedge seating also reduces the friction effect as the tension decreases. This reduction in tension will affect the cable with −1.1 kip/ft (16 kN/m).

Calculations with Results Presented in Figure 11.32

Stressing at Anchor A

Friction losses over the length of the bridge 300 ft $*$ 0.24 kip/ft
= 72 kip (91 m $*$ 3.5 kN/m = 319.5 kN)
Required jacking force in A to achieve 175 kip in the middle of the bridge
= 175 kip + 72/2 kip = 211 kip (778 kN + 319.5/2 kN = 938 kN)

Resulting tension force in B = 211.0 kip − 72 kip = 139 kip (938 kN − 319.5 kN = 618 kN)

Force loss due to wedge seating (3/8 inch draw back) = 0.375 in $*$ 100 kip/in = 37.5 kip (167 kN)

Tension at A after wedge set = 211 kip − 37.5 = 173.5 kip (772 kN)
Since this is smaller than the minimum required force of 175 kip, the jacking force needs to be slightly increased to 213 kip (947 kN), resulting in a final tension at A of 175.5 kip (780 kN)

Length of tension reduction effect = 37.5 kip/1.1 kip/ft = 34 ft (167 kN/16 kN/m = 10.4 m)
Maximum force after wedge is set = 213 kip − (34 ft $*$ 0.24 kip/ft) = 213 kip − 8.2 kip
= 205 kip (947 kN − 3.5 kN/m $*$ 10.4 m = 911 kN)

Stressing at Anchor B

Initial tension = 149.6 kip (1,031 MPa)
Required jacking force = 213 kip (1,469 MPa)
Force loss due to wedge seating = 37.5 kip (258 MPa)
Final tension at anchor B = 175 kip (1,207 MPa)
Maximum tension 34 ft from the end = 205 kip (1,413 MPa)

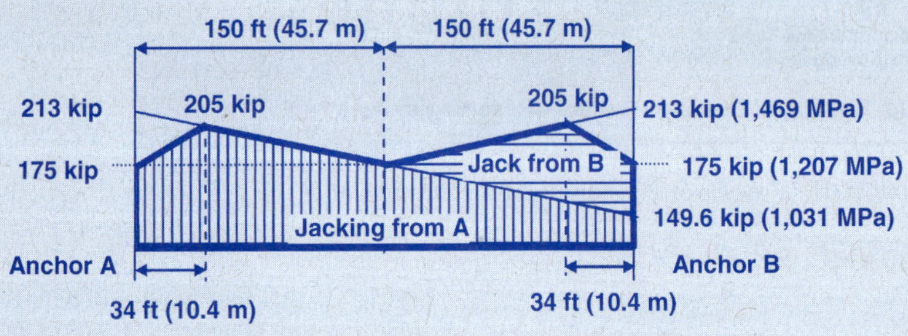

Figure 11.32 Stressing Forces Along the Duct

Locating the post-tensioning cables externally—outside the concrete for the web of the girder—means that no ducts have to be placed and secured during rebar placement, while friction losses during tensioning are eliminated. In addition, because the strands are visible, as compared to being grouted in, visual inspection is made very easy, and bad tendons can be replaced. While these advantages had been well known for a long time, only the development of effective corrosion protection material provided the necessary impetus to apply the external tensioning method to bridge construction.

11.4.7 SUSPENSION BRIDGE CABLES

Suspension bridges owe their elegance to wire cables, which are spun, spliced, and eventually heavily protected against corrosion. The main concept of using steel wire to

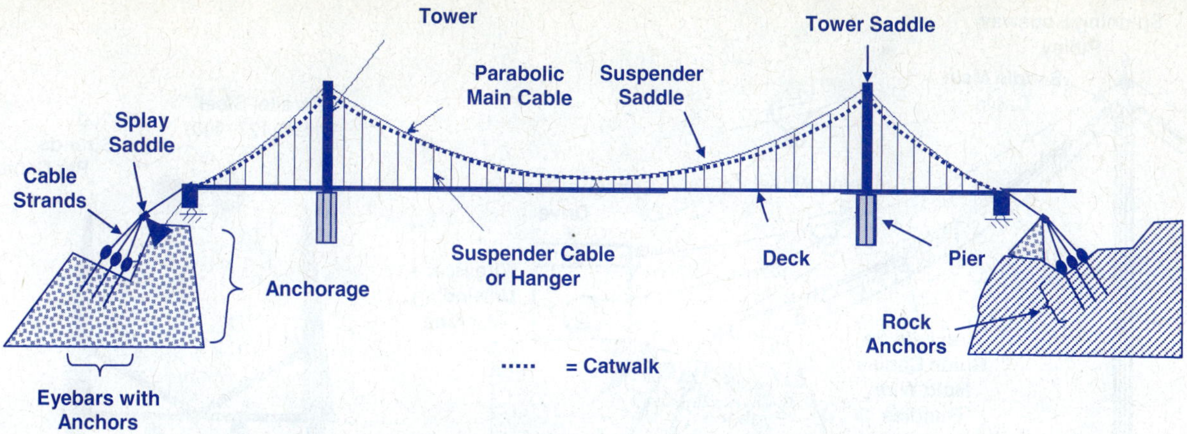

Figure 11.33 The Functions of Steel Cables on Suspension Bridges

build a bridge was developed by John August Roebling in the 1840s and provides the main structure for suspension bridges that reach phenomenal lengths (see Table 11.4). Figure 11.33 shows schematically the main steel wires cables used in building a suspension bridge.

VG11.3 Suspension Bridge Ben Franklin

After studying the theory of prestressed concrete, one can easily recognize the similarities and differences. Most importantly, one of the key functions of tension cable in concrete, preventing the concrete from cracking, is not needed. However, the capability of parabolic cables to carry vertical loads when in tension is fully utilized. Due to the magnitude of the cables, the anchorage increased in size. In the following section, we will take a quick look at the key elements of the cable structure: (1) main cable, (2) hangers, and (3) anchorage.

Installing the Main Cable: There are two methods used to build the main cables. The first is spinning the cable by laying each individual wire into a crate, as John Roebling did in 1850s. In the second method, strands of steel wires are prefabricated in a

John Augustus Roebling (1806–1869)

This German-born civil engineer is known as the "father" of the Brooklyn Bridge, a suspension/cable stay hybrid. His son Washington, together with his lawyer wife, Emily Warren Roebling, opened the bridge in 1883; Roebling had died of tetanus in 1869. During the 1840s John Roebling studied and experimented with steel wires, which he wound helically into strands and cables. Subsequently, he developed methods to "spin" straight wires from anchorage to anchorage, to squeeze the strands into a circle with a movable jack and to protect the cable against corrosion by wrapping it with a wire that is later painted. This basic concept continues to be applied.

Table 11.4 Historical Development of Suspension Bridges

Name of Bridge	Country	Year	Mid Span (m)	Mid Span (ft)
Clifton	England	1860	214	702
Cincinnati	USA	1865	322	1056
Brooklyn	USA	1883	480	1574
Philadelphia-Camden	USA	1926	535	1755
George Washington	USA	1931	1066	3496
Golden Gate	USA	1937	1280	4198
Tacoma Narrows	USA	1940	853	2798
Mackinac	USA	1958	1158	3798
Firth of Forth	Scotland	1964	1002	3287
Verrazano	USA	1964	1298	4257
Severn	England	1966	988	3241
Humber	England	1981	1410	4625
Great Belt	Denmark	1997	1624	5327
Akaski Kaikyo B	Japan	1998	1990	6527
Runyang	China	2005	1490	4887
Xihoumen	China	2007	1650	5412

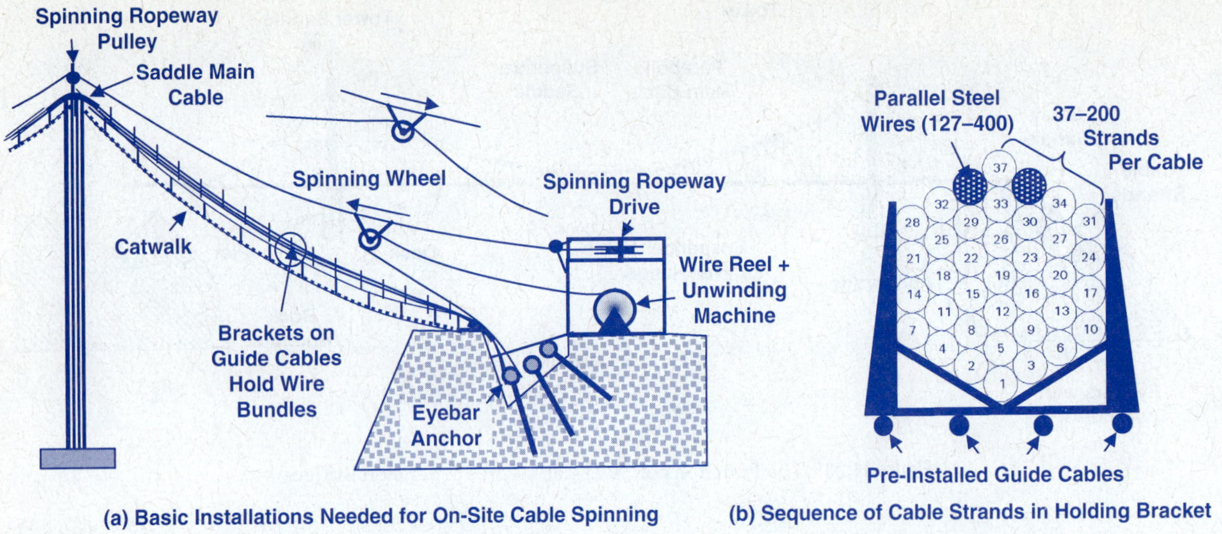

Figure 11.34 Principles of Spinning Main Cable

plant and pulled from one anchorage to the other. Unlike the seven-wire twisted prestressing cables, the main suspension cables are made up of a large number of parallel wires. For example, one main cable for the New San Francisco–Oakland Bay Bridge, with a main span of 385 m (1,174 ft) has a diameter of 78 cm (2.6 ft), consisting of 17,399 galvanized parallel wires bundled into 137 strands with 127 wires each. The minimum strength for a wire is given as 1,800 MPa (265,000 psi). For its construction, the prefabricated wire strand method was used. The following will discuss the aerial spinning process, which has still many benefits.

The spinning process requires completed anchorages, towers, and reels of high-strength steel wires. As Figure 11.34 indicates, however, additional mechanisms and temporary structures are needed.

First, the laborers need a catwalk all the way across to assist the laying process, in particular the straightening out of crossed wires, the bundling of strands, and their proper sequencing as shown in the cross-section view in Figure 11.34b. To support the holding brackets for the wires and to ensure that the final cable follows the predesigned parabolic curvature, guide cables are put in place and surveyed to create the perfect shape. Finally, a ropeway or cableway that carries the spinning wheel back and forth between the anchorage is needed. As depicted in Figure 11.34a, the wheel pulls the loop of an endless wire that is unwound from a reel, called the live line, while the other leg of the wire, called the dead line, drops into the holding brackets. The unwinding machine ensures that the tension is constant during the spinning. As each wheel reaches the opposite anchorage, the loop is transferred to a strand shoe connected to a pair of eyebars that are anchored in either concrete or rock. On the way back, the wheel lays the previous live line, which is now anchored on the other side. To speed up this tedious process, multiple wires are "spun" with the same wheel. Following the placement (of 17,399 wires into 127 strands), the compaction or squeezing of the wires into a circular shape followed by a tight wrapping can begin. The goal of this activity is to protect the wires against corrosion from moisture that could enter the cable. Problems with older suspension bridges have revealed that water already collects inside the wrapping during construction but also penetrates the coating after micro-cracks develop. Hence, new material and procedures are used today. One example is displayed in Figure 11.35.

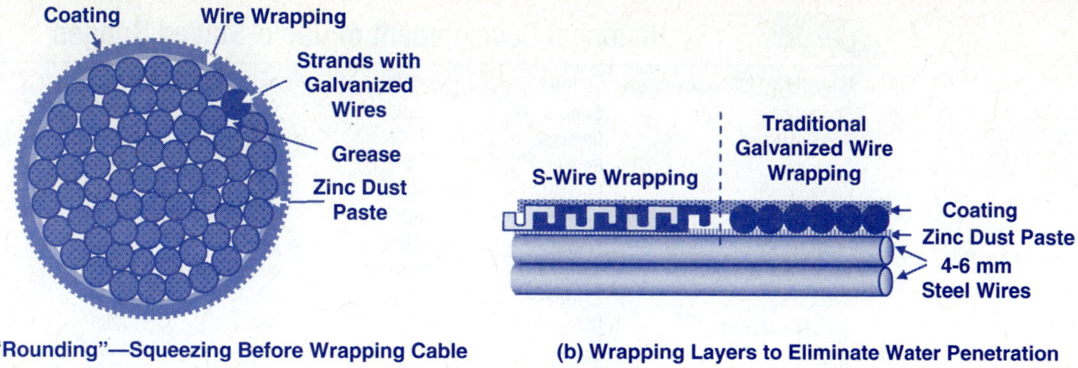

Figure 11.35 Creating a Protected Main Cable

A jacking and wrapping mechanism travels along the cable and compacts the strands into the final circular form while also applying the wrapping. While neoprene wraps and resin coating have been used to replace the "leaky" round wire wraps, the New San Francisco–Oakland Bay Bridge utilized S-wire wrapping developed in Japan. In addition, a dehumidifying system was used to remove all the moisture that had accumulated during construction.

Finally, the suspenders or hangers can be connected to the main cable, an operation that begins with the affixing of a cable band, as depicted in Figure 11.36. The metal band consists of two semi-cylindrical halves, connected by high tensile steel bolts. It goes without saying that the bands have to fit the circumference of the wrapped cable. In particular, they need to be able to create sufficient friction where the main cable is on an incline, without limiting the movement of the wires in a longitudinal direction. Because wires contract slightly when they are stressed during the installation of the heavy bridge deck, the bands need to have sufficient space between the halves so that the bolts can be tightened in the last phase of the project.

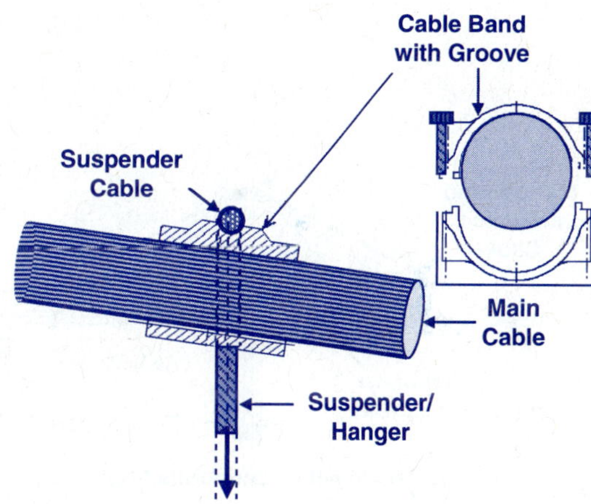

Figure 11.36 Connection of Suspenders to Main Cable

11.4.8 STAY CABLE SYSTEMS

While suspension bridges were still expanding their "reach," in 1955 a "cousin was born" in Strömsund, Sweden, designed by the German engineer Franz Dischinger. The Strömsund bridge is generally considered the first modern cable-stayed bridge. Here, one steel cable was assigned both of the functions that two cables had to fulfill in a suspension bridge. In other words, a stay cable had to serve as a suspender without the benefit of a main cable. Of course, steel wires can be manufactured with a much higher tensile strength (e.g., 1,670 MPa) than structural steel (350 – 500 MPa). The cable-stayed bridge has been used for a span range from approximately 150 m to 900 m, moving into a range that previously has been the domain of suspension bridges (see Table 11.5).

There are two major classes of cable-stayed bridges: in a fan design, shown in Figure 11.37a, the cables all connect to or pass over the top of the tower in a saddle, similar to suspension bridges. From our experiment in Figure 11.14 we learned that the steeper a cable, the more vertical force it is able to carry and the lower the horizontal

Table 11.5 Historical Development of Cable-Stayed Bridges

Name of Bridge	Country	Year	Span (m)	Span (ft)	Material
Strömsund	Sweden	1955	183	600	Steel
Theodor Heuss	Germany	1957	260	853	Steel
Severins	Germany	1961	302	991	Steel
Knie	Germany	1969	319	1046	Steel
Duisburg-Neuenkamp	Germany	1970	350	1148	Steel
St. Nazaire	France	1975	404	1325	Steel
Barrios de Luna	Spain	1983	440	1443	Concrete
Alex Fraser	Canada	1986	465	1525	Composite
Sunshine Skyway	USA	1987	366	1200	Concrete
Iguchi	Japan	1991	490	1607	Steel
Kvarnsund	Norway	1992	530	1738	Concrete
Normandy	France	1995	856	2808	Steel
Tatara	Japan	1999	890	2919	Steel
Rama VIII Bridge	Thailand	2002	300	984	Composite
Rio-Antirrio	Greece	2004	560	1837	Composite
Millau Viaduct	France	2004	342	1122	Composite
Incheon	South Korea	2010	800	2624	Composite

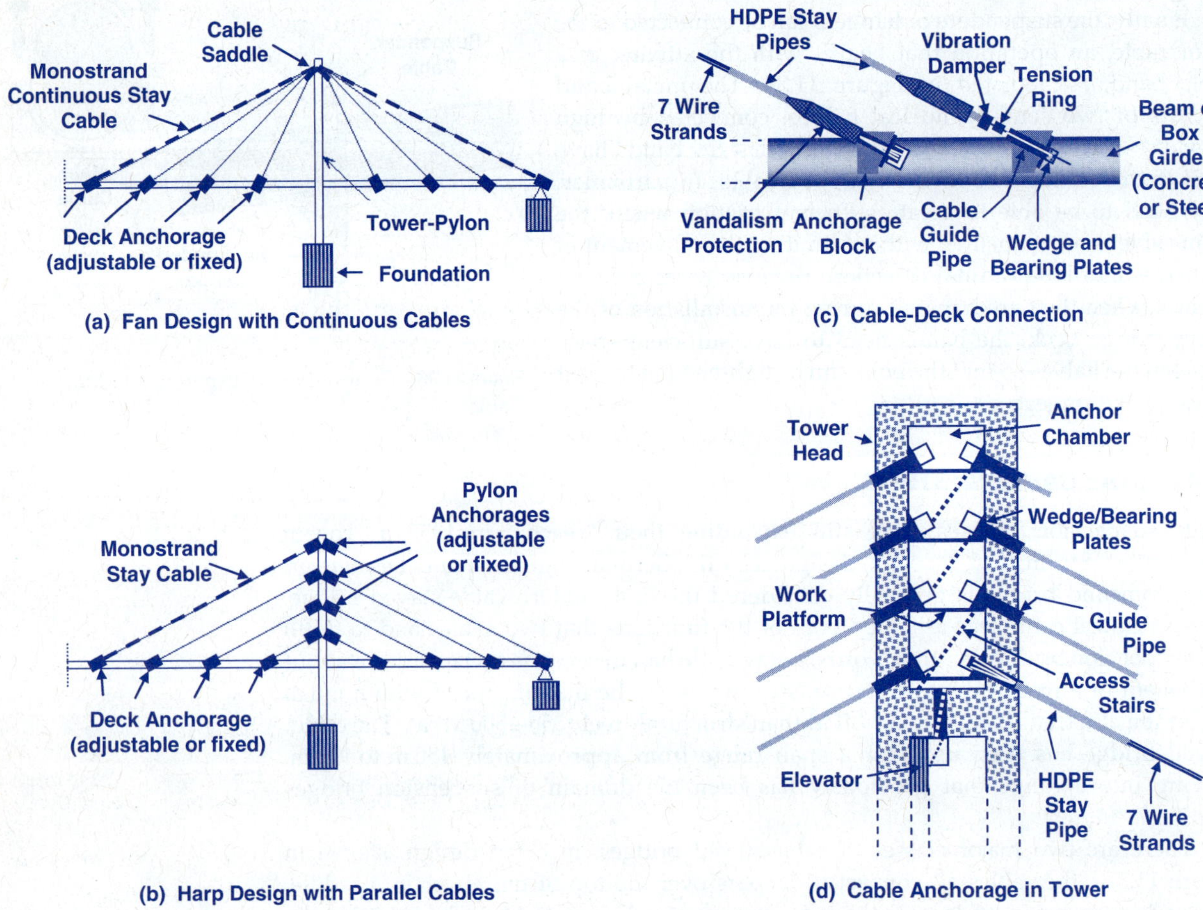

Figure 11.37 Key Elements of a Cable-Stayed Bridge

compression force against the pylon will be. In a harp design, shown in Figure 11.37b, the cables are kept parallel.

VG11.4 Cable-Stayed Incheon Bridge

While the fan-type bridge, like a suspension bridge, depends on a congested cable saddle on the top of the tower, harp-type bridges allow that every stay cable is anchored separately in the tower head, as depicted in Figure 11.37d. The anchorages for the cables, inside a high-density polyethylene (HDPE) pipe, apply similar technologies to those used for suspension bridges. The wires and cable strands are protected by a guide pipe as they pass through an anchor block to reach the wedge plates. Again, welding of the high carbon tension cables is avoided through the use of wedges that are able to grab and hold each wire or strand. Both ends of the stay cables are accessible, either at the level of the deck or up in the tower. Pylons commonly are equipped with a personnel elevator and a crane to reach the anchor chamber for inspection or repair. While one end has an "active" anchor with a simple readjustment mechanism, the other end can be a passive anchorage. The active anchorage can be positioned either at the girder or in the pylon. In order to ease the cost of repairing cables damaged due to wire corrosion or fatigue, today's manufacturers offer means to replace individual strands. This is made possible by small hydraulic jacks that are designed to pull one strand at a time. Naturally, pulling is required to relax the grip of a wedge for removal of the strand.

A critical element, of course, is the anchor point at the girder, which will encounter not only static but also dynamic forces due to wind and earthquakes. Furthermore, acts of vandalism and car accidents have shown that the cable close to the bridge deck is very vulnerable to damage. Figure 11.37c shows two protection measures. One is the use of a solid shield around the cable, while the second is a shock-absorbing damping mechanism designed to suppress vibration in a cable. The tension ring further helps reduce the negative effect of rapidly changing cable angles.

As was learned from past experience, in particular the Tacoma Narrows bridge, long, slender cable-supported bridge decks are susceptible to strong torsional movements that will destroy the deck. On the other hand, extra cost for stiffening the deck should be minimized. This drive to optimize the bridge led to competing concepts that involved the cable stays.

The first solution tries to minimize the number of cable stays by depending on a single plane anchored between the two separated bridge decks. Stiffness is achieved with the use of precast and prestressed concrete box girders, as depicted in Figure 11.38a. The other

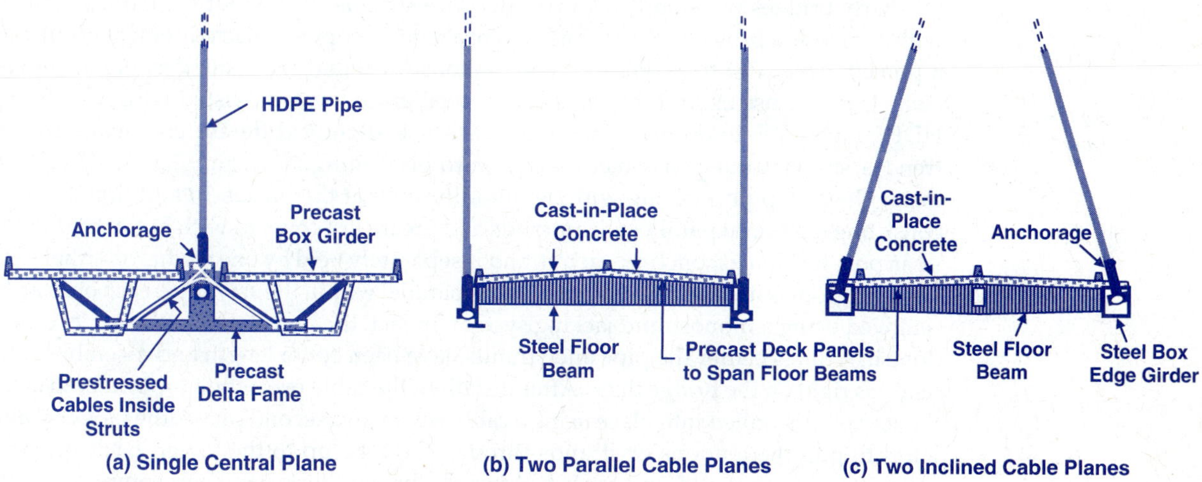

Figure 11.38 Stay Cable Plane Systems

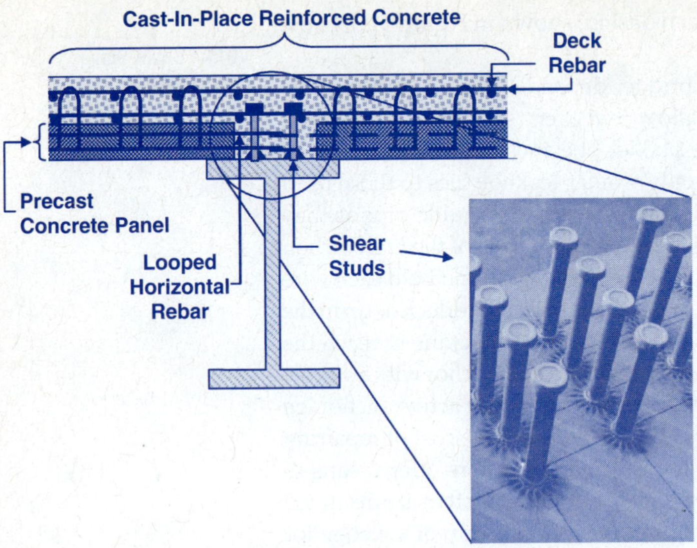

Figure 11.39 Schematic of a Composite Deck

two systems make use of a both steel and concrete, the first to create large spans in support of a quick cantilevered construction and the latter to "forge" a cohesive structure after the assembly of a section is finished. Between a set of two anchors attached to longitudinal girders or I-beams at the edge, two parallel floor beams made of steel create the "foundation" for precast concrete panels to be placed onto the top flanges thus bridging the open span. The desired stiffness of the bridge deck is achieved with the help of strong concrete that is being poured on top. Figure 11.39 indicates schematically the concept of bonding steel and concrete into one composite structure that operates as one solid piece. As shown, two different means are utilized. The first consists of studs with heads that are welded to the upper flange of the beam. They are responsible for bonding the steel with the cast-in-place concrete, both in tension and most importantly in shear. Thus, the studs are often called shear studs. The second connection mechanism is looped rebar that has been inserted in the precast panels. For example, the horizontal rebar extends out and loops back. When a panel is positioned on top of the beam, the loop embraces one or several shear studs as well as rebar coming from the neighboring panel. When concrete is cast, the rebar provides a solid connection in tension. Loops or hooks also extend through the upper panel surface, acting similarly to the shear studs.

Stud connectors are welded not only to the floor beams but, most importantly, to the longitudinal girders, joists, and beams. As a consequence, all the steel and the concrete is bonded into a stiff plate that resists movement in a horizontal direction and torsion along the main axis of the bridge deck.

In the early days of cable-stay bridge construction, the stays were installed as fully prefabricated cables transported in reels. Naturally, the handling of those heavy stays, to be hoisted up and over the high towers, required special equipment. Today, the installation may take a little longer but needs many fewer personnel and can do without tower cranes.

Early bridges were built with parallel wire strands (PWS) that were ungalvanized with a corrosion protection that consisted of a tape wrapping, injected cement grout, and a polyethylene (PE) tube. Figure 11.40 demonstrates that the cable design has moved away from the use of straight wires and cement grout toward twisted wire strands and HDPE pipes. While different manufacturers use slightly different combinations, the trend is, similar to suspension bridges, toward preventing the accumulation of moisture inside the stay pipe, starting with dehumidification at the outset. In addition, tendons consisting of several parallel wire strands and grease are covered with hot-extruded PE. As in post-tensioned concrete, each strand is separately held by one wedge on either side. Since no grout is used and the strands run in parallel within the stay pipe, each one can be removed using a monostrand jacking system. In fact, the installation of the stays is now done in sequence after the pipe and strands have been cut to length and assembled into tendons right on the bridge deck. After installing the cable protection pipes at both ends, the stay pipe is pulled into place using a cable winch. In a second phase, the pipe contains wires to pull the tendons or strands through the tube up to the tower. Each strand is subsequently tensioned and sealed. After the bridge deck has been completed from abutment to abutment, each cable needs to be detensioned. Instead of dealing with every

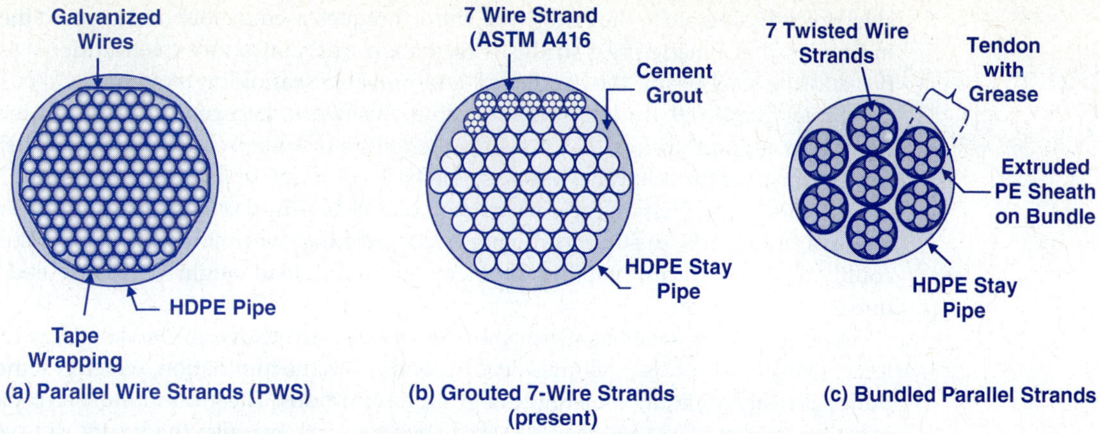

Figure 11.40 Examples of Stay-Cable Construction Past and Present

strand separately, the adjustable anchor provides the mechanism that allows the entire bearing-wedge plate to be moved using a multi-strand jack.

11.5 Conventional Construction of the Superstructure

There is no other sector of construction in which the engineering design, method of construction, environmental conditions, and cost are so closely intertwined. For example, a deep canyon that has to be bridged makes it costly to build piers, but offers the opportunity to build an arch made of steel or concrete. However, steel may be very expensive but extremely cost-efficient when used as a composite with concrete. Other factors to be considered for large undertakings are the availability of sufficient skilled labor and accessibility for large equipment. In many cases, impacting all decision related to design and construction is the temporary support that the contractor has to erect and dismantle when done. In fact, the safe construction of such temporary structures can comprise the largest effort of the entire project.

The economic reality of today's bridges built in public-private partnerships demand high speed and safe construction within a given budget and designs that are easily maintained and repaired when necessary. We have seen that the latter has led to significant technological changes in suspension and stay bridge cable systems. All these factors have changed the make-up of the design team and most often leads to alliances between the designer, consultants, the main construction contractor, and specialty subcontractors such as for foundations, steel, or cable supports.

This chapter so far has covered the characteristics of some of the most important bridge elements. In previous chapters we have looked at some of the main methods for craning heavy loads, excavating caissons and shafts, anchoring temporary walls, and dewatering to the subsurface to build foundations on solid rock. Last, the book has presented falsework technologies and the formwork needed to cast concrete in place. All these methods have to be considered when planning the construction of a bridge substructure requiring deep foundations, piers, arches, and abutments. In this section, we study traditional techniques and methods used to build the superstructure of the most common bridge types.

11.5.1 ON-GROUND SHORING USING FALSEWORK

In ancient times, builders assembled bridges piece by piece or brick by brick in their final location. While some state-of-the-art methods still depend on supplying small elements of steel or fresh concrete to the bridge site, the techniques used to hoist and support the work in progress have made great strides. This section will reintroduce the traditional shoring but add the steel beam, the cantilever, and movable scaffolding system.

From Chapter 9 we know that shoring, falsework, and centering are all terms to describe a support system that prevents the failure or collapse of a permanent structure under construction. Like the Romans, the early users of reinforced concrete in bridge construction depended on the construction of a heavy-duty shoring or centering system made of timber able to support the heavy concrete that was placed in phases. The latter could cause significant stability problems since the load might be distributed in an unbalanced manner.

Bridge shoring systems mirror those used for concrete decks. One key issue that has led to many costly and sometimes deadly accidents is the foundation, which is sometimes built with only minimal knowledge of ground conditions. The ground must be capable of carrying not only the weight of the shoring/formwork but also the heavy wet concrete until the structure can support itself. Thus, a geotechnical engineer should calculate the settlement that the temporary structure will experience and design an adequate foundation that could include special mats, compaction, or even piles. Of course, the ground conditions will change along the axis of the bridge. This requires a comprehensive rather than a "spot" geotechnical investigation. It is common that state and federal agencies require that a contractor, prior to construction, supply plans for monitoring the ground settlements around the temporary shoring and remedial actions, such as jacking of the shoring, that would be initiated if necessary.

Shoring manufacturers offer a variety of designs to fit different situations and loading conditions. Figure 11.41 shows the frame, wood, and steel column systems built on three

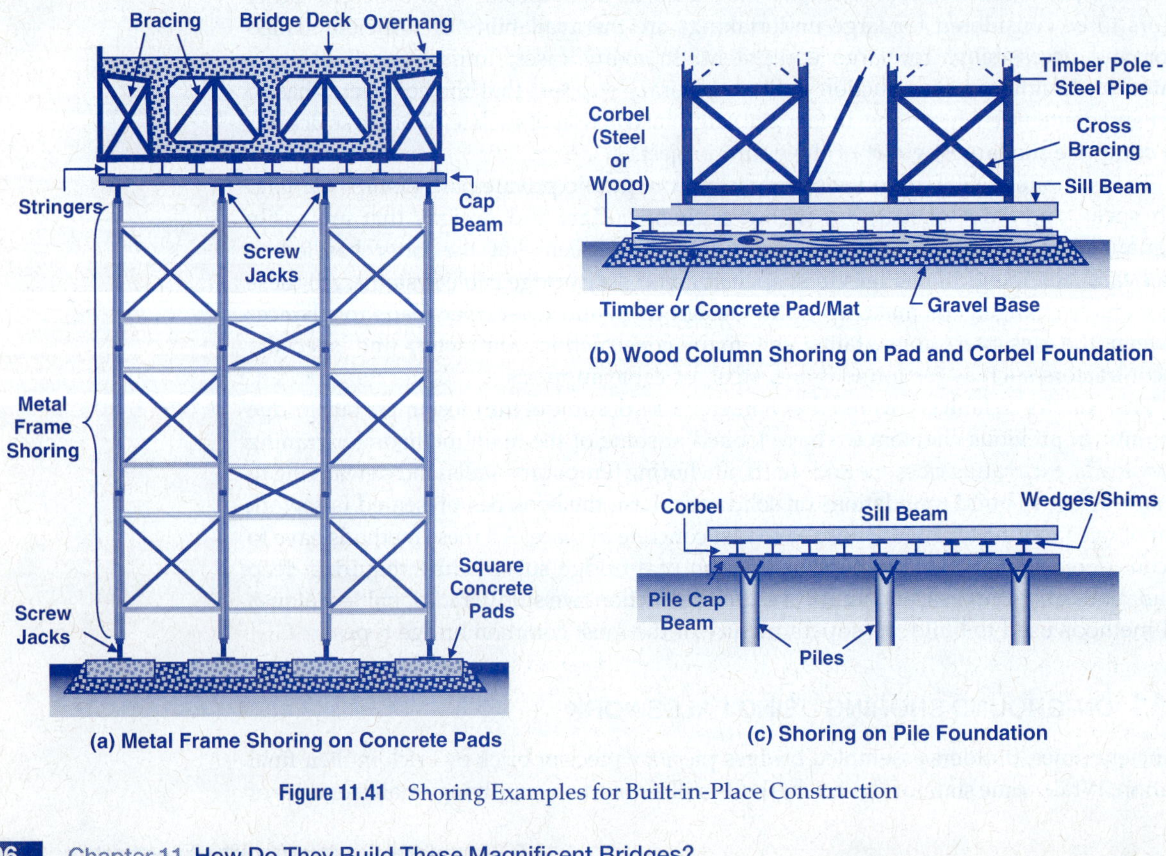

Figure 11.41 Shoring Examples for Built-in-Place Construction

different foundations. It is common that adaptations are needed to cover special conditions. For example, the need to provide an opening for construction or local traffic to pass underneath asks for a raised deck that spans between two rows of reinforced columns. Obstacles or large variations of the surface will require similar adjustments. For further reading about constructing proper shoring, refer to Chapter 9.

11.5.2 UNSHORED GIRDER METHOD

While shored construction requires the erection of temporary supports beneath the future bridge, unshored bridge construction replaces the on-the-ground falsework with structural elements that are able to span the space between piers without any extra support. Of course, this means that those long-span pieces need to be fabricated on the ground and hoisted into position in one piece with the help of a lifting device. The bridge deck is subsequently built on top of those elements. Such an unshored approach eliminates the costly installation and removal of the falsework, and the shores don't interfere with other construction activities. Furthermore, the bulk of the work can be performed in a controlled environment prior to on-site construction, with little or no disruption of traffic. Weather becomes less of a factor, and quality improves. The most efficient structure that is light but strong enough to span the required distances consists of girders fabricated of different materials and built in any size. There are four common girder types used in bridge construction:

VG11.5 Simple Span Girder Bridges

1. I-beam steel girders with web heights of up to 1 m (3 ft), spanning up to 30 m (98 ft)
2. Steel plate girders fabricated with flat steel welded or bolted, spanning up to 100 m (330 ft)
3. Concrete girder pre- or post-tensioned, spanning up to 50 m (164 ft)
4. Steel box girder fabricated as closed or U-shaped boxes.

Before studying the operational process, we need to quickly review composite deck construction one more time. You might remember that welded-on shear studs can be used to bond cast-in-place concrete to steel deck elements, having been installed first. Employing this composite method, the contractor is able to take advantage of prefabricated components that can be quickly put and functions also as falsework that supports the forms for the cast-in-place concrete. With this in mind, let's look at two common situations.

Girder bridges often create highway overpasses with two equally sized main spans. After the two abutments and the three rows of columns with caps and bearings, called bents, are completed, the girders can be set. The contractor of Figure 11.42a decided to rent a truck-crane positioned on the access road to the overpass. While this could be convenient where the access road already exists, limitations in lifting capacity may require the crane to be moved to the other side of the bridge. Reach is not a problem for the option in Figure 11.42b, where the crane, this time a crawler, is positioned at the level of the underpass. This option is most efficient where the same crane has already been used to build the foundation and the columns.

Nowadays most steel beams extend into the main span to where the moment is close to zero. This makes setting them far trickier than precast, because the ends cantilever out, requiring the immediate installation of the cross-bracings. Figure 11.42b shows how the two end-beams are set before the center-segment is hoisted up so that its two ends can be spliced together using pre-drilled holes in the web, flanges, and plates and bolts.

Figure 11.43 displays steel elements needed to splice steel girders together. The pictures were taken in the fabricator's yard where a box-girder bridge in steel was

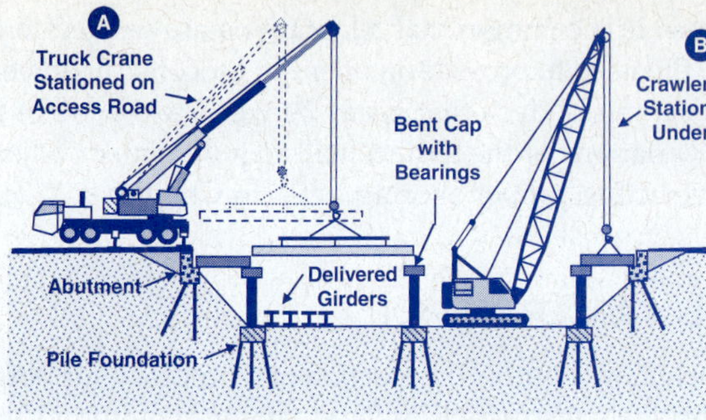

(a) Common Equipment for Setting Single-Span Girders

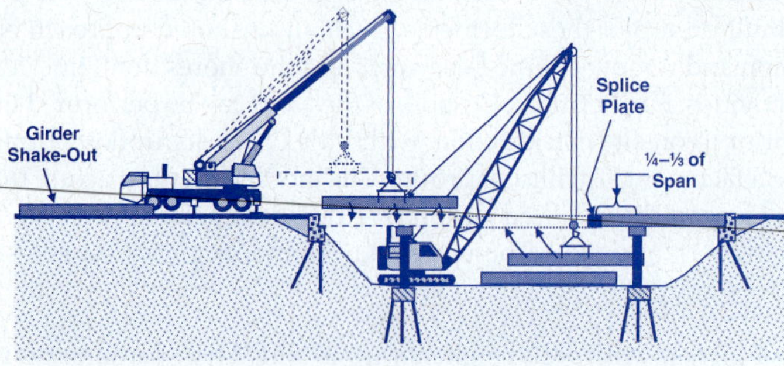

(b) Erection of Continuous Steel Girder with Two Bolt Splices

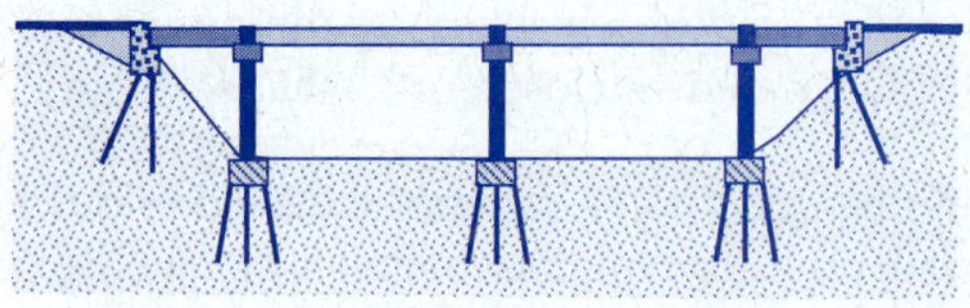

(c) Single-Span Girders Set for Entire Bridge

Figure 11.42 Setting Steel and Precast Girders onto Pier Caps

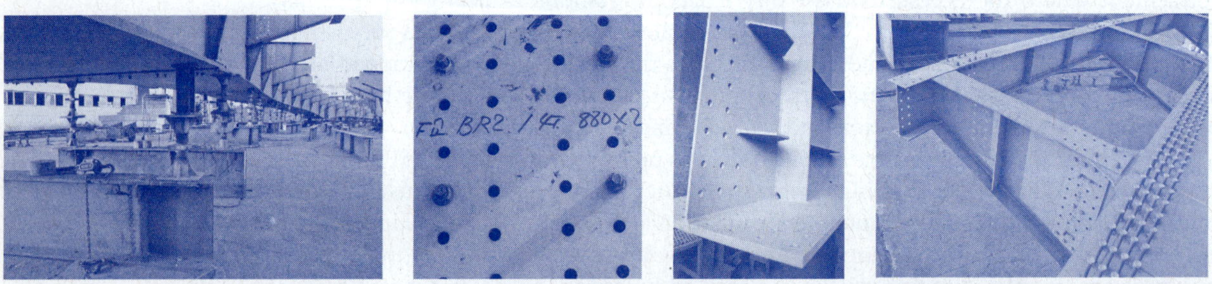

(a) Blocked-Up Box-Girder Bridge (b) Bolt Hole Pattern on Splice Plate (c) Splice with Bolt Holes in Web and Flange (d) Testing the Alignment of Splices with Temporary Bolts

Figure 11.43 Pre-Assembly of Continuous Box-Girder Steel Bridge

pre-assembled on top of blockings that mirror the final elevation of the entire structure, shown in Figure 11.43a.

There are several benefits of pre-assembly that go beyond ensuring that all the pieces have been fabricated. For one, the delivery to the construction site can be planned on an as-needed basis. Testing the perfect alignment of all bolt holes eliminates the risk of creating large muda-waste on site. Depicted in Figures 11.43b–d is the large number of pre-drilled holes and splice plates. Finding out on the construction site that the bolt holes on one plate are misaligned will cause a costly disruption of the erection work.

Three different drawings are commonly used for structural steel: (1) design, (2) detail shop, and (3) erection drawings established by the fabricator. Section 11.4.2, presented the electronic technologies for automated cutting, drilling, and welding that have spurred efficiency and quality of fabrication. All this work is controlled with digital data from the detail design software. Despite these advancements, there is still a lot of hands-on work involved in preparing, transporting, aligning, and spot-welding.

Structural steel erection goes through three distinct processes: (1) assembly, (2) fit-up, and (3) bolting. The process of steel assembly is illustrated in Figure 11.42b. The objective of the fit-up is the proper alignment of the steel members and spice plates that make up the bolted connections with the help of so-called drift pins, cylindrical steel pins that are tapered on both ends. The drift pins are first driven, using only minimal force, into the bolt holes through all the plies at the corners of the web plates in order to align the steel members vertically. Subsequently, the flanges are drifted into correct horizontal alignment using the same principle. Eventually, $^{1}/_{4}$ of all bolt holes need to be filled with drift pins. Subsequently, fit-up or erection bolts are used, to bring all the plies of a splice into contact with each other. As a minimum these bolts should be $^{7}/_{8}''$ diameter and meet ASTM A325 for primary members ($^{3}/_{4}''$ bolts for secondary members) and will be replaced with production bolts. The bolting begins with replacing drift-pins or erection bolts, one by one, with the lubricated final bolt and tightened "snug tight." The correct tension in the production bolts is achieved by the calibrated turn of the nut rotation method and measured by specialized load cell washers called direct tension indicators (DTIs), which compress at a given bolt tension force.

Figure 11.42c shows the bridge with all its girders sitting on the designed abutments and pier bends ready for work on the deck.

Figure 11.44 depicts a barge solution where a bridge crosses a lake or river outside the reach of a land-based crane. Any time a mobile crane is put on a floating barge, a red flag has to go up for anybody who is in involved in the process. Why? For one thing, the load chart attached to the crane is not valid anymore, as it is based on use on a stable support and leveled to within 1% (0.57 degree). A crane mounted on a floating barge will experience continuous changes in trim and list, as shown in Figure 11.44b. **VG11.6 Barge Cranes**

OSHA's construction-related rule, CFR 1926.550(f)(1)(iv), states, "Mobile cranes on barges shall be positively secured." However, this does not mean that a mobile crane will now behave as when stationed on land. Rather, the barge and the crane now behave like one piece of equipment that has not been calibrated as such. As this has to be considered a special operation, the contractor should conduct a stability analysis for the planned crane and barge combination. This analysis should result in a list of radii, boom angles, and loads for each lifts reaching front to back and sideways. Since the crane manufacturer knows the behavior of its cranes beyond their normal usages, the analysis should be reviewed by the manufacturer in order to ensure that lifts are safe and as efficient as possible, which requires up-front planning.

Again, the result of the first phase is a crossing consisting of girders sufficiently strong to carry the deck. It must be added that girders made of steel can obviously be welded or bolted together, either right on top of a bent or in the span area. This would be called a continuous girder and can provide some useful structural benefits. One of the benefits of

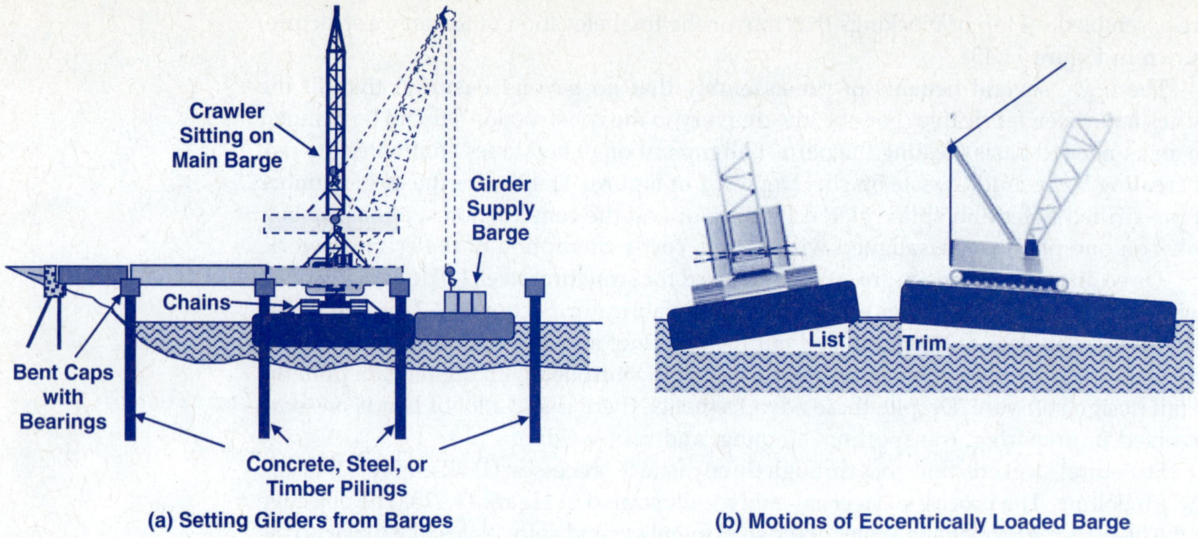

Figure 11.44 Installing Single-Span Girders over Water

composite bridge construction is to make the bridge deck continuous while bonding to single-span girders. The following section reviews some of the combinations that can be found in bridge construction.

Deck Construction Step 1: Installation of Deck Panels (Steel or Precast)

VG11.7 Lost Deck Panels

The bridge girders firmly secured on top of the piers or bents provide the jump-off point to create a safe work platform that will serve, in addition, as the stay-in-place formwork for the concrete to be cast. This timesaving option is the bridge-deck panel system, consisting of panels with the two ends placed on the top flanges of adjacent girders. Made of either metal or prestressed concrete, the panels span the opening and thus create a platform for laborers, tools, and material. Commonly, the panels are secured to the girders and can be adjusted or shimmed in height. The process of placing the panels is illustrated in Figure 11.45a.

Deck Construction Step 2: Cantilever Formwork and Shear Studs

Commonly, bridge designers extend the width of the deck on both sides by cantilevering the concrete. For this reason, removable falsework and the appropriate formwork have to be installed. Figure 11.46b shows how the falsework has to be attached to the edge

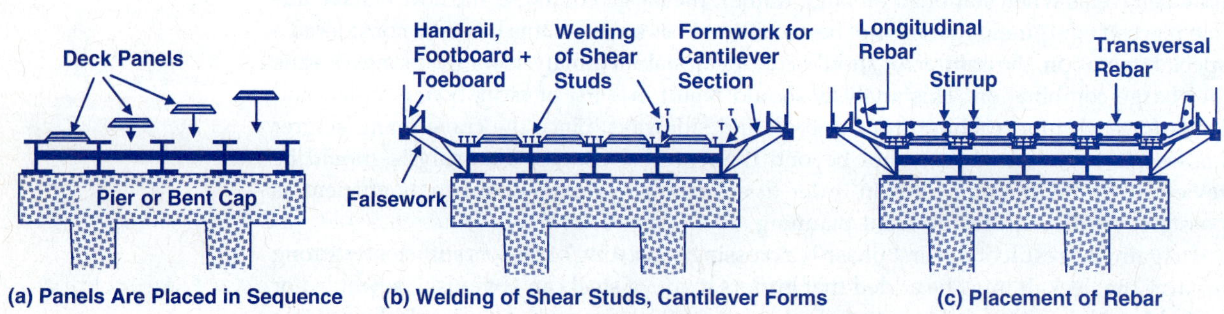

Figure 11.45 Preparing the Bridge Deck for Concrete Placement

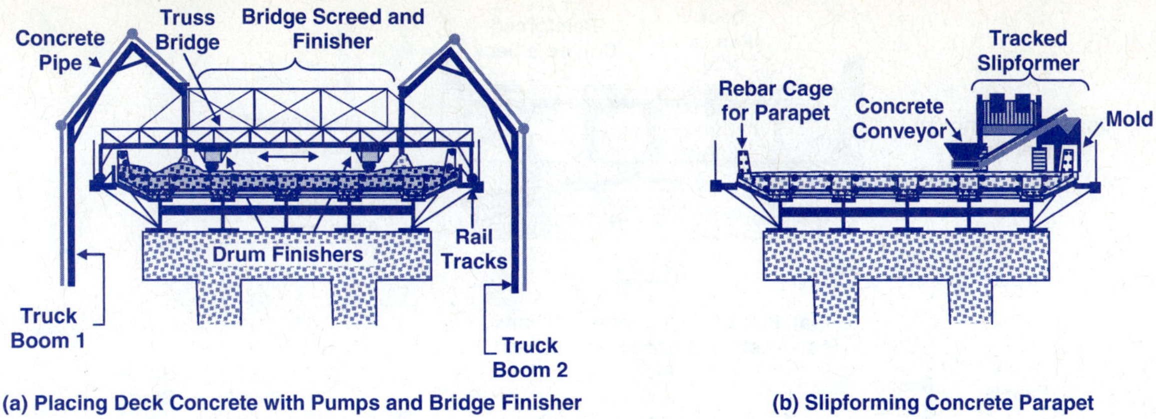

Figure 11.46 Completing the Composite Deck with Cast-in-Place Concrete

girder. Integrated into the falsework are foot- and toeboards as well as handrails preventing tools or laborers from falling to the ground. Occasionally, the cantilevering structure will have to be strengthened in order to carry extra loads during concrete placement (see Figure 11.45b).

Deck Construction Step 3: Rebar Placement and Tying

At this step, the rebar is brought onto the completed deck to be placed and tied according to the design. Commonly, the rebar for bridges is epoxy-coded and needs to be treated with care so as not to damage the coating in the process. In order to prevent corrosion of rebar close to the concrete surface that is exposed to rain and snow, inspectors will check the covering of the rebar with concrete. They will also inspect to see whether the rebar is properly tied and the epoxy is undamaged. At this time, elements for the drainage system have to be installed as well.

Deck Construction Step 4: Concrete Pouring

The first and main pour will establish the flat deck plate on top of the girders and deck panels. Concrete can be supplied along different paths. A widely preferred technology is the mobile truck-mounted boom pump, which depends heavily on the availability of a proper access road for concrete trucks. Figure 11.46a displays a situation where two boom-pumps deliver concrete onto the deck from both sides of the bridge. An alternative route could be one of the abutments if trucks have access. In this case, a pump without the boom is installed there, pushing concrete through pipes placed on the top flange of the girders.

In order to achieve the desired surface quality, this sample equipment setup employs a bridge finisher similar to the one discussed in Chapter 9, section 5.6. The wide span requires the use of two trusses that also support two drum finishers that distribute, screed, and finish the deck via remote control. The concrete is placed in front of the finisher, which moves on two rail tracks, one at each edge of the bridge. Finally, the high wall parapets are slipformed over the rebar cage, using a side-mounted mold.

As mentioned earlier, the use of composite-steel decks has become very popular because of the many benefits during construction, such as providing a work platform without the need for scaffolds and supporting concrete formwork without falsework. In order to optimize the technique for different conditions in terms of span distances or material cost, various alternatives have been tested.

VG11.8 Open Box Girder

Figure 11.47 presents three common composite bridge designs. The first example integrates post-tensioned pier caps with prefabricated concrete beams and a cast-in-place

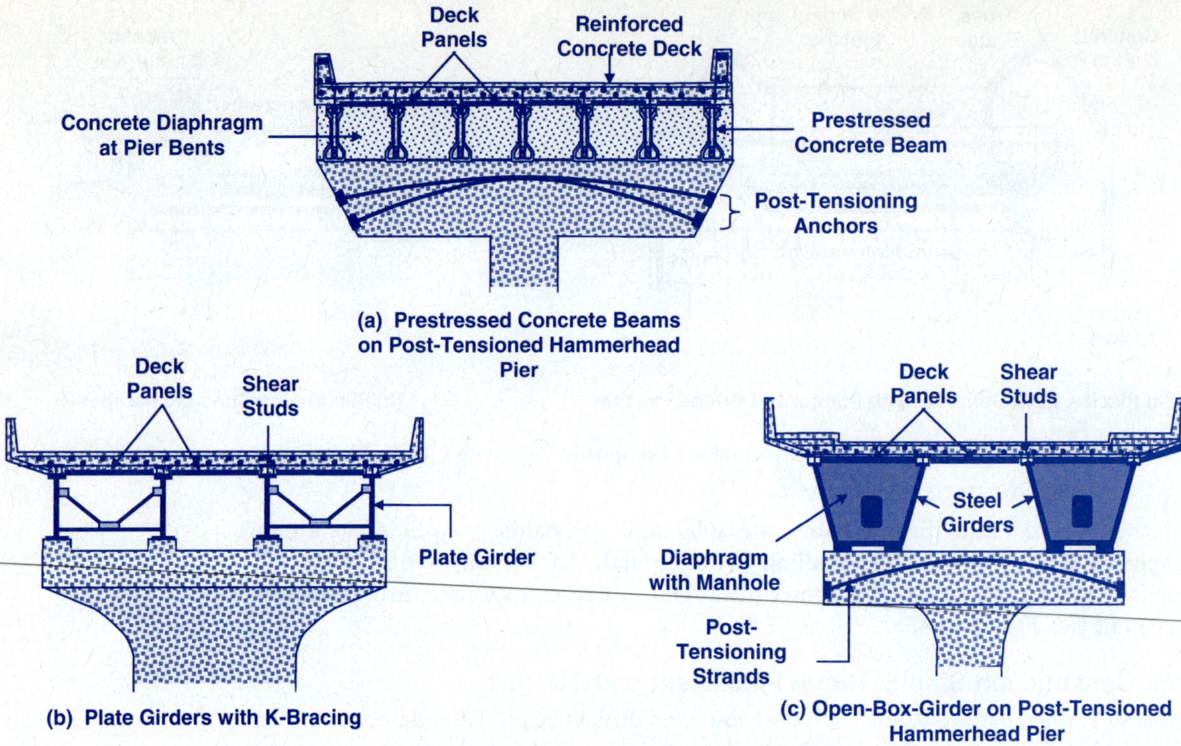

Figure 11.47 Designs of Composite Bridges

concrete deck. Figure 11.47b replaces the many prestressed concrete beams with only four plate girders braced in pairs. The plate girders are fabricated with two flanges welded to a thin web, which usually has transverse stiffening and may have longitudinal stiffening. Shear studs welded to the top flange provide a solid bond between the steel girders and the concrete deck. Plate girder bridges provide a very competitive solution for short and medium span bridges.

The third example shown in Figure 11.47c takes advantage of the rigidity of box girders while integrating the stiffness provided by a cast-in-place concrete slab. The V-shaped box girder is open on the top while equipped with upper flanges with shear studs. This composite takes advantage of the compressive strength of concrete and the tensile strength of steel.

Figure 11.48 compares two common bridge designs for a curved flyover of imaginary highway and exit lanes. The objective of the comparison is not to favor one over the other but to illustrate differences of the two designs. In fact, there are many different solutions for designing the imagined flyover. Nevertheless, the centered box girder offers a very sleek solution that exhibits torsional rigidities that are very useful in curved bridges. Figure 11.48c highlights the fact that box girders offer a convenient way to inspect the superstructure by offering a pathway for a person to walk inside and use manholes to pass from one section to another. The use of a longitudinal orthotropic deck fastened to the cantilevering bracket will further the lightness of the entire structure.

Plate girders provide the opportunity to shape them to a curve, something that is impossible with rolled I-beams. As Figure 11.48b indicates, the curving flyover superstructure built on pier bents requires that the girders be cut, bent, and welded according to a continuously changing shape. While computer-controlled steel cutting and welding

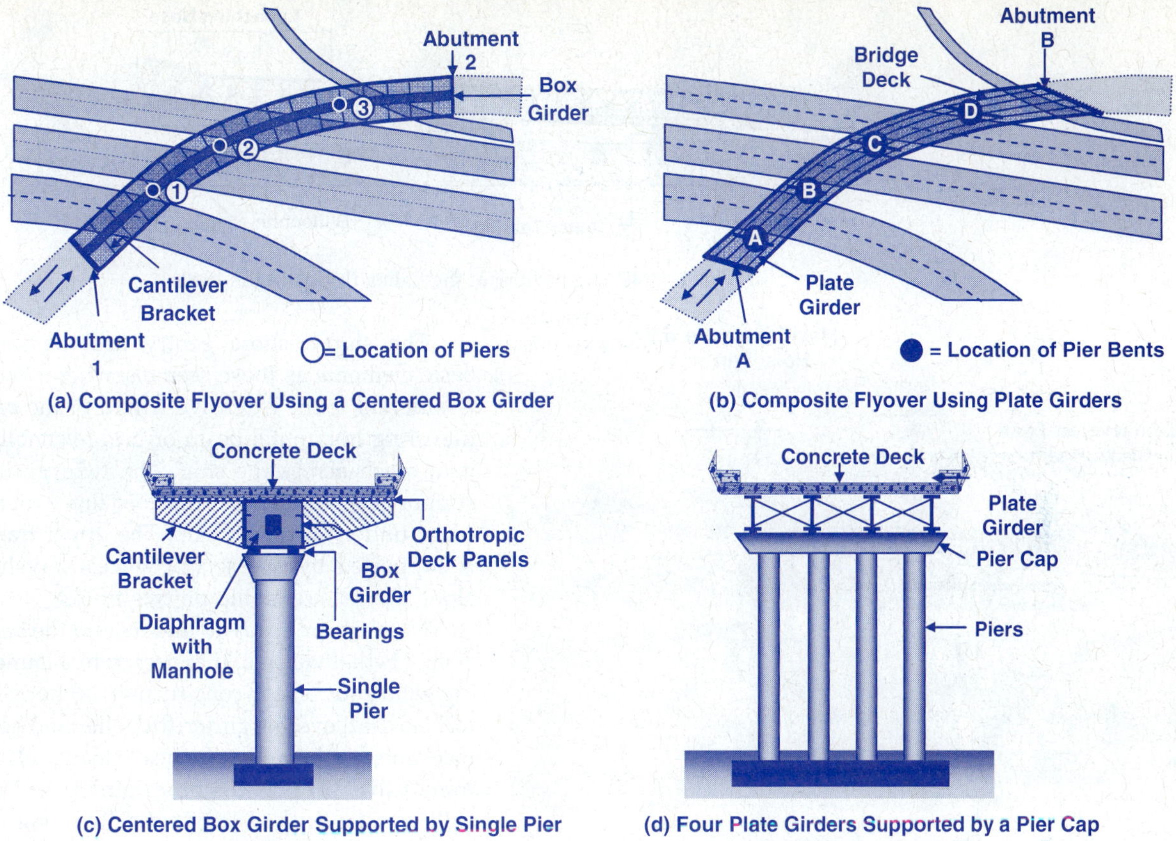

Figure 11.48 Comparison of Flyovers Designed with Centered Box and Plate Girders

machines, fed with the data from the design data files, offer simple solutions to creating the various shapes, the resulting steel scrap and the added problems in handling the many irregular girders as well as the required cross-bracing result in significant added costs over straight girders.

11.6 Bridge Building with Launching Gantry (LG)

How can one launch a bridge similar to ships? Not surprisingly, World War II also made one bridge launching method famous. The bridge got its name from the engineer who patented his design in 1943, Sir Donald Bailey, a British military engineer. The Bailey Bridge was designed to be longitudinally pushed into its final position. Not only had the bridge to be constructible with only six men, but it had to be rapidly constructed. Bailey's concept depended on lightweight and standardized pieces assembled beside or behind the bank of a river and a push and pull launch across rollers, as shown in Figure 11.49. Of course, the pushing and pulling over the rollers had to be supported by machinery such as a tank, bulldozer, or even cables with pulleys and a powered winch.

The second term that seems to be out of place is a gantry. Gantries or their cousin, the overhead crane, are found in factories and shipyards, as depicted in Figure 11.50.

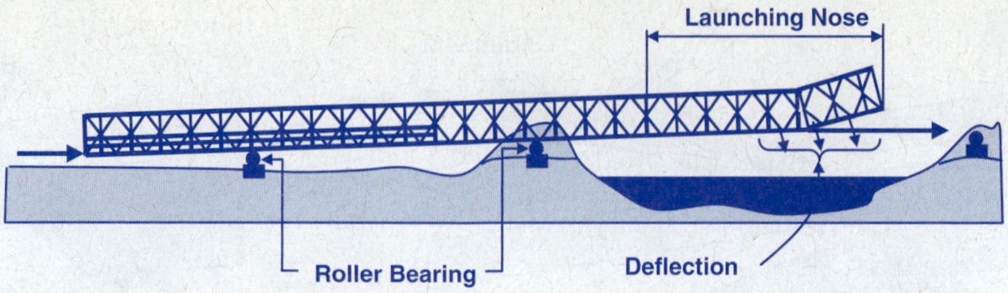

Figure 11.49 Launching of the Bailey Bridge in the 1940s

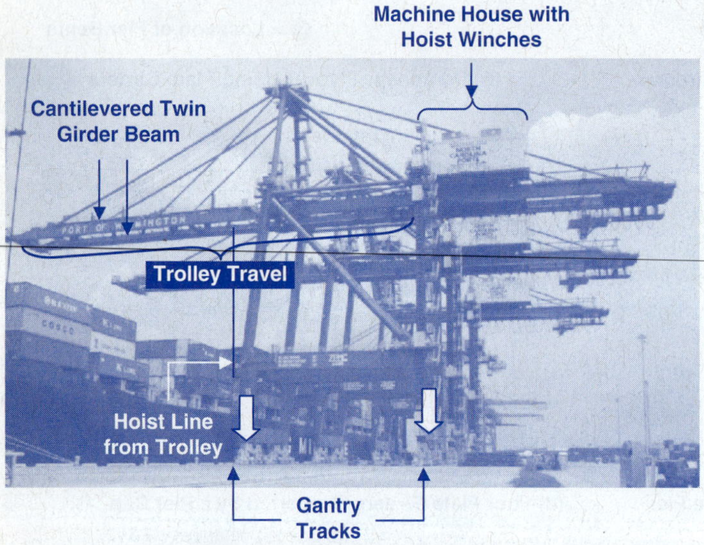

Figure 11.50 Ship-to-Shore Gantry Crane

The ship-to-shore gantry has the same basic elements as those seen elsewhere. What is different is the extensive length of the cantilevering horizontal beam or bridge enabling it to reach across the ship. The twin girders support a trolley to move the hoist line with the cargo underneath the beam. The upper frame is supported by two sets of wheels traveling along the side of a ship on two tracks.

Now we are ready to understand the basic elements that we should expect from a launching gantry for bridge construction: (a) horizontal load cantilevering girder, (b) trolley and hoist mechanism to lift and move heavy loads, (c) two frames that support the beam girder, and (d) ability to be pushed longitudinally along the axis of the bridge. As the following discussion will reveal, there exist several slightly different mechanical bridge building methods that all employ these very basic characteristics.

11.6.1 LG WITH MOVABLE FORMWORK SYSTEM (MFS)

This first example is a logical extension of the in-situ construction of concrete bridges using large shoring (falsework) built from the bottom up. A movable formwork system (MFS) is repeatedly moved longitudinally along the axis of the bridge, supported by a gantry installed underneath, referred to as an underslung launching gantry. Circumventing the need of falsework allows the easy bridging of obstacles in the path of the bridge such as roads, creeks, and even buildings. Nevertheless, the fabrication of the formwork represents a large investment and does not offer any flexibility to make turns. For this reason, this otherwise elegant system is only cost-effective from a minimum distance with mostly straight sections. The following reviews the basic steps of a launching cycle presented in Figure 11.51.

Figure 11.51a depicts the situation when the concrete is curing to 75% of its 28-day strength after a section has been poured. Figure 11.51b presents the main supports of the two gantry girders at the piers. So-called pier brackets have been installed on the top of every pier, providing a platform to install hydraulic jacks and transverse bracings for the gantries. Using the jacks, the entire MFS can be lowered after the superstructure has been post-tensioned so that it is able to support its own weight. The internal formwork, of course, has to be moved separately inside the completed box.

Figure 11.51c highlights the most critical moment of the launching procedure as the transverse bracings have to be removed for the two gantries to pass by the pier head. The

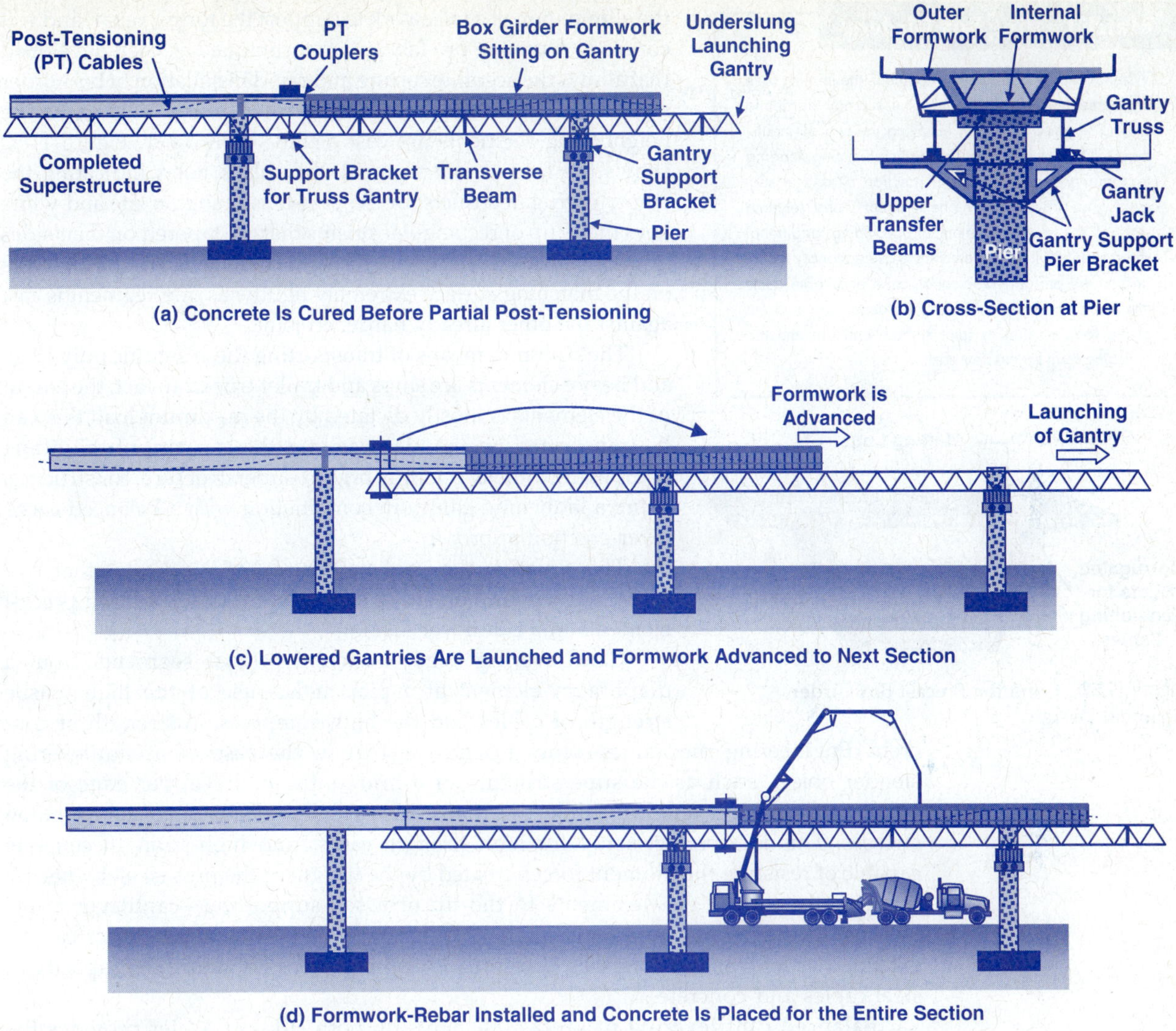

Figure 11.51 Sequence of In-Situ Concrete Casting of Bridge Superstructure with an Underslung LG

trusses are now launched to the next section using horizontal jacks before the cross-beams are reinstalled. At this time, the heavy formwork can be securely moved as well and put in position to be jacked up again and readjusted for concreting the next section. But first, the rebar has to be placed and the post-tensioning cables coupled to the previous section. Finally, the internal formwork can be advanced from the previous section. Figure 11.51d shows the concreting using a concrete pump positioned underneath. It goes without saying that different methods for supplying concrete are possible, including cranes or pumping horizontally along the surface of the completed bridge. Only an analysis incorporating the local constraints and the cost of each option will provide the basis for selecting the best solution.

11.6.2 SEGMENTAL PRECAST BOX GIRDER INSTALLATION WITH LG

In Section 11.4.5, we learned about the concept of precasting and transporting pre-stressed concrete beams onto the site to be lifted onto the pier bends. One of the main benefits was

Match Casting Box Girders

When pre-casting a concrete segment, the jointing face of segment A is cast against the already hardened mating segment B. When installing the two segments, they will fit exactly without requiring shimming, leading to rapid and more accurate erection of bridge segments. The raised keys on one segment fit precisely into slots of the matching element, providing additional shear strength. It is most efficient to use stationary molds and lift the prefabricated rebar cage and cable ducts into and to push the sufficiently hardened segment out of the form so that its joint surface can become the matching form for the next cast.

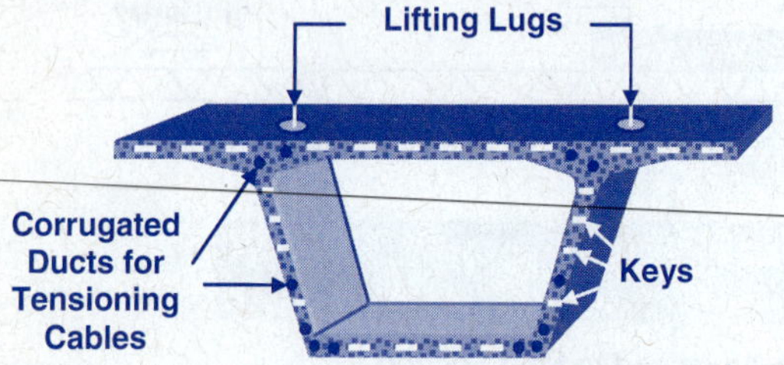

Figure 11.52 Common Precast Box Girder Segment Design

the elimination of a falsework to support the forms, rebar, and wet concrete. A unique pre-fabrication technique has been developed that allows the accurate casting and rapid installation of box girder segments. The method has been termed match casting, as the neighboring segments are cast against each other. Figure 11.52 shows the face of one segment with black holes indicating the empty corrugated ducts for the post-tensioning on site and white keys made up of rectangular prisms that are tapered on their sides extending outward from the surface. These keys will fit into slots on the matching surface extremely tightly, as one segment is cast against the other already hardened joint.

The common means of transporting these significantly large and heavy elements are ships and trailer trucks. In fact, the length of the elements is mostly dictated by the maximum load that can be transported on the highway and the capacity of the lifting equipment. But let's join a bridge superstructure construction using a launching gantry in combination with a balanced cantilever erection approach.

The situation depicted in Figure 11.53 indicates that two crews are working on either side of pier L, both adding precast elements in a cantilever fashion.

Cantilevering heavy concrete precast segments from a diaphragm element at a pier makes use of the high tensile strength of cables and the high compression strength of concrete. Engineering mechanics-statics teaches us that in the case of a cantilevering slender object, such as the superstructure of a bridge, the entire upper edge of the object will develop tension, while the underside will be subjected to compression stresses. This is a perfect setup for high-strength cables and high-strength concrete capable of resisting the moment forces created by the weight of the precast elements. By sequentially adding new elements to the tip of a self-supporting "cantilever arm," holding the segments together with tension cables, the cantilevering object can be extended as long as both stresses (tension and compression) lie within the capacity of steel cables and concrete.

Balanced cantilevering of precast elements on both sides of a pier provides the means to eliminate eccentricity. Balance can be achieved by connecting one segment to the tip of the cantilever on one side of the pier, in parallel to a segment on the other side. When the mid-span is reached, a "closure pour" will connect the two half-span cantilevers from opposite sides of two adjacent piers.

While the lifting of the precast elements can be accomplished with crawler or barge cranes from the bottom, situational constraints don't always make this possible. A technology that has been found useful is a launching gantry installed over the superstructure, the overbridge. As shown in Figure 11.53, the launching gantry system has three legs that are positioned on top of the completed pier caps. Two long truss beams are braced and lifted on top of the three legs that can be actuated both vertically and longitudinally. Crane trolleys are mounted on either the lower or the top flanges of the truss beam. Horizontal hydraulic jacks provide the force to launch the trusses forward on top of the legs. Other jacks permit the legs to be moved individually one at a time while at least two are fixed. For example, the weaker front support leg is used only during launching, positioned on the foremost pier cap, pier M in the figure. For this to occur, the trusses have to be pushed forward to where the leg can be lowered and affixed to the pier. In this position, the gantry rear leg can be released from pier K and pushed forward to take over the pier L from the gantry's front leg, which allows it to be

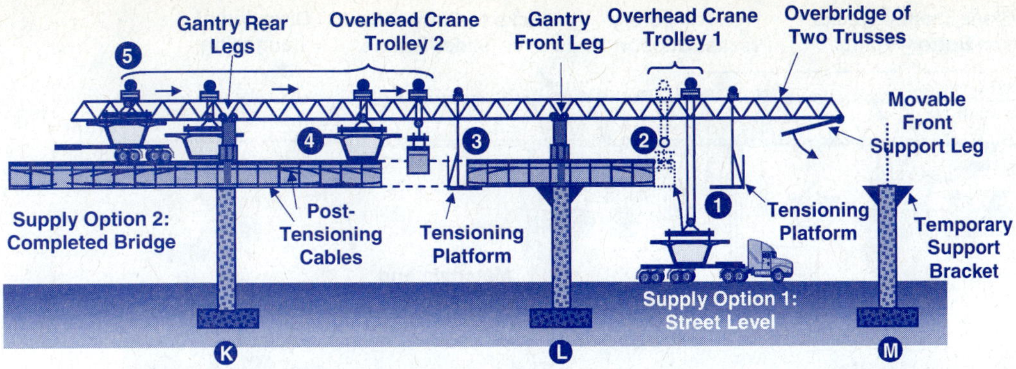

Figure 11.53 Balanced Cantilever Precast Element Installation Using an Overbridge LG

moved to pier Ⓜ. The launch finishes by pushing the truss forward on top of the two main gantry legs.

The precast segments are lifted one at a time via cable-winch and positioned next to the previously connected segment. A gantry should be able to pick up the supplied precast pieces, either from the street level or from the top of the already completed bridge, as at pier Ⓚ in Figure 11.53. Epoxy glue is then applied to the joint surface of the two segments. Temporary rods and jacks are used to connect the new pieces in the inside of the box. Now the tensioning platform can be positioned at the free end to install the hydraulic jacks to stress the wire strands designed for that particular segment.

Let us review of what happens at the construction site shown in Figure 11.53 by following the five labeled focal points.

❶ One precast segment has been transported from the precast plant by a trailer truck and positioned underneath the gantry's crane trolley 1. A spreader beam is attached to the crane hook with two vertical lift rods connected to the precast lifting hooks. As shown, the crane is able to lift the element directly from the truck trailer placed at the street level. Before the start of the lifting, a glue compound is applied to the "face" of the new element, enhancing the aligning process as well as the shear strength of the joint.

❷ The new segment is lifted by the gantry crane to the appropriate height and horizontally aligned with the previous element. By traveling the trolley toward the tip of the cantilevering bridge, the new precast element is joined. A suspended tensioning platform in waiting is brought into position in front of the newest element.

❸ The tensioning platform has been positioned next to the new segment on the other side of the pier. It is used to tightly glue the new element temporarily to the cantilevering segment, using post-tensioning rods that can be removed when the entire span is being tensioned together. It will be lifted up and moved out of the way of a new element.

❹ The second gantry crane trolley is supplying a new precast element from the opposite side of the pier to provide a balance to the raising cantilever.

❺ The second trolley crane is lifting a new precast element off a truck trailer that has reached the site riding on the already completed bridge. Special designs are needed to allow the crane to carry the new element "through" the rear legs of the launching gantry.

In summary, the launching gantry combined with segmental box girders provides several important features. It eliminates the need for cranes or other supports from below the bridge. Even the bridge elements can be brought via the already completed bridge. The balanced cantilever method also allows two or more work fronts to be run in parallel.

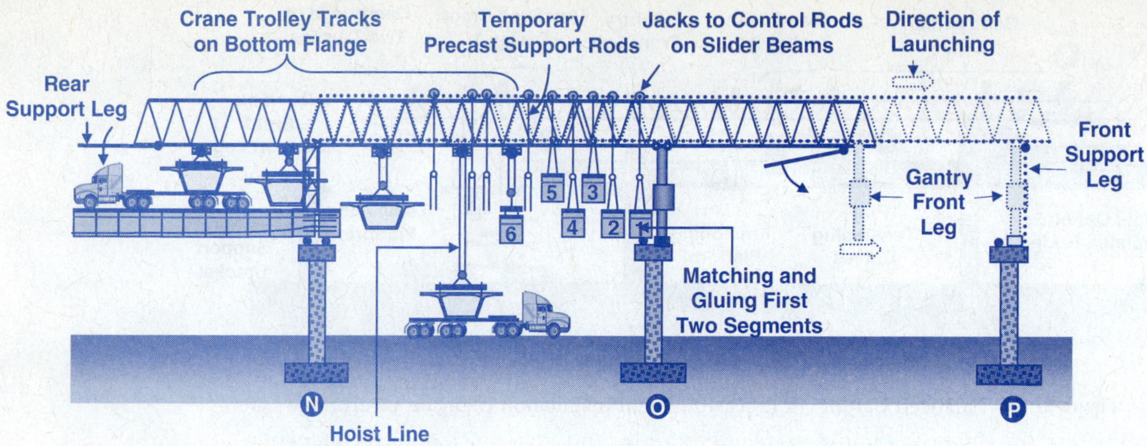

Figure 11.54 Launching Gantry and Movable Lifting Rods Supporting Segments for an Entire Span

11.6.3 SPAN-BY-SPAN BRIDGE ERECTION WITH LG

In this approach, the launching gantry serves as the staging ground to preassemble all the elements for an entire span between two adjacent piers. In fact, all the necessary precast elements are hung onto the overbridge, lined up perfectly, and post-tensioned all together. For example, an LG with a rated lifting capacity of 1,000 tons (907 t) would be able to hold fifteen 10 ft (3 m) long segments each weighing 66 tons (60 t). Such a configuration would require a span to be kept around 150 ft (45.7 m) while the gantry would possibly be 340 ft (104 m) in length. After finishing a span, the LG is traveled forward to the next span similar to the LG for the balanced cantilever method. Modern LGs even include hinges in the overbridge, which allows it to build superstructures with a horizontal curvature. Figure 11.54 presents the mechanical system that enables the span-to-span method of bridge erection.

Again, two supply routes for the segments are possible, one from the street and the other from the already erected bridge section. Depicted in Figure 11.54 is a strategy that allows rapid unloading of the truck-trailers or, if over water, of barges. This is accomplished by suspending the properly sequenced precast pieces at different levels to allow slight overlaps without damaging the joint surfaces. As shown, element #6 is being readied to be attached to the next in line suspension rod at a level below segment #5.

A second crew has already begun lining up the first two segments at pier O, using the rod jacks to raise and lower the piece and the sliding beam holding the jacks to align them

Table 11.6 Possible Task Bar Chart for Span-to-Span Erection with LG

#	Tasks for Installing Segments for One Span and Moving LG	Work Days					
		1	2	3	4	5	6
1	Aligning and preparation of LG	■					
2	Installation and connecting starter segment to last section		■				
3	Hoisting and "parking" elements on suspension rods		■	■			
4	Dry matching of sequenced segments			■			
5	Gluing and connecting with temporary rods			■	■		
6	Inserting and pulling post-tensioning strands/cables				■	■	
7	Tensioning of cables at anchor points possibly at both ends					■	
8	Disconnecting suspension rods, traveling LG to next span						■
9	Grouting of cable ducts						■

on the horizontal plane. Before jointing them, the joint surfaces are also in this method covered with epoxy-glue to "smooth" the joining and to increase the shear strength of the joint. Next, the two segments are held together by temporary rods to ensure a perfect fit. Only when all the segments have been lined up in this manner, each piece still suspended from the LG, does the post-tensioning process start.

The repetitive nature of this bridge construction method can be most effectively demonstrated by sequencing the individual tasks, displayed in Table 11.6, to complete one span, possibly consisting of fifteen 10 ft (3 m) long elements each weighing 60 tons (54 t).

11.6.4 FULL-SPAN-METHOD LAUNCHING WITH TWIN LG

The full-span-method (FSM) bridge construction is able to speed up the erection process by eliminating the time-consuming matching, gluing, and post-tensioning of segments. This feat is accomplished by pre-fabricating and installing entire spans. Such spans are typically 100–160 ft (30.5 m– 48.8 m) long and weigh between 600 and 1,500 tons (544 and 1,360 t). Such large objects require a special transport vehicle that will not be able to use public roads or, alternatively, a barge, or heavy floating crane. This requires, of course, a prefabrication plant next to the bridge or a navigational waterway. After the pier caps are prepared, the heavy transporters supply a full span to a launcher, which places the piece onto two piers. This rapid method of construction is made possible by two different launcher concepts used today. The first is a twin LG with a heavy lifting beam across one span and a lighter launching nose, as shown in Figure 11.55. Its twin is a light-weight underbridge, which is only needed during the travel phase of the overbridge and provides the driving leg to move to the next pier ahead of the gantry front leg.

The transporter of the heavy elements, also called a self-propelled mobile transporter (SPMT), is equipped with rubber tires mounted in pairs that are able to pivot 360° around a vertical center axis. For example, the transporter used to construct the Incheon Bridge, completed in 2009, had to move spans that were 160 ft (48.8 m) long and weighed 1,500 tons (1,360 t). For this, it needed 320 rubber tires. Completely computer-controlled and hydraulically powered, it could be driven by one operator up 8% inclines and change direction on the spot. Smaller units can be coupled together to any length needed, all still operated by one driver.

Recently a second method has been developed that combines the benefits of a launchable underbridge with that of a mobile transporter to create the full-span method using wheeled straddle carriers that are able to travel on a launching underbridge beam as well.

11.6.5 FULL-SPAN-METHOD LAUNCHING AND SELF-PROPELLED STRADDLE CARRIERS

Figure 11.56 depicts the straddle carrier as a transporter split into a front and rear trolley, each with a steel structure to support an elevated beam between them. Instead of operating like a trailer, the heavy object is suspended from the elevated beam via jacks and rods. Similar to the transporter discussed in Section 11.6.4, the two trolleys are highly automated and steerable by one driver.

The key to this technology is the front trolley. As shown, during the stretch from the precast plant it travels on its rubber tires. When the trolley reaches the launching beam resting on top of the last bridge span, it is able to straddle it and continue its transport past pier G, now supported by the launching beam and its legs.

Before the new span can be placed onto the pier caps, the beam has to be pushed forward ❹ and out of the way. But first, the rear leg needs to be moved from pier F to G ❸. Now the mobile leg can be moved, followed by the front beam leg. Now everything is

S11.13 VSL Bridge Construction Systems
S11.14 FHWA Bridge Transporter

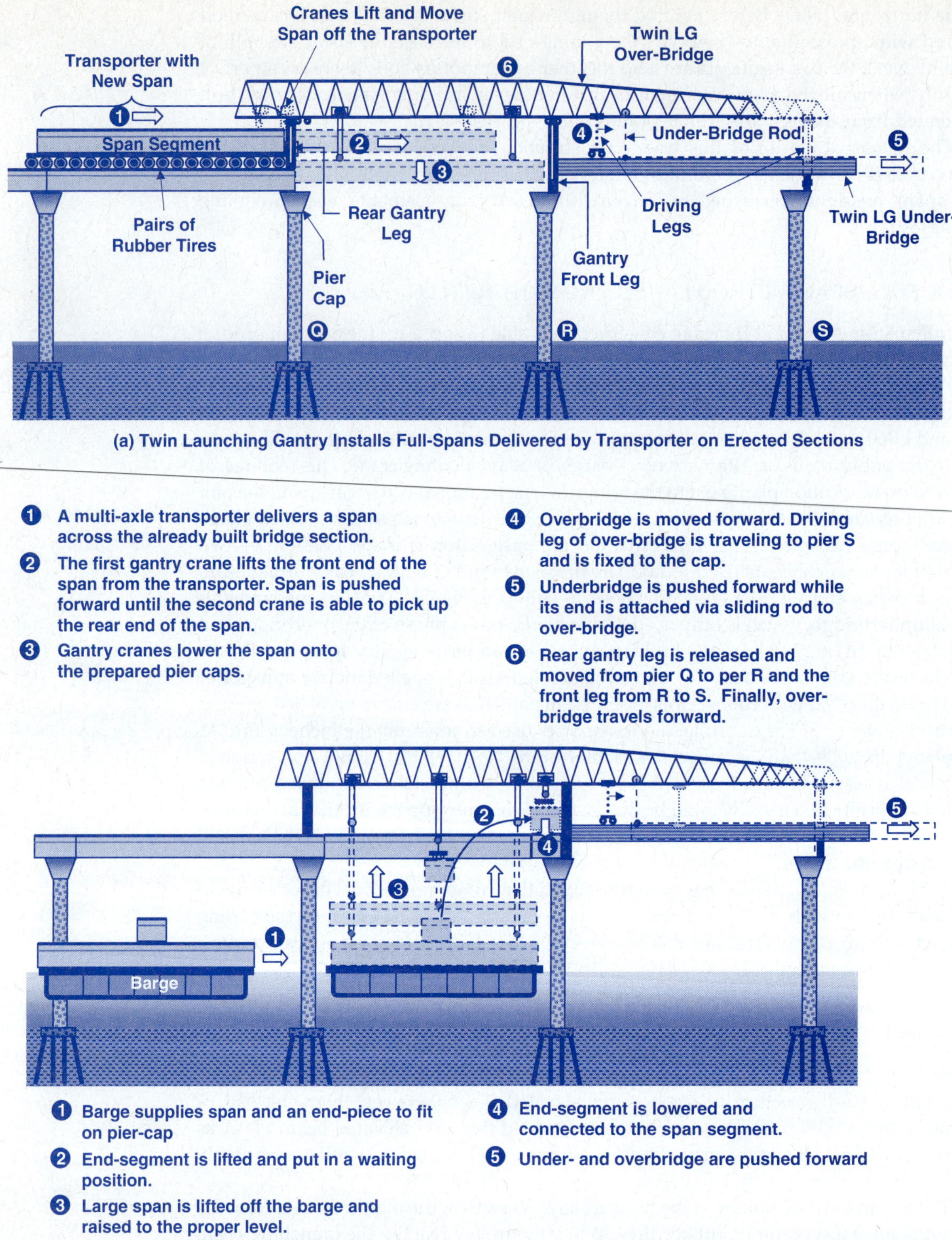

(b) Twin Launching Gantry Installs Full-Spans Delivered by Barge

Figure 11.55 Full-span Installation Methods Using Twin Launching Gantry

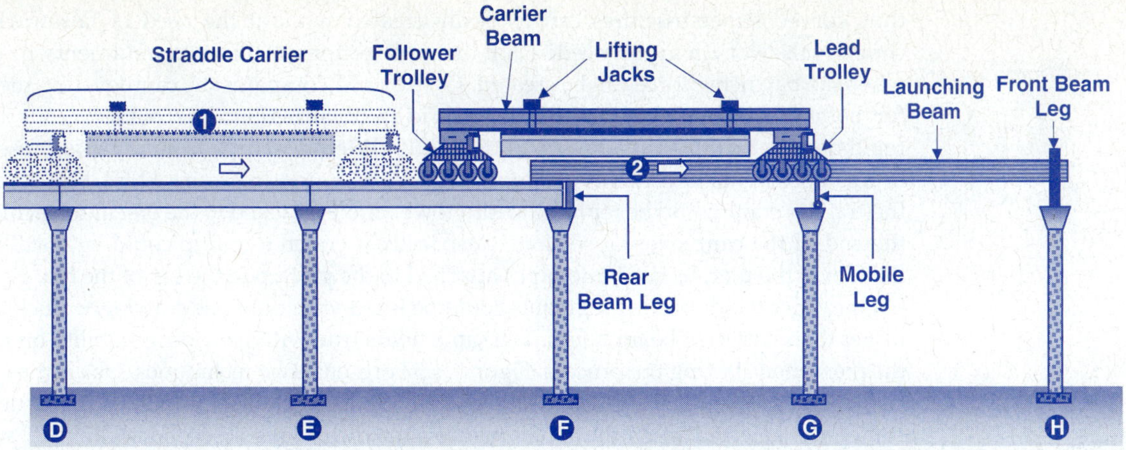

1. Straddle carrier arrives with the span F–G.
2. Lead trolley advances on top of the launching beam.

(a) Lead Trolley Advances on Top of Launching Beam

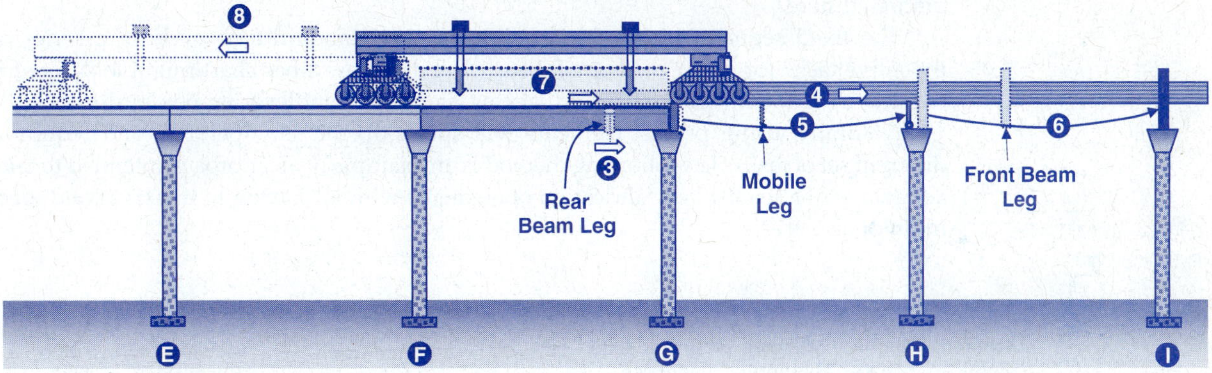

3. After the front trolley reaches the final position beyond pier G the rear beam leg is moved from pier F to G.
4. Launching beam is pushed forward below the front trolley attached to the rear beam leg.
5. The mobile leg on pier G is moved forward to H.
6. Front beam leg is moved from pier H to I.
7. New span is lowered onto piers F and G.
8. Empty carrier drives over the new span back to the storage area.

(b) Beam Launching and Positioning of New Span

Figure 11.56 Placement of a Full-span Segment with a Straddle Carrier

ready for jacking down the span onto piers F and G, providing a traveling surface for the tires of the front trolley to return to fetch the G–H span.

11.6.6 BALANCED CANTILEVER BRIDGE CONSTRUCTION

In 11.6.2 we learned about the balanced cantilever method for erecting precast box girder segments hoisted by an LG. This section discusses how the cantilevering method is also applied to cast the elements in-situ using a traveling form. Similar to the derricks or beam cranes used for the segmental precast erection, a formwork wagon rides on tracks anchored to the tip of the cantilevering bridge superstructure. Tie rods provide the necessary support during rebar placement and the pouring of the concrete. The forms can be readjusted for every box-girder segment with the help of hydraulic jacks;

thus curved superstructures can be easily created without the need of falsework. Of course, falsework may be needed for the end sections at the two abutments in cases where no balancing force can be created. Figure 11.57 compares the cantilevering method for both precast and in-situ box-girder bridge construction. The reader needs to be reminded that contractors always find slightly different setups from what is presented in this book. This is certainly true for what is presented in Figure 11.57. For example, the crawler crane could be replaced with a tower crane attached to the pier and reaching to the end of the cantilevering sections. Also, a fixed concrete pump could be installed at the foot of the pier, feeding into pipes attached to the pier up to the top of the box girders. On the other hand, precast segments could be lifted with only one crawler or truck crane rather than with two beam hoists. The same holds true with providing stability on top of the pier when starting the process. Figure 11.57 presents two techniques. One is the use of two piers to create a stiff connection with the first element of the superstructure, sometimes referred to as the table. Figure 11.57b displays a second technique, which uses two temporary steel towers, one on either side of a slender pier. There are several other techniques, but they are not covered due to space limitations. However, the reader will quickly recognize their unique features after understanding the principles discussed here.

The fixed segment length invites the use of a standard-time cycle of a series of repetitive tasks for either method. Table 11.7 shows how a bar chart might look like for two traveling formwork systems advancing along the cantilevering box girder.

Naturally, using precast elements will speed up the site operation and require a different set of tasks. The following method is an example of an approach intended to take advantage of a fixed-plant fabrication of elements without having to transport and erect individual pieces.

VG11.9 Cantilever Bridge
Svinesund Bridge

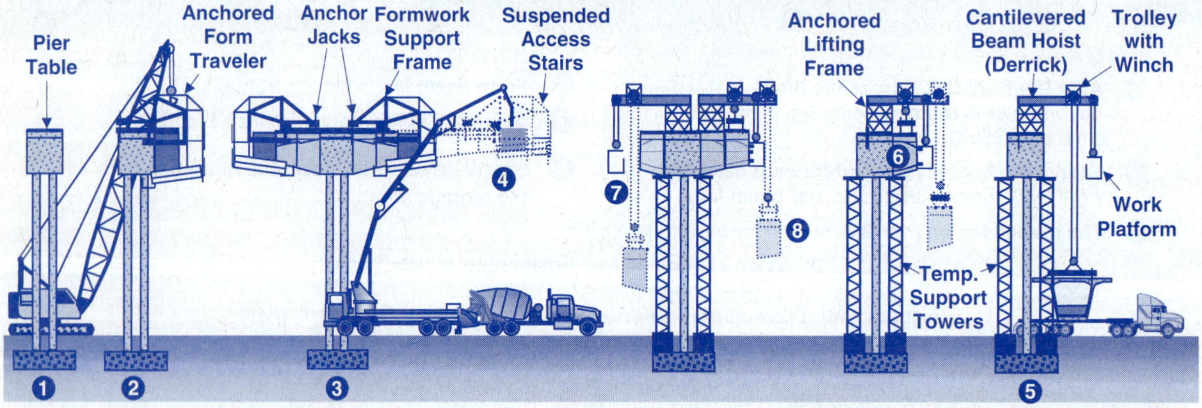

1. Double-pier and pier table have been constructed.
2. Crawler crane is used to install first form traveler anchored to the pier table.
3. Both travelers have been installed and anchored to segments with varying box heights. Post-tensioning cables hold the segments in place.
4. Concrete pump truck delivers to the form.
5. Temporary support towers stabilize the pier table. Beam hoists anchored to the table are used to lift segments from the street level. A working bridge is suspended in a parking position.
6. First pre-cast segment is post-tensioned to the pier table.
7. Second beam hoist is anchored to the table after the first has moved forward, making room available for the hardware and the anchors. First balancing element is attached.
8. First hoist continues with lifting a second segment.

(a) Cast-in-Place Balanced Cantilever with Form Traveler on Rails **(b) Balanced Cantilever with Precast Box Girder Segments**

Figure 11.57 Common Equipment Setups for Balanced Cantilever Bridge Construction

Table 11.7 Seven-Day Cycle for Casting-Curing-Tensioning One Standard Box Segment

#	Task Descriptions for Casting One Segment with a Traveler	Work Days 1	2	3	4	5	6	7
1	Add rails, remove main anchor, travel form, anchor at new position	━					═	
2	Adjust-fasten forms for new segment, place rebar fin bottom slab-web	━	━					═
3	Place tensioning cables into web, advance interior form for top slab	═		━				
4	Install wall ties, complete rebar for top slab, including stressing cables		═		━			
5	Pour concrete for bottom slab, webs, and top slab			═		━		
6	Curing of fresh concrete				═		━	━
7	Post-tension transverse-cantilever tendons, stripping of forms, grouting					═		━

━ = Segment on Cantilever 2
═ = Segment on Cantilever 1

11.6.7 INCREMENTAL LAUNCHING

For this method, a plant is set up behind one or both abutments of a bridge having a consistent cross-section and horizontal curvature. Within the well-organized and protective temporary plant, such as the one shown in Figure 11.58, segments of the bridge are built in sequence and pushed out the "front door" toward and across the already prepared piers. The superstructure can be built with steel, concrete, or composite. The following presents the technique applied to a concrete box-girder bridge.

VG11.10 Svinesund Bridge

The covered area of the casting operation is serviced by an overhead crane. Beside the building is the concrete batch plant providing fresh concrete when needed at a constant quality. In zone 1, rebar is assembled into cages to be lifted into the stationary outer bridge formwork residing in zone 2. It is in this zone that concrete is being match-cast against the previous segment for the bottom slab and the web. After initial curing, the forms for the webs can be pulled back and the uncompleted segment pushed into zone 3 for adding the top slab. Often, however, zones 2 and 3 are combined, meaning that the top slab is added in zone 2. After sufficient curing, the segment can be pushed to zone 4 for continued curing while stressing cables are inserted and prepared for tensioning, to be started when the concrete reaches sufficient strength. Table 11.8 shows that a sufficiently accelerated initial curing might take 2.5 days. The post-tensioning will be done

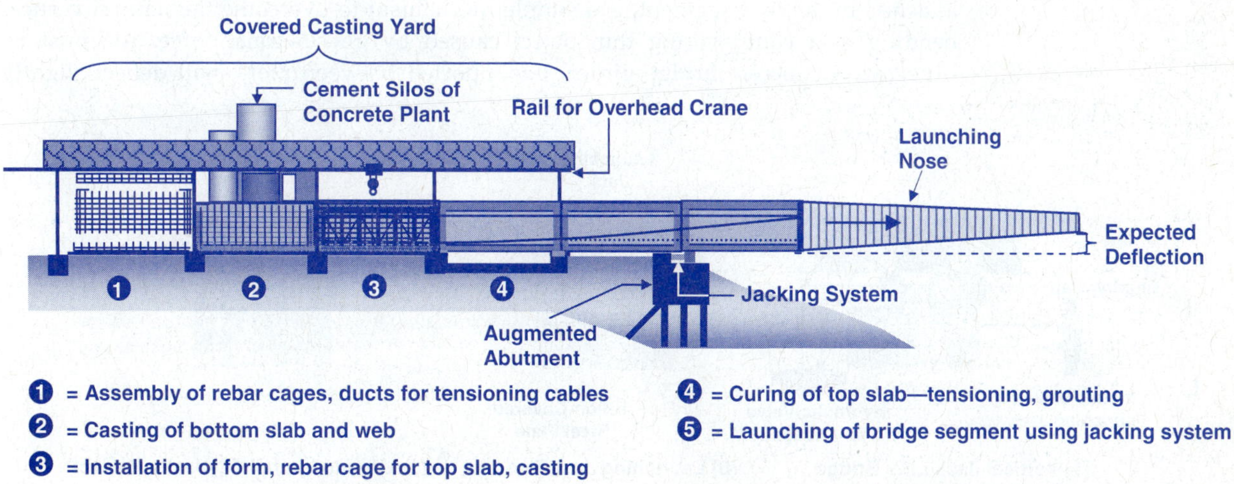

Figure 11.58 Cross-Sectional View of a Casting Yard and Launching System for Box-Girder Bridge

Table 11.8 Example Barchart for One Segment Launch Cycle

#	Task Descriptions for an Incremental Launch Cycle	Work Days						
		1	2	3	4	5	6	7
1	Assembly of rebar for bottom slab and webs in section 1							
2	Insertion of corrugated tubes for longitudinal tension cables into rebar							
3	Cleaning and oiling of formwork for slab and webs							
4	Lifting rebar for bottom slab and webs into forms in section 2							
5	Relocating inside web forms and tying of in- and outside forms							
6	Pouring concrete for bottom slab and webs							
7	Relocating top slab form, placement of rebar, lateral stressing tendons							
8	Placement of concrete for top slab							
9	Curing of concrete (top and bottom slab)							
10	Removing of outer forms							
11	Tensioning of tendons and grouting of tubes							
12	Launching of bridge by one segment							

transversely as well as longitudinally on cables that are coupled to those in the adjacent segment. After the these tasks are accomplished, for example in seven days as laid out in Table 11.8, the segment is freed of all supports and pulled forward toward the abutment.

Zone 5 is the station that is responsible for launching and moving the entire bridge after a section is ready. Figure 11.59 demonstrates the operation of one of the two systems working under each side of the bridge box of either steel or concrete.

For incremental launching, the abutment has been built so that two jacking systems will fit between the top of the abutment and the bottom of the bridge. Two proportional valves control the jacks. The lift jack rests on a Teflon-covered steel plate designed to reduce friction during launching. A cycle begins with the opening of the valve of the lift jack (Figure 11.59a) allowing the high pressure fluid to push the piston upwards until the bridge is off the bearing cap. Now, the horizontal launching jack is activated, causing the piston to slide the lift jack assembly along the Teflon surface for a full stroke, shown in Figures 11.59b and c. At this time the lift jack retracts, setting the bridge back on its abutment bearing, which allows the launching jack to retract as well, pulling the lift cylinder back to the starting position to begin the next cycle.

A unique piece of equipment is mounted to the front tip of the launched bridge—the launching nose. This distinctively shaped lightweight structure, which is firmly attached to the first segment, is a simple mechanism to overcome the natural vertical bending of a cantilevering thin object caused by gravitational force. As must be expected, a concrete bridge girder, unsupported between piers, will deflect slightly

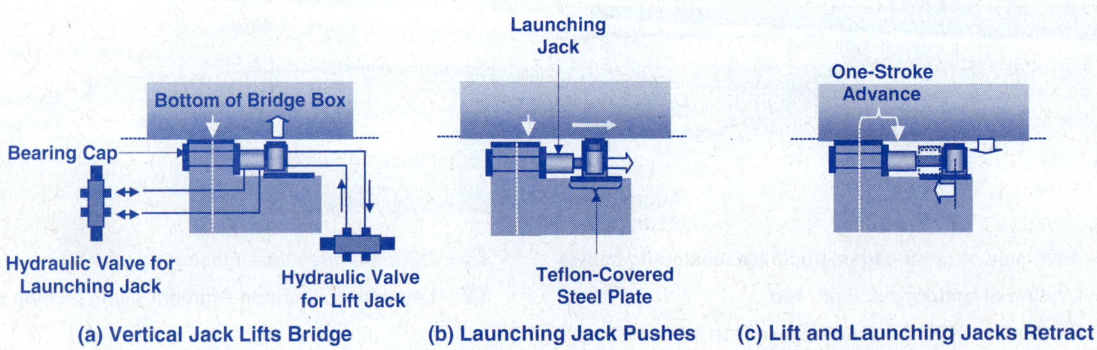

Figure 11.59 Sequence of a Launching Cycle

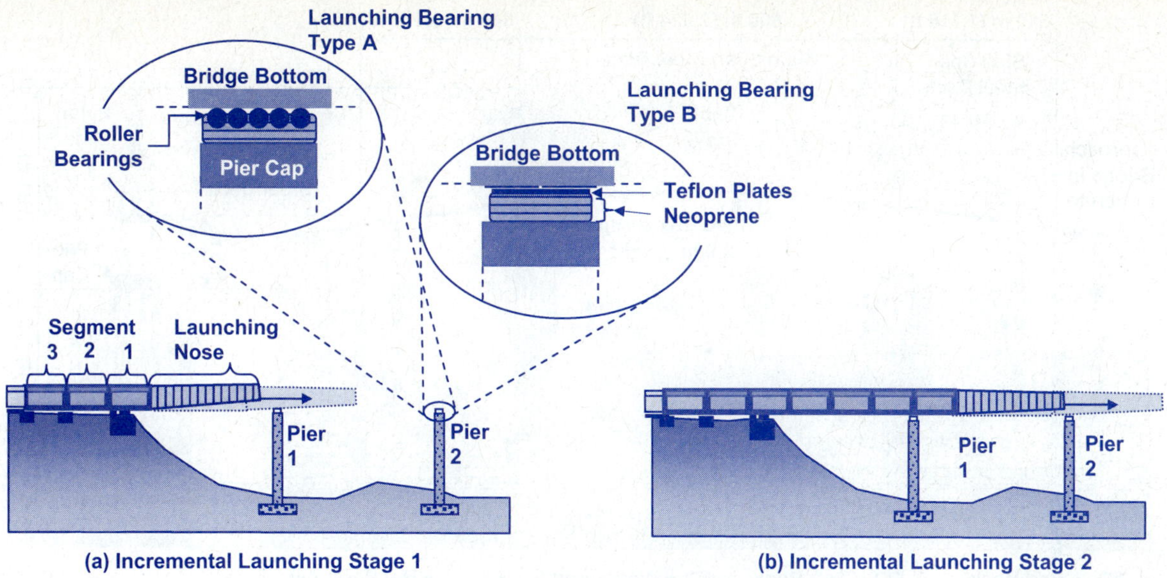

Figure 11.60 Stages of Incremental Bridge Launching Process

despite its significant cross-sectional inertia. The purpose of the launching nose is to end up at the next pier above the pier cap and bearing. For this reason, the nose is shaped upward, making sure that the front edge will make up for the expected deflection and force the first element upward and over the temporary launching bearing as the bridge is launched forward.

As the superstructure is advanced, section by section, it will eventually reach the first pier cap, guided by the launching nose. In order to protect the pier from undue horizontal forces caused by friction, special launching bearings are installed, designed to smoothen the sliding of the box or girders. Figure 11.60a presents two techniques that can be found, one based on rollers and the second on Teflon-neoprene plates with a very low friction coefficient (e.g., 2%). Lateral guides mounted on the side will ensure that the bridge stays on it horizontal track. The bearings and guides will be replaced with permanent bearings after the bridge reaches its final position.

11.7 Cable-Stayed Bridge Construction

In the 1950s, the well-known suspension bridge, with a deck that is suspended by vertical hangers from cables that are anchored on both sides of the bridge, saw the birth of a "cousin," the cable-stayed bridge. This type of bridge has started to increase the range of main spans to over 1000 m, a length that previously could be reached only by suspension bridges. Single-span cable-stayed bridges have one long main span, carried by forestays, and two short-side spans supported by the so-called backstays.

In October 2009, a new 12 km (7.3 ml) long bridge was opened in South Korea, opening a second road to reach the Incheon international airport. It included a 800 m (2, 624 ft) main span cable-stayed bridge that will be used to explain the entire construction sequence. Of course, this will include many construction methods and equipment that have been studied earlier. Figure 11.61 presents an overall view of the bridge when completed. The drawing illustrates that the bridge has to cross the entrance to Incheon, the second-largest harbor in South Korea. Thus, the clearance in the center is 74 m (243 ft) and at the pylons 55 m (180 ft).

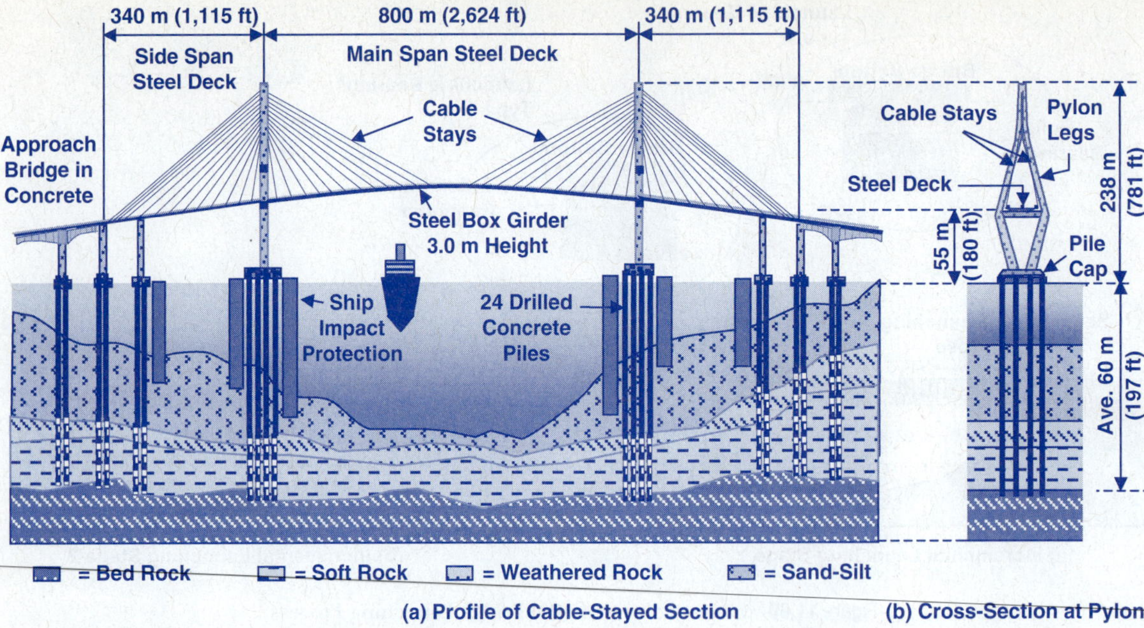

Figure 11.61 Schematic View of the Cable-Stayed Portion of the New Incheon bridge (not to scale)

Figure 11.62 View Down the Centerline of the Completed Bridge

Furthermore, large Ship Impact Protection (SIP) barriers are built to protect the pylons against the force of an off-course ship. The surface of bedrock is at an approximate depth of 60 m, lying below several layers of sand-silt and weak rock. The form of the pylons is an inverted Y, anchoring two rows of cable stays arranged in a semi-fan shape. Steel is used to build the stiffened box girders for the bridge deck. Both approaches are built in concrete. Figure 11.62 offers a view of the bridge from the southern approach at sunset.

Let's go back four years and revisit what construction was needed to erect this massive structure.

11.7.1 BUILDING SOLID FOUNDATIONS

Work on the bridge began in the fall of 2009 with the driving of steel casings 3.0 m (9.8 ft) in diameter through the sandy silt down to the surface of weathered rock. Twenty-four casings were needed for each pylon and 8 casings for piers in the approaches. Figure 11.63 depicts several steps of foundation construction.

A barge-mounted lattice crane is in the process of driving the 24 casings with a hydraulic hammer for pylon A. A platform has been installed as a guide so that each pile head will end up at the desired location to support the cap. As shown, the piling stops above the rock surface so as not to damage the tip of the pipes since the entire diameter will be needed for the next operation to reach the bedrock, work that is proceeding at foundation B.

The goal of step 2 is to drill through the weak rock down to bedrock onto which the concrete pile can be erected. A common practice for drilling large diameters vertically through rock is reverse circulation drilling (RCD). In the case of the Incheon bridge, the drills were clamped onto the top of the casings that provided the reaction forces for the downhole drilling operation. Figure 11.63a shows an RCD system atop a casing, while the upper frame containing the rotary drill motor and the jacks to create down-thrusts has been tilted backward to provide access to the inner part of the casing. Behind the operator cabin, one can see that a new outer pipe is being moved up, as the drilling master is now able to add a new section after having drilled a full stroke length. Also visible are two hoses, one to provide water-mud to be pumped down to the drill bit and the other connected to the drill pipe bringing up the cuttings. Figure 11.64b shows the RCD during a drilling phase when the upper frame sits vertical on the casing, which allows the rotary motor to engage the drill steel rod. Next to the drill on top of a large casing, one can see a rebar cage for a pile suspended by a crane.

In fact, the rebar cage is presently being lowered into the casing of a pile that has already been drilled down to the bedrock. This segment, as shown, is the last of a series that have been welded together, as they have been lowered to a depth of approximately 60 m (197 ft). The indicators that this is the top segment are the white wrappings of the longitudinal rebar at the very top. The reason for protecting the vertical rebar is visible in Figure 11.64c, which shows the top level of a pile casing filled with concrete and the longitudinal rebar protruding through the still wet concrete. In the center of the pile is a suspended pump that removes the water that is still percolating

Reversed Circulation Drilling (RCD) (Dual-Wall)

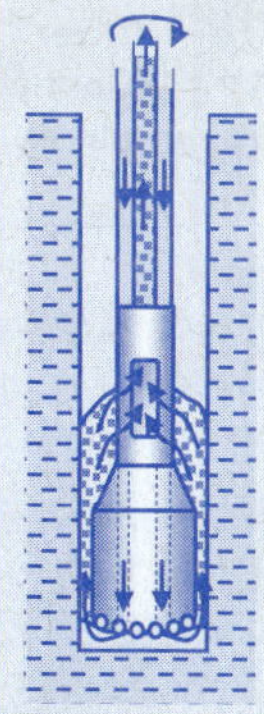

This method received its name from its reversal of the flow of drill mud or water through the drill steel rod. "Clean" water is forced down between an outer and an inner pipe and exits through holes in the rotating drill bit. As it is pushed back upward the drill hole it takes along cuttings from the drill and flows toward the low pressure created by the suction pump of the drill rod. It also cools the bit.

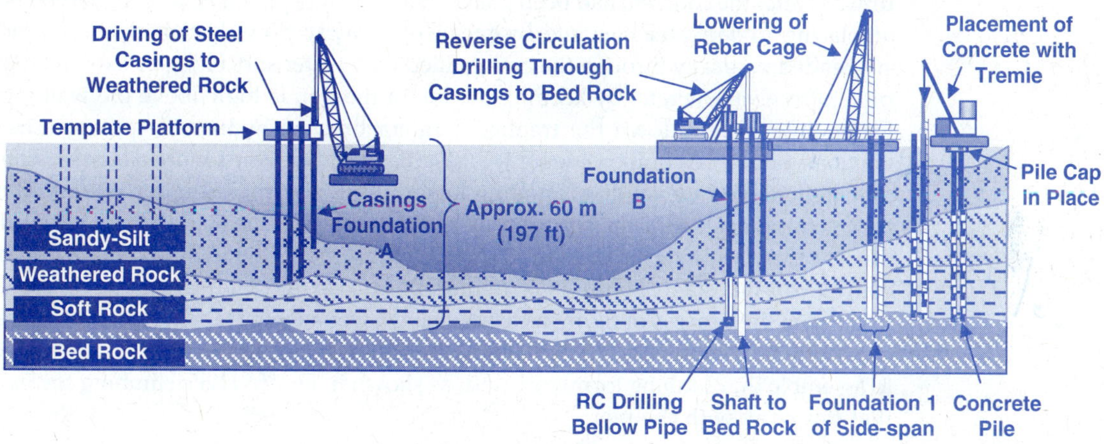

Figure 11.63 Construction of the Foundations for Pylons and Piers

(a) RCD On-the-Pile Readied to Add Drill Steel **(b) RCD Drilling and Rebar Cage is Lowered** **(c) Casing Filled with Concrete, Wrapped Rebar**

Figure 11.64 Construction of Deep Foundations for Bridge Piers and Pylons

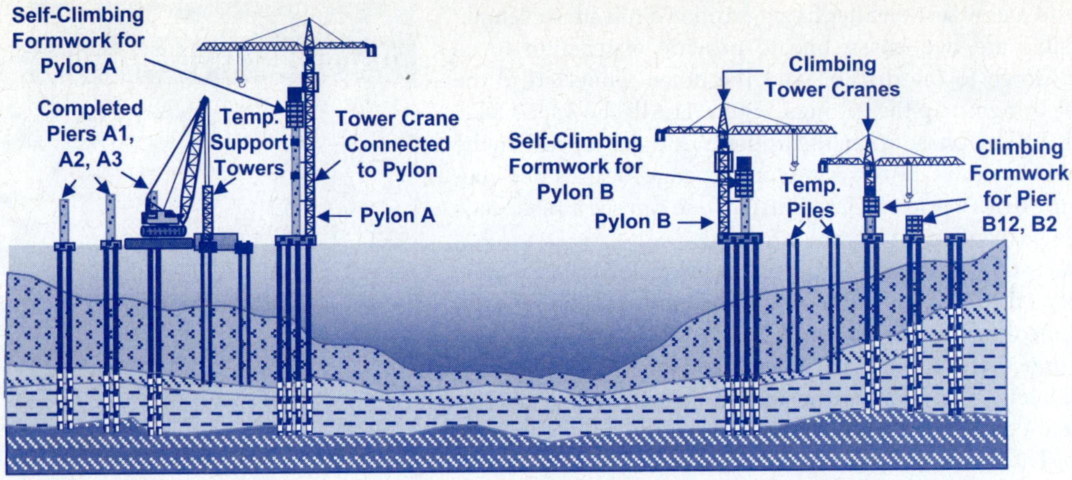

Figure 11.65 Building the Temporary and the Final Supports for the Bridge Superstructure

upward after the concrete had been placed with a tremie pipe. The tremie method is a way of placing underwater concrete by means of gravity flow. It consists of a rigid pipe suspended vertically through the casing filled with water and a hopper fixed on top of the pipe to receive concrete. By keeping the top of the tremie high above the water, gravity forces the concrete down the tremie, counteracting the hydrostatic pressure from the water around it. To compensate for hydrostatic water pressure, more concrete is pumped to a higher level, a level that is getting lower as the water is being forced out. Thus, the rebar wrappings fulfill the function of keeping the steel clear of water and mud, thus offering clean welding surfaces for the cages of the pile caps and perfect bonding with the concrete for that stage. Figure 11.63 depicts this phase. A barge, carrying a concrete batch plant, is in the process of filling a casing using a tremie while anchored to a finished pile cap. With the completion of the pile caps, the hollow concrete piers are getting a solid base to assemble the climbing formwork and, as shown in Figure 11.65, climbing tower cranes that can grow with the pier.

11.7.2 CASTING PYLONS AND PIERS—TEMPORARY SUPPORTS

The installed climbing tower cranes had capacities of 120 m-t and were fixed to the pile-cap next to the pylon. In Chapter 10 we covered some of the different formwork systems for vertical structures. For the Incheon bridge, self-climbing systems used hydraulic power to push the forms upward. Concrete was pumped up in steel pipes from a barge mixing it right there. On average, the cycle time for 4 m vertical advancement was 4 days.

The crawler crane on top of the barge in Figure 11.65 is setting up temporary steel towers on a temporary pile cap. They will be needed for the construction of the superstructure, with the first phase being presented in Figure 11.66.

11.7.3 BUILDING THE SUPERSTRUCTURE

The construction of the superstructure advances at several locations. At pier A1, a floating crane is in the process of setting a pier table onto the cap as the start segment for the balanced cantilever method to erect precast box-girder segments, which is already in

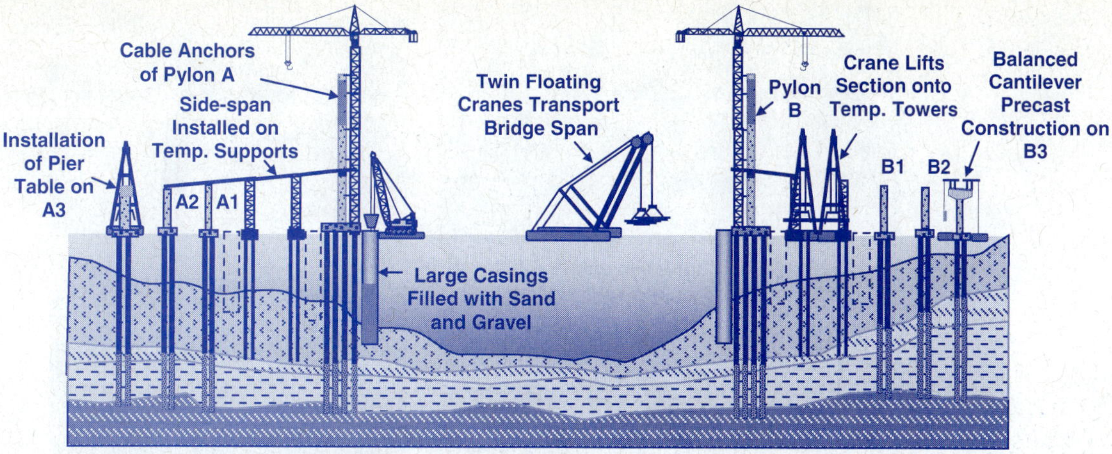

Figure 11.66 Start of the Superstructure and Ship Impact Protection

process on pier B3. There, two beam-cranes lift the elements on both ends of the growing bridge section connection to piers B2 and B4.

The superstructure of the cable-stayed bridge has also already begun. Piers A1 and A2 bear the loads from the steel box-girder deck without help from the stay cables. It is apparent that the side-spans are built without the balancing effect of the cantilevering main span and ahead of its construction. Now the purpose of the two temporary towers becomes clear. They serve as extra piers for the steel deck until the stay cables take over the load in the next phase. The installation of span-to-span sections is presently progressing from pylon B to pier B2. Two twin ship cranes pick up the prefabricated sections from shore and float it toward temporary towers where they are lifted onto the bearings so precisely that they can be welded to the neighboring sections. The large cranes are still standing, as they will assist in the installation of the stay cables, which need to be lifted to the crown of the pylons where they will be anchored.

Three photographs in Figure 11.67 offer views of the actual situation approximately one year before the bridge was opened. Clearly visible is the balanced cantilever construction at pier B3, almost completed for both decks, while the floating crane has already put the pier tables in place on pier B4. The completed side spans to B2 and the two temporary steel towers are shown completed in Figure 11.67b. In fact, work has begun on the main span beyond pylon B, as evidenced in Figure 11.67c.

11.7.4 INSTALLING STAY CABLES

The first stay cables have been put in place on both sides of the pylon, indicating that the weight of the side-span is being used as a balancing force for the growing main span section. The scaffolding around the crown of the pylon provides safe and easy access to the anchoring points during lifting by crane, setup, tensioning and application of the protective covering. The construction of the final elements is shown in Figure 11.68.

Presently, the last of the 3.0 m (9.8 ft) high and 12.3 m (40.3 ft) long steel deck box girder segment has been lifted off the last supply barge. Before this final lift, the last four stayed cables had been pulled up to the very top anchor positions. As indicated, the anchorages had been prefabricated in steel in the precision machine shop. The ever-changing angles both in the vertical and the horizontal dimensions (3-D) make it impossible to construct the anchorages with the necessary precision on site. Thus, strong

(a) Balanced Cantilever with Precast Segment-2 Pier Tables Installed

(b) Steel Box-Girder Section Meets Concrete Box Girder at B2

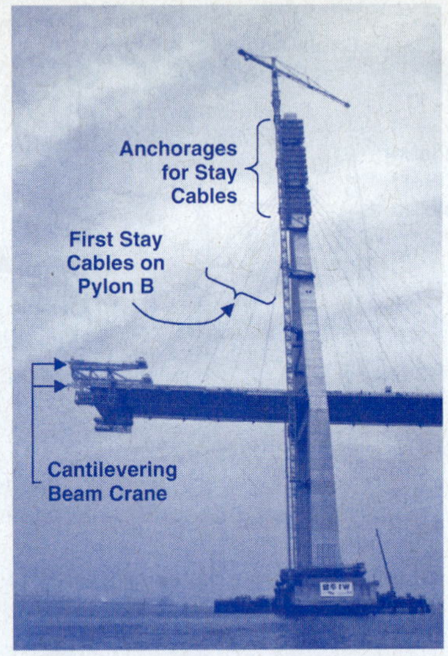

(c) Cable Supported Cantilever Main Span Construction

Figure 11.67 Superstructure Under Construction

and accessible steel boxes containing the individual anchorage frames are embedded in the head section of the pylon. As discussed earlier, one end of a multi-strand cable will be built as fixed and one as an adjustable anchorage connected either to the deck or the pylon. It is important to have the adjustable anchorages aligned on the top and protected from the weather. Of course, this will help in the inspection process and maintenance, especially when winches and a hydraulic pump with hoses can be permanently installed.

Two methods are used for the actual installation of multi-strand cable: (1) prefabricated and (2) strand by strand. In the first, all the strands are pre-cut and put together in advance, while in the second each strand will be pulled up to the top of the pylon

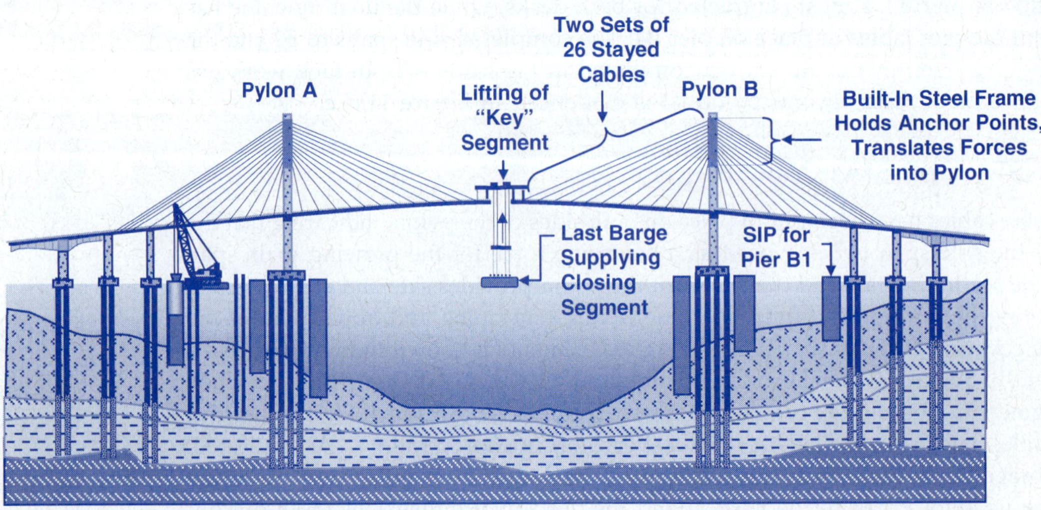

Figure 11.68 Final Phase of Bridge Construction

(a) Deck Anchors and Cable Stays (b) Protections for Anchor and Cable

Figure 11.69 Deck Anchorage at Incheon Bridge

separately. Of course, the first method is quicker, but it requires heavy lifting equipment such as a tower crane. The fact that the Incheon bridge still had one crane at each pylon until the very end tells us that the stay cables were, in fact, installed as heavy complete sets supplied on large reels, ensuring the fastest mode of operation.

After two cables have been properly set into their anchorages on each side of the bridge deck and the pylon, the tensioning at the adjustable end can begin until the desired length or the needed tension force is measured. These settings will normally have to be adjusted when the entire deck is completed, so that each cable will carry a load that creates an even distribution along the bridge. As can be expected, the greatest stress that a cable experiences is right after it has been installed and the next segment is being hoisted. In this situation, the cantilevering beam crane is fastened to the last segment cantilevering from the latest anchorage point. Thus, the weights of the segments being winched up and that of the crane rest on the anchorage alone, in addition to the weight of the segments that create the cantilever effect. The next cable installation will result in a drastic reduction of the tension. Figure 11.69 presents two views of the completed deck anchorage. Clearly visible is the sag of one stay cable caused by its own weight. Also identifiable are the guide pipes for the strands that are separated from the tight packaging so that each individual strand can be tightened at the anchor plate at its end (not visible). The cone-shaped "hat" protects the larger-diameter vibration-damping mechanism against corrosion. At the same time, it is built to guard against vandals. Figure 11.69b highlights the extruded HDPE enclosing the cable itself showing indentation on its surface to reduce the wind drag on the cables.

The final activities are the removal of the temporary supports in the side-spans and the completion of the large filled casings around the pylons and first piers. The latter will provide the SIP for the foundation against ships such as a 100,000 DWT tanker traveling at a speed of 10 knots. The 44 casings, also called dolphins, are between 20 m (65.6 ft) and 25 m (82 ft) in diameter, and were prefabricated with sheet piles on shore. After they are positioned on the sea floor, they are filled with sand and gravel and topped off by a layer of concrete. In case of an impact, the dolphin may be destroyed, but in the process it will absorb up enough energy to stop the ship before it reaches the pile cap.

11.8 The Construction of Arch Bridges

As mentioned earlier, the arch has been used for many centuries to span deep valleys or wide rivers. Its unique structural feature results in that the load-carrying elements are all in compression, making them ideal for building them in stone, brick, concrete, or cast iron. The human eye has long gotten used to looking at arches, considering them aesthetically pleasing independent of the material used.

Unlike in Roman times or later, when the large cathedrals in Europe were built, the spandrels of today's bridges are left almost empty, resulting in significant reductions in weight and in cost savings. Instead, these so-called open-spandrel arch bridges use thin piers or ribs to provide the vertical support for the upper deck.

Spandrel

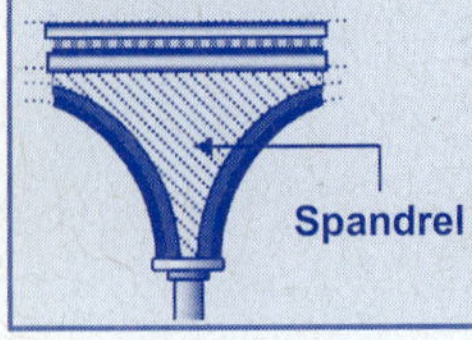

Probably from the Latin *expandere*, meaning to spread out. Today the spandrel defines the filler space between two arches or between an arch and the frame around it. In Baroque and Gothic cathedrals, this space was heavily decorated.

Because arch bridges do not require large piers or pylons, the construction starts with the abutments. Depending on the general type—tied arch or deck arch—the abutments have to be able to sustain different forces. In the following discussion, we will focus mainly on deck arches, where the horizontal reaction forces will be borne by the abutments without the help of ties.

11.8.1 FRAMING THE ARCH

A strong arch is constructed using high-compression materials. For the construction of the arch, heavy falsework, also called the centering, is required to support the individual elements until the keystone is put in place at the crown. Only with this last stone will the arch become self-supporting. As one can easily imagine, setting up such falsework across deep valleys in or across a river was a main engineering feat in itself and a major cost factor. Thus, it is not surprising that alternative methods were sought that did not need heavy falsework. One alternative is cantilevering the arch supported by stayed cables. Initially, this method was only used on arch bridges made of cast-iron and steel. Later the technique was modified to work for concrete and composites as well. For example, two halves of thin-walled steel-arch tubes were assembled and hinged at each abutment. Subsequently, they were lowered from the two abutments using winched cables, to create an arched centering from which the rest of the bridge was built up.

VG11.12 Hoover Dam Bypass
VG11.10 Svinesund Bridge

The last section of this chapter will review recent examples of arches made as a concrete box supported by temporary stay cables until the "key" was cast. Figure 11.70 illustrates the start phase at one of the two abutments that is constructed with the use of a cableway covering the arch span.

The drawing in Figure 11.70 presents three construction systems integrated to work together: (1) a cableway crane covering the entire construction site, (b) a cast-in-place stayed arch under construction, and (c) surveying system.

Chapters 3 and 7 demonstrated the many ways construction utilize ropes, cables, winches, pulleys, sheaves, and booms to move heavy objects to large heights. The construction of stationary and elongated bridges provides yet another opportunity to take advantage of ropes and pulleys. Referred to as cableway or aerial tramway, this type of crane has quietly provided unmatched capabilities in terms of capacities. For example, one cableway for the building of the Hoover Dam lifted 181 t (200 tons) with a stationary span of 383 m (1,256 ft).

Besides the ability to lift heavy weights, the cableway boasts high dependability and excellent safety records when designed and built properly. To provide coverage in the lateral direction, the end towers can either be put on rails, or a mast with a ball-joint bearing can be luffed laterally, which, of course, moves the track cable as well. To this end, the foot of a mast is designed as a swivel, while the top is guyed back while

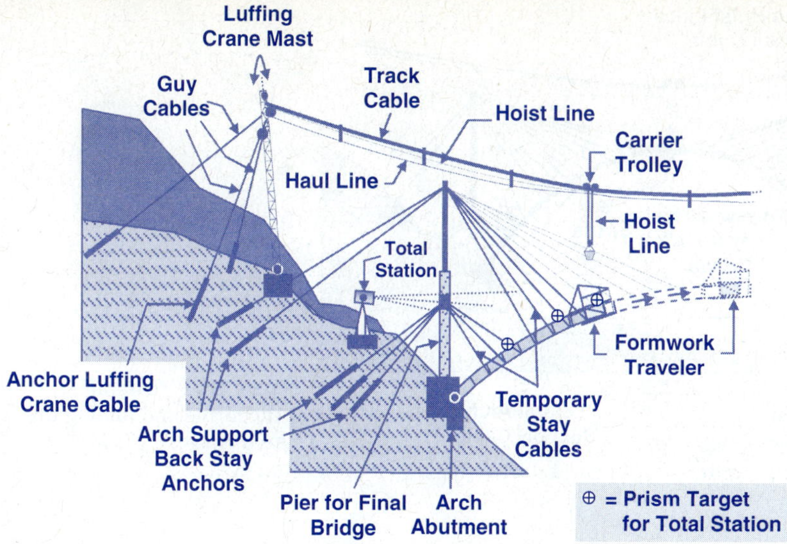

Figure 11.70 Cantilevering Cast-in-Place Arch Construction

side-cables are used to pivot or luff the top of the mast as shown in Figure 11.70. Three ropes and cables span between the two luffing masts on the opposite sides of the divide. Several track cables provide the "way" for the trolley carrier, riding on several pairs of wheel sets to distribute the load along the cable. It is moved across the track by the endless haul line crossing the span twice to be attached to both sides of the carriage and powered by the haul line drive. Finally, the hoist line is guided by pulleys on the trolley carrier down to the hook-block sheaves and dead-ends at the opposite side from the location of the hoist winch. The entire cableway crane is powered by electric winches and controlled from one end or remotely via a wireless control system.

The cantilever construction of the arch reminds us of the balanced cantilever method used for cable-stayed bridge construction. The only difference is that the side-span cable stays are replaced by temporary backstay cables anchored back into the rock. Also, the pylon is replaced with a temporary tower built on top of the final pier resting on the abutment. The formwork traveler, instead of progressing horizontally, is first a mix between a climbing and cantilevering formwork. This will require some additional forms for covering the slabs during the pouring of the concrete and flexible walkways that can be kept horizontal while the traveler slowly reduces its incline. Depending on the shape of the arch, perfectly round or elliptic, the forms themselves may be kept at a fixed position. Nevertheless, the 3-D position of the form traveler needs to be tracked and adjusted in order to ensure the proper shape of the arch and to meet up in the middle with the opposite half. This important operation is performed with very precise computerized surveying equipment.

The center point for keeping the progressing arch on track is a computer-controlled total station that is installed on a fixed and secured foundation behind the abutment, as depicted in Figure 11.70. In Chapter 4 we learned how total stations use a laser beam to hit a prism mounted on a moving grader, thus measuring the distance and the angles of the target several times per second. Built-in software calculates in real time the changing position in 3-D space relative or absolute. The same concept is used but slightly expanded with a programmable target finder. This latter capability is critical in making this system economical in bridge construction, where a growing number of prism targets are affixed

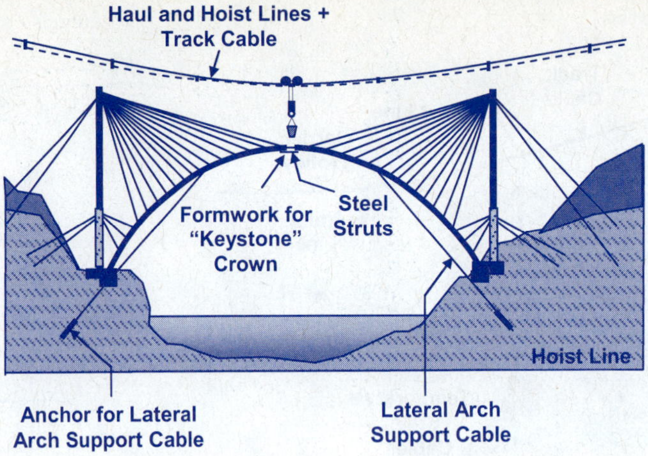

Figure 11.71 Lateral Anchoring and Joining the Arch Halves

to the arch and to the form traveler, as shown in Figure 11.70. For each measurement, the total station has to target its laser to a small prism, a task that requires a great deal of concentration by the surveyor, easily leading to errors. Furthermore, environmental factors, such as temperature changes, wind, and sunshine, cause the cantilevering end to swing or to change its cross-sectional form elastically. These impact factors not only require dynamic 3-D measurements several times per second, but also effective bridge deflection models to calculate the actual center of the moving target. And finally, the target finding needs to be automated in such a way that the system will find each tiny target by its own whenever desired. This is in effect what modern systems do as most targets are being kept at a fixed position after being attached to the bridge. These fixed positions can be "taught" to the system that stores the measurements of its internal motor positions, after which the built-in servo-system is able to return to the exact same readings and start a search path in case the target has moved slightly. Enabled by radio communication, the entire system can be remotely controlled even to follow the form traveler, providing critical data to "steer" the traveler or to tighten or loosen cable stays.

As the arch halves grow, their lateral stability to sustain high wind forces decreases. Thus, at one or several predefined positions, lateral arch supports, in the form of pairs of stay cables, are anchored back into the side of the valley at both sides of the bridge. By equalizing the cable angles and the tension forces, no unwanted side loads or deflections should result.

As indicated in Figure 11.71, the two arch segments reach the "keystone" area where they have to be joined together. Here, the formwork travelers have to be removed and conventional forms have to span the open crown. One special problem is the unequal swinging movements and significant deflection of the two segments due to wind and other environmental factors. Such relative movements, possibly 30 cm (1 ft), are certainly not acceptable during the curing phase of the concrete. One method to eliminate the problem is the use of tubular steel compression struts that are attached to each side, providing a solid connection during casting. Because after closing the crown no vertical cable stays are needed anymore, they are removed as well as the temporary backstays and towers. Figure 11.72 presents the next steps to construct a so-called *through deck* made of two connected steel box girders made of three main beams joined by orthotropic plates and suspended from a centerline row of hangers. This system was used for constructing the 250-m-long arch section of the Svinesund Bridge between Sweden and Norway, completed in June 2005.

VG11.10 Svinesund Bridge

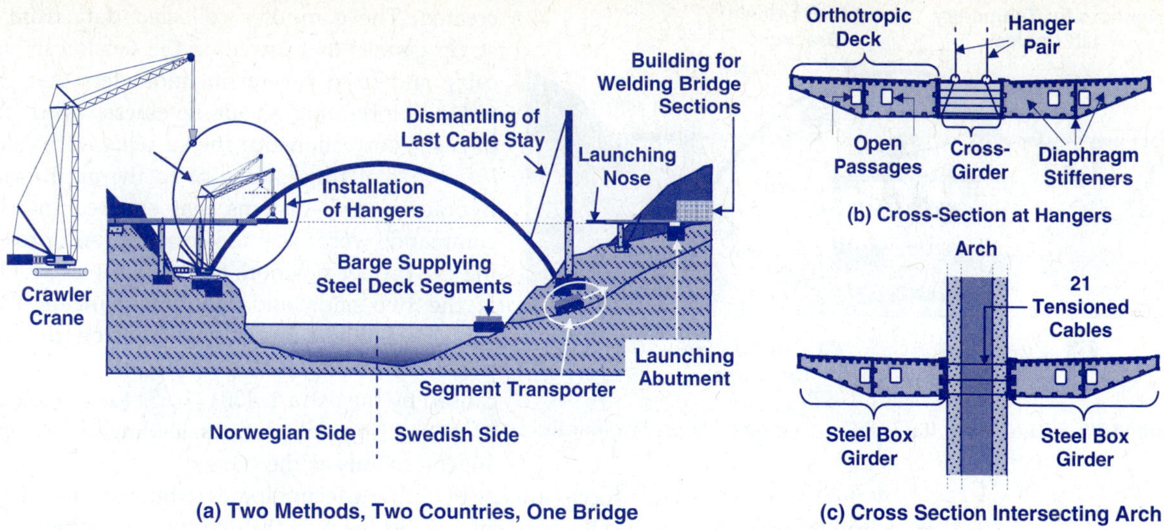

Figure 11.72 Slightly Out of Sync Depiction of Work on the Superstructure

11.8.2 ADDING A THROUGH DECK TO THE ARCH

Figure 11.72 illustrates how two different methods for building the bridge deck are used. On the Swedish side, the bridge is launched incrementally, while on the Norwegian side 75-t (87.8 ton) bridge segments are lifted onto temporary supports by a large crane, with a maximum reach of 86 m (282 ft). The choices were dictated by the available access routes. While the Norwegian side could be reached by an existing road and a bridge strong enough to carry the 75-t (87.8-ton) bridge segments as well as the crane, the same was not possible on the Swedish side. Thus, the 25-m- (82 ft) long, 5-m- (16.4 ft) wide, and 3-m- (9.8 ft) high segments were floated in on barges and hauled up a steep incline by a transporter that delivered them to a welding shop. There, elements were assembled and launched in increments. Figures 11.72b and c show that the bridge actually consists of two separate steel boxes only connected by cross girders at the location of the hangers and by 21 tension cables where they intersected with the hollow concrete arch. To speed up the erection of the middle span, while circumventing the limits of the large mobile crane, a third method of construction was used. Parallel to the welding and incremental launching on the Swedish side and the lifting and welding of segments on the Norwegian side, a third crew was making a large segment to close the gap in a remote welding shop. As welding is a time-consuming process and cannot be shortened without risk of losing quality, the aligning of the three deck construction methods in parallel resulted in considerable time savings. Thus, five 25-m-long sections each with 4 elements were welded together, resulting in a section that weighed 1,450 t (1,598 tons) with a total length of 127.5 m (418 ft). The additional equipment that was needed on top of the bridge to lift the whole section from two barges included a mobile crane and eight cable jacks with related hydraulic equipment amounting to 33 t (36.4 tons). Figure 11.73 depicts the lifting phase that was stopped at night. In total, the entire lift took approximately 14 hours.

A total of 6 pairs or 12 cable jacks were used to lift the section of the barge in 70 cm (2.3 ft) stroke lengths to 55 m (180 ft) above the fjord (= 79 strokes). It was important to coordinate all 12 lifting jacks so that the bridge stayed horizontal and each lift cable experienced an equal load. For this reason, a computer-controlled jacking system was

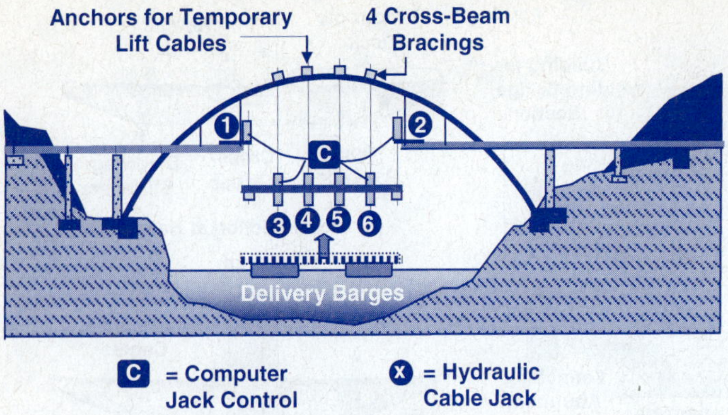

Figure 11.73 Lifting the 1,480 Ton Center Section from Two Barges

created. The computer collected data from 12 strain gauges that provided the tension in each cable and from several inclinometers that provided information about levelness. From this data set, corrections for the 12 jacks were calculated in real time and sent to the appropriate hydraulic servo-systems that ensured that the commands were executed correctly. After reaching the proper position, the section was welded at the two ends and the final hangers were connected to take over the weight from the temporary lift cables. The deformation of the arch caused by the extra 1,480 t (1,631 tons) had been predicted from calculations, leading to appropriate corrections at the piers.

After learning from this successful use of multiple technologies to build a "mid-deck" arch bridge, let's look at a slightly different arch bridge with the deck built over the crown of the arch similar to the one presented in Figures 11.8 and 11.74. The Hoover Dam Bypass Bridge is our example.

11.8.3 ADDING A TOP DECK TO THE ARCH

Several features look very familiar to those of the Svinesund Bridge. For example, cable stays were used to support the forming of the concrete arch as the last ones are being removed by the cableway crane on the right side of the bridge ❸. However, the hangers are replaced by spandrel piers that rest on top of the arch supporting the bridge deck above. They are built as slender hollow concrete boxes using climbing formwork ❷. The crane is presently lowering a precast spandrel pier section. Also different is the design of the superstructure itself. Instead of a continuous steel box girder, the Hoover Dam Bypass Bridge deck is built as a composite of steel and concrete. Figure 11.74 shows the cableway crane in the process of placing one steel girder on top of two adjacent piers ❶. Figures 11.75a and b depict the process of building the superstructure.

Figure 11.75 offers a fascinating view of how the innovative and synergistic use of construction material, structural design, and construction equipment and methods results

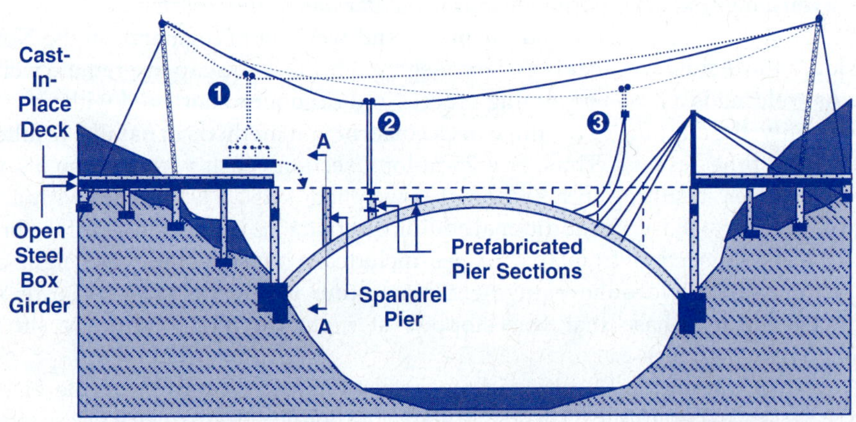

Figure 11.74 Potpourri of Activities to Construct the Upper Deck of an Arch Bridge (Not All Steps Are Simultaneous)

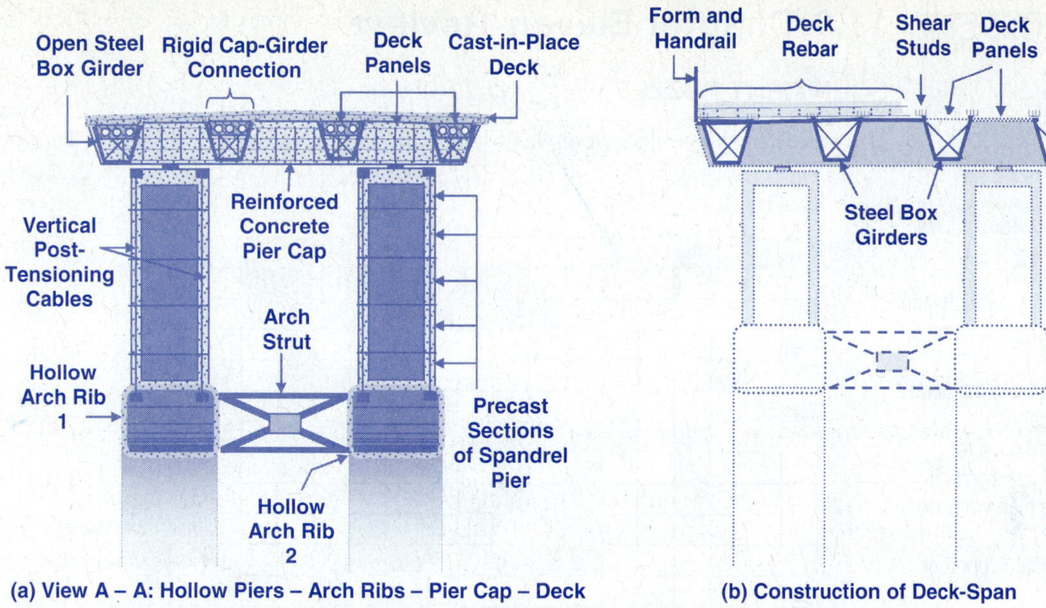

Figure 11.75 A Synergy of Steel, Cast-in-Place Concrete, Precast, and Tensioning Cables

in efficient and aesthetically pleasing results. The two slender arch ribs are connected by strong cross-bracing made of a dampening system of struts. For optimal speed in erecting the spandrel piers prefab, segments were cast off-site, shipped on a truck trailer, and installed via the cableway crane. Similar to segmental bridge construction, the segments were epoxy-glued and post-tensioned to create stiff connections between the segments and with the arch rib depicted in Figure 11.75a. The deck is constructed using four open-box girders that are integrated with the pier cap made of reinforced concrete. This turned the pier cap into a diaphragm for all four steel box girders.

Figure 11.75b illustrates the construction sequence of the deck in the open span from right to left. The box girders serve as platform to first put down and weld the deck panels to close the openings between the upper flanges. Secondly, the deck panels were installed to close the open girders, which provided a flat deck for placing the deck rebar, and the concrete. The shear studs for the upper flanges of the box girder had been affixed in the shop.

VG11.12 Hoover Dam Bypass

11.9 Chapter Eleven Review

EQUIPMENT PUZZLE

The letters in the circled boxes lead to a famous bridge builder (first and last name).

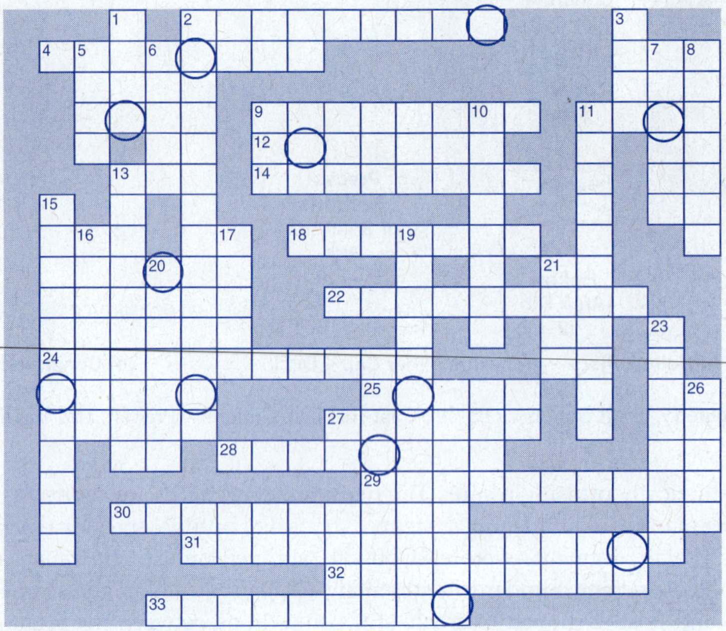

Across

2. A temporary protection for constructing a foundation in a river
4. Leads traffic to the center of a bridge
9. A structure used for medieval cathedrals to support a high walls from buckling
11. Non-vertical support or brace
12. Protects vehicles and pedestrians crossing a bridge
14. Shape of a cable suspended between its two ends
18. Support mechanism the bridge-deck to move
22. Called centering of an arch construction
24. (Two words) A metal piece used to strengthen a truss
25. Center of a road and top of an arch
27. Wire box filled with stones
28. Supports end-span of bridge
29. A structural member only supported at one end
30. Hangers for suspension bridges
31. (Two words) Allows the bridge to move
32. Temporary falsework needed to construct an arch
33. Cross-bracing in the deck at a pier or bridge cap

Down

1. Indicates the horizontal space between two supports
2. (Two words) A structure extending over several piers in support of a bridge deck
3. Soil containing particles between 4 to 61 microns in diameter
5. Column is another name
6. A fastener fitted hot on site to join metal pieces
7. Name of a suspension bridge element that also beds a baby
8. Overpass
10. A beam supporting the deck longitudinally
11. (Two words) A bridge that rotates horizontally
13. (Two words) A span in which the effective length is the same as the length of the spanning structure
15. (Two words) Carries traffic through the interior of the truss
16. Bridge that rotates on one or two trunnions
17. A compressive structural member
19. (Two words) A bearing that allows longitudinal movement
20. Multi-span structure that supports a rails
21. Perpendicular to the bottom chord
23. (Two words) Piles connected by a cap
26. One of at least two main horizontal members of a truss

JOURNALING QUESTIONS

1. Modern bridges are built with basically two materials but in many different combinations requiring slightly different construction methods. Elaborate briefly about some common features of short-span bridges (e.g., flyovers) and long-span bridges (e.g., larger than 600 m) in relation to materials used for the superstructure and methods of construction.

2. Steel (cable strands, box girders, shear studs, bracings, beams, trusses, tension rods, etc.) has found wide use in improving the length and the safety of bridges. Discuss

applications and special benefits of steel in composite bridges in relation to construction (e.g., duration of construction).

3. The Orkin community from Header Problem 9.1 is both weary and excited about the planned Ocean Bay Connector. What proactive measures would you recommend to community leaders so that construction will have minimal impact on Orkin's citizens and businesses?

TRADITIONAL HOMEWORK

1. Develop an I/O process model for a bridge construction with an LG that uses the segmental span-by-span method with precast concrete. Show the material flow from the precast plan to the completion of the span models all the special equipment that is needed for each task.

2. Constructing an overpass in an incline over a train track, as Figure 11.76 shows, creates a series of special problems for the contractors involved. What would be a list of key how-will-you-do it? questions that you would make sure that the contractors address before they start with their site installations?

Figure 11.76 The Inclined Girder Bridge

3. Steel wires used for post-tensioning the box girder, spinning the main cables of suspension, and building cable-stayed bridges have many characteristics in common. However, there are key differences. Divide a piece of paper into three columns (landscape format) and sketch the solutions for each and label the key parts for (a) method of installation, (b) anchoring mechanism, (c) anchorage design, and (d) protection against corrosion.

OPEN-ENDED PROBLEMS

1. The piers of six spans of the Lighthouse Ridge Viaduct as part of the Ocean Bay Connector of the Header Problem require special foundations, as the silty-sandy hillside is slowly sliding downhill. It is proposed to stabilize the hill, which would take fifteen years to take effect. Second, it is proposed to dig caissons down to stable bedrock, providing the footing for the piers. The caissons could further provide opportunities to drain the ground and provide space for the hillside to keep sliding for a while without destroying the piers in the process. Propose a process for the construction of caissons, piers, and a continuous girder bridge for the viaduct. Of course, the instability of the hillside will not allow you to cut a new access road to the caissons for trucks and other equipment or to position heavy equipment directly on the ground in the unstable area.

2. The first accident presented in Section 11.1 involved a launching gantry. Review the file and sketch the exact situation when the rear legs slipped. Establish a force diagram that could explain why the legs slipped exactly at that time (remember the bridge is in a 1% down slope). What would you recommend the LG fabricator change so that such an accident cannot happen again?

BIBLIOGRAPHY

Ancich, E.J., G.J. Chirgwin, and S.C. Brown. "Dynamic Anomalies in a Modular Bridge Expansion Joint." *J. Bridge Engrg* 11, no. 5 (2006).

Braestrup, M.W. "Danish Strait Crossings: Lillebælt, Storebælt, Øresund and Beyond." *Proc. Structures Congress 2005*. New York: April 20–24.

Dean, S. "Design and Construction of the SR 836 Flyover Bridges." *Proc. Structures Congress 2007*. Long Beach, CA: May 16-19, 2007.

Egilmez, O.O., T.A. Helwig, C.A. Jetann, and R. Lowery. "Stiffness and Strength of Metal Bridge Deck Forms." *J. Bridge Engrg.*, 12, no. 4 (2007).

El-Razek, M.E., and I.M. Basha "Constructability Improvement of Bridges Using Stepping Formwork." *J. Constr. Engrg. Mgmt.* 127, no. 3 (2001).

Sawhne, A., and A. Vamadevan. "Petri Net-based Scheduling of a Bridge Project." *Proc. Constr. Congress VI*. Orlando, FL: February 20–22, 2000.

Thomas, H.R., R.E. Minchin, and D. Chen. "Role of Workforce Management in Bridge Superstructure Labor Productivity." *J. Mgmt. Engrg.* 19, no. 1 (2003).

Wallner, M., and M. Pircher. "Kinematics of Movable Bridges." *J. Bridge Engrg*, 12, no. 2 (2007).

Zhou, Z., F. Li, and R.A. Imbsen. "Introduction of the Lateral Post-tension Method for Prestressed Concrete Bridges." *J. Bridge Engrg.* 13, no. 6 (2008).

Glossary

Chapter 1

derrick	Piece of equipment consisting of a mast or equivalent member held at the end by guys or braces, with or without a boom, for use with a hoisting mechanism and operating ropes.
efficiency	The ratio of output to input.
energy	Capacity for doing work.
guy derrick	A fixed derrick consisting of a vertical mast capable of being rotated 360° (not continuous) supported by guys, and a boom that is pivoted at the bottom and capable of moving in a vertical plane; a reeved rope between the head (top) of the mast and the boom harness (at the boom point) allows lifting and lowering of the boom and a reeved rope allows lifting and lowering of a load.
hoist	A system used for lifting or lowering a freely suspended (unguided) load.
hoisting equipment	Includes, but is not limited to, cranes, derricks, tower cranes, barge-mounted derricks or cranes, gin poles, and gantry hoist systems.
hopper	A container having a funnel-shaped bottom, or a bottom reduced in size, narrowed to receive material and direct it to a conveyor, feeder, or chute.
jack	A tool to lift or move a heavy object. Also a tool to turn a flywheel to a desired position.
load block	An assembly of hook or shackle, swivel, pins, and frame.
mast (derrick)	The upright member of the derrick used for support of the boom.
reeving	A system in which a rope winds around drums and sheaves.
resistance	Mechanically, a force opposing the motion of a body measured in pounds. Electrically, that which opposes the flow of an electric current measured in ohms.
simple machine	A mechanism designed to leverage a small into a large force. Key examples are: lever, wheel and axle, pulley, inclined plane, wedge, screw.
stiff leg derrick	A derrick similar to a guy derrick except that the mast is supported or held in place by 2 or more stiff members (stiff legs), which are capable of resisting both tensile or compressive forces. Sills are generally provided to connect the lower ends of the stiff legs to the foot of the mast.
tackle	An assembly of ropes and sheaves arranged for lifting, lowering, and pulling.
torque	The effect that rotates or tends to rotate a body. Torque is the product of force multiplied by the arm, or normal distance from the center of rotation to the force. Torque is measured in lb.-ft or lb.-in.

Chapter 2

account	A record of an asset, liability, owners' equity, revenue, or expense.
amortization	Method of paying back a loan in regular intervals while paying interest on the remainder.
annuity	A series of periodic payments or receipts (e.g., mortgage payments) for a specified amount of time.
assets	Includes all that a company owns. Listing of all valuables presented on left side of the balance sheet, opposite the liabilities.
audit	The inspection of the accounting records and procedures by a trained accountant for the purpose of verifying the accuracy and completeness of the records. A tax audit (IRS in the U.S.) determines whether the appropriate tax was paid.

bank credit The maximum amount of money that is made available by a lending institution to a borrower.

banknote Contract in which the issuer promises that it will pay a certain amount to the bearer on demand.

book value The book value of an asset is its original cost minus all depreciations. This is distinct from market value, the price which the asset might fetch if sold. It is recorded in a company's books, now called the balance sheet.

brake horsepower An engine's horsepower measured immediately at the shaft exiting the engine. Does not include any losses from the transmission, etc.

budget An itemized list of all cost and revenue that a given investment project is expected to have.

cash The liquid form of a currency that can be easily stored (in a cassa), transported, and used for payment.

cash-flow diagram An x-y graph with the x-axis representing time and y-axis cash in (+) and cash out (−). Vertical arrows are used to illustrate the size, sign, and timing of individual cash flows. Subtraction shows the net cash-flow including the lack of sufficient cash.

corporation A legal unit, separate and distinct from its owners and managers, having unlimited life, easy transferability of ownership, and limited liability.

corrective maintenance Maintenance at the time a piece of equipment breaks down.

cost The amount of money that must be paid to take ownership of something (purchase) or to reimburse for work performed by employees or other companies.

cost-benefit analysis A method of assessing the benefits anticipated from a decision when compared to anticipated cost.

CCI Cumulative cost index: function of the cumulative cost of maintenance over time.

declining-balance depreciation A method to accelerate the depreciation of a property during its early life. It results in high depreciation expenses for the first years.

depreciation There are two types of depreciation: (a) for the purpose of tax-savings and (b) for appreciating the loss in value of an asset. Depreciation for tax purposes has to follow fixed rules concerning the life-span of the asset (e.g., 5 years for common construction equipment) and method of depreciation. Common methods for tax purposes are the straight-line (SL), the sum-of-years' digits (SYD), and the double-declining balance (DDB). Accelerated depreciation is selected if the income projections for the next 2 years are very positive, as it is considered cost.

direct costs Costs that can be directly associated with the site operations.

DOT Department of Transportation. A government agency overseeing all projects and facilities related to transportation (highways, ports, etc.).

dozer A heavy crawler or bulldozer with a large metal blade mounted to the front to push soil, gravel, coal, etc. Sometimes equipped with rippers in the back to loosen hard soil.

economic trade-off The trading of one object for another having the same economic values (e.g., price).

equity The amount that is left after assets are sold and all liabilities paid back.

expenses The daily costs incurred in running and maintaining a business.

fixed cost A cost that is fixed for a given period of time and for a given production level.

FOG Filters, oil, and grease.

horsepower A unit of power between 735 and 746 watt (called *Pferdestaerke* in German).

income Excess of revenue over expenses and losses for an accounting period (i.e., net income).

income statement A statement summarizing the firm's revenues and expenses over an accounting period, generally a quarter or a year.

income tax Government levy on the net earnings of an individual, corporation, or other taxable unit. The tax rate is usually graduated as earnings go from one tax bracket to another.

indirect costs Costs of a business that cannot be directly associated with the production of a particular unit or type of operation (e.g., excavation of a foundation).

inflation An increase in the general price level of goods and services; alternatively, a decrease in the purchasing power of the dollar or other currency.

interest rate The rate of interest charged for the use of money, usually expressed as an annual rate.

internal rate of return (IRR) Rate of return (interest rate) that equals out the present value of future cash flows with the initial investment. Also referred to as the yield on investments.

investment tax credit A one-time incentive offered to companies or individuals that make desired investments. The credits can be subtracted from the total taxes owed at that year.

job efficiency A measure that describes how well a job is laid out and run when compared to an average. A highly efficient job has no idleness or other waste.

lease A contract in which the owner (lessor) of an asset (e.g., equipment) provides the asset for use by the lessee for either a fixed or flexible cost and duration. At the end of such period the asset will be returned to the lessor.

liabilities A company's financial obligations to other parties (bank loans, government, persons, etc.). These (borrowed) funds are used to pay for part of the assets. Appear on the right side of the balance sheet opposite the assets.

net present value (NPV) A method used in comparing investments, whereby all future cash out and cash inflows (returns) are estimated and then discounted to the present using a minimum attractive rate of return (MAAR). An investment is acceptable if the NPV is positive, meaning the inflows exceed all outflows.

net worth Total assets less total liabilities. Net worth represents shareholder equity.

money Valuable objects that are used for making payments.

obsolescence Reduction of an asset's value (e.g., scraper), resulting from technological or market changes, wear and tear from use, or natural deterioration.

oil sampling The extraction of an oil samples from an engine. The samples are sent to a laboratory to be analyzed. The measured amounts of contaminants in the oil indicate the wear and tear in the transmission.

operating expenses Expenditure that is incurred in operating and maintaining a production unit from day to day; equipment costs that are linked to its readiness and use such as fuel, maintenance, repair, insurance, etc. Includes depreciation, but not interest.

operating factor A factor used to account for nonstandard conditions of a construction operation such as low or high operator skill.

operator skill A measure that describes the smoothness with which an operator handles a piece of equipment, throttles the engine, avoids unnecessary motions, etc.

overhead cost Any cost that cannot be traced directly to a product or operation is known as an overhead. Examples of such costs include general office salaries and apportioned costs of premises shared by several construction projects.

ownership cost The cost of owning an equipment without operating it.

paver (asphalt or concrete) A piece of mobile paving equipment designed to place a uniform layer of concrete or asphalt onto a prepared base. The paving material is supplied by trucks and placed (into a hopper) in the front of the paver, spreading it out evenly.

present value (PV) Present value represents a series of future cash flows that have been discounted to today using an acceptable interest or discount rate. A given amount of money is more valuable now than in the future (e.g., due to inflation), so present values are generally smaller than corresponding future values.

preventive maintenance Maintenance geared to prevent the breakdown of a tool or piece of equipment.

prime rate An interest rate charged by commercial banks to large and financially solid borrowers.

revenue The inflow of funds from selling assets or constructing a building for a client.

salvage value Realizable value of an asset (e.g., dozer) after deducting costs associated with its sale or pickup as scrap.

sensitivity analysis The analysis of how sensitive outcomes are to changes in the assumptions.

straight-line depreciation Allows an equal amount to be charged as depreciation for each year of the expected use of the asset. It is computed by dividing the purchase cost minus the salvage value by its useful life.

sum-of-years' digits depreciation An alternative to the declining-balance method of depreciation to accelerate the amount of tax depreciation during the early life of a property.

tax Charge imposed by the state to individual or company incomes and to property or transactions to pay for the expenses of government.

tax depreciation Reduction in value of a property considered an (depreciation) expense in a particular tax year.

time value of money The idea that a dollar today is worth more than a dollar in the future, because the dollar received today can build up interest. Inflation will reduce the amount of goods we can buy in the future for the same amount of money.

value depreciation See **depreciation**.

variable costs Expenses that vary in direct proportion to changes in activities, such as machine hours or labor hours within the relevant range. Examples are concrete needed for a foundation or gasoline expense based on mileage driven.

uniform annuity Term used to label a money stream of equal amount of payments made or received at regular time intervals (e.g., yearly) for a predefined period of time.

working capital A firm's short-term or current assets—cash, marketable securities, inventory, and accounts receivable.

Chapter 3

actuating cylinder "Tube" and piston converting fluid pressure into linear force and motion.

administrative control Work practices, work methods, policies, and procedures established by the employer with the goal of reducing exposure to a work-related risk/hazard. Ear plugs are an administrative control to reduce the risk of hearing loss. Some administrative controls designed to reduce the risk of muscular disorders include scheduling rest breaks, job rotation schemes, worker training, etc.

air resistance A measure of the drag force due to the air turbulence produced by a vehicle in motion. Because it varies theoretically as the square of the speed, it affects the ability of a vehicle to reach top speed and its need of fuel to overcome force.

authorized person A person approved or assigned by the employer to perform a specific type of duty or duties or to be at a specific location or locations at the job site.

awkward posture Deviation from the natural or neutral position that puts minimal stress on the body. Awkward postures typically include reaching above or behind and twisting; awkward postures can cause fatigue, pain, and musculoskeletal injury.

biomechanics Biomechanical models calculate the levels of force (compression and shear) and torque that the joints, tendons, muscles, and ligaments are required to withstand and/or generate. The calculated forces and/or torques are compared to recognized limits to evaluate the acceptability of a task, activity, etc.

body harness, full Straps that are secured about a body in a manner that distributes the arresting forces over at least the thighs, waist, chest, shoulders, and pelvis, with provision for attaching a lanyard, lifeline, or deceleration device.

brake horsepower The actual horsepower at the crankshaft measured by a dynamometer.

carpal tunnel syndrome (CTS) A disorder of the hand and wrist that is caused by the compression of the median nerve where the nerve passes from the forearm, through the carpal tunnel in the wrist, and into the hand. Symptoms include tingling and numbness and loss of dexterity and strength in the hand.

cartilage Thick, white connective tissue attached to the surfaces of bones where they contact other bones, forming a low-friction cushion.

cervical vertebrae Seven small irregular bones in the neck that support and allow head movement.

check valve A valve that permits fluid flow in one direction, but prevents flow in the reverse.

circadian rhythms Circadian means "about a day." Cyclic changes in physiological variables or biological rhythms. A large number of biological rhythms take place approximately every 24 hours.

closed-center valve A type of valve that has its pressure passage blocked to fluid flow when the valve is in the OFF position.

certified equipment Equipment that has been tested and found to meet applicable test standards in order to be safe for use in a specified manner. It must bear a record of certification.

competent person Person who is capable of identifying existing and predictable hazards in the surroundings or working area that are hazardous or dangerous to employees, and who has authorization to take prompt corrective measures.

confined space A work space that has limited or restricted means for entry or exit such that a worker's ability to escape in an emergency would be hindered (e.g., tanks, vessels, silos, storage bins, manholes).

contaminant Any material that, by nature of its composition or reaction with other materials, is potentially capable of causing injury, death, illness, damage, loss, or pain.

crane A piece of equipment for lifting or lowering a load and moving it horizontally, with the hoisting mechanism being an integral part.

crankshaft That part of the engine that transmits the reciprocating motion of the pistons to the driven unit in the form of rotary motion.

cumulative trauma disorders Disorders of muscles, tendons, peripheral nerves, or vascular system. These can be caused, precipitated, or aggravated by intense, repeated, or sustained exertions, motions of the body, insufficient recovery, vibration, or cold.

cylinder (hydraulic) Tube with a passive piston rod and a port at each end for hydraulic fluid to pass. One hydraulic valve controls the flow direction of the pressurized oil.

cylinder bore (engine) The inside diameter of an engine cylinder. Also the surface of the cylinder in which the piston slides or moves.

displacement (fluid) Volume of fluid passing through a pump, motor, or cylinder in a single stroke.

deadman control A constant-pressure, hand- or foot-operated control designed so that when released, it automatically returns to a neutral or deactivated position.

decibel (dB) A measure of sound pressure.

dB(A) A weighted measure of sound pressure; the weighting causes the sensitivity of the sound level meter to vary with the frequency and intensity of sound and in doing so duplicates the response of the human ear.

diesel engine A compression-ignition combustion engine first developed by Rudolf Diesel.

differential A gear assembly on the drive axle that will not permit one wheel to spin while the other is motionless—such as when a truck is stuck in snow or mud. Torque is transmitted to both wheels of the driving axle.

displacement (fluid) Volume of fluid that can pass through a pump, motor, or cylinder in a single revolution or stroke.

endurance Endurance is the ability to continue to work, or in the static case, to continue to maintain a posture or position. It is the maximal amount of time a person can perform useful work, and it is related to the amount of effort being exerted.

energy expenditure The amount of energy (power) used by the body at any point of time, whether at work or at rest. It is usually expressed in terms of kilocalories per minute (kcal/min) or in terms of the amount of oxygen used (volume of O_2).

ergonomic force When doing "work" the body uses muscles to generate force to allow the movement of body segments, to resist the "force" of objects being lifted / carried, or to apply force to an object to move it. When muscles contract they also generate "reaction" forces in the body at the joints (e.g., shoulder).

ergonomics Ergonomics (or human factors) is the scientific discipline concerned with the understanding of interactions between a human body and other elements. It applies theory, principles, data, and methods to design in order to optimize human well-being and overall system performance.

ergonomics program A systematic process, often spelled out in writing, for identifying, analyzing, and controlling ergonomic hazards at a particular workplace.

exhaust cam The cam that controls the operation of the exhaust valve.

exhaust valve The valve through which the burnt gases pass to the exhaust manifold.

fatigue Fatigue is a loss of work capacity resulting in a loss of efficiency, capability, and safety. Fatigue is both a physical and a psychological state.

fluid Any liquid, gas, or mixture thereof.

fluid Any liquid, gas, or mixture thereof.

flywheel The wheel on the end of the crankshaft that gives the crankshaft momentum to carry the pistons through the compression stroke.

fulcrum The support on which a lever turns.

gear ratio The number of revolutions a driving gear requires to turn a driven gear through one complete revolution. For a pair of gears, the ratio is found by dividing the number of teeth on the driven gear by the number of teeth on the driving gear.

grasp/grip The way we grip objects makes a big difference in how much effort we must use to hold and manipulate the object. The power grasp/grip is one in which we wrap our fingers and thumb around the object, as when picking up a suitcase handle or grasping a hammer. The other type of grasp is a pinch grasp/grip.

gross combination weight (GCW) Represents the actual weight of a vehicle at the ground with a trailer or trailers including vehicle, equipment, driver, passengers, fuel, and payload.

gross vehicle weight (GVW) Actual weight of the entire vehicle including all equipment, fuel, body, payload, driver, etc. This is for an individual unit only such as a truck or tractor.

guardrail system A rail system erected along the open sides and ends of platforms. The rail system consists of a toprail and midrail and their supports.

hand-arm vibration syndrome (HAVS) A chronic disorder that can develop when the hand is subjected to prolonged exposure to vibration, often resulting in irreversible damage to the blood vessels, nerves, and muscles (e.g., numbness, loss of control of fingers).

hazard A dangerous condition, potential or inherent, that can bring about an interruption or interfere with the expected orderly progress of an activity. A source of potential injury to person or to property.

hazardous environment An environment with an atmosphere that poses a risk of death, incapacitation, injury, or illness due to flammable or explosive hazards; hazardous substances or agents; oxygen concentrations below 19.5% or above 22%; or any other atmospheric condition recognized as IDLH.

hazardous substance A substance that can cause death or injury, by reason of being explosive, flammable, poisonous, corrosive, oxidizing, irritating, or otherwise harmful.

herniated disc A condition where the soft inner part of an intervertebral disc pushes out through a tear in the disc.

horsepower A measure of the amount of work that can be done by an engine in a certain amount of time. One horsepower is equal to 33,000 ft.-lb. of work per minute.

hydraulics A branch of engineering mechanics dealing with the action or use of liquids forced through tubes or lines under pressure to operate various mechanisms.

impeller The rotating part of a centrifugal pump or blower that imparts motion to liquid or air by forcing it outward from the center of the machine.

impulse noise Noise is considered impulse when the variations in sound-pressure level involve peaks at intervals greater than 1 second.

injection pump The pump used to inject fuel oil into the combustion space of a diesel engine.

job enrichment The process of providing workers with greater responsibility for their work output. It is also important to recognize achievement and a job well done.

job satisfaction A multidimensional psychophysical measure that compares a person's opinions about job requirements to individual goals for meaningful work.

lanyard A flexible rope that is used to secure a safety belt or harness to a lifeline or directly to a point of anchorage.

lifeline A rope connecting a secure anchorage with a worker's body belt, body harness, lanyard, or deceleration device; it may be horizontal or vertical in application.

lockout A control using the placement of a lockout device on an energy-isolating device to ensure that a system cannot be operated until the lockout device is removed.

manual material handling (MMH) Any handling task involving the human body as the main power source. MMH includes lifting, lowering, pushing, pulling, carrying, and holding.

material safety data sheet (MSDS) A sheet with information on a substance: ingredients and hazards; physical data; fire and explosion data; reactivity data; health hazard information; spill, leak, and disposal procedures; and special precautions and comments.

maximum rolling grade Greatest grade a vehicle is able to climb while under motion, or the number of feet rise the vehicle attains for each 100 feet of horizontal movement.

musculoskeletal disorders (MSDs) A group of conditions that involve the nerves, tendons, muscles, and supporting structures such as intervertebral discs. The various conditions can differ in severity from mild symptoms to severe chronic and disabling disorders.

musculoskeletal system — The bones, muscles, tendons, ligaments, and joints in the body, and how they interact with each other.

neutral body posture — The natural position of body parts, the best position to minimize stress.

NIOSH — National Institute for Occupational Safety and Health. NIOSH, part of the Centers for Disease Control and Prevention (CDC) under the Department of Health and Human Services, is the federal government agency with a mandate to conduct and fund occupational safety and health research and training.

OSHA — Occupational Safety and Health Administration. OSHA is a federal government agency, part of the U.S. Dept. of Labor, whose mission is to help prevent workplace injuries and protect the health of workers. OSHA adopts and enforces workplace health and safety standards.

orifice — Device to restrict the flow of fluid in order to slow the operation of a component.

performance test — A test to determine the proper operation of a crane and the ability of the crane to safely lift loads within its performance rating. A performance test includes operational performance tests and load performance tests.

personal fall arrest system — An engineered system used to arrest an employee in a fall: consists of an anchorage, connectors, and body harness; it may also include a lanyard, deceleration device, lifeline, or suitable combination of these.

personal fall protection system — An engineered system that protects employees from falls.

pilot valve — Valve used to control the operation of a second valve.

pinch grip — A type of grip that requires an object to be picked up between the thumb and one or more fingers, or the fingers and the palm, or the finger tips and the base.

piston — That part of an actuating cylinder, servo, or motor that the hydraulic fluid works against. In a pump, the pistons work against the fluid.

port — Opening for the intake or release of a fluid.

position hazard analysis (PHA) — A documented process by which the duties (or tasks) of an employee's job position are outlined, the actual or potential hazards of each duty are identified, and measures for the elimination or control of those hazards are developed.

posture — The general position of the whole body (e.g., kneeling) or, more specifically, the position of any body part/joint with respect to adjacent body parts.

power curve — A graphic illustration of maximum output of power and torque at all operating speeds. These curves are established by measuring a running engine.

powertrain — The group of components that transmits engine power to the wheels. It includes clutch, transmission, universal joints, drive shafts, and rear-axle gears.

pressure reducer — Device for lowering the pressure in a hydraulic system to allow a component to operate at a lower pressure than the rest of the system.

pressure relief valve — Valve used to keep system pressure from exceeding predetermined limits.

protective system — A method of protecting employees from cave-ins, from material falling into an excavation, or from the collapse of adjacent structures; includes benching, sloping, shoring, trench shields, underpinning, rock bolting, etc.

pump — Device that converts mechanical energy into fluid energy.

qualified person — "Qualified" means one who, by possession of a recognized degree, certificate, or professional standing, or who by extensive knowledge, training, and experience, has successfully demonstrated his ability to solve or resolve problems relating to the subject matter, the work, or the project.

range of motion The limit of movement defined at a joint of the body. Stresses on the connective tissues at a joint increase as the joint moves towards its range of motion.

recovery time Work periods when task demands are light or when rest breaks are scheduled, permitting a person to recover from heavy-effort work.

repetitive stress injury (RSI) An injury caused by working in the same awkward position, or repeating the same stressful motions, over and over.

reservoir (hydraulic) Container that serves as a supply source of the fluid for a hydraulic pump.

rim pull The force available at the road surface contacting the driving wheels of a truck. It is determined by engine torque, transmission ratio, axle ratio, tire size, and frictional losses in the drive train.

risk factor The probability of a negative outcome. The likelihood of a health problem is greatly increased if a worker is exposed to more than one risk factor at a time.

rolling resistance A measure of the retarding effect of the road surface to forward movement of the vehicle; varies with the type and condition of the road.

rpm Abbreviation for revolutions per minute.

safety factor The ratio of the ultimate breaking strength of a member or piece of material or equipment to the actual working stress or safe load when in use.

safety harness A design of straps that is secured about the employee in a manner to distribute the arresting forces over at least the thighs, shoulders, and pelvis, with provisions for attachment to a lanyard, lifeline, or decelerating device.

safety relief valves Valves that relieve excess pressure or vacuum (depending on their design) that would otherwise damage equipment or cause injury to personnel.

safety tag A notice usually made of card stock, paper, or paperboard with letters, markings, or symbols alerting persons to the presence of a temporary hazard or hazardous condition. The tag is removed when the hazard no longer exists.

safety belt A strap, with means for securing about the waist and attaching to a lanyard, lifeline, or decelerating device that is used to limit the fall of a worker.

servo Device used to convert a small movement into a greater movement or force.

"shall" Mandatory.

stress Demand (or burden) on the human body caused by something outside, such as work, the environment, work-rest schedules, and social relationships.

stroke (engine, cylinder) Distance a piston moves in its bore from one end of its range to the other.

tagout The placement of a tagout on an energy-isolating device is an established procedure to indicate that the energy-isolating device and the system being controlled may not be operated until the tagout device is removed.

task analysis A systematic approach for documenting and assessing the risks associated with a specific task or job.

taut-line hitch A knot used for securing all workers aloft to their climbing rope, and consisting of either one or two wraps over two wraps.

toeboard A vertical barrier at floor level erected along exposed edges of a floor opening, wall opening, platform, runway, or ramp to prevent materials from falling.

tension neck syndrome (TNS) Fatigue, stiffness, tenderness, swelling, weakness, or pain in the neck or shoulder area, or headache radiating from the neck. It is caused by strain on various neck and shoulder muscles.

torque, engine Engine torque is the amount of twisting effort exerted at the crankshaft by an engine. The unit of measure is the pound-foot, which represents a force of one pound acting at right angles at the end of an arm one foot long.

torque, gross The maximum torque developed by an engine without allowing for the power absorbed by the engine's accessory units such as the fan, water pump, generator, and exhaust system.

tractor (highway) Vehicle designed for pulling loads greater than its own weight.

transmission A transmission contains a number of gears that mesh with another, providing a means to change the speed ratio between a driving and an exit shaft.

turbocharger A rotary compressor that pressurizes engine intake air. It raises the pressure in the combustion chamber to increase the power of an engine.

universal joint A coupling that permits a driving shaft to operate between two power train units that are not always in alignment with each other or subject to movement.

vibration The oscillatory motion of a physical body. Localized vibration, such as hand-arm vibration, is produced by contact with powered tools or equipment or with vibrating structures. Whole-body vibration occurs while standing or seated in vibrating environments or objects, such as trucks or heavy machinery.

viscosity Internal resistance to flow in a liquid or gas. In practice, for oils it is measured by the number of seconds required for a defined quantity to flow through a standard orifice under stated test conditions.

wheelbase (WB) The distance between the centerlines of the front and rear axles. For trucks with tandem rear axles, the centerline is midway between the two rear axles.

whole-body vibration Working conditions that involve sitting, standing, or lying on a vibrating surface.

work-related musculo-skeletal disorder (WMSD) A musculoskeletal disorder caused or made worse by the work environment. WMSDs can cause severe symptoms such as pain, numbness, and tingling; reduced productivity; lost time from work; temporary or permanent disability; loss of motion; inability to perform job tasks; and an increase in workers' compensation costs.

Chapter 4

AASHTO American Association of State Highway and Transportation Officials.

aluminum hydraulic shoring A pre-engineered shoring system comprised of aluminum hydraulic cylinders (cross-braces) used in conjunction with vertical rails (uprights) or horizontal rails (wales). Such systems are designed specifically to support the sidewalls of an excavation and prevent cave-ins.

attachments Fitting that mounts onto the stick such as a bucket, breaker, or grapple. Modern attachments include also many kinds of electronics devices that assist its operation or monitor the engine.

auxiliary hydraulics A dedicated source of hydraulic oil, intended to provide oil flow for specific attachments. The oil is sourced from the excavator's pump system and routed to the attachment via tubelines and hoses on the workgroup. Auxiliary hydraulic flow rates and pressures are a key determinant of an excavator's attachment capacity and compatibility.

backfilling Soil that is used as fill around an excavation, against a foundation, or beside and above a pipe installation; it involves filling the excavation with backfilling material in a manner that provides structural integrity to the pipe.

backhoe-loader Tractor that operates as a loader and a backhoe but is limited to a maximum of 180° arm swing, unlike an excavator that can slew 360°. Typically features fold-down stabilizers that often restrict its access to confined areas. Repositioning requires several time-consuming steps.

bedding material Material placed under and around a pipe to develop the required load factor.

benching Method to protect employees from cave-ins by cutting the sides of the excavation in the arrangement of one or more horizontal levels, usually with vertical or near-vertical walls between steps.

boom The primary component of the workgroup that is attached to the house structure via the swing frame. It supports the arm and attachment.

breakout force Vertical upward force (**Fb**) created by the curling action of the bucket.

bucket capacity Actual volume of material that can be carried in an equipment bucket.

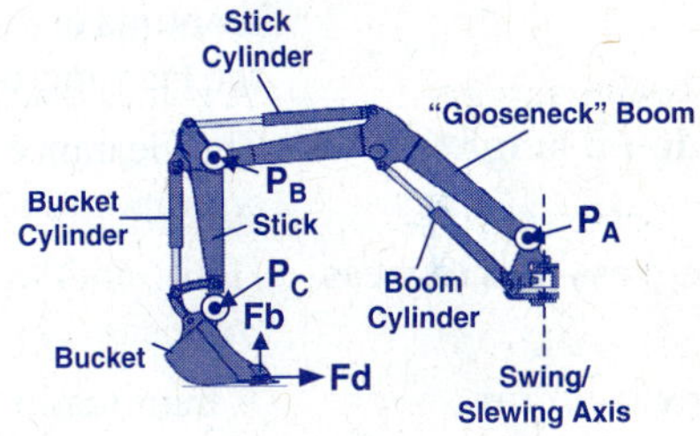

Arm of a Backhoe-Loader

buried facilities Includes utilities (sewer, telephone, fuel, electric, water, and other product lines), tunnels, shafts, vaults, foundations, and other underground fixtures or equipment that may be encountered during excavation or trenching work.

cabin, excavator Located on the left side looking outward along the arm. The operator has good visibility while digging and moderate visibility when loading.

cave-in Separation of a mass of soil or rock material from the side of an excavation, or the loss of soil from under a trench shield or support system, and its sudden movement into the excavation, either by falling or sliding, in sufficient quantity so that it could entrap, bury, or otherwise injure and immobilize a person.

CECE Committee on European Construction Equipment that rates heaped buckets.

cohesive soil A soil consisting of silt and clay where the soil particles tend to stick together, producing a sticky plastic soil that attaches itself to buckets and treads.

compact/mini excavators Generally includes those with operating weights of 14,000 lbs or less and dig depths of 14′ or less.

compaction Hand, mechanical, or water consolidation of backfill.

confined space A space that, by design and/or configuration, has limited openings for entry and exit, as wells as unfavorable natural ventilation, and may contain or produce hazardous substances, and is not intended for continuous employee occupancy.

control lockout A safety system that requires the operator to consciously place a console in an "operational" position in order for predetermined hydraulic functions to be active on the excavator. The purpose of the control lockouts is to prevent unintentional movement of the excavator's components.

counterweight A weight added to the rear of the excavator's house structure in an effort to improve its lifting characteristics. Counterweights are also added to the machine to accommodate variations in arm configurations such as long-arm options or extendable arm options.

crowding force The excavator's ability to produce a pulling force, using the arm hydraulic force. Force that pushes the bucket into the bank. It is a function of hydraulic power and linkage design.

cycle time Cycle time refers to the amount of time, usually in seconds, that a particular function can be cycled (e.g., dumping one bucket load into truck). Cycle time is a means of comparison among excavators and operators.

dig depth The maximum depth that the excavator is able to excavate on flat, level ground. It is a primary consideration when comparing and selecting an excavator.

digging or crowding force The excavator's ability to produce a pulling force (**Fd**).

distress Cave-in is imminent or is likely to occur. Distress is evidenced by such phenomena as the development of fissures in the face of an open excavation or adjacent to it; the slumping of material from the face or the bulging or heaving of material from the bottom of an excavation; the spalling of material from the face of an excavation; and raveling, i.e. small amounts of material such as pebbles or little clumps of material suddenly separating from the face and trickling or rolling down into an excavation.

dump height The clearance required for an excavator to dump the spoil from its bucket. Dump height is an important selection criterion when truck loading will be performed with the excavator.

egress and ingress "Exit" and "entry" respectively used in trenching and excavation. They refer to the provision of safe means for employees to enter or exit an excavation or trench.

excavation A man-made cut, cavity, trench, or depression in an earth surface that is formed by earth removal.

expandable undercarriage A variant of the traditional excavator undercarriage, in that its structure can be widened. This allows increased performance of the excavator when working over the side of the machine, yet when retracted still allows the machine to enter confined areas.

extendable stick/dipper A telescoping stick structure. The extendable arm provides increased reach for trench cleaning or truck loading. It is hydraulically activated and can be used simultaneously with other digging functions. When extra reach is not required, the stick can be retracted to provide maximum digging force.

faces or sides The vertical or inclined earth surfaces formed as a result of excavation work.

flagman A person who warns of construction operations, danger, caution, and general hazardous conditions by the use of flags or signs.

flotation The machine's ability to traverse soils or surfaces which have very little load-bearing capability.

gasket material An industrial adhesive material that is shipped in long coils; it is used to seal pipe sections together.

hazardous atmosphere An environment that by reason of being explosive, flammable, poisonous, corrosive, oxidizing, irritating, oxygen-deficient, toxic, or otherwise harmful may cause death, illness, or injury to persons exposed to it.

heaped bucket Volume in the bucket under the strike-off plane plus the volume of the heaped material above the strike-off plane, having an angle of repose of 1:1, without any consideration for any material supported or carried by the spillgate or bucket teeth.

hydraulic excavator Wheeled or tracked digging machine with a revolving superstructure having an engine mounted at one end and a hydraulically powered arm and stick at the other. A wide range of buckets and other attachments can be mounted on the stick.

initial backfill Native or imported material extending from the top of the bedding material to a height of one foot above the top of the pipe barrel.

kickout The accidental release or failure of a cross-brace.

laser, line-and-grade A device that sends a straight beam of light along the proper line and grade to a target.

lift, backfilling An even quantity of backfilling material placed around the entire perimeter of an assembled pipe; typically, bedding material is installed in 1′ to 2′ lifts as part of the backfilling process.

lift point locations The places on the pipe or manhole from which lifting gear is attached and the piece can be lifted safely; most vaults include four lift points.

log book Daily record of construction activity.

maximum allowable slope The steepest incline of an excavation face that is acceptable for the most favorable site conditions as protected against cave-ins; it is expressed as the ratio of horizontal distance to vertical rise (H:V).

mudsills Wales installed at the toe of the trench side. Wales embedded into mud that prevents the uprights from kicking out.

occupational illness Any abnormal condition or disorder caused by exposure to environmental factors associated with employment. For excavations this might include illnesses caused by the inhalation of toxic vapors.

occupational injury Any injury such as a cut, fracture, sprain, or amputation that results from a work-related accident or from exposure involving a single incident at work.

offset digging The practice of excavating with the boom offset to either the right or the left of the backhoe center. Offset digging allows for simple excavation of soil adjacent to an existing structure. It allows the tracks of the excavator to remain parallel to the trench for efficient repositioning.

oxygen deficiency Oxygen deficiency occurs in atmospheres containing less than 19.5% oxygen.

permitting regulations Procedures defined by a local or state government; they dictate safe and legal practices for projects such as pipe installations.

protective system Refers to a method of protecting employees from cave-ins, from material that could fall or roll from an excavation face or into an excavation, and from the collapse of adjacent structures. Protective systems include support systems, sloping, and benching systems, shield systems, and other systems that provide the necessary protection.

raker A diagonal brace for supporting sheeting of an excavation.

reach at ground level Measured from the excavator's center rotational axis to the tip of the standard bucket tooth, at ground level.

repositioning Moving the excavator to the next position when excavating. As an operator is trenching and has excavated up to the point where the machine is restricting his digging, the machine is moved back along the path of the intended trench. Repositioning with a compact excavator simply involves raising the blade, activating the travel controls in the appropriate direction, and lowering the blade to the ground.

rigging Chains, straps, or other gear connecting a heavy load to a lifting hook, such as the one on an excavator bucket. Most pipes require four-way rigging, or four legs.

SAE Society of Automotive Engineers (standard-setting body).

sheeting Members of a shoring system that retain the earth in position and in turn are supported by other members of the shoring system.

shield A structure that is able to withstand the forces imposed on it by a cave-in and thereby protect employees within the structure. Shields can be permanent structures or can be designed to be portable and moved along as work progresses. Additionally, shields can be either premanufactured or job-built in accordance with OSHA 1926.652(c)(3) or (c)(4). Shields used in trenches are usually referred to as "trench boxes" or "trench shields".

shoring (shoring system) A structure such as a metal hydraulic, mechanical, or timber shoring system that supports the sides of an excavation and that is designed to prevent cave-ins.

short-term exposure A period of time less than or equal to 24 hours that an excavation is open. This applies only to excavations in Type A soil less than 12 feet deep.

site planning One of the most critical steps in the pipe installation process, site planning includes reviewing the site for above- and below-ground obstructions, planning an access path and suitable surface area for the delivery truck, selecting an appropriate crane weight, and evaluating lift conditions.

skip or spot shoring When individual timber uprights or individual hydraulic shores are not placed in contact with the adjacent member.

slewing Rotating the excavator's house assembly. Unlike a conventional backhoe, the operator can "slew" the entire house and workgroup upon the undercarriage for spoil placement. This introduces minimal fatigue to the operator as his body is traveling with the workgroup for maximum visibility and minimal body movement. Compact excavators feature a full 360° of unlimited rotational capability.

slicing The act of consolidating the bedding material in the pipe haunch areas, usually with a shovel.

sloping Protecting employees from cave-ins by excavating to form sides of an excavation that are inclined away from the excavation so as to prevent cave-ins. The angle of incline required to prevent a cave-in varies with differences in such factors as the soil type, environmental conditions of exposure, and application of surcharge loads.

splicing The recommended method for joining one piece of gasket material to another; the two pieces of gasket material are placed end-to-end.

stable base preparation The installation of several inches of material in the bottom of the hole to provide a stable foundation for the pipe.

stable rock Solid mineral material that can be excavated with vertical sides and will remain intact while exposed. Unstable rock is considered to be stable when the rock material on the side or sides of the excavation is secured against caving-in or movement by rock bolts or by another protective system that has been designed by a registered professional engineer.

staging A construction procedure for the storage and or fabrication of material on a job site.

stick Also referred to as a "dipper," it is the element that connects the excavator's attachment to the boom.

stick cylinder This cylinder rotates the stick.

struck Bucket capacity rating. The volume actually enclosed inside the outline of the sideplates and rear and front bucket enclosures without any consideration for any material supported or carried by the spillgate or bucket teeth.

support system Refers to structures such as underpinning, bracing, and shoring that provide support to an adjacent structure or underground installation or to the sides of an excavation or trench.

subsurface encumbrances These include underground utilities, foundations, streams, water tables, transformer vaults, and geological anomalies.

surcharge An extra vertical load or weight caused by spoil, overburden, vehicles, equipment, or activities that may affect trench stability.

swing system The hydraulic motors and mechanical drive transmission mounted on the upperstructure platform enabling the excavator to pivot about a vertical slew axis.

tagout device A warning, such as a tag with a means of attachment, that can be securely attached to an energy-isolating device in accordance with established procedures to indicate that the energy-isolating device and system being controlled may not be operated until the tagout device is removed.

tail swing Refers to the rear overhang of the excavator as it rotates upon the undercarriage. It is measured from the central rotational axis to the furthest rear point of the machine.

teeth Hardened steel cutting edges attached to an excavating device such as a bucket.

tieback A structural tie used to hold sheeting around an excavation by drilling the tie into the soil behind the sheeting and securing it through the sheeting as a tensile member.

trench A narrow excavation (in relation to its length) made below the surface of the ground. In general, the depth of a trench is greater than its width, and the width (measured at the bottom) is not greater than 15 ft (4.6 m).

tongue and groove joint A two-part joint in which a projection on one vault section called a tongue fits into a groove on the other vault section.

TOPS/ROPS TOPS is an acronym for Tip Over Protective Structure. ROPS is an acronym for Roll Over Protective Structure. A TOPS or ROPS rating defines the protection afforded an operator in the event that the machine rolls over or tips over. The structural integrity of an operator's compartment must meet standards prescribed by ISO or SAE to be certified as TOPS or ROPS.

unconfined compressive strength The load per unit area at which soil will fail in compression. This measure can be determined by laboratory testing, or it can be estimated in the field using a pocket penetrometer, by thumb penetration tests, or by other methods.

underground installations Include, but are not limited to, utilities (sewer, telephone, fuel, electric, water, and other product lines), tunnels, shafts, vaults, foundations, and other underground fixtures or equipment that may be encountered during excavation or trenching work.

uprights The vertical members of a trench shoring system placed in contact with the earth and usually positioned so that individual members do not contact each other. Uprights placed so that individual members are closely spaced, in contact with or interconnected to each other, are often called "sheeting".

vibrator, compaction A mechanical or pneumatic piece of equipment that vibrates at a predetermined rate and is used to compact soil or distribute concrete in formwork.

wales The horizontal members of a shoring system placed parallel to the excavation face whose sides bear against the vertical members of the shoring system or earth.

water table The elevation of permanent ground water beneath the surface of the earth.

work area vision system Provides machine operators with views from video cameras to complement exterior mirrors and/or the operators' direct line of sight.

Chapter 5

angle dozer Dozer with a blade that can be pivoted about its center point to cast materials left or right. Ideal for making windrows, digging drainage ditches, backfilling, grading, etc.

apron Hydraulically controlled gate to the bowl. When open soil enters and exits the bowl. During hauling, the apron is down (closed).

automation Term used to indicate a closed-loop and self-correcting operation.

blade (moldboard/moldblade) Curved piece(s) of steel mounted on a dozer or motor grader to scrape, cut, or push soil or rock. In particular hardened cutting edges and wear plates can be added.

blade pitch 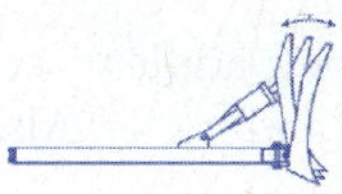 The angle of the blade relative to the push-arm or "C" frame. A back-pitched blade is used for leveling and spreading.

blade tilt  Raising and lowering the opposite ends of the blade. This allows for side slopes along a surface, cutting drainage ditches, or lifting up large or heavy objects for a short distance.

bogie Hinged and cushioned lever arm supporting two or more track rollers providing a flexible backbone for the tracks as they travel over rocks and uneven surfaces.

bowl Steel container similar to Fresno scraper. It has fixed sides, back walls and open front closed with apron.

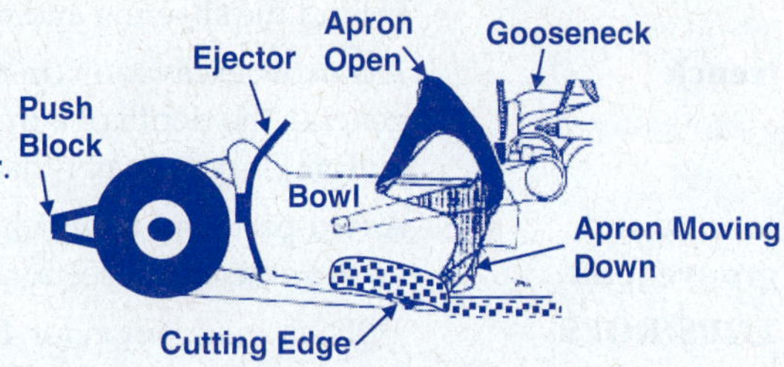

Scraper during loading

bulldozer Another name for a **dozer** or crawler dozer.

"C" frame "C" frame attached to trunnions that wraps around the front of the dozer to support the blade.

caterpillars Term coined by British soldiers to refer to the first crawler tractors.

circle The mechanism on a motor grader that supports the blade and regulates its angle.

coefficient of traction Coefficient of friction between the surface of the road and the tire. Expresses the amount of the normal load on the tire that can be use to accelerate.

crawler/wheeled dozer A tracked/wheeled vehicle that moves earth with a front blade. May have numerous attachments for the rear, including a winch, ripper, or even a backhoe attachment.

crib structure A strong foundation structure constructed of mutually perpendicular, interlocking beams or elements.

culvert A round-shaped conduit connected together with couplers or bands; it differs from a bridge in that it is usually constructed entirely below the road base.

cushion blade Special dozer blade used to push-load scrapers.

cutting edge Piece(s) of wear-resistant hardened steel bolted to the front bottom edge of the bowl.

ditch front slope Side slope of a ditch adjacent to the roadway.

ditch line The top edge of a ditch's side slope from where the street runoff flows into the drain.

dragging Moving loose surface material from high spots and road sides to fill and smooth surface irregularities to restore the crown of an unpaved road without cutting into the crust. Also referred to as "blading".

drawbar pull The available rimpull (gross force) can be turned into a pulling force if enough traction is available. The gross force minus rolling resistance is the drawbar pull.

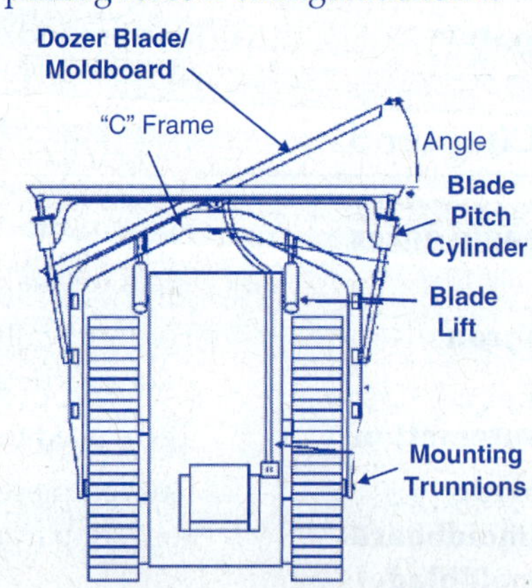

drilling and blasting Time-consuming method to remove rock that is too hard to be ripped and loaded.

drive wheel Wheel that provides power to the tracks.

dozer A tracked vehicle without a front attachment (blade or loader). It may be used for either agricultural or construction purposes.

earthsound Greek for geophone.

elevator Chain mechanism with paddles that pushes the soil into the bowl.

elevating scraper A machine, usually self-propelled, that removes earth by "scraping" it. The earth then travels up the elevator (paddle wheel) into the bowl. It articulates via a gooseneck, similar to a semitractor and trailer. Also known as a pan or self-loading scraper.

ejector Hydraulically operated blade across the bowl. During loading the ejector is at the back of the bowl and pushes forward during unloading.

ejector, scraper Hydraulic mechanism that pushes the soil material out of the scraper bowl.

fresno The name of the first successful scraper design, which was first pulled by horses or mules and later by tractors.

geophone The term geophone derives from the Greek word "geo" meaning "earth" and "phone" meaning "sound".

geotextile Synthetic polyethylene fibers manufactured in a woven or loose nonwoven pattern to form a blanket-like product. Also called geofabric.

gooseneck The mechanical connector/hitch that attaches the scraper to the tractor. It allows the tractor to swivel almost 90 degrees to the left and right.

grade display system Display inside the operator's cabin of a dozer or grader showing the distance between the target and the actual height of the blade. This visual information from the laser-based system allows the operator to adjust the blade very accurately without stakes.

grade resistance A grade (or gradient) is the pitch of a slope and is often expressed as a percent tangent. Grade resistance is caused by the gravitational force.

gross vehicle weight The weight of a piece of equipment that includes its normal (net) weight plus the maximum weights of attachments and loads that could be carried at one time.

grouser Metal ledge or ridge welded to the track plate providing activating the shear strength of the soil below the tracks. Leaves a visible indentation in the soil.

horsepower Power output of internal-combustion engines. 1 hp = 33,000 ft.lbf/min (550 ft.lbf/s) (745.7 watts).

hydroseeding Sowing of seed by distribution in a stream of pressurized water.

hydrostatic transmission A transmission that uses a positive displacement pump to provide high pressure and low flow to a positive-displacement motor in a closed loop. The pump converts a mechanical power input into hydraulic power, and the motor converts the hydraulic power back into mechanical power. Infinite speed selection, direction reversal, and dynamic braking are operational characteristics of this transmission.

idler Wheel supporting the track plates passively (not powered).

laser An amplified beam of light that is detected by a light-sensitive target or is reflected. The first working prototype was demonstrated in May 1960 by Theodore Maiman.

laserplane Invisible plane created by a whirling laser providing a reference plane from where laser sensors mounted on equipment are able to measure how far they are "off" and to correct variations.

leaned front wheels Leaning the wheels makes it possible to keep the grader on a straight line when the material on the moldboard pushes to one side.

LeTourneau, Robert Gilmore Famous engineer who invented the modern scraper and skid-steer wheeled earthwork equipment.

lift cylinders Hydraulic vertical cylinders to raise and lower the blade for efficient dozing action.

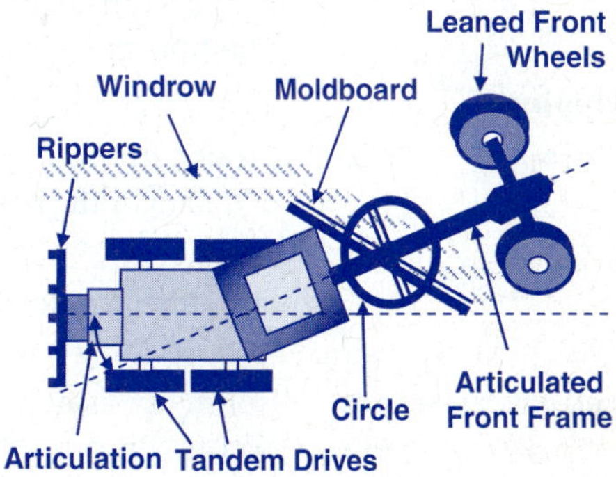

Motor grader

load factor Ratio representing the loosening effect during loading of soil changing from a bank to a loose density (= weight of unit volume of loose soil/weight of unit volume of bank soil).

loose cubic yard (LCY)	A cubic yard (cuyd = $3' \times 3' \times 3'$) after it has been loosened by earthwork equipment.
maximum allowable slope	Steepest incline of an excavation face that is acceptable for the most favorable site conditions as protected against cave-ins; expressed as the ratio of horizontal distance to vertical rise (H:V).
moldboard	Blade mounted on the circle used to cut, push material, and to smooth the surface. Can be rotated, lowered/lifted and tilted.
motor grader	A vehicle used in road construction to maintain dirt and gravel roads. In northern climates they are also used extensively for snow and ice removal. Equipped with a articulated blade (moldboard) in between the wheels and leanable front wheels.
net horsepower	The power produced by an engine after the power losses from the fan and alternator are subtracted.
pads, dozer	Separate parts connected to the track plates of a dozer providing it the flexibility to travel across different surfaces.
pan	Slang for scraper.
power shift transmission	Uses a hydraulic clutch connecting a gear to a shaft to change speed and direction. Or uses a hydraulic brake to stop a gear from turning, thereby forcing other gears to rotate around it to change speed or direction of the machine. Hydraulic transmissions can be shifted on-the-go between speeds or directions without interrupting power flow to the wheels.
pulse, laser	Fixed light burst that is reflected by the laser target and can be detected by the laser station, which allows the calculation of time for the pulse to return at the speed of light.
push block	Fixed structure attached to the back of the bowl that allows a tractor to push the scraper during loading.
retarder, hydraulic	Attachment to the transmission that can be engaged to help in slowing down equipment on a downhill slope in order to protect the mechanical brakes from overheating.
revetment	A retaining wall or other support for a trench wall or embankment, commonly made of heavy material such as stone (riprap) facing, concrete, brick, etc.
rimpull	Available force at the rim of the tire provided by the diesel engine minus all the losses in the drive-train and other power needs (e.g., air-conditioner).
riparian buffer	A strip of undisturbed vegetation between sensitive areas such as rivers, streams, wetlands, ponds, etc., and areas of land disturbance and/or fallow (bare) ground, such as unpaved roads, construction sites, etc.
ripper	A device attached to a crawler tractor designed to loosen the earth/rock, much like a plow. Rippers can come singly (single shank) or in groups of two or more (multishank rippers). The ripper shank is fitted with a replaceable tungsten steel tip.
ripping	Ripping breaks rock into pieces that can be dozed and loaded by an excavator.
riprap	Broken-up rock, cobbles, or boulders placed on earth surfaces, such as the face of a dam or the bank of a stream, for protection against the erosion action of water.
ROPS	Acronym for a safeguard for large equipment operators, a strong Roll Over Protective Structure. It's designed to protect the operator from being crushed in case the equipment rolls over on an incline or when struck by another vehicle.
rolling resistance	Rolling resistance, also called rolling friction or rolling drag, is the resistance that occurs when a wheel or tire rolls on another surface. Energy wasted in the form of heat that is caused by the deformation of the wheel, tire or steel, and/or the ground.

runoff The portion of the precipitation on a drainage area that is discharged from the area in stream channels.

scarifier An attachment for a motor grader used chiefly for loosening hard road surfaces.

scarify To abrade, scratch, or modify the surface; for example, to break the surface of a road with a narrow-bladed implement.

scraper Self-loading earthmoving equipment.

sediment Solid material, both mineral and organic, that is in suspension, is being transported, or has been moved from its site of origin by air, water, gravity, or ice, and has come to rest on the earth's surface either above or below sea level.

servo Short for a control system to achieve and maintain target values very accurately.

shank protectors Attachment to a ripper made of high-strength steel as a protection during highly abrasive or deep ripping.

slope percent The ratio between the vertical rise (or fall) and the horizontal distance in which the rise occurs in percentage terms.

slot dozing A method of moving large quantities of material with a bulldozer, using the same path for each trip so that the spillage from the sides of the blade builds up along each side, creating a slot.

soil bioengineering Use of live, woody vegetative cuttings to repair slope failures and increase slope stability, often combined with inert structures and materials.

spool metering The ability of a hydraulic valve to control oil flow to a given function in a consistent and predictable manner.

spotter Person on foot in cut or fill area who spots (observes) and signals to the scraper operators on completion of loading or spreading (empty bowl).

sprockets "Teeth" on the drive wheel that mesh with the track chain and supply the force to the tracks.

stringline A continuous piece of string attached to stakes on the side of a road to be constructed to indicate (maybe with an offset) the top the final road surface.

subgrade The natural soil material that provides that foundation for the road base.

superelevation Slope change in a road curve that improves driving safety.

swell Increase in soil volume per unit weight during excavation from its natural state.

tandem-powered Scrapers having two engines powering the front and back wheels.

terrace An embankment or combination of an embankment and channel across a slope to control erosion by diverting or storing surface runoff instead of permitting it to flow uninterrupted down the slope.

tilt Blade motion that allows a dozer to lift up heavy objects.

tilt cylinders Coordinated twin cylinders that allow quick change of tilt angle.

toe of a slope Base of a slope.

tire flexure A major source of increased rolling resistance for rubber tires.

tire penetration The depth a tire will sink into a plain ground surface. For each inch the equipment has to overcome, rolling resistance increases 1.5 percent.

TOPS/ROPS TOPS is an acronym for Tip Over Protective Structure. ROPS is an acronym for Roll Over Protective Structure. A TOPS or ROPS rating defines the protection afforded an operator in the event that the machine rolls or tips over. The structural integrity of an operator's compartment must meet standards prescribed by ISO or SAE to be certified as TOPS or ROPS.

torque A turning force generated by an engine that drives all other machine systems. If the torque requirements of the machine systems exceed the torque available from the engine, the engine stalls.

torque converter drive A turbine device utilizing a rotary pump, one or more reactors (stators), and a driven circular turbine or vane whereby power is transmitted from a driving member to a driven member by hydraulic action. With a speed reduction it adds torque during loading or spreading of a scraper.

tracking The process of running a tracked vehicle such as a bulldozer over an earthen area. A common practice is to run a dozer up and down a constructed, dressed slope before and/or after seeding and mulching.

traction Friction force between a wheel and the driving surface. It stresses the "positive" aspect of friction to provide motion to a vehicle.

trunnion A pin-like protrusion from a main object that serves as a pivoting point for the object itself (e.g., a cannon) or an attachment (e.g., C-frame of a dozer).

turn-out Intermittent discharge point in a ditch (most usually a road ditch) where the ditch channel is diverted from its normal profile and tapered out onto an area suitable for the discharge of the ditch water.

underdrain A drain placed beneath the surface of a road.

watershed Area contained within a divide above a specified point on a stream or lake. Oftentimes called drainage areas, drainage basin, or a catchment area.

wetland Land that has a wet and spongy soil, as a marsh, swamp or bog.

wheel tractor Any vehicle that is propelled by wheels rather than tracks. Designed to provide drawbar force/power to implements and attachments for agriculture, landscaping, and earth-moving job sites.

windrow Material that has been pushed into a row to "clean" the area next to it. Examples include gravel, snow and hay.

Chapter 6

aggregate Sized sand, rock, or crushed stone that has to be mixed by weight or volume to meet ASTM standards required in the production of concrete and asphalt.

aggregate gradation The distribution of aggregate particles among various sizes, expressed in terms of cumulative percentages larger or smaller than each of a series of standardized sieve openings.

air ride suspension Suspension that supports the load on air-filled rubber bags rather than steel springs. Compressed air is supplied by the same engine-driven air compressor and reservoir tanks that provide air to the air brake system.

antilock braking system (ABS) Computer, sensors, and solenoid valves that together monitor wheel speed and modulate braking force if wheel lock-up is sensed during braking. Helps to retain control of the vehicle during heavy braking on slippery roads.

articulated dump truck or hauler All-wheel drive articulated dump truck without a traditional steer axle. For steering the one-axle cab is pivoted/articulated at the permanent hinge by hydraulic rams.

automatic vehicle identification (AVI) System combining an on-board transponder with roadside receivers to automate identification of vehicles. Uses include electronic toll collection and stolen vehicle detection.

axle A shaft on which or with which two of more wheels rotate. Axles have designated weight limits according to the overall size and gross vehicle weight of the truck.

- Drive axles are those with powered wheels.
- Front axle is usually called the steer axle.
- Pusher axles are unpowered and go ahead of drive axles.
- Rear axles may be drive, tag or pusher types.
- Tag axles are unpowered and go behind drive axles.

bedrock The solid rock that underlies the soil, overburden, and unconsolidated material.

bench A step excavated into the face of a sloped surface for safety, stability, drainage, and to facilitate drilling operation in the quarry (or pit).

bills of lading Documentation that specifies the weight, nature of commoditiy and other details of shipment of goods that is being transported between a shipper and a receiver.

boulder Pieces of rock larger than 8 in (200 mm).

B–train assembly A rigid frame extension attached to the rear frame of a first semitrailer, which allows for a fifth wheel connection point for the second semitrailer. This combination has one less articulation point than the conventional "A dolly" connected truck–tractor semitrailer–trailer combination.

citizens' band radio (CB) Two-way radio for which no license is required by the Federal Communications Commission (FCC). Used by truckers and motorists for everything from traffic condition reports to emergency calls to idle chatter.

coarse aggregate A general term for aggregate of such size that it is substantially retained on a #4 (4.75 mm) sieve.

commercial driver's license (CDL) License that authorizes an individual to operate commercial motor vehicles and buses over 26,000 pounds gross vehicle weight.

common carrier Freight transportation company that serves the general public. May be regular route service (over designated highways on a regular basis) or irregular route (between various points on an unscheduled basis).

cone crusher A type of crusher that crushes the stone by gyrating an inner cone against a fixed outer cone, thus crushing the stones in between two hardened surfaces.

conveyor Thick rubber belting of various widths, depending on the capacity required, for the movement of aggregates from one part of a process to another.

crush & run aggregate Aggregate that has been broken in a crusher but has not been found useful during the screening process.

drivetrain (powertrain) All the components, excluding engine, that transmit the engine's power to the rear wheels: clutch, transmission, driveline, and drive axle(s).

dry bulk trailer Semitrailer to transport powder or granular material in dry condition. Integrated enclosed aluminum silos/tanks are filled from the top and emptied at the bottom. Unloading and cleaning is assisted with pneumatic pressure.

dump truck The frame of a fixed multiaxle truck is equipped with mountings to hold and raise a dump box made of steel or aluminum helped by a hydraulic ram system. Tailgates can be lifted up remotely via a pneumatic ram.

dust Very fine quarried aggregates passing sieve size 0.63 mm.

face The exposed unbroken rock of the quarry after a blast.

feed The flow of aggregate into a crusher, screen house, or washing plant.

fifth wheel Coupling device attached to a tractor or dolly that supports the front of a semitrailer and locks it to the tractor or dolly. The fifth wheel's center is designed to accept a trailer's kingpin, around which the trailer and tractor or dolly pivot in turns.

fine aggregate Aggregate passing the No. 4 sieve and predominantly retained on the No. 200 sieve.

gradeability	Vehicle's ability to climb a grade at a given speed. Example: A truck with a gradeability of 5% at 60 mph can maintain 60 mph on a grade with a rise of 5%.
gross axle weight rating (GAWR)	The value specified by the vehicle manufacturer as the load-carrying capacity of a single-axle system, as measured at the tire-ground interfaces.
gross combined weight (GCW)	The weight of tractor and trailer and the maximum payload.
gross vehicle weight (GVW)	The weight of the vehicle plus the weight of any load thereon.
heavy truck	A truck weighing in excess of 80,000 lb of gross vehicle weight.
heavy vehicle use tax (HVUT)	Annual federal tax on carriers with registered gross vehicle weight over 55,000lbs. Tax designed to cover the cost of pavement damage caused by heavy vehicles.
horsepower (hp)	Measure of power (the amount of work that can be done over a given amount of time). One horsepower is defined as 33,000 foot-pounds of work in one minute.
impact crusher	Type of crusher in which the breaking process is by impact of fast rotating hammers or a blow bar. Here, the rock is not crushed but rather fragmented by kinetic energy.
inspection levels	Five levels of inspection carried out as part of the Motor Carrier Safety Assistance Program. Level 1 covers both the driver and the vehicle, including the undercarriage.
jaw crusher	The crushing action resulting from an eccentric motion of a vertical moving jaw closing the gap to a fixed jaw. The gap is fed with rock that is crushed moving towards the opening at the bottom.
kingpin (trailer)	Anchor pin at the center of a semitrailer's upper coupler that is captured by the locking jaws of a tractor's fifth wheel to attach the tractor to the semitrailer.
length	Total longitudinal dimension of a single vehicle, trailer, or semitrailer. Length of a trailer or semitrailer is measured from the front of the cargo-carrying unit to its rear, exclusive of all overhand, safety, or energy-efficiency devices.
live bottom trailer	Semitrailer to transport loose bulk material, asphalt, and low-slump concrete without the need to raise the box/tub. Unloading is accomplished via a hydraulically powered conveyor system built into the bottom of the trailer. By activating the conveyor belt, the load is horizontally moved out of the back of the trailer.
longer combination vehicle (LCV)	Any combination of a truck tractor and two or more trailers or semitrailers that operates on the Interstate system at a gross vehicle weight greater than 80,000 pounds.
lowboy	Open flat-bed trailer with a deck height very low to the ground, used to haul construction equipment or bulky or heavy loads.
mesh	The openings in a screen or sieve.
mobile plant	Equipment within the quarry that can be moved about, e.g., crushers, conveyors, loaders, compressors, trucks and drill rigs.
off-road haul truck (dumper, hauler)	Large, mostly rear-back dumping trucks referred to as haulers or dumpers. Largest units used in mining and large earthmoving operation carry 400 tons (364 tonnes).
pit	An open hole in the ground from which sand or gravel is excavated from its natural bed.
primary crusher	The first crushing process of large rock from a quarry.
retarder	Device used to assist brakes in slowing the vehicle. The most common type of retarder on over-the-road trucks manipulates the engine's valves to create engine drag. This type is commonly referred to as a "Jake Brake" because the predominant manufacturer is Jacobs Vehicle Equipment Co. Other types of retarders include exhaust retarders, transmission-mounted hydraulic retarders, and axle-mounted electromagnetic retarders.

riprap Large stone placed along a surface to prevent erosion, scour, or sloughing of a structure or surface such as breakwater, embankment, mountain trail, etc.

secondary crusher The second crusher set up after the surge pile to reduce the size of the larger stones coming from the primary crushers.

screen Large vibrating decks or meshes, of woven wire or polypropylene, with specific hole sizes through which aggregates fall through, and are collected.

semi-bottom-dump (belly dump) A three-axle tractor pulling a two-axle trailer with a clam-shell type dump gate in the belly of the trailer.

semi-end-dump A tractor-trailer combination wherein the trailer itself contains the hydraulic ram. A typical semi-end-dump has a three-axle tractor pulling a two-axle semitrailer.

side dump truck A three-axle tractor pulling a two-axle semitrailer. It has hydraulic rams which tilt the dump body onto its side, spilling the material to either the left or right side of the trailer.

single axle weight The total weight transmitted to the road by all wheels whose centers may be included between two parallel transverse vertical planes 40 inches apart, extending across the full width of the vehicle. The federal single axle weight limit on the Interstate System is 20,000 pounds.

special permits State-issued permits to travel off the Interstate system on state roads. Special loads are greater than 80,000 gross vehicle weight and exceed the axle weights and spacing specified by the federal bridge formula. In many cases, local geography, weather, population, or highway and bridge construction methods impose particular size and weight limits.

stockpile Storage of aggregate sizes in a large mound to be washed and loaded for sale or disposal.

tandem axle weight The total weight transmitted to the road by two or more consecutive axles whose centers may be included between parallel transverse vertical planes spaced more than 40 inches and not more than 96 inches apart, extending across the full width of the vehicle. The federal tandem axle weight limit on the Interstate system is 34,000 pounds.

trailing axle Liftable and load-bearing axle that extends behind the last axle. It stretches the outer "bridge" measurement to the maximum overall length allowed. This increases the gross weight allowed under the federal bridge formula, which sets standards for truck size and weight to a maximum 80,000 pounds GVW.

volumetric concrete mixing truck Mobile concrete production plant. The main materials, aggregates, cement, and water are transported in segregated bins and mixed on site.

weight-in-motion (WIM) Electronic sensor–based technology for measuring and analysis of the dynamic forces applied by a moving vehicle to estimate its gross vehicle weight. System is installed in the pavement or on bridges.

wheelbase Distance between the front and rear axles of a vehicle or the center point of contact of the front and rear wheels with the ground.

Chapter 7

all-terrain crane Mobile crane used on rough construction sites and also allowed to travel on highway.

angle indicator/inclinometer A device measuring the angle of boom or crane base to the horizontal.

anti-two-block A device that, when activated, disengages all crane functions whose movement can cause two-blocking.

articulated jib A tower crane jib that in general has pivot points somewhere in the middle area.

back stay	Guy used to support a boom or mast or that section of a main rope, as on a suspension bridge or cableway leading from the tower to the anchorage.
basket hitch	Loading with the sling passed under the load and both ends on the hook, master link, or lifting device.
block	A term applied to a wire rope sheave (pulley) enclosed inside plates and fitted with some attachment such as a hook or shackle.
block and tackle	A system of two or more pulleys with a rope or cable threaded between them, usually used to lift or pull heavy loads. The block and tackle system was invented by Archimedes. The mechanical advantage is equal to the number of lines running between the two blocks.
boom (crane)	A member, in compression, hinged to the rotating superstructure and used for supporting the hoisting tackle and load.
boom angle	The angle above or below horizontal of the longitudinal axis of the base boom section.
boom base section	The lowermost section of a telescopic boom; it does not telescope but contains the boom-foot-pin mountings and the boom-hoist-cylinder upper end mountings.
boom cylinder	Hydraulic cylinder used instead of a rope boom suspension, the most common means of derricking telescopic booms.
boom head	The portion of a boom that houses the upper load sheaves.
boom hoist	The rope drum(s), drive(s), and reeving controling the luffing motion of the boom.
boom stay	Fixed-length rope forming part of the boom-suspension system; also called boom guy line, hog line, standing line, or stay rope.
boom stop	Device intended to limit the maximum angle to which the boom can be raised.
breaking strength	Load necessary to break a wire rope in tension.
bridge crane	A crane with a single- or multiple-girder movable bridge, carrying movable trolley or fixed hoisting mechanism, and traveling on an overhead fixed runway structure.
cab	Operator's compartment from which a crane is controlled.
cableway	Aerial conveying system for transporting single loads along a suspended track cable.
cantilever gantry crane	A gantry or semigantry crane in which the bridge girders or trusses extend transversely beyond the crane runway on one or both sides.
capstan	A spool-shaped revolving drum, manually or power-operated, used for pulling fiber or synthetic rope. Also called a winch head.
carrier	Rubber-tired highway truck providing the base for operating a crane superstructure.
center of gravity	Defined as the location where an equivalent weight could be concentrated into a single point.
choker	Chain, wire rope, or synthetic fiber rigging assembly that is used to attach a load to a hoisting device. It tightens around its load as it is pulled.
choker hitch	Loading with a sling passed through one eye or choker hook and attached to the hook at the other end.
clearance	Horizontal or vertical distance from any part of the crane to a point of the nearest obstruction.
climbing tower crane	Top slewing tower crane with the capability to insert sections into the mast.
clevis	A U-shaped fitting with holes in each end through which a pin or bolt is run.
come-along	A mechanical device, usually consisting of a chain or cable attached at each end, used to move heavy materials or apply tension in the chain.
competent person	Someone who is capable of identifying existing and predictable hazards in the surroundings or working conditions that are unsanitary, hazardous, or dangerous to employees, and who has authorization to take prompt corrective measures to eliminate them.

core Element of a wire rope around which the strands are helically laid. The core can be made of fiber (cloth), independent wire rope, or wire strands.

counter jib A horizontal member of a tower crane on which the counterweights and usually the hoisting machinery are mounted; also called counterweight jib.

crane A machine for lifting or lowering a load and moving it horizontally, in which the hoisting mechanism is an integral part. It may be driven manually or by power; it may be fixed or mobile.

crawler crane A crane consisting of a rotating superstructure with power plant, operating machinery, and boom, mounted on a base and equipped with crawler treads for travel.

critical lift Any lift that includes more than one crane, where the load is exceeds 75 percent of the rated capacity, is difficult due to the complexity of the situation, or involves a toxic material.

cycle time The time it takes to complete a task or a set of connected tasks from beginning to end.

deadman An object or structure, either existing or built for the purpose, used as anchorage for a guy rope.

derrick An apparatus for lifting or lowering loads, consisting of a mast or equivalent member held at the head by guys or braces, with or without a boom, for use with hoists and ropes.

design safety factor As a function of design, this factor can be based upon the point of equipment failure, such as crane tipping, and brake stopping capacity, or upon strength of materials.

designated leader (DL) A qualified individual assigned to all hoisting and rigging operations to ensure that the lifting operation is properly performed.

double wrap basket hitch Single of pairs of slings forming a basket hitch wrapped around the load to stop it from slipping out of the basket. Will compress the load.

dragging Pulling a load laterally to change its horizontal position.

drive An assembly consisting of motors, couplings, gear, and gear case(s) that is used to propel a bridge, trolley, or hoist.

drum A cylindrical-flanged barrel around which rope is wound for lifting or lowering the load or boom, or swinging the boom-supporting structure.

dynamic loading Loads introduced into the machine or its components by forces in motion.

fiber core Cord or rope made of vegetable or synthetic fiber used in the core of a wire rope.

fiddle block A block consisting of two sheaves in the same plane held in place by the same cheek plates.

fitting Any accessory that serves to attach wire ropes.

floating crane A rotating superstructure, power plant, operating machinery, and boom mounted on a barge or pontoon. The power plant may be installed below deck. The crane's function is to handle loads at various radii.

fly jib An extension attached to the head of a boom to provide added length and flexibility. For example, the luffing fly jib allows a load to be placed on the roof of a building well beyond the outer edge.

gantry crane A crane similar to an overhead bridge crane, except that the bridge for carrying the trolley or trolleys is rigidly supported on two or more legs running on fixed rails or other runway, usually 3 meters (10 feet) or more below the bottom of the bridge.

grade (steel) A classification according to breaking strength.

grooved drum Drum with a grooved surface that accommodates and guides the rope.

guy line Steel wire or rope that braces or supports a structure.

headache ball A weighted hook that is used to attach loads to the hoist load line of the crane.

hoist Apparatus that is used to exert force on an object to be lifted or lowered with a rope or cable.

hook, rigging A hook used as part of tackle. Any hook used in hoisting and rigging that is not the primary hook.

hook latch A mechanical device to bridge the throat opening of a hook.

idler Sheave or roller used to guide or support a rope.

improved plow steel (IPS) A high-carbon steel having a tensile strength of approximately 260,000 psi that is roughly 15 percent stronger than plow steel.

independent wire rope core (IWRC) Wire rope that serves as the core for a greater rope.

jib Extension boom pinned to a base, such as the head of a main boom or a tower. It is supported by guylines either in a fixed position (tower crane) or in a luffing configuration (fly jib).

kink Permanent distortion of wires and strands resulting from sharp bends.

lattice boom A boom constructed of four longitudinal corner members, called chords, assembled with transverse and/or diagonal members, called lacings, to form a space truss.

lay (wire rope) The manner in which the wires are helically wound to form rope. "Lay" refers specifically to the direction of the helical path of the strands in a wire rope; for example, if the helix of the strands is like the threads of a right-hand screw, the lay is known as a right lay or right-hand, but if the strands go to the left, it is a left lay or left-hand.

lead line That part of a rope tackle leading from the first or fast sheave to the drum.

lift (a) Any sequence of operations in which a hoisting device raises an object above the ground, floor, or support, and then places it on the ground, floor, or support; (b) maximum safe vertical distance through which the hook can travel; (c) the hoisting of a load.

line Rope used for supporting and controlling a suspended load.

line pull The pulling force attainable in a rope leading off a rope drum or lagging at a particular pitch diameter (number of layers).

line speed Speed attainable in a rope leading off a rope drum or lagging at a particular pitch diameter (number of layers).

load The total superimposed weight or force to be overcome by the hoisting and rigging equipment.

load radius The horizontal distance from the axis of rotation to the center of gravity of a lifted load. In mobile crane practice, this is more specifically defined as the horizontal distance from the projection to the ground of the axis of rotation before loading to the center of a loaded but vertical hoist line.

load, rated The maximum static vertical load for which a crane or an individual hoist is designed.

load block, lower The assembly of hook or shackle, swivel, sheaves, pins, and frame suspended by hoisting ropes.

load block, upper Assembly of sheaves, pins, and frame suspended from the hoisting platform or from the boom in mobile cranes.

load, safe working The maximum load a piece of equipment (or tackle) can handle without exceeding the rated capacity (the rated capacity of the lowest capacity item used in the lift).

load capacity The lifting capacity established by the certified agent for various angles and positions.

load radius Normally, the horizontal distance from the axis of rotation to the center of gravity of a lifted load. In mobile crane practice, this is defined as the horizontal distance from the projection to the ground of the axis of rotation before loading to the center of a loaded but vertical hoist line.

luffing-boom tower crane Bottom slewing tower cranes with luffing boom.

main hoist The hoist mechanism provided for lifting the machine's maximum-rated load.

mast The upright member of a derrick.

mobile crane Self-propelled crane that is able to travel on its own.

multiple lift rigging Rigging assembly manufactured by wire rope rigging suppliers that facilitates the attachment of up to five independent loads to the hoist rigging of a crane.

nominal strength Nominal wire rope strengths as calculated by a standardized industry-accepted procedure. Minimum acceptance strength is 22 percent lower than nominal strength.

operational safety factor This factor considers all the operational factors that deviate from our assumptions underlying the calculations of the tipping load/load capacity. They include conditions such as uneven soil conditions, light wind, load line not perfectly vertical, operator expertise, etc.

outrigger Extendable arm attached to a crane base mounting, which includes the means for relieving the wheels of crane from carrying the weight; used to achieve stability.

overload Any load in excess of the safe working load or rated capacity of the equipment or tackle.

parts of line A number of running ropes supporting a load or force, also called parts or falls.

pivoted luffing jib A tower crane jib that in general has pivot points somewhere in the middle area; also called articulated jib.

preventive maintenance A periodic or scheduled program that provides lubrication, adjustments, inspection, and testing as required to keep equipment in safe, operable working condition.

qualified person One who, by possession of a recognized degree, certificate, or professional standing, or who by extensive knowledge, training, and experience, has successfully demonstrated the ability to solve or resolve problems relating to the subject matter, the work, or the project.

range diagram A diagram showing an elevation view of a crane with circular arcs marked off to show the luffing path of the tip for all boom and jib lengths and radial lines marking boom angles. A vertical scale indicates height above ground, while a horizontal scale is marked with operating radii. The diagram can be used to determine lift heights, clearance of the load from the boom, and clearances for lifts over obstructions.

rated load (capacity) The maximum load designated by the manufacturer for which a crane, hoist, rigging, or other lifting device is designed and built.

reel The flanged spool on which wire rope or strand is wound for storage or shipment.

reeve The pattern that a rope forms between sheaves in a hoisting system.

reeved blocks Rope passed through a set of blocks, as opposed to laced blocks, and in such a manner that there are no lines crossed or rubbing each other.

reeving (noun) A rope system in which the rope travels around drums and sheaves in a prescribed manner.

revolving superstructure Part of a mobile crane that rotates 360°; also called upper superstructure.

rigging The act of attaching hoisting equipment to the load.

rope core Element of a wire rope around which the strands are helically laid. The core can be made of fiber (cloth), independent wire rope, or wire strand.

saddle jib The horizontal live-load supporting member of a hammerhead-type tower crane having the load falls supported from a trolley that traverses the jib; also called load jib.

shackle A two-piece fastener consisting of a U-shaped piece of metal secured with a pin or bolt across the opening. Round, screw, or bolt-type pin. A simple clevis.

sheave A pulley wheel with a circumferential groove or channel designed for a particular size of wire rope mounted in a frame that guides or alters the direction of a running rope.

shock loading Term describing the sudden, unplanned loading of equipment that would jeopardize the safety of the lift. Typical examples are the fast breaking of a winch, unplanned shifting of the load while suspended, and fracture of a lifting system component.

side loading (side pull) A force applied to the boom at any angle to its vertical plane. Cranes are designed and certified for vertical lifting only.

slewing A crane or derrick function wherein the boom or load-supporting member rotates about a vertical axis (axis of rotation); also called swing.

sling Wire rope, chain, or synthetic fabric, with or without fittings, for handling loads.

spooling (rope) Winding of rope on a cylindrical drum in evenly spaced, uniform layers.

spotter Person(s) whose sole responsibility is to provide a warning or stop signal during vehicle or equipment operation to avoid hazards such as power and communication lines, overhead obstructions, buildings, telephone poles, ground penetrations, etc.

spreader beam A below-the-hook fixture such as a pipe, wide-flange, I-beam, channel, plate, with two or more hooks to assist in lifting a long or wide load. The multiple hooks and slings "spread" the load over more than one lifting point.

stabilizing moment The moment of the dead-weight of the crane or derrick, less boom weight, about the tipping fulcrum; hence, the moment that resists overturning.

strand An assembly of wires that are helically wound around an axis, fiber, or wire center (core) to create a symmetrical portion.

strand grade Classification of strands according to breaking strength. The ranking of increasing breaking strengths is as follows: Common, Siemens Martin, High Strength, and Extra-High Strength; a utility's grade strand is available for certain requirements.

swage The act of fastening a termination to a wire rope through physical deformation of the termination about the rope via a hydraulic press or hammering. The strength is 100 percent of the wire rope rating.

swing Rotation of the superstructure or derrick boom.

tackle Pieces comprising the rigging such as slings, spreader bars, chokers, shackles, thimbles, eyebolts, rings, or other handling fixtures used for attachment of the load to the crane or hoist.

tag line Section of rope used to guide a load that is being lifted into a desired position.

telescoping boom Boom comprised of multiple sections that can be hydraulically extended or retracted as long as the sections stay straight.

thimble Grooved-metal fitting designed to prevent crushing or overstressing wire rope at the terminal end used to protect the eye of a wire rope or sling.

tipping load The load for a particular operating radius that brings the crane or derrick to the point of incipient tipping.

tipping point/axis The line about which a crane or derrick will rotate should it overturn; the axis on which the entire weight of a crane or derrick will be imposed during tipping.

tower crane Crane with a vertical lattice mast (tower) either affixed to a trolley on rails or to the floor/building section. The jibs can be horizontal (saddle), luffing, or articulated, and the slewing ring can either be at the bottom or at the top of the mast. The shorter jib holds the counterweight for the main jib as well as the winches for the hoist lines and trolley.

trolley (crane) A unit that travels on bridge rails or saddle jib consisting of frame, drives supporting the hoisting mechanism, rope, and the load block that holds a load.

truck crane A crane consisting of a rotating superstructure with its own diesel engine and boom on a truck-type carrier equipped with a separate motor for travel.

turnbuckle Device attached to wire rope chain or rods for making limited adjustments in length, which consists of a barrel and right-hand and left-hand threaded bolts.

two-blocking The hoist line hook assembly is winched up but blocked by the boom tip, resulting in the abrupt failure of the overstressed rope and the hook assembly dropping to the ground.

ultimate strength Maximum conventional stress—tensile, compressive, or shear—that a material can stand without failure.

weather the crane Letting the crane rotate with the wind when out of service to expose a minimum area to the wind force.

wheel-mounted crane A crane consisting of a rotating superstructure with power plant, operating machinery, and boom, mounted on a base or platform equipped with axles and rubber-tired wheels for travel. The base may be propelled by an engine in the superstructure, or with a separate engine controlled from the superstructure.

winch Mechanical device used to wind up or let out a rope or wire rope. In its simplest form it consists of a spool and attached crank.

windlass A base-mounted machine, usually power-operated, used for hauling in or paying out rope or chain.

wire A continual span of metal that has been cold drawn from a rod.

wire rope Plurality of strands of wire laid helically around an axis or a core.

work task A segment of a production process or operation that requires a set of resources as inputs, either as active processors, (raw) materials, information, or time, in order to produce a desired output.

wrap (wire) One circumferential turn of wire rope around a rope drum.

Chapter 8

accelerator, grout Chemical admixture that increases the rate of a chemical reaction.

activator, grout Chemical admixture that activates a catalyst to begin a reaction.

aggregate, grout Granular mineral material, such as sand, ground slag, or rock, that is mixed with water and cement to form a grout.

anchor (earth or rock) Anchors are used for excavation support, or as a part of permanent retaining walls, or to resist uplift forces on foundations.

angle of repose The angle that the slope of a bank of loose earth, gravel, or other material makes with the horizontal.

Appold, John Nineteenth-century pioneer of the centrifugal pump.

artesian water Pressurized subsurface water underlying a confining layer of clay.

auger A helical rotary (screw) tool for drilling a cylindrical hole in soil.

boulder A rounded and weathered rock greater than 200 mm in size.

bracing, wall Structural system designed to provide horizontal supports for a vertical wall.

cake (filter cake) A layer of clay built up on the wall of a slurry shaft or wall forming an impermeable lining that maintains pressure against the surrounding soil.

cantilever diaphragm wall A concrete retaining wall for a deep excavation without a horizontal bracing or waling. The wall is prevented from tipping over by burying its base sufficiently far below the bottom of the excavation.

cement grouting The injection of hydraulic cement grouts into soil or rock to add strength.

centrifugal force, pump A rotating impeller pushes a fluid while sucking more with the low pressure on its backside. The position of in- and outtakes takes advantage of both forces.

Chance, Albert Inventor of the first practical earth anchor.

cofferdam A temporary structure built inside an area surrounded by water to create a dry work place.

cold joint Surface where concrete placement was interrupted, then later resumed.

consolidation The gradual reduction in the volume of a soil mass resulting from compaction.

coquina A soft, porous limestone made up largely of shells, coral, and fossils.

data logger Electronic device recording sensor data, such as pressure, tension, or vibration data.

deadman A buried or partially buried structure utilized as an anchorage for tension rods that restrain a retaining wall. Deadman anchorage may be provided by soldier piles, sheet piling, or concrete blocks or walls.

deep well A vertical excavation below the top of the ground-water table.

desander Hydrocyclone that separates sand from the drilling fluid.

dewatering (1) Removing surface water from a construction area, (2) lowering of the groundwater table in order to obtain a "dry" area to be excavated.

diaphragm wall A concrete retaining wall that is constructed in panels from the ground surface. During the excavation, the two faces are supported by a drilling mud such as bentonite. Rebar cages are placed before concrete is tremied into the slot, displacing the mud.

discharge static head The difference in elevation between the liquid level of the discharge tank and the centerline of the pump.

drawdown, dewatering The amount by which the water table has been lowered at a certain distance from the well. Measured inside an observation well or a piezometer.

emulsion A liquid containing a second dispersed phase composed of minute droplets.

earth anchor A rod with a helical screw at one end. The rod is driven and turned into soft ground until a torque is reached that promises to provide the desired tension force needed to stabilize a structure (e.g., earth retaining wall).

extractor An inverted air hammer with yoke to pull piles out of the ground.

fill, construction Any man-made soil deposit. Fills may consist of soils that are carefully compacted to form an extremely dense, incompressible mass.

fines Soils or granular material with a nominal size smaller than 0.075 μm.

fly-ash Fly-ash is 66–68 percent pozzolanic glass resulting from burning pulverized coal and is a common ingredient in modern multi-component, stable grouts.

ground loss Sinking of the ground adjacent or close to a vertical shaft caused by soil moving into the open space laterally during drilling or after.

grout A fluid material injected into a soil or rock to change its strength. Grouts may be cementitious, chemical solutions, or of a variety of resinous origins.

head, pump Specific energy or energy per unit weight of fluid, the unit of head is expressed in feet or meters.

hollow-stem auger An earth auger with an end bit on a hollow center shaft.

impeller The rotating disk of a centrifugal pump with curved vanes.

in-situ soil mixing A soil improvement technique used to construct load-bearing columns, cutoff walls, or retaining walls, and to treat soils in situ. A mixing shaft is pushed into the soil and rotated by a drill mounted on a crane while grout is pumped through the hollow stem and injected at its tip. When the design depth is reached, the stem is withdrawn, but mixing continues on the way to the surface.

internally braced walls Retaining walls supported by struts and wales to keep them from moving horizontally under the pressure of the surrounding soil.

impervious In impervious soil, the close spacing of the particles allows only a slow passage of water. In a typical clay, water is only 1/1,000,000 as fast as in sand.

jet-eductor wellpoint Venturi-based pump creating a low pressure at the bottom of the wellpoint using compressed air that sucks the water from the intake.

lagging Timber boards, planking, or sheathing; reinforced concrete planks; or steel plate secured between adjacent soldier piles.

Larssen, Tryggve German inventor who rammed deformed steel profiles into the ground supporting each other. The product name for steel sheet piles.

Loomis, Thomas Holder of the 1875 patent for steam hammers fabricated by Vulcan, Inc.

mesh Steel wires in a grid that when encased in concrete or shotcrete provide strength.

NPSA Net Positive Suction Available. The head or specific energy at the pump suction flange less the vapor pressure head of the fluid.

operating point The point of intersection between the system and the performance curve of a pump. It corresponds to the flow and head required for the process.

oscillator A device that moves back and forth around a central neutral point, created by changes in energy. The eccentric weights rotated around a central axis oscillate and provide vibrating forces that depend on the masses of the eccentric weights.

packer An inflatable device packed around the injection pipe inside the grout hole to prevent leaking of the grout during injection (mechanic, hydraulic or pneumatic).

packing Steel, wood, concrete, or nonshrink grout used to fill gaps and transfer load between the shoring wall and bracing elements.

performance curve (pump) A plot of total head vs. flow for a specific pump model, impeller diameter, and speed (synonym characteristic curve, water performance curve).

piezometer for water pressure An electronic device embedded in a drill hole to measure the soil pore pressure = pressure head of groundwater. In its simplest form, it may consist of an open hole or well penetrating the water table. In engineering work, it more often refers to a pressure cell placed in a dam embankment, etc., as part of the permanent instrumentation.

pressure The application of external or internal forces to a body producing tension or compression within the body.

pressure cell/sensor Electronic device that can be embedded into a structure and used to monitor changes in pressure and force.

pressure head The specific energy due to pressure.

preloading Placement of initial loads in bracing members by jacking and shimming or wedging to ensure adequate bearing of connected shoring elements and to reduce ground movements.

progressive cavity pump A rotary pump wherein a helical steel rotor rotates inside a softer stator. Spaces containing the grout advance along the rotor in a positive screwing motion.

proportioning pump A pump that, combined with one or several others, is used to apportion separate components of a grout to achieve a mix according to a desired recipe.

pump curve A curve plotting the total head vs. the flow for a specific pump model.

pumpability A relative measure indicating the ease with which a grout mix can be pumped through the pipes, valves, and hoses through which the grout must pass.

rig, drilling rig A piece of mobile equipment for drilling holes in earth or rock.

riser, pipe A vertical suction pipe of a wellpoint system jetted into the ground, feeding into a header pipe on the surface that is connected to a suction pump.

rock anchor A steel bar or strands grouted into a hole drilled in rock. The bars or strands are subsequently tensioned. This provides lateral or vertical force to resist movement of a retaining structure.

secant (pile) wall A method of constructing a concrete retaining wall by means of a continuous row of concrete-filled boreholes. Alternate holes are drilled in two successive series such that adjacent piles overlap.

seepage Small quantities of water percolating through a soil deposit or soil structure.

settlement (1) Downward movement of a foundation when loading is applied; (2) vertical movement of structures or soil surface as the supporting earth compresses.

sheet piling Vertical steel profiles that are driven into the ground and interlocked with one another to form a continuous wall in order to support lateral loads from soil and/or water.

silt A fine-grained nonplastic soil; often mistaken for clay, but quite different in its behavior (particle sizes range from 0.002 to 0.074 mm).

siphon A system of piping or tubing in which the exit point is lower than the entry point.

slurry wall (slurry trench) A narrow, vertical-walled trench of specified width excavated while full of slurry and backfilled with specified materials to form a cutoff wall of low permeability. The slurry prevents the collapse of the trench walls during excavation. A reinforced concrete wall is created via a tremie.

soil nailing A system in which soil nails are typically grouted, and untensioned rebar is installed in drilled holes. Subsequently, steel mesh and shotcrete is applied to the surface of the excavation while the rebar serves as nails holding the mesh.

soldier (beam) A substantial steel or timber section driven into the ground and used to support horizontal lagging.

soldier pile walls Vertical piles set 6 to 10 feet apart with lagging set in between. The soldier piles carry the full earth pressure load transferred by the lagging.

spoil Soil or rock removed from an excavation to be wasted or used as fill.

steel sheet piles Interlocked steel sheets are driven into soft ground to cut off concentrated water flow through pervious layers within or below the excavation and to stabilize the sides of a deep excavation pit with the help of bracings or tiebacks.

strain gauge An instrument or device for measuring relative motion between two points, such as a drilled shaft and the surrounding ground.

strut A brace (compression member) that resists thrusts in the direction of its main axis.

submergence The height between the free surface of a fluid and a submerged pump intake.

subsidence The settlement of the ground, pipelines, or other structures. The effects may not be evenly distributed, and differential settlement may occur.

suction head The height between the free surface of a fluid and the centerline of the pump.

sump and ditch Installation around an excavation to collect surface water.

temporary casing Casing left in place until concrete is placed or as protection for workmen.

test hole A contractor might be asked to demonstrate that his method will work by creating a test hole.

tieback Tensioned strands or rods anchored in the soil behind a retaining wall utilized to keep the wall in place. The strands or rods are grouted into a drilled hole and provide resistance after their ends are tied to the wall.

tied-back walls A retaining wall supported by one or several rows of tiebacks.

total head The difference between the pressure head at the discharge and the suction flange of the pump.

tremie (verb) To place concrete below water level through a pipe with the bottom opening kept immersed in fresh and raising concrete.

tube à manchette (tam) A grout pipe perforated with rings of small holes; each ring of perforations is enclosed by a short rubber sleeve fitting tightly around the pipe so as to act as a one-way valve when used with an inner pipe containing two packer elements that isolate a stage for injection of grout.

turbulent The behavior of fluid particles within a flow stream characterized by rapid movements in many directions.

uncased borehole Any borehole without a lining or pipe inserted; i.e., self-supporting, whether temporary or permanent.

vacuum wellpoint Wellpoint with vacuum to increase the hydraulic gradient. Does not work when raiser pipes are > 15 ft.

vibratory driver/extractor A pile-driving and extracting machine that is mechanically connected to a pile or casing and loosens it while driving or pulling by oscillation. Motors may be either electric or hydraulic.

wale A horizontal member of a bracing system installed parallel to the retaining wall and supported by cross-bracings or struts.

waling A horizontal steel or timber support system to a retaining wall. The waling is restrained at two or more locations by struts back to the other side of the excavation or to vertical soldiers.

walking off Tendency of a rotating bit to deflect laterally when encountering a harder, deflecting layer of rock or an irregular surface.

water content The ratio of the quantity (by weight) of water in a given volume of soil mass to the weight of the soil solids, typically expressed as a percentage.

water table The elevation at which free water will usually be present; groundwater table.

waterproof membrane A skin attached to one side of a retaining wall to improve the water tightness of concrete. It could be a UPV type sheet or tar material.

well screen A system of mesh screening or holes designed to allow water to enter a well or borehole without undue loss of head, but to exclude sand, silt, and other geologic material.

wellpoint System to lower the water table around an excavation pit in sandy-silty soil. Consists of riser, header, and a suction pump. A perforated section is attached to the bottom of the riser pipe preventing the entry of sand during pumping. The riser pipe brings the groundwater to the surface. The pumps are located above the water table and collect the water from the header pipes for discharge.

work, pump The energy required to drive fluid through the system.

Chapter 9

accelerator, grout Chemical admixture that increases the rate of a chemical reaction.

activator, grout Chemical admixture that activates a catalyst to begin a reaction.

adhesion	The cohesive property of soil that makes it "cling," or "adhere" to piers or piles establishing added resistance to movement.
admixture	Additive to a fluid to achieve a desirable characteristic.
ADSC	Association of Drilled Shaft Contractors.
aggregate, grout	Granular mineral material such as sand, ground slag, or rock that is mixed with water and cement to form a grout.
anchor pier	A pier designed to resist uplift or lateral forces.
auger	A helical rotary (screw) tool for drilling a cylindrical hole in soil.
axial load	That portion of the load on a pier or pile that is in the direction of its axis.
bailing bucket	A bucket-like tool for removing water from a shaft before concrete placement.
batch system, grout	Method of injecting where a batch of grout components are mixed at one time.
bearing capacity	The maximum load a soil or rock will sustain without excessive settlement or failure.
bearing stratum	A soil or rock stratum that is expected to carry the drilled shaft load (by end friction, sidewall friction, or a combination of the two).
bell	Enlargement of the bottom of a drilled shaft to increase its bearing area.
belling bucket	A drilling bucket tool with expanding cutters that can enlarge the bottom of the drilled hole, to form a bell or underream.
bends	A disease that was first observed in the 1870s during the construction of the Eads Bridge in St. Louis where 15 workers died.
bentonite	Clay containing 75 percent of a natural substance that swells in water to several times its dry volume.
bond strength, grout	A measure that quantifies the adherence of grout to other material.
boulder	A rounded and weathered rock greater than 256 mm or 10 inch in size.
bracing, wall	Structural system designed to provide horizontal supports for a vertical wall.
cage, pile/shaft	Preassembled and tied rebar for piles or shafts, lowered by crane as a whole.
caisson	A steel box first used in 1851 to build the foundations for the bridge over the Medway at Rochester, England.
calyx (or shot) barrel	A core barrel cutting rock using steel shots (small beads) that grind the rock when forced by the rotating steel edge of the barrel.
catalyst, grout	Compound that increases the speed of a reaction of a main ingredient.
catalyst system	A combination of compounds (initiator and an accelerators) speeding the chemical reaction after initiation.
caving (or sloughing)	Soil falling into an uncased (unsupported) hole or trench during or after the drilling or excavation.
cavitation	The collapse of gas bubbles due to a sudden pressure increase.
cement grouting	The injection of hydraulic cement grouts into soil or rock to add strength.
chemical grouting	Grouting with a pure solution or a grout that contains such fine particles in suspension that it behaves essentially as a Newtonian or near-Newtonian fluid.
cleanout bucket	A cylindrical bucket to remove "cuttings" from the shaft bottom by turning it clockwise, thus opening a bottom door, and counterclockwise to close it.
cold joint	Surface where concrete placement was interrupted and then later resumed.
collar	The surface opening of a grout hole to which the grout delivery line is attached.

continuous flight auger	A long helical auger (screw) with a cutting head to bore a hole in the earth, into which a pile section may be set, concrete cast in place, or tieback grouted.
coquina	A soft, porous limestone made up largely of shells, coral, and fossils.
core barrel	A cylindrical rock-drilling tool cutting around a central cylindrical core of rock, which can then be removed.
cushion, pile	The material placed on a pile head to cushion the impact of a hammer.
cuttings	Particles of soil or rock resulting from the cutting action of drilling or augering.
density meter	Electronic device to measure in real time the density of a fluid passing through.
displacement grouting	Injection of grout into the ground to controllably displace or move a desired part of the subsurface (e.g., push a foundation upwards).
downdrag	A downward force exerted on a drilled shaft, pile, or other structural element by settling soil. Sometimes called "negative skin friction".
drilled shaft	A reinforced or unreinforced concrete foundation formed by drilling a hole in the earth and filling it with concrete.
drilling bucket	A rotary boring tool with a cutting edge at its base. Spoil is emptied from the bucket by swinging it to one side and releasing the hinged bottom door.
drilling mud, mud, or slurry	A fluid mixture of water and clayey soil, or commercial "driller's mud," which may be bentonite or attapulgite.
drawdown, dewatering	The amount by which the water table has been lowered at a certain distance from the well as measured inside an observation well or a piezometer.
emulsion	A liquid containing a second dispersed phase composed of minute droplets.
end bearing	The portion of load-carrying capacity a shaft or pile has due to the end area bearing on the material below.
epoxy resins	Multicomponent resin consisting essentially of epoxide groups that are characterized by very high tensile, compression, and bond strengths.
fill, construction	Any man-made soil deposit. Fills may consist of soils that are carefully compacted to form an extremely dense, incompressible mass.
fines	Soils or granular material with a nominal size smaller than 0.075 μm.
fissure	A tiny fracture in a rock or soil mass.
foam grout	A grout that has been blended with a foaming agent in order to reduce its unit weight and/or to help reduce the cost of filling voids.
fracture	Fissure or break in a rock mass that may be a natural consequence of folding or faulting or artificially produced by pressure injection.
fracturing	High-pressure injection of grout along cracks or fissure to crack them apart.
friction, fluid	Friction produced internally as one layer of fluid moves with respect to another and also at the fluid wall interface.
friction/end-bearing pile	A pile achieving load capacity from both wall friction and end bearing.
friction shaft	A foundation that derives its resistance to load by the friction or bond developed between the side surface and the soil or rock surrounding it.
gel	Condition in which a liquid grout begins to develop strength.
gel time	Time interval between the mixing of a fluid grout and the formation of a gel.
ground freezing	A method to convert soft, flowing material into a strong, watertight structural member by inserting refrigeration pipes.

hardener In a two-component epoxy or resin grout, the chemical component that causes the base component to cure.

hydrofracture grouting Injection of grout at sufficiently high pressures to produce fractures to promote distribution of the grout into fingers, sheets, and/or lenses.

impact force The peak force delivered by a pile driving hammer to the pile head as measured by means of dynamic monitoring.

jet grouting A mixing technology that uses a pipe system equipped with one or more high-pressure jets to erode/cut soils while mixing them with cement grout. The mixing tool is jetted to the final depth prior to starting the mixing action while pulling up.

Kelly bar (Kelly) The Kelly bar is a square shaft attached to the top of the drill steel and turned by a rotary table. Also used to raise the drill column out of the borehole.

load cell An electronic force measuring device. Used in testing pile-bearing load.

matrix The natural material in which any fossil, pebble, crystal, etc., is embedded.

mixer A machine to blend the constituents of grout, mortar, or other components.

mixing cycle One episode of loading the mixer, mixing, and unloading a batch of grout.

mixing speed The revolutions per minute of mixer paddles or pump impeller in a grout mixer.

necking The reduction in diameter in a section of a drilled shaft.

negative pressure Pressure in a vessel that is less than the pressure in the environment.

negative skin friction Effect of settling soil around a pile adding frictional down-forces. Downdrag.

packer An expandable device packed around the injection pipe inside the grout hole to prevent leaking of the grout during injection, (mechanic, hydraulic, or pneumatic).

paddle mixer A grout mixer in which the mixing is accomplished by paddles rotating around a shaft. Vertical or horizontal turning inside large pans.

pan mixer Grout mixer comprised of a horizontal pan or container in which mixing is accomplished by means of a rotating pan, rotating paddles, or both.

particulate grout Any grout characterized by undissolved insoluble particles suspended in the mix. Also referred to as suspension grout or slurry grout.

percussion drilling A drilling process in which a hole is advanced by using a series of impacts to the drill steel and attached bit; the bit is slightly advanced between impacts.

pile A long, slender foundation unit, made of wood, steel, or concrete, or combinations thereof, which is either premanufactured and placed by driving, jacking, jetting, or screwing, or cast-in-situ into a hole.

pile point A special type of pile shoe.

plasticity Term applied to fine-grained soils (such as slays), which when moist can be remolded without raveling or breaking apart.

Pons Sublicius The name of the first Roman bridge, built on timber piles.

reverse circulation Drilling fluid is pumped out of the drill stem at the top, circulated through a pit where cuttings are removed, and returned to the annular space around the drill stem. Circulation is upward inside the drill stem and downward outside it.

rig, drilling rig Mobile equipment for drilling holes in earth or rock.

rock auger An auger-type drilling tool with hard-metal teeth to drill soft or weathered rock.

rotary boring A method of boring, or simple drilling, through soil or hard material.

rotary drill rig Equipment powered hydraulically, pneumatically, electrically, or mechanically to bore holes for drilled shafts, caissons, or in-situ piles (auger or Kelly type).

settlement	(1) Downward movement of a foundation when loading is applied; (2) vertical movement of structures or soil surface as the supporting earth compresses.
SID	This *shaft inspection device* has a video camera that is lowered into a drilled shaft. It measures the thickness of sediment on the bottom and samples soils.
sidewall grooving	The cutting of circular or spiral grooves in the walls of a drilled shaft hole in rock or soil, with the objective of improving sidewall support.
skin friction	Resistance to shearing motion between a pile and soil or rock in contact with it.
sloughing	Soil falling into an uncased shaft during or after the drilling.
spacers	Spacers keep the steel cage centered in the drilled shaft and insure proper concrete cover. They should be of concrete-approved noncorrosive material.
spoil	Soil or rock removed from an excavation to be wasted or used as fill.
standard penetration test (SPT) (N)	A standard ASTM D 1586 method to assess density of soil by driving a 24-inch long soil sampling tool 1 foot vertically into the ground and counting the drops of a 140 lb weight free falling over 30 inch.
strain gauge	An instrument or device for measuring relative motion between two points such as a drilled shaft and the surrounding ground.
template	A guide installation for drilling or piling from a barge to maintain alignment.
temporary casing	Casing left in place until concrete is placed or as protection for workmen.
test hole	A contractor might be asked to demonstrate that his method will work by creating a test hole.
tremie (pipe)	The tremie pipe transports the concrete through the slurry, keeps the concrete from segregating during placement, and keeps the concrete from mixing with the drilling slurry at the slurry/concrete interface.
tremie concrete	Concrete placed under water or slurry by means of a long pipe. The end of the pipe stays immersed in concrete, which is heavier than water or slurry. While the concrete settles at the bottom, the slurry is slowly displaced without mixing.
tube à manchette	A grout pipe perforated with rings of small holes; each ring of perforations is enclosed by a short rubber sleeve fitting tightly around the pipe so as to act as a one-way valve when used with an inner pipe containing two packer elements that isolate a stage for injection of grout.
underream	Enlargement at the bottom of a shaft to increase its bearing area, a "bell".
uplift	An upward force exerted on a pier, pile, or other structural elements, by expanding soil or rock, hydraulic pressure, or structural loading.
vibratory driver/ extractor	A pile-driving and extracting machine that is mechanically connected to a pile or casing and loosens it while driving or pulling by oscillation. Motors may be either electric or hydraulic.
walking off	Tendency of a rotating bit to deflect laterally when encountering a harder, deflecting layer of rock or an irregular surface.
water-bentonite slurry	A stable colloidal suspension of powdered bentonite in water. The terms "slurry" and "water-bentonite slurry" are used interchangeably.
water content	The ratio of the quantity (by weight) of water in a given volume of soil mass to the weight of the soil solids, typically expressed as a percentage.
work tremie	A hopper and drop pipe used to place concrete underwater.

Chapter 10

bamboo	Plant that is widely used to build scaffolds.
bleeding	Term used for water that rises to the surface of fresh concrete.

cleat	Stops a plank from slipping off a scaffold.
competent person	One who can identify existing and predictable hazards in the working environment or working conditions that are dangerous to personnel and who has authorization to take prompt corrective measures to eliminate them.
coupler	A device for locking together the two tube elements of a tubes- and-coupler scaffold.
dowel	A plain round steel bar that extends into two adjoining PCC slabs so as to transfer shear loads.
falsework	Temporary structures used in construction to support spanning or arched structures in order to hold the component in place until construction is sufficiently advanced.
flying	Descriptive adjective for a formwork that is being transported by crane.
form scaffold	A supported scaffold consisting of a platform supported by brackets attached to formwork.
Haglin, Charles	Civil engineer who designed and oversaw the first vertical slip-forming operation.
hoist	A manual or power-operated mechanical device to raise or lower a suspended scaffold.
JPCP	Short for Jointed Reinforced Concrete Pavement that has no reinforcement.
lifeline	Rope connecting a personal fall arrest system to a solid anchor.
mobile scaffold	A powered or unpowered, portable, caster, or wheel-mounted supported scaffold.
Monier, Joseph	Frenchman inventor who patented the first steel reinforced concrete system.
PCC	Short for Portland Cement Concrete.
post shore	Also called a single prop.
preshoring	Also known as backshoring.
raking shore	Consists of one or more timbers sloping between the face of the structure to be supported and the ground.
reshoring	Either single posts or towers, with legs strategically located to transfer construction loads to as many lower floors as required to ensure that the partially completed building is neither overloaded or overstressed.
screwjack	Mechanism to adjust the final position of the formwork.
shoring	Process of supporting a structure in order to prevent collapse.
shoring towers	A common formwork solution with high clearance.
slipform	Formwork that climbs continuously up the rising building during construction.
spalling	The deterioration of the upper 3 to 12 mm (0.125 to 0.5 in) of the concrete surface caused by corroding rebar below. The process includes the continued cracking and falling off of surface concrete, exposing more and more rebar.
tie bar	A deformed steel bar or connector embedded across a longitudinal joint in a PCC slab to prevent separation of abutting slabs.
tipping	Major danger when operating a mobile boom work platform.
transverse	Descriptor for a joint that runs perpendicular to the pavement.
tube	Name for the steel or aluminum pipe used to build scaffolds.
wind	The primary horizontal load onto a structure, such as a vertical formwork, which is a function of wind speed, direction, and surface area. Wind bursts add additional short-term forces.

abutment Part of a structure that supports the end span or accepts the thrust of an arch; often supports and retains the approach embankment.

anchor span The end section of a suspension bridge that counterbalances the span extending in the opposite direction from the tower.

anchorage Location where to the main cables of a suspension or cable-stayed bridge are anchored either to the ground or the pylon.

approach The part of the bridge that carries traffic from the land to the main parts of the bridge.

approach span The span or spans connecting the abutment with the main span or spans.

aqueduct A pipe or channel, open or enclosed, which carries water. May also be used as part of a canal to carry boats. Sometimes carried by a bridge.

arch A curved structure that supports a vertical load mainly by axial compression.

arch bridge A bridge whose main support structure is an arch. A through arch has the deck go through the arch and is partially suspended by hangers.

arch barrel The inner surface of an arch extending the full width of the structure.

arch ring An outer course of stone forming the arch. Made of a series of **voussoirs**. An archivolt is an arch ring with decorating moldings.

balisters A decorative railing, especially one constructed of concrete or stone, including the top and bottom rail and the vertical supports called balusters. May also include larger vertical supports called **stanchions**.

bascule bridge From the French word for "see-saw," a bascule bridge features a movable span (leaf), which rotates on a horizontal hinged axis (trunnion) to raise one end vertically. A large counterweight is used to offset to weight of the raised leaf. May have a single raising leaf or two that meet in the center when closed. Compare to **swing bridge** and **vertical lift bridge**.

beam A horizontal structure member supporting vertical loads by resisting bending. A girder is a larger beam, especially when made of multiple plates. Deeper, longer members are created by using a box structure.

beam bridge A bridge built of beams, classified either as a short-span or long-span beam bridge, which is supported on beams whose ends rest on piers or abutments.

bearing A mechanism placed on top of a pier cap or abutment. It is specially fabricated to allow the bridge to move in a desired manner while blocking other movements.

bedrock The solid rock layer beneath sand or silt.

bent Part of a bridge substructure. A rigid frame commonly made of reinforced concrete or steel that supports a vertical load and is placed transverse to the length of a structure. Bents are commonly used to support beams and girders. An **end bent** is the supporting frame forming part of an abutment. Each vertical member of a bent may be called a **column**, **pier**, or **pile**. The horizontal member resting on top of the columns is a **bent cap**. The columns stand on top of some type of foundation or footer that is usually hidden below. Grade A bent commonly has at least two or more vertical supports. Another term used to describe a bent is **capped pile pier**. A support having a single column with bent cap is sometimes called a **hammerhead pier**.

bowstring truss A truss having a curved top chord and straight bottom chord meeting at each end.

box girder bridge A box girder bridge is a bridge in which the main beams comprise girders in the shape of a hollow box. The box girder normally comprises either prestressed concrete, structural steel, or a composite of steel and reinforced concrete. The box is typically rectangular or trapezoidal in cross-section. Box girder bridges are commonly used for highway flyovers and for modern elevated structures of light rail transport. Although normally the box girder bridge is a form of beam bridge, box girders may also be used on cable-stayed bridges and other forms.

brace A structural support to resist loads.

brace-ribbed arch An arch with parallel chords connected by open webbing (**trussed arch**).

bridge gross weight formula A bridge protection formula used by federal and state governments. The standard specifies the relationship between axle (or groups of axles) spacing and the gross weight that (those) axle(s) may carry. Expressed by:

$$W = 500\left[\frac{LN}{N-1} + 12N + 36\right]$$

W = overall gross weight on any group of two or more consecutive axles to the nearest 500 pounds, L = distance in feet between the extreme of any group of two or more consecutive axles, and N = number of axles in the group under consideration.

brittle Characteristic of a material that fails without warning; brittle materials do not stretch or shorten before failing.

buckle To bend under compression.

buttress A wall projecting perpendicularly from another wall that prevents its outward movement. Usually wider at its base and tapering toward the top.

cable Part of a suspension bridge extending from an anchorage over the tops of the towers and down to the opposite anchorage. Suspenders or hangers are attached along its length to support the deck.

cable-stayed bridge A variation of suspension bridge in which the tension members extend from one or more towers at varying angles to carry the deck. Allowing much more freedom in design form, this type does not use cables draped over towers nor anchorages at each end, as in a traditional suspension bridge.

caisson "Caisson" is the French word for "box." A watertight, dry chamber in which people can work underwater. At the base of the caisson is its "cutting edge" of plate steel. In a suspension bridge the caisson becomes the foundation, the pier, supporting the bridge's towers.

caisson disease An affliction developed by people moving in and out of caissons too quickly; also called the bends and decompression sickness.

camber A positive, upward curve built into a beam that compensates for some of the vertical load and anticipated deflection.

camelback truss A truss having a curved top chord and straight bottom chord meeting at each end, especially when there are more than one used end to end. Compare to **Baltimore truss** and **Pennsylvania truss**.

cantilever A structural member that projects beyond a supporting column or wall and is counterbalanced and/or supported at only one end.

cast in place Concrete poured within formwork on site to create a structural element in its final position.

castellated girder A steel beam fabricated by making a zig-zag cut along its web, then welding the two sides together at their peaks. This creates a beam that has increased depth and therefore greater strength, but is not increased in weight.

cast iron A brittle alloy with high carbon content; iron that has been melted, then poured into a form and cooled; can be made into any shape desired.

catenary Curve formed by a rope or chain hanging freely between two supports. The curved cables or chains used to support suspension bridges may be referred to as catenaries.

catwalks Temporary foot bridges, used by bridge workers to spin the main cables (several feet above each catwalk), and to attach the suspender cables that connect the main cables to the deck.

centering Temporary structure or falsework supporting an arch during construction.

chord Either of the two principal members of a truss extending from end to end, connected by web members.

closed spandrel deck arch One of the oldest and strongest types of bridges, the arch has been used in bridge building since the Roman era. An arch supports weight by compression on an axis or keystone. The axis or keystone transfers the stress of gravity (the traffic load) outward toward the base of the arch. Supports called abutments absorb the pressure and stop the ends of the bridge from spreading apart. Most arch bridges have vertical supports called spandrels that distribute the weight on the deck to the arch below. The first arch bridges were built of stone, but arch bridges are now constructed of concrete or steel for greater strength and durability. Arch bridges are strong because the entire structure is under compression. Therefore, they must be built of materials that are strong under compression. Putting more weight on the arch actually compresses the material and makes the structure more rigid. Most arch bridges span between 200 and 800 feet.

cofferdam A temporary dam built to divert a river around a construction site so the dam can be built on dry ground.

column A vertical structural member used to support compressive loads.

compression member Describes a timber or other truss member that is subjected to compression rather than tension.

continuous beam bridge A simple bridge made by linking one beam bridge to another; some of the longest bridges in the world are continuous span beam bridges.

corbelled arch Masonry built over an opening by progressively overlapping the courses from each side until they meet at the top center. Not a true arch as the structure relies on strictly vertical compression, not axial compression.

counter A truss web member that functions only when a structure is partially loaded.

cradle Part of a suspension bridge that carries the cable over the top of the tower.

cripple A structural member that does not extend the full height of others around it and does not carry as much load.

crown On road surfaces, the crown or center is the highest point and the surface slopes downward in opposite directions, assisting in drainage. Also a point at the top of an arch.

culvert A drain, pipe, or channel that allows water to pass under a road, railroad, or embankment.

damping The action of reducing the vibration of an object. This tends to return the vibrating object to its original position.

deck The top surface of a bridge that carries the traffic.

deck bridge A bridge in which the supporting members are all beneath the roadway.

deck plate girder A plate girder bridge is a bridge supported by two or more plate girders. The plate girders are typically I-beams made up from separate structural steel plates (rather than rolled as a single cross-section), which are welded (or occasionally bolted or riveted) together to form the vertical web and horizontal flanges of the beam. In some cases, the plate girders may be formed in a Z-shape rather than I-shape. Plate girder bridges are suitable for short to medium spans and may support railroads, highways or other traffic. In the deck-type bridge, a steel or reinforced-concrete bridge deck is supported on top of two or more plate girders, and may act compositely with them. Additional beams may span across between the main girders, for example, in the form of bridge known as ladder-deck construction. Also, further elements may be attached to provide cross-bracing and prevent the girders from buckling.

deck truss A truss that carries its deck on its top chord.

deck truss
cantilever bridge A cantilever bridge is a bridge built using cantilevers: structures that project horizontally into space, supported on only one end. For small footbridges, the cantilevers may be simple beams; however, large cantilever bridges designed to handle road or rail traffic use trusses built from structural steel, or box girders built from prestressed concrete. A simple cantilever span is formed by two cantilever arms extending from opposite sides of the obstacle to be crossed, meeting at the center. In a common variant, the suspended span, the cantilever arms do not meet in the center; instead, they support a central truss bridge that rests on the ends of the cantilever arms.

deflection The displacement of a structural member or system under load.

diagonal A sloping structural member of a truss or bracing system.

diaphragm Bracing that spans between the main beams or girders of a bridge or viaduct and assists in the distribution of loads.

embankment Angled grading of the ground.

end post The outward most vertical or angled compression member of a truss.

expansion joint A meeting point between two parts of a structure that is designed to allow for movement of the parts due to thermal or moisture factors while protecting the parts from damage. Commonly visible on a bridge deck as a hinged or movable connection.

extrados The outer exposed curve of an arch; defines the lower arc of a spandrel.

eye bar A structural member having a long body and an enlarged head at each end. Each head has a hole though which a pin is inserted to connect to other members.

falsework Temporary structure used as support during construction. Falsework for arch construction is called "centering".

fill Earth, stone, or other material used to raise the ground level, form an embankment or fill the inside of an abutment, pier, or closed spandrel.

fixed arch A structure anchored in its position. Compare to **hinged arch**.

floor beam Horizontal members that are placed transversely to the major beams, girders, or trusses; used to support the deck.

flyover A bridge formed when one roadway crosses over another at a higher level; an overpass.

footing The enlarged lower portion of the substructure or foundation that rests directly on the soil, bedrock, or piles; usually below grade and not visible.

force Any action that tends to maintain or alter the position of a structure.

forms Temporary structures or molds made of wood, metal, or plastic used when placing concrete to ensure that it is shaped to its desired final form.

formwork A total system of support for freshly placed concrete, including the mold and all supporting members, hardware, and necessary bracing. Formwork must be strong enough to support the considerable weight and pressure of wet concrete without bending or breaking.

gabion A galvanized wire box filled with stones used to form an abutment or retaining wall.

girder A horizontal structure member supporting vertical loads by resisting bending. A girder is a larger beam, especially when made of multiple metal plates. The plates are usually riveted or welded together.

girder bridge A girder bridge is perhaps the most common and most basic bridge. The cross-section of the girder takes the shape of the capital letter "I." The vertical plate in the middle is known as the *web*, and the top and bottom plates are referred to as *flanges*. A box girder is much the same as an I-beam girder except it takes the shape of a box. The typical box girder has two webs and two flanges. However, in some cases there are more than two webs, creating a multiple chamber box girder. Other examples of simple girders include pi girders, named for their likeness to the mathematical symbol for pi, and T-shaped girders.

glu-lam freespan Glu-lam bridges can freespan over 100 ft (30 m). This makes a Glu-lam freespan the ideal choice for road overpasses, bridging steep ravines, and creating a crossing where pile foundations would be prohibited.

gusset plate A metal plate used to unite multiple structural members of a truss.

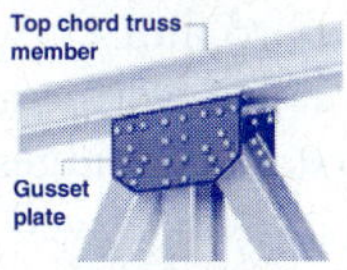

hanger A tension member serving to suspend an attached member.

haunch The enlarged part of a beam near its supported ends, which results in increased strength; visible as the curved or angled bottom edge of a beam.

haunched girder Typical slab-on-beam bridges have space between the bottom of the slab and the top of the top flanges of beams. This space, referred to as the fillet or haunch, typically consists of unreinforced concrete that increases the dead load of the section but is not normally considered to add strength.

hinged arch A two-hinged arch is supported by a pinned connection at each end. A three-hinged arch also includes a third pinned connection at the crown of the arch near the middle of a span.

Howe truss A type of truss in which vertical web members are in tension and diagonal web members in compression. Maybe be recognized by diagonal members that appear to form an "A" shape (without the crossbar) toward the center of the truss when viewed in profile.

impost The surface that receives the vertical weight at the bottom of an arch.

joint A mechanism connecting two or more parts of a structure; a roller joint allows adjacent parts to move controllably past one another; a rigid joint prevents adjacent parts from moving or rotating past one another.

keystone The uppermost wedge-shaped voussoir at the crown of an arch that locks the other voussoirs into place.

knee brace Additional support connecting the deck with the main beam, which keeps the beam from buckling outward. Commonly made from plates and angles.

lag Cross-pieces used to connect the ribs in centering.

lateral bracing Members used to stabilize a structure by introducing diagonal connections.

lattice An assembly of smaller pieces arranged in a gridlike pattern; sometimes used as a decorative element or to form a truss of primarily diagonal members.

lift span The part of a movable span (or drawbridge) that opens to allow river traffic. Also called the draw span.

live load Vehicular traffic, wind, water, and/or earthquakes.

load Weight distribution throughout a structure; loads caused by wind, earthquakes, and the materials that the structure is made of, for example, affect how weight is distributed throughout a structure.

lower chord The bottom horizontal member of a truss.

main beam A beam supporting the spans and bearing directly onto a column or wall.

main span The longest span in a multispan bridge, located between the bridge's main piers or towers (supports). Bridges are typically compared using main-span lengths, which do not account for the length of the entire bridge or its approaches.

member An individual angle, beam plate, or built piece intended to become an integral part of an assembled frame or structure.

open spandrel An arch bridge is based on the ancient concept of spanning an opening with a curved structural member. The arch transmits the load from the bridge deck to the abutments on both sides of the span and thus to the ground below. Early arch bridges were built of stone blocks wedged together to form the arch. Short modern arch bridges may use wood or concrete, while longer arch spans are built of steel. Since the arch requires no central support, it can be used to bridge long open spans. The arch can be either above or below the bridge deck. The arch pushes downward and outward against its abutments, which must be heavy to resist the thrust. Since the abutments transfer both horizontal and vertical forces from the bridge deck, arch bridges can only be used where the ground or foundation is solid and stable. The curved arch structure offers a high resistance to bending forces.

Arch bridges can be constructed with the deck above the arch (a *deck arch bridge*), or the deck can be hung from a segment of the arch that rises above the deck (a *through arch* or *tied arch* bridge). In a deck arch bridge, the space between the bottom of the arch and the deck can be solid (a *closed spandrel deck arch*) or open with supporting vertical members (an *open spandrel deck arch*).

oscillation A periodic movement back and forth between two extreme limits. An example is the string of a guitar that has been plucked. Its vibration back and forth is one oscillation. A vibration is described by its size (amplitude), its oscillation rate (frequency), and its timing (phase). In a suspension bridge, oscillation results from energy collected and stored by the bridge. If a part of the bridge has to store more energy than it is capable of storing, that part could fail.

parabola A form of an arch defined by a moving point that remains equidistant from a fixed point inside the arch and a moving point along a line. This shape when inverted into an arch structure results in a form that allows equal vertical loading along its length.

parapet A low wall along the outside edge of a bridge deck used to protect vehicles and pedestrians.

pier A vertical structure that supports the ends of a multispan superstructure at a location between abutments.

pile A long load-bearing "toothpick" driven deep into the ground to form part of a foundation or substructure.

pile bent A row of driven or placed piles with a pile cap to hold them in their correct positions.

pin A cylindrical bar that is used to connect various members of a truss, such as those inserted through the holes of a meeting pair of eyebars.

plate beam-girder See **deck plate girder**.

pony truss A truss that carries its traffic near its top chord but not low enough to allow cross-bracing between the parallel top chords.

portal The opening at the ends of a **through truss** which forms the entrance. Also the open entrance of a tunnel.

post One of the vertical compression members of a truss that is perpendicular to the bottom chord.

Pratt truss A type of truss in which vertical web members are in compression and diagonal web members in tension. Many possible configurations include pitched, flat, or camelback top chords. May be be recognized by diagonal members that appear to form a "V" shape toward the center of the truss when viewed in profile. Variations include the **Baltimore truss** and **Pennsylvania truss.** Compare to **Warren truss** and **Howe truss.**

pylon A monumental vertical structure marking the entrance to a bridge or forming part of a gateway.

ready-mixed concrete truck Multiaxle truck carrying a large concrete mixer that is charged at the plant. During transit, the drum is slowly turned, which prevents the concrete from setting and disaggregating. To discharge, the drum is reversed which causes the "Archimedes screw" inside the drum to lift the concrete out and onto a chute or conveyor.

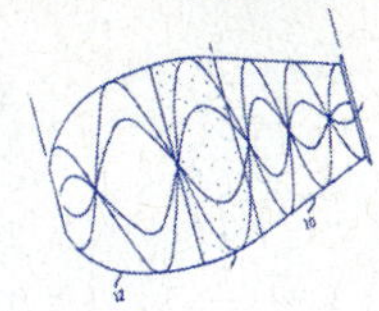

Cross-section through concrete drum

reinforcement Adding strength or bearing capacity to a structural member. Examples include the placing of metal rebar into forms before pouring concrete or attaching gusset plates at the intersection of multiple members of a truss.

retractile draw bridge A bridge with a superstructure designed to move horizontally either longitudinally or diagonally from "closed" to "open" position, the portion acting in cantilever being counterweighted by that supported on rollers; also known as traverse drawbridge.

resonance The regular vibration of an object as it responds in step (at the same frequency) with an external force.

revet The process of covering an embankment with stones.

rib Any one of the arched series of members that is parallel to the length of a bridge, especially those on a metal arch bridge.

rigid frame bridge A type of girder bridge in which the piers and deck girder are fastened to form a single unit. Unlike typical girder bridges that are constructed so that the deck rests on bearings atop the piers, a rigid frame bridge acts as a unit. Pier designs vary.

rise The measure of an arch from the spring line to the highest part, which is to say from its base support to the crown.

rivet A metal fastener used in pre-1970 construction; made with a rounded preformed head at one end and installed hot into a predrilled or punched hole; the other end was hammered into a similarly-shaped head, thereby clamping the adjoining parts together.

rocker bearing A bridge support that accommodates expansion and contraction of the superstructure through a rocking action.

rocker bent A bent hinged or otherwise articulated at one or both ends to provide for the longitudinal movements resulting from temperature changes and superimposed loads.

roller A steel cylinder intended to permit longitudinal movements by rolling contact.

roller bearing A single roller or a group of rollers so housed as to permit longitudinal movement of a structure.

roller nest A group of steel cylinders used to facilitate the longitudinal movements resulting from temperature changes and superimposed loads.

rolling lift bridge A bridge of bascule type devised to roll backward and forward upon supporting girders when operated through an "open and closed" cycle.

segmental arch An arch formed along an arc that is drawn from a point below its spring line, thus forming a less than semicircular arch. The intrados of a Roman arch follows an arc drawn from a point on its spring line, thus forming a semicircle.

shear A force that causes parts of a material to slide past one another in opposite directions.

silt Sediment particles ranging from 0.00016 to 0.0024 inches in diameter.

simple span A span in which the effective length is the same as the length of the spanning structure. The spanning superstructure extends from one vertical support, abutment or pier, to another, without crossing over an intermediate support or creating a cantilever.

skew When the superstructure is not perpendicular to the substructure, a skew angle is created. The skew angle is the acute angle between the alignment of the superstructure and the alignment of the substructure.

span The horizontal space between two supports of a structure. Also refers to the structure itself. May be used as a noun or a verb. The **clear span** is the space between the inside surfaces of piers or other vertical supports. The **effective span** is the distance between the centers of two supports.

spandrel The roughly triangular area above an arch and below a horizontal bridge deck. A **closed spandrel** encloses fill material. An **open spandrel** carries its load using interior walls or columns.

specifications A document that explains all material and construction requirements of the bridge structure to be constructed, usually used by engineers or architects in the planning stages of construction.

splice plate A plate that joins two girders. Commonly riveted or bolted.

springer The first voussoir resting on the impost of an arch.

spring line The place where an arch rises from its support; a line drawn from the impost.

stanchion One of the larger vertical posts supporting a railing. Smaller, closely spaced vertical supports are ballusters. Also see **balustrade**.

stay Diagonal brace installed to minimize structural movement.

steel stringers Load-carrying beams in a viaduct's superstructure that rest on abutments and other intermediate supports.

stiffener On plate girders, structural steel shapes, such as an **angle**, are attached to the **web** to add intermediate strength.

stringer A beam aligned with the length of a span that supports the deck.

strut A compressive member.

substructure The portion of a bridge structure including abutments and piers that supports the superstructure.

superstructure The portion of a bridge structure that carries the traffic load and passes that load to the substructure.

suspended span A simple beam supported by cantilevers of adjacent spans, commonly connected by pins.

suspenders Tension members of a suspension bridge that hang from the main cable to support the deck. Also similar tension members of an arch bridge that features a suspended deck. Also called **hangers**.

suspension bridge A bridge that carries its deck with many tension members attached to cables draped over tower piers.

swing bridge A movable deck bridge that opens by rotating horizontally on an axis. Compare to **bascule bridge** and **vertical lift bridge**.

tension member Any timber or rod of a truss that is subjected to pulling force.

through truss A truss that carries its traffic through the interior of the structure with cross-bracing between the parallel top and bottom chords. Compare to **deck truss** and **pony truss**.

tie A tension member of a truss.

tied arch An arch that has a tension member across the open space underneath the arch connecting one end to the other.

tower A tall pier or frame supporting the cable of a suspension bridge.

trestle While **bridge** is the more general term (which may be a single span or multispan, typically one span is longer than the others), **trestle** is a longer, multispan structure — a series of shorter spans in which most of the spans are of similar length. **Trestle** is a more common term in relation to railroads, while **viaduct** is a similar long, multispan structure for streets. Neither term seems to be exclusive.

truss A structural form that is used in the same way as a beam, but because it is made of an web-like assembly of smaller members it can be made longer, deeper, and therefore stronger than a **beam** or **girder** while being lighter than a beam of similar dimensions.

trussed arch A metal arch bridge that features a curved truss.

two-hinged arch and three-hinged arch Arches use a curved structure that provides a high resistance to bending forces. Unlike girder and truss bridges, both ends of an arch are fixed in the horizontal direction (i.e., no horizontal movement is allowed in the bearing). Thus when a load is placed on the bridge (e.g., a car passes over it), horizontal forces occur in the bearings of the arch. These horizontal forces are unique to the arch and as a result arches can only be used where the ground or foundation is solid and stable. Like the truss, the roadway may pass over or through an arch or in some cases both. Structurally, there are four basic arch types: hinge-less, two-hinged, three-hinged, and tied arches. The three-hinged arch adds an additional hinge at the top or crown of the arch. The three-hinged arch suffers very little if there is movement in either foundation (due to earthquakes, sinking, etc.). However, the three-hinged arch experiences much more deflection, and the hinges are complex and can be difficult to fabricate. The three-hinged arch is rarely used anymore.

upper chord Top chord of a truss.

vault An enclosing structure formed by building a series of adjacent arches.

vertical lift bridge A movable deck bridge in which the deck may be raised vertically by synchronized machinery at each end. Compare to **swing bridge** and **vertical lift bridge**.

viaduct A long, multispan structure, especially one constructed of concrete. More commonly used in relation to structures carrying motor vehicles. **Trestle** is the term for a similar structure when used in relation to railroads.

voussoir Any one of the wedge-shaped blocks used to form an arch.

Warren truss A type of truss in which vertical web members are inclined to form equilateral triangles. May be be recognized by diagonal members that appear to form a series of alternating "V" and "A" shapes (without the crossbar) along the length of the truss when viewed in profile. Often the triangles are bisected by vertical members to reduce the length of the members of the top chord. Compare to **Pratt truss**.

web The system of members connecting the top and bottom chords of a truss. Or the vertical portion of an I-beam or girder.

weld A joint between pieces of metal at faces that have been made plastic by heat or pressure.

welded joint A joint in which the assembled elements and members are united through fusion of metal.

wing walls Extensions of a retaining wall as part of an abutment; used to contain the fill of an approach embankment.

Index

Selected Conversion Tables

LENGTH		
1 foot (ft) =	12	inch (in)
=	0.33	yard (yd)
=	0.305	meter (m)
=	30.5	centimeter (cm)
=	0.0001894	mile (ml)
1 meter (m) =	3.28	ft
=	39.37	in
=	1.09	yd
=	100	cm
=	1,000	millimeter (mm)
1 mile (ml) =	5,280	ft
=	1,760	yd
=	1,609.34	m
=	1.61	kilometer (km)

AREA		
1 square foot (ft^2) =	144	square inch (in^2)
=	0.111	square yard (yd^2)
=	0.093	square meter (m^2)
1 square yard (yd^2) =	8.999	ft^2
=	1296	in^2
=	0.836	m^2
1 square meter (m^2) =	10.76	ft^2
=	1,550	in^2
=	1.196	yd^2
1 square kilometer (km^2) =	1,195,990	yd^2
=	0.386	ml^2
=	1,000,000	m^2

VELOCITY		
1 mile per hour (ml/h) =	17.59	inches per second (in/s)
=	5280	feet per hour (ft/h)
=	1.47	feet per second (ft/s)
=	1.609	kilometer per hour (km/h)
=	0.447	meter per second (m/s)
1 kilometer per hour (km/h) =	3,280.84	ft/h
=	0.911	ft/s
=	10.94	in/s
=	0.621	ml/h

VOLUME		
1 cubic yard (yd^3) =	27	cubic feet (ft^3)
=	46,656	cubic inches (in^3)
=	202.07	gallons (gal)
=	0.7645	cubic meter (m^3)
=	764,554.8	cubic centimeter (cm^3)
=	764.55	liter (l)
1 cubic foot (ft^3) =	1728	in^3
=	0.037	yd^3
=	7.48	gal
=	0.028	m^3
1 cubic meter (m^3) =	1.308	yd^3
=	35.31	ft^3
=	1,000,000	cm^3

WEIGHT		
1 lb =	16	ounces
=	0.4535	kilogram (kg)
1 kg =	2.2	lb
=	35.28	ounces
=	1,000	grams (g)
1 short ton (short) =	2,000	lb
=	907.18	kg
=	0.907	tonnes metric (t)

FORCE

Force of weight (Fg)	**$F_g = m \cdot g$**	
Standard Gravity (g)	**$g = 32.174 ft/s^2 (9.8066\ m/s^2)$**	
Newton is the SI basic unit of force (N)	**$N = 1\ kg \cdot m/s^2$**	
Pound force	**$lb_f = 32.174\ lb_m ft/s^2$**	
1 pound force (lbf or simply lb) =	4.4482	Newton (N)
=	0.0045	kilo Newton (kN)
=	0.453	kilogram force (kg_f or kg)
1 short ton (ton) =	2000	pounds force (lb_f or lb)
=	0.907	metric ton (t)
=	907.18	kgf or kg
=	8,896.44	N
=	8.896	kN
1 kilogram force (kg_f) or kilopond =	2.204	pounds force (lb_f)
=	9.807	N

PRESSURE

Pressure (p)	**p = F/A (Force/Area)**	
Pascal (Pa)	**$Pa = 1\ N/m^2 \equiv 1\ kg/m \cdot s^2$**	
Pound-force per square inch (psi)	**psi = 6,894.757 Pa**	
1 Pound per square inch (psi) =	143.999	lb/ft^2
=	68.05×10^{-3}	atm (standard atmosphere)
=	68.95×10^{-3}	bar
1 kilo-pound per square inch (kips) =	6,894.7	kilo-Pascal (kPa)
=	68.95	bar
=	72	ton per square foot (ton/ft^2)
=	1000	psi
1 Pascal (Pa) =	145.04×10^{-6}	psi
=	9.8692×10^{-6}	atm
=	10^{-5}	bar

DENSITY

Density ρ (rho)	**ρ = m/V (mass/Volume)**		
1 lb/ft^3 =	0.0135	ton/yd^3	
=	27	lb/yd^3	
=	0.134	lb/gal (U.S)	
=	5.787037×10^{-4}	lb/in^3	
=	16.018	kg/m^3	
1 kg/m^3 =	0.0624	lb/ft^3	
=	1.6856	lb/yd^3	
=	0.000843	ton/yd^3	
=	0.008345	lb/gal (U.S)	
Earth/Soil/Dry Sand =	2,700 lb/yd^3	100 lb/ft^3	1,600 kg/m^3
Clay, dry =	2,800 lb/yd^3	103.7 lb/ft^3	1,660 kg/m^3
Sand, moist/packed =	3,200 lb/yd^3	118.5 lb/ft^3	1,900 kg/m^3
Clay, wet/packed =	3,500 lb/yd^3	129.6 lb/ft^3	2,080 kg/m^3
Limestone =	4,400 lb/yd^3	163.0 lb/ft^3	2,610 kg/m^3
Water =	1,685.6 lb/yd^3	62.4 lb/ft^3	1,000 kg/m^3
Ice (at 0 °C) =	1,545.1 lb/yd^3	57.2 lb/ft^3	916.7 kg/m^3
Steel rolled =	13,231.6 lb/yd^3	490.1 lb/ft^3	7,850 kg/m^3
Concrete, conventional =	2,552.5 lb/yd^3	140.0 lb/ft^3	2,242 kg/m^3